Mobility

Mobility

Processes, Computers, and Agents

edited by
Dejan Milojičić,
Frederick Douglis,
and Richard Wheeler

ADDISON-WESLEY
An imprint of Addison Wesley Longman, Inc.
Reading, Massachusetts • Harlow, England • Menlo Park, California
Berkeley, California • Don Mills, Ontario • Sydney
Bonn • Amsterdam • Tokyo • Mexico City

The publisher offers discounts on this book when ordered in quantity for special sales.

For more information, please contact:

Computer & Engineering Publishing Group
Addison Wesley Longman, Inc.
One Jacob Way
Reading, Massachusetts 01867

Visit us on the Web: www.awl.com/cseng/

Library of Congress Cataloging-in-Publication Data
Mobility : processes, computers, and agents / edited by Dejan S. Milojičić, Frederick Douglis, Richard Wheeler.
p. cm.
Includes bibliographical references and index.
ISBN 0-201-37928-7
1. Mobile agents (Computer software) 2. Mobile computing.
I. Milojičić, Dejan S. II. Douglis, Frederick. III. Wheeler, Richard, 1962– .
QA76.76.I58M65 1999
006.3--dc21 99-10175
CIP

Text printed on recycled and acid-free paper.

ISBN 0201379287

2 3 4 5 6 7 CRS 03 02 01 00

2nd Printing February 2000

We dedicate this book to our families.
From Dejan to Maja, Višnja, and Milena.
From Fred to Lisa and Allison.
From Ric to Natalie, Adam, and Joshua.

Contents

Preface

History and Goals

This book grew out of a survey paper on process migration. After spending considerable time and effort collecting and sorting several hundred references for that work, we realized that we had almost enough material for a book. Since then, the scope of the book expanded to encompass broader issues in mobility. Mobile agents descend in many ways from earlier work in process migration, both representing a form of "logical mobility," while mobile computing is "physical mobility." We were struck by some of the similarities and benefits in these seemingly different areas. To the best of our knowledge, these topics are not presented in this combination elsewhere and this book represents a unique perspective on the literature in this field.

After more than a year of selecting papers, and contacting contributing authors and publishers, we settled on the current selection. Each of the editors brings a different perspective: one has long been active in the development of parallel and distributed operating systems, one worked for several years in mobile computing research, and the third is and has been active in the mobile agents research community.

The goals of this book are threefold. First, we provide a distinguished set of papers related to the three types of mobility, focusing on systems and practical experience. In addition, several of the original authors present afterwords that updates and reflects on their original work. Papers and afterwords are presented in Parts II, III and IV. Second, we hope to transfer our experience in using and building mobile systems to a broad audience. This experience is summarized in the Introduction (Part I) of the book, as well as in the individual introductions to Parts II, III and IV. Finally, we want to analyze mobility in these forms by exploring the similarities and differences between these approaches as presented in Summary (Part V).

Who Should Read This Book

This book is intended for three sets of readers. The first set includes software architects and practitioners who can draw from the experiences reflected in this book to guide their own projects. The second set includes computer science professors and graduate students for whom the selection of relevant papers could contribute to an advanced course in operating systems or mobile computing. Our final targeted set of readers are researchers in any of these three areas who can hopefully motivate their own work by an increased exposure to the work of others in these fields.

We organized the book in historical order. First, we present process migration, then mobile computing, and finally mobile agents. Not coincidentally, this parallels the editors' own careers which began with process migration and then dealt with the other areas. Clearly, different readers can take different approaches when reading this book. Those interested in only one form of mobility can focus on their selected part of the book, but we believe that all readers will benefit from reading papers in each of the fields.

Acknowledgments

We would like to acknowledge the work of many people who helped this project. First and foremost, without the support of the authors of the contributed papers included in this collection, this book would not exist. We thank those authors for providing their material, and offer particular thanks to those who contributed afterwords with additional commentary.

Our editors at Addison Wesley Longman, Karen Gettman and Mary Hart, were a continuous source of help and encouragement. David Ruddock and Charles Perkins reviewed our original book proposals and gave us and the publishers encouragement to proceed with this project. AWL enlisted several reviewers, Anurag Acharya, Bob Gray, Mor Harchol-Balter, Danny Lange, Bob Pemberton, M. Satyanarayanan, and Jim White, who provided many useful suggestions related to the technical contents and the organization of the book. We are also indebted to our own set of reviewers which includes Philippe Bernadat, Dan Duchamp, Laura Feeney, Mitsuru Oshima, and Jan Vitek. ACM, Baltzer, IEEE, Springer Verlag, USENIX, and Wiley each allowed us to reprint papers that appeared in their own publications. We would like to thank our employers, AT&T and The Open Group Research Institute, who supported our project. Finally, we would like to acknowledge the patience and support of our families to whom we dedicate this book.

Dejan S. Milojičić, Acton, Massachusetts,
Frederick Douglis, Florham Park, New Jersey, and
Richard Wheeler, Belmont, Massachusetts.

August, 1998.

Part I

Introduction

Chapter 1 Introduction

Mobility has always been an important factor in nature. Animals move from place to place, migrating to find food and shelter. Migration is also important to humans: early nomadic tribes followed sources of food while in contemporary times workers tend to migrate in search of better employment, entertainment, or travel experiences. In its essence, mobility stems from a desire to move either toward resources or away from scarcity.

This book discusses both physical and logical computing entities that move. Physical entities, like mobile computers, change their actual location. Logical entities can be an instance of a running user application or a mobile agent—a network application that migrates in a network and executes on behalf of its owner. While a mobile agent might migrate anywhere on the Internet in pursuit of its specific goals, active applications (known as processes) typically migrate only within a local cluster of computers. This book introduces mobility in chronological order, from its first use as process migration in the late 1970s, to the mobile computers that became common in the 1980s, and finally to the mobile agents that are the products of the 1990s.

Process migration is the act of transferring a process between two computers. A process is an operating system abstraction that encompasses the code, data, and operating system state associated with an instance of a running application. Process migration is traditionally used to enable load distribution or fault resilience. In Part I we discuss how several research operating systems have implemented full blown process migration mechanisms while most commercial migration-related products provide a higher level, checkpoint/restart version of migration. Well-known systems that support process migration include Accent [Rashid, 1981], Charlotte [Artsy, 1987], Chorus [Philippe, 1993], Mach [Accetta, 1986], MOSIX [Barak, 1993], Sprite [Ousterhout, 1994], and the V Kernel [Cheriton, 1988]. In part II we present papers on each of these systems. Checkpoint/restart implementations include Condor [Litzkow, 1988] and Utopia [Zhou, 1993], which run on most commercial operating systems. Techniques used to implement process migration have influenced the subsequent work that has been done on mobile agents [Johansen, 1995; Milojičić, 1998b].

Mobile computing involves the movement of physical devices such as laptop and palmtop computers. In Part III, we focus on three aspects of physical mobility. The first is *weak connectivity*: how a computer can operate when disconnected from the network, intermittently connected, or connected over very slow communication links. The Coda File System [Kistler, 1992] was one of the earliest examples of support for disconnected operation. The second aspect deals with *wireless connectivity* when a computer moves between "cells" in a wireless network. The Mobile IP protocol, which was initially proposed by Ioannidis, Duchamp, and Maguire [Ioannidis, 1993b], has been improved, expanded upon, and evaluated in numerous ways. The third aspect is *ubiquitous computing*, a scenario espoused by Mark Weiser and others at Xerox PARC [Weiser, 1993b] to reflect what could happen when computers are so small and cheap that they fade into the background. Examples of "ubicomp" include active badges (that permit computers to track the whereabouts of humans and interact with them depending on their location), small hand-held computers in the vein of what today is the increasingly popular 3Com PalmPilot, and large-scale interfaces such as interactive whiteboards.

Mobile agents are programs that can move through a network and autonomously execute tasks on behalf of users. There is an important distinction between agents and other user applications: an agent represents and acts on behalf of its owner, and inherits the owner's authority (typically expressed through credentials). Compared to mobile code, such as applets, mobile agents also carry data and possibly the thread of control. Mobile agents require agent environments (or servers), that act like a docking station that accepts agents. They are typically supported on top of a programming environment, such as a Java virtual machine or a Tcl/Tk interpreter. Mobile agents are intended for use in electronic commerce, software distribution, information retrieval, system administration, and network management. They are well-suited for slow and unreliable links where they can improve their performance by improving their locality of reference. Fault resilience is also improved by encapsulating the agent's code and state. Well-known mobile agent systems include Telescript [White, 1996], Aglets [Lange, 1998], Agent Tcl [Kotz, 1997], Concordia [Walsh, 1998], Voyager [Glass], Tacoma [Johansen, 1995], Mole [Baumann, 1998], Ara [Peine, 1997], Sumatra [Ranganathan, 1997], MOA [Milojičić, 1998b], and many others. These systems are discussed in Part IV.

Despite the superficial differences between these three areas, there is much inherent commonality stemming from the nature of mobility. In Section 1.1, we illustrate the common benefits, deployment challenges, and technical issues introduced by process migration, mobile computing, and mobile agents viewed from a common perspective. The introductions of each part of the book include the specifics for each form of mobility. We again compare these three forms of mobility in Part V. An overview of the rest of the book is provided in Section 1.2.

1.1 Benefits, Deployment Challenges, and Technical Issues

Mobility in distributed systems offers benefits, such as enabling movement toward desired resources, the use of computer resources while moving, and improved flexibility. The major challenges in deploying mobility are the lack of applications, widespread infrastructure, and global scalability. Finally, there are technical issues that need to be resolved while supporting mobility: naming, locating, security, reliability, and disconnected behavior. There are certainly many more benefits, deployment challenges, and technical issues, especially those that are specific to each particular form of mobility. We have selected only the most important ones and those that are common for all three forms of mobility for this book. The rest of this chapter addresses these benefits, deployment challenges, and technical issues in more detail.

Benefits

Moving toward a desired resource is the first benefit of mobility. In the case of process migration, processes might move toward an underloaded computer, a specific database, or some rare hardware device. Mobile agents migrate toward the source of information, or a computer that they manage. By locally accessing data or other resources, performance can be dramatically improved. Sometimes, only local access is possible because of the semantics or security requirements. In the case of mobile computing, both the owner and computer move. For example, people often take laptop or palmtop computers along as they move from home to office, or travel on a business trip. In all of these cases, it is the ability to move toward a resource or specific location that provides a qualitative or quantitative benefit.

Using computer resources while moving is another benefit of mobility. It allows users to take a computer away from its usual workplace and still be productive. Process migration and mobile agents provide additional support, by enabling movement of the programming environment and applications along with the mobile computer. The relevance of this benefit will increase as the deployment of computers increases. Computers are used for everyday tasks, and computer access has become more and more important, not only for computer professionals, but also for normal people in their everyday lives. Since professional life typically involves daily commutes and occasional traveling, having continuous access to computers has become a necessity for computer users.

Flexibility can be improved by virtue of mobility. Flexibility can mean easier reconfiguration or improved reliability. If a computer has a partial failure, or if it is about to shut down, a process can migrate to another computer and continue to execute there. If a wireless phone cannot connect from a specific area, moving to a new area may overcome some natural obstacles that prevented the original connection. A mobile agent may not have sufficient resources or connectivity from one host, but it may have the requisite resources after migrating to a new one. The encapsulation of all state, code, or hardware (in the case of mobile computing) into a mobile unit reduces residual dependencies and the chances of failure.

Deployment Challenges

The lack of applications is the biggest challenge for deploying any form of mobility. Despite a need for mobility, application developers typically rely on simple mechanisms, such as remote execution, the reconnection of mobile computers, or remote method invocation. Even though these mechanisms offer less performance, functionality, and flexibility, the benefits have been traded for increased simplicity and the possibility to reuse existing solutions. Nevertheless, as computer technology evolves, more support for the mobility will exist in the underlying infrastructure that makes mobility easier to deploy. For example, object serialization and mobile code provide inherent support for object mobility in Java, which greatly reduces the amount of work needed to implement Java based mobile agents. Similarly, wireless telephony provides increased support for mobile computing as today's mobile phones evolve into powerful, mobile computers.

The lack of widespread infrastructure is another major challenge to the success of mobility. This is represented by the absence of support for connecting computers while or after they migrate, or for programming infrastructure that will allow migrated processes or agents to visit remote computers. Technical solutions do exist for these problems. Solutions, such as Mobile IP, wireless, and the ability to download execution environments, are just a first step on the path to achieving a ubiquitous, global infrastructure that will enable free mobility of computers, mobile agents, and processes.

The infrastructure should support the transparent movement of entities from one location to another. In practice, this means that communication with a migrating entity can proceed as if there had been no mobility. For example, mobile computers can be reconnected after restarting at a new location, migrated processes can be controlled after migration to a new computer, and mobile agents can continue to communicate as they migrate on the Internet.

Performance can suffer because of migration. It can be impacted by the use of slower links, by the overloading of a remote location, or by the increased levels of indirection required to reach a migrated entity. Mobile computers typically communicate over slow and unreliable links, for example, wireless or telephone lines. Mobile agents can migrate away to another continent, and processes can migrate to another subnetwork. Infrastructure needs to address such performance, scalability, and reliability issues.

Global scalability is a requirement for the widespread deployment of mobile systems. This translates into support for process migration at many nodes around the world, unique, worldwide connectivity for mobile computers, and globally interoperable docking stations for mobile agents. Standardization efforts are very important in this respect, such as the proposed Mobile IP standard [Perkins, 1996], or the Mobile Agent System Interoperability Facility (MASIF) standard [Milojičić, 1998a] for mobile agent systems. These standards can result in a wider availability of systems supporting mobility.

Scalability requirements for mobile entities are hard to fulfill since mobility can make the number of abstractions appear larger than the number of abstractions in a non-mobile system. Hierarchical solutions are well suited to handle large numbers of abstractions, but do not lend themselves readily to mobile systems: current hierarchical schemes typically assume a set of relatively stationary abstractions. If mobile agents frequently cross between the sub-trees of a hierarchy, the benefits of using a hierarchical approach are reduced or eliminated.

Technical Issues

Security problems that need to be addressed in mobile environments are similar to those addressed by classic security research: authentication, authorization, integrity, privacy, prevention of the denial of service, and non-repudiation (the act of denying that one has sent a message) [Kaufman, 1995]. Mobility, however, introduces more opportunities for security holes. Stationary places tend to be more secure than mobile ones: central computer systems and cashiers at banks can be protected by firewalls and human guards. Laptop computers and individuals carrying large sums of money are much harder to safeguard. In a similar fashion, it is relatively easy to access data sent from a mobile computer or to interpose between a remote agent and the source of its incoming messages. The integrity of a mobile agent can be attacked by modifying its code or data while it visits a remote node. Furthermore, it is harder to support non-repudiation for mobile entities—especially those that move frequently and are short lived—than it is for static, long-lived objects.

In the world of computing, mobile entities are even more exposed to typical security attacks: unauthorized access, data corruption, denial of access, spoofing, trojan horses, replaying, or eavesdropping. Security mechanisms determine if an entity is allowed to migrate and, once migrated, whether it will be a recognized and trusted entity at the new location. After a mobile computer restarts at a new location, its identity may need to be verified and other computers may need to trust its identity before communication can be established. In the case of mobile agents, security infrastructure supports trust domains; the level of trust in mobile agents and mobile agent systems is determined as a part of migration. Security issues related to process migration mechanisms are typically handled by the underlying operating system. Unlike mobile agents and mobile computing, process migration typically occurs within a single trust domain.

Authentication is the security concern closely related to mobility. If an entity is moved, malicious users could use some other program to masquerade as the migrated entity, for example, during attempts to communicate with the home environment. If an entity moves into non-trusted, potentially hostile domains, it is hard to preserve its integrity and privacy.

Despite an increasing number of researchers who work on possible solutions for security issues raised or emphasized by mobility, the state of the art in mobile security is still significantly inferior when compared to traditional security solutions for non-mobile problems. Security is one of the biggest challenges for mobility. The existing security infrastructure is designed to protect stationary computers and abstractions, leaving support for mobility as an afterthought. Security related research that addresses the needs of mobile code, and applets in particular, has increased. There is also a social aspect: users do not trust foreign computers or mobile entities and are reluctant to allow them access to their local resources. In some cases, mobility can solve security problems. For example, if remote access is forbidden because of security concerns, a combination of mobile agents and controlled local access can be an acceptable alternative solution. In other cases, it could be more difficult to attack a mobile entity because its location is not always known.

Reliability in the presence of disconnection is both a technical issue and a benefit for mobility. Reliability can be improved by mobility; however, it also needs additional support. Mobility can introduce residual dependencies and distributed state, thereby increasing the potential for failure. Due to the increased likelihood of failure during migration, various techniques are supported to improve reliability of mobile entities. Caching and hoarding of state are also important for mobile entities. After an agent or process has moved, some state must be accessed remotely. Furthermore, some of this state is invalidated after each move and it must be reloaded after each migration. Mobile computers typically cache files. Mobile processes cache code and data from their address space and open data files. Mobile agents cache classes representing their code and objects that they need to access. Caching depends on the underlying communication media and distance that the entity migrates. In this book, we present papers dealing with replication and checkpointing for process migration [Litzkow, 1992]; file hoarding for mobile computing [Kistler, 1992; Mummert, 1996]; and transactional message queuing [Zajcew, 1993] and fault tolerance tools, such as Horus [Johansen, 1995; van Renesse, 1996] for mobile agents.

Disconnected behavior is associated with mobility because once an entity moves, it may have only a limited connection to its original environment. This means that mobile entities can lose contact with their owners, as is the case with mobile agents, or with various servers in the case of mobile computing and process migration. This adds a requirement to function autonomously. For example, a mobile computer can access some files locally even while disconnected. However, a mobile agent can act on behalf of its owner even if it is not able to communicate with the owner at that time. This typically implies some amount of self sufficiency, such as locally cached files for mobile computers and intelligence for mobile agents. In the case of process migration, some residual dependency is typically required, such as on the originating (also called home) node.

Naming and locating mobile entities is common for all forms of mobility. Without the support for locating a mobile object, other functionality, like communications or control, is not achievable. When a mobile entity is moved from its original location (as when a laptop computer is disconnected from its original network) and then reconnected at some new location, communication channels must be reestablished with other entities. Similarly, communication channels must be reconstructed after a process or mobile agent migrates to a new node.

Striking similarities occur in the solutions for locating mobile entities in each of the three types of mobility we describe in this book. Typically, mobile agents register their locations with a predefined server and then update this server about subsequent migrations. Servers are selected differently: some schemes use the node where the object was created, or *home node*, while others rely on some other well known server. Other possibilities include either leaving a trail at each node visited, accompanied by a requirement to forward messages, or simply searching for the entity at a set of predefined locations.

Naming mobile entities has similar requirements for different forms of mobility. Naming schemes used in large-scale distributed systems guarantee that each entity has a unique name. In mobile systems, names must be unique and they must allow for secure identification and authentication of the mobile entity in its new environment. Therefore, the naming of mobile entities is typically associated with security. Another problem arises when names get recycled: since entities are migrateable, it is not always known if they still exist, so recycling must be done with care.

Maintaining communication while moving relies on naming and locating. It is of particular importance in collaborative applications. In the case of mobile computing, combinations of active gadgets and palmtops can be used to maintain continuous communication channels within a group, for example military personnel, police, or rescue teams. Large-scale problems can be solved by groups of mobile agents. Controlling migrated processes and agents is even more important. Once an agent is sent out to the Web to perform a task, it must support various control functions, to allow users to check its status, to suspend or kill it, or to recall the agent back to its home node. This requires maintaining communication channels with mobile agents. A message sent to a migrated entity needs to be forwarded to a new location, possibly through a chain of trails, or through a known registry.

1.2 Overview of the Book

The rest of the book is organized in three separate parts: process migration, mobile computing, and mobile agents. Each section begins with an overview of the specific field, followed by a collection of papers from that field and accompanying notes from some of the original authors. Each part is organized thematically, as described in the following three paragraphs.

Part II consists of six chapters. Chapter 2 introduces the field of process migration by describing the benefits and challenges, applications, myths, and facts. Chapter 3 includes two papers describing early work in process migration, the Worms and DEMOS/MP systems. Chapter 4 includes papers describing kernel supported process migration in: MOSIX, Sprite, Charlotte, Accent, V Kernel, and Mach. Chapter 5 includes papers describing user-space process migration: Condor, Emerald, and Tui. While previous chapters addressed migration mechanisms, Chapter 6 includes two papers that address migration policies, and specifically address the lifetime of processes and the impact of migration on load distribution strategies. Finally, Chapter 7 provides pointers to additional sources of information.

Part III consists of five chapters. Chapter 8 introduces mobile computing and includes an overview paper by Forman and Zahorjan on technology challenges. Chapter 9 discusses limits on connectivity: disconnected operation, weakly connected operation, and low bandwidth. It covers applications in the areas of file systems, databases, and Web access. Chapter 10 covers Mobile IP, including the original system, a recent tutorial by Perkins, and performance studies. Chapter 11 covers "ubiquitous computing," the notion that computers become so plentiful that they fade into the background. We include an overview paper on "ubicomp," as it has been called, and a recent paper on support for the 3Com PalmPilot—today's answer to the ParcTab ubicomp device. Chapter 12 provides a summary and a list of other sources of information.

Part IV consists of three chapters. Chapter 13 introduces the area of mobile agents by overviewing the benefits and challenges of mobile agents, describing applications, and analyzing some of the myths and facts about mobile agents. Chapter 14 includes papers describing ten mobile agent systems: Telescript, Aglets, Agent TCL, Concordia, Mole, TACOMA, Sumatra, Ara, MOA, and Voyager; and an overview of the MASIF standard. Chapter 15 contains other sources of information on mobile agents.

Part V, Summary, concludes the book by comparing the three forms of mobility. It attempts to predict the future for each form of mobility and includes an extensive list of references.

Part II

Mobility in a Cluster: Process Migration

Chapter 2 Process Migration

In the late 1970s, people commonly moved data from one computer to another by copying it onto magnetic tapes. As networking grew more popular, researchers began looking in earnest for interesting ways to exploit distributed systems. In this era research in process migration started.

Traditional process migration mechanisms rely on help from the underlying operating system. For example, in MOSIX [Barak, 1993] and Sprite [Ousterhout, 1988], the operating system provides a *single system image* that allows system resources, like files, to be accessed with equal ease regardless of the process's current location. Other researchers provide higher level mechanisms that implement support for migration in user space, such as in Mach [Milojičić,1993b] or by supporting migration at the level of individual applications as discussed in Condor [Litzkow, 1992].

2.1 Benefits and Challenges of Process Migration

Benefits

The basic idea behind all of the implementations is to move an executing process from one node to another. What is this ability good for? Migration provides several key benefits: *load distribution, fault resilience,* and *data access locality.*

Load distribution. This allows a heavily loaded node to migrate away some subset of its active processes to a less loaded node. An example that illustrates the advantages of load distribution is a typical undergraduate computer lab with a few dozen machines connected to a local area network. Students come and go, causing the load on individual nodes to vary at any given time. Load distribution allows this load to be evenly spread among all nodes in the local cluster.

Fault resilience. By enabling processes to migrate away from partially broken machines, migration improves the ability of a process to overcome localized errors or faults. For example, a long-running job could detect a disk failure on its current node and use migration to transfer to a properly working node. In a similar way, migration can be used by system administrators to relocate active processes before shutting down a specific node for maintenance.

Data access locality. Locality of reference can provide significant performance gains. Given a process that needs to perform a series of database queries on a very large data set, migrating to the host node of the data might greatly reduce the access time.

Challenges

Despite the promises of migration, most successful implementations face several challenges that work against their widespread acceptance and popularity. The two most common ones are *social acceptance* and the inherent *complexity* of most migration mechanisms.

Social issues. The most prominent hurdle to process migration has always been a social one: given the ability to move a process from one node to another, to obtain permission to use a new node's resources requires a social contract between the various owners of nodes in a cluster. Researchers implement various mechanisms to help address this issue; for example, by migrating only to idle nodes and then allowing the owner to reclaim the node by rejecting foreign processes. MOSIX allows individual nodes to block incoming migration, even when its own processes are allowed to migrate to other nodes. Sprite requires that any node that enables migration for its local processes must reciprocate by allowing other processes to migrate to it.

Code complexity. Process migration mechanisms, especially those implemented as part of the operating system, tend to be complex and difficult to maintain. Minimizing the impact of process migration and providing a portable mechanism motivates some of the user-space implementations. When code is difficult to maintain, it tends to atrophy and die as the original developers tire of the repetitive burden of reintegrating their work into new releases of the base system.

In addition to these deployment issues, there are a wide range of technical challenges that migration implementations face. The following issues are the most common.

Naming, locating, and controlling migrated processes. In operating systems that support process migration, giving processes unique names is difficult. Process names, typically process identifiers, or PIDs, have to maintain existing single node semantics and yet uniquely identify each process in the cluster namespace. The complexity involved is proportional to the desired degree of transparency. For example, many systems encode the process's creating host identifier in the PID, which has ramifications for scalability.

Process migration uses several schemes to track and maintain up-to-date information about the location of migrated processes. One popular scheme, employed in MOSIX and Sprite, uses the creation site, or *home machine*, of each process to track its location. On migration, the home machine is informed about the process's new location. Charlotte [Artsy, 1989] uses a forwarding scheme that basically leaves a forwarding address at each node visited. This scheme is useful when agents make a small number of migrations. Searching schemes, similar to those used in the V Kernel [Theimer, 1985], can be applied in cases where the number of nodes to be visited is small, even though the number of potential migrations can be high.

Various mechanisms are used to control migrated processes. For example, to terminate or suspend a remote process, a local proxy is used in Locus [Walker, 1989] and OSF/1 AD [Zajcew, 1993]. This virtual process, or *vproc,* acts as a local proxy for a migrated process and keeps track of the current process location. All control requests are forwarded to the process's current node. In Mach [Accetta, 1986], control is achieved by transparently sending messages to any task regardless of location via a distributed IPC service that supports transparent forwarding of messages to remote nodes.

Exporting and transferring process state. A process has several types of data, such as user data, stack, or memory-mapped files. Migrating a process at any arbitrary point in time requires complex support for encapsulating this state. Besides exporting various registers, operating system activity must be cleanly interrupted at a restartable point of execution. This state can either be transferred to another node in a distributed operating system, such as in MOSIX, or exported into user space and then transferred, as in the case of Mach. The latter is required because processes cannot transfer any internal operating system state between nodes.

After migration, the operating system can transparently *page-in* its address space to its current location on demand. This mechanism is provided to various degrees in Accent [Zayas, 1987a], MOSIX, and Mach. This is called *lazy evaluation*. Lazy evaluation of an address space, as well as of any other state that might be too large or costly to transfer eagerly, has proven to be a useful optimization in the case of process migration. Lazy evaluation, however, is very complex to support. This is especially true for complex address space implementations, as shown in the case of the Mach microkernel and its task migration scheme.

Transparent communication. Some systems, such as Charlotte [Artsy, 1987], Mach [Accetta, 1986], and Amoeba [Tanenbaum, 1990; 1992], provide support for transparent communication of processes independent of their location. Messages sent during migration can be discarded or received out of order. In order to prevent such cases, complex algorithms are required to maintain communication in the presence of migration. Typically, these algorithms employ some type of proxy. Maintaining communication in the presence of migration has turned out to be one of the most complex components of migration systems as shown in Charlotte, Sprite, and Mach.

File Systems. In many cases, process migration is supported by the underlying file system (for example, Sprite [Nelson, 1988]). This support consists of transparently moving the open connections for files and other abstractions represented by files, such as pipes. A process can open files on one node, migrate to another node, and still access its opened files. In addition, a distributed file system allows a process to open any file in the system from any location using a universal name.

Security. Operating systems provide a secure trusted environment that can be shared by untrusted abstractions in the local cluster. Security is provided by supporting capabilities, as is done in Amoeba and Mach, or by using access control lists (ACLs), as is done in Locus. These single node abstractions have been extended in various ways to support distributed applications executing on multiple nodes.

Decentralized algorithms. In many cases, there is no central point of control for cluster-wide mechanisms like process migration. This means that migration mechanisms need to make policy decisions locally without the knowledge of other nodes in the network, as is done in the MOSIX [Barak, 1989a] load balancing mechanism.

2.2 Applications

Process migration is used in several different ways. **Parallel applications** are those in which a collection of related processes take part in one computation. Each of these related processes can be migrated to different nodes in a cluster in order to evenly distribute the load and improve performance. Examples of this type of application include a parallel compilation (*make*) tool and parallel computation as described in Sprite [Douglis, 1990], MPVM [Casas, 1995], and in work by Skordos [Skordos, 1995].

Load balancing, as described in the introduction to this section, uses migration to distribute a generic multiuser workload between nodes in a cluster. As described in Harchol-Balter and Downey [Harchol-Balter, 1996; 1997], migrating away long-lived processes can benefit the migrated process and also increase performance for any short-lived processes left on the original node by allowing them a larger share of the local resources, such as the processor or physical memory.

Data locality can be improved by process migration. For example, a process might need to submit queries to a large database that is resident on a remote node. Migration can be used to move the process to the node, improving the performance of future data accesses. This was demonstrated in the case of MOSIX [Barak, 1985a], Mach [Milojičić, 1993a], and Emerald [Jul, 1989], among many others.

Long-running applications can benefit from migration by using migration to avoid getting interrupted by system failures or routine maintenance. For example, an application can be informed of the impending shutdown by a signal. Upon receipt of the signal, migration to any node can be initiated. One of the most successful commercial process migration packages, Load Sharing Facility (LSF) [Platform Computing, 1996], is used heavily by long-running applications.

2.3 Myths and Facts

Despite the evident advantages of process migration, it has been deployed so far mainly in academic and research circles. Several misconceptions contribute to its unpopularity:

Myth: *Migration is too complex.* **Fact:** Most modern operating systems do not provide support for the single system image features that would make migration easier to implement. Early migration implementations, like Sprite and MOSIX, did indeed involve fairly complex modifications of the base operating system to support migration. However, research has addressed the issue of complexity by providing user level migration packages, like Condor, or by implementing migration at a higher level of the system as is done in the most recent MOSIX implementations [Alon, 1987; Barak, 1998]. Both types of efforts produce a much more maintainable and portable mechanism.

Myth: *There is no user demand for migration.* **Fact:** Historically, given the availability of alternatives, like remote invocation and remote data access, full-blown process migration has not proven useful to most users. However, neither of these mechanisms can be used to support long-running jobs as described above or support automatic load balancing of unmodified processes [Harchol-Balter, 1997; Krueger, 1988].

Myth: *Migration is socially impossible.* **Fact:** In the early days of network computers, a typical usage pattern involved one shared file server that provided file service for a set of independent workstations, each of which was owned by a specific person or group who was unwilling to share their machine with others. In many ways, basic human greed will always be a factor. However, the popularity of new programming paradigms, like the Web, have helped users and individual workstation owners change the way they work. Today, many users freely allow Java applets to run on their machines. Also, many migration implementations mitigate the impact of these issues by increasing the amount of control an individual has over migration. For example, MOSIX users can bar foreign processes from migrating into their node while selfishly allowing their own local processes to freely migrate to other nodes in the system [Barak, 1993].

Chapter 3 Early Work

As mentioned in the introduction to this section, the earliest work in process migration was done in the late 1970s and early 1980s. This collection presents two papers from this era. In their **Worm** paper [Shoch, 1982], Shoch and Hupp address several of the issues raised above. The worm program provides a framework for distributed computations that can dynamically detect and take over idle nodes on the local network. The owners of the nodes reclaim their machines by rebooting them back to the original state. This early work addresses some of the security and social acceptance issues by using a voluntary contract: no local disks are accessed and the worm programs "promise" not to prevent the machines from being reclaimed at an arbitrary point. In many ways, this paper echoes the concerns being debated in the mobile agent community.

Powell and Miller describe the **Demos** system, which is one of the earliest operating system based implementations of migration [Powell, 1983]. In Demos, a process communicates with the operating system and other processes by sending messages via *links,* which specify the message's receiver node and process. By encapsulating a process completely behind the link concept, Demos simplifies the process migration mechanism.

Other papers that mention process migration or describe related early work include those on the Distributed Computing System [Farber, 1973], XOS [Miller, 1981], and Butler [Dannenberg, 1982; 1985; Nichols, 1987]. The Distributed Computing System proposes a ring architecture in which processes communicate using a message-passing scheme. XOS is designed as a tool for minimizing the communication between the nodes in an experimental multiprocessor system. The location of a migrated process is found by following hints left at the system's nodes. Early versions of the Butler system support migration and deal with issues surrounding protection, security, and autonomy. Later versions of Butler dropped support for migration in favor of supporting remote invocation.

3.1 Shoch, J. and Hupp, J., The Worm Programs – Early Experience with a Distributed Computation

Shoch, J. and Hupp, J., "The Worm Programs – Early Experience with a Distributed Computation," *Communications of the ACM*, 25(3):172-180, March 1982.

The "Worm" Programs—Early Experience with a Distributed Computation

John F. Shoch and Jon A. Hupp
Xerox Palo Alto Research Center

I guess you all know about tapeworms...? Good. Well, what I turned loose in the net yesterday was the ... father and mother of all tapeworms....
My newest-my masterpiece-breeds by itself....
By now I don't know exactly what there is in the worm. More bits are being added automatically as it works its way to places I never dared guess existed....
And-no, it can't be killed. It's indefinitely self-perpetuating so long as the net exists. Even if one segment of it is inactivated, a counterpart of the missing portion will remain in store at some other station and the worm will automatically subdivide and send a duplicate head to collect the spare groups and restore them to their proper place.
—John Brunner, *The Shockwave Rider*
Ballantine, New York, 1975

SUMMARY: The "worm" programs were an experiment in the development of distributed computations: programs that span machine boundaries and also replicate themselves in idle machines. A "worm" is composed of multiple "segments," each running on a different machine. The underlying worm maintenance mechanisms are responsible for maintaining the worm—finding free machines when needed and replicating the program for each additional segment. These techniques were successfully used to support several real applications, ranging from a simple multimachine test program to a more sophisticated real-time animation system harnessing multiple machines.

1. Introduction

In *The Shockwave Rider*, J. Brunner developed the notion of an omnipotent "tapeworm" program running loose through a network of computers—an idea which may seem rather disturbing, but which is also quite beyond our current capabilities. The basic model, however, remains a very provocative one: a program or a computation that can move from machine to machine, harnessing resources as needed, and replicating itself when necessary.

In a similar vein, we once described a computational model based upon the classic science-fiction film, *The Blob*: a program that started out running in one machine, but as its appetite for computing cycles grew, it could reach out, find unused machines, and grow to encompass those resources. In the middle of the night, such a program could mobilize hundreds of machines in one building; in the morning, as users reclaimed their machines, the "blob" would have to retreat in an orderly manner, gathering up the intermediate results of its computation. Holed up in one or two machines during the day, the program could emerge again later as resources became available, again expanding the computation. (This affinity for nighttime exploration led one researcher to describe these as "vampire programs.")

These kinds of programs represent one of the most interesting and challenging forms of what was once called *distributed computing*. Unfortunately, that particular phrase has already been co-opted by those who market fairly ordinary terminal systems; thus, we prefer to characterize these as *programs which span machine boundaries* or *distributed computations.*

An earlier version of this paper was prepared for the Workshop on Fundamental Issues in Distributed Computing, ACM/SIGOPS and ACM/SIGPLAN, Pala Mesa Resort, December 1980.
Authors' present address: J.F. Shoch and J.A. Hupp, Xerox Corporation, Palo Alto Research Center, 3333 Coyote Hill Road, Palo Alto, CA 94304.

CR Categories and Subject Descriptors: C.2.4 and C.2.5 [**Computer Communication Networks**]: Distributed Systems and Local Networks.
General Terms: Design, Experimentation.
Additional Key Words and Phrases: multimachine programs, Ethernet local network, Pup internetwork architecture.

In recent years, it has become possible to pursue these ideas in newly emerging, richer computing environments: large numbers of powerful computers, connected with a local computer network and a full architecture of internetwork protocols, and supported by a diverse set of specialized network servers. Against this background, we have undertaken the development and operation of several real, multimachine "worm" programs; this paper reports on those efforts

In the following sections, we describe the model for the worm programs, how they can be controlled, and how they were implemented. We then briefly discuss five specific applications which have been built upon these multimachine worms.

The primary focus of this effort has been obtaining real experience with these programs. Our work did not start out specifically addressing formal conceptual models, verifiable control algorithms, or language features for distributed computation, but our experience provides some interesting insights on these questions and helps to focus attention on some fruitful areas for further research.

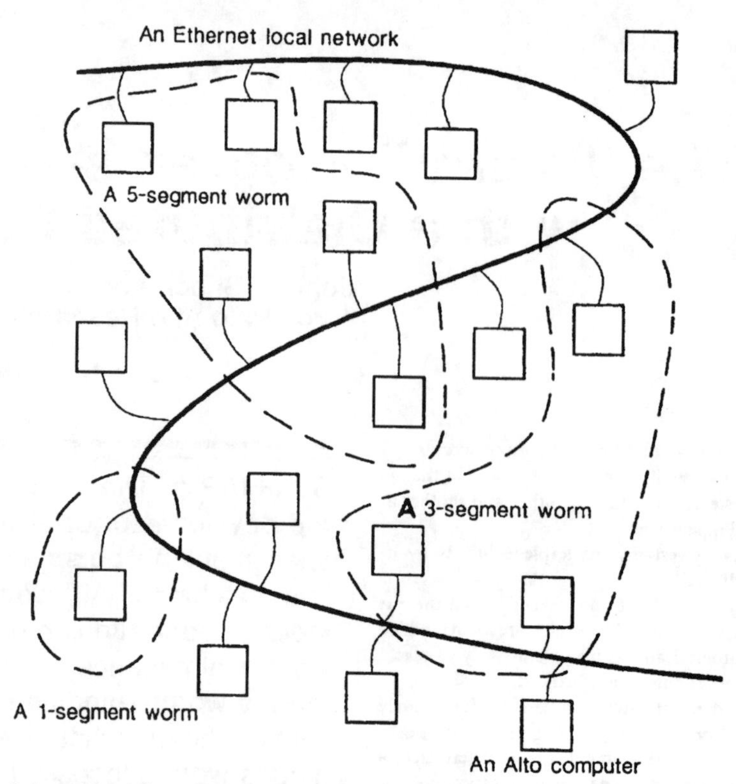

Fig. 1. Schematic of Several Multisegment Worm Programs.

2. Building a Worm

A *worm* is simply a computation which lives on one or more machines (see Figure 1). The programs on individual computers are described as the *segments* of a worm; in the simplest model each segment carries a number indicating how many total machines should be part of the overall worm. The segments in a worm remain in communication with each other; should one segment fail, the remaining pieces must find another free machine, initialize it, and add it to the worm. As segments (machines) join and then leave the computation, the worm itself seems to move through the network. It is important to understand that the worm mechanism is used to gather and maintain the segments of the worm, while actual user programs are then built on top of this mechanism.

Initial construction of the worm programs was simplified by the use of a rich but fairly homogeneous computing environment at the Xerox Palo Alto Research Center. This includes over 100 Alto computers [10], each connected to an Ethernet local network [4, 6]. In addition, there is a diverse set of specialized network servers, including file systems, printers, boot-servers, name-lookup servers, and other utilities. The whole system is held together by the Pup architecture of internetwork protocols [1].

Many of the machines remain idle for lengthy periods, especially at night, when they regularly run a memory diagnostic. Instead of viewing this environment as 100 independent machines connected to a network, we thought of it as a 100-element multiprocessor, in search of a program to run. There is a fairly straightforward set of steps involved in building and running a worm with this set of resources.

2.1 General Issues in Constructing a Worm Program

Almost any program can be modified to incorporate the worm mechanisms; all of the examples described below were written in BCPL for the Alto. There is, however, one very important consideration: since the worm may arrive through the Ethernet at a host with no disk mounted in the drive, the program must not try to access the disk. More important, a user may have left a disk spinning in an otherwise idle machine; writing on such a disk would be viewed as a profoundly antisocial act.

Running a worm depends upon the cooperation of many different machine users, who must have some confidence in the judgment of those writing programs which may enter their machines. In our work with the Alto, we have been able to assure users that there is not even a disk driver included within any of the worm programs; thus, the risk to any spinning disk is no worse than the risk associated with leaving the disk in place while the memory diagnostic runs. We have yet to identify a single case in which a worm program tried to write on a local disk.

It is feasible, of course, for a program to access secondary storage available through the network, on one of the file servers.

2.2 Starting a Worm

A worm program is generally organized with several components: some initialization code to run when it starts on the first machine; some initialization when it starts on any subsequent machine; the main program. The initial program can be started in a machine by any of the standard methods, including loading via the operating system or booting from a network boot-server.

2.3 Locating Other Idle Machines

The first task of a worm is to fill out its full complement of segments; to do that, it must find some number of idle machines. To aid in this process, a very simple protocol was defined: a special packet format is used to inquire if a host is free. If it is, the idle host merely returns a positive reply. These inquiries can be broadcast to all hosts or transmitted to specific destinations. Since multiple worms might be competing for the same idle machines, we have tried to reduce confusion by using a series of specific probes addressed to individual machines. As mentioned above, many of the Altos run a memory diagnostic when otherwise unused; this program responds positively when asked if it is idle.

Various alternative schemes can be used to determine which possible host to probe next when looking for an additional segment. In practice, we have employed a very simple procedure: a segment begins with its own local host number and simply works its way up through the address space. Figure 2, an Ethernet source-destination traffic matrix (similar to the one in [8]), illustrates the use of this procedure. The migrating worm shows up amid the other network traffic with a "staircase" effect. A segment sends packets to successive hosts until finding one which is idle; at that point the program is copied to the new segment, and this host begins probing for the next segment.

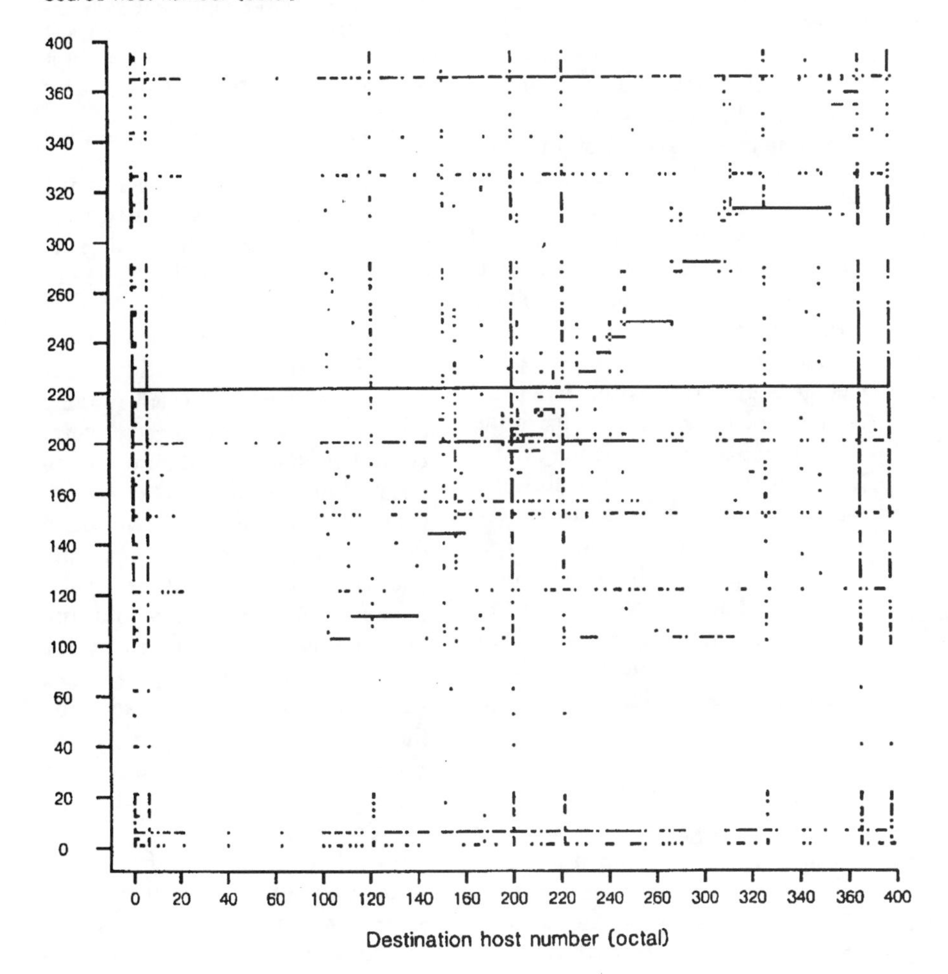

Fig. 2. Ethernet Source-Destination Traffic Matrix with a "Worm" Running. (Note the "staircase" effect as each segment seeks the next one.)

2.4 Booting an Idle Machine

An idle machine can be located through the Ethernet, but there is still no way in which an Alto can be forced to restart through the net-work. By design, it is not possible to reach in and wrench away control from a running program; instead, the machine must willingly accept a request to restart, either by booting from its local disk or through the network.

After finding an idle machine, a worm segment then asks it to go through the standard network boot procedure. In this case, however, the specified source for the new program is the worm segment itself. Thus, we have this sequence:

(1) Existing segment asks if a host is idle.
(2) The host answers that it is.
(3) The existing segment asks the new host to boot through the network, from the segment.
(4) The newcomer uses the standard Pup procedures for requesting a boot file [1].
(5) The file transfer protocol is used to transfer the worm program to the newcomer.

In general, the program sent to a new segment is just a copy of the program currently running in the worm; this makes it easy to transfer any dynamic state information into new segments. But the new segment first executes a piece of initialization code, allowing it to reestablish any important machine-dependent state (for example, the number of the host on which it is running).

2.5 Intra-Worm Communication—The Need for Multidestination Addressing

All segments of the worm must stay in communication, in order to know when one of their members has departed. In our experiments, each segment had a full model of its parent worm—a list of all other segments. This is a classic situation in which one host wants to send some information to a specified collection of hosts—what is known as *multidestination addressing* or *multicasting* (also called *group addressing*) [2, 5]. Unfortunately, the experimental Ethernet design does not directly support any explicit form of multicasting. There are, however, several alternatives available [6]:

(1) *Pseudo-multicast ID*: An unused *physical* host number can be set aside as a special *logical* group address, and all participants in the group set their host ID to this value. This is a workable approach (used in some existing programs), but does require advance coordination. In addition, it consumes one host ID for each worm.

(2) *Brute force multicast*: A copy of the information is sent to each of the group's other members. This is one of the techniques which was used with the worms: each segment periodically sends its status to all other segments.

The latter approch does require sending $n*(n - 1)$ packets for each update; other techniques reduce the total number of packets which must be sent. Many of the worms, however, were actually quite small, requiring only three or four machines to ensure that they would not die when one machine was lost. In these cases, the explicit multicast was very satisfactory. When an application needs a substantial number of machines, they can be obtained with one large worm or with a set of cooperating smaller worms.

This state information being exchanged is used by each independent segment to run an algorithm similar to the one for updating routing tables in store-and-forward packet-switched networks and internetworks: if a host is not heard from after some period of time, it is presumed dead and eliminated from the table. The remaining segments then cooperate to give one machine responsibility for finding a new segment, and the process continues.

2.6 Releasing a Machine

When a segment of a worm is finished with a machine, it needs to return that machine to an idle state. This is very straightforward: the segment invokes the standard network boot procedure to reload the memory diagnostic program, that test is resumed, and the machine is again available as an idle machine for later reuse.

This approach does result in some unfortunate behavior should a machine crash, either while running the segment or while trying to reboot. With no program running, the machine cannot access the network and, as we saw, there is no way to reach in from the net to restart it. The result is a stopped machine, inaccessible to the worm. The machine is still available, of course, to the first user who walks up it it.

3. A Key Problem: Controlling a Worm

> No, Mr. Sullivan, we can't stop it! There's never been a worm with that tough a head or that long a tail! It's building itself, don't you understand? Already it's passed a billion bits and it's still growing. It's the exact inverse of a phage—whatever it takes in, it adds to itself instead of wiping . . . Yes, sir! I'm quite aware that a worm of that type is theoretically impossible! But the fact stands, he's done it; and now it's so goddamn comprehensive that it can't be killed. Not short of demolishing the net!
>
> —John Brunner, *The Shockwave Rider*

We have only briefly mentioned the biggest problem associated with worm management: controlling its growth while maintaining stable behavior.

Early in our experiments, we encountered a rather puzzling situation. A small worm was left running one night, just exercising the worm control mechanism and using a small number of machines. When we returned the next morning, we found dozens of machines dead, apparently crashed. If one restarted the regular memory diagnostic, it would run very briefly, then be seized by the worm. The worm would quickly load its program into this new segment; the program would start to run and promptly crash, leaving the worm incomplete—and still hungrily looking for new segments.

We have speculated that a copy of the program became corrupted at some point in its migration, so that the initialization code would not run properly; this made it impossible for the worm to enlist a new, healthy segment. In any case, some number of worm segments were hidden away, desperately trying to replicate; every machine they touched, however, would crash. Since the building we worked in was quite large, there was no hint of which machines were still running; to complicate matters, some machines available for running worms were physically located in rooms which happened to be locked that morning so we had no way to abort them. At this point, one begins to imagine a scene straight out of Brunner's novel—workers running around the building, fruitlessly trying to chase the worm and stop it before it moves somewhere else.

Fortunately, the situation was not really that grim. Based upon an ill-formed but very real concern about such an occurrence, we had included an emergency escape within the worm mechanism. Using an independent control program, we were able to inject a very special packet into the network, whose sole job was to tell every running worm to stop no matter what else it was doing. All worm behavior ceased. Unfortunately, the embarassing results were left for all to see: 100 dead machines scattered around the building.

This anecdote highlights the need for particular attention to the control algorithm used to maintain the worm. In general, this distributed algorithm involves processing incoming segment status reports and taking actions based upon them. On one hand, you may have a "high strung worm": at the least disturbance or with one lost packet, it may declare a segment gone and seek a new one. If the old segment is still there, it must later be expunged. Alternatively, some control procedures were too slow in responding to changes and were constantly operating at less than full strength. Some worms just withered and died, unable to promptly act to rebuild their resources.

Even worse, however, were the unstable worms, which suddenly seemed to grow out of control, like the one described above. This mechanism is not yet fully understood, but we have identified some circumstances that can make a worm grow improperly. One factor is a classic failure mode in computer communications systems: the *half-up link* (or one-way path) where host *A* can communicate with host *B*, but not the other way around. When information about the state of the worm is being exchanged, this may result in two segments having inconsistent information. One host may think everything is fine, while another insists that a new segment is necessary and goes off to find it.

Should a network be partitioned for some time, a worm may also start to grow. Consider a two-segment worm, with the two segments running on hosts at opposite ends of an Ethernet cable, which has a repeater in the middle. If someone temporarily disconnects the repeater, each segment will assume that the other has died and seek a new partner. Thus, one two-part worm becomes two two-part worms. When the repeater is turned back on, the whole system suddenly has too many hosts committed to worm programs. Similarly, a worm which spans different networks may become partitioned if the intermedite gateway goes down for a while and then comes back up.

In general, the stability of the worm control algorithms was improved by exchanging more information, and by using further checks and error detection as the programs evaluated the information they were receiving. For example, if a segment found that it continually had trouble receiving status reports from other segments, it would conclude that it was the cause of the trouble and thereupon self-destruct.

Furthermore, a special program was developed to serve as a "worm watcher" monitoring the local network. If a worm suddenly started growing beyond certain limits, the worm watcher could automatically take steps to restrict the size of the worm or shut it down altogether. In addition, the worm watcher maintained a running log recording changes in the state of individual segments. This information was invaluable in later analyzing what might have gone wrong with a worm, when, and why.

It should be evident from these comments that the development of distributed worm control algorithms with low delay and stable behavior is a challenging area. Our efforts to understand the control procedures paid off, however: after the initial test period the worms ran flawlessly, until they were deliberately stopped. Some ran for weeks, and one was allowed to run for over a month.

4. Applications Using the Worms

In the previous sections we have described the procedures for starting and maintaining worms; here we look at some real worm programs and applications which have been built.

4.1 The Existential Worm

The simplest worm is one which runs a null program—its sole purpose in life is to stay alive, even in the face of lost machines. There is no substantive application program being run (as a slight embellishment, though, a worm segment can display a message on the machine where it is running).

This simple worm was the first one we constructed, and it was used extensively as the test vehicle for the underlying control mechanisms. After the first segment was started, it would reach out, find additional free machines, copy itself into them, and then just rest. Users were always free to reclaim their machines by booting them; when that happened, the customary worm procedure would find and incorporate a new segment.

As a rule, though, this procedure would only force the worm to change machines at very infrequent intervals. Thus, the program was equipped with an independent self-destruct timer: after a segment ran on a machine for some random interval, it would just allow itself to expire, returning the machine to an idle state. This dramatically increased the segment death rate, and exercised the worm recovery and replication procedures.

4.2 The Billboard Worm

With the fundamental worm mechanism well in hand, we tried to enhance its impact. As we described, the Existential worm could display a small message; the "Billboard worm" advanced this idea one step further, distributing a full-size graphics image to many different machines. Several available graphics programs used a standard representation for an image—pictures either produced from a program or read in with a scanner. These images could then be stored on a network file server and read back through the network for display on a user's machine.

Thus, the initial worm program was modified so that when first started, it could be asked to obtain an image from one of the file servers. From then on, the worm would spread this image, displaying it on screens throughout the building. Two versions of the worm used different methods to obtain the image in each new segment: the full image could be included in the program as it moved, or the new segment could be instructed to read an image directly from one of the network servers.

With a mechanical scanner to capture an image, the Billboard worm was used to distribute a "cartoon of the day"—a greeting for workers as they arrived at their Altos.

4.3 The Alarm Clock Worm

The two examples just described required no application-specific communication among the segments of a worm; with more confidence in the system, we wanted to test this capability, particularly with an application that required high reliability. As a motivating example we chose the development of a computer-based alarm clock which was not tied to a particular machine. This program would accept simple requests through the network and signal a user at some subsequent time; it was important that the service not make a mistake if a single machine should fail.

The alarm clock was built on top of a multimachine worm. A separate user program was written to make contact with a segment of the worm and set the time for a subsequent wake-up. The signalling mechanism from the worm-based alarm clock was convoluted, but effective: the worm could reach out through the network to a server normally used for out-going terminal connections and then place a call to the user's telephone!

This is an interesting application because it needs to maintain in each segment of the worm a copy of the database—the list of wake-up calls to be placed. The strategy was quite simple: each segment was given the current list when it first came up. When a new request arrived, one machine took responsibility for accepting the request and then propagating it to the other segments. When placing the call, one machine notified the others that it was about to make the call, and once completed, notified the others that they could delete the entry. This was, however, primarily a demonstration of a multimachine application, and not an attempt to fully explore the double-commit protocols or other algorithms that maintain the consistency of duplicate databases.

Also note that this was the first application in which it was important for a separate user program to be able to find the worm, in order to schedule a wake-up. In the absence of an effective group-addressing technique, we used two methods: the user program could solicit a response by broadcasting to a well-known socket on all possible machines, or it could monitor all traffic looking for an appropriate status report from a worm segment.

4.4 Multimachine Animation Using a Worm

So far, the examples described have used a distributed worm, with no central control. One alternative way to use a worm, however, is as a robust set of machines supporting a particular application—an application that may itself be tied to a designated machine. An example which we have explored is the development of a multimachine system for real-time animation. In this case, there is a single *control node* or *master* which is controlling the computation and playing back the animation; the multiple machines in the worm are used in parallel to produce successive frames in the sequence, returning them to the control node for display.

The master node initially uses the worm mechanisms to acquire a set of machines. In one approach, the master first determines how many machines are desired and then recruits them with one large worm. As we just discussed, however, a single large worm may be slow to get started as it sequentially looks for idle machines, and it may be unwieldy to maintain. Instead of using one large worm to support the animation, the master spawns one worm with instruction on how many other worms to gather. This starting worm launches some number of secondary worms, which in turn acquire their full complement of segments (in this experiment, three segments per worm). Thus, one can very rapidly collect a set of machines responding to the master; this collection of machines is still maintained by the individual worm procedures.

Each worm segment then becomes a "graphics machine" with a pointer back to the master, and each reports in with an "I'm alive" message after it is created; the master itself is not part of any worm. The master maintains the basic model of the three-dimensional image and controls the steps in the animation. To actually produce each frame, though, it only has to send the coordinates for each object; the "worker" machine then performs the hidden-line elimination and half-tone shading, computing the finished frame. With this approach, all of the worm segments work in parallel, performing the computationally intensive tasks. The master supplies descriptions of the image to the segments and later calls upon them to return their result for display as the next image.

The underlying worm mechanism is used to maintain the collection of graphics workers; if a machine disappears, the worm will find a new one and update the list held by the control program. The worm machines run a fairly simple program, with no specific knowledge about the animation itself. The system was tested with several examples, including a walk through a cave and a collection of bouncing and rotating cubes.

4.5 A Diagnostic Worm for the Ethernet

The combination of a central control machine and a multipart worm is also a useful way to run distributed diagnostics on many machines. We knew, for example, that Alto Ethernet interfaces showed some pair-wise variation in the error rates experienced when communicating with certain other machines. To fully test this, however, would require running a test program in all available machines—a terribly awkward task to start manually.

The worm was the obvious tool. A control program was used to spawn a three-segment worm, which would then find all available machines and load them with a test program; these machines would then check in with the central controller and prepare to run the specified measurements. Tests were conducted with as many as 80, 90, or even 120 machines.

In testing pair-wise error rates, each machine had a list of all other participants already loaded by the worm and registered with the control program. Each host would simply try to exchange packets with each other machine thought to be a part of the test. At the end of the test each machine would report its results to the control host—thus indicating which pairs seemed to have error-prone (or broken) interfaces.

Figure 3 is the Ethernet source-destination traffic matrix produced during this kind of worm-based test. To speed the process of gathering all available machines, a three-segment worm would be spawned, and these segments could then work in parallel. Host 217 was the control Alto, and

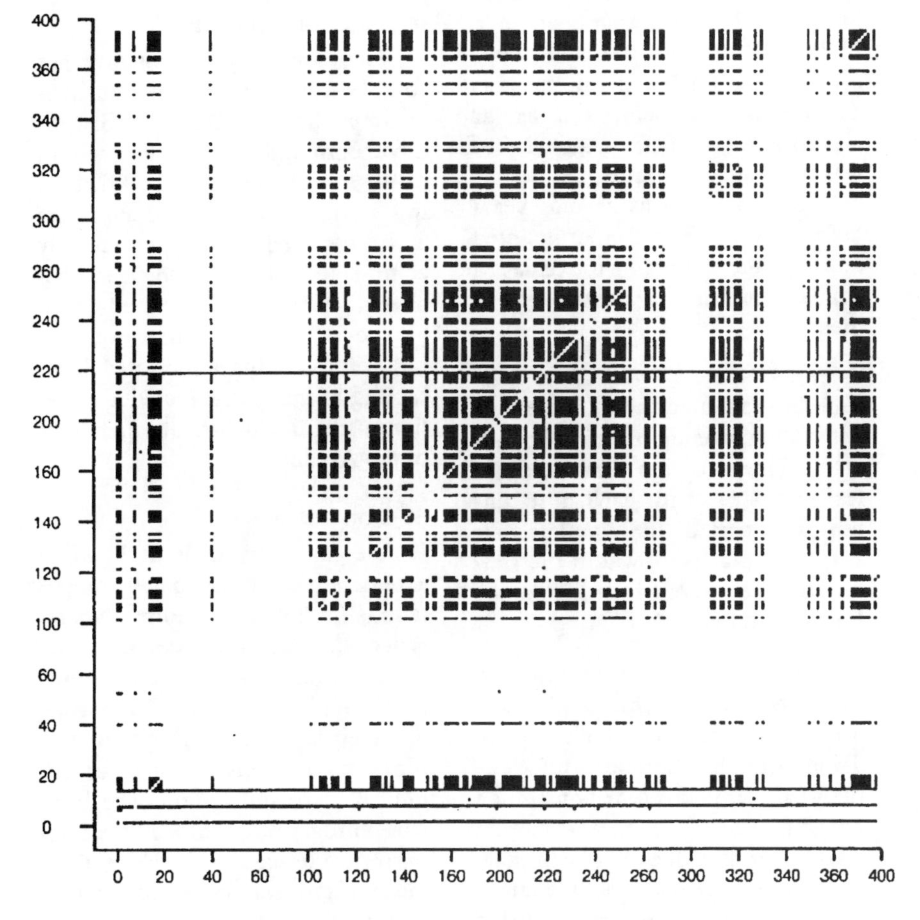

Fig. 3. Ethernet Source-Destination Traffic Matrix When Testing Ethernet Connectivity. (Total number of source-destination pairs = 11,396.)

it found the three segments for its worm on hosts between 0 and 20. Those three segments then located and initialized all of the other participants. As described earlier, a simple linear search through the host address space is used by each segment to identify idle machines. To keep the multiple segments from initially pinging the same hosts, the starting point for each segment could be selected at intervals in the address space. Each segment does make a complete cycle through the address space, however, looking carefully for any idle machines.

To avoid any unusual effects during the course of the test itself, the worm maintenance mechanism was turned off during this period. If hosts had died, the worm could later be reenabled, in an effort to rebuild the collection of hosts for a subsequent test.

At the conclusion of the tests, all of the machines are released and allowed to return to their previous idle state—generally running the memory diagnostic. These machines would boot that diagnostic through the network, from one of the network boot file servers; 120 machines trying to do this at once, however, can cause severe problems. In particular, the boot server becomes a scarce resource that may not be able to handle all of the requests right away, and the error recovery in this very simple network-boot procedure is not foolproof. Thus, all of the participants in the measurements coordinate their departure at the end of a test: each host waits for a quasi-random period before actually attempting to reboot from the network boot server.

5. Some History: Multimachine Programs on the Arpanet

The worm programs, of course, were not the first multimachine experiments. Indeed, some of the worm facilities were suggested by the mechanisms used within the Arpanet or demonstrations built on top of that network:

(1) The Arpanet routing algorithm itself is a large, multimachine distributed computation, as the Interface Message Processors (IMPs) continually exchange information among themselves. The computations continue to run, adapting to the loss or arrival of new IMPs. (Indeed, this is probably one of the longest-running distributed computations.)

(2) In a separate procedure, the Arpanet IMPs can be individually reloaded through the network, from a neighboring IMP. Thus, the IMP program migrates through the Arpanet, as needed.

(3) In late 1970, one of the earliest multimachine applications using the Arpanet took place, sharing resources at both Harvard and MIT to support an aircraft carrier landing simulation. A PDP-10 at Harvard was used to produce the basic simulation program and 3-D graphics data. This material was then shipped to an MIT PDP-10, where the programs could be run using the Evans & Sutherland display processor available at MIT. Final 2-D images produced there were shipped to a PDP-1 at Harvard, for display on a graphics terminal. (All of this was done in the days before the regular Network Control Program (NCP) was running; one participant has remarked that "it was several years before the NCPs were surmounted and we were again able to conduct a similar network graphics experiment.")

(4) "McRoss" was a later multimachine simulation built on top of the NCP, spanning machine boundaries. This program simulated air traffic control, with each host running one part of the simulated air space. As planes moved in the simulation, they were handed from one host to another.

(5) One of the first programs to move by itself through the Arpanet was the "Creeper," built by B. Thomas of Bolt Beranek and Newman (BBN). It was a demonstration program under Tenex that would start to print a file, but then stop, find another Tenex, open a connection, pick itself up and transfer to the other machine (along with its external state, files, etc.), and then start running on the new machine. Thus, this was a relocatable program, using one machine at a time.

(6) The Creeper program led to further work, including a version by R. Tomlinson that not only moved through the net, but also replicated itself at times. To complement this enhanced Creeper, the "Reaper" program moved through the net, trying to find copies of Creeper and log them out.

(7) The idea of moving processes from Creeper was added to the McRoss simulation to make "relocatable McRoss." Not only were planes transferred among air spaces, but entire air space simulators could be moved from one machine to another. Once on the new machine, the simulator had to reestablish communication with the other parts of the simulation. During the move this part of the simulator would be suspended, but there was no loss of simulator functionality.

This summary is probably not complete or fully accurate, but it is an impressive collection of distributed computations, produced within or on top of the Arpanet. Much of this work, however, was done in the early 70s; one participant recently commented, "It's hard for me to believe that this all happened seven years ago." Since that time, we have not witnessed the anticipated blossoming of many distributed applications using the long-haul capabilities of the Arpanet.

6. Conclusions

We have the tools at hand to experiment with distributed computations in their fullest form: dynamically allocating resources and moving from machine to machine. Furthermore, local networks supporting relatively large numbers of hosts now provide a rich environment for this kind of experimentation. The basic worm programs described here demonstrate the ease with which these mechanisms can be explored; they also highlight many areas for further research.

Acknowledgments

This work grew out of some early efforts to control multimachine measurements of Ethernet performance [6, 7, 8]. E. Taft and D. Boggs produced much of the underlying software that made all of these efforts possible. In addition, J. Maleson implemented most of the graphics software needed for the multimachine animation; his imagination helped greatly to focus our effort on a very real, useful, and impressive application. When we first experimented with multimachine migratory programs, it was S. Weyer who pointed out the relevance of John Brunner's novel describing the "tapeworm" programs. (Readers interested in both science fiction and multimachine programs might also wish to read *The Medusa Conspiracy* by Ethan I. Shedley and *The Adolescence of P-1* by Thomas J. Ryan.) Finally, our thanks to the many friends within the Arpanet community who helped piece together our brief review of Arpanet-related experiments, and our apologies to anyone whose work we overlooked.

References

1. Boggs, D.R., Shoch, J.F., Taft, E.A., and Metcalfe, R.M. PUP: An internetwork architecture. *IEEE Trans. Commun. 28*, 4 (April 1980). Describes the Pup internetwork architecture, used to tie together over 1,200 machines on several dozen different networks.

2. Dalal, Y.K. Broadcast protocols in packet switched computer networks. Tech. Rep. 128, Stanford Digital Syst. Lab., Stanford, Calif., April 1977. Discussion of alternative techniques for broadcast addressing.

3. Dalal, Y.K., and Printis, R.S. 48-bit Internet and Ethernet host numbers (to be published in the Proc. 7th Data Comm. Symp., Oct. 1981). Describes the use of broadcast and multicast addresses in an internet design, and how this influenced the development of the Ethernet addressing scheme.

4. Metcalfe, R.M., and Boggs, D.R. Ethernet: Distributed packet switching for local computer networks. *Comm. ACM 19*, 7 (July 1976), 395–404. The original Ethernet paper, describing the principles of operation and experience with the Experimental Ethernet.

5. Shoch, J.F. Internetwork naming, addressing, and routing. Proc. 17th IEEE Comp. Soc. Int. Conf. (Compcon Fall '78), Washington, D.C., Sept. 1978.General discussion of addressing modes, including the use of multicast addressing.

6. Shoch, J.F. *Local Computer Networks.* McGraw-Hill, New York (in press). A survey of alternative local networks and a detailed description of the Ethernet local network.

7. Shoch, J.F., and Hupp, J.A. Performance of an Ethernet local network–a preliminary report. Local Area Comm. Network Symp., Boston, Mass., May 1979 (reprinted in the Proc. 20th IEEE Comp. Soc. Int. Conf. (Compcon Spring '80), San Francisco, Calif., Feb. 1980). Description of the measured performance of the Ethernet.

8. Shoch, J.F., and Hupp, J.A. Measured performance of an Ethernet local network. *Comm. ACM 23*, 12 (Dec. 1980), 711-721. Detailed discussion of the measured performance of the Ethernet, including several source-destination traffic graphs similar to the ones presented here.

9. Shoch, J.F., Dalal, Y.K., Crane, R.C., and Redell, D.D. Evolution of the Ethernet local computer network. Xerox Tech. Rep. OPD–T81–02, Palo Alto, Calif., Sept. 1981. The basic paper on the revised and improved Ethernet Specification, including comparisons with the original Experimental Ethernet.

10. Thacker, C.P., McCreight, E.M., Lampson, B.W., Sproull, R.F., and Boggs, D.R. Alto: A personal computer. In *Computer Structures: Principles and Examples, 2nd edition*, Siewiorek, Bell, and Newell, Eds., McGraw–Hill, New York, 1982, 549–572. Describing the Alto computer—a high-performance, single-user machine—which was used for running the worm programs.

The attention of Computing Practices readers is called to a letter on spelling checkers by Raben in the ACM Forum, pp. 220–221.

3.2 Powell, M. and Miller, B., Process Migration in DEMOS/MP

Powell, M. and Miller, B., "Process Migration in DEMOS/MP," *Proceedings of the 9th ACM Symposium on Operating Systems Principles*, pp. 110-119, October 1983.

Process Migration in DEMOS/MP

Michael L. Powell
Barton P. Miller

Computer Science Division
Department of Electrical Engineering and Computer Sciences
University of California
Berkeley, CA 94720

Abstract

Process migration has been added to the DEMOS/MP operating system. A process can be moved during its execution, and continue on another processor, with continuous access to all its resources. Messages are correctly delivered to the process's new location, and message paths are quickly updated to take advantage of the process's new location. No centralized algorithms are necessary to move a process.

A number of characteristics of DEMOS/MP allowed process migration to be implemented efficiently and with no changes to system services. Among these characteristics are the uniform and location independent communication interface, and the fact that the kernel can participate in message send and receive operations in the same manner as a normal process.

This research was supported by National Science Foundation grant MCS-8010686, the State of California MICRO program, and the Defense Advance Research Projects Agency (DoD) Arpa Order No. 4031 monitored by Naval Electronic System Command under Contract No. N00039-82-C-0235.

1. Introduction

Process migration has been discussed in the operating system literature, and has been among the design goals for a number of systems [Finkel 80][Rashid & Robertson 81]. Theoretical and modeling studies of distributed systems have suggested that performance gains are achievable using relocation of processes [Stone 77, Stone & Bokhari 78, Bokhari 79, Robinson 79, Arora & Rana 80]. Process migration has also been proposed as a tool for building fault tolerant systems [Rennels 80]. Nonetheless, process migration has proved to be a difficult feature to implement in operating systems.

As described here, ***process migration*** is the relocation of a process from the processor on which it is executing (the ***source*** processor) to another processor (the ***destination*** processor) in a distributed (loosely-coupled) system. A loosely-coupled system is one in which the same copy of a process state cannot directly be executed by both processors. Rather, a copy of the state must be moved to a processor before it can run the process. Process migration is normally an involuntary operation that may be initiated without the knowledge of the running process or any processes interacting with it. Ideally, all processes continue execution with no apparent changes in their computation or communications.

One way to improve the overall performance of a distributed system is to distribute the load as evenly as possible across the set of available resources in order to maximize the parallelism in the system. Such resource load balancing is difficult to achieve with static assignment of processes to processors. A balanced execution mix can be disturbed by a process that suddenly requires larger amounts of some resource, or by the creation of a new process with unexpected resource requirements. If it is possible to assess the system load dynamically and to redistribute processes during their lifetimes, a system has the opportunity to achieve better overall throughput, in spite of the communication and computation involved in moving a process to another processor [Stone 77, Bokhard 79]. A smaller relocation cost means that the system has more opportunities to improve performance.

System performance may also be improved by reducing inter-machine communication costs. Accesses to non-local resources require communication, possibly through intermediate processors. Moving a process closer to the resource it is using most heavily may reduce system-wide communication traffic, if the decreased cost of accessing its favorite resource offsets the possible increased cost of accessing its less favored ones.

A static assignment to a processor may not be best even from the perspective of a single program. As a process runs, its resource reference pattern may change, making it profitable to move the process in mid-computation.

The mechanisms used in process migration can also be useful in fault recovery. Process migration provides the ability to stop a process, transport its state to another processor, and restart the process, transparently. If the information necessary to transport a process is saved in stable storage, it may be possible to "migrate" a process from a processor that has crashed to a working one. In failure modes that manifest themselves as gradual degradation of the processor or the failure of some but not all of the software, working processes may be migrated from a dying processor (like rats leaving a sinking ship) before it completely fails.

Process migration has been proposed as a feature in a number of systems [Solomon & Finkel 79, Cheriton 79, Feldman 79, Rashid & Robertson 81], but successful implementations are rare. Some of the problems encountered relate to disconnecting the process from its old environment and connecting it with its new one, not only making the new location of the process transparent to other processes, but performing the transition without affecting operations in progress. In many systems, the state of a process is distributed among a number of tables in the system making it hard to extract that information from the source processor and create corresponding entries on the destination processor. In other systems, the presence of a machine identifier as part of the process identifier used in communication makes continuous transparent interaction with other processes impossible. In most systems, the fact that some parts of the system interact with processes in a location-dependent way has meant that the system is not free to move a process at any point in time.

In the next section, we will discuss some of the structure of DEMOS/MP, which eliminates these impediments to process migration. In subsequent sections, we will describe how a process is moved, how the communication system makes the migration transparent, and the costs involved in moving a process.

2. The Environment: DEMOS/MP

Process migration was added to the DEMOS/MP [Powell, Miller, & Presotto 83] operating system. DEMOS/MP is a version of the DEMOS operating system [Baskett, Howard, & Montague 77, Powell 77] the semantics of which have been extended to operate in a distributed environment. DEMOS/MP has all of the facilities of the original uni-processor implementation, allowing users to access the multi-processor system in the same manner as the uni-processor system.

DEMOS/MP is currently in operation on a network of Z8000 microprocessors, as well as in simulation mode on a DEC VAX running UNIX. Though the processor, I/O, and memory hardware of these two implementations are quite different, essentially the same software runs on both systems. Software can be built and tested using UNIX and subsequently compiled and run in native mode on the microprocessors.

2.1. DEMOS/MP Communications

DEMOS/MP is a message-based operating system, with communication as the most basic mechanism. A kernel implements the primitive objects of the system: executing processes, messages, including inter-processor messages, and message paths, called links. Most of the system functions are implemented in server processes, which are accessed through the communication mechanism.

All interactions between one process and another or between a process and the system are via communication-oriented kernel calls. Most system services are provided by system processes that are accessed by message communication. The kernel implements the message operations and a few special services. Messages are sent to the kernel to access all services except message communication itself.

A copy of the kernel resides on each processor. Although each kernel independently maintains its own resources (CPU, real memory, and I/O ports), all kernels cooperate in providing a location-transparent, reliable, interprocess message facility. In fact, different modules of the kernel on the same processor, as well as kernels on different processors, use the message mechanism to communicate with each other.

In DEMOS/MP, messages are sent using *links* to specify the receiver of the message. Links can be thought of as buffered, one-way message channels, but are essentially protected global process addresses accessed via a local name space. Links may be created, duplicated, passed to other processes, or destroyed. Links are manipulated much like capabilities; that is, the kernel participates in all link operations, but the conceptual control of a link is vested in the process that the link addresses (which is always the process that created it). Addresses in links are context-independent; if a link is passed to a different process, it will still point to the same destination process. A link may also point to a kernel. Messages may be sent to or by a kernel in the same manner as a process.

The most important part of a link is the message process address (see Figure 2-1). This is the field that specifies to which process messages sent over that link are delivered. The address has two components. The first is a system-wide, unique, process identifier. It consists of the identifier of the processor on which the process was created, and a unique local identifier generated by that machine. The second is the last known location of the process. During the lifetime of a link, the first component of its address never changes; the second, however, may.

Changes with process location | *Set on process creation. Does not change*

Last Known Machine	Unique Process ID	
	Creating Machine	Local Unique ID

Structure of a process address
Figure 2-1

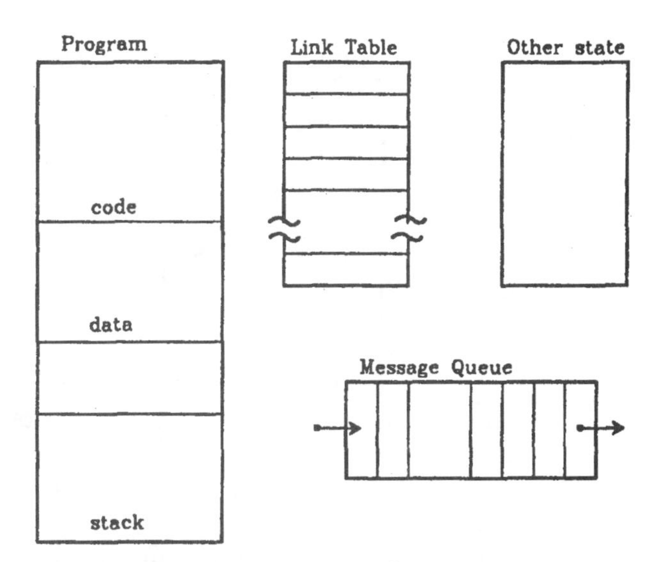

Components of a DEMOS/MP process
Figure 2-2

2.2. Special Kernel Communications

A link attribute, called *DELIVERTOKERNEL*, causes the link to reference the kernel of the processor on which a particular process resides. Except for the DELIVERTOKERNEL flag, a link with this attribute looks the same as a link to the process to which it points. Links with the DELIVERTOKERNEL attribute used to cause the kernel to manipulate the process in ways that system processes cannot do directly.

A message sent over a DELIVERTOKERNEL link follows the normal routing to the process. However, on arrival at the destination process's message queue, the message is received by the kernel on that processor. A link with the DELIVERTOKERNEL attribute allows the system to address control functions to a process without worrying about which processor the process is on (or is moving to).

This mechanism has simplified a number of problems associated with moving a process. It is often the case that some part of the system needs to manipulate the state of a process, for example, the process manager may wish to suspend a process. Using a link with the DELIVERTOKERNEL attribute, the process manager can send a message to the process's kernel asking that the process be stopped. If the process is temporarily unavailable to receive the message (for instance, it is in transit during process migration), the message is held and forwarded for delivery when normal message receiving can continue.

In addition to providing a message path, a link may also provide access to a memory area in another process. When a process creates a link, it may specify in the link read or write access to some part of its address space. The process holding the link may use kernel calls to transfer data to or from the data area defined by the link. This is the mechanism for large data transfers, such as file accesses or data transfer in process migration. The kernel implements the data move operation by sending a sequence of messages containing the data to be transferred. These messages are sent over a DELIVERTOKERNEL link to the kernel of process containing the data area. Using DELIVERTOKERNEL links allows the data to be read from or written to the kernel of the remote process without the kernel that instigated the operation being aware of the process's location.

The inter-machine communication of DEMOS/MP provides reliable delivery of messages. The fundamental guarantee is that any message sent will eventually be delivered.

A DEMOS/MP process is shown in Figure 2-2. A process consists of the program being executed, along with the program's data, stack, and state. The state consists of the execution status, dispatch information, incoming message queue, memory tables, and the process's link table. Links are the only connections a process has to the operating system, system resources, and other processes. Thus, a process's link table provides a complete encapsulation of the execution of the process.

2.3. DEMOS/MP System Processes

DEMOS/MP ***system*** processes are those processes assumed to be present at all times. ***User*** processes are created dynamically to perform computation, usually at the request of some user. A system process will often be a ***server*** process, that is, most other processes will be able to ask it to perform some functions on their behalf. The system processes being used in DEMOS/MP are the switchboard, process manager, memory scheduler, file system (actually, four processes), and command interpreter. The switchboard is a server that distributes links by name. It is used by the system and user processes to connect arbitrary processes together. An example of the system process structure is shown in Figure 2-3.

The process and memory managers handle all the high-level scheduling decisions for processes. These processes allocate and keep track of usage for system resources such as the CPU, real memory, etc. They control processes by sending messages to kernels to manipulate process states. For example, although the kernel implements the mechanisms of migrating a process, the process manager makes the decision of when and to where to migrate a process.

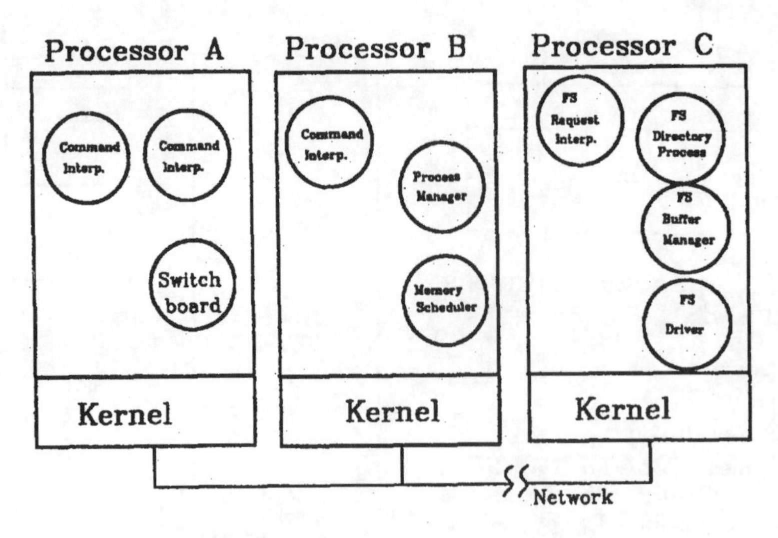

Example of system process lay-out
Figure 2-3

The file system is the same as that implemented for the uni-processor DEMOS [Powell 77], with the added freedom that the file system processes can be located on different processors. The command interpreter allows interactive access to DEMOS/MP programs.

One of our test examples of process migration runs the above processes. It migrates a file system process while several user processes are performing I/O. This is more difficult than moving a user process would be, as we shall see below.

2.4. The Long and Short of Links

It is important to consider all the places where links to a process might be stored when that process is moved, since they contain information specifying the location of the process. Although a link is not useful after the process that it addresses terminates, some links last for relatively long periods of time. For example, a *request* link, which represents a service such as process management, or a *resource* link, which represents an object such as an open file, may exist for as long as the system is up, if they are held by a system process. Other links, such as *reply* links, have short lifetimes, since they are used only once to respond to requests.

Links may be either in some process's link table or in a message that is enroute to a process. Once a link is given out, it may be passed to other processes without the knowledge of the process that created the link (the process to which the link points). There is no way short of a complete system search of finding all links that point to a process. The mechanism for handling messages during and after a process is migrated must provide a way for messages to be directed to the new location, despite out-of-date links. Moreover, for performance reasons, it should eventually bring these links up-to-date.

Moving a *user* process will usually be simple. The only processes likely to have links to a user process are system processes. Such links may be used to send only one message, so the out-of-date link will no longer exist after forwarding the reply message to the new location of the user process.

Moving a *system* process (or, more precisely, a *server* process), is more difficult, since many processes may have links to it, and such links may last a long time, being duplicated and passed to other processes. In fact, the server process may not know how many copies of links there are to it (it is possible, but optional, for a process to keep track of how many, but not where they are). Since such links may be used for many messages, performance considerations will require a method for updating these links.

3. Moving a Process

Most of the low-level mechanisms required to manipulate the process state and move data between kernels were available in the version of DEMOS/MP that existed when this effort began. Process migration was implemented by using those facilities to move the process, and adding the mechanisms for forwarding messages and updating links.

3.1. The Mechanism

A process is moved between two processors called the *source processor* and the *destination processor*. A request to the kernel to move a process is made by the process manager system process. In the absence of an authentic workload for our test cases, the decision to move a particular process and the choice of destination were arbitrary. However, adding a decision rule for when and to where to move a process will be easy. The process manager and memory scheduler already monitor system

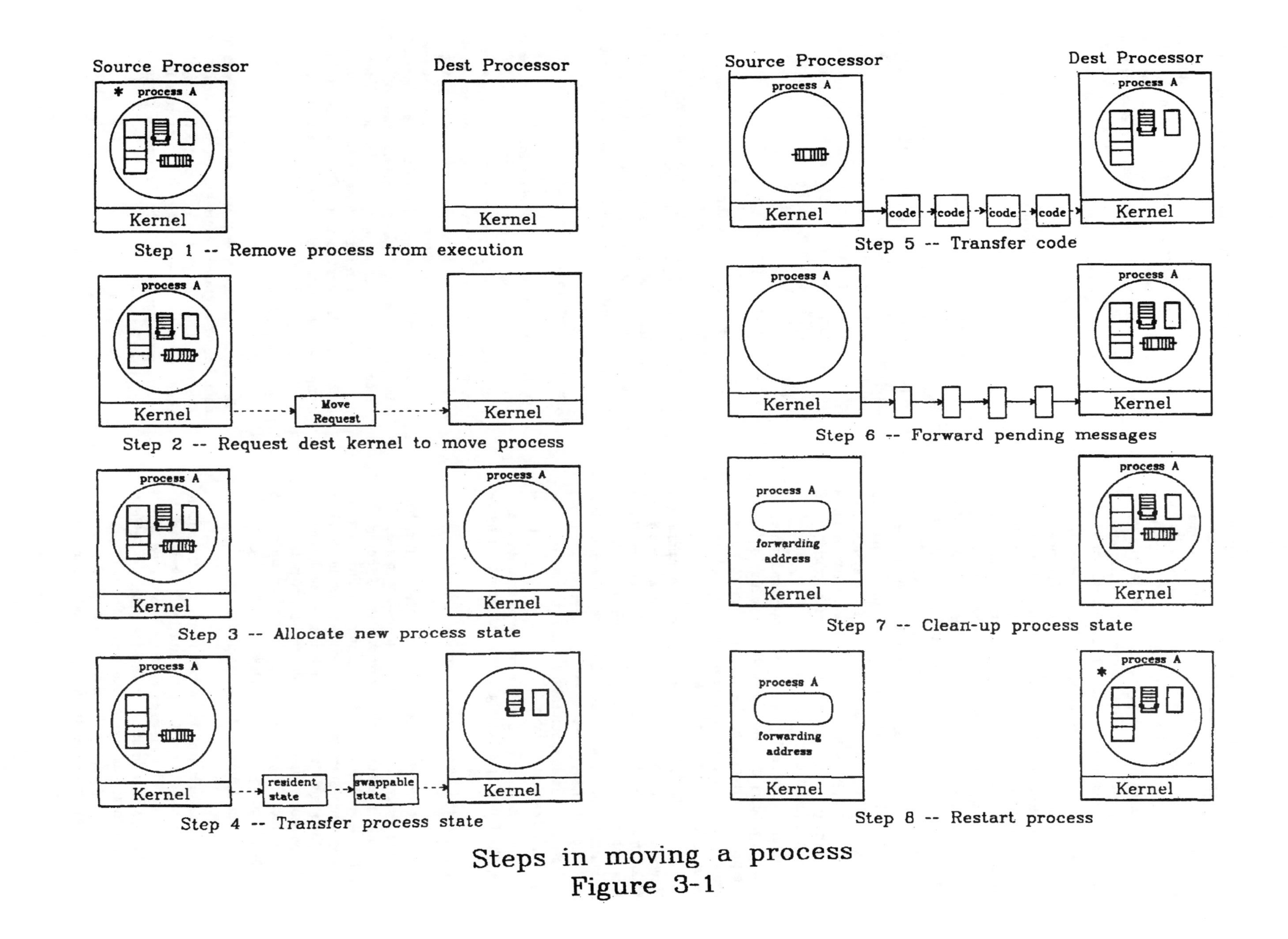

Steps in moving a process
Figure 3-1

activity for memory and cpu scheduling, and can use the same information to make process migration decisions. Information on the communications load is also available. It is of course possible for a process to request its own migration. This request can be thought of as one more piece of information that the process manager can use in making migration decisions. Designing an efficient and effective decision rule is still an open research topic.

There are several features of a decision rule that we have considered in our implementation. The migration scheme depends on the ability to evaluate the resource use patterns of processes. This function is normally available in the accounting or performance monitoring part of the system. There must also be a way to assess the load on individual processors. This function is often available in systems with load-limiting schedulers, which activate or deactivate processes based on overall system load. The three features not usually available are the means to collect the above information in one place, an strategy for improving the operation of the system considering the appropriate costs, and a hysteresis mechanism to keep from incurring the cost of migration more often than justified by the gains.

Information used to determine when and where to move a process involves the state of machine on which the process currently resides, and machines to where the process could move. Processor loading and memory demand for each machine is required.

More difficult is integrating the communications cost incurred by a process. Processes cooperating in a computation may exhibit a great deal of parallelism, and therefore should be on different machines. However, separating them could increase the latency of communication beyond the savings accrued by parallel execution. Collection of the communication data is beyond the ability of most current systems.

Once the decision has been made to migrate a process, the following steps are performed (shown in figure 3-1).

1. Remove the process from execution:

   The process is marked as "in migration". If it had been ready, it is removed from the run queue. No change is made to the recorded state of the process (whether it is suspended, running, waiting for message, etc.), since the process will (at least initially) be in the same state when it reaches its the destination processor. Messages arriving for the migrating process, including DELIVERTOKERNEL messages, will be placed on its message queue.

2. Ask destination kernel to move process:

   A message is sent to the kernel on the destination processor, asking it to migrate the process to its machine. This message contains information about the size and location of the the process's resident state, swappable state, and code. The next part of the migration, up to the forwarding of messages (Step 6), will be controlled by the destination processor kernel.

3. Allocate a process state on the destination processor:

   An empty process state is created on the destination processor. This process state is similar to that allocated during process creation, except that *the newly allocated process state has the same process identifier as the the migrating process.* Resources such as virtual memory swap space are reserved at this time.

4. Transfer the process state:

   Using the move data facility, the destination kernel copies the migrating process's state into the empty process state.

5. Transfer the program:

   Using the move data facility, the destination kernel copies the memory (code, data, and stack) of the process into the destination process. Since the kernel move data operation handles reading or writing of swapped out memory and allocation of new virtual memory, this step will cause definition of memory to take place, if necessary. Control is returned to the source kernel.

6. Forward pending messages:

   Upon being notified that the process is established on the new processor, the source kernel resends all messages that were in the queue when the migration started, or that have arrived since the migration started. Before giving them back to the communication system, the source kernel changes the location part of the process address to reflect the new location of the process.

7. Clean-up process's state:

   On the source processor, all state for the process is removed and space for memory and tables is reclaimed. A *forwarding address* is left on the source processor to forward messages to the process at its new location. The forwarding address is a degenerate process state, whose only contents are the (last known) machine to which the process was migrated. The normal message delivery system tries to find a process when a message arrives for it. When it encounters a forwarding address, it takes the actions described in the next section. The source kernel has completed its work and control is returned to the destination kernel.

8. Restart the process:

   The process is restarted in whatever state it was in before being migrated. Messages may now arrive for the process, although the only part of the system that knows the new location of the process is the source processor kernel. The destination kernel has completed its work.

At this point, the process has been migrated. The links from the migrated process to the rest of the system are all still valid, since links are context-independent. Links created by the process after it has moved will point to the process at its new location. The only problem is what to do with messages sent on links that still point to the old location.

3.2. A Note on Autonomy and Interdomain Migration

The DEMOS/MP kernels trust each other, and thus are not completely autonomous. Moreover, for practical purposes, all DEMOS/MP processors are identical and provide the same services. This makes process migration particularly useful in our environment. However, the process migration mechanism could work even if the kernels were autonomous and had different resources.

The crucial questions for autonomous processors are "Is the process willing to be moved?" and "Will the destination machine accept it?" Any policy to decide which process to migrate could take into account the former question. The second question can be addressed during the migration. Note that the destination machine actually performs most of the steps. In particular, in Step 2, the source machine asks the destination machine to accept the process. If the destination machine refuses, the process cannot be migrated.

It is also possible to migrate processes between domains. By domain, we mean that the destination processor belongs to a collection of machines under a different administrative control than the source processor, and may be suspicious of the source processor and the incoming process. The destination processor may simply refuse to accept any migrations not fitting its criteria. The source processor, once rebuffed, has the option of looking elsewhere.

The source and destination kernels must, of course, be able to communicate with each other in order to accomplish the migration, and the destination machine must be able to handle messages sent over the links held by the process. Since the ability to send and receive messages over links is all a DEMOS process expects of its environment, so long as that continues to be provided, the process can continue to run.

4. Message Forwarding

Since DEMOS/MP guarantees message delivery, in moving a process it must be ensured that all pending, enroute, and future messages arrive at the process's new location. There are three cases to consider: messages sent but not received before the process finished moving, messages sent after the process is moved using an old link, and messages sent using a link created after the process moved.

Messages in the first category were in process's message queue on the source machine, waiting for the process to receive them, when the process restarted on the destination machine. These messages are forwarded immediately as part of the migration procedure.

Messages in the middle category are forwarded as they arrive. After the process has been moved, a forwarding address is left at the source processor pointing toward the destination processor. When a message is received at a given machine, if the receiver is a forwarding address, then the machine address of the message is updated and the message is resubmitted to the message delivery system (see Figure 4-1). As a byproduct of forwarding, an attempt may be made to fix up the link of the sending process (See next section).

The last case, messages sent using links created after the process has moved, is trivial. Links created after the process is moved will contain the same process identifier, and the last known machine identifier in the process address will be that of the new machine.

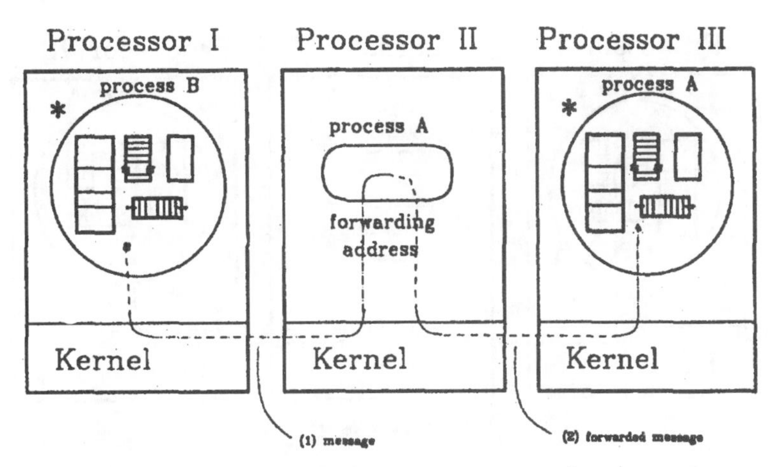

Message sent through a forwarding address
Figure 4-1

Simply forwarding messages is a sufficient mechanism to insure correct operation of the process and processes communicating with it after it has moved. However, the motivation for process migration is often to improve message performance. Routing messages through another processor (with the forwarding address) can defeat possible performance gains and, in many cases, degrade performance. The next section discusses methods for updating links to reduce the cost of forwarding.

An alternative to message forwarding is to return messages to their senders as not deliverable. This method does not require any process state to be left behind on the source processor. The kernel sending the message will receive a response that indicates that the process does not exist on the destination machine. Normally this means that the process the link points to has terminated; in this case, it may mean the process has migrated. The sending kernel can attempt to find the new location of the process, perhaps by notifying the process manager or some system-wide name service, or can notify the sending process that the link is no longer usable, forcing it to take recovery action. The disadvantage of this scheme is that, even if the kernel could redirect the message without impacting the sending process, more of the system would be involved in message forwarding and would have to be aware of process migration. This method also violates the transparency of communications fundamental to DEMOS/MP.

When the forwarding address is no longer needed, it would be desirable to remove it. The optimum time to remove it is when all links that point to the migrated process's old location have been updated. This typically would require a mechanism that makes use of reference counts. An alternative is to remove the forwarding address when the process dies. This can be accomplished by means of pointers backwards along the path of migration.

The forwarding address is compact. In the current implementation, it uses 8 bytes of storage. As a result of the negligible impact on system resources, we have not found it necessary to remove forwarding addresses. Given a long running system, however, some form of garbage collection will eventually have to be used.

It is possible for the processor that is holding forwarding address to crash. Since forwarding addresses are (degenerate) processes, the same recovery mechanism that works for processes works for forwarding addresses. Process migration assumes that reliable message delivery is provided by some lower level mechanism, for example, *published communications* [Powell & Presotto 83].

5. Updating Links

By *updating links*, we mean updating the process address of the link. Recall that a process address contains both a unique identifier and a machine location. The unique identifier is not changed, but the machine location is updated to specify the new machine.

As mentioned above, for performance reasons, it is important to update links that address a process that has been migrated. These links may belong to processes that are resident on the same or different processors, and processes may have more than one link to a given process (including to themselves). Links may also be contained in messages in transit. It is therefore impractical to search the whole system for links that may point to a particular process.

Since race conditions might allow some messages to be in transit while the process is being moved, the message forwarding mechanism is required. As long as it is available, it can also be used for forwarding messages that are sent using links that have not yet been updated to reflect the new location of the process.

The following scheme allows links to be updated as they are used, rather than all at once: As it forwards the

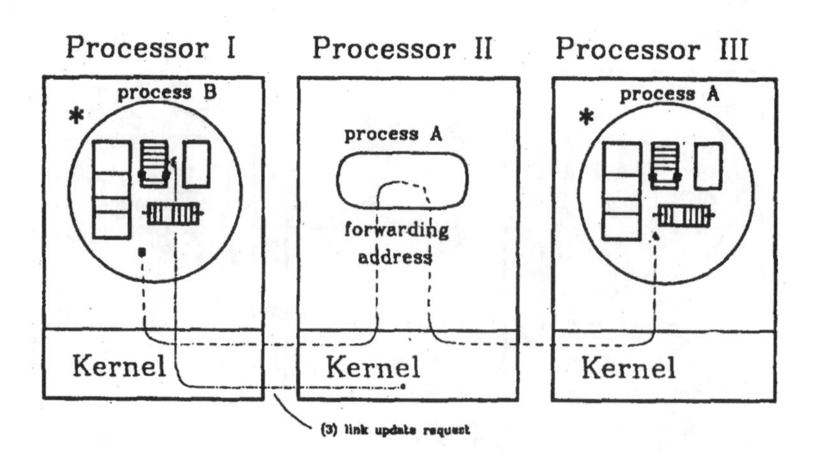

Updating a link after a message forward
Figure 5-1

message, the forwarding machine sends another special message to the kernel of the process that sent the original message (see Figure 5-1). This special message contains the process identifier of the sender of the message, the process identifier of the intended receiver (the migrated process), and the new location of the receiver. All links in the sending process's link table that point to the migrated process are then updated to point to the new location.

Movement of a process should cause only a small perturbation to message communication performance. If the process that has moved is a user process, there will usually be few links that point to it. The links will tend to be either reply links, which will generate only one message and thus not need to be fixed up, or links from other user processes with which it is communicating, which will quickly be updated during the first few message exchanges. As a general rule, system processes do not retain non-reply links to user processes.

The worst case will be when the moving process is a server process. In this case, there may be many links to the process that need to be fixed up. Generally, links to servers are used for more than a few message exchanges, so the overhead of fixing up such a link is traded off against the savings of the cost to forward many messages. Moreover, the likelihood of server processes migrating is lower than for user processes. Servers are often tied to unmovable resources and usually present predictable loads that allow them to be properly located, reducing the need to move them.

6. Cost of Migration

The cost of moving a process dictates how frequently we are willing to move the process. These costs manifest themselves in two areas; the actual cost in moving the process and its related state, and the incremental costs incurred in updating message paths.

The cost of the actual transfer of the process and its state can be separated into *state transfer cost* and *administrative cost*. The state transfer cost includes the messages that contain the process's code, data, state, and message queue. DEMOS/MP uses the data move facility to transfer large blocks of data. This facility is designed to minimize network overhead by sending larger packets (and increasing effective network throughput). The packets are sent to the receiving kernel in a continuous stream. The receiving kernel acknowledges each packet (but the sending kernel does not have to wait for the acknowledgement to send the next packet). Three data moves are involved in moving a process. These are for the program (code and data), the non-swappable (resident) state, and the swappable state. The non-swappable state uses about 250 bytes, and the swappable state uses about 600 bytes (depending on the size of the link table). For non-trivial processes, the size of the program and data overshadow the size of the system information.

In addition, each message that is pending in the queue for the migrating process must be forwarded to the destination machine. The cost for each of these messages is the same as for any other inter-machine message.

The administrative cost includes the message exchanges that are used to initiate and orchestrate the task of moving a process. These costs depend on the internal structures of the system on which it is being implemented. The current DEMOS/MP implementation uses 9 such messages, each message being in the 6-12 byte range. These messages use the standard inter-machine message facility.

The incremental costs for process migration are incurred when a link needs to be updated. Each message that goes through a forwarding address generates two additional messages. The first is the actual message being forwarded to its new destination, and the second is the update message back to the sender. This will occur for each message sent on a given link until the update message reaches the sending process. In current examples, the worst case observed was two messages sent over a link before it was updated. Typically, the link is updated after the first message.

The movement of a process involves a small number of short, control messages, and a large number of block data transfers. The cost of migrating a process depends on the efficiency of both of these types of communications.

7. Conclusion

Process migration has proven to be a reasonable facility to implement in a communication-based distributed operating system. Less than one person-month of time was required to implement and test the mechanism in the current version of DEMOS/MP.

A number of DEMOS/MP design features have made the implementation of process migration possible. DEMOS/MP provides a complete encapsulation of a process, with the only method of access to services and resources being through links. There is no uncontrolled sharing of memory and all contact with the operating system, I/O, and other processes is made through a process's links. DEMOS/MP has a concise process state representation. There is no *process* state hidden in the various functional modules of the operating system. On the other hand, the system servers each maintain their own states, thus no *resource* state (except for links) is in the process state. Once a process is taken out of execution, it is a simple matter to copy its state to another processor. The location transparency and context independence of links make it possible for both the moved process and processes communicating with it to be isolated from the change in venue.

The DELIVERTOKERNEL link attribute allows control operations to be performed without concern for where the process is located. Thus control can follow a process through disturbances in its execution.

The mechanism for moving a process has been implemented, but there is not yet a strategy routine that actually decides when to move a process. The literature contains a few studies of metrics to use for processor and message traffic load optimization. Our continuing work involves implementing process load balancing algorithms, and developing facilities for the measurement and analysis of the performance of communications in distributed programs

References

Arora & Rana 80
Arora, R.K. & Rana, S.P., "Heuristic Algorithms for Process Assignment in Distributed Computing Systems", *Information Processing Letters* **11**, 4-5, December, 1980, pp. 199-203.

Baskett, Howard, & Montague 77
Baskett, F., Howard, J.H., Montague, J.T., "Task Communication in DEMOS", *Proc. of the Sixth Symp. on Operating Sys. Principles*, Purdue, November 1975, pp. 23-32.

Bokhard 79
Bokhari, S.H., "Dual Processor Scheduling with Dynamic Reassignment", *IEEE Trans. on Software Engineering* **SE-5**, 4, July, 1979.

Cheriton 79
Cheriton, D.R., "Process Identification in Thoth", *Technical Report* 79-10, University of British Columbia, October 1979.

Feldman 79
Feldman, J.A., "High-level Programming for Distributive Computing", *CACM* **15**, 4 (April), 1972, pp. 221-230.

Finkel 80
Finkel, R., "The Arachne Kernel", *Technical Report* TR-380, University of Wisconsin, April, 1980.

Powell 77
Powell, M.L., "The DEMOS File System", *Proc. of the Sixth Symp. on Operating Sys. Principles*, Purdue, November 1975, pp. 33-42.

Powell & Presotto 83
Powell, M.L., Presotto, D.L., "Publishing: A Reliable Broadcast Communication Mechanism", *Proc. of the Ninth Symp. on Operating Sys. Principles*, Bretton Woods N.H., October 1983.

Powell, Miller, & Presotto 83
Powell, M.L., Miller, B.P., Presotto, D.L., "DEMOS/MP: A Distributed Operating System", *in preparation*.

Rashid & Robertson 81
Rashid, R.F., Robertson, G.G., "Accent: A Communication Oriented Network Operating System Kernel", *Proc. of the Eighth Symp. on Operating Sys. Principles*, Asilomar, Calif., December 1981, pp. 64-75.

Rennels 80
Rennels, D.A., "Distributed Fault-Tolerant Computer Systems", *Computer*, **13**, 3, March, 1980, pp. 39-46.

Robinson 79
Robinson, J.T., "Some Analysis Techniques for Asynchronous Multiprocessor Algorithms", *IEEE Trans. on Software Engineering* **SE-5**, 1, January, 1979.

Solomon & Finkel 79
Solomon, M.H., Finkel, R.A., "The Roscoe Operating System", *Proc. of the 7th Symp. on Operating Sys. Principles*, Asilomar, Calif., 1979, pp. 108-114.

Stone 77
Stone, H.S., "Multiprocessor Scheduling with the Aid of Network Flow Algorithms", *IEEE Trans. on Software Engineering* **SE-3**, 1, January, 1977, pp. 85-93.

Stone & Bokhari 78
Stone, H.S. & Bokhari, S.H., "Control of Distributed Processes", *Computer*, July, 1978, pp. 97-106.

Chapter 4 Kernel Supported Migration

Most migration implementations are done as modifications to the base operating system. In this section, we start by presenting migration implementations that were done on traditional, monolithic operating systems. Next, we present message-based operating systems that provide migration facilities, and we conclude with a discussion of task migration as implemented in a popular microkernel system.

Both MOSIX [Barak, 1985a, b; 1989a, b; 1993; 1998] and Sprite [Douglis, 1987; 1989; 1990; 1991; Ousterhout, 1994] implement process migration in traditional, full-featured UNIX-like systems. Both systems use remote procedure calls to communicate between the operating system kernels on different nodes and provide a high degree of transparency after migration.

Barak and Wheeler describe the **MOSIX** system in which all of the nodes in a cluster are presented as a *single system image,* that is, a process running on any node can access any other node's resources transparently [Barak, 1989b]. In fact, the user of the MOSIX system described has no knowledge of which node his or her processes are running on. Migration in this system is fully automatic and requires no special actions on the part of running applications: migration takes place at context switch time whenever the system determines that it will be worthwhile. Work on process migration at the MOSIX laboratory has been continuous since the early 1980s, with the latest versions of MOSIX being ported to BSDI [Barak, 1998] and Linux-based systems.

In the next paper, Douglis and Ousterhout describe the migration facility in **Sprite** [Douglis, 1991]. Each Sprite process has a designated home machine, on which the process logically executes regardless of its physical location. A UNIX-style *ps* command on a machine will list the processes that belong to that machine, regardless of where they are physically executing. Therefore, operations that involve process state, such as *fork* and *exit*, are forwarded to the home machine via kernel-to-kernel RPC. They also affect process state on the current machine. Some calls are handled independently, such as memory allocation and those relating to the distributed file systems. In Sprite, migration is typically invoked at *exec*-time by enhanced applications, such as a parallel *make* utility, but nearly any process is capable of invoking migration. The system automatically migrates processes to their home machines when a workstation is no longer idle. Work on Sprite ended in the early 1990s.

Message-based systems are a variation on traditional operating systems. In message-based systems, messages are used to provide access to system services. By forwarding these messages to remote nodes, these systems provide a relatively simple model for supporting higher level distributed mechanisms, like process migration. However, supporting distributed messaging does not eliminate complexity; rather, it moves the complexity down into the distributed messaging implementation itself [Douglis, 1991].

Artsy and Finkel present an early message-based migration implementation in the **Charlotte** system [Artsy, 1987; 1989]. In Charlotte, the migration mechanism and policy are cleanly separated to allow experimentation with different distributed algorithms and load balancing strategies. The system design supports fault resilience and transparent migration.

Next, Zayas describes migration in the **Accent** system [Rashid, 1981; 1986; 1987a, b]. Accent, which is also designed as a distributed operating system, attacks the performance of its migration implementation by using *copy on reference* to transfer the address space. This technique allows a process to migrate to a new node without copying all of its existing address space at the same time. This mechanism speeds the act of migration, but also creates a residual dependency between a process and its source node after migration.

Theimer, Lantz, and Cheriton present the implementation of the **V** kernel's migration scheme, which stresses transparency, minimal interference between program operation and migration, and leaves minimal residual state on the source node [Cheriton, 1988; 1990; Theimer, 1985]. One of their most innovative ideas is to *precopy* the address space of a running process before migration. The system iteratively performs this precopying until it has few pages left to move during the actual migration. The major advantage of this technique is that the amount of time that the process is "frozen" during migration is minimized, which avoids introducing degraded response time to the program's user or triggering time outs for time critical operations as, for example, when responding to network requests.

Microkernel-based systems, like **Mach** [Accetta, 1986], take the separation of policy and mechanism one step further than message-based systems. Milojičić, et al., present a migration mechanism for Mach tasks [Milojičić, 1993a, b; 1994]. In Mach, a task is a low-level abstraction that does not provide all of the higher level semantics associated with a full-fledged UNIX process. File systems and other traditional abstractions are implemented by user-space system daemons. The task migration mechanism proves useful for load distribution at the microkernel level. However, for UNIX process migration, this mechanism needs to be extended to migrate the UNIX process state as was done in OSF1/AD [Paindaveine, 1996].

Other kernel-supported migration systems include Locus [Popek, 1981; 1985; Walker, 1983; 1989], one of the rare systems supporting migration that became a commercial product. The commercial version, known as the Transparent Computing Facility (TNC), runs on IBM's AIX operating system and on commercial derivatives of the OSF1/ AD system [Bryant, 1995; Paindaveine, 1996; Zajcew, 1993]. A second, non-commercial migration mechanism exists for the OSF/1 AD systems that extends the task level migration to support migration of full-blown UNIX processes [Paindaveine, 1996]. Louboutin implemented a version of migration for the MINIX operating system [Louboutin, 1991]. Other message-passing operating systems supporting process migration include Amoeba [Steketee, 1994] and Galaxy [Sinha, 1991]. Several other microkernel-based systems also support migration, including RHODOS [Zhu, 1992], Arcade [Cohn, 1989; Tracey, 1991], Chorus [Philippe, 1993], and Birlix [Lux, 1993]. Other systems supporting process migration include Choices [Rousch, 1996] and work by Dediu and others. [Dediu, 1992]. Eskicioglu [Eskicioglu, 1990], Hac [Hac, 1986], and Nuttal [Nuttal, 1994] survey systems supporting process migration.

4.1 Barak, A. and Wheeler, R., MOSIX: An Integrated Multiprocessor UNIX

Barak, A. and Wheeler, R., "MOSIX: An Integrated Multiprocessor UNIX," *Proceedings of the USENIX Winter 1989 Technical Conference*, pp. 101–112, February 1989.

MOSIX: An Integrated Multiprocessor UNIX

Amnon Barak and Richard Wheeler

Department of Computer Science
The Hebrew University of Jerusalem
Jerusalem 91904, Israel

amnon@humus.huji.ac.il

ABSTRACT

MOSIX is a general-purpose **M**ulticomputer **O**perating **S**ystem which **I**ntegrates a cluster of loosely connected, independent computers (nodes) into a single-machine UNIX* environment. The main properties of MOSIX are its high degree of integration and the possibility of scaling the configuration to a large number of nodes. Developed originally for a network of uniprocessor nodes, it has recently been enhanced to support nodes with multiple processors. In this paper we present the hardware architecture of this multiprocessor workstation and the software architecture of the MOSIX kernel. We then describe the main enhancements made in the multiple-processor version and give some performance measurements of the internal mechanisms of the system.

1. INTRODUCTION

This paper describes a symmetrical multicomputer operating system, called MOSIX, which integrates a cluster of independent computers into a single-machine UNIX environment. By restructuring the UNIX kernel into machine-dependent and machine-independent parts, each MOSIX kernel isolates the user's processes from the specific processor in which they execute, while at the same time providing these processes with the standard UNIX interface [1,2]. In order to provide efficient network-wide services, different MOSIX kernels interact at the system-call level. This means that processes are executed in a site-independent mode and that all system calls are executed in a uniform way, regardless of the initiating site and the site which has the target object. The organization of the MOSIX kernel allows a unified, network-wide interaction among heterogeneous computers, i.e., it can be implemented on any hardware. Another outcome of this organization is the possibility for process migration among (homogeneous subsets of) processors to improve the response time.

The other main property of MOSIX is the possibility of scaling the configuration to a large number of processors. This is achieved by ensuring that all kernel interactions involve at most two processors, and by limiting the network-related activities at each node, using probabilistic algorithms. In other words, the design of the internal management and control algorithms in MOSIX is such that each processor has a fixed amount of overhead, regardless of the size of the network.

The hardware configuration for MOSIX is a cluster of loosely-coupled, independent computers (nodes), which may consist of any combination of uniprocessor nodes, shared-memory multiprocessors, and shared-device multicomputers, interconnected by a communication link. This communication link is used for kernel-to-kernel interaction. In order to provide high reliability and efficiency and to support scaling, MOSIX uses private communication protocols which have been carefully

* UNIX is a trademark of AT&T.

designed to guarantee bus performance and reliability over a local area network (LAN).

The original version of MOSIX , called MOS, was compatible with UNIX Version 7 [3]. It was developed in 1982 for a cluster of PDP-11 computers interconnected by ProNET-10, a 10 MBit/sec LAN. This system was ported in 1984 to a cluster of CADMUS/PCS, MC68000-based computers. Later versions of MOSIX were redeveloped for Digital's VAX and National's 32000 families of computers connected by Ethernet. These versions are compatible with AT&T UNIX Version 5.2. In all of these configurations, each node is a uniprocessor, independent UNIX system, with complete hardware and software facilities, while the whole cluster behaves like a single UNIX machine.

Recently, MOSIX was enhanced to handle nodes with several processors. As in the uniprocessor case, MOSIX can be used to integrate a cluster of several such nodes. However, in this recent implementation, each node is a multiprocessor system with several independent processing elements, each with its own local memory. This paper deals with the architectural and software enhancements of this multiprocessor system. Since in our configuration several processors are placed in the same enclosure, we call this multiprocessor node a "workstation". The term "processor" refers to a single processing element within a workstation.

The main characteristics of the workstation MOSIX are:

- **Replicated kernel**: the system kernel is replicated in each processor
- **Transparency**: the bus and the network are completely invisible at the user level
- **Decentralized control**: each workstation makes all its control decisions independently
- **Autonomy**: each workstation is capable of operating as an independent system
- **A unified file system**: the file system consists of one tree with a replicated 'super-root'
- **Scaling**: probabilistic algorithms are used for system management and network related overhead at each node is limited
- **Adaptive load balancing**: the system initiates process migration within the workstation and among different workstations to improve performance
- **Dynamic configuration**: workstations may be added or removed at any time and processors may fail with minimal effect on the system
- **Compatibility**: the system is compatible with AT&T UNIX Version 5.2

Some details about these properties are given in Section 3. More details can be found in our earlier papers and reports [3,4,5,6].

The motivation behind the MOSIX project has been research and development in distributed operating systems, including the exploration of kernel architectures, internal-system mechanisms, communication protocols, control algorithms, scaling and the support of concurrency, and multiprocessing. The resulting system is intended for large multiuser environments, for users executing a large number of independent tasks, e.g., multiple windows, and for applications that can benefit from concurrent processing with coarse or medium granularity using a large number of nodes [5]. Other projects that address the scaling issues are Grapevine [13], a distributed registry system which serves as a repository of naming information for an electronic mail system, and Andrew [10], a large distributed environment built at Carnegie Mellon University. Other projects with similar goals to MOSIX are Locus [12] developed at UCLA, the V kernel [7] developed at Stanford University, and Sprite [11], developed at UC Berkeley.

This paper is organized as follows: in Section 2 we describe the architecture of the workstation and the specific enhancements made at the board level to support multiprocessing. Due to the internal architecture of the MOSIX kernel, relatively few changes were necessary in order to generalize the uniprocessor version into the workstation system. In Section 3 we describe this architecture and present the mechanism for performing remote operations between different kernels. Some of the important characteristics of MOSIX are also described. In Section 4 we describe the enhancements made to the workstation version of MOSIX, and in Section 5, we give performance measurements of system calls, data transfers and process migration among the processors of a workstation and between workstations. Our conclusions are given in Section 6.

2. THE HARDWARE CONFIGURATION

The hardware configuration for MOSIX consists of a cluster of workstations connected by the ProNET-80, an 80 Mbit/sec token-ring-based LAN. This network is used for internal, kernel-level communication among the workstations, while an Ethernet LAN connected to at least one workstation in the cluster is used for external, TCP/IP communication. The workstation consists of up to eight processors, each with its own local memory and a floating-point unit. All the processors of a workstation share I/O devices and communication controllers over a VME bus, which can also support an optional shared memory. The rationale behind this architecture is an improved cost/performance ratio over the single-processor workstation.

The specific processor used is National's NS32532 microprocessor. The current measurements were carried out on a configuration with two 30MHz (10 MIPS) processors and several 25MHz (8 MIPS) processors. Each processor has 4 MBytes of local memory, a 64-KByte on-board cache, 2 serial lines, a PROM, and an NS32381 floating-point unit. A Weitek WTL3164 floating-point chip and an additional 12 MBytes local memory board are optional. The board level, called the VME532, is a 2-board set incorporating a full VME bus interface, including a master-slave (A32/D32), a location monitor, and a 4-level system controller. In addition, all read/write operations to I/O devices are non-cacheable to both on-chip and external caches. This ensures that all data transfers from/to I/O devices will directly reach the device.

The VME532 supports multiprocessing in configurations that include up to 16 processors in one VME card cage. The specific enhancements that were made at the board level include:

1. A unique ID for each processor in the range 0-F. The ID is set by a thumb-wheel switch that can be read by the software.
2. Each processor has its own cacheable, dual-ported local memory. This memory is mapped into the VME address slot determined by the processor ID. The local memory of each processor can be accessed by other VME masters (processors) in a non-cacheable mode.
3. A location monitor that interrupts the CPU for efficient inter-processor interrupts.
4. A VME bus watcher that maintains consistency between the main memory and the caches.
5. 240-252 MBytes of shared VME address space, cacheable by all the processors.

The multiprocessing support implies that the bus activities, including the I/O interrupts, are available to all the processors. It is the duty of the software level to coordinate the management and synchronization of these activities. One way to organize these activities is by using a software-level master/slave organization. The outcome of such an organization is improved reliability, since any processor can assume the role of the master. If this master crashes, then another processor can take on its role. Note that in this case the workstation must be re-booted. Also note that if a slave processor crashes, the system is not damaged, but processes running at this slave are lost. More details about the software organization are given in Section 4.

3. THE ARCHITECTURE OF THE KERNEL

MOSIX is a symmetrical multiprocessor operating system obtained by restructuring the UNIX kernel while preserving the standard UNIX kernel interface. Each processor running MOSIX is an independent UNIX machine. It has a complete copy of the kernel, with the possible exception of device drivers. In this section we describe the architecture of the MOSIX kernel for a multiprocessor consisting of many uniprocessor independent computers.

The MOSIX kernel consists of three parts: the machine-dependent part, called the *lower-kernel*, the machine-independent part, called the *upper-kernel* and the communication layer, called the *linker* [3]. The *lower-kernel* contains routines that access local resources such as device drivers for local disks, and routines which access file and process structures. The *lower-kernel* is tightly coupled with the local processor, it has complete knowledge of all the local objects, and it can access only these local objects (or objects that have migrated to that processor). The *lower-kernel* does not have any knowledge about processors other than the one in which it executes, and it does not distinguish between requests originating in its local processor or requests originating in other processors. The *upper-kernel* executes in a machine independent mode. On one hand, it provides the standard UNIX system call interface. At the same time, the *upper-kernel* does not know the processor ID in which it executes, but it has a complete knowledge about the locations of all the objects it handles. For example, when a process executes a system call, the *upper-kernel* performs the preliminary processing of the parameters, e.g. calling *namei* to parse a path name into a *universal inode,* or checking that the user has permission to access a certain object. Eventually, the *upper-kernel* calls the relevant remote kernel procedure using the remote procedure call (RPC) mechanism to complete the service. For example, in the following the remote kernel procedure "proc_name" is executed with the parameters specified in "param_list":

Rproc_name (machine_id, param_list).

Next, the call is passed by the *upper-kernel* to the *linker* which decomposes the RPC into the procedure name and the parameter list. The *linker* examines the first parameter of the call to determine where the call needs to be executed. If it is a remote call, the *linker* encapsulates the call into a message and sends the message over the network. If the call can be executed locally, the *linker* invokes the local *lower-kernel* procedure directly. On the target machine, an *ambassador process*, a lightweight kernel process, executes the appropriate *lower-kernel* procedure for the calling process. The result of the system call is then encapsulated into another message and returned to the calling node.

The communication protocol used by the linkers is designed to provide a high degree of reliability. This results from the fact that in MOSIX the information passed between the linkers is critical. For example, a change of one bit could cause the removal of a file rather then closing it. The specific protocol that is used includes a CRC checksum which is validated by the receiver, a software-initiated acknowledgement message in addition to the hardware acknowledgement provided by the token-ring network, a message ordering for each pair of nodes, and a password for each message.

In order to support dynamic reconfiguration, e.g. removing or adding new machines, each remote access generates a new remote kernel call which returns a special error code if the node is currently unreachable. The system does not require any special action to incorporate a new node: the first remote call to the newly joined node is simply sent by the system in the usual way, and the connection is created dynamically.

Remote system calls which need to transmit large amounts of data in addition to the result of the RPC use the *funnel* mechanism to copy data from the memory space of one machine to another. The *linker* handles the implementation of funnels by breaking up large blocks of data into message-size pieces at one end, sending the messages over the network, and reunifying the data in the proper order on the receiving machine. The data is then copied by the *linker* of the receiving node directly to the specified user's address space. For example, for the *read* system call, the *upper-kernel* sets up an input funnel on the local machine before calling the remote *Rread* system call. If the system call accesses a remote file, the *linker* routes the call to the target machine, and an ambassador process there calls the appropriate kernel procedure for reading a file. As each logical file block is read, the data is placed into the remote end of the funnel and passed back to the initiating machine's *linker*. After the call is completed, the returned status of the system call is encapsulated by the remote *linker* and passed back to the calling machine.

The net result of this kernel architecture is that the user's process is completely isolated from the processor in which it is currently running. This isolation requires all kernel references to be made in a machine-independent (universal) mode, a unified naming scheme, and good performance from the RPC mechanism. Note that this organization can be applied to any version of UNIX and can easily support heterogeneous hardware. The only parts which has to be implemented for each machine type is the *lower-kernel*. Another outcome of this organization is a simple process-migration mechanism.

The main characteristics of the uniprocessor MOSIX include network transparency, decentralized control, dynamic configuration, a distributed file system, and load balancing by dynamic process migration. Details about these characteristics can be found in [3]. For completeness, we give below a short description of the file system [4], the load-balancing mechanism [6], and some of the probabilistic algorithms.

3.1. The file system

As in UNIX, the file system of MOSIX is a forest, with several disjoint trees. Each tree is a complete UNIX file system, with regular files and devices as leaves, and directory files as internal nodes. Each user is assigned to one of these trees, and the root of that tree is the root directory of this user. In this scheme, a user always has the same root (home) directory at all login sessions, but different users may have different root directories [4].

The MOSIX file system uses a super-root, **"/..."**, as a network-wide root. The super-root is replicated in all the nodes, and it is supported by the kernel. When addressing a file using an absolute path, mainly on a tree which differs from the user's root directory, the user prefixes the super-root to the machine name and then adds the usual UNIX path. For example, **"/.../m2/etc/passwd"** is the absolute path name for the password file on machine number 2. One version of MOSIX uses a special file type which stores the remote-machine number in its inode. When such an inode is accessed by the kernel, the inode of the special file is automatically replaced with the inode of the root directory of the remote node indicated. For example, if the special file **"/usr/systems/bert"** is created as a remote escape to the file system on machine m2, the path **"/usr/systems/bert/etc/passwd"** refers to the password file on machine m2. Note that this change eliminates the need for a special, non-standard UNIX pathname syntax. Also, it allows the MOSIX file system to have arbitrarily placed links between nodes.

In MOSIX, the *inode* of an open file is held by the site at which the file resides. All remote *opens* are returned with a universal pointer to the file. This *universal file pointer* includes the identity of the machine in which the file resides, and it is used for future file accesses. We note that MOSIX does not support file migration. The garbage-collection algorithm is used to clean up allocated inode structures in case a failure occurs in the calling node.

3.2. Load balancing

In MOSIX, load-balancing is carried out by dynamic process migration [6]. As a result of the system architecture, a process running under MOSIX is not sensitive to its physical location: system calls which access resources not located on its current node are automatically forwarded by the *linker* to the remote node. In order to support process migration, the hardware configuration must consist of clusters of homogeneous processors. Process migration is allowed only among processors with the same instruction set, because application tasks may be assigned or migrate arbitrarily among the nodes of the cluster. Note that all of the other functions of MOSIX may be supported among heterogeneous processors.

In MOSIX, each processor exchanges its own local load estimate with that of a set of randomly selected processors every unit of time (one second in the current implementations). Load estimates received from other processors are kept in a *load vector* and are "aged" to reflect their decreasing relevancy. The algorithms used for load balancing are probabilistic, and are intended to provide each node with the latest, up-to-date information about the loads of other nodes. As proved in [9] this is achieved in $O(\log N)$ units of time for an N-processor system. Another goal of these algorithms is to overcome a node failure and to be responsive to dynamic configuration.

The decision to migrate a process is based on many parameters, including the past profile of the process, the amount of local-versus-remote I/O, the relevant locations of this I/O, the relative load of the sending and receiving nodes, the size of the process, etc. After a decision has been made to migrate a process, the target processor may still refuse to accept the migrating process if it so desires (due to local circumstances). Processes which have a history of "forking" new child processes are given migration preference by the algorithm to further speed up the even distribution of the load. Also, processes with a history of I/O operations to some specific set of nodes are given priority for migration to one of these nodes.

Several heuristics are used to avoid over-migration. To begin with, the load-balancing algorithm does not migrate a process unless it has accumulated a minimal amount of CPU time on the current processor. Also, the load estimate sent out to other processors, or *export load*, is slightly higher than the true local load. Each process that has been allowed to migrate to a processor is immediately counted in the receiving processor's *export load* estimate and removed from the local load estimate of the sending processor. New requests for migration are accepted only if the difference between the revised load estimates is still greater than a known threshold value. The actual migration starts by passing the necessary information required to re-build the page table, followed by the data pages which were changed, i.e., the "dirty pages", including in-core or swapped-out pages. The rest is either mapped to the original file, or defined as "all zeroes". Note that after the process migration is completed, no "tail" remains in the sending site.

3.3. Distributed probabilistic algorithms

In order to limit the network-related management activities performed by each node, the internal control algorithms of MOSIX are distributed. These probabilistic algorithms attempt to achieve near optimal performance at the fraction of the cost required by a centralized algorithm. For example, the load-balancing algorithm described above ensures that each processor has sufficient, but limited amount of information about other processors. This method allows load-balancing interactions between many different pairs of processors while at the same time it eliminates flooding of a single processor, for example, immediately after it joins the network.

Another example of a distributed algorithm is the remnant-collection mechanism. Unlike the more traditional garbage-collection mechanism, the MOSIX algorithm is intended to remove any remnants of a process that lost its objects or communication links due to a failure in a remote node. For this purpose, all the resources allocated in MOSIX include a timer and a *"keep-alive"* mechanism which is supported by the kernel. The timer is reset at regular intervals and upon each interaction between the process and the object. If the timer expires, then the object is removed, since the process that uses this object does not exist any longer.

4. THE WORKSTATION VERSION

The workstation version of MOSIX, like all previous versions, is a symmetric system, in the sense that each workstation is an independent computer with full UNIX capability. This symmetry can be extended to the processor level if sufficient I/O and communication controllers are available. In order to simplify the organization of the software level and to use shared I/O and communication devices, a master/slave organization of the workstation software was chosen. This implies a point of asymmetry since one processor (processor 0) assumes the role of a **master** while the remaining processors are **slaves**, i.e., they need the master processor to coordinate some of their I/O and communication requests. We note that in the implementation of the workstation kernel a great effort was made to minimize this dependency. For example, slave-to-slave (within the same workstation) communication and load balancing is done without the interference of the master processor. Also, in swapping, data is directly transferred between the slave processor and the disk, thus reducing the amount of overhead required from the master processor.

The main enhancements of the workstation kernel were made in the communication interface at the *linker* level and in the boot procedure. Recall that the local memory of each processor can be addressed directly by all the other processors (in the same workstation). For example, when the workstation is rebooted, the master processor loads the kernel from the disk, then copies the kernel to each slave processor using the direct-addressing mode. Another mechanism that was modified in the workstation version is the handling of the file-system buffer cache. In the current version, the master processor handles all of the file system operations since the slave processors do not have a file system of their own. Similarly, a new process ID (PID) is assigned only by the master processor.

Only minor modifications were necessary to enhance the single-machine kernel to the multiprocessor version. In fact, due to the architecture of the kernel, most of the kernel routines were used without any change. The only modifications that were made resulted from the hardware architecture of the workstation. More specifically, changes were made to the I/O and communication routines, and to the boot function described earlier, and to the system administration. The other main changes that were made were:

1. Remote swapping allows slave processors to swap directly to the disk in the same workstation. In this case the master processor serves only as coordinator, but the actual addressing and block allocation is done by the slave. For this purpose each slave processor has an allocated swap space on the disk.
2. A multi-level load-balancing scheme between master and slave processors. The algorithms of the load-balancing mechanisms were tuned to include master-to-master, master-to-local-slave, master-to-remote-slave, and slave-to-slave migration (in the same workstation and in different workstations). Each case requires a different set of migration parameters. These parameters were provided by the automatic tuning mechanisms developed for this purpose.
3. A two-level *linker* for master and slave-oriented commands. The linker was enhanced to include all possible combinations of data transfers and commands between local and remote master and slave processors.

A TCP/IP package for external communication, including remote terminal access, file transfer, and remote command execution was also added. The network level supports the Internet Protocol (IP) and the Internet Control Message Protocol (ICMP). The host level supports the Transmission Control Protocol (TCP) and the User Datagram Protocol (UDP). Applications supported include the TELNET protocol for connecting remote terminals (RFC), and the File Transfer Protocol (FTP). Also supported are the UNIX special services such as *rcp* (remote file copy), *rlogin* (remote login), and *rsh* (remote execution of commands). The application programmer has access to the host level protocol using a BSD UNIX 4.2-like socket software library.

5. PERFORMANCE MEASUREMENTS

In this section we start by comparing the time required to perform local vs. remote system calls, both between different processors of the same workstation and between different workstations. The system calls were measured by a set of benchmarking programs which execute each particular call on a remote object and a local object 100,000 times. We also present measurements of the rate of data transfers and process migration. The section concludes with a performance analysis of a distributed implementation of the traveling-salesman problem.

First, we note that due to the architecture of the MOSIX kernel, many system calls are performed locally by each processor without any slowdown penalty. Examples of system calls of this type are *getpid, signal* and *time.* File-related system calls are always performed by a master processor because of the architecture of the workstation. File-related system calls which access a file on some remote workstation are handled by the remote workstation's master. The times required to perform the most frequently used file-related system calls are given in Table 1. All times are given in milliseconds. The first column shows the time required by a process running on the master processor to access a local file. The second column shows the time required by a process running on a slave processor in the same workstation to access the same file, i.e., via the VME bus. The third column shows the time required by a process running on one master processor to access a file located in another workstation, i.e., master-to-master, using the ProNET-80. The fourth column shows the slowdown factor associated with the VME bus (master-slave in the same workstation). It is computed as the slave time (column 2) divided by the master time (column 1). The fifth column is the remote-master time (column 3), divided by the slave time (column 2). In both cases, the involved processors ran at 30MHz.

System call	Local call Master	Slave to local Master	Master to remote Master	VME Slowdown Ratio	ProNET Slowdown Ratio
read (1KByte)	0.34	1.36	1.93	4.00	1.42
write (1KByte)	0.68	1.65	2.33	2.43	1.41
open & close	2.06	4.31	5.19	2.09	1.20
fork (256KByte)	7.80	21.60	23.10	2.77	1.07
exec (256KByte)	25.30	51.50	53.80	2.04	1.04

Table 1: Local vs. remote system call execution times (ms)

One useful measure of the overall effect of remote access on a process is a *weighted average* of the slowdown factors. Using the frequencies of system calls measured in [8], the frequency of each call is multiplied by the slowdown factor associated with remote execution. The results show a weighted slowdown factor of 2.8 for system calls executed by the slave versus the same set of system calls executed by the master. Comparison of the time required to perform system calls between a master and slave in the same workstation, using the VME bus, and the time required between two masters in different workstations, using the ProNET LAN, shows a slowdown of only 32%. Recall that the LAN protocol uses a software-generated CRC at both the sending and receiving nodes, as well as an acknowledgement message. Thus it is estimated that in a network that supports this protocol at the hardware level there would be no slowdown factor, other then the one generated by the VME bus. This indicates that the configuration can be scaled up with only a slight degradation in performance.

In Table 2 we give the speeds of several data-transfer mechanisms of MOSIX. First, we give the rate of a memory-to-memory copy using the *funnel* mechanism between processes running on the same processor. The next figure shows the throughput of a memory copy between a process that runs on a master processor and a process that runs on one of its slave processors. The last figures show the speed of process migration.

System Throughput	KByte/S
Funnel Copy (**Same processor**)	11,299
Memory Copy (**Master to Local Slave**)	1,820
Process Migration (**Master to Local Slave**)	1,235
Process Migration (*Slave to Local Slave*)	1,045
Process Migration (**Master to Remote Master**)	794
Process Migration (**Master** to *Remote Slave*)	638
Process Migration (*Remote Slave to Remote Slave*)	530

Table 2: System data throughput (KByte/S)

In the above chart, 30MHz nodes are printed in boldface while 25MHz nodes are printed in italics. In every case, the involved masters were running at 30MHz. Note that a migration between two slaves not in the same workstation involves four processors: two masters and the slaves themselves. We note that the speed of data transfer between a processor and a common memory (on the VME bus) is expected to be around 3MByte/sec. As indicated earlier, the reliability of the network is essential, since in MOSIX the LAN is used for kernel-to-kernel communication. The results of Table 2 imply that the speed of process migration between different workstations (master to remote master) using the LAN is about 35% slower than the speed of local migration (master to local slave). This difference in migration speeds is expected to diminish when faster LANs become available. All of these results have ramifications in scaling MOSIX, since the different migration speeds are taken into account by the load-balancing mechanism when a process is considered for migration to a remote processor.

The next two tables show the relation between the granularity of the computation, the frequency of communication, and the speedup. Using a 6-processor configuration in 2 identical workstations, the computation included one master process assigning identical computational tasks to slave processes. Upon the completion of each task, the slave process communicates with the master and is assigned a new task. The amount of computation done in each iteration was increased from 0.1 to 10 seconds, resulting in a decrease in the communication and management overhead of the master process. In each case we allowed two different initial process assignment schemes. The first uses MOSIX load balancing, and the second uses static assignment by the master process. Note that in the first case, each process was executed for at least one second in the processor in which it was created before it was migrated. For comparison, we note that the total execution time of the benchmark when executed as a single process with no communication overhead was exactly 60 minutes.

In Table 3 we list the results of executing a master and six slave processes. The first column lists the amount of work assigned to each slave by the master process. Column 2 is the elapsed execution time for the whole benchmark. Column 3 lists the total overhead (user + system) time which resulted from the work distribution and the communication overhead. Column 4 is the speedup, defined as the total execution time using a single machine divided by the elapsed time using six processors. Utilization is defined as the ratio between the single-machine execution time and the measured overhead (column 3), divided by the product of the elapsed time and the number of processes. Note that utilization refers to CPU utilization for all work done (including overhead time), while speedup refers to actual work done (not including overhead).

Comparison of the task granularity shows that even with a small unit of work, the speedup and utilization obtained are quite good. Comparison of MOSIX load balancing and the optimal static assignment shows the efficiency of the MOSIX scheme. This is also reflected in the high degree of utilization obtained. The corresponding results for a master and five slave processes, each executing on a different processor are given in Table 4. Comparison between the two tables shows a consistent increase in the speedup and utilization ratios, with only 1-2% decrease in the utilization when the number of processes is increased from 6 to 7.

	Automatic Process Migration				Static Process Assignment			
Work Unit *(sec)*	**Elapsed** *min:sec*	**Overhead** *user+sys (sec)*	**Speedup**	**Utiliz.**	**Elapsed** *min:sec*	**Overhead** *user+sys (sec)*	**Speedup**	**Utiliz.**
0.1	11:50	14.2+159.2	5.08	88%	11:38	12.0+157.2	5.16	90%
0.2	11:00	6.5+80.9	5.45	93%	10:50	9.5+80.1	5.54	94%
0.3	10:43	5.2+56.5	5.61	94%	10:34	5.0+55.8	5.68	96%
0.4	10:40	6.7+39.7	5.63	94%	10:26	5.7+40.7	5.75	97%
0.5	10:30	3.8+33.8	5.71	96%	10:22	5.0+32.9	5.80	97%
0.7	10:29	5.3+24.0	5.72	96%	10:16	2.3+25.6	5.84	98%
1.0	10:20	2.3+17.4	5.81	97%	10:12	2.4+18.0	5.89	98%
1.5	10:17	5.0+12.7	5.83	97%	10:09	3.1+13.0	5.92	98%
2.0	10:14	4.3+9.9	5.86	98%	10:08	3.8+10.3	5.94	99%
3.0	10:14	4.9+6.5	5.88	98%	10:06	3.2+6.9	5.94	99%
5.0	10:16	5.0+4.9	5.85	97%	10:06	4.4+4.7	5.95	99%
10.0	10:17	1.0+5.4	5.84	97%	10:04	1.0+9.3	5.96	99%

Table 3: Task granularity benchmark, master + 6 slave processes

	Automatic Process Migration				Static Process Assignment			
Work Unit *(sec)*	**Elapsed** *min:sec*	**Overhead** *user+sys (sec)*	**Speedup**	**Utiliz.**	**Elapsed** *min:sec*	**Overhead** *user+sys (sec)*	**Speedup**	**Utiliz.**
0.1	13:49	6.6+127.5	4.35	90%	13:43	7.8+130.4	4.37	90%
0.2	12:57	1.2+67.5	4.63	94%	12:53	4.7+66.2	4.66	94%
0.3	12:41	4.7+41.6	4.74	95%	12:36	5.3+45.1	4.77	96%
0.4	12:35	3.7+35.3	4.77	96%	12:28	5.7+33.9	4.82	97%
0.5	12:27	2.7+29.1	4.82	97%	12:23	3.0+28.0	4.85	97%
0.7	12:22	3.2+19.5	4.86	97%	12:16	5.1+18.3	4.89	98%
1.0	12:19	2.0+16.9	4.88	97%	12:13	0.8+15.3	4.92	98%
1.5	12:23	4.2+10.2	4.85	97%	12:09	3.6+10.4	4.94	99%
2.0	12:13	4.8+7.9	4.91	98%	12:08	3.3+8.3	4.95	99%
3.0	12:13	4.2+7.0	4.92	98%	12:06	2.2+6.2	4.96	99%
5.0	12:15	2.4+5.3	4.90	98%	12:06	1.8+5.0	4.97	99%
10.0	12:15	1.0+8.4	4.90	98%	12:05	2.7+3.4	4.97	99%

Table 4: Task granularity benchmark, master + 5 slave processes

The last table shows how effectively some CPU-bound applications can utilize the load balancing algorithm of the system. A distributed implementation of the traveling-salesman problem was developed. It uses a master process which forks several subprocesses. The processes use System V messages for interprocess communication. The example shown is fairly large, requiring almost 40 minutes of CPU time. Each of the executions involves one master process and four slave processes.

Number of Processors	Total CPU Time	Elapsed Time	Speedup
1	2344.8	2348.4	1.00
2	2293.2	1178.6	1.99
3	2265.1	790.0	2.97
4	2264.8	598.1	3.93

Table 5: The traveling-salesman problem

The total CPU time given in Table 5 is the number of CPU seconds spent in user mode by all of the processes. As expected, this number is approximately the same regardless of the number of processors working on the problem. The elapsed time given is the real time required to execute the program. The speedups gained are quite impressive: all are linear. These speedups are obtained through the use of the automatic load-balancing mechanism of the MOSIX kernel.

6. CONCLUSIONS

In this paper we describe a symmetrical multiprocessor operating system which integrates a set of processors that share a common VME bus into a single-machine UNIX workstation. This system can further integrates several such workstations, via a local area communication network into a single-machine UNIX environment. With the present tuning, and the specific hardware described, the maximal configuration is 496 processors, configured into 62 workstations, with 8 processors each. Each workstation is expected to deliver a theoretical computing power of 64 MIPS, with as much as 8 MIPS for each process. Measurements carried out with a few 30MHz (10 MIPS) processors indicate that the system's performance will improve by 20 percent when all of the 25 MHz processors are replaced. Thus the expected total computing power of the maximal configuration is 4 - 5 GIPS. The cost/performance ratio of this workstation is expected to be $1,000/MIP. We note that the restriction on the number of workstations is imposed by the size of the UNIX process ID, which was preserved as a *short* integer. If this restriction were removed, then the number of workstations may be further increased.

As of the summer of 1988, a system of 3 workstations, each with 4 processors was operational. A workstation with 8 processors has also been tested. Our target configuration is 8 workstations, with 64 processors.

The software system is fully compatible with AT&T UNIX Version 5.2 including virtual demand paging for both local and remote processes (migrated to another processor) as well as the TCP/IP communication software. The development of the multiprocessor version kernel (from the uniprocessor version) required 4 man/months. In addition to the system kernel, an extensive set of software development tools and a powerful kernel debugger were also developed.

When used for a CPU-bound application, the system has shown a linear speedup with the number of processors. As can be expected, in I/O and communication-bound processes a lower speedup was obtained. In all cases, the load-balancing algorithm was a major contributor to the high utilization of the system when multiple processes were executed.

ACKNOWLEDGEMENT

The authors wish to thank A. Shiloh for the implementation, National Semiconductor Corp. for the technical support and G. Shwed for performing the granularity benchmarks.

REFERENCES

[1] *AT&T System V Interface Definition,* AT&T Ed., 1985.

[2] Bach M. J., *The Design of the UNIX Operating System,* Prentice-Hall, 1986.

[3] Barak A. and Litman A., "MOS: A Multicomputer Distributed Operating System," *Software Practice & Experience,* Vol. 15, No. 8, pp. 725-737, Aug. 1985.

[4] Barak A., Malki, D. and Wheeler R., "AFS, BFS, CFS... or Distributed File Systems for UNIX," *Proc. EUUG Conference on Dist. Systems,* pp. 461-472, Manchester, Sept. 1986.

[5] Barak A. and Paradise O. G., "MOS - Scaling Up UNIX," *Proc. Summer 1986 USENIX Conference,* pp. 414-418, Atlanta, GA, June 1986.

[6] Barak A. and Shiloh A., "A Distributed Load-balancing Policy for a Multicomputer," *Software Practice & Experience,* Vol. 15, No. 9, pp. 901-913, Sept. 1985.

[7] Cheriton D.R. and Zwaenepoel W., "The Distributed V Kernel and its Performance for Diskless Workstations," *Proc. of the Ninth Symposium on Operating Systems Principles,* pp. 129-140, Oct. 1983.

[8] Douglis F. and Ousterhout J., " Process Migration in the Sprite Operating System," *Proc. 7-th International Conf. on Distributed Computing Systems,* pp. 18-25, Berlin, Sept. 1987.

[9] Drezner Z. and Barak A., "An Asynchronous Algorithm for Scattering Information Between the Active Nodes of a Multicomputer System," *Journal of Parallel and Distributed Computing,* Vol. 3, No. 3, pp. 344-351, 1986.

[10] Morris J.H., Satyanarayanan M, Conner M.H., Howard J.H., Rosenthal D.S. and Smith F.D., "Andrew: A Distributed Personal Computing Environment," *Communication of the ACM,* Vol. 29, No. 3, March 1986.

[11] Ousterhout J., Cherenson A.R., Douglis F., Nelson M.N. and Welch B.B., "The Sprite Network Operating System," *IEEE Computers,* Vol. 21, No. 2, Feb. 1988.

[12] Popek G.J. and Walker B.J., *The LOCUS Distributed System Architecture,* MIT Press, 1985.

[13] Schroeder M.D., Birrell A.D. and Needham, R.M., "Experience with Grapevine: The Growth of a Distributed System," *ACM Transactions on Computer Systems,* Vol. 2, No. 1, pp. 3-23, Feb. 1984.

MOSIX for Scalable Computing Clusters

Amnon Barak and Richard Wheeler

Both the MOSIX team and the MOSIX system continue to evolve and change [1]. The version of MOSIX described in this book documents the system as it stood almost ten years ago: a full featured system that requires extensive modifications to the existing operating system code.

As others describe in this book, maintaining this amount of changes is a never-ending task. The latest release of MOSIX tackles this problem by redesigning the dynamic process migration support to be less intrusive. It is designed for Networks of Workstations (NOW) [2] and Scalable Computing Clusters (SCC) [6] systems. SCC MOSIX supports preemptive process migration, dynamic load and memory balancing, network transparency and decentralized control. The system is built on the latest release of BSDI's BSD/OS. The release adds 10 new files, modifies 20 of the existing files and adds a few fields to the kernel's data structures. Clearly, this is a more manageable configuration.

Despite this more limited arena, SCC MOSIX [3, 4] continues to provide most of the features of the original MOSIX system. Each process is presented with the illusion that it is still running on its home machine, the node in the network where it was created. Each process is broken into a deputy process that contains system context and runs on the home node and a body process that contains user specific context information and migrates freely around the cluster. Since the system call interface is well defined, each system call that depends on site specific information is intercepted and routed to the deputy process at the home node. A special channel is established to optimize the communication between the deputy and the body.

This approach has the major advantage of limiting the impact of base operating system changes on the process migration mechanism. Effort can be invested in making the most frequent system calls non-node-specific, for example system calls for a distributed file system. The disadvantages of the deputy scheme are added overhead in the execution time of system calls and overhead added to store the local state information at the home site.

We contend that the trade off is beneficial: we have demonstrated performance gains both in a generic load balancing case [3] and in MOSIX versus PVM executions [5]. Readers are invited to visit the MOSIX home page for an up-to-date list of publications and updates on the evolution of the MOSIX system. The BSDI implementation is freely available to BSDI source code licensees.

References.

1. **http://www.mosix.cs.huji.ac.il**
2. Anderson, T.E., Culler, D.E., Patterson, D.A., "A Case for NOW (Networks of Workstations)," *IEEE Micro,* 159(1), pp. 54-64, February, 1995.
3. Barak A. and La'adan O., The MOSIX Multicomputer Operating System for High Performance Cluster Computing, *Journal of Future Generation Computer Systems,* 13(4-5), pp. 361-372, March 1998.
4. Barak A. and Braverman A., Memory Ushering in a Scalable Computing Cluster, *Proceedings of the IEEE Third Int. Conference on Algorithms and Architecture for Parallel Processing,* Melbourne, December 1997.
5. Barak A., Braverman A., Gilderman I. and La'adan O., Performance of PVM with the MOSIX Preemptive Process Migration, *Proceedings of the 7th Israeli Conference on Computer Systems and Software Engineering,* Herzliya, pp. 38-45, June 1996.
6. Barak A. and La'adan O., Experience with a Scalable PC Cluster for HPC, *Proceedings of the First Cluster Computing Conference,* Emory University, Atlanta, GA, March 1997.

4.2 Douglis, F. and Ousterhout, J., Transparent Process Migration: Design Alternatives and the Sprite Implementation

Douglis, F. and Ousterhout, J., "Transparent Process Migration: Design Alternatives and the Sprite Implementation," *Software-Practice and Experience*, 21(8):757–785, August 1991. Reproduced by Permission of John Wiley & Sons Limited.

Transparent Process Migration: Design Alternatives and the Sprite Implementation*

Fred Douglis[†]
John Ousterhout

Computer Science Division
Electrical Engineering and Computer Sciences
University of California
Berkeley, CA 94720

Summary

The Sprite operating system allows executing processes to be moved between hosts at any time. We use this *process migration* mechanism to offload work onto idle machines, and also to *evict* migrated processes when idle workstations are reclaimed by their owners. Sprite's migration mechanism provides a high degree of transparency both for migrated processes and for users. Idle machines are identified, and eviction is invoked, automatically by daemon processes. On Sprite it takes up to a few hundred milliseconds on SPARCstation 1 workstations to perform a remote *exec*, while evictions typically occur in a few seconds. The *pmake* program uses remote invocation to invoke tasks concurrently. Compilations commonly obtain speedup factors in the range of three to six; they are limited primarily by contention for centralized resources such as file servers. CPU-bound tasks such as simulations can make more effective use of idle hosts, obtaining as much as eight-fold speedup over a period of hours. Process migration has been in regular service for over two years.

Keywords: Process migration, Load sharing, Operating systems, Distributed Systems, Experience

Introduction

In a network of personal workstations, many machines are typically idle at any given time. These idle hosts represent a substantial pool of processing power, many times greater than what is available on any user's personal machine in isolation. In recent years a number of mechanisms have been proposed or implemented to harness idle processors (*e.g.*, References 1, 2, 3, 4). We

*This work was supported in part by the Defense Advanced Research Projects Agency under contract N00039-85-C-0269 and in part by the National Science Foundation under grant ECS-8351961. This paper appeared in Software–Practice & Experience, 21(8):757–785, August 1991.

[†]Author's present address: MITL, Panasonic Technologies Inc.; 2 Research Way, Third Floor; Princeton, NJ 08540-6628. Internet: douglis@research.panasonic.com.

have implemented process migration in the Sprite operating system for this purpose; this paper is a description of our implementation and our experiences using it.

By "process migration" we mean the ability to move a process's execution site at any time from a *source* machine to a *destination* (or *target*) machine of the same architecture. In practice, process migration in Sprite usually occurs at two particular times. Most often, migration happens as part of the *exec* system call when a resource-intensive program is about to be initiated. *Exec*-time migration is particularly convenient because the process's virtual memory is reinitialized by the *exec* system call and thus need not be transferred from the source to the target machine. The second common occurrence of migration is when a user returns to a workstation when processes have been migrated to it. At that time all the foreign processes are automatically *evicted* back to their home machines to minimize their impact on the returning user's interactive response.

Sprite's process migration mechanism provides an unusual degree of transparency. Process migration is almost completely invisible both to processes and to users. In Sprite, transparency is defined relative to the *home machine* for a process, which is the machine where the process would have executed if there had been no migration at all. A *remote process* (one that has been migrated to a machine other than its home) has exactly the same access to virtual memory, files, devices, and nearly all other system resources that it would have if it were executing on its home machine. Furthermore, the process appears to users as if it were still executing on its home machine: its process identifier does not change, it appears in process listings on the home machine, and it may be stopped, restarted, and killed just like other processes. The only obvious sign that a process has migrated is that the load on the source machine suddenly drops and the load on the destination machine suddenly increases.

Although many experimental process migration mechanisms have been implemented, Sprite's is one of only a few to receive extensive practical use (other notable examples are LOCUS [5] and MOSIX [6]). Sprite's migration facility has been in regular use for over two years. Our version of the *make* utility [7] uses process migration automatically so that compilations of different files, and other activities controlled by *make*, are performed concurrently. The speed-up from migration depends on the number of idle machines and the amount of parallelism in the task to be performed, but we commonly see speed-up factors of two or three in compilations and we occasionally obtain speed-ups as high as five or six. In our environment, about 30% of all user activity is performed by processes that are not executing on their home machine.

In designing Sprite's migration mechanism, many alternatives were available to us. Our choice among those alternatives consisted of a tradeoff among four factors: transparency, residual dependencies, performance, and complexity. A high degree of transparency implies that processes and users need not act differently after migration occurs than before. If a migration mechanism leaves *residual dependencies* (also known as "residual host dependencies" [3, 8]), the source machine must continue to provide some services for a process even after the process has migrated away from it. Residual dependencies are generally undesirable, since they impact the performance of the source machine and make the process vulnerable to failures of the source. By performance, we mean that the act of migration should be efficient and that remote processes should (ideally) execute with the same efficiency as if they hadn't migrated. Lastly, complexity is an important factor because process migration tends to affect virtually every major piece of an operating system kernel. If the migration mechanism is to be maintainable, it is important to limit this impact as much as possible.

Unfortunately, these four factors are in conflict with each other. For example, highly-transparent migration mechanisms are likely to be more complicated and cause residual dependencies. High-performance migration mechanisms may transfer processes quickly at the cost of residual dependencies that degrade the performance of remote processes. A practical implementation of migration must make tradeoffs among the factors to fit the needs of its particular environment. As will be seen in the sections below, we emphasized transparency and performance, but accepted residual dependencies in some situations. (See Reference 9 for another discussion of the tradeoffs in migration, with a somewhat different result.)

A broad spectrum of alternatives also exists for the policy decisions that determine what, when, and where to migrate. For Sprite we chose a semi-automatic approach. The system helps to identify idle hosts, but it does not automatically migrate processes except for eviction. Instead, a few application programs like *pmake* identify long-running processes (perhaps with user assistance) and arrange for them to be migrated to idle machines. When users return to their machines, a system program automatically evicts any processes that had been migrated onto those machines.

The Sprite Environment

Sprite is an operating system for a collection of personal workstations and file servers on a local area network.[10] Sprite's kernel-call interface is much like that of 4.3 BSD UNIX,[†] but Sprite's implementation is a new one that provides a high degree of network integration. For example, all the hosts on the network share a common high-performance file system. Processes may access files or devices on any host, and Sprite allows file data to be cached around the network while guaranteeing the consistency of shared access to files.[11] Each host runs a distinct copy of the Sprite kernel, but the kernels work closely together using a remote-procedure-call (RPC) mechanism similar to that described by Birrell and Nelson.[12]

Four aspects of our environment were particularly important in the design of Sprite's process migration facility:

Idle hosts are plentiful. Since our environment consists primarily of personal machines, it seemed likely to us that many machines would be idle at any given time. For example, Theimer reported that one-third of all machines were typically idle in a similar environment;[3] Nichols reported that 50-70 workstations were typically idle during the day in an environment with 350 workstations total;[1] and our own measurements below show 66–78% of all workstations idle on average. The availability of many idle machines suggests that simple algorithms can be used for selecting where to migrate: there is no need to make complex choices among partially-loaded machines.

Users "own" their workstations. A user who is sitting in front of a workstation expects to receive the full resources of that workstation. For migration to be accepted by our users, it seemed essential that migrated processes not degrade interactive response. This suggests that a machine should only be used as a target for migration if it is known to be idle, and that foreign processes should be evicted if the user returns before they finish.

Sprite uses kernel calls. Most other implementations of process migration are in message-passing systems where all communication between a process and the rest of the world occurs through message channels. In these systems, many of the transparency aspects of migration can be handled simply by redirecting message communication to follow processes as they migrate. In contrast, Sprite processes are like UNIX processes in that system calls and other forms of interprocess communication are invoked by making protected procedure calls into the kernel. In such a system the solution to the transparency problem is not as obvious; in

[†]UNIX is a registered trademark of AT&T.

the worst case, every kernel call might have to be specially coded to handle remote processes differently than local ones. We consider this issue in greater depth below.

Sprite already provides network support. We were able to capitalize on existing mechanisms in Sprite to simplify the implementation of process migration. For example, Sprite already provided remote access to files and devices, and it has a single network-wide space of process identifiers; these features and others made it much easier to provide transparency in the migration mechanism. In addition, process migration was able to use the same kernel-to-kernel remote procedure call facility that is used for the network file system and many other purposes. On SPARCstation 1 workstations (roughly 10 MIPS) running on a 10 megabits/second Ethernet, the minimum round-trip latency of a remote procedure call is about 1.6 milliseconds and the throughput is 480-660 Kbytes/second. Much of the efficiency of our migration mechanism can be attributed to the efficiency of the underlying RPC mechanism.

To summarize our environmental considerations, we wished to offload work to machines whose users are gone, and to do it in a way that would not be noticed by those users when they returned. We also wanted the migration mechanism to work within the existing Sprite kernel structure, which had one potential disadvantage (kernel calls) and several potential advantages (network-transparent facilities and a fast RPC mechanism).

Why Migration?

Much simpler mechanisms than migration are already available for invoking operations on other machines. In order to understand why migration might be useful, consider the *rsh* command, which provides an extremely simple form of remote invocation under the BSD versions of UNIX. *Rsh* takes as arguments the name of a machine and a command, and causes the given command to be executed on the given remote machine.[13]

Rsh has the advantages of being simple and readily available, but it lacks four important features: transparency, eviction, performance, and automatic selection. First, a process created by *rsh* does not run in the same environment as the parent process: the current directory may be different, environment variables are not transmitted to the remote process, and in many systems the remote process will not have access to the same files and devices as the parent process. In addition, the user has no direct access to remote processes created by *rsh*: the processes do not appear in listings of the user's processes and they cannot be manipulated unless the user logs in to the remote machine. We felt that a mechanism with greater transparency than *rsh* would be easier to use.

The second problem with *rsh* is that it does not permit eviction. A process started by *rsh* cannot be moved once it has begun execution. If a user returns to a machine with *rsh*-generated processes, then either the user must tolerate degraded response until the foreign processes complete, or the foreign processes must be killed, which causes work to be lost and annoyance to the user who owns the foreign processes. Nichols' *butler* system terminates foreign processes after warning the user and providing the processes with the opportunity to save their state, but Nichols noted that the ability to migrate existing processes would make *butler* "much more pleasant to use." [1] Another option is to run foreign processes at low priority so that a returning user receives acceptable interactive response, but this would slow down the execution of the foreign processes. It seemed us to that several opportunities for annoyance could be eliminated, both for the user whose jobs are offloaded and for the user whose workstation is borrowed, by evicting foreign processes when the workstation's user returns.

The third problem with *rsh* is performance. *Rsh* uses standard network protocols with no particular kernel support; the overhead of establishing connections, checking access permissions, and establishing an execution environment may result in delays of several seconds. This makes *rsh* impractical for short-lived jobs and limits the speed-ups that can be obtained using it.

The final problem with *rsh* is that it requires the user to pick a suitable destination machine for offloading. In order to make offloading as convenient as possible for users, we decided to provide an automatic mechanism to keep track of idle machines and select destinations for migration.

Of course, it is unfair to make comparisons with *rsh*, since some of its disadvantages could be eliminated without resorting to full-fledged process migration. For example, Nichols' *butler* layers an automatic selection mechanism on top of a *rsh*-like remote execution facility. Several remote execution mechanisms, including *butler*, preserve the current directory and environment variables. Some UNIX systems even provide a "checkpoint/restart" facility that permits a process to be terminated and later recreated as a different process with the same address space and open files.[14] A combination of these approaches, providing remote invocation and checkpointing but not process migration, would offer significant functionality without the complexity of a full-fledged process migration facility.

The justification for process migration, above and beyond remote invocation, is two-fold. First, process migration provides additional flexibility that a system with only remote invocation lacks. Checkpointing and restarting a long-running process is not always possible, especially if the process interacts with other processes; ultimately, the user would have to decide whether a process can be checkpointed or not. With transparent process migration, the system need not restrict which processes make use of load-sharing. Second, migration is only moderately more complicated than transparent remote invocation. Much of the complexity in remote execution arises even if processes can only move in conjunction with program invocation. In particular, if remote execution is transparent it turns shared state into *distributed* shared state, which is much more difficult to manage. The access position of a file is one example of this effect, as described below in the section on transferring open files. Many of the other issues about maintaining transparency during remote execution would also remain. Permitting a process to migrate at other times during its lifetime requires the system to transfer additional state, such as the process's address space, but is not significantly more complicated.

Thus we decided to take an extreme approach and implement a migration mechanism that allows processes to be moved at any time, to make that mechanism as transparent as possible, and to automate the selection of idle machines. We felt that this combination of features would encourage the use of migration. We also recognized that our mechanism would probably be much more complex than *rsh*. As a result, one of our key criteria in choosing among implementation alternatives was simplicity.

The Overall Problem: Managing State

The techniques used to migrate a process depend on the *state* associated with the process being migrated. If there existed such a thing as a stateless process, then migrating such a process would be trivial. In reality processes have large amounts of state, and both the amount and variety of state seem to be increasing as operating systems evolve. The more state, the more complex the migration mechanism is likely to be. Process state typically includes the following:

- **Virtual memory**. In terms of bytes, the greatest amount of state associated with a process is likely to be the memory that it accesses. Thus the time to migrate a process is limited by the speed of transferring virtual memory.

- **Open files.** If the process is manipulating files or devices, then there will be state associated with these open channels, both in the virtual memory of the process and also in the operating system kernel's memory. The state for an open file includes the internal identifier for the file, the current access position, and possibly cached file blocks. The cached file blocks may represent a substantial amount of storage, in some cases greater than the process's virtual memory.

- **Message channels.** In a message-based operating system such as Mach [15] or V,[16] state of this form would exist in place of open files. (In such a system message channels would be used to access files, whereas in Sprite, file-like channels are used for interprocess communication.) The state associated with a message channel includes buffered messages plus information about senders and receivers.

- **Execution state.** This consists of information that the kernel saves and restores during a context switch, such as register values and condition codes.

- **Other kernel state.** Operating systems typically store other data associated with a process, such as the process's identifier, a user identifier, a current working directory, signal masks and handlers, resource usage statistics, references to the process's parent and children, and so on.

The overall problem in migration is to maintain a process's access to its state after it migrates. For each portion of state, the system must do one of three things during migration: transfer the state, arrange for forwarding, or ignore the state and sacrifice transparency. To transfer a piece of state, it must be extracted from its environment on the source machine, transmitted to the destination machine, and reinstated in the process's new environment on that machine. For state that is private to the process, such as its execution state, state transfer is relatively straightforward. Other state, such as internal kernel state distributed among complex data structures, may be much more difficult to extract and reinstate. An example of "difficult" state in Sprite is information about open files—particularly those being accessed on remote file servers—as described below. Lastly, some state may be impossible to transfer. Such state is usually associated with physical devices on the source machine. For example, the frame buffer associated with a display must remain on the machine containing the display; if a process with access to the frame buffer migrates, it will not be possible to transfer the frame buffer.

The second option for each piece of state is to arrange for forwarding. Rather than transfer the state to stay with the process, the system may leave the state where it is and forward operations back and forth between the state and the process. For example, I/O devices cannot be transferred, but the operating system can arrange for output requests to be passed back from the process to the device, and for input data to be forwarded from the device's machine to the process. In the case of message channels, arranging for forwarding might consist of changing sender and receiver addresses so that messages to and from the channel can find their way from and to the process. Ideally, forwarding should be implemented transparently, so that it is not obvious outside the operating system whether the state was transferred or forwarding was arranged.

The third option, sacrificing transparency, is a last resort: if neither state transfer nor forwarding is feasible, then one can ignore the state on the source machine and simply use the corresponding state on the target machine. The only situation in Sprite where neither state transfer nor forwarding seemed reasonable is for memory-mapped I/O devices such as frame buffers, as alluded to above. In our current implementation, we disallow migration for processes using these devices.

In a few rare cases, lack of transparency may be desirable. For example, a process that requests the amount of physical memory available should obtain information about its current host rather than its home machine. For Sprite, a few special-purpose kernel calls, such as to read instrumentation counters in the kernel, are also intentionally non-transparent with respect to migration. In general, though, it would be unfortunate if a process behaved differently after migration than before.

On the surface, it might appear that message-based systems like Accent,[17] Charlotte,[9] or V [16] simplify many of the state-management problems. In these systems all of a process's interactions with the rest of the world occur in a uniform fashion through message channels. Once the basic execution state of a process has been migrated, it would seem that all of the remaining issues could be solved simply by forwarding messages on the process's message channels. The message forwarding could be done in a uniform fashion, independent of the servers being communicated with or their state about the migrated process.

In contrast, state management might seem more difficult in a system like Sprite that is based on kernel calls. In such a system most of a process's services must be provided by the kernel of the machine where the process executes. This requires that the state for each service be transferred during migration. The state for each service will be different, so this approach would seem to be much more complicated than the uniform message-forwarding approach.

It turns out that neither of these initial impressions is correct. For example, it would be possible to implement forwarding in a kernel-call-based system by leaving *all* of the kernel state on the home machine and using remote procedure call to forward home every kernel call.[14] This would result in something very similar to forwarding messages, and we initially used an approach like this in Sprite.

Unfortunately, an approach based entirely on forwarding kernel calls or forwarding messages will not work in practice, for two reasons. The first problem is that some services must necessarily be provided on the machine where a process is executing. If a process invokes a kernel call to allocate virtual memory (or if it sends a message to a memory server to allocate virtual memory), the request *must* be processed by the kernel or server on the machine where the process executes, since only that kernel or server has control over the machine's page tables. Forwarding is not a viable option for such machine-specific functions: state for these operations must be migrated with processes. The second problem with forwarding is cost. It will often be much more expensive to forward an operation to some other machine than to process it locally. If a service is available locally on a migrated process's new machine, it will be more efficient to use the local service than to forward operations back to the service on the process's old machine.

Thus, in practice all systems must transfer substantial amounts of state as part of process migration. Message-based systems make migration somewhat easier than kernel-call-based systems, because some of the state that is maintained by the kernel in a kernel-call-based system is maintained in a process's address space in a message-based system. This state is transferred implicitly with the address space of the process. For other state, both types of system must address the same issues.

Mechanics of Migration

This section describes how Sprite deals with the various components of process state during migration. The solution for each component consists of some combination of transferring state and arranging for forwarding.

Virtual Memory Transfer

Virtual memory transfer is the aspect of migration that has been discussed the most in the literature, perhaps because it is generally believed to be the limiting factor in the speed of migration.[17] One simple method for transferring virtual memory is to send the process's entire memory image to the target machine at migration time, as in Charlotte[9] and LOCUS.[5] This approach is simple but it has two disadvantages. First, the transfer can take many seconds, during which time the process is *frozen*: it cannot execute on either the source or destination machine. For some processes, particularly those with real-time needs, long freeze times may be unacceptable. The second disadvantage of a monolithic virtual memory transfer is that it may result in wasted work for portions of the virtual memory that are not used by the process after it migrates. The extra work is particularly unfortunate (and costly) if it requires old pages to be read from secondary storage. For these reasons, several other approaches have been used to reduce the overhead of virtual memory transfer; the mechanisms are diagrammed in Figure 1 and described in the paragraphs below.

In the V System, long freeze times could have resulted in timeouts for processes trying to communicate with a migrating process. To address this problem, Theimer used a method called *pre-copying*.[3, 8] Rather than freezing a process at the beginning of migration, V allows the process to continue executing while its address space is transferred. In the original implementation of migration in V, the entire memory of the process was transferred directly to the target; Theimer also proposed an implementation that would use virtual memory to write modified pages to a shared "backing storage server" on the network. In either case, some pages could be modified on the source machine after they have been copied elsewhere, so V then freezes the process and copies the pages that have been modified. Theimer showed that pre-copying reduces freeze times substantially. However, it has the disadvantage of copying some pages twice, which increases the total amount of work to migrate a process. Pre-copying seems most useful in an environment like V where processes have real-time response requirements.

The Accent system uses a *lazy copying* approach to reduce the cost of process migration.[4, 17] When a process migrates in Accent, its virtual memory pages are left on the source machine until they are actually referenced on the target machine. Pages are copied to the target when they are referenced for the first time. This approach allows a process to begin execution on the target with minimal freeze time but introduces many short delays later as pages are retrieved from the source machine. Overall, lazy copying reduces the cost of migration because pages that are not used are never copied at all. Zayas found that for typical programs only one-quarter to one-half of a process's allocated memory needed to be transferred. One disadvantage of lazy copying is that it leaves residual dependencies on the source machine: the source must store the unreferenced pages and provide them on demand to the target. In the worst case, a process that migrates several times could leave virtual memory dependencies on any or all of the hosts on which it ever executed.

Sprite's migration facility uses a different form of lazy copying that takes advantage of our existing network services while providing some of the advantages of lazy copying. In Sprite, as in the proposed implementation for V, backing storage for virtual memory is implemented using ordinary files. Since these backing files are stored in the network file system, they are accessible throughout the network. During migration the source machine freezes the process, flushes its dirty pages to backing files, and discards its address space. On the target machine, the process starts executing with no resident pages and uses the standard paging mechanisms to load pages from backing files as they are needed.

In most cases no disk operations are required to flush dirty pages in Sprite. This is because the backing files are stored on network file servers and the file servers use their memories to cache

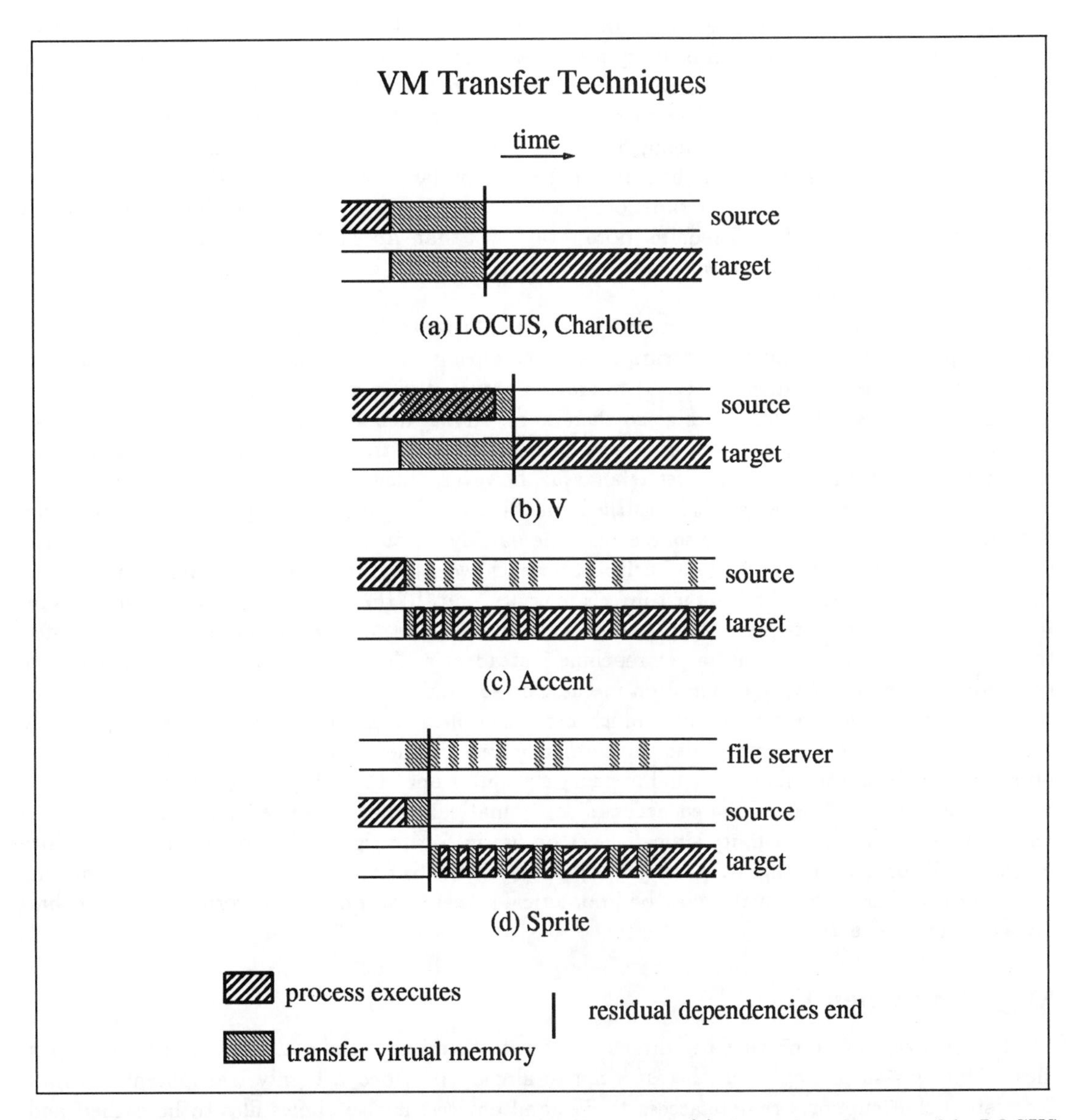

Figure 1: *Different techniques for transferring virtual memory.* (a) shows the scheme used in LOCUS and Charlotte, where the entire address space is copied at the time a process migrates. (b) shows the pre-copying scheme used in V, where the virtual memory is transferred during migration but the process continues to execute during most of the transfer. (c) shows Accent's lazy-copying approach, where pages are retrieved from the source machine as they are referenced on the target. Residual dependencies in Accent can last for the life of the migrated process. (d) shows Sprite's approach, where dirty pages are flushed to a file server during migration and the target retrieves pages from the file server as they are referenced. In the case of eviction, there are no residual dependencies on the source after migration. When a process migrates away from its home machine, it has residual dependencies on its home throughout its lifetime.

recently-used file data. When the source machine flushes a dirty page it is simply transferred over the network to the server's main-memory file cache. If the destination machine accesses the page then it is retrieved from the cache. Disk operations will only occur if the server's cache overflows.

Sprite's virtual memory transfer mechanism was simple to implement because it uses pre-existing mechanisms both for flushing dirty pages on the source and for handling page faults on the target. It has some of the benefits of the Accent lazy-copying approach since only dirty pages incur overhead at migration time; other pages are sent to the target machine when they are referenced. Our approach will require more total work than Accent's, though, since dirty pages may be transferred over the network twice: once to a file server during flushing, and once later to the destination machine.

The Sprite approach to virtual memory transfer fits well with the way migration is typically used in Sprite. Process migration occurs most often during an *exec* system call, which completely replaces the process's address space. If migration occurs during an *exec*, the new address space is created on the destination machine so there is no virtual memory to transfer. As others have observed (*e.g.*, LOCUS [5]), the performance of virtual memory transfer for *exec*-time migration is not an issue. Virtual memory transfer *is* an issue, however, when migration is used to evict a process from a machine whose user has returned. In this situation the most important consideration is to remove the process from its source machine quickly, in order to minimize any performance degradation for the returning user. Sprite's approach works well in this regard since (a) it does the least possible work to free up the source's memory, and (b) the source need not retain pages or respond to later paging requests as in Accent. It would have been more efficient overall to transfer the dirty pages directly to the target machine instead of a file server, but this approach would have added complexity to the migration mechanism so we decided against it.

Virtual memory transfer becomes much more complicated if the process to be migrated is sharing writable virtual memory with some other process on the source machine. In principle, it is possible to maintain the shared virtual memory even after one of the sharing processes migrates,[18] but this changes the cost of shared accesses so dramatically that it seemed unreasonable to us. Shared writable virtual memory almost never occurs in Sprite right now, so we simply disallow migration for processes using it. A better long-term solution is probably to migrate all the sharing processes together, but even this may be impractical if there are complex patterns of sharing that involve many processes.

Migrating Open Files

It turned out to be particularly difficult in Sprite to migrate the state associated with open files. This was surprising to us, because Sprite already provided a highly transparent network file system that supports remote access to files and devices; it also allows files to be cached and to be accessed concurrently on different workstations. Thus, we expected that the migration of file-related information would mostly be a matter of reusing existing mechanisms. Unfortunately, process migration introduced new problems in managing the distributed state of open files. Migration also made it possible for a file's current access position to become shared among several machines.

The migration mechanism would have been much simpler if we had chosen the "arrange for forwarding" approach for open files instead of the "transfer state" approach. This would have implied that all file-related kernel calls be forwarded back to the machine where the file was opened, so that the state associated with the file could have stayed on that machine. Because of the frequency of file-related kernel calls and the cost of forwarding a kernel call over the network, we felt that this approach would be unacceptable both because it would slow down the remote process and because it would load the machine that stores the file state. Sprite workstations are typically diskless and files are accessed remotely from file servers, so the forwarding approach would have meant that each file request would be passed over the network once to the machine where the file was opened, and possibly a second time to the server. Instead, we decided to transfer open-file state along with a migrating process and then use the normal mechanisms to access the file (*i.e.*, communicate directly with the file's server).

There are three main components of the state associated with an open file: a file reference, caching information, and an access position. Each of these components introduced problems for migration. The file reference indicates where the file is stored, and also provides a guarantee that the file exists (as required by UNIX semantics): if a file is deleted while open then the deletion is deferred until the file is closed. Our first attempt at migrating files simply closed the file on the source machine and reopened it on the target. Unfortunately, this approach caused files to disappear if they were deleted before the reopen completed. This is such a common occurrence in UNIX programs that file transfer had to be changed to move the reference from source to target without ever closing the file.

The second component of the state of an open file is caching information. Sprite permits the data of a file to be cached in the memory of one or more machines, with file servers responsible for guaranteeing "consistent access" to the cached data.[11] The server for a file keeps track of which hosts have the file open for reading and writing. If a file is open on more than one host and at least one of them is writing it, then caching is disabled: all hosts must forward their read and write requests for that file to the server so they can be serialized. In our second attempt at migrating files, the server was notified of the file's use on the target machine before being told that the file was no longer in use on the source; this made the file appear to be write-shared and caused the server to disable caching for the file unnecessarily. To solve both this problem and the reference problem above we built special server code just for migrating files, so that the transfer from source to destination is made atomically. Migration can still cause caching to be disabled for a file, but only if the file is also in use by some other process on the source machine; if the only use is by the migrating process, then the file will be cacheable on the target machine. In the current implementation, once caching is disabled for a file, it remains disabled until no process has the file open (even if all processes accessing the file migrate to the same machine); however, in practice, caching is disabled infrequently enough that an optimization to reenable caching of uncacheable files has not been a high priority.

When an open file is transferred during migration, the file cache on the source machine may contain modified blocks for the file. These blocks are flushed to the file's server machine during migration, so that after migration the target machine can retrieve the blocks from the file server without involving the source. This approach is similar to the mechanism for virtual memory transfer and thus has the same advantages and disadvantages. It is also similar to what happens in Sprite for shared file access without migration: if a file is opened, modified, and closed on one machine, then opened on another machine, the modified blocks are flushed from the first machine's cache to the server at the time of the second open.

The third component of the state of an open file is an access position, which indicates where in the file the next read or write operation will occur. Unfortunately the access position for a file may be shared between two or more processes. This happens, for example, when a process opens a file and then forks a child process: the child inherits both the open file and the access position. Under normal circumstances all of the processes sharing a single access position will reside on the same machine, but migration can move one of the processes without the others, so that the access position becomes shared between machines. After several false starts we eventually dealt with this problem in a fashion similar to caching: if an access position becomes shared between

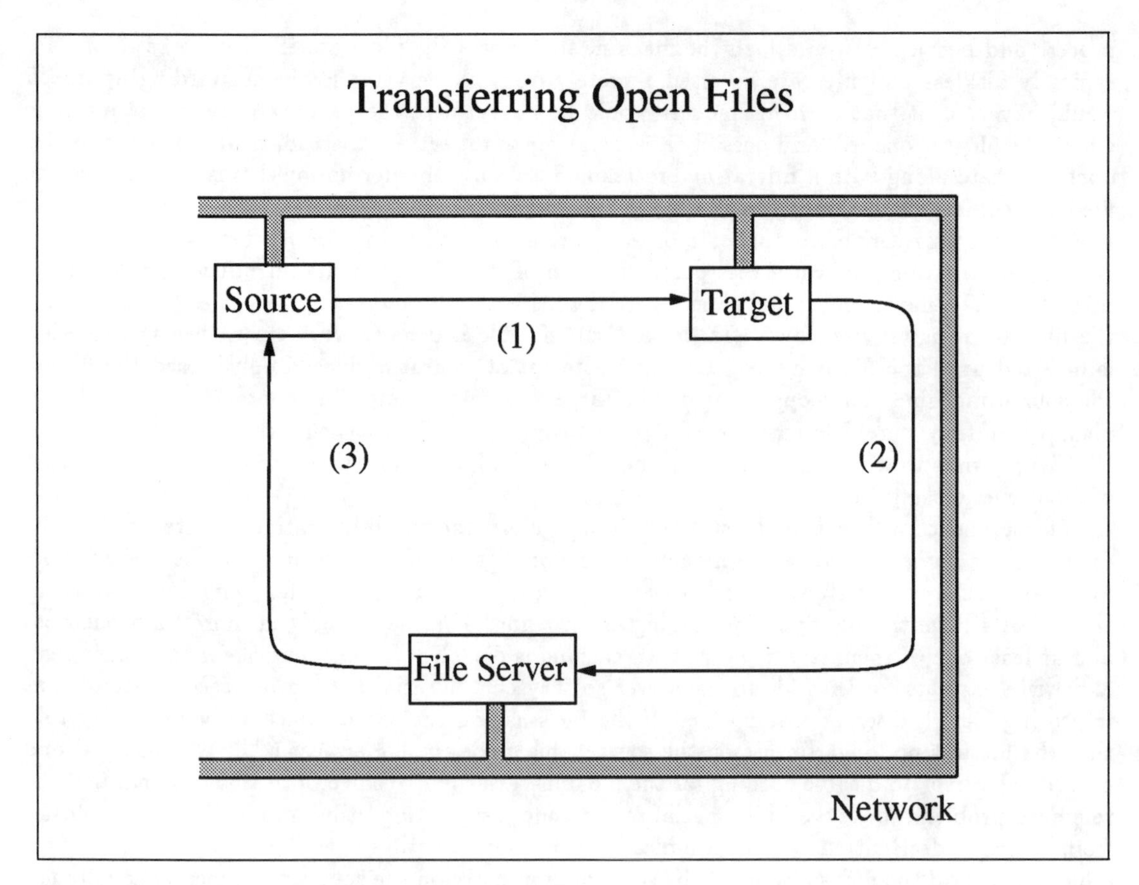

Figure 2: *Transferring open files.* (1) The source passes information about all open files to the target. (2) For each file, the target notifies the server that the open file has been moved; (3) during this call the server communicates again with the source to release its state associated with the file and to obtain the most recent state associated with the file.

machines, then neither machine stores the access position (nor do they cache the file); instead, the file's server maintains the access position and all operations on the file are forwarded to the server.

Another possible approach to shared file offsets is the one used in LOCUS.[5] If process migration causes a file access position to be shared between machines, LOCUS lets the sharing machines take turns managing the access position. In order to perform I/O on a file with a shared access position, a machine must acquire the "access position token" for the file. While a machine has the access position token it caches the access position and no other machine may access the file. The token rotates among machines as needed to give each machine access to the file in turn. This approach is similar to the approach LOCUS uses for managing a shared file, where clients take turns caching the file and pass read and write tokens around to ensure cache consistency. We chose not to use the Locus approach because the token-passing approach is more complex than the disable-caching approach, and because the disable-caching approach meshed better with the existing Sprite file system.

Figure 2 shows the mechanism currently used by Sprite for migrating open files. The key part of this mechanism occurs in a late phase of migration when the target machine requests that the server update its internal tables to reflect that the file is now in use on the target instead of the source. The server in turn calls the source machine to retrieve information about the file, such as the file's access position and whether the file is in use by other processes on the source machine. This two-level remote procedure call synchronizes the three machines (source, target, and server) and provides a convenient point for updating state about the open file.

The Process Control Block

Aside from virtual memory and open files, the main remaining issue is how to deal with the process control block (PCB) for the migrating process: should it be left on the source machine or transferred with the migrating process? For Sprite we use a combination of both approaches. The home machine for a process (the one where it would execute if there were no migration) must assist in some operations on the process, so it always maintains a PCB for the process. The details of this interaction are described in the next section. In addition, the current machine for a process also has a PCB for it. If a process is migrated, then most of the information about the process is kept in the PCB on its current machine; the PCB on the home machine serves primarily to locate the process and most of its fields are unused.

The other elements of process state besides virtual memory and open files are much easier to transfer than virtual memory and open files, since they are not as bulky as virtual memory and they don't involve distributed state like open files. At present the other state consists almost entirely of fields from the process control block. In general, all that needs to be done is to transfer these fields to the target machine and reinstate them in the process control block on the target.

Supporting Transparency: Home Machines

As was mentioned previously, transparency was one of our most important goals in implementing migration. By "transparency" we mean two things in particular. First, a process's behavior should not be affected by migration. Its execution environment should appear the same, it should have the same access to system resources such as files and devices, and it should produce exactly the same results as if it hadn't migrated. Second, a process's appearance to the rest of the world should not be affected by migration. To the rest of the world the process should appear as if it never left its original machine, and any operation that is possible on an unmigrated process (such as stopping or signalling) should be possible on a migrated process. Sprite provides both of these forms of transparency; we know of no other implementation of process migration that provides transparency to the same degree.

In Sprite the two aspects of transparency are defined with respect to a process's *home machine*, which is the machine where it would execute if there were no migration at all. Even after migration, everything should appear as if the process were still executing on its home machine. In order to achieve transparency, Sprite uses four different techniques, which are described in the paragraphs below.

The most desirable approach is to make kernel calls location-independent; Sprite has been gradually evolving in this direction. For example, in the early versions of the system we permitted different machines to have different views of the file system name space. This required *open* and several other kernel calls to be forwarded home after migration, imposing about a 20% penalty on the performance of remote compilations. In order to simplify migration (and for several other good reasons also), we changed the file system so that every machine in the network sees the same name space. This made the *open* kernel call location-independent, so no extra effort was necessary to make *open* work transparently for remote processes.

Our second technique was to transfer state from the source machine to the target at migration time as described above, so that normal kernel calls may be used after migration. We used the state-transfer approach for virtual memory, open files, process and user identifiers, resource usage statistics, and a variety of other things.

Our third technique was to forward kernel calls home. This technique was originally used for a large number of kernel calls, but we have gradually replaced most uses of forwarding with transparency or state transfer. At present there are only a few kernel calls that cannot be implemented transparently and for which we cannot easily transfer state. For example, clocks are not synchronized between Sprite machines, so for remote processes Sprite forwards the *gettimeofday* kernel call back to the home machine. This guarantees that time advances monotonically even for remote processes, but incurs a performance penalty for processes that read the time frequently. Another example is the *getpgrp* kernel call, which obtains state about the "process group" of a process. The home machine maintains the state that groups collections of processes together, since they may physically execute on different machines.

Forwarding also occurs from the home machine to a remote process's current machine. For example, when a process is signalled (*e.g.*, when some other process specifies its identifier in the *kill* kernel call), the signal operation is sent initially to the process's home machine. If the process is not executing on the home machine, then the home machine forwards the operation on to the process's current machine. The performance of such operations could be improved by retaining a cache on each machine of recently-used process identifiers and their last known execution sites. This approach is used in LOCUS and V and allows many operations to be sent directly to a remote process without passing through another host. An incorrect execution site is detected the next time it is used and correct information is found by sending a message to the host on which the process was created (LOCUS) or by multi-casting (V).

The fourth "approach" is really just a set of *ad hoc* techniques for a few kernel calls that must update state on both a process's current execution site and its home machine. One example of such a kernel call is *fork*, which creates a new process. Process identifiers in Sprite consist of a home machine identifier and an index of a process within that machine. Management of process identifiers, including allocation and deallocation, is the responsibility of the home machines named in the identifiers. If a remote process *fork*s, the child process must have the same home machine as the parent, which requires that the home machine allocate the new process identifier. Furthermore, the home machine must initialize its own copy of the process control block for the process, as described previously. Thus, even though the child process will execute remotely on the same machine as its parent, both its current machine and its home machine must update state. Similar kinds of cooperation occur for *exit*, which is invoked by a process to terminate itself, and *wait*, which is used by a parent to wait for one of its children to terminate. There are several potential race conditions between a process exiting, its parent waiting for it to exit, and one or both processes migrating; we found it easier to synchronize these operations by keeping all the state for the *wait-exit* rendezvous on a single machine (the home). LOCUS similarly uses the site on which a process is created to synchronize operations on the process.

Residual Dependencies

We define a *residual dependency* as an on-going need for a host to maintain data structures or provide functionality for a process even after the process migrates away from the host. One example of a residual dependency occurs in Accent, where a process's virtual memory pages are left on the source machine until they are referenced on the target. Another example occurs in Sprite, where the home machine must participate whenever a remote process forks or exits.

Residual dependencies are undesirable for three reasons: reliability, performance, and complexity. Residual dependencies decrease reliability by allowing the failure of one host to affect processes on other hosts. Residual dependencies decrease performance for the remote process because they require remote operations where local ones would otherwise have sufficed. Residual dependencies also add to the load of the host that is depended upon, thereby reducing the performance of other processes executing on that host. Lastly, residual dependencies complicate the system by distributing a process's state around the network instead of concentrating it on a single host; a particularly bad scenario is one where a process can migrate several times, leaving residual dependencies on every host it has visited.

Despite the disadvantages of residual dependencies, it may be impractical to eliminate them all. In some cases dependencies are inherent, such as when a process is using a device on a specific host; these dependencies cannot be eliminated without changing the behavior of the process. In other cases, dependencies are necessary or convenient to maintain transparency, such as the home machine knowing about all process creations and terminations. Lastly, residual dependencies may actually improve performance in some cases, such as lazy copying in Accent, by deferring state transfer until it is absolutely necessary.

In Sprite we were much more concerned about transparency than about reliability, so we permitted some residual dependencies on the home machine where those dependencies made it easier to implement transparency. As described above in the section on transparency, there are only a few situations where the home machine must participate so the performance impact is minimal. Measurements of the overhead of remote execution are reported below.

Although Sprite permits residual dependencies on the home machine, it does not leave dependencies on any other machines. If a process migrates to a machine and is then evicted or migrates away for any other reason, there will be no residual dependencies on that machine. This provides yet another assurance that process migration will not impact users' response when they return to their workstations. The only noticeable long-term effect of foreign processes is the resources they may have utilized during their execution: in particular, the user's virtual memory working set may have to be demand-paged back into memory upon the user's return.

The greatest drawback of residual dependencies on the home machine is the inability of users to migrate processes in order to survive the failure of their home machine. We are considering a nontransparent variant of process migration, which would change the home machine of a process when it migrates and break all dependencies on its previous host.

Migration Policies

Until now we have focussed our discussion on the *mechanisms* for transferring processes and supporting remote execution. This section considers the *policies* that determine how migration is used. Migration policy decisions fall into four categories:

What. Which processes should be migrated? Should all processes be considered candidates for migration, or only a few particularly CPU-intensive processes? How are CPU-intensive processes to be identified?

When. Should processes only be migrated at the time they are initiated, or may processes also be migrated after they have been running?

Where. What criteria should be used to select the machines that will be the targets of migration?

Who. Who makes all of the above decisions? How much should be decided by the user and how much should be automated in system software?

At one end of the policy spectrum lies the *pool of processors* model. In this model the processors of the system are treated as a shared pool and all of the above decisions are made automatically by system software. Users submit jobs to the system without any idea of where they will execute. The system assigns jobs to processors dynamically, and if process migration is available it may move processes during execution to balance the loads of the processors in the pool. MOSIX [6] is one example of the "pool of processors" model: processors are shared equally by all processes and the system dynamically balances the load throughout the system, using process migration.

At the other end of the policy spectrum lies *rsh*, which provides no policy support whatsoever. In this model individual users are responsible for locating idle machines, negotiating with other users over the use of those machines, and deciding which processes to offload.

For Sprite we chose an intermediate approach where the selection of idle hosts is fully automated but the other policy decisions are only partially automated. There were two reasons for this decision. First, our environment consists of personal workstations. Users are happy running almost all of their processes locally on their own personal workstations, and they expect to have complete control of their workstations. Users do not think of their workstations as "shared". Second, the dynamic pool-of-processors approach appeared to us to involve considerable additional complexity, and we were not convinced that the benefits would justify the implementation difficulties. For example, most processes in a UNIX-like environment are so short-lived that migration will not produce a noticeable benefit and may even slow things down. Eager et al. provide additional evidence that migration is only useful under particular conditions.[19] Thus, for Sprite we decided to make migration a special case rather than the normal case.

The Sprite kernels provide no particular support for any of the migration policy decisions, but user-level applications provide assistance in four forms: idle-host selection, the *pmake* program, a *mig* shell command, and eviction. These are discussed in the following subsections.

Selecting Idle Hosts

Each Sprite machine runs a background process called the *load-average daemon*, which monitors the usage of that machine. When the workstation appears to be idle, the load-average daemon notifies the *central migration server* that the machine is ready to accept migrated processes. Programs that invoke migration, such as *pmake* and *mig* described below, call a standard library procedure *Mig_RequestIdleHosts* to obtain the identifiers for one or more idle hosts, which they then pass to the kernel when they invoke migration. Normally only one process may be assigned to any host at any one time, in order to avoid contention for processor time; however, processes that request idle hosts can indicate that they will be executing long-running processes and the central server will permit shorter tasks to execute on those hosts as well.

Maintaining the database of idle hosts can be a challenging problem in a distributed system, particularly if the system is very large in size or if there are no shared facilities available for storing load information. A number of distributed algorithms have been proposed to solve this problem, such as disseminating load information among hosts periodically,[6] querying other hosts at random to find an idle one,[20] or multicasting and accepting a response from any host that indicates availability.[8]

In Sprite we have used centralized approaches for storing the idle-host database. Centralized techniques are generally simpler, they permit better decisions by keeping all the information up-to-date in a single place, and they can scale to systems with hundreds of workstations without contention problems for the centralized database.

We initially stored the database in a single file in the file system. The load-average daemons set flags in the file when their hosts became idle, and the *Mig_RequestIdleHosts* library procedure selected idle hosts at random from the file, marking the selected hosts so that no one else would select them. Standard file-locking primitives were used to synchronize access to the file.

We later switched to a server-based approach, where a single server process keeps the database in its virtual memory. The load-average daemons and the *Mig_RequestIdleHosts* procedure communicate with the server using a message protocol. The server approach has a number of advantages over the file-based approach. It is more efficient, because only a single remote operation is required to select an idle machine; the file-based approach required several remote operations to open the file, lock it, read it, etc. The server approach makes it easy to retain state from request to request; we use this, for example, to provide fair allocation of idle hosts when there are more would-be users than idle machines. Although some of these features could have been implemented with a shared file, they would incur a high overhead from repeated communication with a file server. Lastly, the server approach provides better protection of the database information (in the shared-file approach the file had to be readable and writable by all users).

We initially chose a conservative set of criteria for determining whether a machine is "idle". The load-average daemon originally considered a host to be idle only if (a) it had had no keyboard or mouse input for at least five minutes, and (b) there were fewer runnable processes than processors, on average. In choosing these criteria we wanted to be certain not to inconvenience active users or delay background processes they might have left running. We assumed that there would usually be plenty of idle machines to go around, so we were less concerned about using them efficiently. After experience with the five-minute threshold, we reduced the threshold for input to 30 seconds; this increased the pool of available machines without any noticeable impact on the owners of those machines.

Pmake and Mig

Sprite provides two convenient ways to use migration. The most common use of process migration is by the *pmake* program. *Pmake* is similar in function to the *make* UNIX utility [7] and is used, for example, to detect when source files have changed and recompile the corresponding object files. *Make* performs its compilations and other actions serially; in contrast, *pmake* uses process migration to invoke as many commands in parallel as there are idle hosts available. This use of process migration is completely transparent to users and results in substantial speed-ups in many situations, as shown below. Other systems besides Sprite have also benefitted from parallel make facilities; see References 21 and 2 for examples.

The approach used by *pmake* has at least one advantage over a fully-automatic "processor pool" approach where all the migration decisions are made centrally. Because *pmake* makes the choice of processes to offload, and knows how many hosts are available, it can scale its parallelism to match the number of idle hosts. If the offloading choice were made by some other agent, *pmake* might overload the system by creating more processes than could be accommodated efficiently. *Pmake* also provides a degree of flexibility by permitting the user to specify that certain tasks should not be offloaded if they are poorly suited for remote execution.

The second easy way to use migration is with a program called *mig*, which takes as argument a shell command. *Mig* will select an idle machine using the mechanism described above and use process migration to execute the given command on that machine. *Mig* may also be used to migrate an existing process.

Eviction

The final form of system support for migration is eviction. The load-average daemons detect when a user returns. On the first keystroke or mouse-motion invoked by the user, the load-average daemon will check for foreign processes and evict them. When an eviction occurs, foreign processes are migrated back to their home machines, and the process that obtained the host is notified that the host has been reclaimed. That process is free to remigrate the evicted processes or to suspend them if there is no new host available. To date, *pmake* is the only application that automatically remigrates processes, but other applications (such as *mig*) could remigrate processes as well.

Evictions also occur when a host is reclaimed from one process in order to allocate it to another. If the centralized server receives a request for an idle host when no idle hosts are available, and one process has been allocated more than its fair share of hosts, the server reclaims one of the hosts being used by that process. It grants that host to the process that had received less than its fair share. The process that lost the host must reduce its parallelism until it can obtain additional hosts again.

A possible optimization for evictions would be to permit an evicted process to migrate directly to a new idle host rather than to its home machine. In practice, half of the evictions that occur in the system take place due to fairness considerations rather than because a user has returned to an idle workstation.[22] Permitting direct migration between two remote hosts would benefit the other half of the evictions that occur, but would complicate the implementation: it would require a three-way communication between the two remote hosts and the home machine, which always knows where its processes execute. Thus far, this optimization has not seemed to be warranted.

Performance and Usage Patterns

We evaluated process migration in Sprite by taking three sets of measurements. The next subsections discuss particular operations in isolation, such as the time to migrate a trivial process or invoke a remote command; the performance improvement of *pmake* using parallel remote execution; and empirical measurements of Sprite's process migration facility over a period of several weeks, including the extent to which migration is used, the cost and frequency of eviction, and the availability of idle hosts.

Migration Overhead

Table 1 summarizes the costs associated with migration. Host selection on SPARCstation 1 workstations takes an average of 36 milliseconds. Process transfer is a function of some fixed overhead, plus variable overhead in proportion to the number of modified virtual memory pages and file blocks copied over the network and the number of files the process has open. If a process *execs* at the time of migration, as is normally the case, no virtual memory is transferred.

The costs in Table 1 reflect the latency and bandwidth of Sprite's remote procedure call mechanism. For example, the cost of transferring open files is dominated by RPC latency (3 RPC's at 1 ms latency each), and the speed of transferring virtual memory pages and file blocks is determined by RPC bandwidth (480-660 Kbytes/second). All things considered, it takes about a tenth of a second to select an idle host and start a new process on it, not counting any time needed to transfer open files or flush modified file blocks to servers. Empirically, the average time to perform an *exec*-time migration in our system is about 330 milliseconds.[22] This latency may be too great to warrant running trivial programs remotely, but it is substantially less than the

Action	Time/Rate
Select & release idle host	36 milliseconds
Migrate "null" process	76 milliseconds
Transfer info for open files	9.4 milliseconds/file
Flush modified file blocks	480 Kbytes/second
Flush modified pages	660 Kbytes/second
Transfer *exec* arguments	480 Kbytes/second
Fork, *exec* null process with migration, wait for child to exit	81 milliseconds
Fork, *exec* null process locally, wait for child to exit	46 milliseconds

Table 1: *Costs associated with process migration.* All measurements were performed on SPARCstation 1 workstations. Host selection may be amortized across several migrations if applications such as *pmake* reuse idle hosts. The time to migrate a process depends on how many open files the process has and how many modified blocks for those files are cached locally (these must be flushed to the server). If the migration is not done at *exec*-time, modified virtual memory pages must be flushed as well. If done at *exec*-time, the process's arguments and environment variables are transferred. The *execs* were performed with no open files. The bandwidth of the RPC system is 480 Kbytes/second using a single channel, and 660 Kbytes/second using multiple RPC connections in parallel for the virtual memory system.

time needed to compile typical source programs, run text formatters, or do any number of other CPU-bound tasks.

After a process migrates away from its home machine, it may suffer from the overhead of forwarding system calls. The degradation due to remote execution depends on the ratio of location-dependent system calls to other operations, such as computation and file I/O. Figure 3 shows the total execution time to run several programs, listed in Table 2, both entirely locally and entirely on a single remote host. Applications that communicate frequently with the home machine suffered considerable degradation. Two of the benchmarks, *fork* and *gettime*, are contrived examples of the type of degradation a process might experience if it performed many location-dependent system calls without much user-level computation. The *rcp* benchmark is a more realistic example of the penalties processes can encounter: it copies data using TCP, and TCP operations are sent to a user-level TCP server on the home machine. Forwarding these TCP operations causes *rcp* to perform about 40% more slowly when run remotely than locally. As may be seen in Figure 3, however, applications such as compilations and text formatting show little degradation due to remote execution.

Application Performance

The benchmarks in the previous section measured the component costs of migration. This section measures the overall benefits of migration using *pmake*. We measured the performance improvements obtained by parallel compilations and simulations.

The first benchmark consists of compiling 276 Sprite kernel source files, then linking the resulting object files into a single file. Each *pmake* command (compiling or linking) is performed on a remote host using *exec*-time migration. Once a host is obtained from the pool of available hosts, it is reused until *pmake* finishes or the host is no longer available.

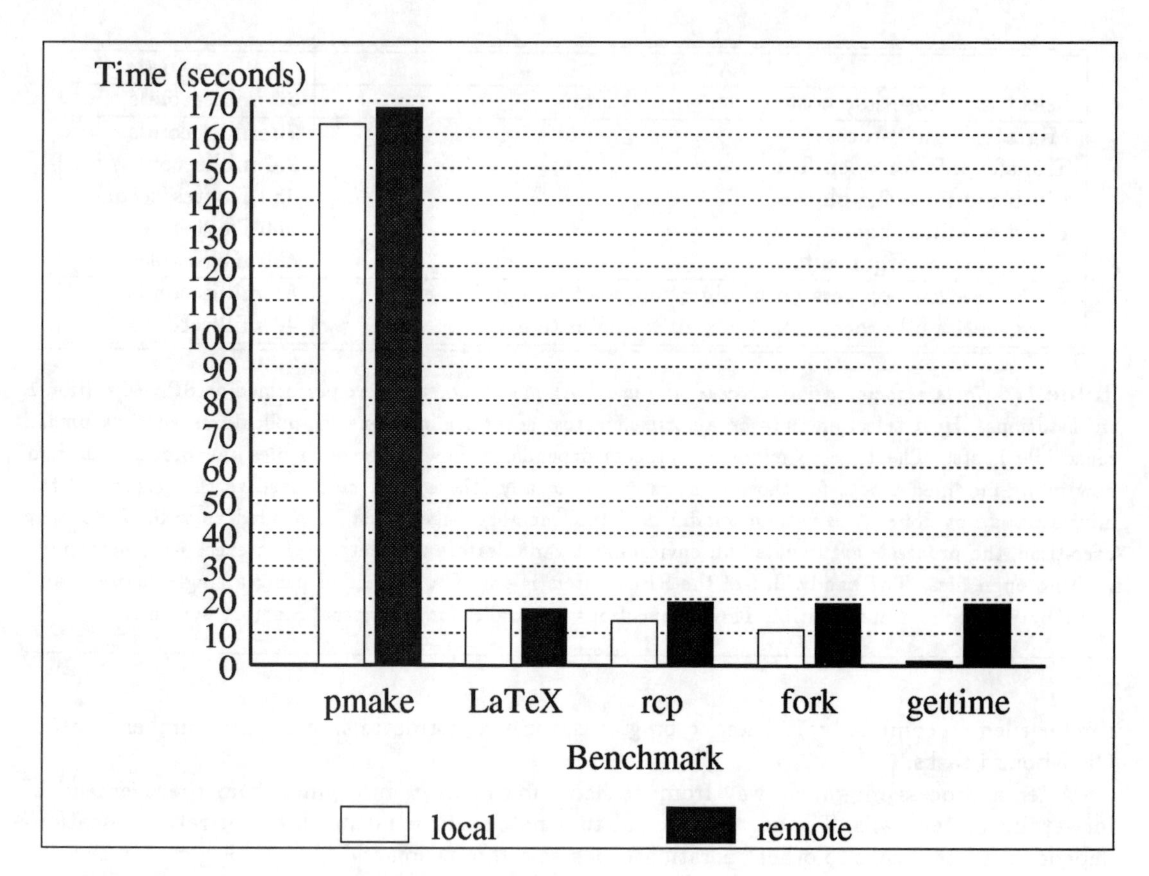

Figure 3: *Comparison between local and remote execution of programs.* The elapsed time to execute CPU-intensive and file-intensive applications such as *pmake* and LaTeX showed negligible effects from remote execution (3% and 1% degradation, respectively). Other applications suffered performance penalties ranging from 42% (*rcp*), to 73% (*fork*), to 3200% (*gettime*).

Name	Description
pmake	recompile *pmake* source sequentially using *pmake*
LaTeX	run LaTeX on a draft of this article
rcp	copy a 1 Mbyte file to another host using TCP
fork	fork and wait for child, 1000 times
gettime	get the time of day 10000 times

Table 2: *Workload for comparisons between local and remote execution.*

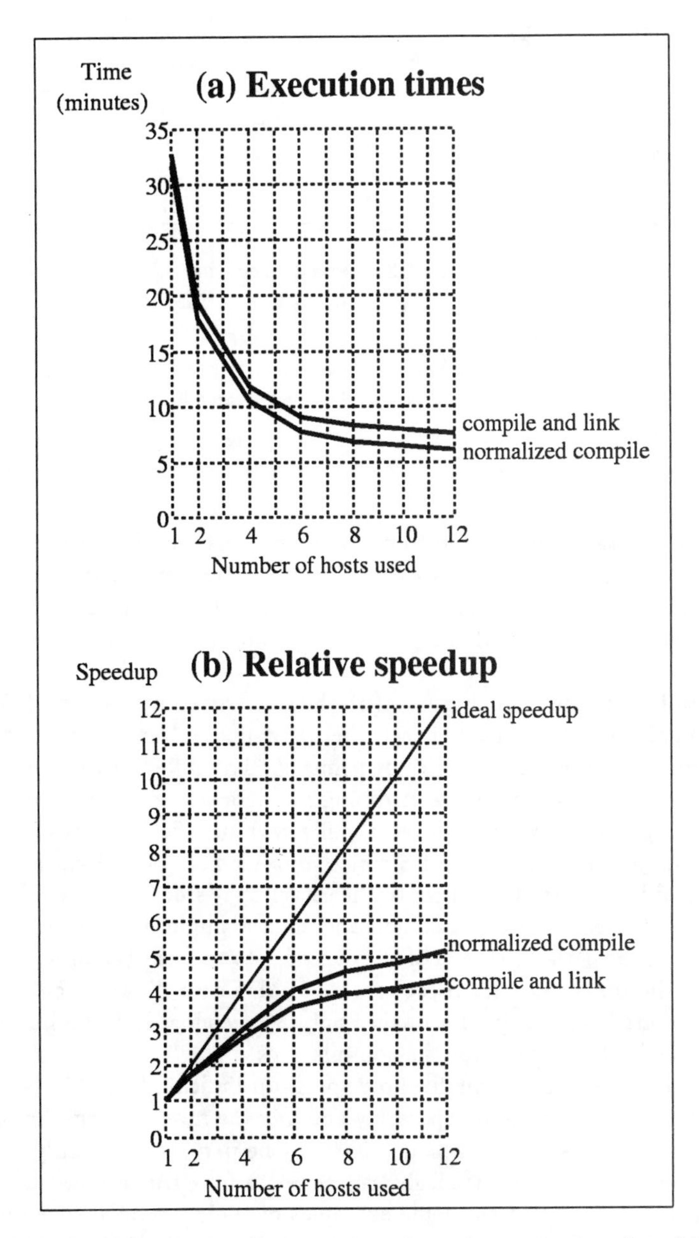

Figure 4: *Performance of recompiling the Sprite kernel using a varying number of hosts and the* pmake *program.* Graph (a) shows the time to compile all the input files and then link the resulting object files into a single file. In addition, it shows a "normalized" curve that shows the time taken for the compilation only, deducting as well the *pmake* startup overhead of 19 seconds to determine dependencies; this curve represents the parallelizable portion of the *pmake* benchmark. Graph (b) shows the speedup obtained for each point in (a), which is the ratio between the time taken on a single host and the time using multiple hosts in parallel.

Program	Number of Files	Number of Links	Sequential Time	Parallel Time	Speed-Up
gremlin	24	1	180	41	4.43
$\TeX$	36	1	259	48	5.42
pmake	49	3	162	55	2.95
kernel	276	1	1971	453	4.35

Table 3: *Examples of* pmake *performance.* Sequential execution is done on a single host; parallel execution uses migration to execute up to 12 tasks in parallel. Each measurement gives the time to compile the indicated number of files and link the resulting object files together in one or more steps. When multiple steps are required, their sequentiality reduces the speed-up that may be obtained; *pmake*, for example, is organized into two directories that are compiled and linked separately, and then the two linked object files are linked together.

Figure 4 shows the total elapsed time to compile and link the Sprite kernel using a varying number of machines in parallel, as well as the performance improvement obtained. In this environment, *pmake* is able to make effective use of about three-fourths of each host it uses up to a point (4-6 hosts), but it uses only half the processing power available to it once additional hosts are used.

The "compile and link" curve in Figure 4(b) shows a speed-up factor of 5 using 12 hosts. Clearly, there is a significant difference between the speed-ups obtained for the "normalized compile" benchmark and the "compile and link" benchmark. The difference is partly attributable to the sequential parts of running *pmake*: determining file dependencies and linking object files all must be done on a single host. More importantly, file caching affects speed-up substantially. As described above, when a host opens a file for which another host is caching modified blocks, the host with the modified blocks transfers them to the server that stores the file. Thus, if *pmake* uses many hosts to compile different files in parallel, and then a single host links the resulting object files together, that host must wait for each of the other hosts to flush the object files they created. It then must obtain the object files from the server. In this case, linking the files together when they have all been created on a single host takes only 56 seconds, but the link step takes 65–69 seconds when multiple hosts are used for the compilations.

In practice, we don't even obtain the five-fold speed-up indicated by this benchmark, because we compile and link each kernel module separately and link the modules together afterwards. Each link step is an additional synchronization point that may be performed by only one host at a time. In our development environment, we typically see three to four times speed-up when rebuilding a kernel from scratch. Table 3 presents some examples of typical *pmake* speed-ups. These times are representative of the performance improvements seen in day-to-day use. Figure 5 shows the corresponding speedup curves for each set of compilations when the number of hosts used varies from 1 to 12. In each case, the marginal improvement of additional hosts decreases as more hosts are added.

The speedup curves in Figure 4(b) and Figure 5 show that the marginal improvement from using additional hosts is significantly less than the processing power of the hosts would suggest. The poor improvement is due to bottlenecks on both the file server and the workstation running *pmake*. Figure 6 shows the utilization of the processors on the file server and client workstation over 5-second intervals during the 12-way kernel *pmake*. It shows that the *pmake* process uses

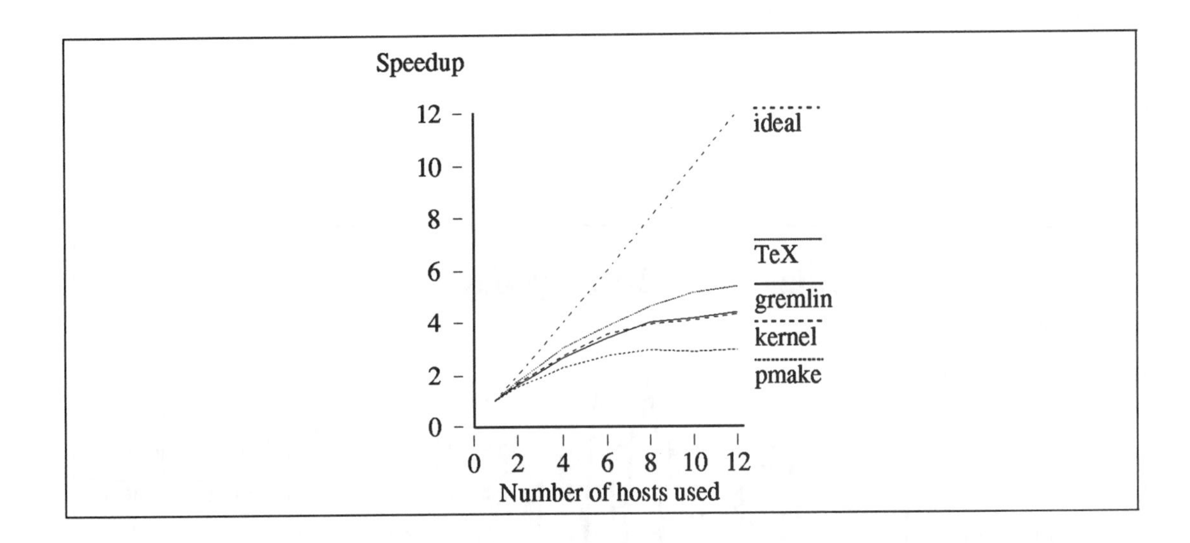

Figure 5: *Speedup of compilations using a variable number of hosts.* This graph shows the speedup relative to running *pmake* on one host (*i.e.*, without migration). The speedup obtained depends on the extent that hosts can be kept busy, the amount of parallelization available to *pmake*, and system bottlenecks.

nearly 100% of a SPARCstation processor while it determines dependencies and starts to migrate processes to perform compilations. Then the Sun-4/280 file server's processor becomes a bottleneck as the 12 hosts performing compilations open files and write back cached object files. The network utilization, also shown in Figure 6, averaged around 20% and is thus not yet a problem. However, as the server and client processors get faster, the network may easily become the next bottleneck.

Though migration has been used in Sprite to perform compilations for nearly two years, it has only recently been used for more wide-ranging applications. Excluding compilations, simulations are the primary application for Sprite's process migration facility. It is now common for users to use *pmake* to run up to one hundred simulations, letting *pmake* control the parallelism. The length and parallelism of simulations results in more frequent evictions than occur with most compilations, and *pmake* automatically remigrates or suspends processes subsequent to eviction.

In addition to having a longer average execution time, simulations also sometimes differ from compilations in their use of the file system. While some simulators are quite I/O intensive, others are completely limited by processor time. Because they perform minimal interaction with file servers and use little network bandwidth, they can scale better than parallel compilations do. One set of simulations obtained over 800% effective processor utilization—eight minutes of processing time per minute of elapsed time—over the course of an hour, using all idle hosts on the system (up to 10–15 hosts of the same architecture).

Usage Patterns

We instrumented Sprite to keep track of remote execution, migrations, evictions, and the availability of idle hosts. First, when a process exited, the total time during which it executed was added to a global counter; if the process had been executing remotely, its time was added

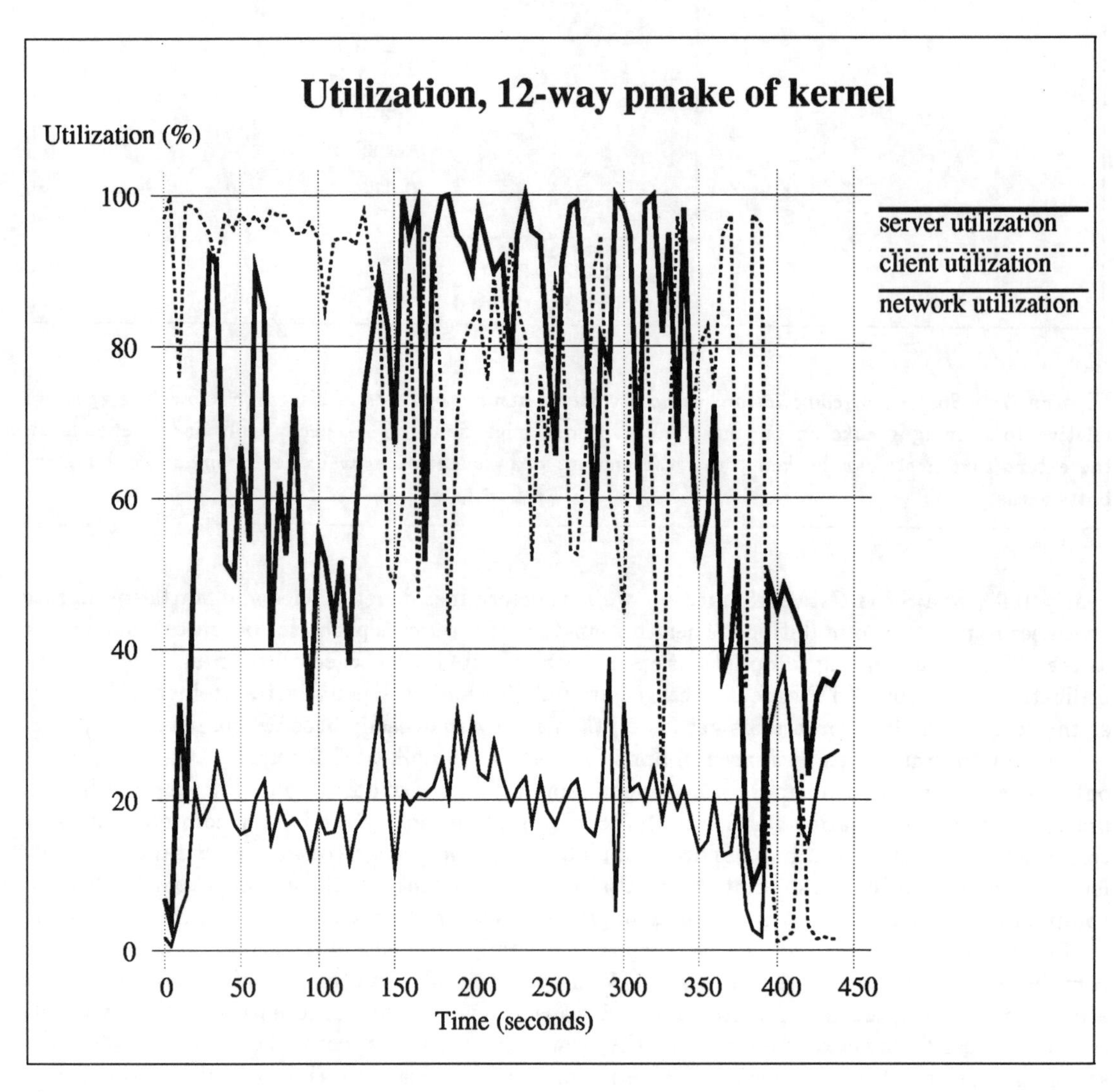

Figure 6: *Processor and network utilization during the 12-way* pmake. Both the file server and the client workstation running *pmake* were saturated.

Host	Total CPU Time	Remote CPU Time	Fraction Remote
garlic	314,218 secs	228,641 secs	72.77 %
crackle	172,355	14,451	8.38 %
sassafras	158,515	138,821	87.58 %
burble	151,117	2,352	1.56 %
vagrancy	107,853	81,343	75.42 %
buzz	96,402	260	0.27 %
sage	92,063	32,525	35.33 %
kvetching	91,611	26,765	29.22 %
jaywalk	75,394	24,017	31.86 %
joyride	58,231	6,233	10.70 %
Others	857,532	120,727	14.1 %
Total	2,175,291	676,135	31.08 %

Table 4: *Remote processing use over a one-month period.* The ten hosts with the greatest total processor usage are shown individually. Sprite hosts performed roughly 30% of user activity using process migration. The standard deviation of the fraction of remote use was 25%.

to a separate counter as well. (These counters therefore excluded some long-running processes that did not exit before a host rebooted; however, these processes were daemons, display servers, and other processes that would normally be unsuitable for migration.) Over a typical one-month period, remote processes accounted for about 31% of all processing done on Sprite. One host ran applications that made much greater use of remote execution, executing as much as 88% of user cycles on other hosts. Table 4 lists some sample processor usage over this period.

During the same time frame, we recorded the frequency of *exec*-time migrations and full migrations in order to determine the most common usage of the migration facility. Since full migrations require that virtual memory be copied, the choice of a virtual memory transfer method would be important if full migrations occurred relatively often. In the one-month period studied, *exec*-time migrations occurred at a rate of 1.76/hour/host over that period, constituting 86% of all migrations.

Second, we recorded each time a host changed from *idle* to *active*, indicating that foreign processes would be evicted if they exist, and we counted the number of times evictions actually occured. To date, evictions have been extremely rare. On the average, each host changed to the *active* state only once every 26 minutes, and very few of these transitions actually resulted in processes being evicted (0.12 processes per hour per host in a collection of more than 25 hosts). The infrequency of evictions has been due primarily to the policy used for allocating hosts: hosts are assigned in decreasing order of idle time, so that the hosts that have been idle the longest are used most often for migration. The average time that hosts had been idle prior to being allocated for remote execution was 17 hours, but the average idle time of those hosts that later evicted processes was only 4 minutes. (One may therefore assume that if hosts were allocated randomly, rather than in order of idle time, evictions would be considerably more frequent.) Finally, when evictions did occur, the time needed to evict varied considerably, with a mean of 3.0 seconds and a standard deviation of 3.1 seconds to migrate an average of 3.3 processes. An average of 37 4-Kbyte pages were written per process that migrated, with a standard deviation of 6.5 from host to host.

Third, over the course of over a year, we periodically recorded the state of every host (active,

Time Frame	In Use	Idle	In Use for Migration
weekdays	31 %	66 %	3 %
off-hours	20 %	78 %	2 %
total	23 %	75 %	2 %

Table 5: *Host availability.* Weekdays are Monday through Friday from 9:00 A.M. to 5:00 P.M. Off-hours are all other times.

idle, or hosting foreign processes) in a log file. A surprisingly large number (66–78%) of hosts are available for migration at any time, even during the day on weekdays. This is partly due to our environment, in which several users own both a Sun and a DECstation and use only one or the other at a time. Some workstations are available for public use and are not used on a regular basis. However, after discounting for extra workstations, we still find a sizable fraction of hosts available, concurring with Theimer, Nichols, and others. Table 5 summarizes the availability of hosts in Sprite over this period.

To further study the availability of idle hosts, we recorded information about requests for idle hosts over a 25-day period. During this period, over 17,000 processes requested one or more idle hosts, and 86% of those processes obtained as many hosts as they requested. Only 2% of processes were unable to obtain any hosts at all. Processes requested an average of 2.6 hosts, with a standard deviation of 4.58 hosts and 76% of processes requesting at most one host at a time. Since there were typically 10 or more idle machines available for each machine type, one would expect processes that request few hosts to be able to obtain them; more interestingly, however, over 80% of those hosts requesting at least 10 hosts were able to obtain 10 hosts. Figure 7 shows the fraction of processes during this period that received as many hosts as requested, as a cumulative function of the number of hosts requested.

Observations

Based on our experience, as well as those of others (V,[8] Charlotte,[9] and Accent [17]), we have observed the following:

- The overall improvement from using idle hosts can be substantial, depending upon the degree of parallelism in an application.
- Remote execution currently accounts for a sizable fraction of all processing on Sprite. Even so, idle hosts are plentiful. Our use of idle hosts is currently limited more by a lack of applications (other than *pmake*) than by a lack of hosts.
- The cost of *exec*-time migration is high by comparison to the cost of local process creation, but it is relatively small compared to times that are noticeable by humans. Furthermore, the overhead of providing transparent remote execution in Sprite is negligible for most classes of processes. The system may therefore be liberal about placing processes on other hosts at *exec* time, as long as the likelihood of eviction is relatively low.
- The cost of transferring a process's address space and flushing modified file blocks dominates the cost of migrating long-running processes, thereby limiting the effectiveness of a

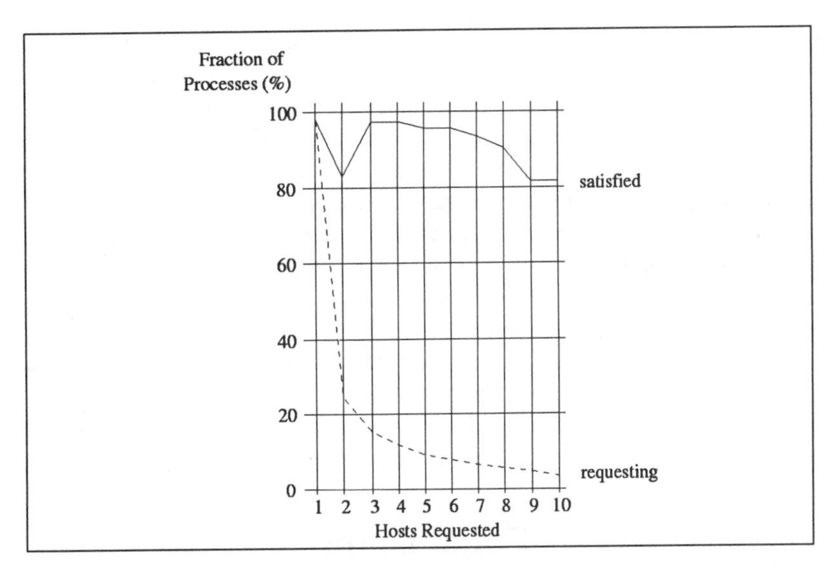

Figure 7: ***Distribution of host requests and satisfaction rates.*** For a given number of hosts, shown on the X-axis, the line labeled ***requesting*** shows the fraction of processes that requested at least that many hosts. The line labeled ***satisfied*** shows, out of those processes that requested at least that number of hosts, the fraction of processes that successfully obtained that many hosts. Thus, 98% of all processes were able to obtain at least one host, and over 80% of processes that requested at least ten hosts obtained 10 hosts. Only 24% of processes requested more than one host.

dynamic "pool of processors" approach. Although there are other environments in which such an approach could have many favorable aspects, given our assumptions above about host availability and workstation "ownership", using process migration to balance the load among all Sprite hosts would likely be both unnecessary and undesirable.

History and Experience

The greatest lesson we have learned from our experience with process migration is the old adage "use it or lose it." Although an experimental version of migration was operational in 1986,[23] it took another two years to make migration a useful utility. Part of the problem was that a few important mechanisms weren't implemented initially (*e.g.*, there was no automatic host selection, migration was not integrated with *pmake*, and process migration did not deal gracefully with machine crashes). But the main problem was that migration continually broke due to other changes in the Sprite kernel. Without regular use, problems with migration weren't noticed and tended to accumulate. As a result, migration was only used for occasional experiments. Before each experiment a major effort was required to fix the accumulated problems, and migration quickly broke again after the experiment was finished.

By the fall of 1988 we were beginning to suspect that migration was too fragile to be maintainable. Before abandoning it we decided to make one last push to make process migration completely usable, integrate it with the *pmake* program, and use it for long enough to understand its benefits as well as its drawbacks. This was a fortunate decision. Within one week after migration became available in *pmake*, other members of the Sprite project were happily using it and achieving speed-up factors of two to five in compilations. Because of its complex interactions with the rest of the kernel, migration is still more fragile than we would like and it occasionally breaks in response to other changes in the kernel. However, it is used so frequently that problems are detected immediately and they can usually be fixed quickly. The maintenance load is still higher for migration than for many other parts of the kernel, but only slightly. Today we consider migration to be an indispensable part of the Sprite system.

We are not the only ones to have had difficulties keeping process migration running: for example, Theimer reported similar experiences with his implementation in V.[8] The problem seems to be inherent in migration, since it interacts with many other parts of the kernel. In Sprite the most complicated aspects of migration were those related to migrating open files. In particular, locking and updating the data structures for an open file on multiple hosts provided numerous opportunities for distributed deadlocks, race conditions, and inconsistent reference counts. It is worth reiterating that these problems would have been present even if we had chosen to implement a "simpler" remote invocation facility without process migration.

Conclusions

Process migration is now taken for granted as an essential part of the Sprite system. It is used hundreds of times daily and provides substantial speed-ups for applications that are amenable to coarse-grain parallel processing, such as compilation and simulation. The transparency provided by the migration mechanism makes it easy to use migration, and eviction keeps migration from bothering the people whose machines are borrowed. Collectively, remote execution accounts for a sizable portion of all user activity on Sprite.

We were originally very conservative in our use of migration, in order to gain acceptance among our users. As time has passed, our users have become accustomed to their workstations being used for migration and they have gained confidence in the eviction mechanism. We have gradually become more liberal about using idle machines, and we are experimenting with new system-wide migration tools, such as command shells that automatically migrate some tasks (*e.g.*, jobs run in background). So far our users have appreciated the additional opportunities for migration and have not perceived any degradation in their interactive response.

From the outset we expected migration to be difficult to build and maintain. Even so, we were surprised at the complexity of the interactions between process migration and the rest of the kernel, particularly where distributed state was involved as with open files. It was interesting that Sprite's network file system both simplified migration (by providing transparent remote access to files and devices) and complicated it (because of the file system's complex distributed state). We believe that our implementation has now reached a stable and maintainable state, but it has taken us a long time to get there.

For us, the bottom line is that process migration is too useful to pass up. We encourage others to make process migration available in their systems, but to beware of the implementation pitfalls.

Acknowledgments

In addition to acting as guinea pigs for the early unstable implementations of process migration, other members of the Sprite project have made significant contributions to Sprite's process migration facility. Mike Nelson and Brent Welch implemented most of the mechanism for migrating open files, and Adam de Boor wrote the *pmake* program. Lisa Bahler, Thorsten von Eicken, John Hartman, Darrell Long, Mendel Rosenblum, and Ken Shirriff provided comments on early drafts of this paper, which improved the presentation substantially. We are also grateful to the anonymous referees of *Software Practice & Experience*, who provided valuable feedback and suggestions. Of course, any errors in this article are our responsibility alone.

References

1. D. Nichols. Using idle workstations in a shared computing environment. In *Proceedings of the Eleventh ACM Symposium on Operating Systems Principles*, pages 5–12, Austin, TX, November 1987. ACM.

2. E. Roberts and J. Ellis. *parmake* and *dp*: Experience with a distributed, parallel implementation of make. In *Proceedings from the Second Workshop on Large-Grained Parallelism*. Software Engineering Institute, Carnegie-Mellon University, November 1987. Report CMU/SEI-87-SR-5.

3. M. Theimer, K. Lantz, and D. Cheriton. Preemptable remote execution facilities for the V-System. In *Proceedings of the 10th Symposium on Operating System Principles*, pages 2–12, December 1985.

4. E. Zayas. Attacking the process migration bottleneck. In *Proceedings of the Eleventh ACM Symposium on Operating Systems Principles*, pages 13–22, Austin, TX, November 1987.

5. G. J. Popek and B. J. Walker, editors. *The LOCUS Distributed System Architecture.* Computer Systems Series. The MIT Press, 1985.

6. A. Barak, A. Shiloh, and R. Wheeler. Flood prevention in the MOSIX load-balancing scheme. *IEEE Computer Society Technical Committee on Operating Systems Newsletter*, 3(1):23–27, Winter 1989.

7. S. I. Feldman. Make — a program for maintaining computer programs. *Software—Practice and Experience*, 9(4):255–265, April 1979.

8. M. Theimer. *Preemptable Remote Execution Facilities for Loosely-Coupled Distributed Systems.* PhD thesis, Stanford University, 1986.

9. Y. Artsy and R. Finkel. Designing a process migration facility: The Charlotte experience. *IEEE Computer*, 22(9):47–56, September 1989.

10. J. K. Ousterhout, A. R. Cherenson, F. Douglis, M. N. Nelson, and B. B. Welch. The Sprite network operating system. *IEEE Computer*, 21(2):23–36, February 1988.

11. M. Nelson, B. Welch, and J. Ousterhout. Caching in the Sprite network file system. *ACM Transactions on Computer Systems*, 6(1):134–154, February 1988.

12. A. D. Birrell and B. J. Nelson. Implementing remote procedure calls. *ACM Transactions on Computer Systems*, 2(1):39–59, February 1984.

13. Computer Science Division, University of California, Berkeley. *UNIX User's Reference Manual, 4.3 Berkeley Software Distribution, Virtual VAX-11 Version*, April 1986.

14. M. Litzkow. Remote UNIX. In *Proceedings of the USENIX 1987 Summer Conference*, June 1987.

15. M. Accetta, R. Baron, W. Bolosky, D. Golub, R. Rashid, A. Tevanian, and M. Young. Mach: A new kernel foundation for UNIX development. In *Proceedings of the USENIX 1986 Summer Conference*, July 1986.

16. D. R. Cheriton. The V distributed system. *Communications of the ACM*, 31(3):314–333, March 1988.

17. E. Zayas. *The Use of Copy-On-Reference in a Process Migration System.* PhD thesis, Carnegie Mellon University, Pittsburgh, PA, April 1987. Report No. CMU-CS-87-121.

18. K. Li and P. Hudak. Memory coherence in shared virtual memory systems. *ACM Transactions on Computer Systems*, 7(4):321–359, November 1989.

19. D. L. Eager, E. D. Lazowska, and J. Zahorjan. The limited performance benefits of migrating active processes for load sharing. In *ACM SIGMETRICS 1988*, May 1988.

20. D. L. Eager, E. D. Lazowska, and J. Zahorjan. Adaptive load sharing in homogeneous distributed systems. *IEEE Transactions on Software Engineering*, SE-12(5):662–675, May 1986.

21. E. H. Baalbergen. Parallel and distributed compilations in loosely-coupled systems: A case study. In *Proceedings of Workshop on Large Grain Parallelism*, Providence, RI, October 1986.

22. F. Douglis. *Transparent Process Migration in the Sprite Operating System.* PhD thesis, University of California, Berkeley, CA 94720, September 1990. Available as Technical Report UCB/CSD 90/598.

23. F. Douglis and J. Ousterhout. Process migration in the Sprite operating system. In *Proceedings of the 7th International Conference on Distribu ted Computing Systems*, pages 18–25, Berlin, West Germany, September 1987. IEEE.

Sprite Process Migration: a Retrospective

Frederick Douglis

August, 1998

As I write this, it has been about ten years since process migration moved into general use in Sprite at U.C. Berkeley, and about five years since the entire Sprite system was last used.

Within the context of the Sprite system [5], and its users and designers, migration was an unqualified success once it was stable enough to be used on a regular basis. It provided an enormous boost in the computational facilities available, both for compilations (a task that most of us performed many times each day) and computationally-intensive applications such as simulations. The ability to "evict" foreign processes was an important sociological factor in gaining acceptance of process migration even among those who were not benefitting from it directly.

Unfortunately, Sprite as a whole did not achieve a long-lasting success, so its process migration facility suffered with it. Sprite's failure to expand significantly beyond U.C. Berkeley was due to a conscious decision among its designers not to invest the enormous effort that would have been required to support a large external community. Instead, individual ideas from Sprite, particularly in the areas of file systems [4, 6] and virtual memory [3, 4], migrated into commercial systems (e.g., Network Appliance's Write Anywhere File System [1]) over time.

The failure of Sprite's process migration facility to similarly influence the commercial marketplace has come as a surprise. Ten years ago I would have predicted that process migration in UNIX would be commonplace today, despite the difficulties in supporting it. Instead, user-level load distribution is commonplace [2, 7], but it is commonly limited to applications that can run on different hosts without ill effects, and relies either on explicit checkpointing or the ability to run to completion.

One reason for the failure of full migration to become commonplace commercially is no doubt the rapid improvement of processing capacities and other resources. Years ago, interactive performance was much more likely to suffer because of the intrusion of cpu-intensive or memory-intensive processes, and evicting them seemed important. Today, offloading work to idle machines is still important, but reclaiming those idle machines is less so.

Migration in Sprite had other benefits besides eviction, of course: its degree of transparency was much greater than many other systems before or since. Supporting fully transparent remote execution, rather than a limited set of applications, may still prove useful in the future. And, given that degree of transparency, full-fledged process migration may be just a step away.

References

[1] D. Hitz, J. Lau, and M. Malcolm. File system design for an NFS file server appliance. In *Proceedings of the USENIX 1994 Winter Conference*, pages 235–245, January 1994.

[2] M. Litzkow and M. Solomon. Supporting checkpointing and process migration outside the UNIX kernel. In *Proceedings of the USENIX Winter 1992 Technical Conference*, pages 283–290, January 1991.

[3] M. Nelson and J. Ousterhout. Copy-on-write for Sprite. In *USENIX 1988 Summer Conference*, pages 187–201, San Francisco, CA, June 1988.

[4] M. Nelson, B. Welch, and J. Ousterhout. Caching in the Sprite network file system. *ACM Transactions on Computer Systems*, 6(1):134–154, February 1988.

[5] J. Ousterhout, A. Cherenson, F. Douglis, M. Nelson, and B. Welch. The Sprite network operating system. *IEEE Computer*, 21(2):23–36, February 1988.

[6] M. Rosenblum and J. Ousterhout. The design and implementation of a log-structured file system. *ACM Transactions on Computer Systems*, 10(1):26–52, February 1992.

[7] S. Zhou, X. Zheng, J. Wang, and P. Delisle. Utopia: A load sharing facility for large, heterogeneous distributed computer systems. *Software—Practice and Experience*, 23(12):1305–1336, December 1993.

4.3 Artsy, Y. and Finkel, R., Designing a Process Migration Facility: The Charlotte Experience

Artsy, Y. and Finkel, R., "Designing a Process Migration Facility: The Charlotte Experience," *Computer*, 22(9):47-56, September 1989.

Designing a process migration facility: The Charlotte experience

Yeshayahu Artsy
Digital Equipment Corporation
550 King Street
Littleton, Massachusetts 01460

Raphael Finkel
Computer Science Department
University of Kentucky
Lexington, KY 40506-0027

1998 addresses
raphael@cs.uky.edu y-artsy@msn.com

This paper was published in *IEEE Computer* **22:** 9, pp. 47-56, September 1989.

Key words: Operating Systems, Distributed Systems, Process Migration, System Design

Abstract

Our goal in this paper is to discuss our experience with process migration in the Charlotte distributed operating system. We also draw upon the experience of other operating systems in which migration has been implemented. A process migration facility in a distributed operating system dynamically relocates processes among the component machines. A successful process migration facility is not easy to design and implement. Foremost, a general-purpose migration mechanism should be able to support a range of policies to meet various goals, such as load distribution, and improved concurrency, and reduced communication. We discuss how Charlotte's migration mechanism detaches a running process from its source environment, transfers it, and attaches it into a new environment on the destination machine. Our mechanism succeeds in handling communication and machine failures that occur during the transfer. Migration does not affect the course of execution of the migrant nor that of any process communicating with it. The migration facility adds negligible overhead to the distributed operating system and provides fast process transfer.

1. Introduction

Our goal in this paper is to discuss our experience with process migration in the Charlotte distributed operating system. We identify major design issues and explain our implementation choices. We also contrast these choices with other migration implementations in the literature. This paper provides insights both into the specific task of implementing process migration for distributed operating systems and into the more general task of designing such systems.

A **preemptive process migration** facility dynamically relocates running processes between peer machines in a distributed system. Such relocation has many advantages. Studies have shown that it can be used to cope with dynamic fluctuations in loads and service needs[1], to meet real-time scheduling deadlines, to bring a process to a special device, or to improve the fault-tolerance of the system. Yet successful process migration facilities are not commonplace in distributed operating systems[2,3,4,5,6,7]. The reason for this paucity is the inherent complexity of such a facility and the potential execution penalty if the migration policy and mechanism are not tuned correctly. It is not surprising that some operating systems prefer to terminate remote processes rather than rescue them by migration.

We can identify several reasons why migration is hard to design and implement. The mechanism for moving processes must be able to detach a migrant process from its source environment, transfer it with its context (the per-process data structures held in the kernel) and attach it in a new environment on the destination machine. These actions should complete reliably and efficiently. Migration may fail in case of machine and communication failures, but it should do so completely. That is, the effect should be as if the process was never migrated at all, or at worst as if the process had terminated due to machine failure. A wide range of migration policies might be needed, depending on whether the main concern is load sharing (avoiding idle time on one machine when another has a non-trivial work queue), load balancing (such as keeping the work queues of similar length), or application concurrency (mapping application processes to machines in order to achieve high parallelism). Policies may need elaborate and timely state information, since otherwise unnecessary process relocations may inflict performance degradation on both the migrant process and the entire system. The mechanisms to support different policies might differ significantly. If several policies are used under different circumstances, the migration mechanism must be flexible enough to allow policy modules to switch policies. The migration mechanism cannot be completely separated from process scheduling, memory management, and interprocess communication. Nevertheless, one would prefer to keep mechanisms for these activities as separate from each other as possible, to allow more freedom in testing and upgrading them. The fact that a process has moved should be invisible both to it and its peers, while at the same time interested users or processes should be able to advise the system about desired process distribution.

The process migration facility implemented for Charlotte is a fairly elaborate addition to the underlying Charlotte kernel and utility-process base. It separates policy (when to migrate which process to what destination) from mechanism (how to detach, transfer, and reattach the migrant process). While the mechanism is fixed in the kernel and one of the utilities, the policy is relegated to a utility and can be replaced easily. The kernel provides elaborate state information to that utility. The mechanism allows concurrent multiple migrations and premature cancellation of migration. It leaves no residual dependency on the source machine for the migrant process. The mechanism copes with all conceivable crash and termination scenarios, rescuing the migrant process in most cases.

The next section presents an overview of Charlotte, and Section 3 presents its process migration facility. In Section 4 we discuss the issues encountered in building Charlotte's migration facility that have general application for any such facility. In discussing each issue, we present alternative design approaches adopted by other process migration facilities. We leave out the discussion of specific migration policies, as they are beyond the scope of this paper. We hope this account will give assistance to others contemplating adding a process migration facility, as well as advice to those designing other operating system facilities that may interact with later addition of process migration. We conclude with a brief list of concrete lessons and suggestions.

2. Charlotte overview

Charlotte is a message-based distributed operating system developed at the University of Wisconsin for a multicomputer composed of 20 VAX-11/750 computers connected by a token ring[9]. Each machine in Charlotte runs a kernel responsible for simple short-term process scheduling and a message-based inter-process communication (IPC) protocol. Processes are not swapped to backing store. A battery of privileged processes (**utilities**) runs at the user level to provide additional operating systems services and policies. The kernel and some utilities are multithreaded.

Processes communicate via **links**, which are capabilities for duplex communication channels. (The high-level language Lynx[10] actually hides this low-level mechanism and provides a remote procedure call (RPC) interface.) The processes at the two ends of a link may both send and receive messages by using non-blocking service calls. A process may post several such requests and await their completion later; it may cancel a pending request before that request completes. A link may be destroyed or given away to another process even during communication. In particular, a link is automatically destroyed when the process holding its other end terminates or its machine crashes; the process holding the local link end is so notified by the kernel. The protocol that implements communication semantics is efficient but quite complex[11]. It depends on full, up-to-date link information in the kernels of both ends of each link. Processes are completely unaware of the location of their communicating partners. Instead, they establish links to servers by having other processes (their parents or a name server utility) provide them.

Utility processes are distributed throughout the multicomputer, cooperating to allocate resources, provide file and connection services, and set policy. In particular, the **KernJob (KJ)** utility runs on each machine to provide a communication path between the local kernel and non-local processes. The **Starter** utility creates processes, allocates memory, and dictates medium-term scheduling policy. Each Starter process controls a subset of the machines; it communicates with their kernels (directly or via their KJs) to receive state information and specify its decisions.

3. Process migration in Charlotte

Charlotte was designed as a platform for experimentation with distributed algorithms and load distribution strategies. We added the process migration facility in order to better support such experiments. Equally important, we wanted to explore the design issues that process migration raises in a message-based operating system. Figure 1 shows the effect of process migration. For convenience, throughout the paper we call the kernel on the source and destination machines **S** and **D**, respectively, and use **P** to represent the migrant process. During transfer, **P**'s process identifier changes, and the kernel data structures for it are completely removed from **S**, but the transfer is invisible to both **P** and its communication partners.

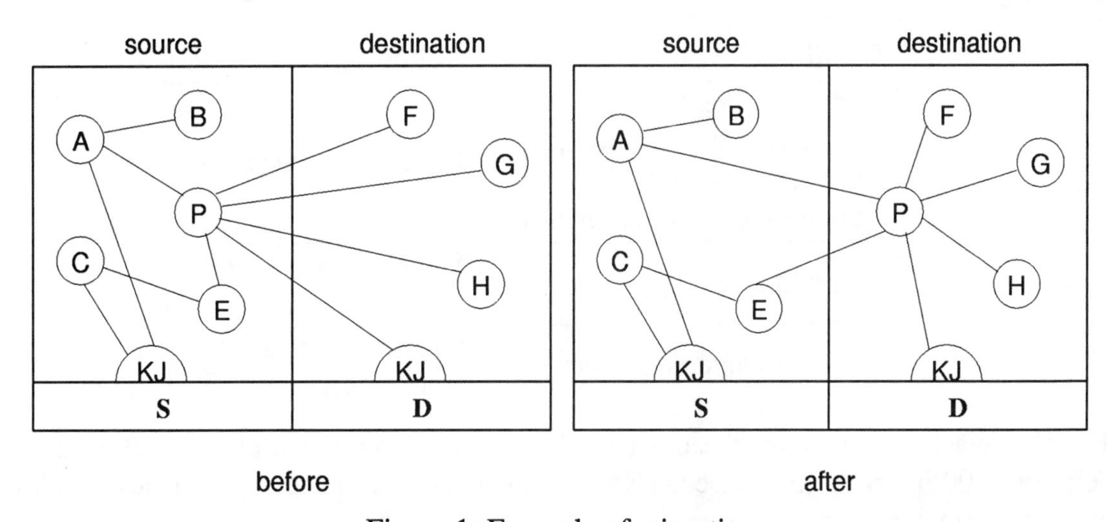

Figure 1: Example of migration

As shown, **P**'s links are relocated to the new machine. All processes continue to name their links the same after migration; they are unaware that link descriptors have moved sites and that local communication (performed in shared memory) has become remote communication (sent over the wire) and vice versa, and they see no change in message flow.

3.1. Policy

Migration policy is dictated by Starter utility processes. They base their decisions on statistical information provided by the kernels they control and on summary information they exchange among themselves. In addition, Starters accept advice from privileged utilities (to allow manual direction of migration and to enable or disable automatic control). When messages carrying statistics, advice, or notice of process creation or termination arrive, the Starter executes a policy procedure. (Introducing migration into the Starter only required writing that policy procedure and invoking it at the right times.) The policy procedure may choose to send messages to other Starters or to request some source kernel to undertake migration. Such requests are sent to the

KernJob residing on the source machine to relay to its kernel. As discussed later, this approach adds insignificantly to the cost of migration (a few procedure calls and perhaps a round-trip message), while it allows policies that integrate scheduling and memory allocation as well as local, clustered, or global policies.

3.2. Mechanism

The migration mechanism has two independent parts: collecting statistics and transferring processes. Both parts are implemented in the kernel.

Statistics include data on machine load (number of processes, links, CPU and network loads), individual processes (age, state, CPU utilization, communication rate), and selected active links (packets sent and received). These statistics are intended to be comprehensive enough to support most conceivable policies. We collect statistics in the following way.

Condition	**Action**
Significant event: message sent or received, data structure freed process created or terminated	Increment associated count
Interval passes	Sample process states and CPU, network loads
Period of *n* intervals passes	Summarize data, Send to starter

To balance accuracy with overhead, we used in our tests an interval of 50 to 80 ms and a period of 100 intervals (5 to 8 seconds). The overhead for collecting statistics was less than 1% of total cpu time.

Transferring processes occurs in three phases.

(1) **Negotiation**. After being told by their controlling Starter processes to migrate **P**, **S** and **D** agree to the transfer and reserve required resources. If agreement cannot be reached, for example because resources are not available, migration is aborted and the Starter that requested it is notified.

(2) **Transfer**. **P**'s address space is moved from the source to the destination machine. Meanwhile, separate messages are sent to each kernel controlling a process with a link to **P** informing that kernel of the link's new address.

(3) **Establishment**. Kernel data structures pertaining to the migrant process are marshaled, transferred, and demarshaled. (Marshaling requires copying the structure to a byte-stream buffer, and converting some data types, particularly pointer types.) No information related to the migrant is retained at the source machine.

Process-kernel interface

We added four kernel calls to the process-kernel interface.

`Statistics(What : action; Where : address)`

The KernJob invokes this call (on behalf of a Starter) so that the kernel will start collecting statistics and placing them in the given address (in the KernJob virtual space). The call can also be used to stop statistics collection.

`MigrateOut(Which : process; WhereTo : machine)`

This call enables the Starter (or its KernJob proxy, if the Starter resides on another machine) to initiate a migration episode.

`MigrateIn(Which : process; WhereFrom : machine; Accept : Boolean; Memory : list of physical regions)`

The Starter (or its KernJob proxy) uses this call to approve or refuse a migration from the given machine to the machine on which the call is performed. If Starters have negotiated among themselves, the Starter controlling the destination machine may approve a migration even before the one controlling the source machine calls `MigrateOut`. The Memory parameter tells the kernel where in physical store to place the segments that constitute the new process. (The Starter learns the segment sizes either through negotiation with its peer or from **D**'s request to approve a migration offer received from **S**.)

`CancelMigration(Which : process; Where : machine)`

The Starter invokes this call to abort an active `MigrateIn` or `MigrateOut` request. This call is rejected if the migration has already reached a commit point.

None of these calls blocks the caller. The kernel reports the eventual success or failure of the request by a message back to the caller.

Mechanism details

Three new modules were created in the kernel to implement the migration mechanism. The migration interface module deals with the new service calls from processes. The migration protocol module performs the three phases listed above. The statistics module collects and reports statistics. These modules are invoked by two new kernel threads. The statistician thread awakens at each interval to sample, or average and report statistics to the Starter. A process-receiver thread starts in **D** for each incoming migrant process. It uses a simpler and faster communication protocol than that used by ordinary IPC.* However, negotiation and other control messages use the ordinary communication protocol and are funneled through the IPC queues in order to synchronize process and link activities.

Figure 2 shows both high- and low- level negotiation messages. In our example, the left Starter process controls machine 1, and its peer controls machine 3. The first two

*The standard protocol must expect extremely complex scenarios that cannot arise in this conversation and must employ link data structures that are not germane here. The cost of introducing a streamlined protocol was slight in comparison to the speed it achieved.

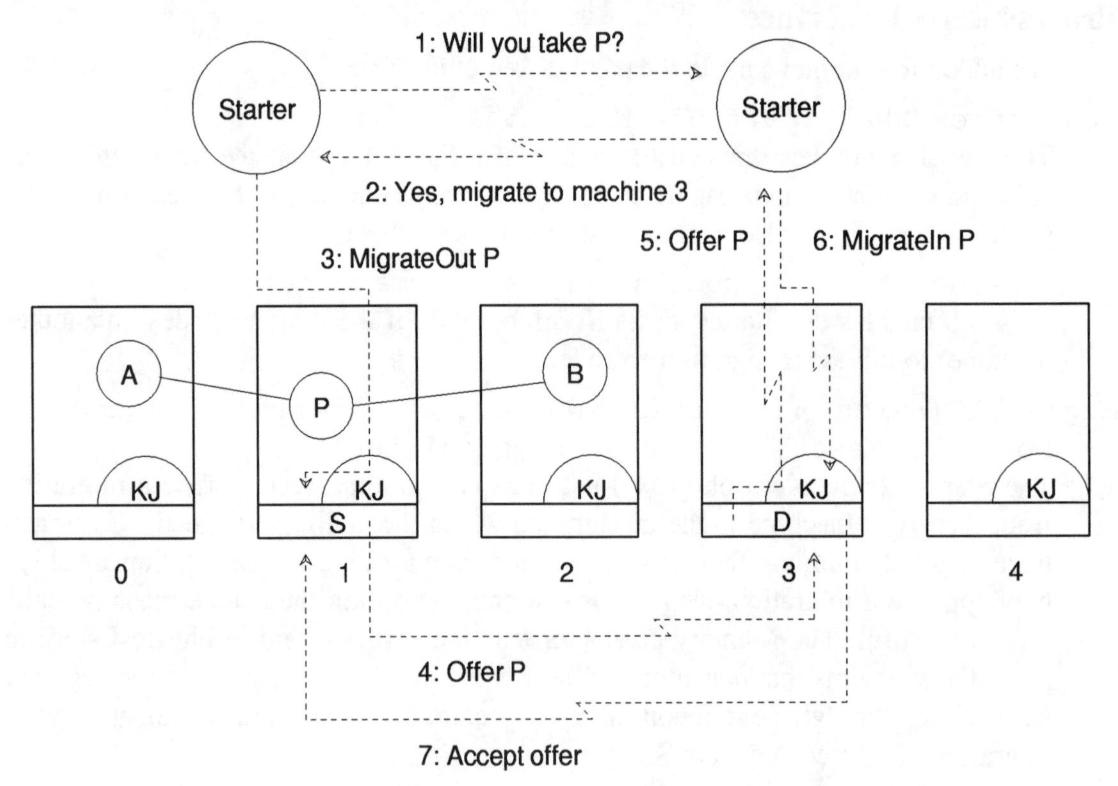

Figure 2: Negotiation phase

messages represent a Starter-to-Starter negotiation that results in deciding to migrate process **P** from machine 1 to 3. Their decision is communicated to **S** in message 3, which is either a direct service call (if the Starter runs on machine 1) or a message to the KernJob on machine 1 to be translated into a service call. **S** then offers to send the process to **D**. The offer includes **P**'s memory requirements, its age, its recent CPU and network use, and information about its links. If **D** is short of the resources needed for **P**, or if too many migrations are in progress, it may reject the offer outright. Otherwise, **D** relays the offer to its controlling Starter (message 5). The relay includes the same information as the offer from **S**. We relay the offer to let the policy module reject a migrant at this point. Although that Starter may have already agreed to accept **P** (in message 2), it may now need to reject the offer due to an increase in actual or anticipated load or lack of memory. Furthermore, the Starter must be asked because the kernel has no way to know if it has even been consulted by its peer Starter, and the Starter must allocate memory for the migrant. The Starter's decision is communicated to **D** by a `MigrateIn` call (message 6). No relay occurs if the Starter has already called `MigrateIn` to preapprove the migration. Before responding to **S** (message 7), **D** reserves necessary resources to avoid deadlock and flow-control problems. Preallocation is conservative; it guarantees successful completion of multiple migrations at the expense of reducing the number of concurrent incoming migrations.

After message 7 is sent, **D** has committed itself to the migration. If **P** fails to arrive and the migration has not been cancelled by **S** (see next), then the machine of **S** must be down or unreachable. **D** discovers this condition through the standard mechanism by which kernels exchange "heart-beat" messages and reclaims resources and cleans up its state.

When message 7 is received, **S** is also committed and starts the transfer. Before each kernel commits itself, its Starter can successfully cancel the migration, in which case **D** replies *Rejected* to **S** (in message 7), or **S** sends *Regretted* to **D** (not shown). The latter also occurs if **P** dies abruptly during negotiation. To separate policy from mechanism, **S** does not retry a rejected migration unless so ordered by its Starter.

Figure 3 shows the transfer phase. **S** concurrently sends **P**'s virtual space to **D** (message 8) and link update messages (9) to the kernels controlling all of **P**'s peers. Message 8 is broken into packets as required by the network. **D** has already reserved physical store for them, so the packets are copied directly into the correct place. Message 9 indicates the new address of the link; it is acknowledged (not shown) for synchronization purposes. After this point, messages sent to **P** will be directed to the new address and buffered there until **P** is reattached. Kernels that have not received message 9 yet may still continue to send messages for **P** to **S**. Failure of either the source or the destination machine during this interval leaves the state of **P** very unclear. Since it would require a very complex protocol (sensitive to further machine failures) to recover **P**'s state, we opted to terminate **P** if one of these machines crashes at this stage.

Finally, **S** collects all of **P**'s context into a single message and sends it to **D** (message 10). This message includes control information, the state of all of **P**'s links, and details of communication requests that have arrived for **P** since transfer began. Pointers in **S**'s data structures are tracked down, and all relevant data are marshaled together. **D**

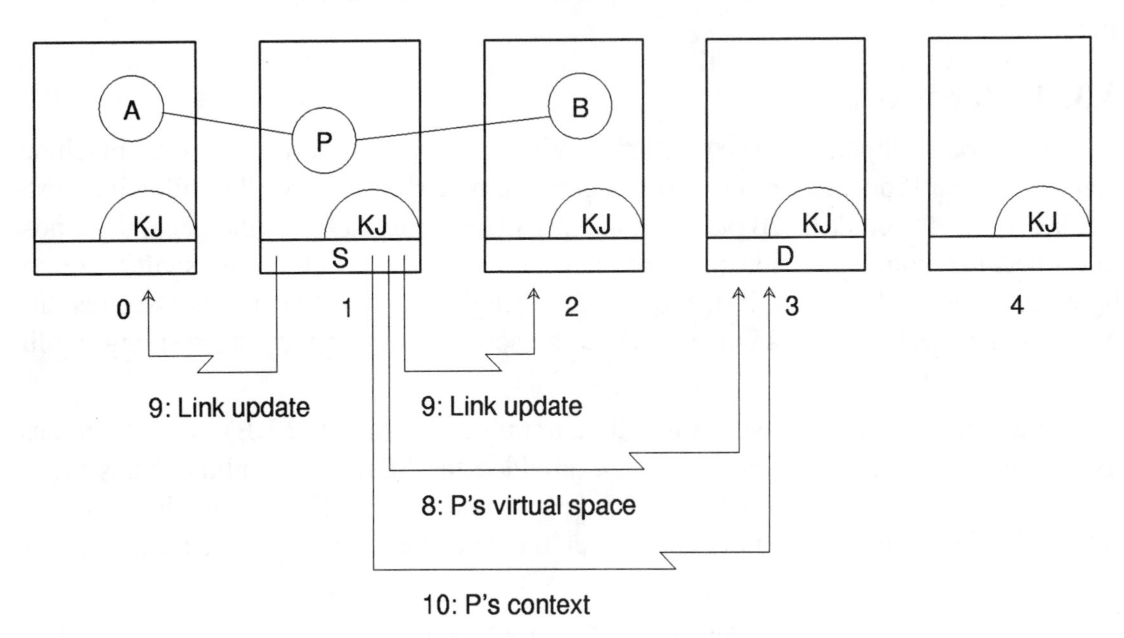

Figure 3: Transfer phase

demarshals the message into its own data structures.

Although it is conceptually simple, the transfer stage is actually quite complex and time-consuming, mainly because Charlotte IPC is rich in context. Luckily, our design saves the migration mechanism from dealing with messages in transit to or from **P**. Since the kernel provides message caching but no buffering, a message remains in its sender's virtual address space until the receiver is ready to accept it or until the send is cancelled. Hence, **S** does not need to be concerned with **P**'s outgoing messages; likewise, it may drop from its cache any message received for **P** that **P** has not yet received. Such a message will be requested by **D** from the sender's kernel when **P** requests to receive it. The link structures sent in message 9 clearly indicate which links have pending sent or received messages. Another advantage in our design is that we do not have to alter or transfer **P**'s context maintained by distributed utilites, such as open files, which are accessed via location-independent links.

The establishment phase is interleaved with transfer. Data structures are deallocated as part of marshaling, and the reserved ones are filled during demarshaling. After transfer has finished, **D** adjusts **P**'s links and pending events and inserts **P** into the appropriate scheduling queue. Those communication requests that were postponed while **P** was moving have been buffered by **S** and **D**; they are now directed to the IPC kernel thread in **D** in their order of arrival. (For each link, all those buffered at **S** precede those buffered at **D**.) Their effect on **P** is not influenced by the fact that it has moved. Finally, the Starter and KernJob processes for both the source and destination machine are informed that migration has completed so they can update their data structures appropriately. A failure of either of the two machines at the transfer phase is detected by the remaining one, which will abort the migration, terminate the migrant, and clean up its state.

3.3. Performance

We measured performance of migration in Charlotte on our VAX/11-750 machines connected by a Pronet token ring. The underlying mechanisms have the following costs. It takes 11 ms to send a 2 KB packet to another machine reliably via the general-purpose inter-machine communication package that Charlotte uses, 0.4 ms to switch context between kernel and process, 10 ms to transfer a single packet between processes residing on the same machine, and 23 ms to transfer a packet between processes residing on different machines.

We measured the average elapsed time to migrate a small (32 KB), linkless process as 242 ms (standard deviation $\sigma = 2$ ms), provided the Starter controlling **D** has preapproved the migration. Each additional 2 KB of image adds 12.2 ms to the migration time. The following formula fits our measurements of the average elapsed time spent in migration.

$$\text{Charlotte time} = 45 + 78p + 12.2s + 9.9r + 1.7q$$

$p = 0$ if **D**'s Starter has approved migration in advance;
else 1 if **D**'s Starter is not on the destination machine;
else about 0.2.
s = size of the virtual space in 2-KB blocks
$r = 0$ if all links are local; 1 otherwise
q = number of non-local links (1 if none).

These measures deviate by about 5% with different locations of the Starter and the overall load. This formula shows that it takes about 750ms to migrate a typical process of 100KB and 6 links (or 670ms if **D**'s Starter is local), and about 6 seconds for a large process of 1MB. Actual CPU time spent on the migration effort for a 32 KB process with no links is about 60 ms for **S** and about 32 ms for **D**. Table 1 shows how this time is spent.

	S	**D**
5.0	Handle an offer	5.4 Handle an offer
2.6	Prepare 2 KB image to transfer	1.2 Install 2 KB of image
1.8	Marshal context	1.2 Demarshal context
6.9	Other (mostly kernel context switching)	4.7 Other

Table 1: Kernel time spent migrating a linkless 32 KB process

Each link costs **S** an additional 1.6 to 2.8 ms of CPU time to prepare link-update messages and to marshal relevant data structures. Collecting statistics requires about 1% of overall elapsed time, and another 2% of all time is spent delivering the statistics to the Starter. A production version of Charlotte, optimized and stripped of debugging code, could exhibit a significant speed improvement.

It is hard to compare Charlotte's migration performance with results published for other implementations, because each uses a different underlying computer, and each operating system dictates its own process structure. Nonetheless, to give the reader some form of comparison, we present formulas for migration speed under Sprite (Sun-3 workstations, about 4 times faster than our VAX-11/750 machines), V (Sun-2, about 2 times faster than our machines), and Accent (Perq workstations). These formulas are extrapolations from a few measurement points reported elsewhere[4,6,7].

$$\text{Sprite time} = 200 + 3.6s + 14f$$

$$\text{V time} = 80 + 6s$$

$$\text{Accent time} = 1180 + 115s$$

s = size of the virtual space in KB
f = number of open files

In particular, a "typical" 100KB process would be transferred in about 560ms in Sprite, 680ms in V, and perhaps 12.7 seconds in Accent. To migrate a large, 1MB process would take at least 3.8 seconds in Sprite, 6 seconds in V, and 116 seconds in Accent. In Accent, sending the context of a process occupies about 1 full second. The virtual space is sent later on demand, so the full cost of transfer is spread over a long period, but part of this cost is saved if not all the pages are referenced. V precopies the address space while the process is still running, so the lost time suffered by the process is quite short.

4. Design issues

Designing a process migration facility requires that one consider many complex issues. We will discuss the separation of policy and mechanism, the interplay between migration and other mechanisms, reliability, concurrency, the nature of context transfer, and to what extent processes should be independent of their location. These issues interrelate to one another, so the following discussion will occasionally need to postpone details until later sections. Moreover, the approaches that we and others adopt to various problems depend somewhat on the design of other components of the operating system. Due to space limitations, we do not discuss these dependencies in detail.

4.1. Structure

The first step in designing a process migration facility is to decide where the policy-making and mechanism modules should reside. We believe that this decision is of major importance since it cannot be easily reversed, unlike most of the design of the migration protocol. Communication-kernel operating systems tend to put mechanism in the kernel and policy in trusted utility processes. In the case of process migration, mechanism is intertwined with both short-term scheduling and IPC, so it fits best in the kernel. Policy, on the other hand, is associated with long-term scheduling and resource management, so it fits well in a utility process. Several considerations affect the success of separation: how efficient the result is, how adequately it provides the needed function, and how conceptually simple are the interfaces and the implementation.

Efficiency and Simplicity

The principal reason one might place policy in the kernel instead of in a utility is to simplify and speed up the interface between policy and mechanism. Any reasonable policy depends on statistics that are maintained primarily in the kernel. High quality decisions may well require large amounts of accurate and comprehensive data. Placing policy outside the kernel incurs execution overhead and latency in passing these statistics in one direction and decisions in the other.

Our experience with Charlotte, however, shows that placing the policy in a utility results in a net efficiency *gain*. Although separation incurs the extra cost of one message for statistics reporting and one kernel call (and perhaps another message round-trip) for decision reporting, it allows reduction of communication and more global policy due to the fact that each Starter process decides policy for a set of machines. As to the latency in passing statistics and decisions, studies have found that good policies tend to depend mostly on aggregate and medium-term conditions, ignoring short-term conditions or small delays.

The designer may choose to support only simple policies, in which case they may well be put in the kernel. For example, the migration policy in V[6] and Sprite[4] is mostly manual, choosing a remote idle workstation for a process or evicting it when the station's owner so requires. In systems where migration is used to meet real-time scheduling deadlines, policy tends to be simple or very sensitive to even small delays, and hence could or should be placed in the kernel. Integrating the policy in the kernel, however, might obstruct later expansion or generalization.

We can achieve conceptual separation of policy and mechanism without incurring a large interface cost by assigning them to separate layers that share memory. MOS[2] adopts this approach by dividing the kernel to two layers, one to implement migration mechanism and other low-level functions, and the other layer to provide policy. These layers share data structures and communicate by procedure calls. Although such sharing improves efficiency, it becomes harder to modify policy, since changes require kernel recompilation, and inadvertent errors are more serious.

Function and flexibility

Placing policy outside the kernel facilitates testing diverse policies and choosing among policies tuned for different goals, such as load sharing, load balancing, improving responsiveness, communication-load reduction, and placing processes close to special devices. Being able to modify policy is especially important in an experimental environment. Our students needed only a few hours to learn the interface and major components of the Starter in order to start trying different policies; they did not need to learn peculiarities of the kernel or of the migration mechanism. This flexibility would be impossible if policy were embedded in the kernel.

In various distributed systems, such as Demos/MP[5], Accent[12], and Charlotte, resource-management policies are often relegated to utilities. Putting migration policy in those same processes can allow more integration and coordination of the policies governing the system.

The designer of process migration should be aware of the danger of separating policy and mechanism too far. Letting policy escape from trusted utility servers into application programs may result in performance degradation or even thrashing. This problem occurs, for example, if applications may decide the initial placement and later relocation of their processes, as in Locus, without getting any assistance from the kernel in the form of timely state and load information.

4.2. Interplay between migration and other mechanisms

The process migration mechanism can be designed independently of other mechanisms, such as IPC and memory management. The actual implementation is likely to see interactions among these mechanisms. However, design separation means that the migration protocol should not change when the IPC protocol does. In Charlotte, for example, we did not change the IPC to add process migration, nor did migration change when we later modified the semantics of two IPC primitives. In contrast, we had to change the marshaling routines when an IPC data structure changed.

We feel that ease of implementation is a dominant motivation for separating mechanisms from each other when process migration is added to an existing operating system, such as was the case in Demos/MP, Charlotte, V, and Accent. A secondary motivation is that the migration code can be deactivated without interfering with other parts of the kernel. In Charlotte, for instance, we can easily remove all the code and structures of process migration at compile time or dynamically turn the mechanism on and off. In contrast, efficiency arguments would favor integrating all mechanisms. Accent, for example, uses a transfer-on-reference approach to transmitting the virtual space of the migrant process that is based on its copy-on-write memory management. If process migration is intended from the start, as in MOS and Sprite, integration can reduce redundancy of mechanisms. In retrospect, Charlotte would have used a different implementation for IPC if the two mechanisms had been integrated from the start. We would have used **hints** for link addresses, which are inaccurate but can be readily checked and inexpensively maintained, rather than using **absolutes**, whose complete accuracy is achieved at a high maintenance cost.

Some interactions seem to be necessary. In Charlotte, for instance, we chose to simplify the migration protocol by refusing to migrate a process engaged in multi-packet message transfer. We therefore depend slightly on knowledge of the IPC mechanism to avoid complex protocols. Similarly, both MOS and Sprite refuse to migrate a process engaged in RPC until it reaches a convenient point, which may not happen for a long time. Other interactions make sense in order for process migration to take advantage of existing facilities. For example, Locus uses existing process-creation code to assist in process migration.

4.3. Reliability

Migration failures can occur due to network or machine failure. The migration mechanism can simply ignore these possibilities (as does Demos/MP) in order to streamline protocols. The Charlotte implementation is able to rescue the migrant from many failures by several means. First, it transfers responsibility for the migrant as late as possible, to survive failure of the destination or the network. Second, it detaches the migrant completely from its source, to survive later failures there. Third, the migrant is protected from failures of other machines; at most, some of its links are automatically destroyed if the machine where their other ends reside has crashed. Rescuing migrating processes under all failure circumstances requires complex recovery protocols, and most likely large overhead for maintaining process replicas, checkpoints, or communication logs. We were unwilling to pay that cost in Charlotte. Instead, we terminate the migrant if either the source or destination machine crashes during the sensitive time of transfer when messages for the migrant may have arrived at either machine, as discussed earlier. Modifying our IPC to use hints for link addresses, as mentioned above, would have made this step less fragile.

4.4. Concurrency

Various levels of concurrency are conceivable:

- Only one migration in the network at a time

- Only one migration affecting a given machine at a time
- No constraints on the number of simultaneous migrations

The Charlotte mechanism puts no constraint on concurrency. Restricting process migration can make the mechanism simpler, especially in operating systems using a connection-based IPC. The most restrictive alternative guarantees that the peers of the migrant process are stationary, so redirection of messages is straightforward. It also tends to mitigate policy problems of migration thrashing, flooding a lightly-loaded machine with immigrants, and completely emptying a loaded machine.

Enforcing such a constraint, on the other hand, requires arbitrating contention, which can be expensive. In addition, limiting concurrency constrains policies that otherwise would be able to evacuate a failing machine quickly or react immediately to a severe load imbalance. We therefore believe that the policy problems alluded to above should be solved by policy algorithms, not by a limitation imposed by the mechanism.

Allowing simultaneous migrations introduces the peculiar problem of name and address consistency: ensuring that all processes and kernels have a consistent view of the world. The problem is manifest in operating systems like Charlotte, in which communication is carried out over established channels and kernels require up-to-date location information. If two processes connected by a channel migrate at the same time, their kernels may have false conception of the remote channel ends. The problem is not critical in operating systems that treat communication addresses as hints, such as V, because communication encountering a hint fault will restore the hint by invoking a process-finding algorithm. This solution incurs execution and latency costs as messages are transmitted. Where absolutes are used, forwarding pointers, such as those used in Demos/MP, may solve the problem, but they introduce long-lived residual dependencies. In Charlotte, we send link-address updates before migration completes, and we buffer notifications for messages arriving during the transfer. The immediate acknowledgement of the updates, even when the other link end is simultaneously given away or migrating, prevents deadlock. When migration completes, **D** processes the notifications buffered by the two kernels and regains a consistent view of **P**'s links, even if their remote ends have moved meanwhile.

Within a single source or destination, we could restrict concurrency to one migration attempt at a time. This restriction simplifies the kernel state and again reduces risks of thrashing. However, complexity can be reduced by creating a new kernel thread for each migration in progress, executing a finite-state protocol independently of other migration efforts. Using these techniques, we found that allowing concurrent migrations in the same machine incurs only a small space overhead and minor execution costs.

4.5. Context transfer and residual dependency

At some point during migration, the process must be frozen to ensure a consistent transfer.

What and when to freeze

Three activities need to be frozen: (1) process execution, (2) outgoing communication, and (3) incoming communication. The first two activities are trivial to freeze. Freezing incoming communication can be accomplished by (a) telling all peers to stop sending, (b) delaying incoming messages, or (c) rejecting incoming messages. Option (a) requires a complex protocol if concurrent migrations are supported or if crashes must be tolerated. Option (c) requires that the IPC be able to resend rejected messages, as in V. In Charlotte, we chose option (b) because it seems the simplest and because it does not interfere with other mechanisms.

Very early freezing (for example, when a process is considered as a migration candidate) has the advantage that the process does not change state between the decision and migration. Otherwise, the migration decision may be worthless, since the process could terminate or start using resources differently. However, freezing a process hurts its response time, which flies in the face of one of the goals of migration. Less conservatively, we can freeze a process when it is selected as a candidate, but before the destination machine has accepted the offer. Even less conservative alternatives include freezing at the point migration is agreed upon, or even when it is completed. Each more liberal choice increases the process' responsiveness at the cost of protocol complexity.

In Charlotte, we chose to balance responsiveness and protocol simplicity by freezing both execution and communication only when context is marshaled and transferred. We delay incoming communication by buffering input notifications at **S** and both notifications and data at **D** until **P** is established. We verified (by exhaustive enumeration of states in our automata that drive the IPC protocol) that the ensuing delays could not cause deadlock or flow control problems[11]. In this way, a minimal context is transferred during negotiation (such as how many links **P** has and where their ends are); the final transfer reflects any change in **P**'s state during migration.

MOS and Locus freeze the migrant earlier, when it is selected for migration. V, in contrast, freezes a process for a minuscule interval near the end of transfer. While transfer is in progress, the migrant continues to execute; pages dirtied during that episode are sent again in another transfer pass, and so forth until a final pass. Incoming messages are rejected during the short freeze, with the understanding that the IPC mechanism will timeout and retransmit them. The result is that the migrant suffers a delay comparable to that required to load a process into memory[6].

Redirecting communication

Redirecting communication requires that state information relevant to the communication channels be updated and that peer kernels discover the migrant's new location. In a connectionless IPC mechanism, a process holds the names of its communication peers. For example, V processes use process identifiers as destinations[13]. To redirect communications in such an environment, a kernel may broadcast the new location. Broadcast can be expensive for large networks with frequent migrations. Alternatively, peers can be left with incorrect data that can be resolved on hint faults. Another alternative is to assign a home machine to each process; the home machine always knows where the process is. Locus uses this method to find the target of a signal. Sprite is similar; the home machine manages signals and other location-dependent operations on behalf of the migrant. Of course, resorting to a home machine makes communication failures more likely and sharply increases the cost of certain kernel calls.

In a connection-based IPC environment with simplex connections, such as Accent and Demos/MP, the kernel of the receiving end of a connection does not know where the senders are. That means that **S** cannot tell which kernels to inform about **P**'s migration. Instead, a forwarding pointer may be left on **S** to redirect new messages as they arrive. Demos/MP uses this strategy. Another approach is to introduce a stationary "middle-man" between two or more mobile ends of a connection. In Locus, cross-machine pipes may have several readers and writers, but they have only one fixed storage site. When a reader or writer migrates, the kernel managing the storage site is informed. In Charlotte, the duplex nature of links suggests maintaining information at both ends about each other, so **S** can tell all peers that **P** has moved. Transferring these link data along with **P**, though, incurs marshaling, transmission, and demarshaling overhead.

Residual dependency

The migrant process can start working on the destination machine faster if it can leave some of its state temporarily on the source machine. When it needs to refer to that state, it can access it with some penalty. To reduce the penalty, state can be gradually transferred during idle moments. State can also be pulled upon demand. The choice between moving the entire address space or only a part is reminiscent of the controversy in network file systems whether entire files should be transferred or only pages for remote file access. Locality of execution suggests transferring at least the working set of **P** during migration, and the rest when needed. On the other hand, the objective of residual independence suggests removing any trace of **P** from the source machine.

In MOS, virtually the entire state of **P** could remain in the source machine, since **D** can make remote calls on **S** for anything it needs. For efficiency reasons, however, MOS transfers most of **P**'s context when it migrates. In Sprite, part of **P**'s context always resides in its home machine, but none is left on the source machine when it is evicted. This approach costs about 15 ms to demand-load a page and perhaps 4 ms to execute some of the kernel calls remotely (about 9-fold increase). In Accent, processes do not make kernel calls directly, but rather send messages to a kernel port. Therefore, no state needs to be moved with a process; it can all remain with **S** and be accessed as needed by kernel calls to the old port. In addition, Accent implements a lazy transfer of data pages on demand. Similarly, in Sprite, **S** acts as a paging device for **D**. These approaches trade efficiency of address-space transfer for risks of machine unavailability, protocol complexity, and later access penalties.

4.6. Location independence

Many distributed operating systems adhere to the principle of location transparency. In particular, process names are independent of their location, processes can request identical kernel services wherever they reside, and they can communicate with their peers equally well (except for speed) wherever they might be. The principle of location transparency must be followed carefully to enable migration. Migration requires that naming schemes be uniform for local and remote communication and that resource references not depend on the host machine. For example, Charlotte objects are all named by the links that connect a client to them. When a process moves, the names it uses for its links are unchanged, even though **D** remaps them to different internal names. The fact that local communication is treated differently from remote communication is localized in a few places in the kernel. Processes may have pointers or indices to kernel data structures, but those are maintained by the kernel. The actual data structures, pointers and indices are remapped invisibly during migration. If such values were buried inside the processes' address spaces, migration would be impossible or extremely complicated. Sprite maintains location transparency throughout multiple migrations by keeping location-dependent information on **P**'s home machine and by directing some of **P**'s kernel calls there.

Transaction management and multithreading also pose transparency problems. A transaction manager must not depend on the location of its clients. Multithreaded processes must be moved *in toto*. If threads may cross address spaces, the identity of one thread may be recorded in several address spaces, leading to location dependencies.

Of course, any policy setter, such as the Charlotte Starter, needs to know the location of all processes and perhaps the endpoints of their heavily-used links. Making this information available need not compromise the principle of transparency. The policy module does not use this information to send messages, only to inform itself about decisions it needs to make. Likewise, for the sake of openness, a design may allow processes willing to participate in migration decisions to receive location information and contribute migration advice.

5. Conclusions

Our experience with Charlotte and others' experience with Sprite, V, MOS, and Demos/MP, show that process migration is possible, if not always pleasant. We found that separating the modules that implement mechanism from those responsible for policy allows more efficient and flexible policies and simplifies the design. Migration interact with other parts of the kernel. In particular, the implementation shares structures and low-level functions with other mechanisms. Nonetheless, we found it possible to keep the mechanisms fairly independent of each other, gaining high code modularity and ease of maintenance.

Software and hardware failures are a fact of life. Our migration protocol can rescue the migrant in most failure situations and restore the state in all of them, despite the fact that the migrant continues its interaction with other processes at early stages of migration. In some cases, though, we opt to kill the migrant even if rescue is dimly conceivable. We chose to postpone committing migration until late during the transfer itself (to deal with early destination crash), while removing any dependency of the migrant on the source as soon as migration completes (to deal with late source crash).

Except for potential confusion suffered by policy modules, it is not particularly hard to achieve simultaneous migrations, even those involving a single machine. The Charlotte IPC requires absolute state information, so we could not try to reduce the cost of migration by sacrificing accuracy. IPC mechanisms that use hints or are connectionless can shorten the elapsed time for migration but then probably pay more during communication. Designs that require previous hosts to retain forwarding information for an arbitrary period after migration are overly susceptible to machine failure. Forwarding data structures, although small, tend to build up over time.

6. Acknowledgements

The design of process migration in Charlotte was inspired by discussions with Amnon Barak of the Hebrew University of Jerusalem in 1984. The authors are indebted to Cui-Qing Yang for modifying Charlotte utilities to support process migration and to Hung-Yang Chang for many fruitful discussions about the design. Andrew Black and Marvin Theimer provided helpful comments on an early draft, and the referees suggested many stylistic improvements. The Charlotte project was supported by NSF grant MCS-8105904 and DARPA contracts N00014-82-C-2087 and N00014-85-K-0788.

References

1. P. Krueger and M. Livny, "When is the best load sharing algorithm a load balancing algorithm?," Computer Sciences Technical Report #694, University of Wisconsin–Madison (April 1987).
2. A. B. Barak and A. Litman, "MOS: A Multicomputer Distributed Operating System," *Software — Practice and Experience* **15**(8) pp. 725-737 (August 1985).
3. D. A. Butterfield and G. J. Popek, "Network tasking in the Locus distributed UNIX system," *Proc. of the Summer USENIX conference*, pp. 62-71 USENIX Association, (June 1984).
4. F. Douglis and J. Ousterhout, "Process migration in the Sprite Operating System," *Proc. of the 7th Int'l Conf. on Distributed Computing Systems*, pp. 18-25 IEEE Computer Press, (September 1987).
5. M. L. Powell and B. P. Miller, "Process migration in DEMOS/MP," *Proc. of the Ninth ACM Symp. on Operating Systems Principles*, pp. 110-118 ACM SIGOPS, (October 1983). In *Operating Systems Review* 17:5
6. M. M. Theimer, K. A. Lantz, and D. R. Cheriton, "Preemptable Remote Execution Facilities for the V-System," *Proc. of the Tenth Symp. on Operating Systems Principles*, pp. 2-12 ACM SIGOPS, (December 1985).
7. E. R. Zayas, "Attacking the process migration bottleneck," *Proc. of the Eleventh ACM Symp. on Operating Systems Principles*, pp. 13-24 ACM SIGOPS, (November 1987). In *Operating Systems Review* 21:5

8. D. A. Nichols, "Using idle workstations in a shared computing environment," *Proc. of the Eleventh ACM Symp. on Operating Systems Principles*, pp. 5-12 ACM SIGOPS, (November 1987). In *Operating Systems Review* 21:5

9. Y. Artsy, H-Y. Chang, and R. Finkel, "Interprocess communication in Charlotte," *IEEE Software* **4**(1) pp. 22-28 IEEE Computer Society, (January 1987).

10. M. L. Scott, "Language support for loosely coupled distributed programs," *IEEE Trans. on Software Eng.* **SE-13**(1) pp. 88-103 IEEE, (January 1987).

11. Y. Artsy, H-Y. Chang, and R. Finkel, "Charlotte: design and implementation of a distributed kernel," Computer Sciences Technical Report #554, University of Wisconsin–Madison (August 1984).

12. R. F. Rashid and G. G. Robertson, "Accent: A communication oriented network operating system kernel," *Proc. of the Eighth ACM Symp. on Operating Systems Principles*, pp. 64-75 ACM SIGOPS, (December 1981).

13. D. Cheriton, "The V Kernel: A software base for distributed systems," *IEEE Software* **1**(2) pp. 19-42 (April 1984).

4.4 Zayas, E., Attacking the Process Migration Bottleneck

Zayas, E., "Attacking the Process Migration Bottleneck," *Proceedings of the 11th ACM Symposium on Operating Systems Principles*, pp. 13-24, November 1987.

Attacking the Process Migration Bottleneck

Edward R. Zayas
Computer Science Department
Carnegie Mellon University
Pittsburgh, PA 15213

(Currently at the Information Technology Center, Carnegie Mellon University)

Abstract

Moving the contents of a large virtual address space stands out as the bottleneck in process migration, dominating all other costs and growing with the size of the program. Copy-on-reference shipment is shown to successfully attack this problem in the Accent distributed computing environment. *Logical* memory transfers at migration time with individual on-demand page fetches during remote execution allows relocations to occur up to one thousand times faster than with standard techniques. While the amount of allocated memory varies by four orders of magnitude across the processes studied, their transfer times are practically constant. The number of bytes exchanged between machines as a result of migration and remote execution drops by an average of 58% in the representative processes studied, and message-handling costs are cut by over 47% on average. The assumption that processes touch a relatively small part of their memory while executing is shown to be correct, helping to account for these figures. Accent's copy-on-reference facility can be used by *any* application wishing to take advantage of lazy shipment of data.

1. Introduction

Process migration is a valuable resource management tool in a distributed computing environment. However, very few migration facilities exist for such systems. Part of the problem lies in providing an efficient method for naming resources that is completely independent of their location. The major difficulty, though, is the cost of transferring a computation's context from one system node to another. This context, which consists primarily of the process virtual address space, is typically large in proportion to the usable bandwidth of the interconnection medium. Moving the contents of a large virtual address space thus stands out as the bottleneck in process migration, dominating all other costs. As programs continue to grow, the cost of migrating them by direct copy will also grow in a linear fashion.

This research was supported by the AT&T Cooperative Research Fellowship Program. It was also supported by the Defense Advanced Research Projects Agency (DoD), ARPA Order No. 3597, monitored by the Air Force Avionics Laboratory under contract F33615-84-K-1520.

Any attempt to make process migration a more usable and attractive facility in the presence of large address spaces must focus on this basic bottleneck. One approach is to perform a *logical* transfer, which in reality requires only portions of the address space to be *physically* transmitted. Instead of shipping the entire contents at migration time, an *IOU* for all or part of of the data can be sent. As the relocated process executes on the new host, attempts to reference "owed" memory pages will result in the generation of requests to copy in the desired blocks from their remote locations. Context transmission times during migration are greatly reduced with this demand-driven *copy-on-reference* approach, and are virtually independent of the size of the address space. Processes are assumed to touch relatively small portions of their address spaces, justifying the higher cost of accessing each page during remote execution.

This paper describes the process migration facility built for the SPICE [12] environment at Carnegie Mellon University, which demonstrates the validity of using copy-on-reference transfer to attack the migration bottleneck. Section 2 describes the design of the Accent copy-on-reference mechanism, available to *any* application wishing to lazy-evaluate its data transfers. Accent's organization and abstractions not only provide the transparency needed to support migration, but lend themselves to the natural construction of such a mechanism. Section 3 show how the migration system capitalizes on copy-on-reference data delivery. Section 4 presents performance measurements taken on a set of representative processes that were migrated using different transmission strategies. Process relocations occur up to one thousand times faster using copy-on-reference transfers. While the amount of allocated data varies by four orders of magnitude across the processes studied, their transfer times are practically constant. The number of bytes exchanged between machines as a result of migration and remote execution drops by 58.2% on average, and message-handling costs are cut by 47.8%. The assumption that processes touch a relatively small part of their memory while executing is shown to be correct, helping to account for these figures. The detailed measurements are used to assess the effect of such copy-on-reference variations as prefetching in response to remote page requests and migration-time transfer of the address space portions resident in main memory. Section 5 compares the Accent migration work to other activity in the field. Finally, Section 6 summarizes the lessons learned from the Accent migration system and considers future research directions suggested by this work.

2. The Accent Copy-On-Reference Mechanism

Accent's design and organization allows such intelligent virtual memory techniques as copy-on-write to be applied to data passed through the IPC system. It is this feature which aids in the construction of another intelligent strategy, copy-on-reference. This section begins by providing a quick overview of the Accent features that contribute to the natural construction of a transparent, generic copy-on-reference facility. Accent's *imaginary segment* abstraction serves as the basis for lazy data delivery, and is described next. The consequences of permitting imaginary objects to exist are explored, along with the method of shipping imaginary areas between machine boundaries.

2.1. Accent Features

The Accent IPC and virtual memory facilities are closely integrated, operating symbiotically. Unlike most message-based systems, a single Accent IPC message can hold *all* of the memory addressible by a process. Message contents are conceptually copied by value directly from the sender's address space into the receiver's. In reality, a message is first copied into the kernel's memory, buffered there until the recipient decides to accept it, and then copied out again. Accent provides the advantages of double-copy semantics for transferring message data between address spaces while still achieving the performance expected of a system that passes data by reference. This is possible through the use of a *copy-on-write* virtual memory mechanism by the IPC facility. If the amount of message data falls below a certain threshold, it is physically copied to the receiver. However, the kernel uses much faster memory-mapping techniques for messages exceeding this threshold. The receiver's virtual memory map is modified to provide access to the message data, and the region is marked copy-on-write for both parties. The two processes share this single copy of the data until either one tries to modify it. The deferred copy operation is then carried out, but *only* for the 512-byte page(s) affected. Files are accessed through an IPC interface and mapped in their entirety into process memory, allowing these techniques to be applied to their data as well. Since large amounts of data are often transferred through IPC messages and only rarely modified to any degree, this lazy strategy realizes performance that approaches by-reference transfer. Fitzgerald's study [3] reveals that up to 99.98% of data passed between processes in a system-building application did not have to be physically copied.

2.2. Imaginary Segments

Accent's copy-on-reference mechanism is based on a new segment class, the imaginary segment. Imaginary segment data is accessed not by direct reference to physical memory or a hard disk, but rather through the IPC system. Each imaginary segment is associated with a *backing IPC port* which provides memory management services for the object. When a process touches a page mapped to an imaginary segment, the high-level *Pager/Scheduler* process sends an *Imaginary Read Request* message to the region's backing port. The process with Receive rights for this port interprets the request and returns the required page in an *Imaginary Read Reply* message. The *Pager/Scheduler* completes the handling of the imaginary "fault" by mapping in the page and resuming the process attempting the access. Currently, page-outs for imaginary data are performed to the local disk at the site that touched the page. Any process may create an imaginary segment based on one of its ports, map all or part of it into its address space and pass this memory to another process via an IPC message. In effect, it transmits an "IOU" for the region's data, promising to deliver it as needed. The backing process continues to field page request messages aimed at the imaginary object until all references to it die out. At this point, Accent informs the backer of the object's demise by sending it an *Imaginary Segment Death* message.

2.3. Accessibility Maps

The existence of imaginary objects forces the operating system to provide a facility for determining the accessibility of any given virtual address range. Carelessly touching imaginary regions can result in deadlock. For example, an Accent process executing in the kernel context deadlocks if it touches a page with port-based backing. The faulter is caught holding the system critical section, preventing the backing process from executing the protected *Receive* operation needed to respond to the fault.

Accessibility Maps (or *AMaps*) were created to supply the necessary addressing information in Accent. Four different memory "distances" have been defined for AMaps:

1. **RealZeroMem**: This is a region that has been validated (allocated) by a process but has never been accessed. When memory is validated, it is conceptually filled with zeros. Accent postpones these filling operations until the pages are first touched. A special fault condition, the *FillZero* fault, is realized for this case. The only action the *Pager/Scheduler* process takes is to reserve a page of physical memory, fill it with zeros and create the appropriate virtual memory mappings. The disk is never consulted while handling this type of fault. In practice, Accent processes validate large amounts of virtual memory and only touch a small percentage. Lazy initialization of address space regions and the use of a special inexpensive fault-handling operation combine to make creation and maintenance of large virtual memory regions affordable. These *RealZeroMem* pages are considered immediately accessible to the process.

2. **RealMem**: The data in this type of region is either already present in physical memory or accessible by fetching the corresponding local disk page. The distinction between disk address mappings owned by the kernel and process mappings for the same data allows a disk page image to be resident without being visible to a user process. In this eventuality, the

Pager/Scheduler again simply fills in the missing user mapping and promotes the faulted process to a runnable state. If neither the disk nor the process mapping are available for the page, the matching disk block is determined. The page is brought in, and disk and process mappings for it are entered. *RealMem* pages are rated "moderately" accessible, since the system may have to go out to disk to get them.

3. **ImagMem:** The contents of memory regions mapped to imaginary segments have *ImagMem* accessibility. Touching a page in this accessibility class results in the the generation and processing of an imaginary fault, as described in Section 2.2. *ImagMem* pages are considered distantly accessible, since it may take an arbitrarily long time to complete a page fetch. The network state, the load on the machines involved and the amount of work being performed by the backing process all contribute variables to the service time.

4. **BadMem:** Attempting to touch a page in a region that hasn't been validated causes a true addressing error. Referencing a *BadMem* page invokes a debugger so the human user can analyze and properly terminate the delinquent process. Since referencing a *BadMem* page is illegal, its accessibility is considered infinitely distant.

2.4. Extending Imaginary Segments

As with the port abstraction, copy-on-reference access via imaginary segments depends on a user-level server for transparent extension across the network. The *NetMsgServer* process, running on each host, provides this service by changing its message fragmentation and reassembly algorithms to account for imaginary subranges. Using an AMap as a guide on both sides, the *RealMem* portions are physically transmitted to the remote location and placed in the corresponding locations in the reassembly buffer. The receiving *NetMsgServer* creates its own local ports and imaginary object(s) to stand in for the originals. Messages generated in response to faults on the remote imaginary objects are automatically channeled to the correct backing site.

On its own initiative, a *NetMsgServer* may cache the *RealMem* portions of a message destined to a remote site and instead pass IOUs for them, becoming the manager for that data. Senders can inhibit this behavior by setting the *NoIOUs* bit in the message header, which is inspected by the *NetMsgServer*. This action guarantees that non-imaginary message data is physically copied to the remote site.

3. Migration Using Copy-On-Reference

The SPICE migration facility is designed to take advantage of the copy-on-reference mechanism described in the previous section. This is done by special migration primitives which automatically separate out the context portions eligible for copy-on-reference shipment. Using these operations, the *MigrationManager* process on each machine has several options for context delivery to the new execution site.

3.1. *ExciseProcess* and *InsertProcess*

The *ExciseProcess* kernel trap allows the complete context of an active process to be removed from its current host. Accent contexts are divided into five components: the state of the Perq[1] microengine, the kernel stack if the process is executing in supervisor mode, the PCB, the set of port rights owned by the process and the virtual address space contents. While the first four parts combined only account for roughly 1 Kbyte, the address space contributes up to 4 gigabytes. Once a context is excised, the process ceases to exist. Since all port rights are passed transparently to the caller, there is no disruption to the set of processes capable of naming these ports.

ExciseProcess delivers a process context in two separate IPC messages, ready for shipment to the new execution site. The *Core* message contains the first four context pieces, which must be physically copied to the remote site. It also carries an AMap describing the entire process address space. The *RIMAS*[2] message contains all of the *RealMem* and *ImagMem* portions of the address space, collapsed into a contiguous area. This allows the caller to fit one or more excised address spaces into its own memory at one time. It also allows the bearer to cache the *RealMem* portions and substitute its own imaginary objects in the *RIMAS* message. If the migration agent doesn't wish to actively manage the excised address space, it simply turns off the *NoIOUs* bit in the *RIMAS* message header as described in Section 2.4, prompting the local *NetMsgServer* to assume backing services for the memory.

The counterpart for *ExciseProcess* is *InsertProcess*, which uses the two context messages to recreate the process. Since the messages are self-contained, they do not have to be preprocessed in any way. The embedded port rights are passed to the new incarnation. Using the AMap for guidance and the *RIMAS* data for ammunition, the process address space mappings are restored. The reconstituted process is finally placed into the kernel queue representing the original execution status.

3.2. The *MigrationManager* Process

Each SPICE machine wishing to participate in process migration runs a simple *MigrationManager* process. This server accepts and executes commands to perform migrations. Given a process name, it uses the *ExciseProcess* primitive to acquire the process context. The two context messages are then simply sent to the *MigrationManager* at the new execution site, which uses *InsertProcess* to reconstruct the target process.

The current *MigrationManager* doesn't attempt sophisticated address space management for the processes it extracts. If asked to use copy-on-reference transfer for process memory, the *MigrationManager* allows the intermediary *NetMsgServers* to cache the data and become its backer.

---

[1] Designed by Perq Systems, Inc., the Perq workstation has a microcoded CPU, 16-bit words and a 150 nanosecond cycle time. It's rated at between 1/5 and 1/2 the speed of a Vax-11/780, depending on the instruction set used. More detailed specs are in [3].

[2] *RIMAS* stands for *R*eal and *I*maginary *M*emory *A*ddress *S*pace.

4. Evaluation

This section summarizes the results of experiments carried out on the augmented Accent testbed system to determine the effectiveness of the copy-on-reference technique in reducing the dominant cost of migration: transfer of large process address spaces. Representative processes were chosen and monitored as they were migrated with the different strategies of interest. These programs are implemented in a variety of languages, perform widely different tasks and differ greatly in memory requirements and access patterns. Figures on their address space composition and utilization are presented, along with the basic costs of the migration primitives used to extract and insert process contexts on a host. Based on such metrics as the quantity and distribution of byte traffic, message processing costs and end-to-end elapsed times, copy-on-reference is shown to be superior to the brute-force method. Two variations on the basic lazy-transfer theme were simulated using the detailed performance figures and also evaluated. While prefetch of between 1 and 15 nearby pages in response to imaginary faults proved to be a valuable optimization, the shipment of process resident sets (as an approximation to their working sets [2]) was found to be generally detrimental. Overall, the experiments show that this lazy transfer technique significantly reduces the dominant context transmission costs by exploiting the fact that processes tend to use only small portions of their address spaces during execution.

4.1. Representative Processes

Several processes were chosen to undergo relocation, each representing a class of programs sharing similar attributes. The results obtained for these representatives should be characteristic of other programs in their class.

1. **Minprog**: This program is used to judge the effects of the various transmission strategies on a "minimal" program. Written in Perq Pascal, Minprog prints a message on the standard output, waits for user input and terminates. Measurement of this program is the equivalent of timing the "null trap" when exploring operating system performance.

2. **Lisp-T**: Accent supports the SPICE Lisp dialect, complete with a customizable screen editor and compiler. The Lisp-T trial resembles Minprog in that the minimum computation is performed. After migration, the Lisp interpreter is simply asked to evaluate T. This process represents simple Lisp programs, or larger Lisp jobs migrated late in life. The primary difference between Lisp-T and Minprog is the amount of address space used. Lisp processes validate their entire 4 gigabyte address spaces at birth, compared to Minprog's use of only 330 Kbytes.

3. **Lisp-Del**: This Lisp process performs a significant amount of computation and I/O. Immediately after migration, a Delaunay triangulation package written at Carnegie Mellon by Rex Dwyer is loaded. Utilizing a divide-and-conquer algorithm on a random set of points, this package displays its actions graphically on the screen as the triangulation is built.

4. **PM-Start**: The Pasmac macro processor for Perq Pascal represents the class of programs whose primary duty is to read files from the disk, process them in some way and write the results back out. In this instance, a 164 Kbyte file containing the program with macro references imports five definition files totaling 114 Kbytes. Migration takes place at the point the first definition file is being accessed.

5. **PM-Mid**: This trial postpones migration of the above macro processor until all of the definition files have been read in. Thus, the file images have become part of the process context and are carried along by the migration. The relocated program doesn't perform any more file accesses until it writes out the expanded program text.

6. **PM-End**: The final trial involving the Perq Pascal macro processor further postpones migration until the original file has almost been completely expanded. With little computation left to perform, this trial reveals the performance of the various migration strategies on processes near the end of their lifetimes.

7. **Chess**: A chess program written by Charly Drechsler at Siemens rounds out the group. It performs a large amount of computation to evaluate board positions and generate moves, but doesn't use a lot of its address space. A graphical representation of the chess board is displayed on the screen along with a game clock. The game clock ticks every second, so screen updates occur at least that often. Migration takes place as soon as the program initializes itself and draws the first screen image.

4.2. Address Space Analysis

4.2.1. Composition

Table 4-1 expresses the address space sizes and breakdowns of the representative processes at migration time.

	Real	RealZ	Total	% RealZ
Minprog	142,336	187,904	330,240	56.9
Lisp-T	2,203,136	4,225,926,144	4,228,129,280	99.9
Lisp-Del	2,200,064	4,225,929,216	4,228,129,280	99.9
PM-Start	449,024	501,760	950,784	52.8
PM-Mid	446,464	466,432	912,896	51.1
PM-End	492,032	398,848	890,880	44.8
Chess	195,584	305,152	500,736	60.9

Table 4-1: Representative Address Space Sizes in Bytes

Listed for each representative process is the amount of non-zero data it addresses (*Real*), the allocated but untouched zero-filled memory (*RealZ*), the total memory addressed (*Total*) and the percentage of the overall process memory taken up by allocated, untouched zero-filled regions (*% RealZ*). Memory quantities are in bytes.

There is wide variance in the amount of validated memory in

the representative Accent processes. The space utilized by the biggest process is a factor of 12,803 larger than that of the smallest. This is the consequence of the way Lisp processes manage their address spaces. The amount of *RealMem* mapped into processes doesn't vary nearly as much, only by a factor of 15 for these samples. Notice that *RealZeroMem* forms a significant part of all process address spaces, more than half even in most non-Lisp examples.

4.2.2. Resident Set Analysis

The process resident set sizes at migration time and their relationships to their host address spaces are shown in Table 4-2.

	RS Size	% of Real	% of Total
Minprog	71,680	50.4	21.7
Lisp-T	190,464	8.6	0.005
Lisp-Del	190,464	8.7	0.005
PM-Start	132,096	29.4	13.9
PM-Mid	190,976	42.8	20.9
PM-End	302,080	61.4	33.9
Chess	110,080	56.3	22.0

Table 4-2: Representative Resident Sets

Listed is the resident set size in bytes at migration time (column *RS Size*) for each representative, as well as the relative size compared to the process non-zero data (*% of Real*) and total allocated space (*% of Total*).

The range of resident set sizes is even narrower than that of the *RealMem* figures in Section 4.2.1, a factor of only 4. With the unrealistic Minprog process excluded, the factor drops to 2.7. This implies that the transfer of a process resident set will contribute a relatively consistent delay to the migration operation. Because of the amount of memory involved, resident set transfers are a significant expense. Viewing resident set transfer as a middle ground between a pure-copy transfer and a pure-IOU strategy appears reasonable, since the resident sets are roughly half as large as the *RealMem* in most cases. However, Section 4.3.4 demonstrates that this added expense at migration time doesn't translate into better overall performance.

4.2.3. Address Space Utilization

As postulated, Accent processes reference a small portion of their address spaces on average in their lifetimes. Table 4-3 reveals the amount of data transferred between machines during the trials in relation to address space size. Percentages are listed for the pure-IOU and resident set strategies without prefetching (pure-copy transmits 100% of *RealMem* by definition). Pure-IOU figures (the first column) indicate the portions actually touched by the process at the remote site.

	IOU		RS	
Minprog	8.6	[3.7]	50.4	[21.7]
Lisp-T	3.0	[0.002]	9.0	[0.005]
Lisp-Del	16.5	[0.009]	17.4	[0.009]
PM-Start	58.0	[27.4]	76.0	[35.9]
PM-Mid	51.5	[25.2]	77.5	[37.9]
PM-End	26.9	[14.8]	72.5	[40.1]
Chess	35.6	[13.9]	66.0	[25.8]

Table 4-3: Percent of Address Space Accessed

For each representative process, the portion of the address space transferred to the new site is given for the pure copy-on-reference (*IOU*) and resident set (*RS*) strategies. The first number in each column represents the percent of the allocated, non-zero (*RealMem*) memory shipped, while the number in square brackets reports the percent of the total allocated address space. By definition, the pure-copy technique transfers 100% of non-zero data.

The Lisp representatives, while they have the largest address spaces, touch the smallest percentage in the course of execution. This applies even when performing a considerable amount of computation and I/O, as in the case of Lisp-Del. The Pasmac macro processor trials showed the highest address space utilization, as their mapped disk files are touched sequentially and in their entirety. In all cases, the resident set transfer method accessed larger portions of the address space, bringing over pages that are never used. This is especially acute for Pasmac. Since physical memory under Accent tends to act as a disk cache, old file pages that have ***already*** been processed are still sent to the new execution site. This explains why the pure-IOU method references significantly less of the Pasmac process address space the later in life it is migrated while the resident set approach results in nearly constant utilization.

4.3. Migration Phase Timings

Migration under Accent may be broken down into three phases:

1. Packaging and unpackaging the process context at the source and destination hosts.
2. Transferring the context between the sites.
3. Running the program at its new location.

This section examines how the migration strategies and their variations perform in each of these phases, and also presents an end-to-end analysis. Copy-on-reference transfers are shown to greatly reduce the time spent in the transfer phase while only moderately increasing remote execution times, resulting in significant overall performance improvements. While the first phase is insensitive to the migration strategy chosen, the experiments reveal some interesting facts about Accent's virtual memory system.

4.3.1. Process Excision and Insertion

Two operations dominate the removal and packaging of a process context, as revealed by Table 4-4: AMap construction for the target address space and the collapse of process memory into a contiguous chunk.

	AMap	RIMAS	Overall
Minprog	.37	.36	.82
Lisp-T	2.12	.59	2.79
Lisp-Del	2.46	.73	3.38
PM-Start	.98	.63	1.67
PM-Mid	1.01	.68	1.74
PM-End	1.4	.94	2.45
Chess	.37	.43	1.00

Table 4-4: Process Excision Times in Seconds

The rightmost column of this table lists the amount of elapsed time used by the *ExciseProcess* kernel trap on each of the representatives (*Overall*). Also listed are the individual timings for the two dominant activities carried out during extraction: AMap construction (*AMap*) and creation of the IPC message containing the condensed process address space (*RIMAS*).

There are two reasons why AMap construction is an expensive operation under Accent. The complex process map organization chosen to support sparse address spaces and copy-on-write makes it difficult to determine accessibility for *ranges* of addresses. Also, the lazy update algorithm employed for process maps often forces a costly search of system virtual memory tables. The Lisp processes take the longest to service, as might be expected. The Minprog and Chess programs have small, uncomplicated address spaces and hence require the shortest amount of time.

While process memory is rearranged into a compact form and delivered to the migration agent via memory-mapping techniques instead of physical copies, it is still an important part of the excision activity. Address space collapses contribute a much smaller variation to excision times than does AMap construction. Overall, excision times vary only by a factor of 4, compared to the 4 *orders of magnitude* difference in the address space contents.

Process reincarnation given the two context messages involves reestablishing the microcode and port state of the process, along with setting up its address space to correspond to the original structure. The times required to insert the transferred contexts into the new site ranged from 263 milliseconds for Minprog to 853 milliseconds for Lisp-Del. Address space reconstruction is the major factor in the insertion operation, and times are very similar to the *RIMAS* creation times during context extraction. As with other portions of the migration mechanism, this insertion costs grow much more slowly than the address spaces involved, only a factor of 3.3.

4.3.2. Context Transfer Times

Approximately one second is required to transmit the *Core* context message (microstate, PCB, port rights) in all cases. These messages differ by a small number of bytes, since some AMaps are slightly larger than others. The real variation involves the delivery of the *RIMAS* message (valid, non-zero address space) under the different transfer strategies. Table 4-5 provides these timings.

	Pure-IOU	RS	Copy
Minprog	.16	5.0	8.5
Lisp-T	.16	25.8	157.0
Lisp-Del	.17	25.8	168.5
PM-Start	.15	9.0	30.8
PM-Mid	.16	13.0	28.1
PM-End	.19	20.5	31.0
Chess	.21	7.7	11.7

Table 4-5: Address Space Transfer Times in Seconds

Address space transfer times are closely clustered for the copy-on-reference approach (*IOU*), but vary considerably for the resident set (*RS*) and pure-copy (*Copy*) techniques.

Times required to ship process address spaces pure-IOU are nearly independent of the amount of memory involved. Use of pure-copy doesn't fare nearly as well, where *RIMAS* trans-

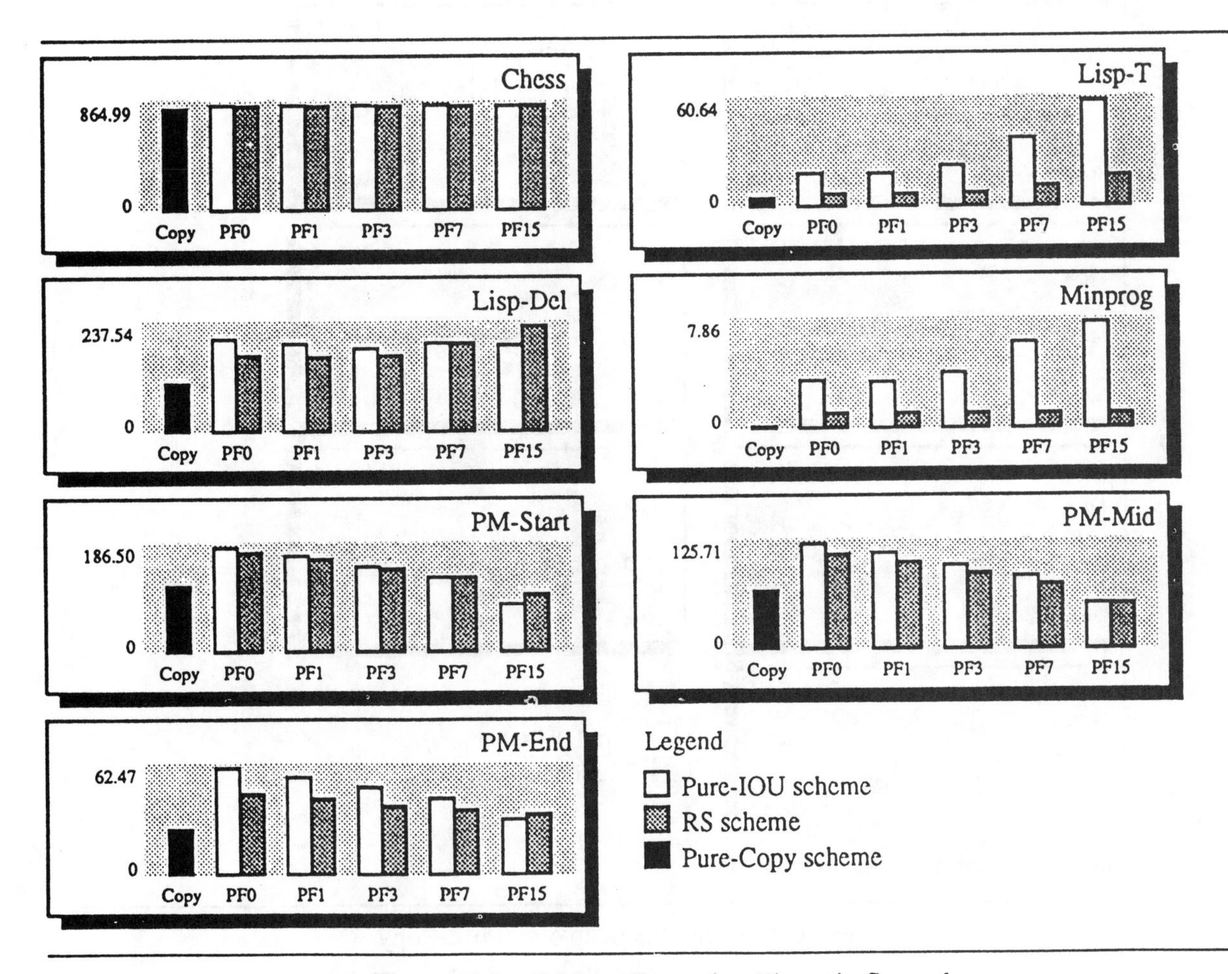

Figure 4-1: Remote Execution Times in Seconds

The measurement interval starts when the relocated program is restarted at its new location and ends when remote execution completes. Column PF*n* describes a trial where *n* pages were prefetched in response to an imaginary fault. Note: each chart is scaled individually.

mission times vary by a factor of 20. Pure-IOU allows the address space transmission to complete in significantly less time. Lisp-Del is the most extreme example, where a physical copy is almost 1,000 times more expensive. Resident set transfers once again display intermediate performance.

4.3.3. Remote Execution Times

Figure 4-1 shows the remote execution times of the representative processes, namely the elapsed time in seconds from the first instruction executed at the new host up to the program's termination. These figures show the effects of the different migration strategies, combined with differing prefetch values for the pure-IOU and resident set approaches.

Part of the effort saved in the lazy transfer of an address space must be expended as the process accesses its memory remotely. Referencing imaginary memory through the intermediary *Scheduler* and *NetMsgServer* processes on both testbed machines is roughly 2.8 times more expensive than accessing data backed by a local disk (115 milliseconds vs. 40.8 milliseconds). The most glaring effect of this cost differential on remote execution time is seen in the Minprog case, which executes 44 times slower under the pure-IOU strategy. The majority of this time is spent collecting its working set as it attempts to execute the few instructions before it terminates. The long-lived, compute-bound Chess program suffers a much smaller execution penalty, running only about 3% longer.

Dependent on the memory access patterns exhibited, the effect of prefetch varies considerably among the representatives. The Lisp family, which doesn't display memory locality, suffered from increased prefetch. The additional pages were rarely used and did not justify the larger fault-handling time. Hit ratios on these extra Lisp pages dropped from around 40% to 20% as prefetch increased. On the other hand, programs such as Pasmac, which access large tracts of memory in a sequential fashion, benefitted greatly from large prefetch. Pasmac tallied a steady 78% hit ratio across all prefetch values used, and improved its IOU remote execution times by up to a factor of 2 across this range.

Transferring process resident sets to the new execution site only had a significant impact on the extremely short-lived processes (Lisp-T, Minprog). This implies that the underlying working sets change quickly for Accent processes, in turn suggesting that resident set transfers are not a useful optimization in this setting.

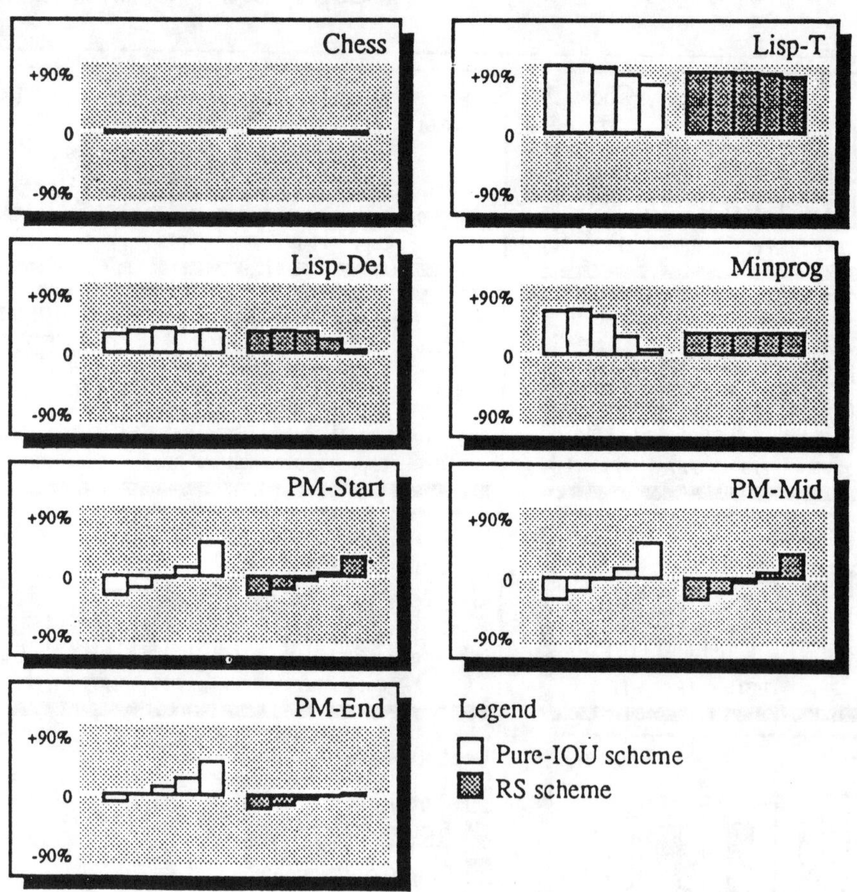

Figure 4-2: Percent Overall Speedup over Pure-Copy

Shown are the end-to-end speedups resulting from use of pure-IOU and RS transmissions. Elapsed times for address space transfer and remote execution are summed for each representative process and prefetch value and compared to the pure-copy results. From left to right in each group, bars indicate percent speedup over pure-copy for prefetch values of 0, 1, 3, 7 and 15 pages. Negative values (bars drawn in the bottom half of each gray area) represent slowdowns in relation to pure-copy.

4.3.4. Overall Migration Speedup

As demonstrated above, the pure-IOU and RS*schemes hold a clear advantage in the address space transfer phase of migration yet generally cause processes to execute longer at the remote site. In order to get overall or end-to-end performance figures, elapsed times for context transfer and remote execution are summed for these strategies and compared to the pure-copy results. The percent speedups over the straightforward pure-copy technique are displayed in Figure 4-2 for the pure-IOU and RS approaches as different amounts of prefetch are performed. The pure-IOU results (white bars) are grouped together on the left-hand side of each chart; similarly, the resident set results (dark gray bars) are placed on the right-hand side. From left to right in each group, the bars show the percent speedup for prefetch values of 0, 1, 3, 7 and 15. Negative values indicate slowdowns in relation to pure-copy.

As expected, processes that access the smallest portion of their address spaces at the new site are best suited to use the copy-on-reference technique when overall elapsed time for migration and remote execution is the metric. In the current implementation, the breakeven point is around one-quarter of the process *RealMem*. Once past this percentage, as in the Pasmac family of processes, the higher cost of fetching individual pages during remote execution in the pure-IOU system outweighs the savings achieved during migration itself. The exception to this observation is the Chess program, which is insensitive to the transfer method used. In that case, the differences imposed the various strategies were drowned out by the program's longevity.

With its strong influence on remote execution times, the amount of prefetch performed is a critical factor in end-to-end performance. Pasmac, as a representative for processes past the breakeven point and demonstrating strong sequential access patterns, went from an overall 21% slowdown on average to a 44% *speedup* as prefetch increased. In all cases, the results demonstrate that returning one additional contiguous page per remote fault improves performance. With intelligent use of prefetch, copy-on-reference migration is significantly faster than pure-copy transfer for the representatives (except the long-lived Chess process) when overall timings are considered. On the other hand, process resident sets didn't "pay their way" by cutting remote faulting activity enough to offset their shipment costs.

4.4. Cost Analysis

Section 4.3 reports that copy-on-reference treatment of ad-

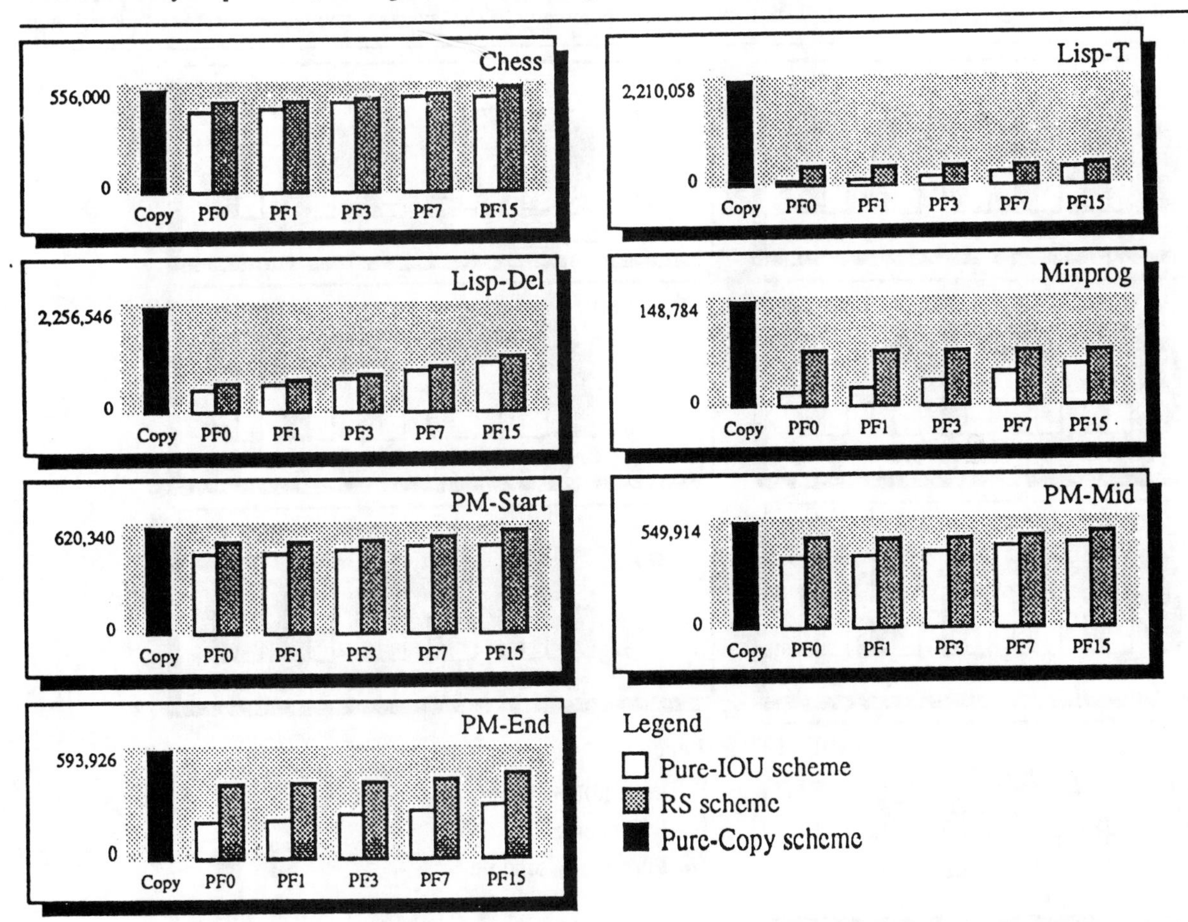

Figure 4-3: Bytes Transferred During Trials

Number of bytes transferred for each program, transmission strategy and page prefetch value during the migration trials. The measurement interval starts when the migration request is received by the *MigrationManager* and ends when the program completes its remote execution. Column PF*n* describes a trial where *n* pages were prefetched in response to an imaginary fault. Note: each chart is scaled individually.

dress space transfers significantly improves the time required to migrate a process to a new site and complete its execution there. This section supports these results by examining the specific costs incurred by the different migration strategies, and how these costs are distributed across the migration phases. Experiments reveal that copy-on-reference reduces the number of bytes transferred between the hosts as well as the cost of handling messages related to migration activities. Not only are the overall costs lowered by this approach, but they are also more evenly distributed across the context transfer and remote execution phases.

4.4.1. Bytes Transferred

Figure 4-3 reports the number of bytes exchanged between machines due to migration and remote execution of the representatives under the different strategies. Note that a single value is reported for each pure-copy trial, since prefetch doesn't apply in these cases.

The pure copy-on-reference strategy was superior to pure-copy across all prefetch settings. This technique reduced byte traffic by an average of 58.2% over pure-copy when no prefetch was used. As a rule, more data was exchanged as the number of contiguous pages prefetched grew. This is reasonable, since not all the extra pages were referenced. Shipping resident sets cut into the savings realized by the IOU strategy, again implying that very little of this data was actually used at the remote site.

4.4.2. Message Costs

Pure-copy is the clear winner when evaluated by the *number* of messages processed by the test systems. However, it does not fare nearly as well in a more important metric, the amount of time required to process and deliver these messages. Each second of execution time spent by the *NetMsgServer* to handle message traffic is not only a second stolen from the migrated process but from *all* processes in *both* systems. Figure 4-4 displays the amount of time spent by each node in message manipulation.

These figures further confirm the utility of a lazy approach to address space access. By putting off the apparent work that needs to be performed until the last moment, a significant portion does not need to be done at all. Although the bulk transfer of the process context when the pure-copy strategy is employed allows a higher throughput than the page-by-page access imposed by the pure-IOU and resident set approaches, the majority of pages sent by the pure-copy approach are never used. The pure-IOU strategy only performs work that is productive and necessary.

In every case, the IOU and resident set strategies outperform pure-copy. The average savings in message processing is

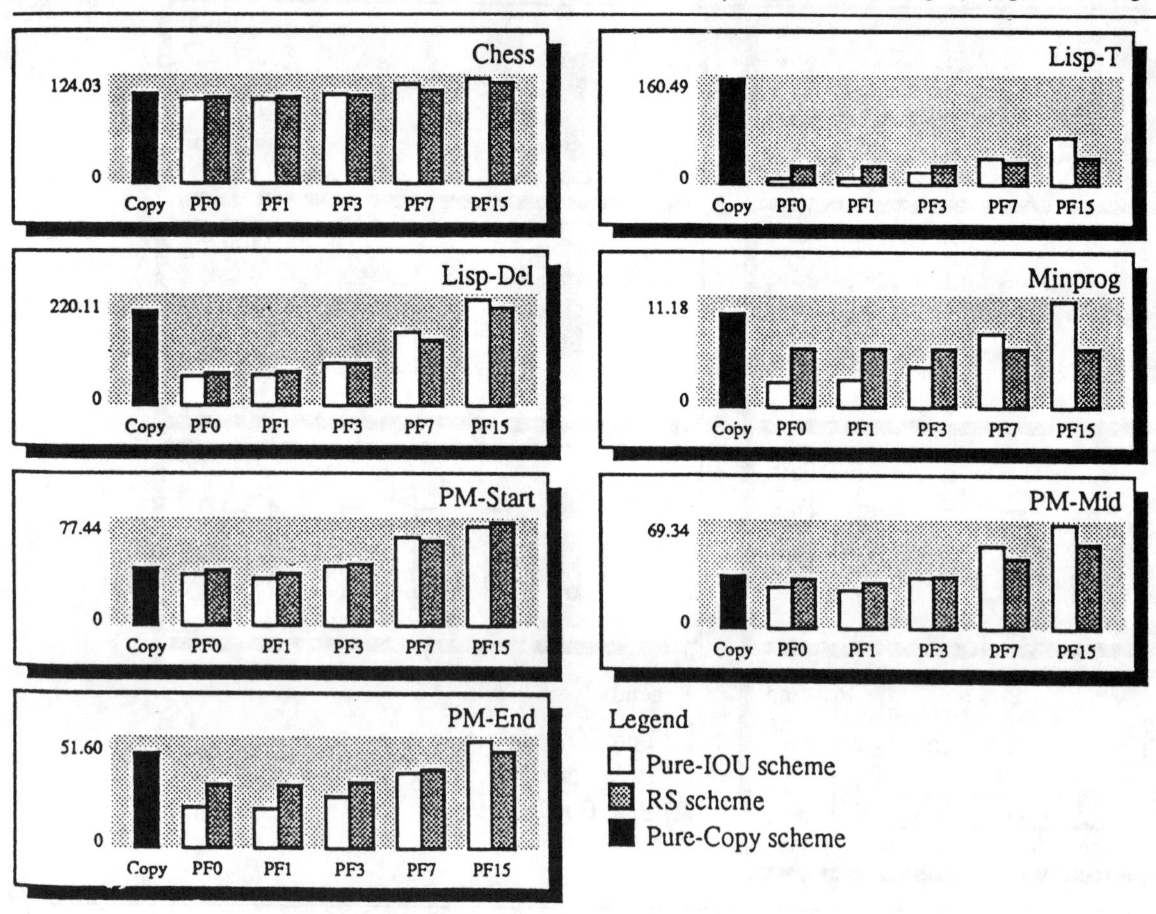

Figure 4-4: Total Message Times in Seconds

Displayed are the elapsed time in seconds required to process the IPC messages generated for each migration trial. Column PF*n* describes a trial where *n* pages were prefetched in response to an imaginary fault. Note: each chart is scaled individually.

47.8% for IOU trials without prefetch. The effect of prefetch is an interesting one. When only a single additional page is prefetched in response to an imaginary fault, the time spent processing messages drops slightly. As we increase the number of pages prefetched, the system spends more and more time in message handling. Although the prefetching eliminates many of the imaginary faults, it also transfers some "dead weight" pages that are never used. Also, since each message carries more data, the time to process each imaginary reply message grows.

Combined with the results on end-to-end costs, these figures suggest that one page should be prefetched regardless of the transfer strategy chosen.

4.4.3. Distribution of Costs

The vast majority of migration costs charged to the pure-copy strategy are incurred during the transfer phase of process migration. On the other hand, the copy-on-reference approaches radically reduce the cost of context shipment and instead incurs its expenses across the remote lifetime of the process involved. Thus, not only are costs reduced overall, but they are also more evenly distributed. Pure IOU transfers don't experience the same magnitudes and bursts of activity required by the pure-copy strategy. Instead, a lower, more constant rate of work is exhibited. The trials demonstrate that sustained network transmission speeds are reduced up to 66%.

Figure 4-5 presents the data transfer rates caused by the migration and remote execution of the Lisp-Del case under the different strategies, starting at the time of migration and ending with the execution of the final remote instruction.

These panels depict the results of a full-IOU transfer of Lisp-Del, a resident set approach and finally the full-copy method. The areas in white represent bytes exchanged in support of imaginary fault activity. Full-copy transfers have a characteristic signature, with a large bulk data transfer early on. The resident set panel illustrates that a sizable amount of data is still physically shipped during the migration phase, but does not improve the overall time significantly from the pure-

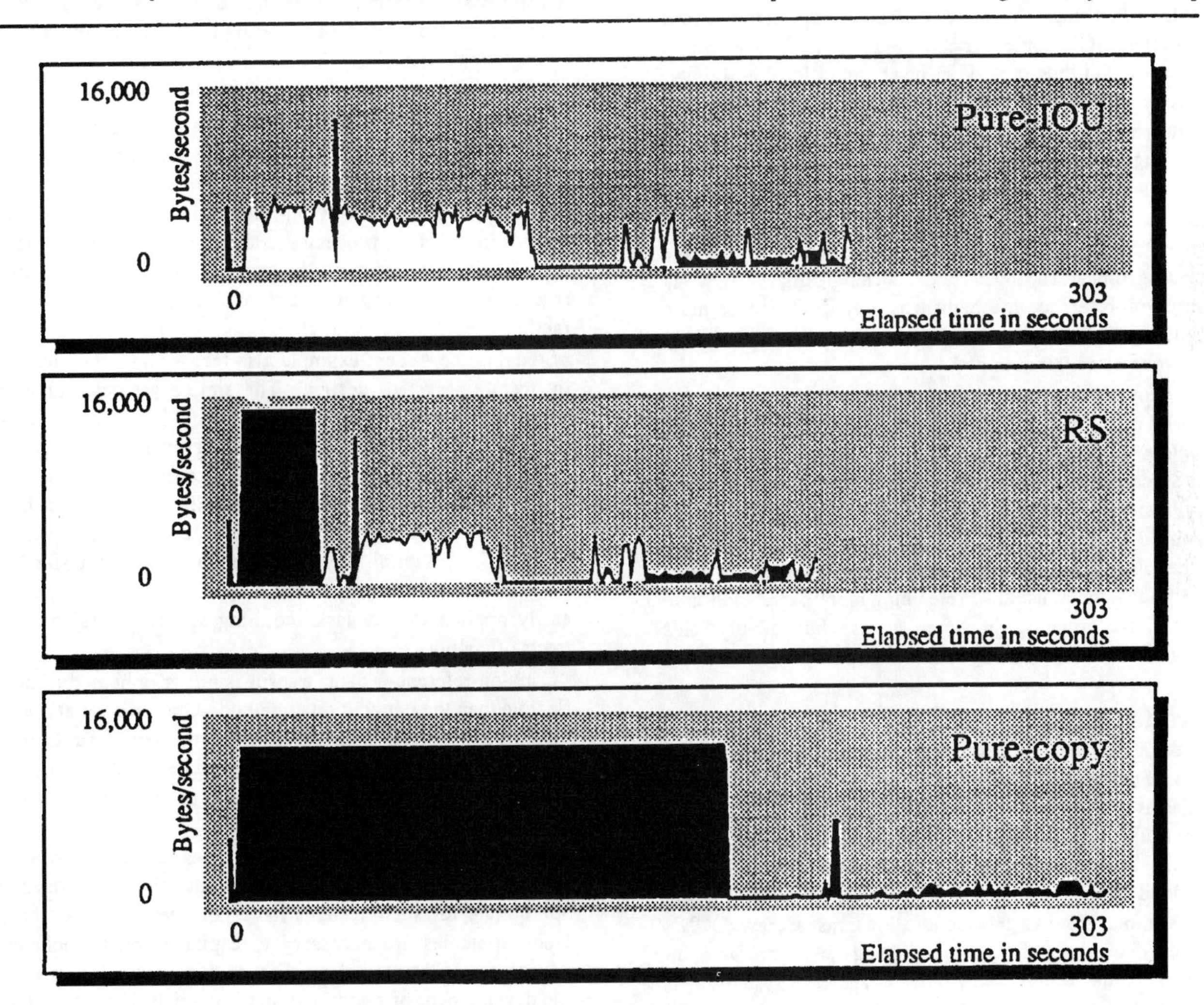

Figure 4-5: Byte Transfer Rates for Lisp-Del

Network data transfer rates during the migration and remote execution of the Lisp-Del (Delaunay triangulation) program. No prefetch is being performed. White areas show data transferred in support of imaginary faults, black areas show all other transfers.

IOU approach. Copy-on-reference allows the process to resume execution very quickly. In this case, Lisp-Del finishes its work shortly after the full-copy trial *begins* its remote execution.

4.5. Summary

The trial data collected for the Accent migration facility reveals several interesting facts about process composition and behavior. While address space size varies by as much as a factor of 12,803 in the representative processes, the amount of *RealMem* only differs by a factor of 15. *RealZeroMem* forms a significant portion of every process address space, more than half in most cases and 99.9% in the Lisp examples. These representatives touched between 0.002% and 27.4% of their validated address spaces, and between 3% and 58% of the *RealMem* portions. This verifies the assumption that processes access relatively small parts of their addressible data.

Process excision and insertion times are also much less variable in this study, factors of 4 and 3.3 respectively across the samples. IOU context transfers take roughly one second in all cases, and thus provide a lower bound for this activity. Pure-copy transfers vary by a factor of 20, and in the most extreme case are 1,000 times more expensive than the corresponding pure-IOU transfer.

Much less data needs to be communicated between machines when copy-on-reference tactics are used. On average, 58% fewer bytes are transferred and message processing times drop by 47%. Touching remote pages via the copy-on-reference mechanism is roughly 2.8 times more expensive than local disk accesses, and this figure can likely be improved through tuning.

The copy-on-reference variations studied in this system produced mixed results. Resident sets were found to be poor predictors of the data required by the process at its remote site. Since Accent uses its physical memory as a disk cache, many resident pages are sometimes guaranteed *not* to be referenced again, especially by the Pasmac class of processes. On the other hand, small amounts of page prefetch were found to always be useful. Prefetching more pages each time degrades performance in some cases, but greatly aids programs performing mostly sequential accesses.

5. Related Work

Investigation into process migration began in the early 1970's. Such efforts as the "Creeper" program [11] by Bob Thomas at BBN and the "Relocatable McRoss" [14] air traffic controller demonstrated migration's feasibility. However, they did little to address the transparency issues. DCN [6] added name transparency by associating resources with processes, but failed to provide location transparency. DCN's resource names specified the supplying host, and were invalidated if the resource was moved. The RIG system [4] is Accent's direct ancestor and shared many of the same concepts. RIG's ports were visibly tied with the process owning them, so it suffered from DCN's problem. The DEMOS/MP operating system [9] was among the first to offer full transparency. Link names contained *hints* to the location of the service, and were not invalidated by resource relocation. The University of Washington's object-oriented Eden [5] system provided full transparency and migration services, but could not take advantage of a copy-on-reference mechanism. Eden's objects were forced to reside entirely on a single host. Dannenberg's Butler [1] made use of an older version of Accent which did not provide copy-on-reference data shipment, but demonstrated Accent's suitability for transparent migration support.

Various systems have attempted different attacks on the cost of context transfer. The LOCUS [8] remote invocation facility exploits shared code present at the target site, cutting down the amount of data that must flow to the new site. This approach doesn't address the data portions of a process context, including memory-mapped files. Marvin Theimer's migration facility for the V system [13] tried to hide transmission costs from processes by *pre-copying* the context in an iterative fashion before moving the process. Process downtime was thus reduced, but both hosts still paid the transfer costs. Theimer's measurements reveal that this technique suffers from network buffering problems and overruns.

6. Conclusions

The Accent testbed's use of copy-on-reference address space transfer has demonstrated its effectiveness in tackling process migration's dominant cost. Unlike the conventional transmission technique, copy-on-reference avoids the linear growth in costs as processes address more and more data. Any distributed system in the same class can expect similar results in the construction and use of a copy-on-reference facility.

Studying the Accent example also teaches important lessons in operating system design. The simple yet powerful port abstraction and the close integration of IPC and virtual memory facilities give Accent the transparency needed to cleanly support migration without sacrificing performance. These features, along with extensibility through user-level processes, allows a generic copy-on-reference mechanism to be built in a natural way. This mode of data transfer has proven useful in the migration domain, but may be just as easily applied to *any* task requiring sparse access to large tracts of memory.

Copy-on-reference data transmission is inherently more flexible than the conventional method. Only two variations of actual data delivery have been explored here. Tasks with special knowledge of the data requirements they will encounter may apply that knowledge to optimize the physical shipment of data.

This investigation opens many avenues for future research. The creation and evaluation of automatic migration strategies appropriate for such systems have not been addressed here. Good strategies are necessary to capitalize on the inherent advantages of lazy transfers. Part of this activity will involve the development of good load metrics which specifically take into account the fact that a process virtual address space may be physically dispersed among several computational hosts. Copy-on-reference may be proven useful in remote file and database accesses, remote invocation facilities and intelligent RPCs. It would be interesting to attempt to extend this work to systems allowing shared memory, and to evaluate the application of copy-on-reference techniques to a shared centralized file system such as Andrew [7].

Although Accent is no longer actively in use at Carnegie Mellon University, the lessons learned from this work are being applied to the Mach environment [10] currently being developed there. A successor to Accent aimed at supporting a wide range of hardware configurations, Mach allows *external pager* processes which provide copy-on-reference administration of data. Study of copy-on-reference behavior in this new facility will provide further insights on the basic mechanism in a more modern computing system.

References

1. Roger B. Dannenberg. *Resource Sharing in a Network of Personal Computers.* Ph.D. Th., Carnegie Mellon University, December 1982.

2. Peter J. Denning. "The Working Set Model for Program Behavior". *Communications of the ACM 11*, 5 (May 1968), 323-333.

3. Robert P. Fitzgerald. *A Performance Evaluation of the Integration of Virtual Memory Management and Inter-Process Communication in Accent.* Ph.D. Th., Carnegie Mellon University, October 1986. Available as CMU technical report CMU-CS-86-158.

4. Keith A. Lantz, Klaus D. Gradischnig, Jerome A. Feldman and Richard F. Rashid. "Rochester's Intelligent Gateway". *Computer* (October 1982), 54-68.

5. E.D. Lazowska, H.M. Levy, G.T. Almes, M.J. Fischer, R.J. Fowler, S.C. Vestal. The Architecture of the Eden System. Tech. Rept. 81-04-01, Department of Computer Science, University of Washington, April, 1981.

6. David. L. Mills. An Overview of the Distributed Computer Network. National Computer Conference, University of Maryland, 1976, pp. 523-531.

7. James H. Morris, Mahadev Satyanarayanan, Michael E. Conner, John H. Howard, David S. H. Rosenthal and Donelson Smith. "Andrew: A Distributed Personal Computing Environment". *Communications of the ACM 19*, 3 (March 1986), 184-201.

8. G. Popek, B. Walker, J. Chow, D. Edwards, C. Kline, G. Rudisin, G. Thiel. LOCUS: A Network Transparent, High Reliability Distributed System. Joint Conference on Computer Performance Modelling, Measurement and Evaluation, ACM, 1986.

9. Michael L. Powell and Barton P. Miller. Process Migration in DEMOS/MP. Proceedings of the Sixth Symposium of Operating System Principles, ACM, November, 1983, pp. 110-119.

10. Richard F. Rashid. "Threads of a New System". *Unix Review 4*, 8 (August 1986), 37-49.

11. John F. Shoch and Jon A. Hupp. "The 'Worm' Programs - Early Experience with a Distributed Computation". *Communications of the ACM 25*, 3 (March 1982), 172-180.

12. CMU Computer Science Department. Proposal for a Joint Effort in Personal Scientific Computing. Carnegie Mellon University, August, 1979.

13. Marvin M. Theimer, Keith A. Lantz and David R. Cheriton. Preemptable Remote Execution Facilities for the V-System. Proceedings of the Tenth Symposium on Operating System Principles, ACM SIGOPS, 1985, pp. 2-12.

14. Robert H. Thomas and D. Austin Henderson. McRoss - A Multi-Computer Programming System. Proceedings, Spring Joint Conference, 1972.

4.5 Theimer, M., Lantz, K., and Cheriton, D., Preemptable Remote Execution Facilities for the V System

This research was supported by the Defense Advanced Research Projects Agency under contracts MDA 903-80-L-0102 and N00039-83-K-0431.

Theimer, M., Lantz, K., and Cheriton, D., "Preemptable Remote Execution Facilities for the V System," *Proceedings of the 10th ACM Symposium on Operating System Principles,* pp. 2–12, December 1985.

Preemptable Remote Execution Facilities for the V-System

Marvin M. Theimer, Keith A. Lantz, and David R. Cheriton
Computer Science Department
Stanford University
Stanford, CA 94305

Abstract

A remote execution facility allows a user of a workstation-based distributed system to offload programs onto idle workstations, thereby providing the user with access to computational resources beyond that provided by his personal workstation. In this paper, we describe the design and performance of the remote execution facility in the V distributed system, as well as several implementation issues of interest. In particular, we focus on network transparency of the execution environment, preemption and migration of remotely executed programs, and avoidance of residual dependencies on the original host. We argue that preemptable remote execution allows idle workstations to be used as a "pool of processors" without interfering with use by their owners and without significant overhead for the normal execution of programs. In general, we conclude that the cost of providing preemption is modest compared to providing a similar amount of computation service by dedicated "computation engines".

1. Introduction

A distributed computer system consisting of a cluster of workstations and server machines represents a large amount of computational power, much of which is frequently idle. For example, our research system consists of about 25 workstations and server machines, providing a total of about 25 MIPS. With a personal workstation per project member, we observe over one third of our workstations idle, even at the busiest times of the day.

There are many circumstances in which the user can make use of this idle processing power. For example, a user may wish to compile a program and reformat the documentation after fixing a program error, while continuing to read mail. In general, a user may have batch jobs to run concurrently with, but unrelated to, some interactive activity. Although any one of these programs may perform satisfactorily in isolation on a workstation, forcing them to share a single workstation degrades interactive response and increases the running time of non-interactive programs.

Use of idle workstations as computation servers increases the processing power available to users and improves the utilization of the hardware base. However, this use must not compromise a workstation owner's claim to his machine: A user must be able to quickly reclaim his workstation to avoid interference with personal activities, implying removal of remotely executed programs within a few seconds time. In addition, use of workstations as computation servers should not require programs to be written with special provisions for executing remotely. That is, remote execution should be *preemptable* and *transparent*. By preemptable, we mean that a remotely executed program can be migrated elsewhere on demand.

In this paper, we describe the preemptable remote execution facilities of the V-system [4, 2] and examine several issues of interest. We argue that preemptable remote execution allows idle workstations to be used as a "pool of processors" without interfering with use by their owners and without significant overhead for the normal execution of programs. In general, we conclude that the cost of providing preemption is modest compared to providing a similar amount of computation service by dedicated "computation engines". Our facilities also support truly distributed programs in that a program may be decomposed into subprograms, each of which can be run on a separate host.

There are three basic issues we address in our design. First, programs should be provided with a network-transparent execution environment so that execution on a remote machine is the same execution on the local machine. By *execution environment*, we mean the names, operations and data with which the program can interact during execution. As an example of a problem that can arise here, programs that directly access hardware devices, such as a graphics frame buffer, may be inefficient if not impossible to execute remotely.

Second. migration of a program should result in minimal interference with the execution of the program and the rest of the system, even though migration requires atomic transfer of a copy of the program state to another host. Atomic transfer is required so that the rest of the system at no time detects there being other than one copy. However, suspending the execution of the migrating program or the interactions with the program for the entire time required for migration may cause interference with system execution for several seconds and may even result in failure of the program. Such long "freeze times" must be avoided.

Finally, a migrated program should not continue to depend on its previous host once it is executing on a new host, that is, it should have no *residual dependencies* on the previous host. For example, a program either should not create temporary files local to its current computation server or else those files should be migrated along with the program. Otherwise, the migrated program continues to impose a load on its previous host, thus diminishing some of the benefits of migrating the program. Also, a failure or reboot of the previous host causes the program to fail because of these inter-host dependencies.

The paper presents our design with particular focus on how we have addressed these problems. The next section describes the remote execution facility. Section 3 describes migration. Section 4 describes performance and experience to date with the use of these facilities. Section 5 compares this work to that in some other distributed systems. Finally, we close with conclusions and indications of problems for further study.

2. Remote Execution

A V program is executed on another machine at the command interpreter level by typing:

<program> <arguments> @ *<machine-name>*

Using the meta-machine name *:

<program> <arguments> @ *

executes the specified program at a random idle machine on the network. A standard library routine provides a similar facility that can be directly invoked by arbitrary programs. Any program can be executed remotely providing that it does not require low-level access to the hardware devices of the machine from which it originated. Hardware devices include disks, frame buffers, network interfaces, and serial lines.

A suite of programs and library functions are provided for querying and managing program execution on a particular workstation as well as all workstations in the system. Facilities for terminating, suspending and debugging programs work independent of whether the program is executing locally or remotely.

It is often feasible for a user to use his workstation simultaneously with its use as a computation server. Because of priority scheduling for locally invoked programs, a text-editting user need not notice the presence of background jobs providing they are not contending for memory with locally executing programs.

2.1. Implementation

The V-system consists of a distributed kernel and a distributed collection of server processes. A functionally identical copy of the kernel resides on each host and provides address spaces, processes that run within these address spaces, and network-transparent interprocess communication (IPC). Low-level process and memory management functions are provided by a *kernel server* executing inside the kernel. All other services provided by the system are implemented by processes running outside the kernel. In particular, there is a *program manager* on each workstation that provides program management for programs executing on that workstation.

V address spaces and their associated processes are grouped into *logical hosts*. A V process identifier is structured as a *(logical-host-id, local-index)* pair. In the extreme, each program can be run in its own logical host. There may be multiple logical hosts associated with a single workstation, however, a logical host is local to a single workstation.

Initiating local execution of a program involves sending a request to the local program manager to create a new address space and load a specified program image file into this address space. The program manager uses the kernel server to set up the address space and create an initial process that is awaiting reply from its creator. The program manager then turns over control of the newly created process to the requester by forwarding the newly created process to it. The requester initializes the new program space with program arguments, default I/O, and various "environment variables", including a name cache for commonly used global names. Finally, it starts the program in execution by replying to its initial process. The communication paths between programs and servers are illustrated in Figure 2-1.

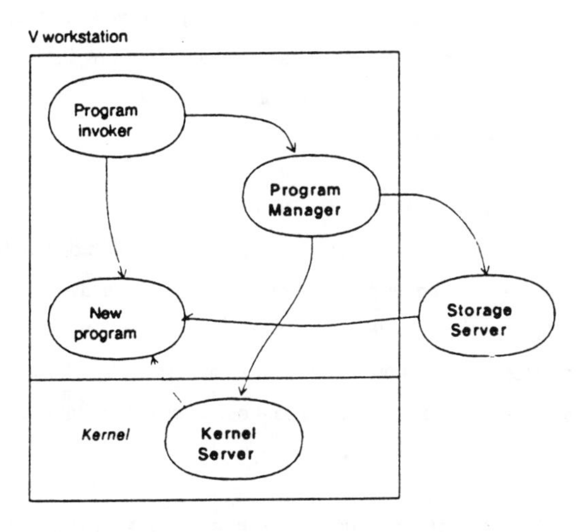

Figure 2-1: Communication paths for program creation.

A program is executed on another workstation by addressing the program creation request to the program manager on the other workstation. The appropriate program manager is selected using the process group mechanism in V, which allows a message to be sent to a group of processes rather than just individual processes [5]. Every program manager belongs to the well-known program manager group. When the user specifies a particular machine, a query is sent to this group with the specified host name, requesting that the associated program manager respond. The response indicates the program manager to which to send the program creation request. When the user specifies "*", a query is sent requesting a response from those hosts with a reasonable amount of processor and memory resources available for remotely executing programs. Typically, the client receives several responses to the request. Currently, it simply selects the program manager that responds first since that is generally the least loaded host. This simple mechanism provides a decentralized implementation of scheduling that performs well at minimal cost for reasonably small systems.

Beyond selection of a program manager, remote program execution appears the same as local program execution because programs are provided with a network-transparent execution environment, assuming they do not directly access hardware devices. In particular:

- The program address space is initialized the same as when the program is executed locally. For example, arguments and environment variables are passed in the same manner.

- All references by the program outside its address space are performed using network-transparent IPC primitives and globally unique identifiers, with the exceptions of the host-specific kernel server and program manager. For example, standard input and output are specified by global identifiers for the server processes implementing them.

- The kernel server and program manager on a remote workstation provide identical services for remotely executing programs as the local kernel server and program manager provide for locally executing programs.[1] Access to the kernel server and program manager of the workstation on which a program is running is obtained through *well-known local process groups*, which in this case contain only a single process. For example, the kernel server can be accessed by constructing a process-group-id consisting of the program's logical-host-id concatenated with the index value for the kernel server.[2] Thus, host-specific servers can be referenced in a location-independent manner.

- We assume that remotely executed programs do not directly access the device server[3] The limitation on device access is not a problem in V since most programs access physical devices through server processes that remain co-resident with the devices that they manage. In particular, programs perform all "terminal output" via a display server that remains co-resident with the frame buffer it manages [10, 14].

---

[1] It only makes sense to use the kernel server and program manager local to the executing program since their services are intrinsically bound to the machine on which the program is executing, namely management of processor and memory resources.

[2] A process-group-id is identical in format to a process-id.

[3] Actually, references to devices bind to devices on the workstation on which they execute, which is useful in many circumstances. However, we are not able to migrate these programs.

3. Migration

A program is migrated by invoking:

migrateprog [-n] [<program>]

to remove the specified program from the workstation. If no other host can be found for the program, the program is not removed unless the "-n" flag is present, in which case it is simply destroyed. If no program is specified, **migrateprog** removes all remotely executed programs.

A program may create sub-programs, all of which typically execute within a single logical host. Migration of a program is actually migration of the logical host containing the program. Thus, typically, all sub-programs of a program are migrated when the program is migrated. One exception is when a sub-program is executed remotely from its parent program.

3.1. Implementation

The simplest approach to migrating a logical host is to freeze its state while the migration is in progress. By *freezing* the state, we mean that execution of processes in the logical host is suspended and all external interactions with those processes are deferred.

The problem with this simple approach is that it may suspend the execution of the programs in the logical host and programs that are executing IPC operations on processes in the logical host for too long. In fact, various operations may abort because their timeout periods are exceeded. Although aborts can be prevented via "operation pending" packets, this effectively suspends the operation until the migration is complete. Suspension implies that operations that normally take a few milliseconds could take several seconds to complete. For example, the time to copy address spaces is roughly 3 seconds per megabyte in V using 10 Mb Ethernet. A 2 megabyte logical host state would therefore be frozen for over 6 seconds. Moreover, significant overhead may be incurred by retransmissions during an extended suspension period. For instance, V routinely transfers 32 kilobytes or more as a unit over the network.

We reduce the effect of these problems by copying the bulk of the logical host state before freezing it, thereby reducing the time during which it is frozen. We refer to this operation as *pre-copying*. Thus, the complete procedure to migrate a logical host is:

1. Locate another workstation (via the program manager group) that is willing and able to accommodate the logical host to be migrated.
2. Initialize the new host to accept the logical host.
3. Pre-copy the state of the logical host.
4. Freeze the logical host and complete the copy of its state.
5. Unfreeze the new copy, delete the old copy, and rebind references.

The first step of migration is accomplished by the same mechanisms employed when the program was executed remotely in the first place. These mechanisms were discussed in Section 2. The remainder of this section discusses the remaining steps.

3.1.1. Initialization on the New Host

Once a new host is located, it is initialized with descriptors for the new copy of the logical host. To allow it to be referenced before the transfer of control, the new copy is created with a different logical-host-id. The identifier is then changed to the original logical-host-id in a subsequent step (Section 3.1.3).

The technique of creating the new copy as a logical host with a different identifier allows both the old copy and the new copy to exist and be accessible at the same time. In particular, this allows the standard interprocess copy operations, ***CopyTo*** and ***CopyFrom***, to be used to copy the bulk of the program state.

3.1.2. Pre-copying the State

Once the new host is initialized, we pre-copy the state of the migrating logical host to the new logical host. Pre-copying is done as an initial copy of the complete address spaces followed by repeated copies of the pages modified during the previous copy until the number of modified pages is relatively small or until no significant reduction in the number of modified pages is achieved.[4] The remaining modified pages are recopied after the logical host is frozen.

The first copy operation moves most of the state and takes the longest time, therefore providing the longest time for modifications to the program state to occur. The second copy moves only that state modified during the first copy, therefore

[4] Modified pages are detected using dirty bits.

taking less time and presumably allowing fewer modifications to occur during its execution time. In a non-virtual memory system, a major benefit of this approach is moving the code and initialized data of a logical host, portions that are never modified, while the logical host continues to execute. For example, consider a logical host consisting of 1 megabyte of code, .25 megabytes of initialized (unmodified data) and .75 megabytes of "active" data. The first copy operation takes roughly 6 seconds. If, during those 6 seconds, .1 megabytes of memory were modified, the second copy operation should take roughly .3 seconds. If during those .3 seconds, .01 megabytes of memory were modified, so the third copy operation should take about 0.03 seconds. At this point, we might freeze the logical host state, completing the copy and transferring the logical host to the next machine. Thus, the logical host is frozen for about .03 seconds (assuming no packet loss), rather than about 6 seconds.

The pre-copy operation is executed at a higher priority than all other programs on the originating host to prevent these other programs from interfering with the progress of the pre-copy operation.

3.1.3. Completing the Copy

After the pre-copy, the logical host is frozen and the copy of its state is completed. Freezing the logical host state, even if for a relatively short time, requires some care. Although we can suspend execution of all processes within a logical host, we must still deal with external IPC interactions. In V, interprocess communication primitives can change the state of a process in three basic ways: by sending a request message, by sending a reply message, or by executing a kernel server or program manager operation on the process.[5] When a process or logical host is frozen, the kernel server and program manager defer handling requests that modify this logical host until it is unfrozen. When the logical host is unfrozen, the requests are forwarded to the new program manager or kernel server, assuming the logical host is successfully migrated at this point.

In the case of a request message, the message is queued for the recipient process. (The recipient is modified slightly to indicate that it is not prepared to immediately receive the message.) A "reply-pending" packet is sent to the sender on each retransmission, as is done in the normal case. When the logical host is deleted after the transfer of logical host has taken place, all queued messages are discarded and the remote senders are prompted to retransmit to the new host running these processes. For local senders, this entails restarting the send operation. The normal Send then maps to a remote Send operation to the new host, given that the recipient process is no longer recorded as local. For remote senders, the next retransmission (or at least a subsequent one) uses the new binding of logical host to host address that is broadcast when a logical host is migrated. Therefore, the retransmission delivers the message to the new copy of the logical host.

Reply messages are handled by discarding them and relying on the retransmission capabilities of the IPC facilities. A process on a frozen logical host that is awaiting reply continues to retransmit to its replier periodically, even if a reply has been received. This basically resets the replier's timeout for retaining the reply message so that the reply message is still be available once the migration is complete.

The last part of copying the original logical host's state consists of copying its state in the kernel server and program manager. Copying the kernel state consists of replacing the kernel state of the newly created logical host with that of the migrating one. This includes changing the logical-host-id of the new logical host to be the same as that of the original logical host.

Once this operation has succeeded, there exist two frozen identical copies of the logical host. The rest of the system cannot detect the existence of two copies because operations on both of them are suspended. The the kernel server on the original machine continues to respond with reply-pending packets to any processes sending to the logical host as well as retransmit Send requests, thereby preventing timeout.

If the copy operation fails due to lack of acknowledgement, we assume that the new host failed and that the logical host has not been transferred. The logical host is unfrozen to avoid timeouts, another host is selected for this logical host and the migration process is retried. Care must be taken in retrying this migration that we do not exceed the amount of time the user is willing to wait. In our current implementation, we simply give up if the first attempt at migration fails.

[5] We treat a *CopyTo* operation to a process as a request message.

3.1.4. Unfreezing the New Copy and Rebinding References

Once all state has been transferred, the new copy of the logical host is unfrozen and the old copy is deleted. References to the logical host are rebound as discussed next.

The only way to refer to a process in V is to use its globally unique process identifier. As defined in Section 2, a process identifier is bound to a logical host, which is in turn bound to a physical host via a cache of mappings in each kernel. Rebinding a logical host to a different physical host effectively rebinds the identifiers for all processes on that logical host. When a reference to a process fails to get a response after a small number of retransmissions, the cache entry for the associated logical host is invalidated and the reference is broadcast. A correct cache entry is derived from the response. The cache is also updated based on incoming requests. Thus, when a logical host is migrated, these mechanisms automatically update the logical host cache, thereby rebinding references to the associated process identifiers.

Various optimizations are possible, including broadcasting the new binding at the time the new copy is unfrozen.

3.2. Effect of Virtual Memory

Work is underway to provide demand paged virtual memory in V, such that workstations may page to network file servers. In this configuration, it suffices to flush modified virtual memory pages to the network file server rather than explicitly copy the address space of processes in the migrating logical host between workstations. Then, the new host can fault in the pages from the file server on demand. This is illustrated in Figure 3-1.

Just as with the pre-copy approach described above, one can repeatedly flush dirty pages out without suspending the processes until there are relatively few dirty pages and then suspend the logical host.

This approach to migration takes two network transfers instead of just one for pages that are dirty on the original host and then referenced on the new host. However, we expect this technique to allow us to move programs off of the original host faster, which is an important consideration. Also, the number of pages that require two copies should be small.

Finally, paging to a local disk might be handled by flushing the program state to disk, as above, and then doing a disk-to-disk transfer of program state over the network to the new host or the file server it uses for paging. The same techniques for minimizing freeze time appear to carry over to this case, although we can only speculate at this time.

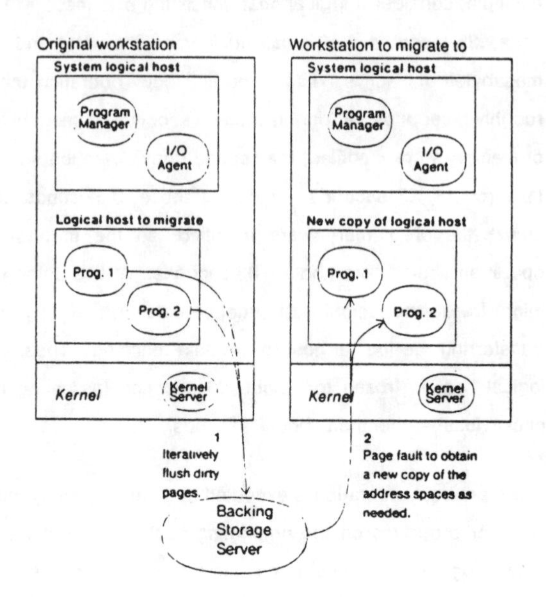

Figure 3-1: Migration with virtual memory.

3.3. Residual Host Dependencies

The design, as described, only deals with migrating programs whose state is contained in the kernel, the program manager and the address space(s) of the program being migrated. However, there are several situations where relevant state is stored elsewhere. This is not a problem if the state is stored in a globally accessible service, such as a network file server. However, extraneous state that is created in the original host workstation may lead to residual dependencies on this host after the program has been migrated. For example, the program may have accessed files on the original host workstation. After the program has been migrated, the program continues to have access to those files, by virtue of V's network-transparent IPC. However, this use imposes a continued load on the original host and results in failure of the program should the original host fail or be rebooted.

To remove this dependency it is necessary to identify and migrate (copies of) all files associated with the program being migrated. These files are effectively extensions of the program state, as considered in the above discussion, and could be moved in a similar fashion. We note, however, a few complications. First, a file may be arbitrarily large, introducing unbounded delay to the program migration. Second, a file that is being read may be a copy of some standard header file that also exists on the machine to which the process is being moved. Recognizing this case would save on file copying as well as reduce the final problem, namely, a program may have the symbolic name of the file stored internally, precluding changing the symbolic name when the file is migrated. However, this name may conflict with an existing file on the machine to which the file is moving. Related to this, a program may have written and closed a temporary file, yet be planning to read the file at some later point. There is no reliable way of detecting such files, so there is no reliable way to ensure they are migrated. With our current use of diskless workstations, file migration is not required and, moreover, file access cost is essentially the same for all workstations (and quite acceptable) [4, 11].

Similar problems can arise with other servers that store state specific to a particular program during its execution. Again, our approach has been to avoid local servers and thereby circumvent the problem of residual host dependencies.

4. Evaluation

At the time of writing, the remote execution facility has been in general use for one year and the migration facility is operational on an experimental basis. This section gives an indication of the cost of remote execution and migration, in both time and space, as well as general remarks about the usage we have observed.

4.1. Time for Remote Execution and Migration

The performance figures presented are for the SUN workstation with a 10 MHz 68010 processor and 2 Mbytes of local memory. Workstations are connected via a 10 Mbit Ethernet local area network.

The cost of remotely executing a program can be split into three parts: Selecting a host to use, setting up and later destroying a new execution environment, and actually loading the program file to run. The latter considerably dominates the first two. The cost of selecting a remote host has been measured to be 23 milliseconds, this being the time required to receive the first response from a multicast request for candidate hosts. Some additional overhead arises from receiving additional responses. (That is, the response time and the processing overhead are different.) The cost of setting up and later destroying a new execution environment on a specific remote host is 40 milliseconds. For diskless workstations, program files are loaded from network file servers so the cost of program loading is independent of whether a program is executed locally or remotely. This cost is a function of the size of the program to load, and is typically 330 milliseconds per 100 Kbytes of program.

The cost of migration is similar to that of remote execution: A host must be selected, a new copy of the logical host's state in the kernel server and program manager must be made, and the logical host's address spaces must be copied. The time required to create a copy of the logical host's kernel server and program manager state depends on the number of processes and address spaces in the logical host. 14 milliseconds plus an additional 9 milliseconds for each process and address space are required for this operation. The time required to copy 1 Mbyte of an address space between two physical hosts is 3 *seconds*.

Measurements for our C-compiler[6] and ***Tex*** text formatter programs indicated that usually 2 precopy iterations were useful (i.e. one initial copy of an address space and one copy of subsequently modified pages). The resulting amount of address space that must be copied, on average, while a program is frozen was between 0.5 and 70 Kbytes in size, implying program suspension times between 5 and 210 milliseconds (in addition to the time needed to copy the kernel server and program manager state). Table 4-1 shows the average rates at which dirty pages are generated by the programs measured.

Time interval (secs)	0.2	1	3
make	0.8	1.8	4.2
cc68	0.6	2.2	6.2
preprocessor	25.0	40.2	59.6
parser	50.0	76.8	109.4
optimizer	19.8	32.2	41.0
assembler	21.6	33.4	48.4
linking loader	25.0	39.2	37.8
tex	68.6	111.6	142.8

Table 4-1: Dirty page generation rates (in Kbytes).

[6] Which consists of 5 separate subprograms: a preprocessor, a parser front-end, an optimizer, an assembler, a linking loader, and a control program.

The execution time overhead of remote execution and migration facilities on the rest of the system is small:

- The overhead of identifying the team servers and kernel servers by local group identifiers adds about 100 microseconds to every kernel server or team server operation.

- The mechanism for binding logical hosts to network addresses predates the migration facility since it is necessary for mapping 32 bit process-ids to 48 bit physical Ethernet host addresses anyway. Thus, no extra time cost is incurred here for the ability to rebind logical hosts to physical hosts. The actual cost of rebinding a logical host to physical host is only incurred when a logical host is migrated.

- 13 microseconds is added to several kernel operations to test whether a process (as part of a logical host) is frozen.

4.2. Space Cost

There is no space cost in the kernel server and program manager attributable to remote execution since the kernel provides network-transparent operation and the program manager uses the kernel primitives. Migration, however, introduces a non-trivial amount of code. Several new kernel operations, an additional module in the program manager and the new command `migrateprog` had to be added to the standard system. These added 8 Kbytes to the code and data space of the kernel and 4 Kbytes to the permanently resident program manager.

4.3. Observations on Usage

When remote program execution initially became available, it was necessary to specify the exact machine on which to execute the program. In this form, there was limited use of the facility. Subsequently, we added the "@ *" facility, allowing the user to effectively specify "some other lightly loaded machine". With this change, the use increased significantly. Most of the use is for remotely executing compilations. However, there has also been considerable use for running simulations. In general, users tend to remotely execute non-interactive programs with non-trivial running times. Since most of our workstations are over 80% idle even during the peek usage hours of the day (the most common activity is editing files), almost all remote execution requests are honored. In fact, the only users that have complained about the remote execution facilities are those doing experiments in parallel distributed execution where the remotely executed programs want to commandeer 10 or more workstations at a time.

Very limited experience is available with preemption and migration at this time. The ability to preempt has to date proven most useful for allowing very long running simulation jobs to run on the idle workstations in the system and then migrate elsewhere when their users want to use them. There is also the potential of migrating "floating" server processes such as a transaction manager that are not tied to a particular hardware device. We expect that preemption will receive greater use as the amount of remote and distributed execution grows to match or exceed the available machine resources.

5. Related Work

Demos/MP provides preemption and migration for individual processes [15]. However, hosts are assumed to be connected by a reliable network, so Demos-MP does not deal with packet loss when pending messages are forwarded. Moreover, Demos/MP relies on a *forwarding address* remaining on the machine from which the process was migrated, in order to handle the update of outstanding references to the process. Without sophisticated error recovery procedures, this leads to failure when this machine is subsequently rebooted and an old reference is still outstanding. In contrast, our use of logical hosts allows a simple rebinding that works without forwarding addresses.

Locus provides preemptable remote execution as well [3]. However, it emphasizes load balancing across a cluster of multi-user machines rather than on sharing access to idle personal workstations. Perhaps as a consequence, there appear to have been no attempts to reduce the "freeze time" as discussed above. Moreover, in contrast to Locus, virtually all the mechanisms in V have been implemented outside the kernel, thus providing some insight into the requirements for a reduced kernel interface for this type of facility.

The Cambridge distributed system provides remote execution as a simple extension of the way in which processors are allocated to individual users [13]. Specifically, a user can allocate additional processors from the pool of processors from which his initial processor is allocated. Because no user owns a particular processor, there is less need for (and no provision for) preemption and migration, assuming the pool of processors is not exhausted. However, this processor allocation is feasible because the Cambridge user display is coupled to a processor by a byte stream providing conventional terminal service. In contrast, with high-performance bitmap displays, each workstation processor is closely bound to a particular display, for instance, when the frame buffer is part of its addressable memory. With this architecture, personal claim on a display implies claim on the attached processor. Personal claim on a processor that is currently in use for another user's programs requires that these programs are either destroyed or migrated to another processor.

Several problem-specific examples of preemptable remote execution also exist. Limited facilities for remote "down-loading", of ARPANET IMPs, for example, have been available for some time. Early experiments with preemptable execution facilities included the "Creeper" and "relocatable McRoss" programs at BBN and the "Worms" programs at Xerox [17]. Finally, a number of application-specific distributed systems have been built, which support preemptable remote executives (see, for example, [1, 9]). However, in each case, the facilities provided were suitable for only a very narrow range of applications (often only one). In contrast, the facilities presented here are invisible to applications and, therefore, available to all applications.

Our work is complementary to the extant literature on task decomposition, host selection, and load balancing issues (see, for example [6, 7, 8, 12]).

6. Concluding Remarks

We have described the design and performance of the remote execution facilities in the V distributed system. From our experience, transparent preemptable remote program execution is feasible to provide, allowing idle workstations to be used as a "pool of processors" without interfering with personal use by their owners. The facility has reduced our need for dedicated "computation engines" and higher-performance workstations, resulting in considerable economic and administrative savings. However, it does not preclude having such machines for floating point-intensive programs and other specialized computations. Such a dedicated server machine would simply appear as an additional free processor and would not, in general, require the migration mechanism.

The presentation has focused on three major issues, namely: network-transparent execution environment, minimizing interference due to migration and avoiding residual host dependencies. We review below the treatment of each of these issues.

The provision of a network-transparent execution environment was facilitated by several aspects of the V kernel. First, the network-transparency of the V IPC primitives and the use of global naming, both at the process identifier and symbolic name level, provide a means of communication between a program and the operating system (as well as other programs) that is network-transparent. In fact, even the V debugger can debug local and remote programs with no change using the conventional V IPC primitives for interaction with the process being debugged.

Second, a process is encapsulated in an address space so that it is restricted to only using the IPC primitives to access outside of its address space. For example, a process cannot directly examine kernel data structures but must send a message to the kernel to query, for example, its processor utilization, current-time and so on. This prevents uncontrolled sharing of memory, either between applications or between an application and the operating system. In contrast, a so-called *open system* [16] cannot prevent applications from circumventing proper interfaces, thereby rendering remote execution impossible, not to mention migration.

Finally, this encapsulation also provides protection between programs executing on behalf of different (remote) users on the same workstation, as well as protection of the operating system from remotely executed programs. In fact, the V kernel provides most of the facilities of a multi-user operating system kernel. Consequently, remotely executed programs can be prevented from accessing, for example, a local file system, from interfering with other programs and from crashing the local operating system. Without these protection facilities, each idle workstation could only serve one user at a time and might have to reboot after each such use to ensure it is starting in a clean state. In particular, a user returning to his workstation would be safest rebooting if the machine had been used by remote programs.

Thus, we conclude that many aspects of multi-user timesharing system design should be applied to the design of a workstation operating system if it is to support multi-user resource sharing of the nature provided by the V remote execution facility.

Interference with the execution of the migrating program and the system as a whole is minimized by our use of a technique we call *pre-copying*. With pre-copying, program execution and operations on this program by other processes are suspended only for the last portion of the copying of the program state, rather than for the entire copy time. In particular, critical system servers, such as file servers, are not subjected to inordinate delays when communicating with a migrating program.

The suspension of an operation depends on the reliable communication facilities provided by the kernel. Any packet arriving for the migrating program can be dropped with assurance that either the sender of the packet is storing a copy of the data and is prepared to retransmit or else receipt of the packet is not critical to the execution of the migrating program. Effectively, reliable communication provides time-limited storage of packets, thereby relaxing the real-time constraints of communication.

To avoid residual dependencies arising from migration, we espouse the principle that one should, to the degree possible, place the state of a program's execution environment either in its address space or in global servers. That way, the state is either migrated to the new host as part of copying the address space or else the state does not need to move. For example, name bindings in V are stored in a cache in the program's address space as well in global servers. Similarly, files are typically accessed from network file servers. The exceptions to this principle in V are state information stored in the kernel server and the program manager. These exceptions seem unavoidable in our design, but are handled easily by the migration mechanism.

Violating this principle in V does not prevent a program from migrating but may lead to continued dependence on the previous host after the program has migrated. While by convention we avoid such problems, there is currently no mechanism for detecting or handling these dependencies. Although this deficiency has not been a problem in practice, it is a possible area for future work.

There are several other issues that we have yet to address in our work. One issue is failure recovery. If the system provided full recovery from workstation crashes, a process could be migrated by simply destroying it. The recovery mechanism would presumably recreate it on another workstation, thereby effectively migrating the process. However, it appears to be quite expensive to provide application-independent checkpointing and restart facilities. Even application-specific checkpointing can introduce significant overhead if provided only for migration. Consequently, we are handling migration and recovery as separate facilities, although the two facilities might use some common operations in their implementations.

Second, we have not used the preemption facility to balance the load across multiple workstations. At the current level of workstation utilization and use of remote execution, load balancing has not been a problem. However, increasing use of *distributed execution*, in which one program executes subprograms in parallel on multiple host, may provide motivation to address this issue.

Finally, this remote execution facility is currently only available within a workstation cluster connected by one (logical) local network, as is the V-system in general. Efforts are currently under way to provide a version of the system that *can* run in an internet environment. We expect the internet version to present new issues of scale, protection, reliability and performance.

In conclusion, we view preemptable remote execution as an important facility for workstation-based distributed systems. With the increasing prevalence of powerful personal workstations in the computing environments of many organizations, idle workstations will continue to represent a large source of computing cycles. With system facilities such as we have described, a user has access to computational power far in excess of that provided by his personal workstation. And with this step, traditional timesharing systems lose much of the strength of one of their claimed advantages over workstation-based distributed systems, namely, resource sharing at the processor and memory level.

Acknowledgements

Michael Stumm implemented the decentralized scheduler and exercised many aspects of the remote execution early in its availability. Other members of the Stanford Distributed Systems Group have also contributed to the ideas and their implementation presented in this paper. Comments of the referees inspired a significant revision and improvement to the paper.

This research was supported by the Defense Advanced Research Projects Agency under contracts MDA903-80-C-0102 and N00039-83-K-0431.

References

1. J.-M. Ayache, J.-P. Courtiat, and M. Diaz. "REBUS, a fault-tolerate distributed system for industrial real-time control." *IEEE Transactions on Computers C-31*, 7 (July 1982), 637-647.

2. E.J. Berglund, K.P. Brooks, D.R. Cheriton, D.R. Kaelbling, K.A. Lantz, T.P. Mann, R.J. Nagler, W.I. Nowicki, M. M. Theimer, and W. Zwaenepoel. *V-System Reference Manual.* Distributed Systems Group, Department of Computer Science, Stanford University, 1983.

3. D.A. Butterfield and G.J. Popek. Network tasking in the Locus distributed UNIX system. Proc. Summer USENIX Conference, USENIX, June, 1984, pp. 62-71.

4. D.R. Cheriton. "The V Kernel: A software base for distributed systems." *IEEE Software 1*, 2 (April 1984), 19-42.

5. D.R. Cheriton and W. Zwaenepoel. "Distributed process groups in the V kernel." *ACM Transactions on Computer Systems 3*, 2 (May 1985), 77-107. Presented at the SIGCOMM '84 Symposium on Communications Architectures and Protocols, ACM, June 1984.

6. T.C.K. Chou and J.A. Abraham. "Load balancing in distributed systems." *IEEE Transactions on Software Engineering SE-8*, 4 (July 1982), 401-412.

7. D.H. Craft. Resource management in a decentralized system. Proc. 9th Symposium on Operating Systems Principles, ACM, October, 1983, pp. 11-19. Published as *Operating Systems Review* 17(5).

8. E.J. Gilbert. *Algorithm partitioning tools for a high-performance multiprocessor*. Ph.D. Th., Stanford University, 1983. Technical Report STAN-CS-83-946, Department of Computer Science.

9. H.D. Kirrmann and F. Kaufmann. "Poolpo: A pool of processors for process control applications." *IEEE Transactions on Computers C-33*, 10 (October 1984), 869-878.

10. K.A. Lantz and W.I. Nowicki. "Structured graphics for distributed systems." *ACM Transactions on Graphics 3*, 1 (January 1984), 23-51.

11. E.D. Lazowska, J. Zahorjan, D.R. Cheriton, and W. Zwaenepoel. File access performance of diskless workstations. Tech. Rept. STAN-CS-84-1010, Department of Computer Science, Stanford University, June, 1984.

12. R. Marcogliese and R. Novarese. Module and data allocation methods in distributed systems. Proc. 2nd International Conference on Distributed Computing Systems, INRIA/LRI, April, 1981, pp. 50-59.

13. R.M. Needham and A.J. Herbert. *The Cambridge Distributed Computing System.* Addison-Wesley, 1982.

14. W.I. Nowicki. *Partitioning of Function in a Distributed Graphics System*. Ph.D. Th., Stanford University, 1985.

15. M.L. Powell and B.P. Miller. Process migration in DEMOS/MP. Proc. 9th Symposium on Operating Systems Principles, ACM, October, 1983, pp. 110-119. Published as *Operating Systems Review* 17(5).

16. D.D. Redell, Y.K. Dalal, T.R. Horsley, H.C. Lauer, W.C. Lynch, P.R. McJones, H.G. Murray, and S.C. Purcell. "Pilot: An operating system for a personal computer." *Comm. ACM 23*, 2 (February 1980), 81-92. Presented at the 7th Symposium on Operating Systems Principles, ACM, December 1979.

17. J.F. Shoch and J.A. Hupp. "Worms." *Comm. ACM 25*, 3 (March 1982), .

4.6 Milojičić, D.S, Zint, W., Dangel, A., and Giese, P., Task Migration on the top of the Mach Microkernel

Milojičić, D.S, Zint, W., Dangel, A., and Giese, P., "Task Migration on the top of the Mach Microkernel," *Proceedings of the 3rd USENIX Mach Symposium*, pp. 273–290, April 1993.

Task Migration on the top of the Mach Microkernel*

Dejan S. Milojičić[†] Wolfgang Zint, Andreas Dangel and Peter Giese

University of Kaiserslautern, Informatik, Geb. 36, Zim. 436
Erwin-Schrödingerstraße, 6750 Kaiserslautern, Germany
tel (49 631) 205 3293 fax (49 631) 205 3558
e-mail: [dejan,zint,dangel,giese]@informatik.uni-kl.de
Paper presented at the third USENIX Mach Symposium, Santa Fe, April 1993.

Abstract

This paper presents initial results in the design and implementation of task migration on the top of the Mach μkernel. The presented work is part of a broader project concerning research on load distribution. Our task migration is implemented in user space in order to improve portability, maintainability and flexibility. At the same time we paid attention not to sacrifice performance, transparency and functionality. Although we have implemented task migration in user space, some modifications to the kernel were necessary. We have designed on the μkernel abstraction level, unaware of issues such as files, signals or other UNIXisms which were main obstacles to simple and transparent process migration so far. We expect to benefit much more when we start dealing with load distribution decisions. Our design allows us to make scheduling decisions entirely based on Mach virtual memory, interprocess communication and processor load.

1 Introduction

The field of Load Distribution (LD), encompassing load balancing, load sharing and related mechanisms, such as Task Migration (TM) and remote execution, has been explored for quite some time [Bokh79, Eage86, Milo91]. Many practical and theoretical results have evolved, but none has achieved wide acceptance. There have been many obstacles, such as a relatively old operating system design, artificial support for distributed computing by extending stand-alone operating systems in a network, the lack of distributed applications, dependence on the particular modified version of the underlying operating system, limited functionality and transparency etc. We believe that contemporary μkernels provide a convenient base for LD and particularly for TM. In a μkernel, basic abstractions are supported within the kernel, while other functionality is provided within user space. This allows access to data and functionality formerly hidden inside the kernel, and thereby improves opportunities for a user level TM implementation. This and other characteristics, such as network transparency, modern virtual memory design and modularity, that modern μkernels possess (particularly Mach [Blac92]), are a promising base for yet another effort in the field of TM. Therefore, we have set as our goal to demonstrate that the Mach μkernel is a suitable environment for the implementation of TM. Our implementation aims at a transparent and portable migration in user space (possibly with minor modifications to the kernel), without paying significant penalties in performance and functionality.

The paper describes two user-space TM servers, necessary modifications to the kernel and a preliminary performance evaluation. Besides serving as a base for future work, the goal of this phase was

*Research is supported by DAAD, University of Kaiserslautern (Germany) and Institute "Mihajlo Pupin", Belgrade (Yugoslavia).

†Currently on a leave from Institute "Mihajlo Pupin", Belgrade, (Yugoslavia).

to provide insight into the complexities of TM implementation, level of transparency, functionality and performance penalties paid for the user level implementation.

The rest of the paper is organized in the following manner. In this section we present an overview of the field and the relevant Mach characteristics. The design of our task migration scheme is presented in section 2. The implementation is described in section 3 by surveying the modifications applied to Mach, and two versions of the TM server. In section 4, we present some preliminary performance measurements. Related research is described in section 5. Our conclusions and future research are presented in section 6.

1.1 Overview of the Field

The field of TM has often been surveyed and described, e.g. in [Gosc91, Arts89]. Here, we briefly mention the motivations for TM, its issues and important implementations.

Computing history has shown that technology continuously provides more resources, but also increases demand for these resources. No matter how much is provided, there will always be an application that needs more, be it computing power, primary memory, huge paging space or specialized hardware. One of the convenient ways to satisfy this need is TM. Other applications of TM include the areas of fault-tolerance, real-time, system administration etc. Mobile computers may become yet another important area to apply TM [Doug92].

One of the intriguing issues of TM is **transparency**. It implies that a task can transparently execute after its migration, without recognizing that it has been migrated, except possibly for the performance difference. **User v. kernel space implementation** concerns the tradeoff between efficiency and simplicity. A user space implementation, while sacrificing speed to some extent, can provide for a simple, robust solution [Arts89]. A kernel implementation may be necessary in the case of real-time requirements. **Residual dependency** is related to the amount of the task state that may be left at the originating or some other node. It is directly related to the reliability of task migration. Task migration should be recoverable, i.e., it should be possible to recreate a task in the case that its migration has failed. **Autonomy and security** are important with respect to the environment as well as for the migrated task itself. A task should in no way influence the integrity of the environment that it is migrated to. Reclaiming of resources, if requested, should happen in a limited time. Finally, **performance** consists of immediate transfer and freeze time (the time during task is suspended), as well as the transfer time of lazy migrated state, such as applied for "Copy-On-Reference" (COR).

The history of TM is relatively old. It runs from early multiprocessors and distributed systems [Bokh79, Mill81], to the modern distributed environments and massively-parallel-processors [Zajc93]. It is an area where implementation dominates. Many prototypes have been running, but only a few have emerged as working products, e.g. Locus [Walk89]. We have chosen to mention the following work. **MOS(IX)** [Bara85] was one of the first to demonstrate μkernel appropriateness for LD, splitting the operating system into two layers. **Accent** [Zaya87], the predecessor of Mach, was the first to introduce the COR technique. The COR technique allows for faster address space transfer and achieves performance improvement by lazy copy of pages, thereby avoiding unnecessary page transfers. The **V kernel** introduced a technique known as "precopying" of the task address space [Thei85]. Precopying significantly improves freeze time, the period while the task is suspended during migration. **Charlotte's** TM scheme [Arts89] extensively relies on the underlying operating system, its communication mechanisms and language. Interprocess communication has been modified in order to support transparent network communication. **Locus** [Walk89] is an exception to most TM implementations because it was the only one to achieve a product stage. It has been ported to the AIX operating system on the IBM 370 and PS/2 computers under the name of Transparent Computing Facility. Locus has achieved a high level of functionality and transparency, paying the cost in significant kernel modifications. It has been recently ported to OSF/1 operating system, running on top of Mach [Zajc93]. **Sprite** [Doug91], is one of the recent and successful process migration implementations. It will be covered in more detail in section 5.

1.2 Relevant Mach Characteristics

The purpose of this subsection is not to give a Mach description, but rather to mention particular Mach features relevant to our TM scheme. More details about Mach can be obtained in [Blac92, Golu90]. The following issues are relevant to our scheme:

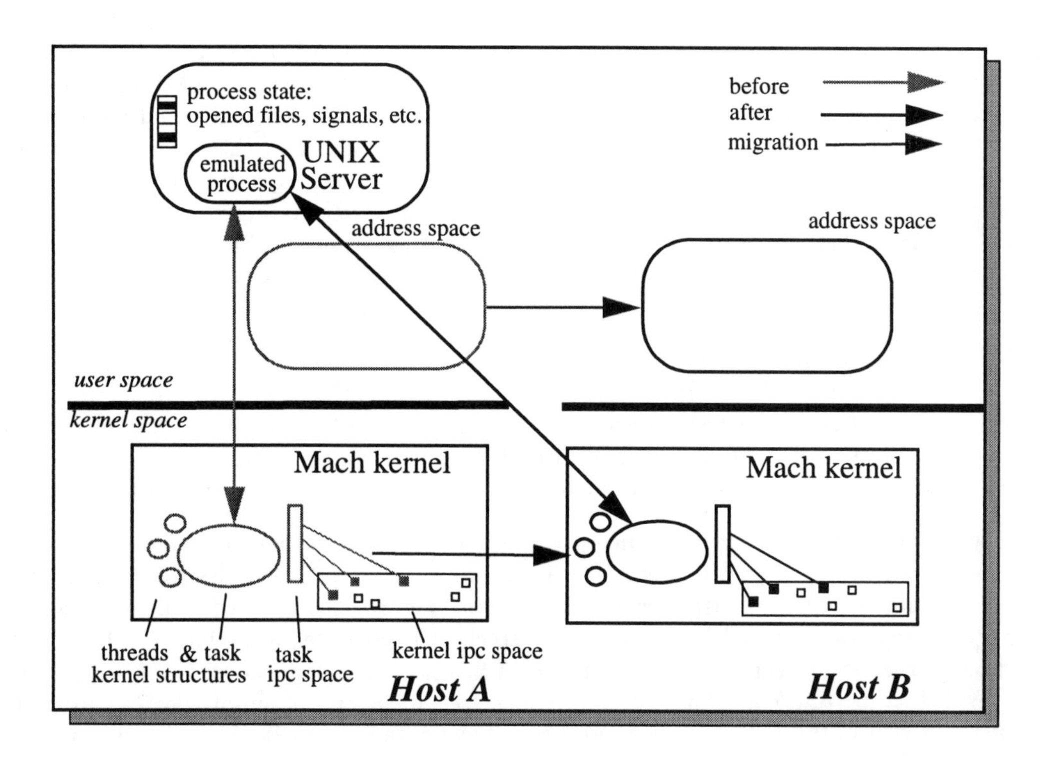

Figure 1: *The transferred task state consists of the address space, IPC space, thread states, and other state, such as task and thread kernel ports, exception ports, task bootstrap port etc. The UNIX server state, contained within the task, is transferred implicitly, as a part of the address space (shared pages) and IPC space (rights through which the emulator communicates with the UNIX server).*

- **Multicomputer support,** as provided by the Mach NORMA version, made our task much easier. The support for network Interprocess Communication (IPC) [Barr91], and Distributed Shared Memory (DSM) [Barr93] is of particular significance. Network IPC has been implemented within the kernel in order to improve performance. It transparently extends local IPC over the network. DSM overcomes current pagers' limitations to support only a single kernel. DSM code is transparently inserted between kernel and pager code, allowing existing pagers to support multiple kernels.
- **UNIX emulation** [Golu90] relieved us from handling UNIX abstractions in early design phases. However, this does not preclude us from considering modifications later, in the case of low performance. We have particularly observed the unacceptable performance of the emulator and the current implementation of mapped files, which have been optimized for parallel, but not for distributed, systems.
- **Device access,** based on IPC, provided a further level of transparency [Fori91]. Formerly, it used to be very hard, if not impossible, to extract the kernel state that remained in device drivers [Free91].

The Mach μkernel has the potential to become a standard basis for building operating systems. There are a lot of projects that have Mach as an underlying environment. Finally, its predecessor Accent was the base for TM experiments [Zaya87]. Mach inherited its design from Accent, particularly in some aspects of memory management. Therefore, Mach is a promising environment for TM experiments.

2 Task Migration Design

TM is implemented on top of Mach, but outside of any operating system emulation server. This implies that we migrate Mach tasks, while the UNIX process remains on the source machine, as presented in Figure 1. The obvious disadvantage is home dependency, a residual dependency with regard to the originating machine [Doug91]. The advantage is that the same TM scheme may be used for the applications running on any operating system emulated on the top of Mach (e.g. BSD UNIX [Golu90], VMS [Wiec92] etc.), as well as for the applications running on the bare Mach. As in most TM implementations, our scheme consists of extracting the task state, transferring it across the network and reestablishing it at a new site. The world notices no difference while communicating with the migrated task, except for performance. The Mach task state consists of the address space, the capability space, the thread states and some other state. We are taking advantage of the Mach NORMA version which supports in-kernel network IPC and DSM. We do not implement capability migration and message forwarding; NORMA code takes care of that, as well as of DSM. However, not all of the task state can be transferred from within user space. We have faced the following problems in the design of TM.

- Mach objects are represented by kernel ports. Actions on behalf of the objects are performed by sending messages to their kernel ports. Therefore, it is necessary to preserve the task and thread kernel ports after migration. However, the kernel ports are not handled the same way as normal ports whose receive capabilities reside within the task IPC space. There are no receive capabilities for the kernel ports, and therefore there is no way to extract them using the existing Mach interface.

- Any shared or externally mapped memory causes a consistency problem for TM. Common examples are memory that is shared through inheritance and memory mapped files. Such memory areas are non-eligible for user space transfer with the current pagers, such as the default pager and the inode pager, which are designed to see only one kernel. The consistency problem could be overcome if Mach pagers would support multiple kernels. Basically, the same consistency support should be provided as in NORMA DSM. Internal, non-shared memory can simply be read from the source side and written into the task instance on the destination side. We call such areas "eligible" for user space transfer.

These were the only two cases where we did not have kernel support. As a solution to the first problem we provide two system calls to interpose the task and thread kernel ports. Both calls return the kernel port and leave the supplied port as a new kernel port. After that, we are the only ones who have the capability for the task kernel port and, therefore, only we can control it. All messages sent to a task end up at the extracted kernel port, whose receive capability is now in the TM server's IPC space. In the meantime, we perform actions on behalf of the task and transfer its state to the destination side. Once the state is transferred, we make the interpose call on the destination side, inserting the original task kernel port into the new task instance.

Shared memory and memory mapped files are a complex problem. However, this is not inherent in TM. Shared memory is not supported by most TM implementations, and files are handled by a distributed file system, e.g., in Sprite [Doug91], MOS(IX) [Bara85] and Locus [Walk89]. We adopt a similar approach. Shared memory and mapped files are supported by NORMA distributed shared memory. We currently handle it by exporting the pager port that represents a region of memory and then remapping it on the remote side. As soon as Mach exports the needed functionality, we may switch to it.

Except for the aforementioned two cases, everything else is implemented in user space. Our TM algorithm can be described with the following steps:

- suspend the task and abort the threads in order to clear the kernel state,

- interpose the task/thread kernel ports,

- transfer the address space, using various user and NORMA memory transfer strategies,

- transfer the threads by getting the state and setting it on the remote side,

- transfer the capabilities (we extract, transport and insert capabilities; NORMA does the actual port transfer),

- transfer the other task/thread state,

- interpose back the task/thread kernel ports at the destination site and

- resume the task.

Before migration, it is necessary to clear the thread kernel state, since it is almost impossible to transfer it. Typically, this case arises when threads are waiting within the kernel, e.g. for a message to be sent or received or a faulted page to be paged-in. Mach provides a particular call *thread_abort*, which aborts thread execution within the kernel and leaves the thread at a clear point, containing no kernel state inconvenient for migration. The execution is restarted after migration, unless instructed otherwise. Lux et al. describe how kernel state complicates migration in BirliX [Lux92].

The adopted design relieves us from issues, such as signals and distributed file systems, which have traditionally been a nightmare in TM and remote execution implementations. We expect major benefits later, when we are to make LD decisions. We shall make decisions only in terms of Virtual Memory (VM and XMM), IPC and processor load, while previous systems separately dealt with files, network traffic, local and remote IPC. Of particular interest for us are XMM and network IPC which, as we presume, dominate the costs.

3 Implementation

The underlying environment in our TM research consists of 3 PCs, interconnected via Ethernet. The PCs are based on a 33 MHz i80486 processor with 8 MB RAM and 400MB SCSI disk. We have been using the Mach NORMA version (MK7, MK12, MK13) and the UNIX server UX28. TM has been accomplished by designing and implementing two versions of migration server, a simple and an optimized one. They have been implemented in a relatively short period. We installed our first PC at the end of November 1991, and the current environment at the end of March 1992. At the end of May we migrated a task for the first time. Polishing it up took more time, which is partially due to the parallel NORMA development. The task migration is currently stable for test applications.

The routines concerning the kernel modifications fit in a file of 300 lines of C code, most of which are comments, debugging code and assertions. The simple migration server has about 600 lines of code, and is actually a library routine. It could be used either by linking it to the task that is going to initiate migration, or by providing a server interface, in which case the migration code is linked to the server. In the former case it is not possible for a task to migrate itself, since one of the first actions on behalf of the migrating task is to suspend it, causing a deadlock. In such a case, it is necessary to provide a server which will indirectly start the migration. The optimized migration server has about 13000 lines. It is based on cooperating servers running on all nodes and has a few threads of control.

In the following subsections we describe the kernel modifications and the simple and optimized TM server.

3.1 Necessary Modifications to the Mach Microkernel

As mentioned in the previous section, the kernel modifications consist of the kernel port interpose, which is a necessary, permanent modification, and a temporary extension to export the memory object port until the NORMA code exports the needed functionality.

The kernel port interpose modification introduced two new system calls, one for the task and one for the thread kernel port interposition. Like most other calls, these are actually messages, sent to the host on which the task executes. Unfortunately, it can not be sent to the task or thread. The interpose routine takes care that all messages accumulated in the original kernel port (while being interposed) are handled properly and the threads blocked in the full port message queue are woken up and their requests handled. In order to avoid unnecessary blocking, the queue limit for the interposed kernel ports is increased to its maximum. Once the ports are interposed back on the destination side, the default queue limit is restored. Interposing, as presented in Figure 2, consists of exchanging the pointer to the IPC space the port belongs to, as well as modifying some other internal task state. On the return from the call, the receive capability for the task kernel port resides within the server IPC space and the receive capability for the interposing port is extracted from the server IPC space into the kernel IPC space.

Exporting the pager port is a temporary modification. The existing NORMA version does not support distributed memory shared through inheritance, although DSM functionality exists. Therefore, we adopted a temporary solution. We export the memory object for the particular region. For security reasons, the Mach interface does not export the memory object port capability, but only the related

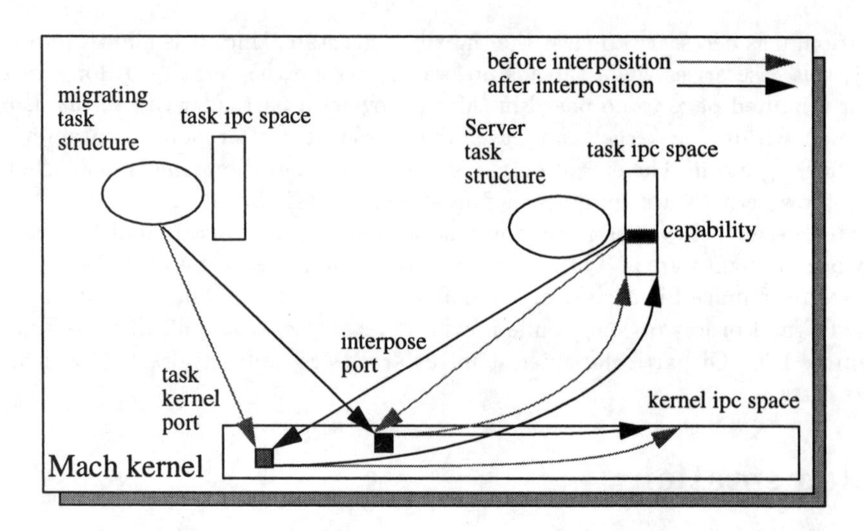

Figure 2: *Interposing the task kernel port - internal perspective: the task_interpose routine exchanges the port that represents the task with an interposing port; it prerequisites exchanging pointers to ports and the IPC spaces the ports belong to, namely kernel IPC space with the IPC space of the task that initiates migration.*

name capability [Youn89]. We extract the capability for the memory object port and map it onto the remote side. This provided the necessary functionality in the early design phases. The added system call looks up the memory object port which represents the memory region and exports a send right out of the kernel.

3.2 Simple Migration Server

At the very beginning the Simple Migration Server (SMS) was implemented in order to verify design feasibility and to get the initial insight into performance. Later on, we completely switched to it. It relies entirely on the NORMA memory transfer and is deliberately unoptimized. From the memory transfer point of view, the SMS has better performance than the Optimized Migration Server (OMS), which relies on user level address space transfer and is more flexible regarding the choice of the memory transfer strategy. Regarding capabilities, threads and other state transfer, the SMS has worse performance than the OMS, since each capability, thread state, etc., is transferred separately, resulting in a new message sent across the network. The SMS consists of a few parts that migrate capabilities, memory, thread state and the other task state. Before migration, the task is suspended and its kernel port is interposed. From now on, we are sure that no unprivileged user has access to the task. A privileged user can always obtain a send capability for the current task kernel port, regardless of the fact that we have exchanged it.

The **thread migration** consists of transferring the thread state. There are a few different flavors of the thread state which should be transferred, such as the contents of CPU and FPU registers etc. The thread states are extracted on the source side and inserted into the newly created threads on the destination side. We have never had an interest in any particular state *per se*, we just copied all of them. This improves the portability of our TM scheme across various processor architectures.

The **memory migration** is based on the default NORMA COR strategy, as opposed to the optimized server where different strategies are implemented in user space. We wanted to explore different approaches in order to get more insight into performance and functionality tradeoffs. In both approaches, the task address space is analyzed, and areas not eligible for user-space migration are transferred using NORMA, as discussed in section 2. These areas are mapped in the destination task, using NORMA DSM. The rest of the memory is either copied in user space (OMS) or using NORMA support (SMS). In either case, non-initialized areas are simply re-allocated.

The **capability migration** is performed by extracting the capabilities on the source node and inserting them on the destination node. We migrate send, send-once and receive capabilities. Port sets are migrated by extracting all receive capabilities from the port set, migrating them and reconstructing the port set on the remote side. Capabilities are migrated one by one in SMS. This has detrimental effects on performance, due to the high costs involved in capability migration. Therefore, we may switch to the same method as in the OMS: extract all capabilities, transfer them in a message to the destination node, and insert them. This requires servers on both sides.

The **other state** consists of suspend counts for task and threads, bootstrap ports, exception ports etc. Each of these states is appropriately extracted and later inserted into the migrated instance of the task. Some of the state is transferred as part of creating the remote instance of the migrated task.

3.3 Optimized Migration Server

The Optimized Migration Server (OMS) has functionality similar to SMS, except that it supports user-space memory transfer and various optimizations. Only memory eligible for user-space migration is transferred across the network by the OMS, while shared and externally mapped memory is still transferred by NORMA. The OMS treats error recovery in the case of an erroneous migration, and potentially retries migration on the same or a different destination node. The following data transfer strategies are supported:

- **Flushing residual dependency** causes the transfer of all memory that remained either on the source or on other nodes due to the previous migrations.
- **Precopy** is implemented similarly to the V kernel; however, its performance rendered it useless: it was too slow to switch from the default pager to a new pager. The basic idea is to remap the address space to the new pager and then to extract the information about modified pages. We are investigating improvements.
- **Copy on reference** is supported in the classical way, similar to Accent or NORMA, but in user space. There is also an extended version that transfers a few pages surrounding the faulting page.
- **Read ahead** is meant to transfer the whole task address space within a given time. It periodically transfers some of the remaining pages to the destination node, until all pages are transferred.
- **Direct copy** is a simple transfer of all memory from the source node to the destination node.

The applied optimizations consist of **packing all capabilities and other relevant information in messages**, requiring servers on both machines; **overlapping of various state transfer**, e.g. memory with capability and thread transfer; **truncating** zero ended memory areas and **substitution of emulator and emulation vectors** by local instances. OMS uses an External Memory Manager (EMM) to provide the functionality for user-level memory transfer strategies. The address space of the migrating task is reconstructed on the destination side by remapping memory objects. After that, the EMM receives paging requests, and either transports referenced pages over the network or retrieves them from the local paging-file. The basic COR-mechanism is modified for COR-extend to fetch some remote pages surrounding the faulting page and transport the requested pages in parallel. Extra pages are saved locally to resolve succeeding page faults. Flushing the entire address space to backing store, as done in Sprite, doesn't make much sense in our environment with local paging files. Hence OMS uses a variant of the strategy. It transfers all eligible pages in parallel with the thread state and capability migration. Network and paging file access are serviced by separate threads to increase performance. The paging file is accessed by a stand-alone file system (the same as the default memory manager uses) to avoid deadlock with the BSD server.

OMS has been useful during development, however, despite many opportunities, currently we have abandoned the use of OMS due to the following reasons:

- Task memory regions not eligible for user-space transfer, such as shared memory or memory backed up by specific pagers, such as the inode pager, still need support of XMM code or the support of new distributed pagers. We have considered the latter case, but we found that it has not been preferred by other main Mach developers. Choosing this way would mean a departure from the mainstream development. In the former case, we duplicate the functionality.

- We have found SMS more robust for further research on LD. Its simplicity, despite somewhat slower performance, e.g. for port transfer, allows us to trace the problems more easily. OMS requires few threads of control, a separate pager, etc., making debugging harder.
- Our main goal was not TM itself, but rather LD, where TM is one of its important mechanisms. Migration candidates are usually long lived tasks (range of tenths of seconds); therefore the performance difference of a few hundred milliseconds between SMS and OMS is of secondary importance.

If we later, for any reason, find SMS inappropriate, we can always re-investigate the choice of migration techniques once again. One of the most likely reasons for reviving the OMS would be the comparison of various migration strategies. This would become an issue once we have a working LD scheme and other Mach applications, such as a distributed file system. As a short term improvement, we may merge some of the good characteristics of OMS, such as packing rights and other state, in a few messages, avoiding many short messages over the network.

4 Performance Measurements

Performance measurements are an important part of any TM implementation. Unfortunately it is hard to find adequate applications for measurements. There are a few true distributed applications, particularly for the Mach μkernel. Most researchers in the field of LD have made artificial loads or used some benchmarks; they rarely used real applications. We used the following: Artificial Load Task (ALT), WPI benchmark suite, parallel make support; matrix multiplication, simulation programs, and some other applications.

All results presented should be accepted with caution. There are many influences on its accuracy, since TM and the Mach NORMA version are continuously being changed. We present performance measurements for the sake of completeness and to give more insight into the order of magnitude. All results presented in Figures 3 to 10 are obtained as a mean of the five consecutive measurements. Only the results for Figures 11 and 12 are the mean of two out of the five measurements, since the experiment failed for some input values due to a known bug. If not otherwise indicated, SMS has been used. All measurements have been performed on normal3[1] and [UX,emulator]28.

4.1 Migration Server Measurements

In this subsection we present some low-level measurement results for both migration servers. In order to gain insight into the influence of particular migration parameters on the overall performance, we designed an Artificial Load Task (ALT). We can tune the following ALT parameters: ratios between computation, IPC activity and memory access, amount of memory (internally mapped by the default memory manager, locally and remotely shared memory), the number of threads, and the number of capabilities (receive, send for local and remote ports). ALT loops for a given number of iterations and in each loop it performs a number of remote procedure calls to the local and remote server, accesses locally and remotely mapped shared memory, followed by a computation part, currently represented by the Linpack benchmark. The main idea behind ALT is to represent the task VM, IPC and processor load behavior. We plan to experiment with various applications and thereby obtain realistic values for ALT parameters and use it as an artificial load for distributed scheduling experiments.

TM consists of three phases which transfer virtual memory, threads and capabilities. Each of these phases was measured using ALT. In Figure 3, we present measurements of transfer time v. memory size, performed using SMS, i.e. using NORMA default COR strategy. Transfer time as a function of the thread number is presented in Figure 4. Figure 5 shows transfer time as a function of the number

[1] In the meanwhile we have also ported SMS to normal4, however, we shall still present results only for normal3 version for a few reasons. First, although normal4 has been more stable for many previous problems, we did observe some new ones, preventing stable measurements. Second, we have switched within SMS to a new kernel supported address space transfer. This could favor SMS over OMS, which hasn't been ported to normal4. Finally, except for XMM, we have not observed any significant difference in relative performance behavior between the two versions. Despite some changes, the curves presented in Appendix A remained similar. We observed faster address space transfer (above 30%), but slower lazy copying.

of receive capabilities. Comparison of receive and send capabilities transfer is given in Figure 6. In Figure 7, we present an interesting side-effect of extending a VM layer to a distributed environment. While experimenting with memory transfers, we noticed unacceptable values for a particular test task. The test task address space regions are created by its parent using a *vm_write* system call on a page-by-page basis. As we closely inspected the address space, we saw that it consisted of many entries, which could be, but have not been, compacted into one. In local cases this does not represent a bottleneck, however, once that task address space is being transferred across the network, entry transfer cost is significantly increased. As we modified the program slightly to write all the memory in a big chunk instead of page-by-page, thereby reducing the number of entries, we significantly improved transfer time. This anomaly could be easily overcome by using the existing kernel function *vm_map_simplify*. From aforementioned measurements we can conclude that migration transfer time is independent of task address space size. It is a linear function of the number of internal memory regions, but they are usually limited in number, except for the unoptimized cases, such as the one presented in Figure 7. The number of threads and rights are usually small for tasks representing UNIX process, except for the servers, which are not good candidates for migrations, anyway. Except for address space transfer which is characterized by fluctuations, all curves are linear. Address space transfer has been subject to change with various norma versions, therefore we didn't further inspect the reasons for its fluctuation.

The presented measurements characterize the transfer phase. Equally important is performance during the task execution after migration, which depends on the transfer of residual state. Figure 8 presents an interesting and unexpected behavior of COR for OMS and SMS. While we expected better capability transfer performance for OMS, we did not expect better address space transfer. According to the measurements, it seems like OMS also has better address space transfer. The measurements are performed with an ALT version in which only the first integer of the page is written. Since OMS introduces an optimization of transferring the data up to the last non-null character in an area, just one integer is transferred instead of the page. As soon as the last integer within the page is written, performance of OMS decreased below the SMS performance. Performance of SMS is the same in both cases, since it has page granularity. This simple experiment demonstrates the benefits of implementing user-space TM, since it is rather easy to substitute various strategies and parameters. In Figure 9, we demonstrate the expected difference in transferring receive capabilities for SMS and OMS. The performance for send capabilities is similar and is not presented. Finally, the obvious benefit of migration towards a server with which the task communicates is demonstrated in Figure 10, where we present the task execution time as a function of the number of remote procedure calls with a server to/from which we migrate the task.

Based on above results we can conclude that the performance of our TM implementation is comparable to other kernel supported implementations, while it significantly outperforms user space implementations. For example, reported process migration in Sprite [Doug91] is few hundred milliseconds for a standard process. We have measured similar performance for our TM implementation[2]. In Condor [Litz92], it takes two minutes to migrate a 6MB process. In our servers it wouldn't cost much more to migrate any other average size task (assuming that memory regions are not chopped in many entries, which is not likely to be). Of course, the actual transfer is done lazy, preventing direct comparison. Similar results of user level implementations have been reported by other implementors, e.g. [Alon88]. It should be noted, though, that we have not aimed at optimizing performance in this phase of the project. For completeness, we shall compare our implementation with the performance of other systems. We derived formulas similar to those that the implementors of Charlotte have presented[3].

Accent time (ms) $= 1180 + 115*s$
V time (ms) $= 80 + 6*s$
Charlotte time (ms) $= 45 + 78*p + 12.2*s + 9.9*r + 1.7*q$
Sprite time (ms) $= 76 + 9.4*f + 0.48*fs + 0.66*s$
SMS time (ms) $\approx 150 + 48*n + 22.8*r + 5.5*s + 5.5*so + 58*t$
OMS time (ms) $\approx 50 + 2.4*n + 7.9*r + 1.9*s + 1.1*so + 5.4*t$

In the above formulas the following parameters have been used. In Charlotte: p is process migration management parameter with values [0,1,0.2], s is the virtual memory size in 2kB blocks, r is 0 if links are

[2] Our measurements are related to task, not process, migration, therefore, additional time should be accounted for. The same is true for most other examples.

[3] Results on Accent, V and Charlotte origin from [Arts89] and for Sprite from [Doug91], since they are more recent.

local, 1 otherwise, and q is the number of nonlocal links. In V, Sprite and Accent, s is VM size in kB, f is the number of open files and fs is file size in kB. In SMS and OMS, r is the number of receive rights, s is the number of send rights, so is the number of send-once rights, and n is the number of regions. According to these formulas, a typical UNIX process (or the corresponding task in relevant cases) would migrate in 330ms in Sprite (exec-time process), 680ms in V, 13 sec in Accent, 750ms in Charlotte, 500ms in SMS and 250ms in OMS. It should be noted that the presented performance measurements have been performed on different computers, therefore direct comparison is inappropriate.

4.2 WPI Benchmarks

We obtained interesting results while running the Jigsaw benchmark [Fink90]. The Jigsaw benchmark stresses the memory management activities of the operating system. It consists of making, scrambling, and solving a puzzle. During each phase, there is an allocation of huge amounts of memory. We repeat similar experiments enhanced by TM in four cases. The first case consists of locally executed Jigsaw. In the second case, we migrate the task after some amount of memory allocation, and thereby distribute memory on both machines. In the third case, the task is migrated after all memory is allocated on the source node. Finally, in the fourth case, the task is executed remotely, resulting in memory allocation mostly on the destination node. These four cases are presented in Figures 11 and 12. The worst performance is achieved in the third case where the task has to access all of its memory remotely. The first and the fourth case have similar performance, since all of the task's memory is on the node where the task executes. The second case exhibits the best performance for the tile sizes which cause excessive paging. In this case, the task has memory on both machines, so it starts thrashing later and achieves better performance. From this example we do not benefit too much, unless we apply similar reasoning as in memory overloaded systems. Distributing the memory across the network could improve performance in much the same way as distributing the load in an overloaded machine.

Finally, in Figure 13 we present the difference between the execution of a real make program and the WPI synthetic gcc compilation, *scomp*. For the real make, there is a performance penalty of more than 500% for the remote execution, while the *scomp* exhibits about a 50% degradation. This simple example demonstrates the inappropriateness of the existing emulator and the current implementation of a mapped file system for TM. The emulator has been designed to support UNIX emulation in a non-distributed environment. The task communicates through the emulator with the server by using shared memory and IPC. In the local case using shared memory for emulator/server communication does not impose a significant overhead. However, it turns out to be inappropriate for a distributed system, due to the unnecessary transfers of a complete page across the network, even for small information exchange. Emulator issues have also been investigated in [Pati93]. Similar reasoning is valid for mapped files. It is not the technique of memory mapped files itself that is a bottleneck; rather it is the current implementation which needs to be optimized for a distributed environment. We expect significant improvements when the existing emulator is replaced and file access optimized. This is, however, related to the particular operating system personality. Therefore, we plan to upgrade to OSF/1 server with a new distributed file system that is currently being developed within the OSF cluster project [Roga92].

4.3 Parallel Make and other Applications

One of the traditional TM applications is a parallel make. We are using *pmake*, written by Adam de Boor for Sprite. We modified it by inserting our migration routines. Due to the emulator problems, we are still not using it extensively. We expect major results to evolve when we switch to a new distributed file system. With current file system support, performance is significantly degraded. We are also running other applications, such as matrix multiplication, simulations, various benchmarks, etc. Most of them have been ported straightforwardly, e.g. simulations, which have been used as common UNIX programs. For matrix multiplication and the traveling salesman problem we have put additional effort to port them to Mach instead of UNIX. These programs run smoothly for small problem sizes, however, for big problem sizes we still encounter problems, which are likely to stem from some current limitations in Norma versions. We expect to have more results on these applications once we start making LD decisions.

5 Related Research

There is a lot of related research in the area of TM and LD. We have chosen some that represents TM on μkernels, monolithic kernels and in user-space.

- **Chorus** was expanded to support TM [?], as a basis for load balancing experiments. The TM consists of elements similar to our scheme, but is applied on the process, instead of the Actor, level (Actors correspond to Mach tasks). It is more oriented towards a hypercube implementation. Some limitations arise from the fact that Chorus currently doesn't support port migration. It is too early, though, to draw deeper comparison with our research, since both projects are at the early stage of development.

- Closely related to our research are University of Utah - **"Schizo"** project [Swan93] and the **TCN** project, performed by Locus for Intel Paragon and OSF/1 AD [Zajc93]. Both projects are Mach based. "Schizo" is targeted at a distributed environment consisting of autonomous workstations. It investigates issues such as autonomy and privacy. "Schizo" inherited some aspects of the Stealth project [Krue91], namely prioritizing VM, file and CPU scheduling, extending it with prioritizing IPC, more robust failure handling, etc. We consider a cluster environment in our work, and therefore do not concern such issues. Our work on task migration has been used as one of the mechanisms for the "Schizo" project. The **TCN** project is concerned with traditional issues of monolithical systems, providing process migration and explicitly considering files, signals and sockets. Currently, TCN is only partially concerned with μkernel TM issues. In the OSF/1 environment, the Mach interface is not exposed to the user, and therefore atomicity of process migration is not affected. A complete solution would be a combination of our work and the work on TCN, i.e. a process migration that makes use of task migration.

- **Sprite** supports one of the important TM implementations [Doug91]. It is tightly coupled with the distributed file system. The address space transfer is optimized by using various caching techniques and by flushing the state onto the server. Compared to Sprite, we have acted on the lower level. We have never dealt with issues such as files or signals. In our scheme, these abstractions are transparently supported by the network IPC and DSM, which correspond to handling open I/O streams and caching in Sprite [?, Doug92]. Therefore, there is an illusion that we have achieved a simpler design by retaining the existing file system instead of introducing the distributed one. However, for performance reasons, the current implementation of signals, files etc., needs to be optimized for distributed systems. This is, however, the flavor of the operating system personality emulated on top of Mach and is related neither to Mach nor to our scheme. It will be hard to compete with Sprite in performance, since their approach allowed for various optimizations regarding caching on the server and client side. Sprite is not only a challenge to our TM, but to Mach itself. All μkernel advantages, such as portability, flexibility, maintainability etc., should be confirmed with performance comparable to monolithic kernels, such as Sprite [Welc91]. Sprite has recently been ported to Mach [Kupf93], unfortunately process migration is not supported, preventing us from interesting comparison.

- **Condor** TM [Litz92] is dedicated to long running jobs that do not need transparency and allow for limitations on system calls the migrated task may issue. Performance penalties in Condor are much higher because it is necessary to dump the core, combine it with executables and return it to the submitting machine. The Condor approach is acceptable for long-running, computation-intensive jobs; for other classes of jobs, it is too expensive. Due to the less expensive migration, our scheme pays off for short running tasks.

6 Conclusion

We have presented a TM design and implementation on top of the Mach μkernel. We explained the reasons why we believe that our choice of a state-of-the-art message passing μkernel for TM implementation is at least comparable to other platforms, while in some aspects it has advantages. We have described minor modifications we had to apply to the Mach μkernel in order to transparently migrate tasks, and two implementations of migration server. Afterwards, we presented some measurements of TM performance, as well as performance of the migrated applications. Finally, we compared our research to the related projects in this area.

As we base our research on Mach, we have tried to use existing Mach features as much as possible. Most of the issues, e.g. network IPC, DSM, and exporting the task/kernel state, have already been provided by Mach.

The work we presented in this paper is a part of wider research on load distribution. Task migration is a problem that we have to solve efficiently, but we expect to achieve major contributions later, when we start using task migration in the framework of load distribution. In the present phase of our work we can summarize the following contributions.

- We showed that it is possible to implement TM on top of the Mach μkernel without much effort, and that μkernels are a suitable level to implement TM on. We have acted in user space with minimal modifications to the kernel. The user space implementation significantly improved maintainability and extensibility.
- Our implementation has achieved high transparency without paying the price in performance comparable to other user space implementations, such as Condor. TM is completely transparent to the migrating task and to the tasks it communicates with. There is no need to link the task with any special libraries, nor are there any limitations to the calls that the task may issue. Since we do not depend on hardware, TM is also portable across the various hardware architectures. We act on the Mach level, transparently supporting various operating system emulations, e.g. UNIX or VMS, as well as applications running on bare Mach.
- We supported memory transfer strategies, such as precopy, flushing and COR in user space. Formerly, only a single strategy per TM implementation has been provided, implemented in the kernel.

OSF Grenoble Research Institute and the University of Utah are currently investigating our TM scheme as a possible technology for use in their "OSF/1 Cluster" and "Schizo" projects respectively. A collaboration with the OSF Grenoble Research Institute is underway, which should involve continuation of our work, namely load information and distributed scheduling.

Our further research is related to TM improvement, as a short-term goal, and to LD, as a long-term goal. In the first phase of TM we have provided functionality; now it is necessary to improve performance. We are going to combine the migration servers into one unique server that will merge the best characteristics of both. Then we need to modify or completely eliminate the emulator, since it has been designed for local execution and as such is not acceptable for distributed systems. It would also be necessary to redesign the mapped file system in order to improve performance. We shall continue to profile various applications and try to learn more about the task characteristics as a function of IPC, VM and processor load. In particular, we shall stress our experiments on IPC and VM dependency. Processor load behavior has been much more researched and existing results could be readily applied, while VM and IPC are less investigated. As a part of this research, we are currently implementing a load information management scheme for Mach and a user level IPC profiler.

As a long-term goal, we have started to work on distributed scheduling. We are interested in the relation of global and local scheduling as suggested by Krueger [Krue87]. Krueger shows that not every local scheduling algorithm matches the distributed one and vice versa. In traditional systems it was hard to access and modify local schedulers, while Mach may be suitable for such experiments [Blac90].

7 Availability

All mentioned programs are available upon request. They have still not been put for anonymous ftp access due to the continuous adjustments to new norma versions. It should be noted that OMS server is not supported anymore, although the last version is available. Beside TM related programs, we have available preliminary versions of load information manager (including user space IPC profiler) and simple distributed scheduler, supporting few basic strategies. Complete LD scheme with numerous strategies is expected to be finished in the second half of 1993.

8 Acknowledgements

We would like to thank: CMU and OSF people for providing us with Mach and with continuous support; Joe Barrera for important hints during development; Fred Douglis and Brent Welch for discussing Mach and Sprite peculiarities; Prof. Dušan Velašević for initial encouragement and support; Prof. Amnon Barak and Prof. Andzej Goscinski for continuous support; Prof. Nehmer for making all of this happen. Special thanks are due to Fred Douglis, Mike Kupfer, Jelena Vučetić, Nikola Šerbedžija, Sean O'Neill and Holger Assenmacher for proofreading the paper and giving many useful suggestions. Finally we would like to thank the program committee, the chairman David Black, and our liaison to program committee, Jay Lepreau, for many useful suggestions and for giving us the opportunity to present our work.

References

[Alon88] ALONSO, R. and KYRIMIS, K. (February 1988) *A Process Migration Implementation for a Unix System.* Proceedings of the USENIX Winter Conference, pages 365–372.

[Arts89] ARTSY, Y. and FINKEL, R. (September 1989) *Designing a Process Migration Facility: The Charlotte Experience.* IEEE Computer, pages 47–56.

[Bara85] BARAK, A. and SHILOH, A. (September 1985) *A Distributed Load-Balancing Policy for a Multicomputer.* Software-Practice and Experience, 15:901–913.

[Barr91] BARRERA, J. (November 1991) *A Fast Mach Network IPC Implementation.* Proceedings of the Second USENIX Mach Symposium, pages 1–12.

[Barr93] BARRERA, J. (1993) *Operating System Support for Multicomputers.* Technical Report CMU-CS-93-?, PhD Thesis in Preparation, Carnegie Mellon University.

[Blac90] BLACK, D. (October 1990) *The Mach Timing Facility: An Implementation of Accurate Low-Overhead Usage Timing.* Proceedings of the USENIX Mach Workshop, pages 53–73.

[Blac92] BLACK, D., GOLUB, D., JULIN, D., RASHID, R., DRAVES, R., DEAN, R., FORIN, A., BARRERA, J., TOKUDA, H., MALAN, G., and BOHMAN, D. (April 1992) *Microkernel Operating System Architecture and Mach.* Proceedings of the USENIX Workshop on Micro-Kernels and Other Kernel Architectures, pages 11–30.

[Bokh79] BOKHARI, S. H. (July 1979) *Dual Processor Scheduling with Dynamic Reassignment.* IEEE Transactions on Software Engineering, SE-5:326–334.

[Doug91] DOUGLIS, F. and OUSTERHOUT, J. (August 1991) *Transparent Process Migration: Design Alternatives and the Sprite Implementation.* Software-Practice and Experience, 21:757–785.

[Doug92] DOUGLIS, F. Personal communication 1992.

[Eage86] EAGER, D., LAZOWSKA, E., and ZAHORJAN, J. (May 1986) *Dynamic Load Sharing in Homogeneous Distributed Systems.* IEEE Transactions on Software Engineering, 12:662–675.

[Fink90] FINKEL, D., KINICKI, R. E., AJU, J., NICHOLS, B., and RAO, S. (October 1990) *Developing Benchmarks to Measure the Performance of the Mach Operating System.* Proceedings of the USENIX Mach Workshop, pages 93–100.

[Fori91] FORIN, A., GOLUB, D., and BERSHAD, B. (November 1991) *An I/O System for Mach 3.0.* Proceedings of the Second USENIX Mach Symposium, pages 163–176.

[Free91] FREEDMAN, D. (January, 1991) *Experience Building a Process Migration Subsystem for UNIX.* Winter USENIX Conference, pages 349–355.

[Golu90] GOLUB, D., DEAN, R., FORIN, A., and RASHID, R. (June 1990) *UNIX as an Application Program.* Proceedings of the Summer USENIX Conference, pages 87–95.

[Gosc91] GOSCINSKI, A. (1991) *Distributed Operating Systems The Logical Design.* Addison-Wesley.

[Krue87] KRUEGER, P. and LIVNY, M. (August 1987) *When is the Best Load Sharing Algorithm a Load Balancing Algorithm.* Technical Report 694, CS Department, University of Wisconsin-Madison.

[Krue91] KRUEGER, P. and CHAWLA, R. (June 1991) *The Stealth Distributed Scheduler.* Proceedings of the 11th International Conference on Distributed Computing Systems, pages 336–343.

[Kupf93] KUPFER, M. (April 1993) *Sprite on Mach.* Proceedings of the third USENIX Mach Symposium, pages 307–322.

[Litz92] LITZKOW, M. and SOLOMON, M. (January 1992) *Supporting Checkpointing and Process Migration outside the UNIX Kernel.* Proceedings of the USENIX Winter Conference, pages 283–290.

[Lux92] LUX, W., KUHNHAUSER, W. E., and HARTIG, H. (June 1992) *The BirliX Migration Mechanism.* Proceedings of the Workshop on Dynamic Object Placement and Load Balancing in Parallel and Distributed Systems Programs, pages 83–90.

[Mill81] MILLER, B. and PRESOTTO, D. (1981) *XOS: an Operating System for the XTREE Architecture.* Operating Systems Review, 2:21–32.

[Milo91] MILOJICIC, D., PJEVAC, M., and VELASEVIC, D. (September 1991) *Load Balancing Survey.* Proceedings of the Summer EurOpen Conference, pages 157–172.

[Pati93] PATIENCE, S. (April 1993) *Redirecting System Calls in Mach 3.0: An Alternative to the Emulator.* Proceedings of the third USENIX Mach Symposium, pages 57–74.

[Roga92] ROGADO, J. (October 1992) *A Strawman Proposal for the Cluster Project.* Technical Report, OSF RI, Grenoble.

[Swan93] SWANSON, M., STOLLER, L., CRITCHLOW, T., and KESSLER, R. (April 1993) *The Design of the Schizophrenic Workstation System.* Proceedings of the third USENIX Mach Symposium, pages 291–306.

[Thei85] THEIMER, M., LANTZ, K., and CHERITON, D. (December 1985) *Preemptable Remote Execution Facilities for the V System.* Proceedings of the 10th ACM Symposium on OS Principles, pages 2–12.

[Walk89] WALKER, B. J. and MATHEWS, R. M. (Winter 1989) *Process Migration in AIX's Transparent Computing Facility.* IEEE TCOS Newsletter, 3(1):5–7.

[Welc91] WELCH, B. (November 1991) *The File System Belongs in the Kernel.* Proceedings of the Second USENIX Mach Symposium, pages 233–250.

[Wiec92] WIECEK, C. A. (April 1992) *A Model and Prototype of VMS Using the Mach 3.0 Kernel.* Proceedings of the USENIX Workshop on Micro-Kernels and Other Kernel Architectures, pages 187–204.

[Youn89] YOUNG, M. W. (November 1989) *Exporting a User Interface to Memory Management from a Communication-Oriented Operating System.* Technical Report CMU-CS-89-202, PhD Thesis, Carnegie Mellon University.

[Zajc93] ZAJCEW, R., ET AL. (January 1993) *An OSF/1 UNIX for Massively Parallel Multicomputers.* Proceedings of the Winter USENIX Conference, pages 449–468.

[Zaya87] ZAYAS, E. (November 1987) *Attacking the Process Migration Bottleneck.* Proceedings of the 11th Symposium on Operating Systems Principles, pages 13–24.

A Measurements

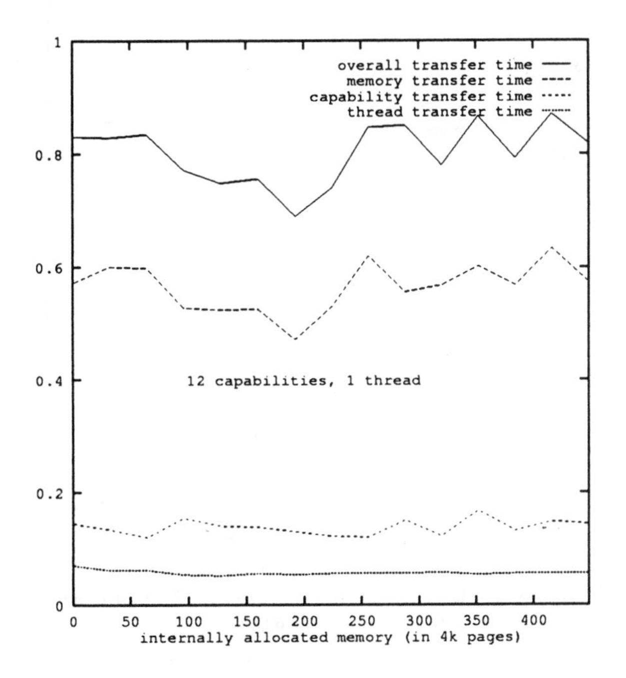

Figure 3. Transfer Time (in sec) v. Memory Size

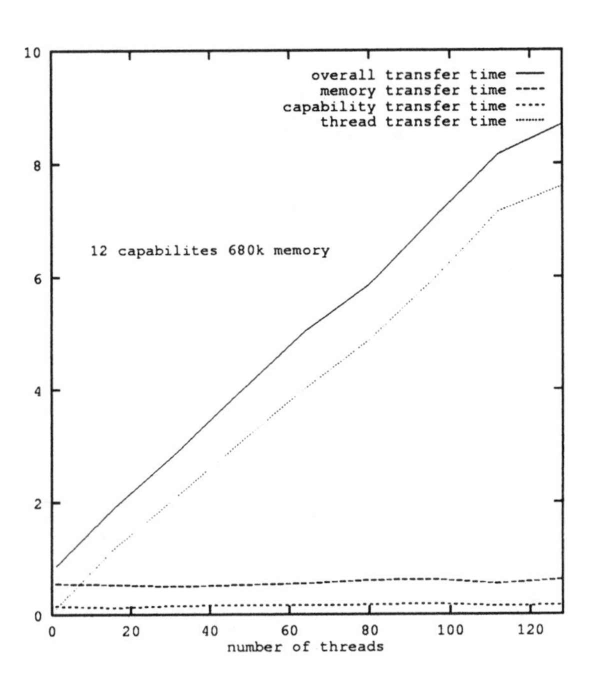

Figure 4. Transfer Time (in sec) v. Thread Number

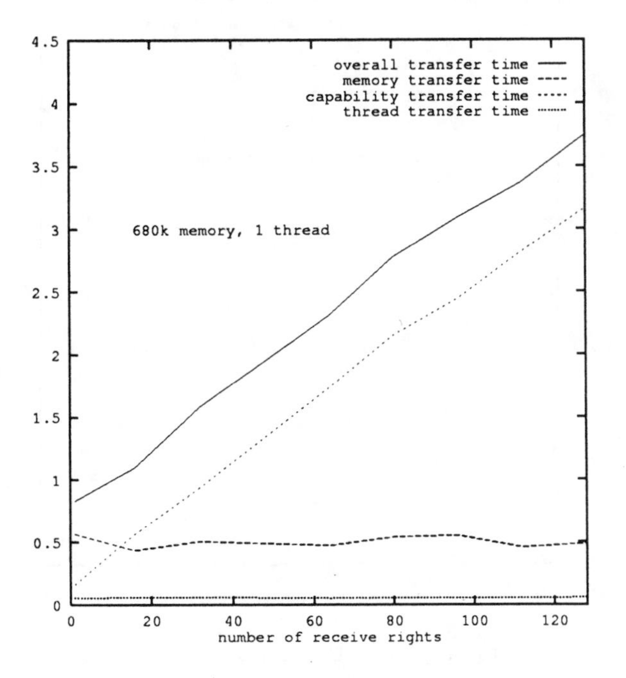

Figure 5. Transfer Time (in sec) v. Number of Receive Rights

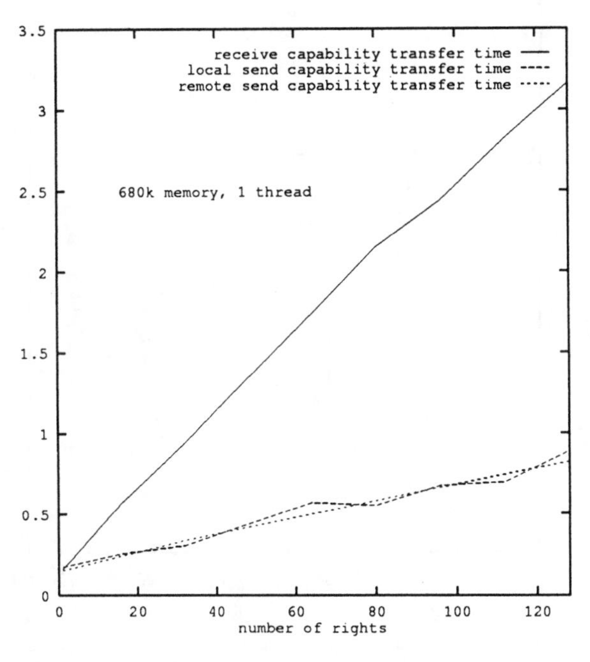

Figure 6. Rights Transfer Time (in sec) v. Number of Rights

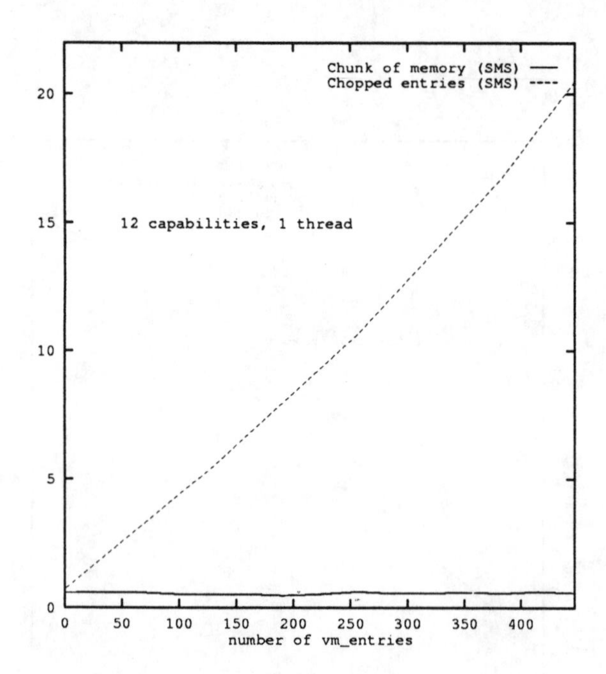

Figure 7. Transfer Time (in sec) v. Memory Contiguity

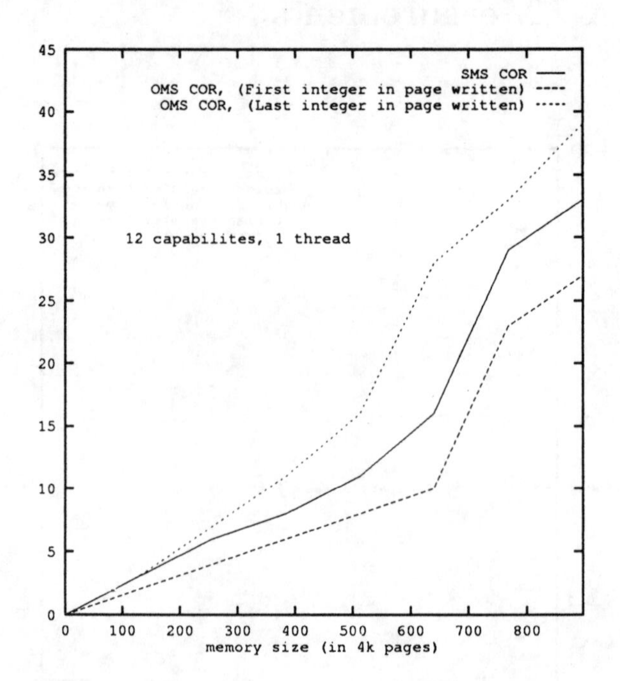

Figure 8. Execution Time (in sec) v. Memory Size, (First/Last Integer in Page Written)

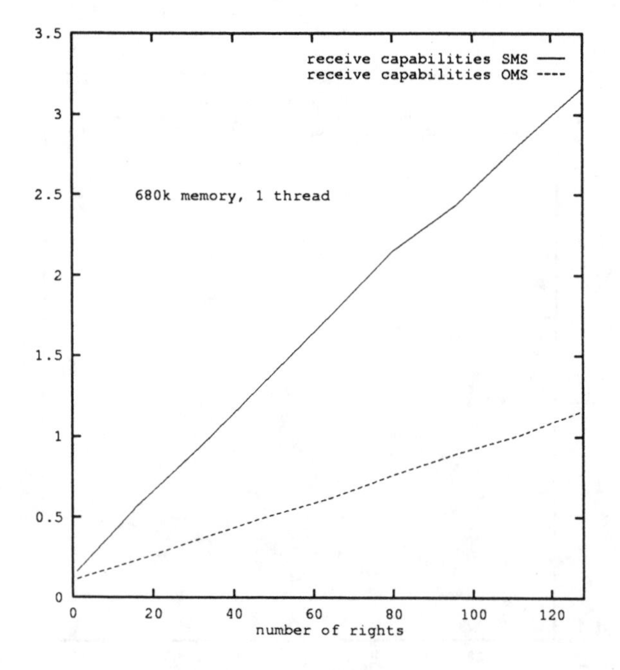

Figure 9. Transfer Time (in sec) v. Number of Receive Rights

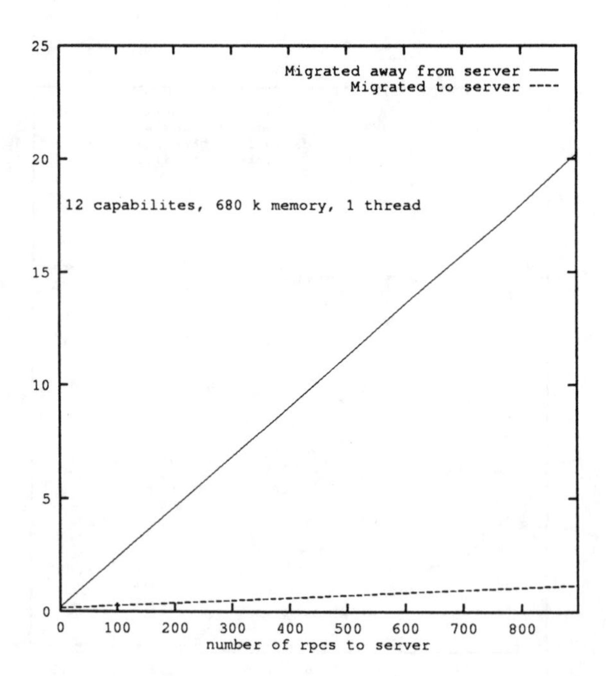

Figure 10. Execution Time (in sec) v. Remote/Local IPC

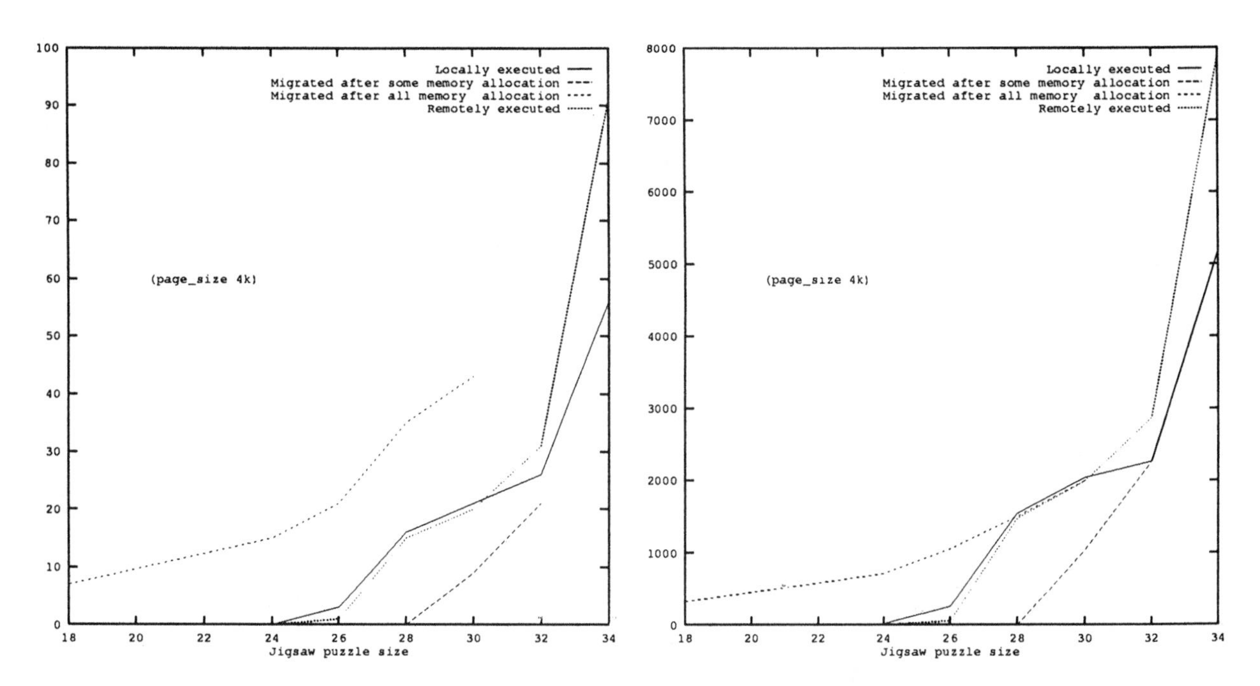

Figure 11. Solving Time (in sec) v. Jigsaw Puzzle Size

Figure 12. Paging (Number of Pageins & Pageouts) v. Jigsaw Puzzle Size

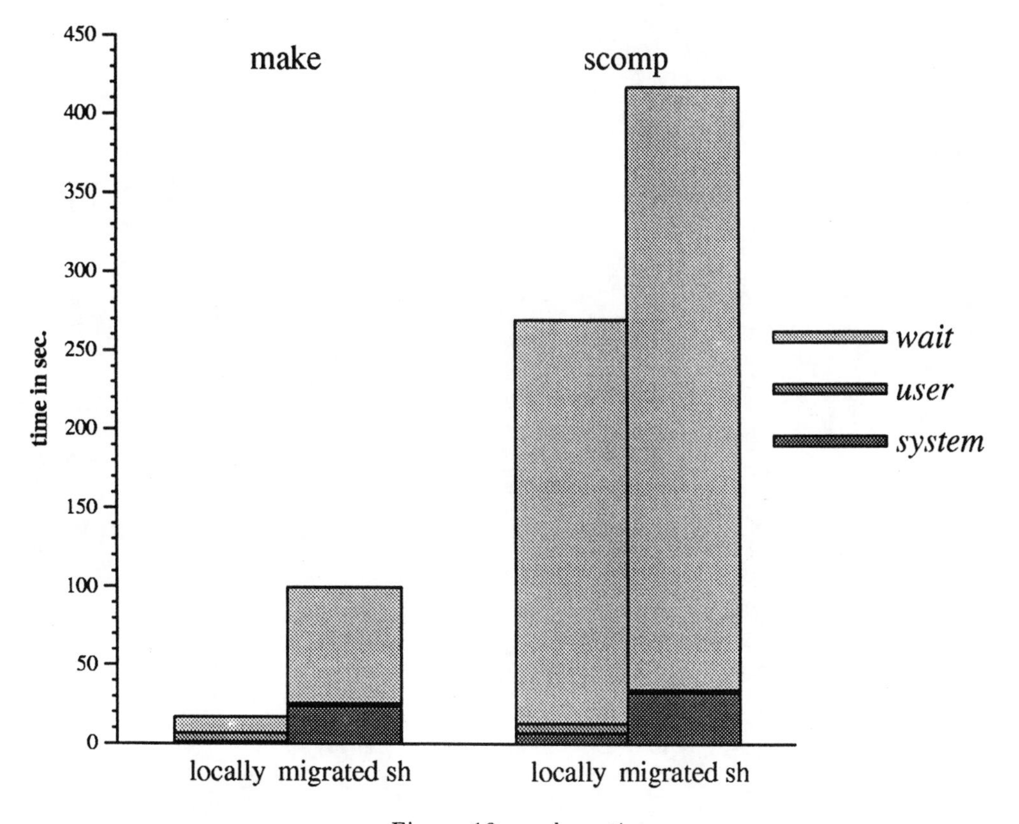

Figure 13: *make ratio.*

Chapter 5 User-space Migration

One way to avoid the complexity of kernel-based migration policies is to implement migration entirely as a user-space mechanism. Frequently, this trades the relatively high performance of kernel-based implementations away in favor of reduced complexity and eased portability.

In our first user-space migration scheme, Litzkow and Solomon implement a user-level migration mechanism in their **Condor** package [Litzkow, 1987; 1988; 1992]. When a process is about to migrate, Condor produces a core file for the processes, which is sent as part of the migration to the new node. A shadow process is left on the source node to execute system calls for the migrated entity. Condor does not support migration for all types of applications: processes are not eligible for migration if they use signals, timers, memory-mapped files or inter-process communication (IPC). Its user-space mechanism uses a brute force copy of the address space, by dumping process state in a core file. The main advantage of Condor is that it is much easier to maintain implementation and to port to new systems, and that it provides a process migration mechanism for systems that would not otherwise have one.

Emerald takes another approach to mobility: it is a programming language and an environment that supports mobile objects in a distributed environment [Jul, 1988]. An Emerald process is a thread of control within an Emerald object that is initiated upon object creation. As an object migrates, so do its associated processes. The advantages of this fine-grained mobility include the ability of the language to hide the mobile nature of an object from the programmer, which allows the same object definition to be used to create either a local or distributed object. The system manages the migration automatically.

Most systems that support process migration require that all of the nodes in the migration domain are homogeneous, that is, they must have the same operating system and hardware architecture. Smith and Hutchinson present a system that overcomes this limitation. In **Tui**, processes can migrate between non-homogeneous nodes [Smith, 1998]. Before migration can take place, the entire image of the migrating process must be translated into a process image the target node can execute.

Other systems supporting user-space migration mechanisms include those done by Alonso [Alonso, 1988], Mandelberg [Mandelberg, 1988], Petri and Langendoerfer [Petri, 1995], MPVM [Casas, 1995], and LSF [Zhou, 1993]. In addition, there are application specific migrations, such as in the work by Freedman [Freedman, 1991], Skordos [Skordos, 1995], and Bharat and Cardelli [Bharat, 1995]. Other examples of mobile object systems supporting (or describing) migration include KNOS [Tsichritzis, 1987], SOS [Shapiro, 1989], COOL [Amaral, 1992], and DCE [Schill, 1993]. Other examples of heterogeneous migration include work by Dubach [Dubach, 1989], Shub [Shub, 1990] and the follow-up work done on Emerald [Steensgaard, 1995]. Mobility of objects is also addressed by two recent mobile object systems, Legion [Grimshaw, 1997] and Globe [Hamburg, 1996].

5.1 Litzkow, M. and Solomon, M., Supporting Checkpointing and Process Migration outside the UNIX Kernel

Litzkow, M. and Solomon, M., "Supporting Checkpointing and Process Migration outside the UNIX Kernel," *Proceedings of the USENIX Winter Conference*, pp. 283-290, January 1992.

SUPPORTING CHECKPOINTING AND PROCESS MIGRATION OUTSIDE THE UNIX KERNEL

Michael Litzkow
Marvin Solomon

Computer Sciences Department
University of Wisconsin—Madison

Abstract

We have implemented both checkpointing and migration of processes under UNIX as a part of the Condor package. Checkpointing, remote execution, and process migration are different, but closely related ideas; the relationship between these ideas is explored. A unique feature of the Condor implementation of these items is that they are accomplished entirely at user level. Costs and benefits of implementing these features without kernel support are presented. Portability issues, and the mechanisms we have devised to deal with these issues, are discussed in concrete terms. The limitations of our implementation, and possible avenues to relieve some of these limitations, are presented.

1. Introduction

Condor is a software package for executing long-running, computation-intensive jobs on workstations which would otherwise be idle. Idle workstations are located and allocated to users automatically. Condor preserves a large measure of the originating machine's execution environment on the execution machine, even if the originating and execution machines do not share a common file system. Condor jobs are automatically checkpointed and migrated between workstations as needed to ensure eventual completion. This paper describes the checkpointing and remote execution mechanisms.

The Condor package as a whole has been documented elsewhere [1,2,3]. This paper focuses on the actual mechanisms for remote execution and process migration, limitations on migrating processes without kernel support, portability issues, and the evolution of our implementation of these mechanisms.

In contrast to the process migration mechanisms offered by such systems as the V-system[4], Sprite[5], and Charlotte[6], Condor supports remote execution and process migration on a variety of UNIX® platforms, and is implemented completely outside the kernel. Because it requires no changes to the operating system, Condor is portable and can be used in environments where access to the internals of the system is not possible. Condor does pay a price for this flexibility in both the speed and completeness of its process migration.

The combination of remote execution and checkpointing means that a process may migrate among processors during its lifetime, and thus must see the same file system view wherever it is run. "In kernel" process migration systems such as Sprite and Charlotte were designed for environments that ensure such uniformity. While careful administration of NFS® or AFS®, can ensure a homogeneous environment, heterogeneous environments are common, and requiring a uniform name space would severely limit the scope of Condor's usefulness. Therefore Condor uses a technique called "remote system calls" in which requests for file-system access are trapped and forwarded to a "shadow" process on the submitting machine.

Several systems have been implemented which offer process migration, but not checkpointing. Offering both mechanisms in combination has two advantages. First, jobs are immune to machine or network crashes. Because

---

Authors' address: Computer Sciences Department, 1210 W. Dayton St., Madison, WI 53706; mike@cs.wisc.edu, solomon@cs.wisc.edu.

This paper was originally presented at the Usenix Winter Conference, San Francisco Ca., Jan 1992

® UNIX is a registered trademark of AT&T Bell Laboratories

® NFS is a registered trademark of Sun Microsystems.

® AFS is a registered trademark of Transarc Corporation.

Condor jobs aren't killed by these events, users can submit large batches of jobs, and then go on to other work, (or leave for the weekend), without worrying about all their jobs being aborted by a sudden power outage or other disaster. Second, checkpointing allows a job to remain idle during periods when all workstations are busy with interactive work. In UNIX, a process, even if stopped, consumes resources (such as swap space). Because the most important benefit of having a workstation is immediate response, our policy is to ensure absolute priority for interactive use.

2. Remote System Calls

To understand Condor's remote system calls, one must first consider normal UNIX system calls. Every UNIX program, whether or not written in the C language, is linked with the "C" library. This library provides a large number of functions, including the standard I/O library (traditionally described in section 3 of the manual), as well as interfaces to the kernel facilities described in section 2. These latter functions are generally implemented as "stubs," which push their arguments and a "call number" identifying the facility onto the user stack, and execute a machine-defined *supervisor call* instruction. In some newer implementations, the mechanism is a bit different, but the general idea is the same; each system call has a corresponding function in the C library which comprises a very thin layer between the user code and the system. Figure 1 illustrates the normal UNIX system call mechanism.

Figure 2 shows how we have altered the system call mechanism by providing a special version of the C library which performs system calls remotely. This library, like the normal C library, has a stub for each UNIX system call. These stubs either execute a request locally by mimicking the normal stubs or package it into a message which is sent to the *shadow* process. The *shadow* executes the system call on the initiating machine, packages the results, and sends them back to the stub. The stub then returns to the application program in exactly the same way the normal system call stub would have, had the call been done locally. The shadow runs with the same user and group ids, and in the same directory as the user process would have had it been executing on the submitting machine. This scheme ensures a uniform view of the file system, as well as avoiding certain security problems.

3. Checkpointing

Ideally, checkpointing and restarting a process means storing the process state and later restoring it in such a way that the process can continue where it left off. In other words, as far as the user code is concerned, the checkpoint never happened. In the most general case, the state of a UNIX process may include pieces of information which are known only to the kernel, or which may not be possible to recreate. For example if a process is communicating with other processes at the time of the checkpoint, and those processes are no longer extant at the time of the restoration, that part of the state cannot be reproduced. While some UNIX processes include state which cannot be

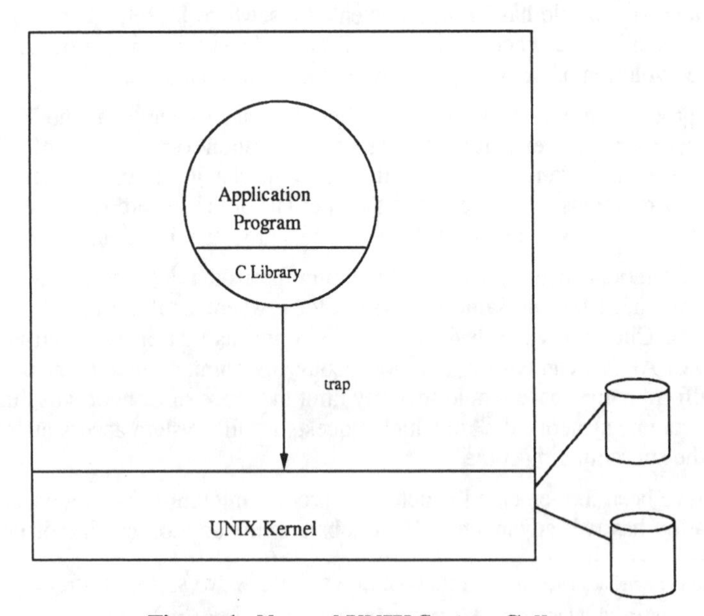

Figure 1. Normal UNIX System Calls

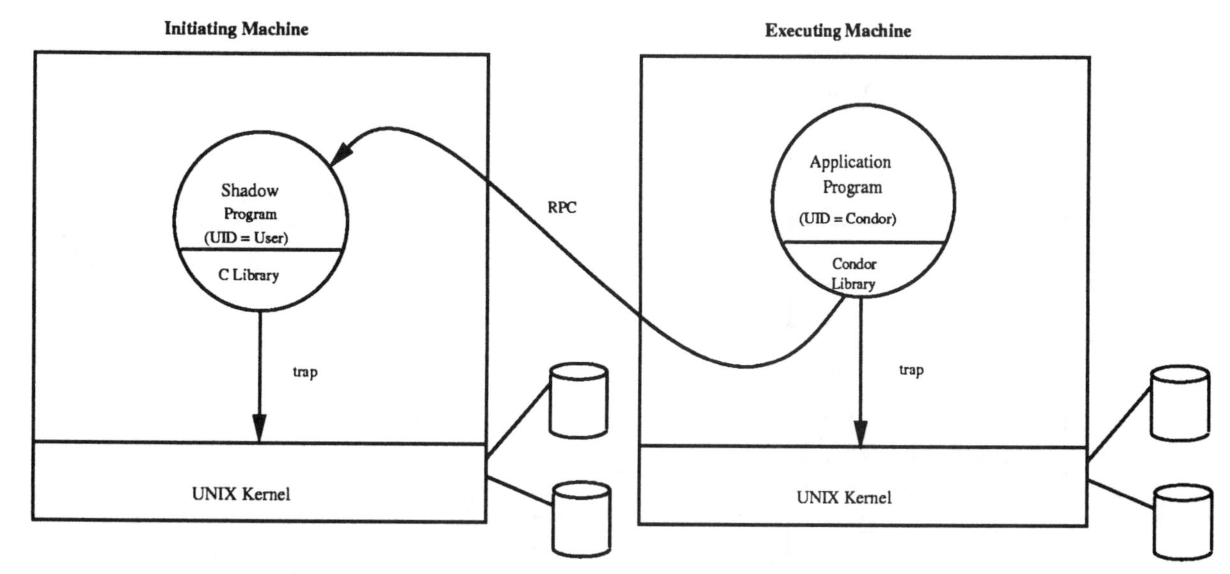

Figure 2. Remote System Calls

saved and restored, there are a large number of jobs whose state is simple enough that they can be checkpointed and restarted without harm to their original mission. In the Condor project we have concentrated on providing practical means for remote execution, checkpointing, and migration of such jobs.

The state of a UNIX process includes the contents of memory (the text, data, and stack segments), processor registers, and the status of open files. Restoring the text segment is easy since it does not change and is available as part of the original executable file. Our approaches to saving and restoring each of the other items have changed over the years, mainly in response to portability issues. In the earliest versions of Condor, the data and stack were saved by writing them directly into a file, and the register contents were saved by architecture-specific assembler routines. Each time we ported Condor to a new hardware platform or a different version of UNIX we were "bitten" by some part of the checkpointing mechanism which was either very difficult or impossible to port. Thus we gradually changed our methods to rely on basic UNIX mechanisms rather than on specific implementations. In other words, we use standard mechanisms that have already been ported by the provider of the kernel and C library.

Our current approach is to create a new checkpoint file from pieces of the previous checkpoint and a core image. The checkpoint is itself a UNIX executable file ("a.out"). While core files are generally intended to aid in debugging a process which has committed an error, such as attempting an illegal memory access, they also serve as a portable mechanism for saving the state of a process at a given point in time. Not surprisingly, the information needed to debug a process and the information needed to restart it are almost exactly the same. The text for the new executable is of course an exact copy of the text from the original. The data area is copied directly from the core file to the initialized data area of the new executable file. The saved stack area is also copied into the new executable in a section which is not normally used by the UNIX process initialization mechanism. Restoring some of the other items in the new instantiation of the process is trickier. For example, although volatile state (such as register contents and the program counter) is recorded in the core image, restoring it directly would require machine-dependent assembler routines. Instead, we use the Unix signal-handling machinery and the setjmp/longjmp facilities in the C library.

Figures 3 through 7 illustrate various steps in checkpointing and restoring a typical UNIX process. Figure 3 depicts the virtual address space of an application program that has been linked with the Condor versions of the C library and startup routine. Routines written by the application programmer as well as those provided by the condor version of the C library co-exist at various locations in the Text segment. The data area consists of both the initialized and uninitialized data from the original executable as well as any data space allocated at run time, e.g. by the `sbrk` system call. The stack area consists of the per-process kernel data, (the "u_area", followed by stack frames for each function in the currently active execution stack. Since Condor must do its own initialization before any of the the user's code is called, the first frame on the stack is for the Condor routine `MAIN`. `MAIN` calls the user's `main` with the correct `argc`, `argv`, and `envp`.

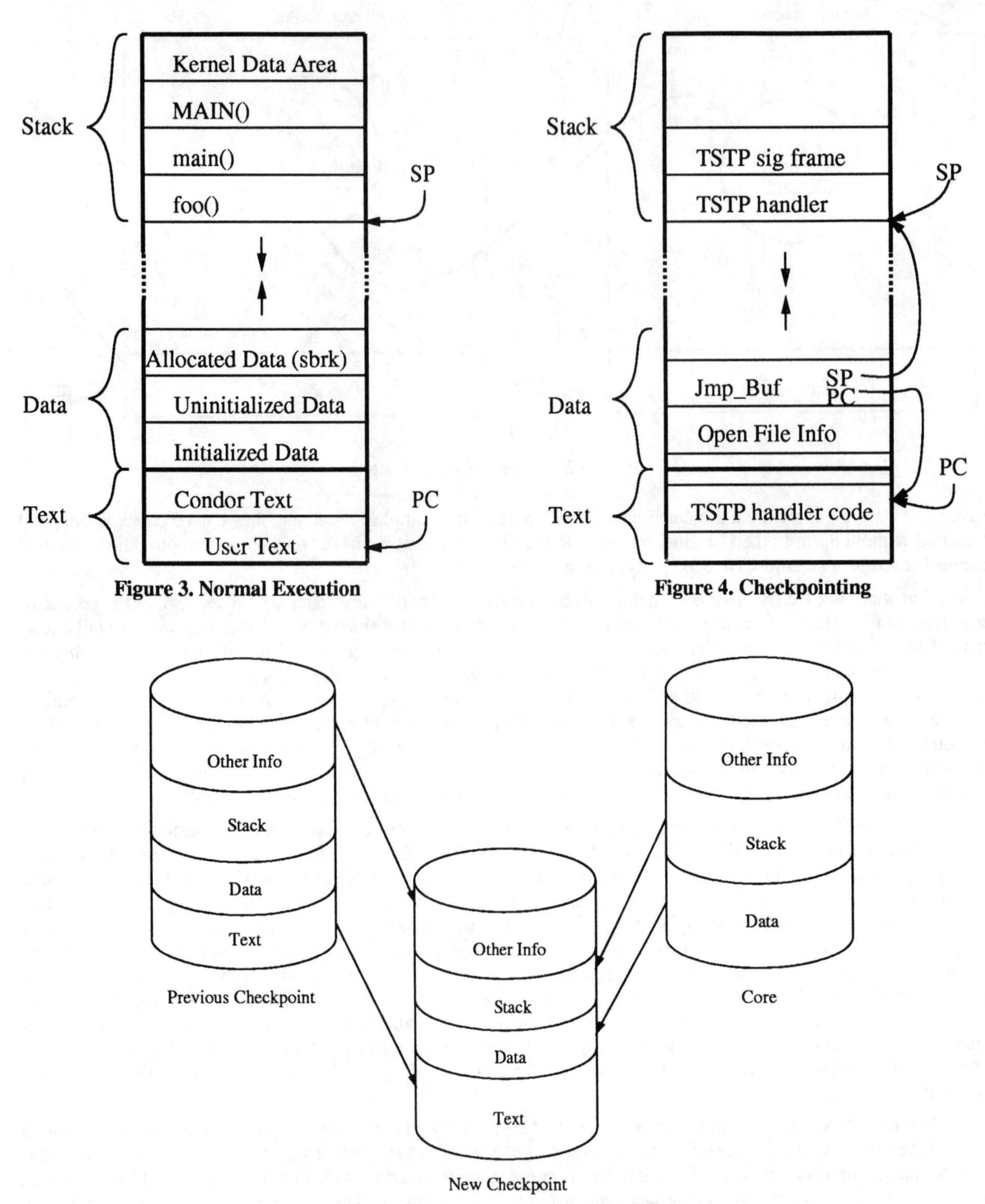

Figure 3. Normal Execution

Figure 4. Checkpointing

Figure 5. Creating a New Checkpoint

The initialization code in `MAIN` establishes a signal handler for the `SIGTSTP` signal, which is used to inform a Condor job that it should checkpoint itself. Information about all files which the process currently has open is kept in a table by the Condor version of the `open` system call routine. The `TSTP` handler updates this information with the current file pointer location for each file using the `lseek` system call and uses the C library's `setjmp` call to record its state (including the program counter) in a buffer. Finally the handler sends itself a

SIGQUIT signal to cause a core-dump and terminate. Figure 4 depicts the state of the process just before the SIGQUIT is received.

Figure 5 depicts the new checkpoint file being created from pieces of the previous checkpoint file and the core file. The data and stack areas come from the core file, while the text comes from the previous checkpoint file. The "other info" referred to is generally symbol table information, which may be preserved for the benefit of debuggers. Before a Condor process is executed for the first time, its executable file is modified to look exactly like a checkpoint file with a zero size stack area, so that every checkpoint is done the same way.

When the new process starts, UNIX will initialize it with the data saved in the executable file, thus the process is born with the data area already restored. Besides the data needed by the user code, the data area contains the jump buffer and the open file table which were saved by the TSTP handler in the previous execution. UNIX will initialize the process with a minimal stack, which must be replaced by the stack information saved in the checkpoint file. The code that restores the stack needs a special stack of its own so that it doesn't interfere with the "real" stack being restored. To move its stack pointer into the data segment, it establishes a handler for the SIGUSR2 signal with a "signal stack" in the data segment, and then sends itself that signal. The USR2 handler uses the values stored in the file table to reopen all files which were open at the time of the checkpoint, and seek them to their correct offsets. Figure 6 shows the state of the process just before the stack is overwritten with the saved stack information.

After the stack is restored, the USR2 handler calls longjmp with the state saved by the setjmp in the TSTP handler of the original process. The stack pointer and program counter are restored to their values as of the setjmp call, and the SIGTSTP handler simply returns, restoring other volatile state (such as processor registers) as it was before the SIGTSTP signal. Figure 7 shows the state of the process after the longjmp, and just before the TSTP handler returns.

Condor uses inherently portable mechanisms to restore a process's stack and registers from their checkpointed values, and does not resort to any assembler code whatsoever. Since individual write calls are not traced, the file recovery scheme requires that all file updates be *idempotent* (repeating operations does no harm). For typical UNIX file usage, this restriction seems not be much of a problem.

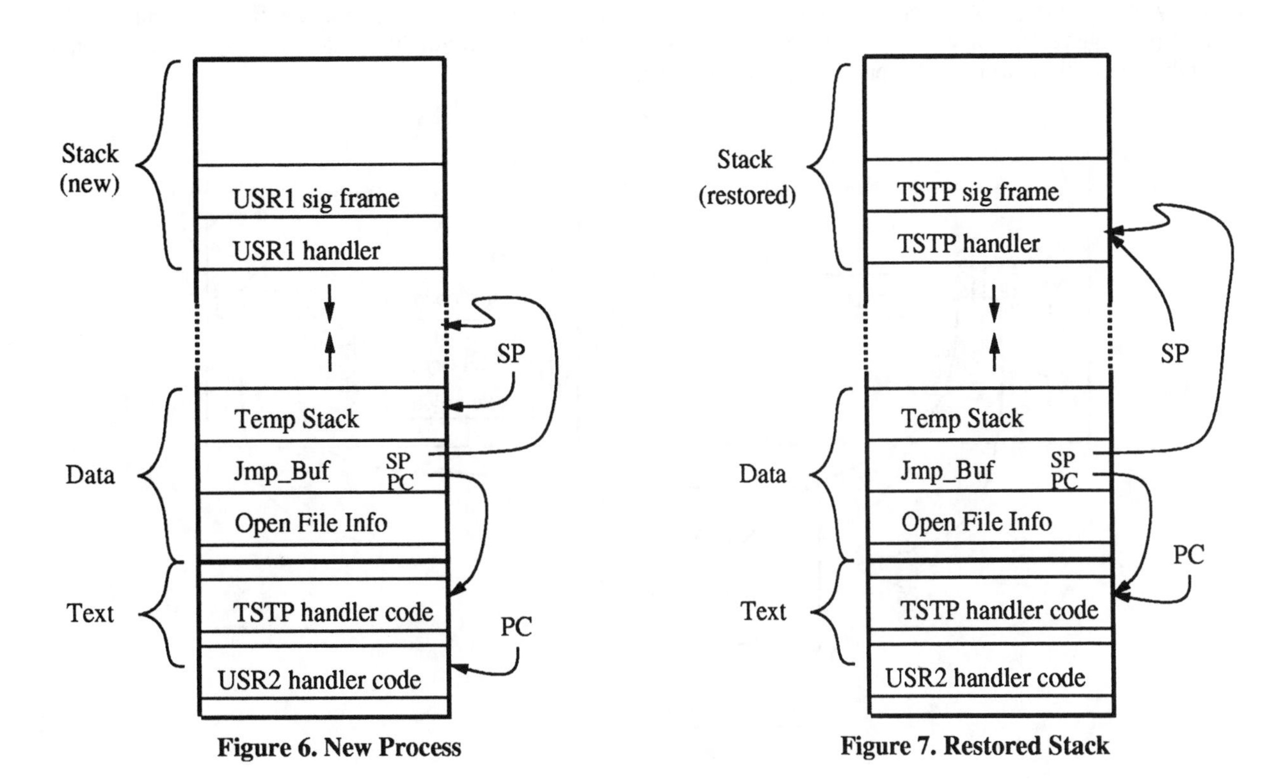

Figure 6. New Process **Figure 7. Restored Stack**

4. Network File Systems

The remote system call scheme described above assures correct operation in inhomogeneous file-naming environments. If, however, a network file system such as NFS is in use, an important optimization is possible. For example, consider the common situation in which a single file server serves an entire local-area network. Each I/O request is forwarded from the "worker" machine to the shadow, which then accesses the file over the network using NFS. If the stub executed the I/O request directly, the results would be the same, but one network round-trip would be avoided. The situation is even worse if the file is physically located on the worker machine: Two network round-trips are used when none are needed. To improve the performance in these cases, Condor incorporates a mechanism to avoid the use of remote system calls when a file is more directly accessible. At the time of an `open` request, the stub sends a name translation request to the submitting machine. The shadow process responds with a translated pathname in the form of *hostname:pathname*, where the *hostname* may refer to a file server, and the *pathname* is the name by which the file is known on the server (which may be different from the pathname on the submitting machine, because of mount points and symbolic links). The stub then examines the mount table on the machine where it is executing, and if possible accesses the file without using the shadow process. Whenever a process is checkpointed and restarted on another machine, the name translation process is repeated, since access to remotely mounted files may vary among the execution machines. Figure 8 illustrates an example where the pathnames of the target file on the initiating and executing machines are different. The `open` routine sends a request to the shadow for a translation of the pathname "/u2/john", and the shadow responds with the external name "fileserver:/staff/john". The `open` routine then translates the external name to the name by which the file is known on the executing machine, namely "/usr1/john", and opens the file using normal system calls, e.g. via NFS.

5. Limitations

Condor has been found to be an extremely useful tool for a particular class of application: A single-process, computation-intensive, long-running job. Other kinds of application are currently hindered by Condor's limitations, some of which could be relatively easily lifted (and probably will be in future releases), and others of which appear to be inherent. Condor currently does not support applications that use signals, timers, memory-mapped files, or shared libraries. We expect some of these restrictions will be relaxed in future versions of Condor. The most painful limitation is that Condor does not support any sort of inter-process communication. Aside from the well-known difficulty of achieving a consistent snapshot of a multi-process program, there is the problem of hidden state in the form of messages inside pipes. Implementing multi-process Condor jobs would also greatly complicate the Condor scheduling algorithm (which is beyond the scope of this paper).

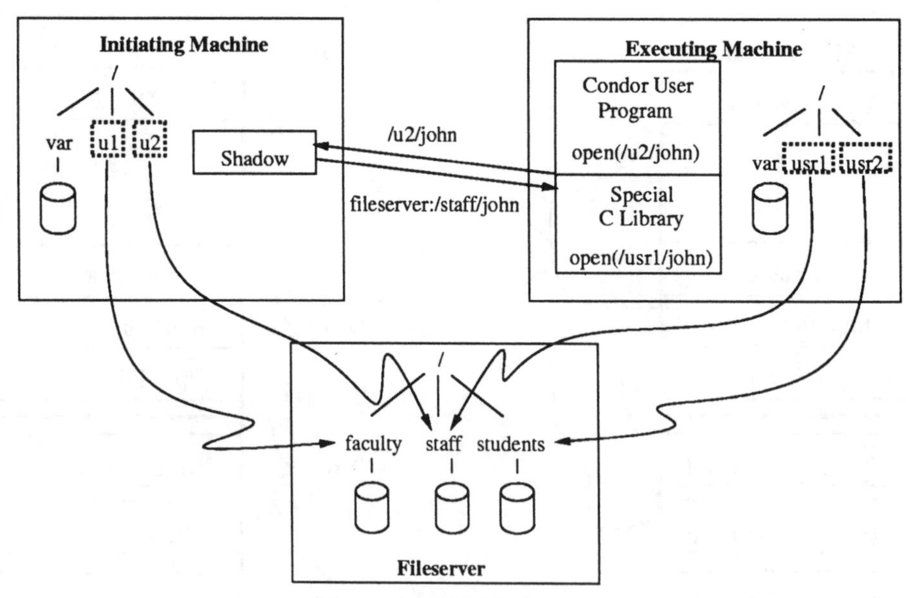

Figure 8. NFS File Access

The high cost of process migration limits the usefulness of Condor for small jobs. Process migration involves many steps: First the process causes a core dump. The core file and executable module are combined to produce a new executable, which is returned to the submitting machine, where it resides until a new execution site becomes available. The executable is then transferred over the network, and execution resumes. Transferring a Condor job with a 6 megabyte address space between two DECstation 3100's on an Ethernet we obtained rates of about 250K Bps (bytes/sec) for dumping core, 130K Bps for updating the checkpoint file, 250K Bps for transferring the new checkpoint file from the execution site to the originating machine, and 250K Bps for transferring the new checkpoint file from the originating machine to a new execution site. Altogether, it takes about two minutes to migrate such a process. In practice the job must wait on the originating machine until the scheduler can locate a new execution site, and the scheduler only runs once every 10 minutes.

There are however some significant benefits to an implementation which does not depend on access to the kernel code. Most obvious of these is the portability which may be obtained. Condor currently runs on ten different hardware/software platforms: the IBM R6000 running AIX, the IBM RT/PC running AOS, Sun 3 and Sun 4 workstations running SunOS, the Silicon Graphics 4D workstation running IRIX, the Hewlett-Packard 9000 under 4.3 BSD UNIX, the Digital Equipment DECstation and VAXstation running Ultrix, and the Sequent Symmetry running Dynix. A related but somewhat different issue is availability in the real world. If our process migration and checkpointing facilities were embedded in an operating system, then it is unlikely many others would run that system even if it were ported to their hardware and freely available. If Condor required a homogeneous server-based file environment, many sites would not be able to use it. As it is, Condor is available via anonymous ftp (contact the first author for details) and is running at many sites throughout the United States and Europe.

6. Future Work

Work with Condor is ongoing at Wisconsin. We are particularly interested in moving Condor to larger groups of machines, including groups which are connected by wide-area rather than local-area networks. We are also interested in integrating constrained forms of inter process communication with Condor such as Linda[7]. The three areas of continuing effort relating to the mechanisms described here are efficiency, functionality, and portability. With regard to efficiency, many methods are possible for saving, migrating, and restoring the necessary pieces of a UNIX process. In some circumstances, we could transfer processes directly between execution sites rather than always sending a checkpoint file back to the originating site. We could use data compression to reduce the volume of data transferred and stored. We could read the stack and data directly from a core file into a new instantiation of a process rather than converting the core file to an executable module.

In the area of functionality, the most often requested item is support for signals. This feature is non-trivial, because much of the information regarding user defined signal handlers, the handling of signals on special stacks, the blocking and unblocking of signals, and so forth is maintained by the kernel in ways which vary among UNIX implementations. Nonetheless, a restricted form of signal facility could probably be provided that would significantly increase the number of applications supported. Unfortunately both optimizations and more general UNIX semantics work against portability. It is often easy to find a solution to one of the optimization or functionality problems which works on some platforms, but not others.

7. Conclusions

Condor has accomplished checkpointing and process migration on "vanilla" UNIX systems. Both facilities are generally only implemented inside an operating system, if at all. Although this design decision incurs a cost, both in efficiency and generality of application programs supported, it makes Condor available to far wider user community,

8. Acknowledgements

Parts of this research have been supported by the National Science Foundation under grant DCR-8521228, by a Digital Equipment Corporation External Research Grant, and by an International Business Machines Joint Study agreement. The port to the Silicon Graphics 4D workstation was funded by NRL/SFA Inc.

A great many people have contributed time, guidance, and ideas to Condor; a chosen few have also contributed code. We wish to particularly thank our users who have been very patient and supportive throughout the development of Condor, and most especially those brave souls who are not users of Condor, but have allowed their machines to be candidates for remote execution anyway. The original idea for Condor was suggested by David DeWitt, based on a suggestion from Maurice Wilkes. It was Dewitt who insisted that Condor must be implemented without any changes whatever to the UNIX kernel. Miron Livny and Matt Mutka first convinced us that checkpointing was both possible and necessary. Livny and Mutka also provided much guidance in the overall philosophy and structure behind the design of Condor, as well as the name "Condor" itself. Allan Bricker implemented portions of Condor and contributed many useful ideas. It was Allan who first suggested using the core file for saving the state of a process and the `signal` mechanism for getting the registers restored.

Availability

Condor is available via anonymous ftp from "shorty.cs.wisc.edu", (128.105.2.8). Mail "condor-request@cs.wisc.edu" if you would like to be on our mailing list.

References

[1]. M. Litzkow, "Remote Unix—Turning Idle Workstations Into Cycle Servers," *Proceedings of the Usenix Summer Conference*, Phoenix, Arizona, June 1987.

[2]. M. Litzkow, M. Livny, and M Mutka, "Condor—A Hunter of Idle Workstations," *8th International Conference on Distributed Computing Systems*, Jan Jose, Calif, June 1988.

[3]. M. Litzkow and M. Livny, "Experience With the Condor Distributed Batch System," *Proceedings of the IEEE Workshop on Experimental Distributed Systems*, Hunstville, AL Oct. 1990.

[4]. M. Theimer, K. Lantz, and D. Cheriton, "Preemptable remote execution facilities for the V-System," *Proceedings of the 10th Symposium on Operating System Principles*, December 1985.

[5]. F. Douglis and J. Ousterhout, "Transparent Process Migration: Design Alternatives and the Sprite Implementation," *Software Practice and Experience*, to appear.

[6]. Y. Artsy and R. Finkel, "Designing a process migration facility: The Charlotte experience," *IEEE Computer*, September 1988.

[7]. N. Carriero and D. Gelernter, "The S/Net's Linda Kernel," *ACM Transactions on Computer Systems*, Vol.4, No. 2, May 1986.

Michael Litzkow has been a member of the technical staff at the University of Wisconsin—Madison since 1983. Past projects include the Csnet Nameserver and a fileserver for the Charlotte distributed operating system. He received his B.S. in computer sciences from the University of Wisconsin in 1983.

Marvin Solomon received a B.S. degree in Mathematics from the University of Chicago and M.S. and Ph.D. degrees in Computer Science from Cornell University. In 1976, he joined the Department of Computer Sciences at the University of Wisconsin—Madison, where he is currently Professor. Dr. Solomon was Visiting Lecturer at Aarhus (Denmark) University during 1975-76 and Visiting Scientist at IBM Research in San Jose California during 1984-85. His research interests include programming languages, program development environments, operating systems, and computer networks. He is a member of the IEEE Computer Society and the Association for Computing Machinery.

The Evolution of Condor Checkpointing 1991–1998

Michael Litzkow
Marvin Solomon

The checkpointing mechanism described in this paper was successfully used by Condor for several years. By eliminating assembly language in favor of "standard" Unix facilities such as core dumps and `setjmp/longjmp`, it greatly eased the difficulty of porting Condor to new platforms. However, the mechanism was found wanting for a variety of reasons some of which were anticipated in the original paper, and evolved into the algorithm described below. This note describes Version 6 of Condor, released in early 1998. A more detailed description of Condor checkpointing as of Version 5 has been published elsewhere [1].

Capturing the state of the program with a core dump, although supported on all Unix platforms, presented several difficulties.

- The format of the `core` file varies widely from platform to platform.
- Since core dumps are expected to be exceptional events, implementors put little effort into making them efficient. As a result, dumping core—particularly for a large program—tends to be far slower than creating a file of comparable size with standard I/O facilities.
- Once started, a core dump cannot be delayed or suspended.
- Data is always written to a file named `core` in the current directory. There is no way to send it directly to an alternative destination such as a pipe or network connection.
- The program receives no indication if the core dump fails (e.g. because of inadequate permissions or space to create the `core` file).

Another obstacle to portability was the need to interpret and create executable (`a.out` format) files. The introduction of the *Common Object File Format* (coff) raised hopes that the industry would settle on a standard format for executable files, but these hopes were not realized. Current Unix systems still use a wide variety of formats (including *coff*, *ecoff*, *elf*, and *dwarf*), and most formats only describe how the file is divided into sections, with the contents of each section left up to the implementation.

The current technique creates a checkpoint file that is idiosyncratic to Condor, but uniform across all platforms. The checkpoint file contains

- the contents of the writable data and stack segments of the checkpointed process,
- information about dynamically linked libraries that have been loaded,
- a table of open files, indicating the file name, mode, and seek pointer for each one,
- volatile CPU state, such as the contents of registers, stored by `setjmp` in a `JMP_BUF` structure, and
- a table of information about signal handlers, as described below.

The format is designed to allow the file to allow sequential access, so that it can be sent directly across a pipe or network connection.

The Condor startup routine `MAIN` linked into the job's executable file performs some Condor-specific setup activities and then determines whether the job should continue from a checkpoint. If not, it simply calls `main`. If so, it uses information from the checkpoint file to restore the process state: signal masks and handlers are set, files are re-opened and positioned, the data and stack segments are reloaded, dynamic libraries are restored, and the remaining context is re-established by a call to `setjmp`.

The new scheme is highly portable. Moving Condor to a new platform only requires a few simple platform-specific primitives that determine the ends of the stack and data segments, the direction of stack growth, and the offset within the `JMP_BUF` where the stack pointer is stored. For HPUX, there is one additional routine to patch the segment ("space") registers. All other code is system-independent.

Three limitations cited in the original paper—no signals, no shared libraries, and no user-defined signal handlers—have been eliminated or eased. Unix signals are now fully supported. The checkpointing code uses the `sigprocmask` system call to determine which signals are blocked or ignored `sigaction` to find out which which signals have custom system handlers installed, and `sigpending` to learn about pending signals (signals that have been sent to the process but not delivered because they are blocked). When a process is restarted from a checkpoint, it restores the state of all signal handlers and masks and re-sends pending signals to itself by calling the `kill` or `sigsend` system call.

Shared (dynamically linked) libraries are also fully supported. In the years since 1991, there has been significant increase in the number of dialects of Unix offering shared libraries, with a corresponding increase in dependence on them, to the extent that some dialects do not even provide static versions of all standard libraries. Support for shared libraries has therefore become a necessity. The checkpoint code determines which library routines have been loaded (with the aid of the `/proc` file system on versions of Unix that support it) and copies them to the checkpoint file. The code and data of shared libraries are restored from the checkpoint file to the same virtual addresses where they appeared when the checkpoint was created.

Multi-process Condor jobs are still not fully supported, but Condor has added support for PVM [2]. A prototype implementation of checkpoints for PVM jobs has been written, but it not yet deployed in the production release of Condor.

See `http://wwwbode.informatik.tu-muenchen.de/~stellner/CoCheck/` for more details.

Acknowledgements

Todd Tannenbaum added support for checkpointing signals, Jim Basney added the code to handle shared libraries, and Jim Pruyne wrote the Condor/PVM implementation. Miron Livny has been the guiding force and leader of the Condor team for the past decade. We are grateful to all of them for reviewing a draft of this update. Responsibility for any remaining errors of course remains with the authors.

References

[1] Todd Tannenbaum and Michael Litzkow, "Checkpointing and migration of Unix processes in the Condor distributed processing system," *Dr. Dobbs Journal,* pp 40-48, Feb. 1995.

[2] Al Geist, Adam Beguelin, Jack Dongarra, Weicheng Jiang, Robert Mancheck, and Vaidy Sunderam, *PVM: Parallel Virtual Machine,* Cambridge, MA: MIT Press, 1994. See also `http://www.epm.ornl.gov/pvm/`.

5.2 Jul, E., Levy, H., Hutchinson, N., and Black, A., Fine-Grained Mobility in the Emerald System

Jul, E., Levy, H., Hutchinson, N., and Black, A., "Fine-Grained Mobility in the Emerald System," *ACM Transactions on Computer Systems*, 6(1):109–133, February 1988.

Fine-Grained Mobility in the Emerald System*

Eric Jul, Henry Levy, Norman Hutchinson, and Andrew Black †

Abstract

Emerald is an object-based language and system designed for the construction of distributed programs. An explicit goal of Emerald is support for object mobility; objects in Emerald can freely move within the system to take advantage of distribution and dynamically changing environments. We say that Emerald has *fine-grained* mobility because Emerald objects can be small data objects as well as process objects. Fine-grained mobility allows us to apply mobility in new ways but presents implementation problems as well. This paper discusses the benefits of fine-grained mobility, the Emerald language and run-time mechanisms that support mobility, and techniques for implementing mobility that do not degrade the performance of local operations. Performance measurements of the current implementation are included.

1 Introduction

Process migration has been implemented or described as a goal of several distributed systems [24, 15, 20, 23, 8, 28, 10]. In these systems, entire address spaces are moved from node to node. For example, a process manager might initiate a move to share processor load more evenly, or users might initiate remote execution explicitly. In either case, the running process is typically ignorant of its location and unaffected by the move.

During the last three years, we have designed and implemented Emerald [5, 6], a distributed object-based language and system. A principal goal of Emerald is to experiment with the use of mobility in distributed programming. Mobility in the Emerald system differs from existing process migration schemes in two important respects. First, Emerald is object-based and the unit of distribution and mobility is the object. While some Emerald objects contain processes, others contain only data: arrays, records, and single integers are all objects. Thus, the unit of mobility can be much smaller than in process migration systems. Object mobility in Emerald subsumes both process migration and data transfer. Second, Emerald has language support for mobility. Not only does the Emerald language explicitly recognize the notions of location and mobility, but the design of conventional parts of the language (e.g., parameter passing) is affected by mobility.

The advantages of process migration, which have been noted in previous work, include:

1. *Load sharing* – By moving objects around the system, one can take advantage of lightly used processors.

2. *Communications performance* – Active objects that interact intensively can be moved to the same node to reduce the communications cost for the duration of their interaction.

3. *Availability* – Objects can be moved to different nodes to provide better failure coverage.

4. *Reconfiguration* – Objects can be moved following either a failure or a recovery, or prior to scheduled down-time.

5. *Utilizing special capabilities* – An object can move to take advantage of unique hardware or software capabilities on a particular node.

Along with these advantages, fine-grained mobility provides three additional benefits:

*This work was supported in part by the National Science Foundation under Grants No. MCS-8004111 and DCR-8420945, by Kbenhavns Universitet (the University of Copenhagen), Denmark under Grant J.nr. 574-2,2, by a Digital Equipment Corporation External Research Grant, and by an IBM Graduate Fellowship. This paper was first published in Transactions on Computer Systems, vol 6 no. 1 1988, pages 109-133. A preliminary version of the paper was presented at the 11th ACM Symposium on Operating Systems Principles, Proceedings of the 11th ACM Symposium on Operating System Principles, vol 21, no. 5 1987, pages 105-106.

†Authors' current addresses: Eric Jul, DIKU, Dept. of Computer Science, University of Copenhagen, Universitetsparken 1, DK-2100 Copenhagen, Denmark. Henry Levy, Dept. of Computer Science, University of Washington, Seattle, WA 98195. Norman Hutchinson, Dept. of Computer Science, University of British Columbia, 201-2366 Main Mall, Vancouver, B.C. V6T 1Z4. Andrew Black, Dept. of Computer Science and Engineering, Oregon Graduate Institute of Science and Technology, 20000 NW Walker Road, Beaverton, Oregon 97006.

1. *Data Movement* – Mobility provides a simple way for the programmer to move data from node to node without having to explicitly package data. No separate message passing or file transfer mechanism is required.

2. *Invocation Performance* – Mobility has the potential to improve the performance of remote invocation by moving parameter objects to the remote site for the duration of the invocation.

3. *Garbage Collection* – Mobility can help simplify distributed garbage collection by moving objects to sites where references exist [15, 29].

To our knowledge, the only other system that implements object mobility in a style similar to Emerald is a recent implementation of distributed Smalltalk [4].

In addition to mobility and distribution, Emerald is intended to provide efficient execution. We wanted to achieve performance competitive with standard procedural languages in the local case and standard remote procedure call systems in the remote case. These goals are not trivial in a location-independent object-based environment. To meet them, we have relied heavily on an appropriate choice of language semantics, a tight coupling between the compiler and run-time kernel, and careful attention to implementation.

Emerald is not intended to run in large, long-haul networks. We assume a local area network with a modest number of nodes (e.g., 100). In addition, we assume that nodes are homogeneous in the sense that they all run the same instruction set, and that they are trusted.

In this paper we concentrate primarily on the language and run-time mechanisms that support fine-grained mobility while retaining efficient intra-node operation. First, we present a brief overview of the Emerald language and system, and its mobility and location primitives. A more detailed description of object structure in Emerald can be found in [5], and of the type system in [6]. Second, we discuss the implementation of fine-grained mobility in Emerald and new problems that arise from providing such support. Third, we present measurements of the implementation and draw implications from the measurements and our design experience.

2 Overview of Emerald

As previously stated, an important goal of Emerald was explicit support for mobility. From a conceptual viewpoint, a more important goal was a single object model. Object-based systems typically lie at the ends of a spectrum: object-based languages such as Smalltalk [12] and CLU [22] provide small, local, data objects; object-based operating systems, like Hydra [30] and Clouds [1], provide large, active objects. Distributed systems such as Argus [21] and Eden [3] that support both kinds of object have a separate object definition mechanism for each. Choosing the right mechanism requires that the programmer know ahead of time all uses to which an object will be put; the alternative is to accept the inefficiency and inconvenience of using the "wrong" mechanism, or to reprogram the object later as needs change. For example, while programming a Collaborative Editing System in Argus, Greif et al. have observed that a designer can be forced to use a Guardian where a cluster might be more appropriate [13].

The motivation for two distinct definition mechanisms is the need for two distinct implementations. In distributed object-based systems such as Clouds and Eden, a *local* execution of the general invocation mechanism can take milliseconds or tens of milliseconds [26]. A more restrictive and efficient implementation is appropriate for objects that are known to be always local; for example, shared store can be used in preference to messages.

While we believe in the importance of multiple implementations, we do not believe that these need to be visible to the programmer. In Emerald, programmers use a single object definition mechanism with a single semantics for defining all objects. This includes small, local, data-only objects and active, mobile, distributed objects. However, the Emerald compiler is capable of analyzing the needs of each object and generating an appropriate implementation. For example, an array object whose use is entirely local to another object will be implemented differently from an array that is shared globally. The compiler produces different implementations from the same piece of code, depending on the context in which it is compiled [17].

The motivation for designing a new language, rather than applying these ideas to an existing language, is that the semantics of a language often preclude efficient implementation in either the local or remote case. In designing Emerald, we kept both implementations in mind. Moreover, Emerald's unique type system allows the programmer to state either nothing or a great deal about the use of a variable; in general, the more information the compiler has, the better the code that it generates.

We believe that the current Emerald implementation demonstrates the viability of this approach and meets our goal of local performance commensurate with procedural languages. Table 1 shows the performance of several local Emerald operations executed on a MicroVAX II[1]; more details on the compiler and its implementation can be found in [17]. The "resident global invocation" time is for a global object (i.e., one that can move around the network) when invoked by another object resident on the same node.

For comparison with procedural languages, a C procedure call takes 13.4 microseconds, while a Concurrent Euclid procedure call takes 16.4 microseconds. Concurrent Euclid is slower because, like Emerald, it must make explicit stack overflow checks on each call.

2.1 Emerald Objects

Each Emerald object has four components:

1. A unique network-wide name.
2. A representation, i.e., the data local to the object, which consists of primitive data and references to other objects.
3. A set of operations that can be invoked on the object.
4. An optional process.

Emerald objects that contain a process are active; objects without a process are passive data structures. Objects with processes make invocations on other objects, which in turn invoke other objects, and so on to any depth. As a consequence, a thread of control originating in one object may span other objects, both locally and on remote machines. Multiple threads of control may be active concurrently within a single object; synchronization is provided by monitors.

Figure 1 shows an example definition of an Emerald object – in this case a simple directory object called *aDirectory*. The representation of the object consists of an array, *a*, of directory elements. The object exports three operations: Add, Lookup, and Delete. The array *a* and the operations are defined within a monitor to guarantee exclusive access to the array.

2.2 Types in Emerald

The Emerald language supports the concept of *abstract type* [6]. The abstract type of an object defines its interface: the number of operations that it exports, their names, and the number and abstract types of the parameters to each operation. For example, consider the abstract type definition for *SimpleDirectoryType* below:

[1] MicroVAX is a trademark of Digital Equipment Corporation.

```
object aDirectory
  export Add, Lookup, Delete
  monitor
    const DirElement == record DirElement
      var name : String
      var obj : Any
    end DirElement
    const a == Array.of[DirElement].empty
    function Lookup[n : String] → [o : Any]
      var element : DirElement
      var i : Integer ← a.lowerbound
      loop
        exit when i > a.upperbound
        element ← a.getelement[i]
        if element.getname = n then
          o ← element.getobj
          return
        end if
        i ← i + 1
      end loop
      o ← nil
    end Lookup
    % Implementation of Add and Delete
  end monitor
end aDirectory
```

Figure 1: An Emerald Directory Object Definition

```
const SimpleDirType == type SimpleDirType
  operation Lookup[ String ] → [ Any ]
  operation Add[ String, Any ]
end SimpleDirType
```

This defines an abstract type with two operations, *Lookup* and *Add*. *Lookup* has an input parameter of abstract type **String** and returns an object of abstract type **Any**. We say that an object *conforms* to an abstract type if it implements at least the operations of that abstract type, and if the abstract types of the parameters conform in the proper way. When an object is assigned to a variable, the abstract type of that object must conform to the declared abstract type of the variable. All objects conform to type **Any**, since **Any** has no operation.

Abstract types permit new implementations of an object to be added to an executing system. To use a new object in place of another, the abstract type of the new object must conform to the required abstract type. For example, we could assign the object *aDirectory* in Figure 1 to a variable declared to have abstract type *SimpleDirType* because *aDirectory* conforms to *SimpleDirType*. Note that each object can implement a number of different abstract types, and an abstract type can be implemented by a number of different objects.

Emerald Operation	Example	Time/μs
primitive integer invocation	$i \leftarrow i + 23$	0.4
primitive real invocation	$x \leftarrow x + 23.0$	3.4
local invocation	localobject.no-op	16.6
resident global invocation	globalobject.no-op	19.4

Table 1: Timings of Local Emerald Invocations

Emerald has no class/instance hierarchy, in contrast to Smalltalk. Objects are not members of a class; conceptually, each object carries its own code. This distinction is important in a distributed environment where separating an object from its code would be costly. However, identically implemented Emerald objects on each node do share code. In the implementation, the code is stored in a *concrete type object*. Because concrete type objects are immutable, they can be freely copied. When an object is moved to another node, only its data is moved. If the object contains a process, part of that data will include the process' stack, but no code is transferred.

When a kernel receives an object, it determines whether a copy of the concrete type object implementing that object already exists locally; if it does not, the kernel obtains a copy of it by finding one on another node using the location algorithm (described in Section 3.2). Typically the concrete type will be available from the node that sent the object. When a concrete type object arrives, it is dynamically linked into the kernel – the compiler generates relocatable code and sufficient symbol table information to make such dynamic linking possible. This scheme makes it possible to add dynamically new concrete types that implement existing abstract types. Concrete type objects are kept on a node for as long as there are objects referencing them, after which they are garbage collected.

2.3 Primitives for Mobility

Object mobility in Emerald is provided by a small set of language primitives. An Emerald object can:

- *Locate* an object (e.g., "**locate** X" returns the node where X resides).
- *Move* an object to another node (e.g., "**move** X **to** Y" co-locates X with Y).
- *Fix* an object at a particular node (e.g., "**fix** X **at** Y").
- *Unfix* an object, making it mobile again following a fix (e.g., "**unfix** X").
- *Refix* an object, atomically performing an Unfix, Move, and Fix at a new node (e.g., "**refix** X **at** Z").

The move primitive is actually a hint; the kernel is not obliged to perform the move and the object is not obliged to remain at the destination site. Fix and refix have stronger semantics; if the primitives succeed the object will stay at the destination until it is explicitly unfixed.

Central to these primitives is the concept of location, which is encapsulated in a ***node*** object. A node object is an abstraction of a physical machine. Location may be specified by naming either a node object or any other object. If the programmer specifies a non-node object, the location implied is the node on which that object resides. These concepts are similar to the location dependent primitives in Eden [7].

A crucial issue when moving objects containing references is deciding how much to move [25]. An object is part of a graph of references, and one could move a single object, several levels of objects, or the entire graph. The simplest approach — moving the specified object alone — may be inappropriate. Depending on how the object is implemented, invocations of the moved object may require remote references that would have been avoided if other related objects had been moved as well.

The Emerald programmer may wish to specify explicitly which objects move together. For this purpose, the Emerald language allows the programmer to ***attach*** objects to other objects. When a variable is declared, the programmer can specify the variable to be an "attached variable".

For example, in the Emerald mail system, mail messages have four fields: a sender, an array of destination mailboxes, a subject line and a text string. It makes sense for the array of destination mailboxes to be attached to the mail message, and this could be specified as:

attached var *ToList* : **Array**.*of*[*Mailbox*]

When the mail message is moved, the array pointed to at that time by *ToList* is moved with it. This may effect the performance of invocations on *ToList*, but not their semantics.

Attachment is transitive: any object attached to *a* will also be moved. For example, linked structures may be moved as a whole by attaching the link fields. Attachment is not symmetric; the object named by *a* can itself be moved, perhaps before it is invoked, and no attempt will be made to move *directory* with it.

2.4 Parameter Passing

An important issue in the design of distributed, object-based systems (as well as remote procedure call systems) is the choice of parameter passing semantics. In an object-based system, all variables refer to other objects. The natural parameter passing method is therefore call-by-object-reference, where a reference to the argument object is passed. This is in fact the semantics chosen by CLU (where it is called *call by sharing*) [22] and Smalltalk [12].

In a distributed object-oriented system, the desire to treat local and remote operations identically leads one to use the same semantics. However, such a choice could cause serious performance problems: on a remote invocation, access by the remote operation to an argument is likely to cause an additional remote invocation. For this reason, systems such as Argus have required that arguments to remote calls be passed by value, not by object-reference [14]. Similarly, remote procedure call systems require call-by-value since addresses are context-dependent and have no meaning in the remote environment.

The Emerald language uses call-by-object-reference parameter passing semantics for all invocations, local or remote. In both cases, the invoking code constructs an activation record that contains references to the argument objects. In the local case, the invoked object is called directly and receives a pointer to the activation record for the invocation. In the remote case, the activation record must be reconstructed on the remote system, but the basic operation and semantics are identical.

Because Emerald objects are mobile, it may be possible to avoid many remote references by moving argument objects to the site of a remote invocation. Whether this is worthwhile depends on (1) the size of an argument object, (2) other current or future invocations of the argument, (3) the number of invocations that will be issued by the remote object to the argument, and (4) the relative costs of mobility and local and remote invocation.

In the current Emerald prototype, arguments are moved in two cases. First, based on compile-time information, the Emerald compiler may decide to move an object along with an invocation. For example, small immutable objects are obvious candidates for moving because they can be copied cheaply. Obviously, it makes little sense to send a remote reference to a small string or integer. Second, the Emerald programmer may decide that an object should be moved based on knowledge about the application. To make this possible, Emerald provides a parameter passing mode that we call *call-by-move*. Call-by-move does not change the semantics, which is still call-by-object-reference, but at invocation time the argument object is relocated to the destination site. Following the call the argument object may either return to the source of the call or remain at the destination site (we call these two modes *call-by-visit* and *call-by-move*, respectively).

Call-by-move is a convenience and a performance optimization. Arguments could instead be moved by explicit move statements. However, providing call-by-move as a parameter passing mode allows packaging of the argument objects in the same network packet as the invocation message.

As an example, consider another mail system example. After composing a mail message (whose fields were described previously), the user invokes the message's *Deliver* operation:

```
operation Deliver
   var aMailbox : Mailbox
   if ToList.length = 1 then
      aMailbox ← Tolist.getelement[ ToList.lowerbound ]
      aMailbox.Deliver[ move self ]
   else
      var i : Integer ← ToList.lowerbound
      loop
         exit when i > ToList.upperbound
         aMailbox ← ToList.getelement[ i ]
         aMailbox.Deliver[ self ]
         i ← i + 1
      end loop
   end if
end Deliver
```

This operation delivers the message to all the mailboxes on the *ToList*. However, in the common case where there is only one destination, call-by-move is used to co-locate the mail message with the (single) destination mailbox.

2.5 Processes, Objects, and Mobility

An Emerald process is a thread of control that is initiated when an object with a process is created. A process can invoke operations on its object or on any object that it can reference. We think of a process

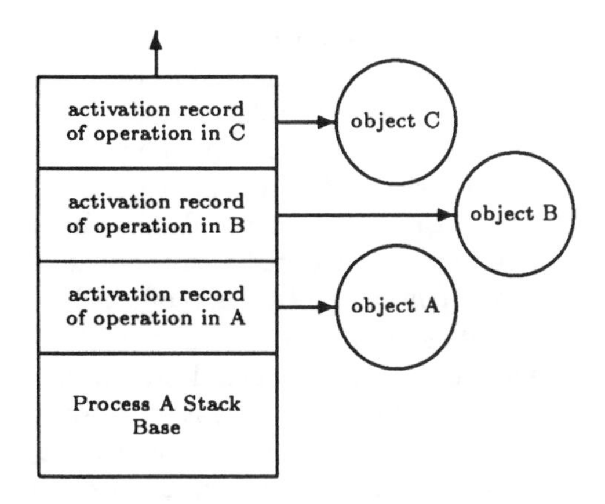

Figure 2: Process Stack and Activation Records

as being a stack of activation records, as shown in Figure 2. The thread of control of one object's process may pass through other objects; in the case of Figure 2, the process owned by object A invokes operations in objects A, B, and C.

One can think of remote invocations in several ways. In the traditional remote operation model [27], the sending process blocks and an existing remote process executes the operation, possibly returning a value to the caller, which then continues execution. In Emerald, when a remote invocation occurs, we think of the process moving to the destination node and invoking the object there. Or, alternatively, the new activation record moves to the destination node to become the base of a new segment of the process stack on that node. The invocation stack of a single Emerald process can therefore be distributed across several nodes.

Mobility presents a special problem to this process structure. For example, given the process activation stack in Figure 2, suppose that object B is moved to another node. In that case, the part of the thread that is executing in B must move along with B; that is, the activation record must move. Furthermore, when the operation in C terminates it must now return to a different node. If object C were to move to a different node from B, we would have three parts of the process stack on three different nodes. Invocation returns would propagate control back from node to node.

One could imagine a different scheme that left the stack intact, with invocations always returning to the node on which the root process resides; at that point the situation could be analyzed and control passed to the proper location. The problem with this design is that it leaves *residual dependencies*. In the situation where objects C and B have moved to different nodes, it should be possible for control to return from C to B even if A is temporarily unreachable. Depending on B's behavior, it may in fact be some time before a return to A or its node is actually required. Moving invocation frames along with the objects in which they execute ensures that execution can continue as long as possible, and removes the computational burden from nodes that do not need to be involved in a communication.

3 Implementing Mobility in Emerald

Adding process mobility to existing systems often proves to be a difficult task. One problem is extracting the entire state of a process, which may be distributed through numerous operating system data structures. Second, the process may have variables that directly index those operating system data structures, such as open file descriptors, window numbers, etc.

In a distributed object-based system, this problem may be somewhat simplified. Objects cleanly define the boundaries of all system entities. Furthermore, since all resources are objects, addressing is standardized and location-independent. All objects, whether user-implemented or kernel-implemented, are addressed indirectly using an object ID. Operations are performed through a standard invocation interface.

While distribution and mobility increase the generality of a system they often reduce its performance. Anyone building an object-based system must be sensitive to performance because of the generally poor performance of such systems. The implementation of mobility in Emerald involved tradeoffs between the performance of mobility and that of more fundamental mechanisms, such as local invocation. Where possible, we have made these tradeoffs in favor of the performance of frequent operations, and we would typically be willing to increase the complexity of mobility to save a microsecond or two on local invocation. Furthermore, it takes a hundred times longer to move an object than to perform a local invocation; adding 5 microseconds to the object move time makes little relative difference, while 5 microseconds is 25 percent of the local invocation time. The result of this philosophy is that, to a great extent, the existence of mobility and distribution in Emerald do not interfere with the performance of objects on a single node.

In the following sections, we describe some of the implementation of the Emerald kernel that is relevant to mobility, and some of the tradeoffs that we have made in this design.

3.1 Object Implementation and Addressing

To meet our goal of building a distributed object-based system with efficient local execution, the Emerald implementation relies heavily on shared memory. We have implemented a prototype of Emerald on top of DEC's Ultrix system (which is based on Unix 4.2BSD) running on five DEC MicroVAX II workstations[2]. The Emerald kernel and all Emerald objects on a single node execute within a single Ultrix address space. Emerald processes are lightweight threads scheduled within that address space. Protection among objects is guaranteed by the compiler both through type checking and through run-time checks inserted into the code. Objects that are resident on the same node address each other directly — an implementation style that has implications for mobility.

As previously stated, all objects are coded using a single object definition mechanism. However, based on its knowledge of an object's use, the compiler is free to choose an appropriate addressing mechanism, storage strategy, and invocation protocol [17]. The Emerald compiler uses three different styles of object implementation:

- A *global* object can be moved independently, can be referenced globally in the network, and can be invoked by objects not known at compile time. Global objects are heap allocated. An invocation of such an object may require a remote invocation. In Figure 1, the object *aDirectory* is implemented as a global object.

- A *local* object is completely contained within another object; that is, a reference to the local object is never exported outside the boundary of the enclosing object. Such objects cannot move independently; they always move along with their enclosing object. Local objects are heap allocated. An invocation is implemented by a local procedure call or inline code. The array *a* in Figure 1 is not used outside of the directory and can thus be implemented as a local object.

- A *direct* object is a local object whose data area is allocated directly in the representation of the enclosing object. Direct objects are used mainly for primitive built-in types, structures of primitive types, and other simple objects whose organization can be deduced at compile time. For example, all integers are direct objects.

[2]Unix is a trademark of AT&T. Ultrix is a trademark of Digital Equipment Corporation.

Figure 3 shows the various implementation and addressing options used by Emerald. Variable X names a global object and the value stored in X is the address of a local *object descriptor*. Each node contains an object descriptor for every global object for which references exist on that node. When the last reference to object m is deleted from node k, k's object descriptor for m can be garbage collected.

An object descriptor contains information about the state and location of a global object. The first word of the object descriptor identifies it as a descriptor and contains control bits indicating whether the object is local or global (the G bit) and whether or not the object is resident (the R bit). If the resident bit is set, the object descriptor contains the memory address of the object's data area; otherwise, the descriptor contains a forwarding address to the object as described in Section 3.2.

Variable Y in Figure 3 names a local object. The value stored in Y is the address of the object's data area. The first word of this data area, like the first word of an object descriptor, contains fields identifying the area and indicating that this is a local object, i.e., the data area acts as its own descriptor. Finally, variable Z names a direct object that was allocated within the variable itself.

Notice that within a single node, all objects can be addressed directly without kernel intervention. Emerald variables contain references that are location *dependent*, that is, they have meaning only within the context of a particular node. For invocation of global objects, compiled code first checks the *resident* bit to see if a local invocation can be performed directly. If the target object is not resident, the compiled code will trap to the kernel so that a remote invocation can be performed. In this way, global objects can be invoked locally in time comparable to a local procedure call.

3.2 Finding Objects

Since objects are allowed to move freely, it is not always possible to know the location of a given object, e.g., when invoking it. The run-time system must keep track of objects or at least be able to find them when needed. Keeping every node in the system up-to-date on the current location of every object is expensive and unnecessary. Instead, we use a scheme based on the concept of *forwarding addresses* as described by Fowler [11].

Each global object is assigned a unique network-wide *Object Identifier (OID)*, and each node has a hashed *access table* mapping OIDs to object descrip-

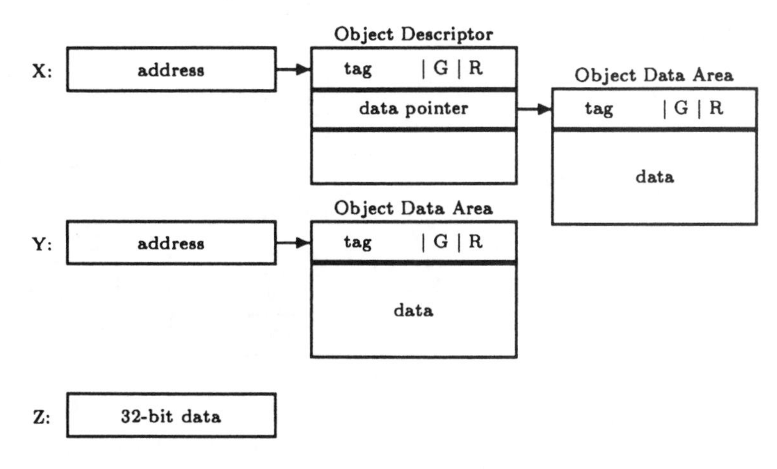

Figure 3: Emerald Addressing Structures

tors. The access table contains an entry for each local object for which a remote reference exists and each remote object for which a local reference exists.

As previously described, an object descriptor contains a *forwarding address* as well as the object's OID. A forwarding address is a tuple < *timestamp, node* > where the node is the last known location of the object and the timestamp specifies the age of the forwarding address. Fowler [11] has shown that it is sufficient to maintain the timestamp as a counter incremented every time the object moves. Given conflicting forwarding addresses for the same object, it is simple to determine which one is most recent. Every reference sent across a node boundary contains the OID of the referenced object and the latest available forwarding address. The receiving node may then update its forwarding address for the referenced object, if required.

If an object is moved from node A to node B, both A and B will update their forwarding addresses for the object. No action is taken to inform other nodes. Should node C try to invoke the object at A, A will forward the invocation message to B. When the invocation completes, B will send the reply to C with the new forwarding address piggybacked onto the reply message.

An alternative strategy, which we did not adopt, would be to keep track of all nodes that have references to a particular object. Should that object move, update messages could be sent to those nodes. However, these extra messages could significantly increase the cost of move and of passing references. For example, when an object reference is passed to a node for the first time, that node would have to register with the node responsible for the object. The DEMOS/MP system used a forwarding address update scheme, and updating forwarding addresses was shown to incur significant overhead [23]. In addition, sending update messages on every move will not avoid the need for invocation forwarding, since update messages do not arrive immediately at all destinations. Our scheme places the cost of forwarding address maintenance on the current users of a forwarding address.

When it is necessary to locate an object, for example when the *locate* primitive is used, we apply the following algorithm. If the kernel has a forwarding address for the object, it asks the specified node whether the object is resident there; if it is we are done. Otherwise, if that node has a newer forwarding address, then we start over using that forwarding address. However, if that node is unreachable or has no better information, we resort to a broadcast protocol.

The broadcast protocol is used whenever the previous step has failed to find the object. The searching kernel sends a first broadcast message to all other nodes seeking the location of the object. To reduce message traffic, only a kernel that has the specified object responds to the broadcast. If the searching kernel receives no response within a timelimit, it sends a second broadcast, this time requesting a positive or negative reply from all other nodes. All nodes not responding within a short time are sent a reliable, point-to-point message with the location request. If every node responds negatively we conclude that the object is unavailable.

When performing remote invocations, the invocation message is sent without locating the target object first. Only if there is a lost forwarding address somewhere along the path will the location algorithm be used. This optimizes for the common case where the object has not moved or where a valid forwarding address exists.

```
const simpleobject == object simpleobject
   monitor
      var myself : Any ← simpleobject
      var name : String ← "Emerald"
      var i : Integer ← 17
      operation GetMyName → [n : String]
         n ← name
      end GetMyName
         :
   end monitor
end simpleobject
```

Figure 4: Simple Emerald Object Definition

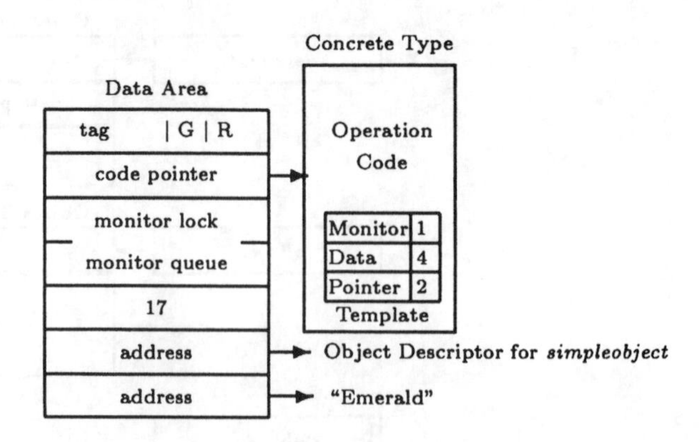

Figure 5: Data Area and Template Structure

3.3 Finding and Translating Pointers

The use of direct memory addresses in Emerald (as opposed to indirect references, such as those used in the standard Smalltalk implementation [12]) increases the performance of local invocations. Consequently, movement of an object involves finding and modifying all of the direct addresses, increasing the cost of mobility. We feel that this is reasonable, since motion is less frequent than invocation. This design places the price of mobility on those who use it.

Finding and translating references could be done in several ways. For example, a *tag* bit in each word could indicate whether or not the word contains an object reference. Smalltalk 80 uses such bits to distinguish integers from references, but using tags increases the overhead of arithmetic operations and complicates the implementation in general.

Instead, the Emerald compiler generates *templates* for object data areas describing the layout of the area. The template is stored with the code in the concrete type object that defines the object's operations. Each object data area contains a reference to the concrete type object so that the code and the template can be found given only the data area. In addition to their use for mobility, templates are used for garbage collection and debugging, as these tasks must also understand an object's data area.

As an example, consider the Emerald program shown in Figure 4 which defines a single object containing three variables inside a monitor. The variable *myself* contains a pointer to its own object descriptor. The variable *name* is initialized to point to a local string object. The variable *i* does not contain a pointer since integers are implemented as direct objects. The corresponding object data area and template are shown in Figure 5.

The data area for *simpleobject* contains:

- control information as described earlier.
- a pointer to the code for *simpleobject.*
- a lock for the monitor.
- the variable *i* allocated as 4 bytes of data.
- the variables *myself* and *name*, each allocated as a pointer to an object.

The template does not describe the first two items since every data area contains them. Each template entry contains a count of the number of items described and the types of the items (called *template-types*). Typical template-types are:

- *Pointer*, which is the address of an object; pointers must be translated if the object is moved.
- *Data*, which is direct data (e.g., integers) stored as a number of bytes; these are not translated.
- *MonitorLock*, which controls access to the object's monitor. Monitors are implemented as a Boolean and a queue of processes awaiting entry to the monitor. A monitor must be translated if the object is moved.

Attached objects, which must move along with an object being moved, are indicated simply by a bit in the template entry. The compiler contiguously allocates variables that can be described by identical template entries. Therefore, the average template contains only two or three entries.

In addition to data areas, the compiler must produce templates to describe activation records so that active invocations can be moved along with objects. A template for an activation record describes three things: the parameters to the operation, the local variables used by the operation, and the contents of the CPU registers.

To simplify activation record templates the Emerald compiler does not permit registers to change their template-type during an operation. A register that contains a pointer must contain a pointer for the lifetime of the invocation; however, the pointer register can point to different objects during its lifetime. This restriction is similar to the segregation of address and data registers in some architectures, but is more dynamic since the division is made for each specific operation. Without this restriction, we would need to have different templates at different points in an operation's execution – a design considered early in the project, but later abandoned as unnecessary.

3.4 Moving Objects

Using the addressing and implementation structure described above, the actual moving of an object is rather straightforward. Although some systems precopy objects to be moved for performance reasons [28], we do not believe this is necessary in the Emerald environment for several reasons. First, unlike process mobility systems, we do not copy entire address spaces. Second, many objects contain only a small amount of data. Third, even when an object with an active process is moved, we may not need to copy any code.

3.4.1 Moving Data Objects

Objects without active invocations are the simplest ones to move. For these, the Emerald kernel builds a message to be transmitted to the destination node. At the head of this message is the data area of the object to be moved. As we previously described, this data area is likely to contain pointers to both global and local objects. Following the data area is translation information to aid the destination kernel in mapping location-dependent addresses. For global object pointers, the kernel sends the OID, the forwarding address, and the address of the object's descriptor on the source node. For local objects, the data area is sent along with its address.

On receipt of this information, the destination kernel allocates space for the moved objects, copies the data areas into the newly allocated space, and builds a translation table that maps the original addresses into addresses in the newly allocated space. OIDs are used to locate object descriptors for existing global objects, or new object descriptors are created where necessary. The kernel then locates the template for each moved object, traverses its data area, and replaces any pointers with their corresponding addresses found in the translation table.

3.4.2 Moving Process Activation Records

As previously described in Section 2.5, when an object is moved the activation records for processes executing its operations must also move. This presents a particularly difficult problem: given an object to move, how do we know which activation records need to move with it? Finding the correct activation records requires a list of all active invocations for a particular object.

Several solutions are possible, but all have potentially serious performance implications. The simplest solution is to link each activation record to the object on each invocation and unlink it on invocation exit. Unfortunately, this would increase our invocation overhead by 50 percent in the current implementation. On the other hand, finding the invocations to move would require only a simple list traversal.

A second solution is to create the list only at move time. This would eliminate the invocation-time cost but would require a search of all activation records on the node. While we believe that mobility should not increase the cost of invocation, exhaustive search seems to be an unacceptable price to pay on every move.

We have therefore adopted an intermediate solution. We do maintain a list of all activation records executing in each object, as in the first solution above. However, on invocation the activation record is not actually linked into this structure. Instead, space is left for the links and the activation record is marked as "not linked", which is an inexpensive operation. When an Emerald process is preempted, its activation stack is searched for "not linked" activation records, and these are then linked to the object descriptors of their respective objects.

The search stops as soon as an activation record is found that has been linked previously. In this way, the work is only done at preemption time and its cost is related to *the difference* in stack depth between the start and end of the execution interval, not to the number of invocations performed.

An operation must still unlink its activation record when it terminates. Each return must check for a queued activation record and dequeue it before freeing the record. However, most returns will find a "not linked" activation record, in which case no work need be done.

Therefore, when an object moves we can find all activation records that must move with it merely by traversing the linked list associated with the object. These activation records are moved in a manner similar to moving data areas as described above.

If necessary, the activation records are removed

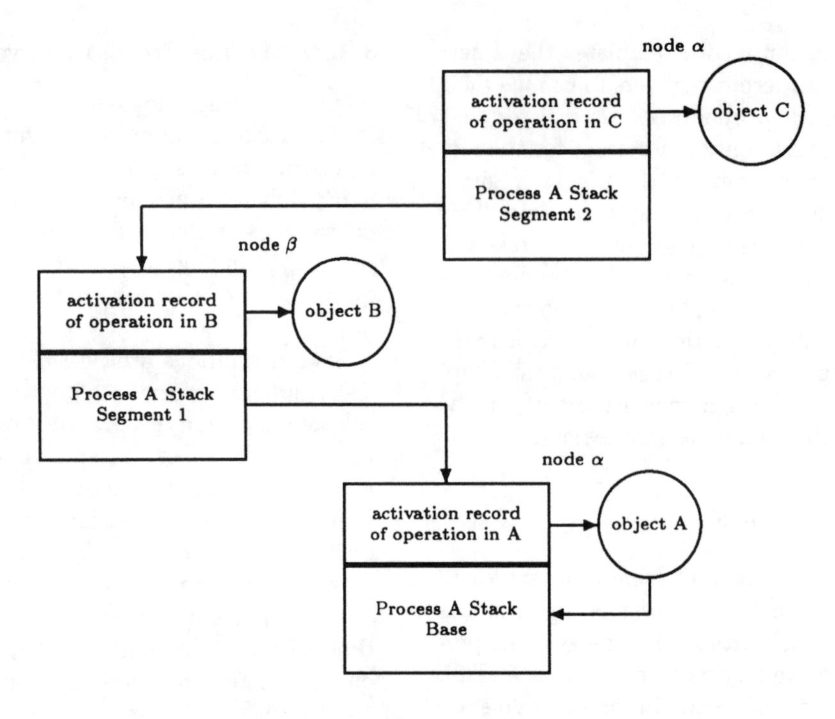

Figure 6: Process Stack after Object Move

from the stack containing them. This is accomplished by splitting the stack into (at most) three parts: the "bottom" part which remains on the source node, the "middle" part which is moved to the destination node, and the "top" part which is copied onto a new stack segment on the source node. The stack break points are found using the templates for the activation records. At each of the two stack breaks, invocation frames are modified to appear as if remote invocations had been performed instead of local invocations. Figure 6 shows the structure that would exist if object B from Figure 2 were moved from node α to node β.

3.4.3 Handling Processor Registers

An additional complexity in moving Emerald processes and activation records is the management of processor registers. The Emerald compiler attempts to optimize the addressing of objects by storing local variables in registers instead of in the activation record. In this way, some of the processor registers may contain machine-dependent pointers and these must be translated when the activation record moves.

Unfortunately, the registers for a given activation record are not kept in one place. Each invocation saves in its activation record a copy of registers that will be modified by that invocation. Referring back to Figure 2, suppose that the first invocation (of A) and the third invocation (of C) both use register 5. In this case, a copy of A's register 5 is saved in C's activation record, as it would be in any conventional stack-based language implementation.

If object A moves, its activation record will move with it. The stack will be segmented, and the rest of the stack will be left behind. Furthermore, the copy of A's register stored in C's activation record will be incorrect when C returns, because the data that it refers to will be at a different location on a different node.

To handle this situation, the kernel sends a copy of the registers used in an invocation along with the moving activation record. First, the kernel finds the template for the activation record in the concrete type object of the invoked object. Second, it determines which registers are used as pointers in an activation record by looking at its template. The templates for activation records have special entries for registers and for the area of the activation record where registers are saved. Third, the kernel scans the invocation stack, looking for the next activation record that has saved each of the registers. In this way, copies of the current values of the registers can be sent along with the record. On the destination node, the registers are modified using the translation table (as described in Section 3.4.1) and stored with the newly created stack segment.

For each stack segment of an Emerald process, there is a separate image of the registers. When an invocation return crosses a stack segment boundary, the registers used are those stored with the stack segment receiving control. These are the possibly translated values of registers that were computed when the stack was segmented.

3.5 Garbage Collection

As with any object-based system, Emerald must rely on garbage collection to recover memory occupied by objects that are no longer reachable. Furthermore, Emerald must deal with the problems of garbage collection in a distributed environment. While our garbage collector is not yet fully implemented, in this section we describe its general design.

The principal problem with distributed garbage collection is that object references can cross node boundaries. The system must ensure that it does not delete an object that can still be referenced. In a distributed system, a reference to an object could be on a different node than the object, on a node that is unavailable, or "on the wire" in a message. While Emerald has mobile objects, this presents no special difficulty; other distributed object-based systems may not have mobile *objects*, but they all have mobile *references*, which are the root of the problem. In fact, if an Emerald object moves, we know implicitly that it cannot be garbage, since either the object is actively executing or someone with a reference to that object must have requested the move. Furthermore, garbage collection is simplified by the presence of object descriptors. Each node retains a descriptor for every non-resident object that it has referenced since the last collection.

The Emerald garbage collection design calls for two collectors: a node-local collector that can be run at any time independently of other nodes, and a distributed collector that requires the nodes to cooperate to collect distributed garbage. Both are mark-and-sweep collectors modified to operate in parallel with executing Emerald processes.

We expect most garbage to consist of objects that are created and disposed of on a single node, with no reference ever leaving that node. To know which objects can be collected by the node-local collector, each object descriptor has a flag called the *RefGivenOut* bit. The kernel sets this bit in an object's descriptor whenever a reference to the object is passed to another node. The kernel also sets this bit in the descriptor for a moving object that has arrived at its destination, since the source node retains a reference to the object. When the node-local collector finds an object with the *RefGivenOut* bit set, it considers the object to be reachable. The node-local collector ignores every reference to a non-resident object.

Distributed collection is performed using a modified mark-and-sweep collection algorithm. In the conventional mark-and-sweep, all objects are initially marked as *white*, indicating that they are not yet known to be reachable. Then, objects that are known to be reachable, e.g., objects containing executable processes, are marked *gray*. A gray object is reachable but its references need to be scanned to mark gray all objects reachable from that object; once this is done the original object is marked *black*. When all gray objects have been scanned, the system consists of black objects that are reachable and white objects that are garbage and can be deleted.

To perform a distributed collection in Emerald, a collecting process is started on each node, and all global collectors proceed in parallel. All global objects, i.e., all objects with *RefGivenOut* set, are first marked as being unreachable as in the traditional mark-and-sweep scheme. Each global collector marks all of its explicitly reachable global objects gray. When attempting to scan a gray object, a global collector may find that the object resides on another node. In that case it sends a mark-gray message to the node where the object resides. The collector on the receiving node adds the object to its gray set and sends back an object-is-black message when the object has been traversed and marked. Upon receiving an object-is-black message, a collector removes the object from its gray set and marks it black. The collection is complete when all nodes have exhausted their gray sets.

To prevent an object from "outrunning" traverse-and-mark requests by moving often, objects are traversed and marked black when moved. This is done even for objects currently marked white since any moved object is a priori reachable – the object would eventually be marked anyway.

If a node is currently unavailable (e.g., has crashed) when a mark-gray message is sent to it then the reference is ignored for the moment. Eventually the only gray references left are to objects on unreachable nodes. At this point, the collectors exchange information about the remaining gray objects so that every collector knows which objects still need to be scanned.

When an unavailable node becomes available again, the collectors continue marking gray objects until either the collection is done or there is a gray reference to an object on an unavailable node. The collectors again exchange gray sets and wait for a node to become available. This process is repeated until the collection completes, at which point garbage objects and object descriptors can be collected. Note that it is not necessary for all nodes to be up simultaneously — it is only necessary for each node to be available long enough for the collection to make progress over time.

Finally, a major problem with the traditional mark-and-sweep scheme is that all other activity must be suspended while collecting. In a distributed system, this is obviously not acceptable. There have been several suggestions for making mark-and-sweep collectors operate in parallel with the garbage generating processes [9, 19], some of which have been implemented [18, 4]. Typically, parallel mark-and-sweep requires processes to cooperate with the collector by setting coloring bits of referenced objects when performing assignments.

Emerald avoids this extra work on assignment by using a scheme proposed by Hewitt [16]. At the start of the marking phase, each executable process is marked before being allowed to run again. Marking a process means marking the objects reachable from the activation records of the process and, transitively, any object reachable from such objects. After an individual process has been marked, it can proceed in parallel with the rest of the collection even though not all objects and processes have been marked. Should a process become executable (e.g., after waiting for entry to a monitor) then that process must be marked before being allowed to execute.

This scheme allows our collectors to proceed in parallel with executing processes but there is a high initial cost when making a process executable: all objects reachable from the process must be marked. To reduce the number of objects traversed before a process may be restarted, we have developed a ***faulting garbage collection*** scheme. Reachable global objects are marked but are not traversed. Instead, they are ***frozen*** by setting a bit in the object descriptor. When a process subsequently attempts to invoke a frozen object, it will fault to the kernel exactly as if the object had been remote. The kernel lets the collector traverse the object, unfreezes the object, and allows the invocation to continue. Thus, only the global objects immediately reachable from the process need be traversed. This replaces one large delay at the start of garbage collection by a number of smaller delays spread throughout a process' execution.

4 Performance

We measured the performance of Emerald's mobility primitives on four MicroVAX II workstations connected by a 10 megabit/second Ethernet. These primitives have been operational for only a short time and no effort has yet been made to optimize their implementation. In addition, we measured the impact of mobility on network message traffic using the Emerald mail system driven by a synthetic workload. The results of these measurements are reported in the following sections.

4.1 Emerald Mobility Primitives

Table 2 shows the elapsed time cost of various Emerald operations. The measured performance figures are averages of repeated measurements. For the simplest remote invocation, the time spent in the Emerald kernel is 3.4 milliseconds. For historical reasons, we currently use a set of network communications routines that provide reliable, flow-controlled message passing on top of UDP datagrams. These routines are slow: the time to transmit 128 bytes of data and receive a reply is about 24.5 milliseconds. Hence, the total elapsed time to send the invocation message and receive the reply is 27.9 milliseconds.

Table 3 shows the benefit of call-by-move for a simple argument object. The table compares the incremental cost of call-by-move and call-by-visit with the incremental cost of call-by-object-reference. The additional cost of call-by-move was 2 milliseconds while call-by-visit cost 6.4 milliseconds. These are computed by subtracting the time for a remote invocation with an argument reference that is local to the destination. The call-by-visit time includes sending the invocation message and the argument object, performing the remote invocation (which then invokes its argument), and returning the argument object with the reply. Had the argument been a reference to a remote object (i.e., had the object not been moved), the incremental cost would have been 30.8 milliseconds. These measurements are somewhat of a lower bound because the cost of moving an object depends on the complexity of the object and the types of objects it names.

Compared with the cost of a remote invocation, call-by-move and call-by-visit are worthwhile for even a single invocation of the argument object. As previously stated, the advantage of call-by-move depends on the size of the argument object, the number of invocations of the argument object, and the local and remote invocation costs. Emerald's fast local invocation time, about 20 *micro*seconds, easily recaptures the time for the move. Even with the current unoptimized implementation, call-by-move and call-by-visit would be worthwhile for a remote invocation cost of under 10 milliseconds.

Moving a simple data object, such as the object in Figure 4, takes about 12 milliseconds. This time is less than the round-trip message time because the reply messages are "piggybacked" on other messages (i.e.,

Operation Type	Time/ms
local invocation	0.019
kernel CPU time, remote invocation	3.4
elapsed time, remote invocation	27.9
remote invocation, local reference parameter	31.0
remote invocation, call-by-move parameter	33.0
remote invocation, call-by-visit parameter	37.4
remote invocation, remote reference parameter	61.8

Table 2: Remote Operation Timing

Parameter Passing Mode	Time/ms
call-by-move	2.0
call-by-visit	6.4
call-by-remote-reference	30.8

Table 3: Incremental Cost of Remote Invocation Parameters

each move does not require a unique reply). Moving an object with a process is more complex; as previously stated, while Emerald does not need to move an entire address space, it must send translation data so that the object can be linked into the address space on the destination node. The time to move a small process object with 6 variables is 40 milliseconds. In this case, the Emerald kernel constructs a message consisting of about 600 bytes of information, including object references, immediate data for replicated objects, a stack segment, and general process control information. The process control information and stack segment together consume about 180 bytes.

4.2 Message Traffic in the Emerald Mail System

The elapsed time benefit of call-by-move, as shown in Table 3, is due primarily to the reduction in network message traffic. We have measured the effect of this traffic reduction in the Emerald mail system, an experimental application modeled after the Eden mail system [2]. Mailboxes and mail messages are both implemented as Emerald objects. In contrast to traditional mail systems, a message addressed to multiple recipients is not copied into each mailbox. Rather, the single mail message is shared between the multiple mailboxes to which it is addressed.

In a workstation environment, we would expect each person's mailbox normally to remain on its owner's private workstation. Only when a person changes workstations or reads his mail from another workstation would the mailbox be moved. However, we expect mail messages to be more mobile. When a message is composed, it will be invoked heavily by the sender (in order to define the contents of its fields) and should reside on the sender's node. In section 2.4 we discussed how mail messages may utilize call-by-move to co-locate themselves with a single destination mailbox upon delivery. If there are multiple destinations it is reasonable for the message to stay at the sender's node, but when the message is read it may be profitable to co-locate the message with the reader's mailbox.

To measure the impact of mobility in the mail system, we have implemented two versions: one which does not use mobility, and one which uses mobility in an attempt to decrease message traffic. In the Emerald mail system, the reading of a mail message takes five invocations: one to get the mail message from a mailbox, and four to read the four fields. If the mail message is remote then reading the message will take four remote invocations. By moving the mail message, these four remote invocations are replaced by a move followed by four local invocations. However, additional effort may be required by other mailboxes to find the message once it has moved.

To facilitate comparison, a synthetic workload was used to drive each of the mail system implementations. Ten short messages (about one hundred bytes) and ten long messages (several thousand bytes) were sent from a user on each of four nodes to various combinations of users on other nodes; the recipients then read the mail that they received.

Table 4 shows some of the measurement data collected by the Emerald kernel. As the table shows, the use of mobility more than halved the number of remote invocations, reduced the number of network packets by 34 percent, and cut the total elapsed time by 22 percent. The number of network messages sent

	Without mobility	With mobility
Total Elapsed Time	71 sec.	55 sec.
Remote Invocations	1386	666
Network Messages Sent	2772	1312
Network Packets Sent	2940	1954
Total Bytes Transferred	568716	528696
Total Bytes Moved	0	382848

Table 4: Mail system traffic

is exactly twice the number of invocations; each invocation requires a send and a reply. The number of packets is slightly higher than the number of network messages because the long mail messages require two packets. Note that the number of packets required per invocation is higher with mobility because mobile mail messages cause subsequent message readers to follow forwarding addresses.

Moving the mail messages reduces the total number of bytes transferred only slightly — by seven percent. Although the same data must eventually arrive at the remote site, whether by remote invocation or by move, the per-byte overhead of move is slightly less than that of invocation. In applications where only a small portion of the data in an object is required at the remote site, invocation might still be more efficient than move.

Finally, it is interesting to note that the 22 percent execution time difference was achieved by simply adding the word "move" in two places in the application.

5 Summary

We have designed and implemented Emerald, an object-based language and system for distributed programming. Emerald is operational on a small network of VAX computers and has recently been ported to the SUN 3[3]. Several applications have been implemented including a hierarchical directory system, a replicated name server, a load sharing application, a shared appointment calendar system, and a mail system.

The goals of Emerald included:

- support for fine-grained object mobility,
- efficient local execution, and
- a single object model, suitable for programming both small, local data-only objects and active, mobile, distributed objects.

[3]SUN is a trademark of SUN Microsystems, Inc.

This paper has described the language features and run-time mechanisms that support fine-grained mobility. While *process* mobility (i.e., the movement of complete address spaces) has been previously demonstrated in distributed systems, we believe that *object* mobility, as implemented in Emerald, has additional benefits. Because the overhead of an Emerald object is commensurate with its complexity, mobility provides a relatively efficient way to transfer fine grained data from node to node.

The need for semantic support for mobility, distribution, and abstract types led us to design a new language, and language support is a crucial part of mobility in Emerald. While invocation is location-independent, language primitives can be used to find and manipulate the location of objects. The programmer can declare "attached" variables; the objects named by attached variables move along with the objects to which they are attached. More important, on remote invocations a parameter passing mode called call-by-move permits an invocation's argument object to be moved along with the invocation request. Our measurements demonstrate the potential of this facility to improve remote invocation performance while retaining the advantages of call-by-reference semantics.

Implementing fine-grained mobility while minimizing its impact on local performance presents significant problems. In Emerald, all objects on a node share a single address space and objects are addressed directly. Invocations are implemented through procedure call or in-line code where possible. The result is that pointers must be translated when an object is moved. Addresses can appear in an object's representation, in activation records, and in registers. The Emerald run-time system relies on compiler-produced templates to describe the format of these structures. A combination of compiled invocation code and run-time support is responsible for maintaining data structures linking activation records to the objects they invoke. A lazy evaluation of this structure helps to reduce the cost of its maintenance.

Through the use of language support and a tightly-coupled compiler and kernel, we believe that our design has been successful in providing generalized mobility without much degradation of local performance.

6 Acknowledgements

We would like to thank Edward Lazowska and Richard Pattis for extensive reviews of early versions of this paper. We also thank Brian Bershad, Carl Binding, Kevin Jeffay, Rajendra Raj, and the referees for their helpful comments.

7 1998 Note

An earlier version of this paper was presented at the 11th ACM Symposium on Operating Systems Principles in Austin. Emerald represents the first language and system that implements type conformity and full, unrestrained, on-the-fly object mobility. Besides object mobility, processes executing native machine code can also be moved on the fly, but only between computers that have the same architecture. This restriction was later lifted when full heteorogeneous mobility was implemented; a paper describing this was presented at the 15th ACM Symposium on Operating System Principles in Copper Mountain Resort (Proceedings of the 15th ACM Symposium on Operating System Principles, vol 29, no. 5 1995, pages 68-78. Further information concerning Emerald is available at the Emerald web site: http://www.cs.ubc.ca/nest/dsg/emerald.html.

References

[1] James E. Allchin and Martin S. McKendry. Synchronization and recovery of actions. In *Proceedings of the 2nd Annual Symposium on Principles of Distributed Computing*, pages 31–44, August 1983.

[2] Guy T. Almes, Andrew P. Black, Carl Bunje, and Douglas Wiebe. Edmas: A locally distributed mail system. In *Proceedings of the Seventh International Conference on Software Engineering*, pages 56–66, Orlando, Florida, March 1984.

[3] Guy T. Almes, Andrew P. Black, Edward D. Lazowska, and Jerre D. Noe. The Eden System: A Technical Review. *IEEE Transactions on Software Engineering*, SE-11(1):43–59, January 1985.

[4] John K. Bennett. Distributed Smalltalk. In *Proceedings of the 2nd Conference on Object-Oriented Programming Systems, Languages, and Applications*, October 1987.

[5] Andrew Black, Norman Hutchinson, Eric Jul, and Henry Levy. Object structure in the Emerald system. In *Proceedings of the Conference on Object-Oriented Programming Systems, Languages, and Applications*, pages 78–86, October 1986.

[6] Andrew Black, Norman Hutchinson, Eric Jul, Henry Levy, and Larry Carter. Distribution and abstract types in Emerald. *IEEE Transactions on Software Engineering*, 13(1), January 1987.

[7] Andrew P. Black. Supporting distributed applications: Experience with Eden. In *Proceedings of the Tenth ACM Symposium on Operating System Principles*, pages 181–93. ACM, December 1985.

[8] David A. Butterfield and Gerald J. Popek. Network tasking in the Locus distributed Unix system. In *USENIX Summer 1984 Conference Proceedings*, pages 62–71, 1984.

[9] Edsger W. Dijkstra, Leslie Lamport, A. J. Martin, C. S. Scholten, and E. F. M. Steffens. On-the-fly garbage collection: An exercise in cooperation. *Communications of the ACM*, 21(11):966–975, November 1978.

[10] Fred Douglis. Process migration in the sprite operating system. Technical Report UCB/CSD 87/343, Computer Science Division, University of California, Berkeley, February 1987.

[11] Robert J. Fowler. *Decentralized Object Finding Using Forwarding Addresses*. PhD thesis, University of Washington, December 1985. Department of Computer Science technical report 85-12-1.

[12] Adele Goldberg and David Robson. *Smalltalk-80: The Language and Its Implementation*. Addison-Wesley Publishing Company, Reading, Massachusetts, 1983.

[13] Irene Greif, Robert Seliger, and William Weihl. Atomic data abstractions in a distributed collaborative editing system. In *Proceedings of the Thirteenth Symposium on Principles of Distributed Computing*, January 1986.

[14] M. Herlihy and B. Liskov. A value transmission method for abstract data types. *ACM Transactions on Programming Languages and Systems*, 4(4):527–551, October 1982.

[15] Carl Hewitt. The Apiary network architecture for knowledgeable systems. In *Conference Record of the 1980 Lisp Conference*, pages 107–118, Palo Alto, California, August 1980. Stanford University.

[16] Carl Hewitt and Henry Baker. Actors and continuous functionals. In *IFIP Working Conference on Formal Description of Programming Concepts*, pages 16.1 - 16.21, August 1977.

[17] Norman C. Hutchinson. *Emerald: An Object-Based Language for Distributed Programming.* PhD thesis, University of Washington, January 1987. Department of Computer Science technical report 87-01-01.

[18] Robert J. Chansler Jr. *Coupling in Systems with Many Processors.* PhD thesis, Department of Computer Science, Carnegie Mellon University, August 1982.

[19] H. T. Kung and S. W. Song. An efficient parallel garbage collection system and its correctness proof. In *Proceedings of the Eighth Annual Symposium on the Foundations of Computer Science*, pages 120–131, October 1977.

[20] Edward D. Lazowska, Henry M. Levy, Guy T. Almes, Michael J. Fischer, Robert J. Fowler, and Stephen C. Vestal. The architecture of the Eden system. In *Proceedings of the 8th Symposium on Operating Systems Principles*, pages 148–159, December 1981.

[21] Barbara Liskov. Overview of the argus language and system. Programming Methodology Group Memo 40, M.I.T. Laboratory for Computer Science, February 1984.

[22] Barbara Liskov, Russ Atkinson, Toby Bloom, Eliot Moss, Craig Schaffert, Bob Scheifler, and Alan Snyder. CLU reference manual. Technical Report MIT/LCS/TR-225, MIT Laboratory for Computer Science, October 1979.

[23] Michael L. Powell and Barton P. Miller. Process migration in DEMOS/MP. In *Proceedings of the Ninth ACM Symposium on Operating Systems Principles*, pages 110–119. ACM/SIGOPS, October 1983.

[24] Richard F. Rashid and George G. Robertson. Accent: A communication oriented network operating system kernel. In *Proceedings of the Eighth Symposium on Operating System Principles*, pages 64–75, December 1981.

[25] Karen R. Sollins. Copying complex structures in a distributed system. Master's thesis, MIT, May 1979. MIT/LCS/TR-219.

[26] Eugene H. Spafford. *Kernel Structures for a Distributed Operating System.* PhD thesis, School of Information and Computer Science, Georgia Institute of Technology, May 1986. Also Georgia Institute of Technology Technical Report GIT-ICS-86/16.

[27] Alfred Z. Spector. Performing remote operations efficiently on a local computer network. *CACM*, 25(4):246–260, April 1982.

[28] Marvin M. Theimer, Keith A. Lantz, and David R. Cheriton. Preemptable remote execution facilities for the V-system. In *Proceedings of the Tenth ACM Symposium on Operating Systems Principles*, pages 2–12. ACM/SIGOPS, December 1985.

[29] Stephen Vestal. *Garbage Collection: An Exercise in Distributed, Fault-Tolerant Programming.* PhD thesis, University of Washington, January 1987. Department of Computer Science technical report 87-01-03.

[30] William A. Wulf, Roy Levin, and Samuel P. Harbison. *HYDRA/C.mmp: An Experimental Computer System.* McGraw-Hill Book Company, 1981.

5.3 Smith, P. and Hutchinson, N.C., Heterogeneous Process Migration: The Tui System

Smith, P. and Hutchinson, N.C., "Heterogeneous Process Migration: The Tui System," *Software—Practice and Experience*, 28(6):611–639, May 1998.

Heterogeneous Process Migration: The Tui System

PETER SMITH AND NORMAN C. HUTCHINSON
Department of Computer Science, University of British Columbia, Vancouver, B.C., Canada V6T 1Z4
(email: norm@cs.ubc.ca)

SUMMARY

Heterogeneous process migration is a technique whereby an active process is moved from one machine to another. It must then continue normal execution and communication. The source and destination processors can have a different architecture, that is, different instruction sets and data formats. Because of this heterogeneity, the entire process memory image must be translated during the migration. Tui is a migration system that is able to translate the memory image of a program (written in ANSI-C) between four common architectures (m68000, SPARC, i486 and PowerPC). This requires detailed knowledge of all data types and variables used with the program. This is not always possible in non-type-safe (but popular) languages such as ANSI-C, Pascal and Fortran. The important features of the Tui algorithm are discussed in great detail. This includes the method by which a program's entire set of data values can be located, and eventually reconstructed on the target processor. Performance figures demonstrating the viability of using Tui to migrate real applications are given. ©1998 John Wiley & Sons, Ltd.

KEY WORDS: process migration; heterogeneous; type-safety

INTRODUCTION

What is heterogeneous process migration?

Process migration can be defined as the ability to move a currently executing process between different processors which are connected only by a network (that is, not using locally shared memory). The operating system of the originating machine must package the entire state of the process so that the destination machine may continue its execution. The process should not normally be concerned by any changes in its environment, other than in obtaining better performance.

Research into the field of process migration has concentrated on efficient exchange of the state information. For example, moving the memory pages of a process from the source machine to the destination, correctly capturing and restoring the state of the process (such as register contents), and ensuring that the communication links to and from the process are maintained. Careful design of an operating system's IPC mechanism can ease the migration of a process.

Most process migration systems make the assumption that the source and destination hosts have the same architecture. That is, their CPUs understand the same instruction set, and their operating systems have the same set of system calls and the same memory conventions. This allows state information to be copied verbatim between the hosts, so that no changes need to be made to the memory image.

Received 16 April 1997
Revised 10 December 1997
Accepted 15 December 1997

CCC 0038–0644/98/060611–29 $17·50

Heterogeneous process migration removes this assumption, allowing the source and destination hosts to differ in architecture. In addition to the homogeneous migration issues, the mechanism must translate the entire state of the process so it may be understood by the destination machine. This requires knowledge of the type and location of all data values (in global variables, stack frames and on the heap).

This paper examines an experimental heterogeneous migration system known as *Tui*. An implementation has revealed the issues involved in translating the data component of a migrating process. Tui does not address the issues normally associated with homogeneous migration, nor does it address the translation of a program's instructions between different architectures.

MOTIVATIONS

The traditional reasons for using process migration have been identified[1] as:

(a) *Load sharing among a pool of processors*. For a process to obtain as much CPU time as possible, it must be executed on the processor that will provide the most instructions and I/O operations in the smallest amount of time. Often this will mean that the fastest processors as well as those executing a small number of jobs will be the most attractive. Migration allows a process to take advantage of underutilized resources in the system, by moving it to a suitable machine.

(b) *Improving communication performance*. If a process requires frequent communication with other processes, the cost of this communication can be reduced by bringing the processes closer together. This is done by moving one of the communicating partners to the same CPU as the other (or perhaps to a nearby CPU).

(c) *Availability*. As machines in the network become unavailable, users would like their jobs to continue functioning correctly. Processes should be moved away from machines that are expected to be removed from service. In most situations, the loss of a process is simply an annoyance, but at other times it can be disastrous (such as an air traffic control system).

(d) *Reconfiguration*. While administering a network of computers, it is often necessary to move services from one place to another (for example, a name server). It is undesirable to halt the system for a large amount of time in order to move a service. A transparent migration system will make changes of this kind be unnoticeable.

(e) *Utilizing special capabilities*. If a process will benefit from the special capabilities of a particular machine, it should be executed on that machine. For example, a mathematics program could benefit from the use of a special math coprocessor, or an array of processors in a supercomputer. Without some type of migration system, the user will be required to make their own decision of where to execute a process, without the ability to change the location during the lifetime of the process. Often users will not even be aware of their program's special needs.

Although process migration has successfully been implemented in several experimental operating systems, it has not become widely accepted. One reason is that the mainstream platforms (such as MSDOS, MS Windows and most variants of Unix), do not have sufficient operating system support for migration. Secondly, the benefits of using process migration are generally not great enough to justify the cost, i.e. moving a process to another machine may be more costly than not moving it.

Recently, two new areas of computing have created new motivations for the use of process migration. Both these issues, *mobile computing* and *wide area computing,* will now be discussed in more detail. In both cases, heterogeneity plays a significant role.

Mobile computing

Mobile computing is a term used to describe the use of small personal computers that can easily be carried by a person, for example, a laptop or a hand-held computer. To make full use of these systems, the user needs to be able to communicate with larger machines without being physically connected to them, normally done via wireless LANs or cellular telephones.

It has been proposed[2] that process migration is important in this area. For example, a user may activate a program on their laptop, but in order to save battery power or to speed up processing, may later choose to transfer the running process onto a larger compute server. The process would be returned to the smaller machine to display results.

These concepts can be extended to allow a program to move between workstations as its owner moves. A person may be using a home computer, with a large number of windows on their screen. By remotely connecting to the computers at their place of work, they will be able to continue executing those programs in their office. If they choose to move between offices, the window system (and programs) could potentially follow them.

Wide area computing

For a computer to be part of the Internet, it must understand the internet communication protocols. Since there are no constraints on other software, such as operating systems and programming languages, an enormous amount of heterogeneity exists.

The one limitation of global computing which will never be resolved is the propagation delay that is suffered over wide area networks. At best, data can only be transmitted at the speed of light, causing noticeable delays. If a program makes frequent use of remote data, its performance will suffer.

Process migration can help alleviate this problem by moving the program closer to the data, rather than moving the data to the program.[3] Typically, a program would start executing on the user's local machine. If it later makes frequent accesses to remote data, the migration system will reduce the delay by moving the process to a machine that is physically closer to the data. This makes the most sense in the case where the program is smaller than the data.

Wide area processing is a topic that has already been addressed in the Java[4] and Telescript[5] languages. Java is most commonly used for transmission of programs using the World Wide Web. Although it supports remote method invocation, it does not currently support migration of active code. On the other hand, Telescript allows migration, as its primary purpose is for 'agent' programs to move between sites. Because of migration, a Telescript program may complete its tasks while minimizing long distance communication costs.

HETEROGENEOUS MIGRATION AND THE TUI SYSTEM

Existing systems

Before discussing the purpose of this research, it is necessary to look at the various classes of heterogeneous migration or mobility systems already in existence. The discussion focusses on the unit of information being migrated and describes how that information can be moved. Further references are given in a later section.

Heterogeneous migration systems can be classified into the following categories:

1. *Passive object.* The process (or object) contains only passive data. There is no executable code to be moved. This situation requires that data can be converted from the source machine's format to that of the destination machine.
2. *Active object, migrate when inactive.* The process has executable code as well as data. Migration may only occur when the code is not active. For example, in an object based system, objects will remain inactive unless an outside agent requests some action. Assuming that migration only occurs during these idle periods, moving a process is simply a matter of translating data. It is assumed that the executable code is available on the destination machine.
3. *Active object, interpreted code.* If a process is currently executing code by using an interpreter, moving the process involves translating the state of the interpreter and all the data values it may access. If these values (that is, variables, parameters, temporaries and other miscellaneous values on the call stack) are stored in a machine independent fashion, then migration is straightforward.
4. *Active object, native code.* If the active program is compiled into native machine code, then fetching the active state is more difficult. Each machine has its own method of storing a program's values. Differences are obvious in the layout of each stack frame, the usage of registers and the structure of the executable code.

Other issues that should be considered when designing a heterogeneous migration system include:

1. *Process originated migration.* The process being migrated makes the decision of when to migrate, rather than leaving the choice to an external agent such as the operating system. This ensures that the process is in a well known state (for example, during a call to a 'migrate now' procedure), rather than in relatively random place within the code. Migrating in a known state will reduce the amount of information to be migrated, and simplify its retrieval.
2. *Additional code within the migrating process.* The task of collecting data values from a process, and then restoring them, can often be simplified by adding code to the migratible program (explicitly by the programmer, or implicitly by the compiler). This can be in the form of a special library that must be linked with the program, or perhaps in the form of extra code at the beginning and end of each procedure. The problem of adding this code is the overhead of the additional execution time.
3. *Type-safety.* All existing heterogeneous migration systems known to the authors require that the migratible program be implemented in either a totally type-safe language or in a type-safe subset of a language. The migration algorithm requires complete knowledge of type information, usually generated by a compiler, to correctly marshall data for the destination machine. If the type data is inconsistent due to deficiencies in the implementation language, migration become more difficult or even impossible.

The purpose of Tui

The approach taken by the Tui system focusses on supplying a migration mechanism suitable for general purpose use. By far the majority of existing software, and programmer experience, is in traditional (and non-type-safe) languages such as C, Pascal, COBOL and Fortran. Having the ability to migrate programs written in more common languages will make migration much more widely available.

For these languages, the data conversion component of the migration algorithm becomes more complex. It must allow for difficulties such as the misuse of pointers, type casting and lack of explicit type information. The less type-safe the language, the more difficult it becomes to locate and assign a type to the data. These problems do not appear in type-safe languages.

Although it is possible to say that non-type-safe programming languages tend to generate non-migratible programs, it is useful to approach each problem on an individual basis. For example, a program written using C may be non-migratible due to the way that one small part of the program has been written. Rewriting this section of code in a different way will ensure that migration is possible. Alternatively, the language compiler and run-time system could generate extra type information to clarify the type of a piece of data.

The following example of C code demonstrates this:

```
main()
{
        union {
                int a;
                float b;
        } u;

        u.a = 1;
        u.b = 23.45;
}
```

Upon migrating this program, the migration system must be aware of the most recent assignment to the union. Either the integer 1 or the floating point number 23.45 is translated, but since they share the same memory location, the migration system does not know how to interpret the data. In this case, several solutions are possible:

(a) Modify the compiler to maintain a 'union tag' that records which element of the union was most recently accessed.
(b) Internally convert unions into structs, so elements have distinct memory locations.
(c) The programmer must refrain from using the union type.

The aim of Tui has been to discover and attempt to solve the problems that make common languages unattractive for heterogeneous migration. The following goals have been followed as closely as possible:

1. To provide a general heterogeneous migration package capable of functioning on a wide range of common operating systems, CPU types, and programming languages. The major limitation being that only operating systems that already supply homogeneous migration will allow totally correct migration. Other systems (such as UNIX) will have limited functionality.
2. To minimize the run time overhead of the process being migrated. It is preferable that the additional overhead due to migration is limited to compile time and migration time, rather than reducing the efficiency of the program during its normal execution period. However, it may be considered worthwhile to sacrifice a small amount of program efficiency if it allows a program to migrate, even though it was previously considered to be non-migratible.

3. To not restrict the implementation language, unless totally necessary. In the previous example of the 'union' declaration in ANSI C, converting the union definition to a similar struct definition is not permitted by the ANSI standard. However, performing this conversion will allow a much larger number of existing programs to be migratible.
4. To not require the user to write extra code or directives to help the migration system. It is considered undesirable to ask the user to register the data types and values that need to be migrated. The determination of this information should be done automatically.
5. To not require the process to initiate the migration. As well as reducing the execution overhead, this also allows an external agent (for example, the operating system) to request that migration take place at any time. However, allowing the compiler to suggest suitable places to preempt the process will reduce the complexity of the migration algorithm.
6. To be as efficient as possible so that the migration cost is considered to be negligible.

Ideally, the final migration algorithm, in conjunction with the language compiler, must be able to successfully migrate a process with no extra intervention from the programmer. That is, any existing software should be migratible without the need to alter the source code. If this is not possible, the compiler or migration system must warn the programmer of features in the program that are not migratible. As a last resort, a written document will describe any non-migratible language features that are not detectable by the compiler.

The eventual aim is to have a complete system where the programmer can take their existing programs and use Tui to migrate their software, or use Tui's suggestions to improve its migratibility.

THE TUI MIGRATION ALGORITHM

In its current form, the Tui Heterogeneous Process Migration System is able to migrate *type-safe* ANSI-C programs between four different architectures: Solaris executing on a SPARC processor (i.e. Sun 4), SunOS on an m68020 (i.e. Sun 3), Linux on an i486 and AIX on a PowerPC. A program is considered to be type-safe if it is possible to uniquely determine the type of each data value within the program. Some work has been done to relax the type-safety restriction, although this is mostly left as future research.

The Tui software has existed in both a prototype and final form. It is important to note that several lessons were learned from the prototype, and led to the creation of the final implementation. The prototype version made use of a garbage collection style algorithm for locating blocks of data to be migrated. Practical studies and performance tests indicated that this method was not sufficient in many situations, and hence motivated a second version.

This section gives a complete description of the revised Tui algorithm, with focus placed on the interesting features. First, an overview of the algorithm is given, with a details of how the four major components interact. Next, each of these components is described in greater detail. Even though this section only discusses the second version, a comparison between the two systems will be made in a later section.

Overview of Tui

Figure 1 shows how a process is migrated within the Tui environment. The following sequence of steps must occur for a program to be compiled, executed on the source machine, then migrated to a destination machine of a different architecture:

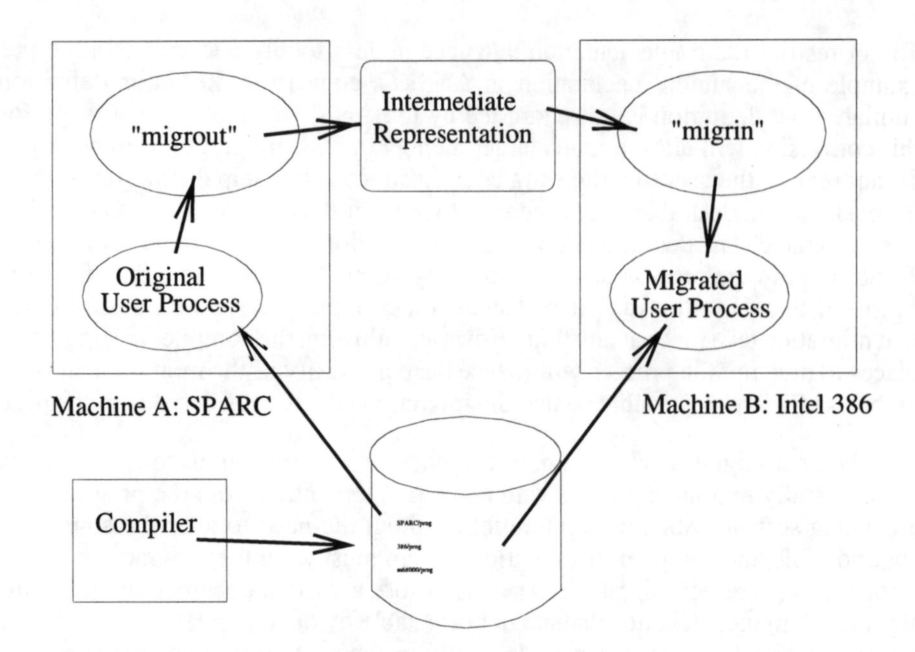

Figure 1. The Tui migration system

1. A program (written in ANSI-C) is compiled, once for each architecture. A modified version of the Amsterdam Compiler Kit (ACK)[6] is able to produce binaries for each of the four machine types supported by Tui.
2. The program is executed on the source machine, in the standard way (such as from the command line).
3. When the process has been selected for migration, the `migrout` program is called upon to checkpoint that process. Given the Process ID and the name of the executable file (containing type information), `migrout` will fetch the memory of the process and scan the global variables, stack and heap to locate all data values. Finally, all these values are converted into an intermediate form and sent to the target machine.
4. On the destination machine, the `migrin` program takes the intermediate representation and creates a new process. It is assumed that the program has been compiled for the target architecture so that the complete text segment, and type information for the data segment is available. After reconstructing the global variables, heap and stack, the process is restarted from the same point of execution as when it was checkpointed.

To make migration in Tui useful, an ANSI-C run time environment exists. Since each of the four architectures runs a different version of Unix, this library hides any inconsistencies. It was not possible to use the standard set of libraries, as Tui requires that processes have the same view of the operating system on both the source and destination machines. The Tui ANSI-C library operates by directly accessing the machine's system calls. This library has not been modified in any way that would slow down the execution of a program, other than what was needed to make the code type-safe.

Most variants of Unix do not allow migration, so movement of communication links and files (other than stdin and stdout) is not easy. However, a simple remote file server that allows migratible clients has been constructed.

The following sections describe the compiler and the executable files it produces, the `migrout` program, the `migrin` program, and the intermediate file format.

Compiler requirements and changes

To create programs that can be migrated by Tui, the compiler must ensure that sufficient type and location information is available to the other components of the system (`migrin` and `migrout`). Also, it must avoid generating code that is inherently non-migratible.

There were two main criteria for choosing a suitable compilation system. First, the compiler must support a wide range of target architectures, and hopefully more than one source language. Secondly, the entire source code for the compiler, assembler and linker had to be available (for all architectures), so that modifications to their output could be made.

Three different compilers were considered. The `gcc` compiler[7] was the obvious choice as it can generate code for most common architectures. However, modifying the compiler and its related tools was considered too difficult due to the complexity of the source code. The `lcc` compiler[8] was considered, due to its wide range of target architectures and its ease of modification. However, it became obvious that important changes had to be made to the assembler and linker, which were not supplied as part of the package.

The compiler that was eventually chosen was ACK (Amsterdam Compiler Kit).[6] This system is very easy to modify, and contains source code for all components. It has frontends for languages such as C, Pascal, Modula-2 and Fortran, as well as backends for architectures such as SPARC, m68020, i386 and PowerPC. The major drawback of ACK is that it is only available at a cost.

Features of ACK generated code

The structure of ACK has proven to be well suited to generating migratible code. It is desirable that an executable program has exactly the same structure on all target machines. That is, each program is compiled to contain the same set of symbols (procedure and variable names), and each procedure contains the same set of local variables and temporaries. The storage location and size of these entities may differ widely between machines, for example, local variables may be stored on the stack or in registers.

Since ACK frontends generate intermediate code,[9] the differences between the various executable files is minimal. The majority of optimizations are performed on intermediate code, with the backends being primarily responsible for performing target instruction selection, as well as a small amount of peephole optimization. Since more complex machine dependent optimizations are not performed, the problems of code motion[10] are not relevant here.

ACK front ends generate *stabs* format[11] debugging information. These describe the type and location of all data values, using a compact ASCII encoding. Also, the mapping between source code line numbers and target machine addresses is recorded. Normally this information is used by debugging tools to allow the programmer to study an active program's data values. Tui uses these values in a similar, but more automatic fashion.

Modifications to ACK

The basic type information used by debuggers is not sufficient to correctly migrate a program. There are several important additions to the stabs format that Tui requires to successfully translate all data values. Aside from these additions, there are several other trivial modifications that were made (for example, the ACK backends were altered to correctly indicate which machine registers were used to store local variables).

The three major additions will now be discussed in more detail.

(a) **Preemption points.** When a process is migrated to a machine of a different architecture, we must deal with the fact that the corresponding point of execution (program counter) will have a different location within the text segment. To solve this, we select a set of logical points within the program at which migration is allowable. When performing the `migrout` operation to checkpoint a process, we must ensure execution stops at one of these *preemption points*. Upon restarting the process, the correct program counter value can be determined. Clearly, the program must have an identical set of preemption points on each target architecture.
Placing preemption points within a program is an interesting issue. Points must be placed often enough so that the process will stop within an insignificant amount of time (excluding the possibility of system calls that could block). However, having too many preemption points will require an excessive amount of information, or may even lead to a situation where the process can not be started at an equivalent point. For example, if a preemption point for a SPARC processor is placed within a sequence of instructions that perform a multiply operation, there is no way of locating the corresponding point within the program on a VAX processor, since it only requires one instruction to perform multiplication.
With these limitations in mind, it was decided that it is sufficient to place preemption points at the beginning of a loop, and at the end of each compound statement. The program will be halted within a very small amount time since no loop can repeat without passing through a preemption point (assuming the process was not blocked inside the operating system). Also, each machine's target optimizer is permitted to manipulate any code within a basic block, but it must not move code across preemption points.

(b) **Call points.** Although careful placement of preemption points can minimize the number of temporary values (partial results of a computation) that we must know about when the program is checkpointed, there is still the possibility that temporaries might exist across procedure calls. The following example illustrates this:

```
x = foo(y) + bar(y)
```

In this code fragment, the result of `foo(y)` needs to be saved somewhere while `bar(y)` is being calculated. However, if the process is preempted during the call to `bar`, it is necessary to retrieve the value of `foo(y)` from its temporary location (on the stack or in a register). Upon reconstructing the process at the target machine, the temporary is restored so that the calculation will complete correctly.
This is achieved by generating a *call point* stabs at each procedure call. This specifies the address of the call instruction, the number of temporaries (partially evaluated expressions), the number of parameters being passed, and the type and location details of each of these values. Although the information about parameters is already specified as part of the callee's stabs information, there are some procedures (such as `printf`), where only the caller is aware of how many parameters are being passed and what their types are.

(c) **Stack frame details**. During the `migrin` process, Tui must reconstruct each stack frame that existed before migration occurred. At compile time, a special stabs string is output at the beginning of each procedure. This specifies the size of the stack frame (that is, how many bytes are used for information such as local variables) as well as which registers were saved on the stack upon entry to that procedure.

Migrout: Checkpointing the process

The following description of the `migrout` process is divided into four main phases. Firstly, the type and location information (generated by the compiler), is entered into Tui's internal data structures. Next, the migrating process is halted, and its memory image is copied into Tui's address space for easy access. Thirdly, the type information is used as a guide for scanning this memory, and locating all data values. Finally, these values are translated into an intermediate format for transmission to the `migrin` component of Tui.

Reading the type information

The *stabs* debugging information associated with a program is specified in a manner that follows the structure of that program. The executable file's symbol table contains a section for each *object file* (`.o` file) that makes up the executable. Within each section, the global variables and procedures are listed, with their appropriate type and location information. For procedures, the same type of information is given for parameters, locals and temporaries. Although the type information is specified in a one dimensional format within the file, Tui creates a multidimensional structure for internal use.

The stabs debugging format strings are converted (when the source code is compiled) into more appropriate type structures. These structures, known as *type trees*, are similar to those used inside most compilers. They are able to represent all of the basic types as well as pointers, arrays and structures. To prevent name clashes, each symbol is prepended with the name of its enclosing file, and for local values, the procedure name.

Figure 2 shows the ASCII stabs strings for the given set of C declarations. It then shows the corresponding type tree entries.

In addition, two extra tables are required. The first table records the preemption points, each entry containing a single address for that point. The second table performs a similar operation, but for call points. In both cases, the table index is used as machine independent representation of the point's address.

Halting the process

Halting a program is more complex than in homogeneous migration. The Unix `ptrace` system call is used to place the process into the `trace` state. `migrout` may now make a copy of the memory and registers. However, we must ensure that the process is in a consistent state (at a preemption point). The exact code for implementing this is machine dependent.

The current version of Tui stops the process, places a breakpoint instruction at *every* preemption point, then continues execution of the process until a breakpoint trap occurs. For

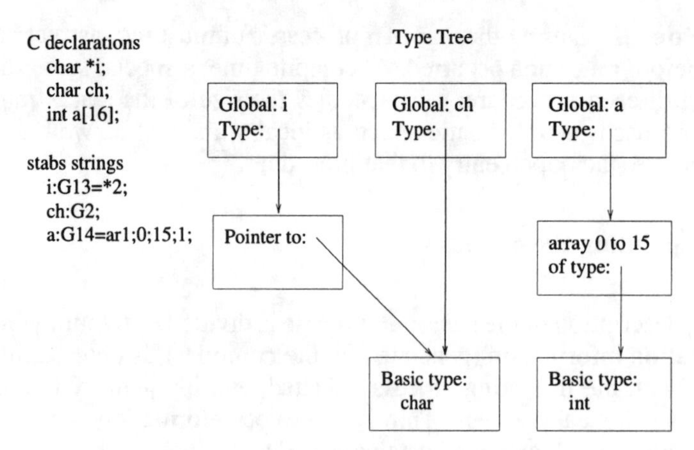

Figure 2. Stabs strings and the type tree

large processes, it would be more efficient to insert only one breakpoint, but it is not always easy to determine which preemption point will be reached next.

As a final step, Tui fetches copies of the stack and data segments of the process, which includes the heap segment, into its own address space. The process can now be killed.

It is probable that altering the Unix kernel would allow Tui to have faster access to the information it needs, rather than using the `ptrace` system call. However, we have performed all of our research without modifying the operating system.

Scanning the memory

While searching the memory of the process in order to locate all the data values, we must ensure that each value is detected exactly once. This is done by maintaining a *value table* that records the starting address, size and type of each piece of data. The value table is implemented as an expandable data structure where the only way to add a new value is to append it to the end. Therefore, the memory is scanned in a linear fashion, so that values are appended to the table in the correct order.

First, the procedure entry points and global variables are scanned, and their details are entered into the value table. Global variables are very simple to deal with since their locations are fixed and their types are well defined.

For the heap data to be scanned in linear order, it was necessary to alter the `malloc` and `free` memory allocation procedures so they would record all the blocks (empty and used) in linear order. This addition costs one extra pointer per memory block. Also, the compiler must generate a small amount of extra code for recording the data type of each block that is allocated. This issue will be discussed further in a later section.

Local variables (contained within stack frames) are scanned in a similar way. The frames are examined, starting at the most recent procedure activation. At each point, Tui queries the program's type information to obtain a list of the procedure's stack or register based values. Since stack based values are specified as offsets from the procedure's frame pointer, the absolute addresses must be calculated. Special care is also taken to maintain a correct idea of the current register set, especially since they are often saved on the stack across procedure calls.

At each point where a procedure call was made, Tui locates the associated call point information to determine which temporaries and arguments were stored on the stack for the duration of that call. A procedure's arguments will be scanned from the caller's perspective to correctly handle procedures that allow a variable number of arguments.

Finally, the command line arguments and environment variables are scanned. This must be done separately from the stack frames since this information is not always described by an explicit variable name as would a normal stack variable.

Marshalling to the intermediate form

The final stage of `migrout` is to traverse the value table and encode all data values from the memory of the process into the intermediate file. This potentially requires that data format conversion take place (for example, little endian to big endian integer formats). Full details of the intermediate file format are given later.

The only difficulty of this phase is that we must represent the relationship between the different data items. That is, some data values will be (or will contain) pointers to other data values. When marshalling a pointer value, Tui performs a binary search on the value table to locate the information about the object being pointed to.

Each entry in the value table is assigned a unique number. When a reference is made to a data item, the pointer is encoded by specifying this machine independent number, rather than the machine specific address. Also, in the case where a pointer refers to a location that is part-way through a composite data item, an *offset* states how many indivisible subelements must be skipped in order to locate the correct value.

The following C code demonstrates this:

```
{
        struct {
            int a;
            int b;
        } c[10];

        int *p = &c[2].b;
}
```

In this case, the offset for the pointer `p` would be 5, since the structure contains two subelements, and `p` refers to the second element of the third instance of that struct within the array `c`.

Migrin: Reconstructing the process

To restart a process on the destination machine, the `migrin` algorithm must obtain the program's type and location information in the same manner as for `migrout`. Next, it reads through the intermediate file and places all the data values in their appropriate locations. This phase reads the intermediate file sequentially, and therefore can mostly be done in parallel with the `migrout` phase.

Global variables are placed directly into their absolute memory locations. Virtual stack and heap pointers are maintained, with all new values being added to the end of the appropriate segment. Clearly, it is vital that the data items on the stack are restored in the correct order. Also, due to the linear fashion in which the value table is constructed during the `migrout` phase, the heap must maintain its correct ordering as well.

Pointers also cause problems when placing data values into memory. It is not possible to determine the final value of a pointer until the object it refers to has been assigned a memory location. Consequently, a table is used to record all pointers, and once all data values have been dealt with, the pointers are converted from their (*Object ID*, *offset*) pairs into machine addresses.

As a last step, the process is restarted by loading the program's binary file into memory, then writing the newly constructed data and stack segments into the address space (using `ptrace`). The preemption point number that represents the continuation address of the process is converted into the correct machine dependent address. Finally, the correct register values are given to the process, and it continues execution.

The intermediate representation

The intermediate file is a machine independent representation of the value table. It lists all data values in a well defined storage format, and if necessary, states the type of the values and the relationship between them. The file format has not yet been optimized to any great extent.

All data values (`int` and `float`) are encoded using the native storage format for Sun 4 machines. That is, big endian two's complement integers and IEEE floating point values. Since Sun 3 and PowerPC machines also use this format, the Intel 386 is the only machine that needs to perform any format conversion.

The data items are listed in the order: *procedures, global variables, heap values* and *stack*. This is the order in which they appear within the address space of all architectures currently supported by Tui:

(a) *Procedures*. The name of each procedure is listed, since it is possible for a pointer to refer to a procedure. No other information is given about the text segment.
(b) *Global variables*. The variable's name and value are specified. It is necessary to include the name, since variables may appear in a different order on different architectures. Also, some symbols may exist on one machine, but not on the other; these will typically be machine dependent values and are not normally meaningful to migrate.
(c) *Heap values*. These do not have names, and the destination machine can not determine the type of the data in advance. Therefore, values are listed alongside their stabs type number. It is necessary that all architectures use a common type numbering system.
(d) *Stack values*. These are listed within their respective frames. Each frame is identified by the name of the procedure and the number of the call point that created the frame. Parameters, local variables, temporaries and arguments are listed in an order that is consistent among all machines. No variable names are needed.

One interesting optimization has been made to the way in which integers are encoded. The number of bytes used to represent integers depends entirely on the value of the number, and not the size that it had on the source machine, or will have on the destination machine. That is, small values (such as 5) can be encoded in one byte, whereas larger values require more.

COMPARISON: THE PROTOTYPE OF TUI AND THE CURRENT VERSION

The implementation of Tui, as described in the previous sections, is now considered to be complete. However, it is worth discussing the earlier prototype version to show the important discoveries that were made, as well as the trade-offs between the two different systems.

The original version of Tui used a different approach to scanning the memory of a process. To locate the data stored on the heap, a traversal algorithm (similar to those used in garbage collection systems) was used. While scanning the global and stack data area, any pointers that refer to data in the heap area were followed and the details added to the value table. If the heap data itself contained pointer references, the traversal process continued until all reachable heap data had been located.

Although the prototype algorithm functioned correctly for most programs, there were two major limitations identified. First, it was possible that when migrating non-type-safe programs, a 'type conflict' could occur. Secondly, the performance of the value table was not satisfactory. These limitations will now be examined in more detail.

Problems due to lack of type-safety

When a pointer was followed to locate a data item in the heap, the base type of that pointer was used to determine the type of that heap data. After much analysis, this approach appeared insufficient given that Tui should function correctly for a non-type-safe language. The following examples will clarify this problem.

Example: Determining array sizes

If an array is dynamically allocated on the heap, there is often only a pointer to the beginning of the array. In ANSI C, there is no way of automatically determining its length. As an estimate, the total size of the heap block could be divided by the size of a single array element. However, this is not totally reliable, since the programmer may not intend to use the entire heap block for storing the array.

Example: Type conflicts

If a pointer refers to a heap block that has already been discovered (by following a previous pointer), both pointers must agree on the data type. If the pointer types differ, there is no way for Tui to ensure that it will correctly interpret the data values.

The following fragment of code is non-migratible since it violates this property. Pointer `a` refers to an integer value (or array of integers) while point `b` suggests that this same area of memory stores characters:

```
{
    int *a = malloc(100);
    char *b = (char *)a;
}
```

When this program is migrated, a 'migrate-time' error would be reported.

A similar difficulty appears if we vary the ordering in which values are discovered. In the following fragment of code, the pointer b refers to a element of the array a:

```
{
    int a[10];
    int *b = &a[5];
}
```

If a is entered into the value table first, b will refer to a known element of the array a. On the other hand, if b is discovered first, the value table must be carefully rearranged to record that a is in fact the most significant data value. This functionality is not impossible to deal with, but the complexity of the necessary code has proved to negatively affect its performance.

Solutions

The solution to these two problems is to require the programmer to more carefully specify the type and size of each malloc block at their time of creation. In each call to the `malloc` library function, the programmer must use the form:

```
malloc(size * sizeof(type))
```

The C compiler will incorporate this size and type into a call to a special version of `malloc` that will record the information for later use by Tui.

Now that extra information is available, there can be no ambiguity over the type of heap data. A pointer of any type may refer to heap data of any other type, as long as it refers to the beginning of an atomic data value. For example, a character pointer may refer to the first byte of a four byte integer, but not to any of the remaining three bytes.

With the new (current) implementation, 'type conflicts' have been reduced to 'alignment conflicts', and it is always possible to determine the size of an array. It is expected that these changes will greatly expand the number of programs that are migratible. However, future work is needed to prove this.

Aside from the extra cost of storing type information with each heap block, there is a requirement that all calls to `malloc` be put in the correct form. Experience has shown that this is often a simple matter of including a suitable `sizeof` expression, but occasionally more work must be done. For example, the following structure definition may occur:

```
struct foo {
    int len;
    char buf[1]
}
```

The programmer's intention is that at run time they will know the length of `buf` and will then be able to allocate appropriately sized storage. However, this violates Tui's rules on using `malloc`. The solution is to rewrite the definition as follows:

```
struct foo {
    int len;
    char *buf;
}
```

With these changes, two calls to `malloc` are required (one for `struct foo` and one for the `buf` array), in each case, the size and type of each object are correctly stored.

Performance

The second limitation of the original garbage collection style algorithm was that due to the potentially random ordering of insertions into the value table, it was not possible to use a linearly expanding data structure. The prototype version used a splay tree[12] that allows randomly ordered insertions. Although a splay tree will typically give excellent performance for random accesses, there were circumstances where the performance was less than satisfactory.

As an example, when migrating a program that contained a large number of stack frames, each new data item that was added to the value table was guaranteed to be inserted at the end of the table. At the same time, this value was being 'splayed' to the root of the splay tree, leading to a very unbalanced structure. It was clear that using the linear table of the new version would give better performance for many programs.

The revised scanning algorithm will always give linear performance (based on the number of data items), but introduces the restriction that the memory of the process must be scanned in order of increasing (or decreasing) address. This is not a significant restriction since all the architectures supported by Tui have their text, data, heap and stack segments layed out in the same order, although at different memory locations. The introduction of a new architecture may cause a minor problem.

PERFORMANCE TESTS

To fully test the performance of the Tui algorithms, three different test programs (for Tui to migrate) have been created. Each is designed to test the complexity of the various components of Tui. The three programs are:

(a) `fibonacci`. An inefficient recursive implementation of the Fibonacci algorithm that creates a large number of stack frames, each with a small number of local variables and temporaries. A single preemption point is placed so that migration will occur when n stack frames are active (n is the input parameter). This program tests `migrout`'s efficiency when scanning the stack.

(b) `tree`. Builds a binary tree of n nodes (n is a command line parameter). Numeric values are selected randomly and then inserted into the tree. Once construction of the tree has completed, migration will occur. This program tests Tui's ability to scan the heap space in a random order.

(c) `arrays`. 50 character arrays (of user specified size) are dynamically allocated on the heap and then filled with characters. This test demonstrates the efficiency of encoding and reconstructing large areas of memory.

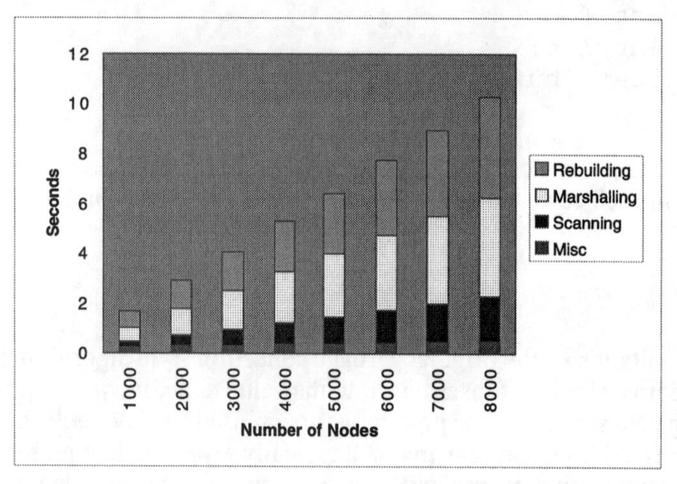

Figure 3. `tree` *on Sun 4*

Components of the `migrin` and `migrout` algorithms

To demonstrate that Tui can correctly function on the four supported architectures, each of the programs was migrated. Figures 3–6 show the time taken by the main components of both the `migrin` and `migrout` algorithms for the `tree` program. In these tests, the number of tree nodes varies from 1000 to 8000. Although only the `tree` program is analyzed, `fibonacci` and `arrays` were similar. The exact machines are:

(a) Sun 4/75 (SPARCStation 2).
(b) Sun 3/60.
(c) i486 running at 50Mhz.
(d) PowerPC 601 running at 66 Mhz.

All measurements are averaged over five runs on an otherwise idle CPU. The machines have sufficient memory to avoid paging.

In this analysis, the total execution time is divided into the major components of both `migrout` and `migrin`. We must pay attention to the relative costs between the components and the growth of each component as the problem size increases. The following list gives an explanation of each cost.

(a) *Miscellaneous*. The time required to read the migrating program's memory image into Tui's address space, as well as the time to read the program's type information from disk. The memory image size will vary depending on the size of the program, but the amount of type information will remain constant.
(b) *Scanning*. The scanning of the memory segments and the construction of the value table. This cost depends on the number of individual data values that are located, not the size of those values.
(c) *Encoding*. The data values must be marshalled into the intermediate file. This cost depends on the total size of all data values, as well as the operating system's performance when writing to files.

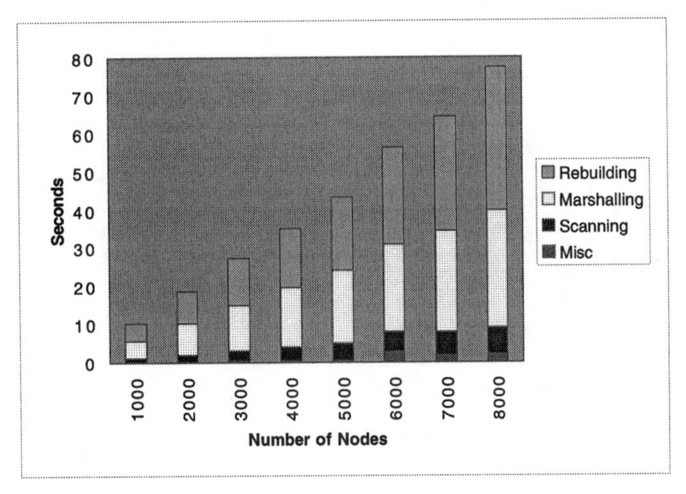

Figure 4. `tree` *on Sun 3*

(d) *Rebuilding*. This is the only component of the `migrin` algorithm that has been analyzed. Given the intermediate file, the new data and stack segments are constructed. The other components of `migrin`, such as reading the type information and reading/writing core memory is the same as for `migrout`.
Note that the final rebuilding of `migrin` can almost entirely occur in parallel with the scanning and encoding of the `migrout` phase. Therefore, the total migration time will be less than the total time required over all components.

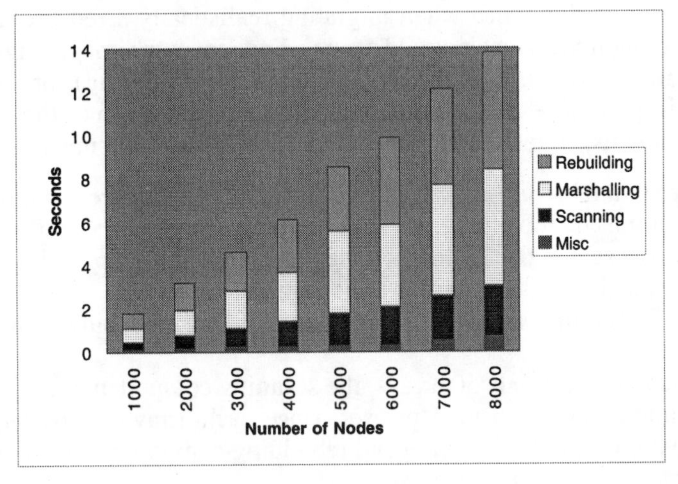

Figure 5. `tree` *on i486*

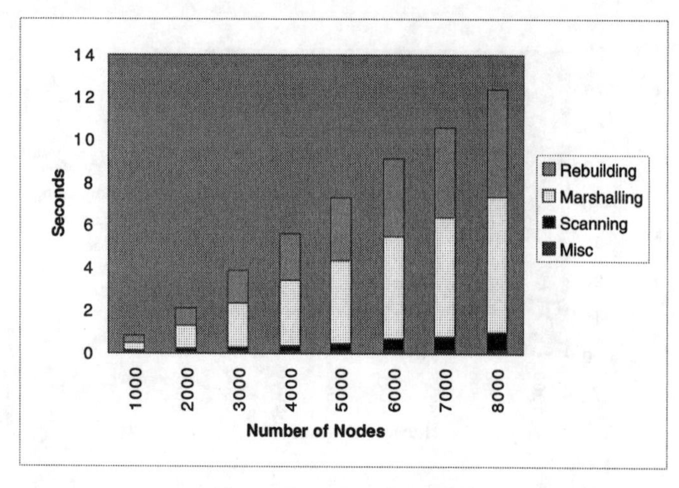

Figure 6. `tree` *on PowerPC*

It can be seen that scanning, encoding and rebuilding are the major components of the migration cost. The miscellaneous costs of reading type information and the memory image is small enough to ignore. We next study how the cost of the three major components increases as the input size becomes extremely large.

Asymptotic growth in migration time

To examine Tui's performance when migrating realistically sized programs, each of the three tests was configured so that it would create a large memory image. Figures 7–9 show the contribution of the major costs (scanning, encoding and rebuilding) for various input sizes. The following list gives an explanation of the performance for each of the three programs. To avoid paging problems, all tests were performed on the same large machine (the PowerPC).

(a) `tree`. This program has close to linear performance for all three components. The makes sense for scanning and rebuilding, but for encoding we expect a slightly higher complexity due to the binary search that is done on the value table for every pointer. In these results, this extra complexity does not appear to be significant.
(b) `fibonacci`. The complexity is roughly the same as for `tree`, but the overall running time is lower.
(c) `arrays`. Since there are only 50 arrays, the scanning component requires an insignificant amount of time to locate them. However, since each array can be large (up to 50,000 characters in our case), the encoding and rebuilding components are significant, although they will always have linear complexity.

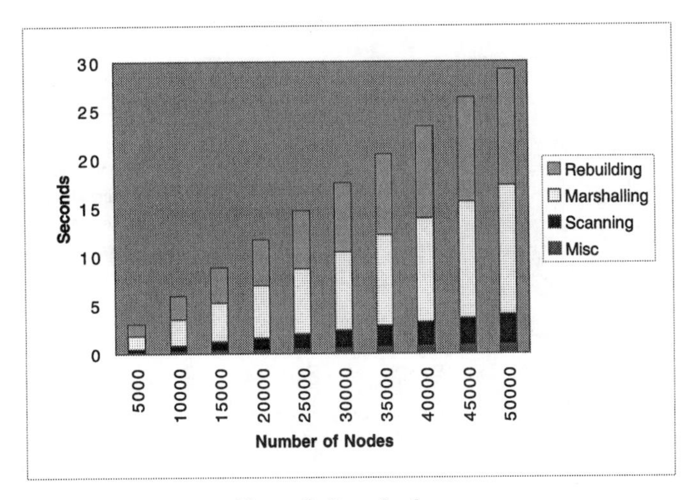

Figure 7. Growth of `tree`

Migrating realistic programs

The three test programs discussed so far were designed to determine the performance of the major components of Tui. However, to demonstrate that Tui is not limited to a small set of contrived programs, two applications that were not designed for migration have been used. A matrix multiplication package and a popular text editor demonstrate how easily 'real' programs can be migrated.

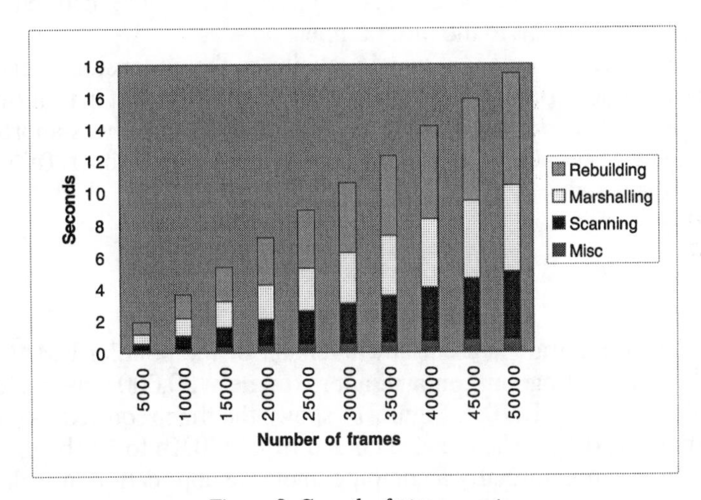

Figure 8. Growth of `fibonacci`

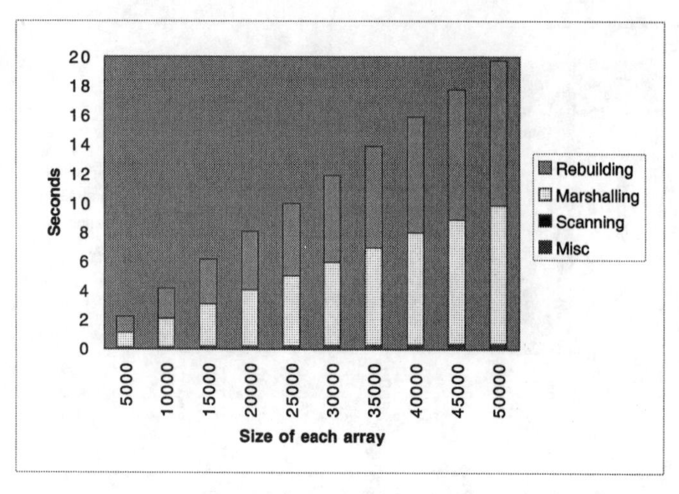

Figure 9. Growth of `arrays`

Matrix multiplication

A matrix manipulation package is a very practical piece of software that often requires extensive amounts of memory and CPU time. The particular package that was used[13] contains approximately 1000 lines of code written in ANSI-C. The intended purpose of this software is to execute and compare the performance of several different matrix multiplication algorithms. Other than altering the program's main loop (to allow for user selected matrix sizes), no modifications to the code were necessary. This is clearly a desirable feature as our goal is that migration should be transparent to the programmer.

In Figure 10, the migration time is shown in relation to the dimension of each matrix. Since the matrices are two dimensional, a doubling in dimension will result in four times the memory usage. In the largest test shown, the memory usage of the program was approximately 4 Mb, requiring a total of 20 seconds to migrate. The asymptotic growth of Tui's execution time remains linear, as expected.

The MicroEMACS editor

The second realistic program is a cut-down version of the popular EMACS editor, called MicroEMACS.[14] This C program contains approximately 20,000 lines of code (that were relevant when running under UNIX). Figure 11 shows the time required to migrate the editor while it was editing a text file. The file size ranged from 200 Kb to 1 Mb.

Although an editor is not normally a compute intensive application, it is desirable to move the program between different machines in a mobile environment. As an example, a user may wish to migrate a set of desktop applications between their workstation and their laptop. For this purpose, a transfer delay of up to a minute would be acceptable.

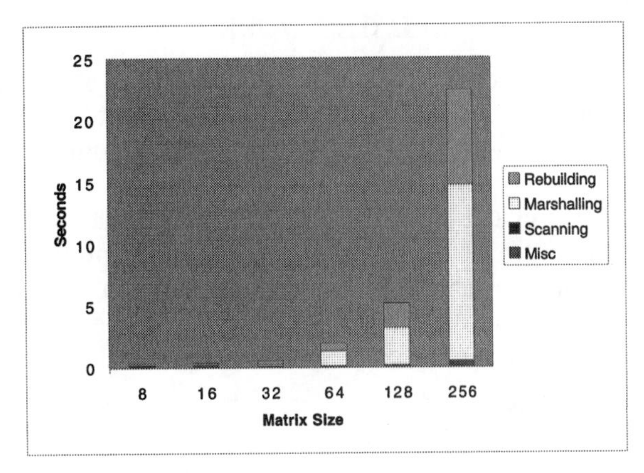

Figure 10. Growth of `matrix`

Unlike the matrix program, several minor changes were required. First, the program was converted into ANSI-C by altering the function headings and adding function prototypes. Secondly, there were 10 uses of the 'malloc' function that were missing the 'sizeof' operator. In all cases, the compiler warned about the missing type information. Finally, there were three additional situations where a heap block was being allocated as a variable sized structure. Each of these occurences was fixed by using the technique discussed earlier (that is, adding a second call to 'malloc').

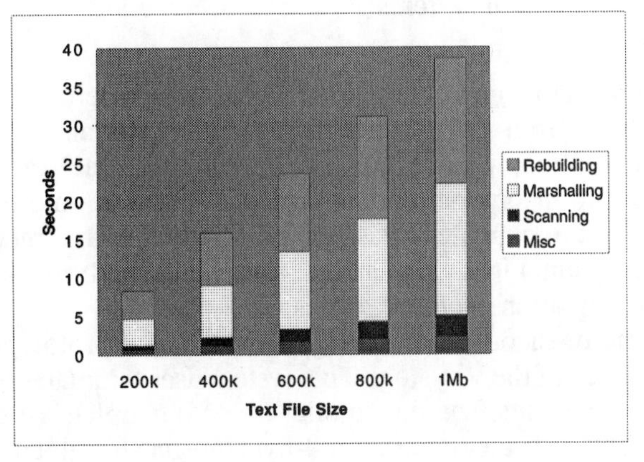

Figure 11. Growth of `uemacs`

RELATED WORK

Until recently, heterogeneous process migration was considered as an interesting topic, but no mature implementation had been developed. However, current interest in world-wide and mobile computing has lead to several implementations, although only with limited use within research environments.

This section briefly lists previous work in homogeneous migration, both on a per-process and per-object basis. Next, a discussion of other heterogeneous migration systems will show the different the range of approaches, in particular, how they compare to Tui. Finally, other supporting areas such as garbage collection and debugging will be surveyed.

Traditional migration systems

Process migration (in its homogeneous form) is not a new topic, and has been studied extensively since the late 1970s. Much of the previous research has involved finding new and improved methods of transferring the state of the process from one machine to another. Examples of homogeneous process migration systems are V,[15,16] Charlotte,[17] DEMOS/MP,[18] Sprite,[19] Condor[20] and Accent.[21] A good summary of these and other systems is given in Nuttall[22] as well as a more recent survey in Milojicic *et al.*[23]

Object mobility

The idea of process migration has been incorporated into distributed object-oriented systems. However, it has become more relevant to migrate on a per-object basis (or in groups of objects), rather than moving a whole program. Migration in this form is more commonly known as *mobility*, that is, the object is mobile. Examples of such systems are Emerald,[24,1,25] DOWL,[26] DCE++,[27] COMET[28] and COOL.[29]

Other heterogeneous migration systems

All of the following systems support migration of native code across different architectures, however, they each differ from Tui in some significant way. Many of them solely support 'process-originated' migration, whereas Tui also allows an external agent to make a request. Secondly, it is common to incorporate the data marshalling code into the migrating process itself (either created by the compiler, or specified by the programmer), whereas Tui is a completely separate program. Finally, Tui has addressed and resolved some of the type-safety issues that could limit migration in other systems.

Possibly the first heterogeneous migration system[30,31] was a prototype built on top of the existing migration features of the V system. This system uses templates to describe the layout of the various memory segments. These (compiler created) templates specify the size of each data element (for example, whether it is a 2- or 4-byte integer) for global data, stack and heap blocks. The type templates are similar to, but simpler than, Tui's type tree.

The primary limitation of this system is the assumption that data will reside at exactly the same address on all architectures (possible in the V system). This simplifies migration, since there is no requirement to adjust pointer values. However, data types must be of the same size, and data structures must be padded to the largest size required by any of the architectures.

Later work (related to this system)[32] has lead to a more formal analysis of the points at which a process may be migrated (known in Tui as preemption points). 'Pointwise equivalence' is required between two computations (that is, executable programs on different architectures) if migrating between the computations is to be guaranteed correct. They discuss issues relating to the granularity of migration points, especially in respect to optimization of program's code. Placement of preemption points in Tui could benefit from this type of analysis.

A second system,[33] that was never implemented, introduces the idea of migration by recompilation. At migration time, a source level program is automatically created, transferred to the destination machine, and then recompiled. When the program is executed, it restores the state of the process and then continues execution at the correct location.

The motivation for this method was that the machine dependent knowledge (such as register usage and stack frame layout) is already embedded into debuggers and compilers. Therefore, a process can be migrated to and from any architecture that supports commonly available debugging and compilation tools. The main disadvantage is that migration time is greatly increased because of the need for source code compilation.

Another approach[34] requires that the migratible program check a state variable at various points throughout its execution (specifically, at the beginning and end of each procedure). If the state variable indicates that normal execution should occur, no migration code will be executed. However, when migration is requested, the variable is set to indicate that the contents of the currently active procedure should be saved to, or restored from, an alternate address space. A source-to-source C language translator is used to insert the additional code.

The Emerald system[24,1,10] is an object-oriented language and environment that permits fine-grained migration of native code objects. The Emerald compiler creates a template describing the internal structure of an object as well as the format of each method's stack frames. Whereas most heterogeneous migration systems make use of the C language, Emerald is itself a type-safe language, so correct migration will always be possible.

The HMF (Heterogeneous Migration Facility) system[35] requires the programmer to explicitly register the data to be migrated. The migration library is linked with the migrating program and provides procedures for registering data values (given their address and a type description), for initiating migration, and for converting to external data formats.

Checkpointing

Checkpointing and migration are very similar. The main difference is that checkpointing requires that a process can be restarted after a long period of time, whereas migration assumes that the current external state will not change. For example, a checkpointing system may need to rollback any files that were being written to. A migration system would assume that the files remained consistent.

In most cases, a checkpointing algorithm assumes that the process will be restarted on exactly the same machine that it started on. This implies that heterogeneity is not an issue. However, if we wish to restart it on a different machine, with a different architecture, then the problem is identical to that of heterogeneous process migration.

Several checkpointing systems have been created for Unix systems,[20,36], but they only function in a homogeneous environment. The recent concept of 'Memory exclusion'[37] demonstrates that careful selection of data values to be saved can reduce the cost of checkpointing. Another system[38] divides programs into modules that can be individually checkpointed. Each module is initialized by supplying it with either a fresh (empty) checkpoint file, or a checkpoint file from a previous execution.

Debugging

Source level debugging is also closely related to heterogeneous process migration. Compilers generate extensive amounts of information describing such things as the type and location of all variables, and the location of each source code statement. A debugger such as `dbx` or `gdb` uses this information to aid the programmer in studying a process.

Recently, work has progressed in the field of debugging optimized code.[39,40] Whereas traditional debuggers have only been able to correctly debug unoptimized programs, it has been recognized that many errors do not become obvious in this situation. The *DWARF* debugging format[41] is a newly developed format that is capable of expressing the structure of an optimized program.

Data marshalling packages

For software that is expected to function correctly in a distributed environment, it is vital that the heterogeneity present in the data storage formats be taken into account. Any data that is externally visible must be in a form that all consumers can interpret.

Several general packages are available to automate the data translation process. Given some form of data description, these systems will generate suitable functions for translating between a machine's native data format and some intermediate format. Two of the most common systems are Sun's XDR[42] and ISO's ASN.1.[43]

Tui does not take advantage of any standard system, since the packaging of the whole data structure is handled as part of `migrout`, and the translation of single data values is trivial in the four machines supported by Tui.

One solution[44] has addressed the issue of transmitting cyclic data structures within the CLU programming environment (XDR and ASN.1 cannot correctly deal with cycles). This problem has also been solved by Tui through the use of the value table and the method of encoding pointers.

Garbage collection

A garbage collector is capable of scanning through the program's memory, searching for, and freeing areas that are no longer being used. A good overview of uniprocessor garbage collection methods is given in Wilson.[45]

The prototype version of Tui made use of garbage collection techniques to help locate data. However, most existing garbage collection algorithms are not accurate enough to correctly migrate programs written in non-type-safe languages. For example, it is often assumed that all data items are distinct (as in object-oriented programming), and that marking the data, to indicate that it has already been scanned, is somehow possible. Also, it is necessary for pointers to be clearly identified in some manner (such as tagging), so they are not confused with other data values.

One system[46] allows garbage collection to function within C programs, but without proper type information, an educated guess must be made to identify pointers. Any pointer sized data value in a register or on the stack is considered to potentially be a pointer. The memory allocator is used to decide whether the value points to a valid memory block or not. The limitation of this system is that we can never be totally sure of whether a data item is a pointer, or simply an integer. Although an incorrect guess is not fatal for a garbage collection system, it will not suffice for a migrator.

In considering type-safety, Diwan *et al.*[47] discuss garbage collection for Modula-3. They introduce the idea that although the source code for a program is type-safe, the compiled executable code may not be. For example, in an array that is not zero based, a 'virtual array origin' pointer often refers to a memory location before the start of the array. This improves performance when calculating array offsets, but also gives a misleading indicator of which memory is in use.

A second important issue raised in this paper is that of determining how to locate the 'derived' pointer variables at garbage collection time. That is, if one pointer variable is derived from a second pointer variable (for example, it may point to a field within an object), that derived pointer must be updated if the object is relocated. This requirement is only an issue if objects are moved individually, as opposed to moving an entire program.

Heterogeneous distributed shared memory

The *Mermaid* system[48,49] allows Distributed Shared Memory (DSM) to function between heterogeneous machines. That is, a group of processes residing on different machines are able to share a consistent view of a segment of memory. Unlike traditional DSM, the machines may have different data formats, requiring that the segment is translated as it is moved between machines.

This system uses information provided by the compiler to determine the types of the data being shared. It then generates stubs to perform the necessary conversion. Using customized conversion code is said to be more efficient than using general conversion facilities such as XDR and ASN.1. Problems with unconvertible data values, pointer correctness and variations in data sizes are raised, but not addressed.

The methods used in Mermaid will be useful for process migration, although they are for a rather simplified environment. Primarily, Mermaid does not address the vital aspect of converting the active components of the process (such as registers and stack). Secondly, it is limited to a well defined segment of memory, rather than the whole process image.

Binary translation

Binary translation is a technique that is used to convert machine code from one architecture to another. For example, one of its main uses was in the introduction of DEC's Alpha processor.[50] There was a desire to convert existing VAX software to the Alpha platform, without using the original source code. Another system[51] talks about emulating complex instruction set machines by using binary translation within a RISC environment.

In the context of heterogeneous process migration, binary translation could be used to migrate the executable program code to a different architecture. Even though the simple solution of recompiling the program from the source code has been chosen, Tui could also take advantage of binary translation.

ONGOING WORK

Even though the current implementation of Tui has been successful, the following topics are seen as necessary additions to the work:

(a) A more detailed survey of non-migratible languages features will be made. We have already shown that for a program to be successfully migrated, type-safety is not always a requirement. Since we are not totally sure of the extent of 'migratibility', a survey of common programs (for example, compilers, editors and scientific software) will be made. This work will act as a guide for those who wish to use Tui to migrate their own programs.

(b) The current implementation of Tui will be extended to deal with a wider range of programming languages. ACK already supports Pascal, Modula-2 and Fortran, so the coding requirements should be minimal. Since C is generally considered to be one of the least type-safe languages, it is expected that other languages will present fewer problems for migration.

(c) Realistic performance tests will be performed to demonstrate that migration is beneficial in wide area and mobile computing. A single program will be used to compare the cost of data access across the internet (while keeping the program at a fixed location), versus the cost of migrating that program closer to the data.

SUMMARY

The Tui Heterogeneous Process Migration system is able to move a process between machines of different architecture, with minimal programmer intervention. It uses type information that is generated when the migratible program is compiled. The bulk of the algorithm's work involves locating and determining the type of each data value within the process. This data is marshalled into an intermediate form so that the process may be reconstructed on the destination machine.

The algorithms for checkpointing and reconstructing a process image have been described, with most focus placed on those features that are unique to heterogeneous migration. Performance measurements of several programs, including two real applications, have shown that migration is possible within an acceptable amount of time. This is especially true if we consider the potential cost of not migrating a program across a wide area or mobile link.

Tui has been designed to migrate ANSI-C programs, although other languages such as Pascal and Fortran could be supported. The required compiler changes are fairly minimal, so new languages can easily be incorporated into the system. A common problem is that most practical languages are not type-safe, so determining the correct type of each data item, therefore migrating successfully, is not always possible.

Experience with two versions of Tui has shown that some type-unsafe programs can be migrated, as long as the programmer is required to be more specific about the type details of dynamically allocated memory. Proposed future work will involve categorizing non-migratible language features, and attempting to eradicate them.

REFERENCES

1. E. Jul, H. Levy, N. Hutchinson and A. Black, 'Fine-grained mobility in the Emerald system', *ACM Transactions on Computer Systems* (1988).
2. F. Douglis and B. Marsh, 'The workstation as a waystation: Integrating mobility into computing environments', *The Third Workshop on Workstation Operating Systems (IEEE)*, 1992.
3. W. C. Hsieh, P. Wang and W. E. Weihl, 'Computation migration: Enhancing locality for distributed memory parallel systems', *SIGPLAN Notices*, **28**(7), 239–248 (1993).
4. J. Gosling, B. Joy and G. Steele, *The Java Language Specification*, Addison-Wesley, May 1996.
5. General Magic. Telescript technology, 1996.
6. A. S. Tanenbaum, H. van Staveren, E. G. Keizer and J. W. Stevenson, 'A practical toolkit for making portable compilers', *Communications of the ACM*, **26**(9), 654–660 (1983).
7. R. M. Stallman, 'Using and Porting GNU CC', 1995.
8. C. Fraser and D. Hanson, *A Retargetable C Compiler: Design and Implementation*, Benjamin/Cummings, 1995.
9. A. S. Tanenbaum, H. van Staveren, E. G. Keizer and J. W. Stevenson, 'Description of a machine architecture for use with block structure languages', *Technical report*, Vrije Universiteit Amsterdam, 1983.
10. B. Steensgaard and E. Jul, 'Object and native code process mobility among heterogeneous computers', *Symposium on Operating System Principles*, 1995.
11. J. Menapace, J. Kingdon and D. MacKenzie, 'The "stabs" debug format', *Technical report*, Cygnus support.
12. D. D. Sleator and R. E. Tarjan, 'Self-adusting binary search trees', *Journal of the ACM*, **32**(3) (1985).
13. Matrix multiplication code. Available by anonymous ftp from: usc.edu/pub/C-numanal/matmult.tar.gz.
14. The microemacs editor. Available by anonymous ftp from: ftp.agt.net/pub/Simtel/msdos/uemacs/ue312src.zip.
15. D. R. Cheriton, 'The V distributed system', *Communications of the ACM*, **31**(3), 314–333 (1988).
16. M. M. Theimer, K. A. Lantz and D. R. Cheriton, 'Preemptable remote execution facilities for the V-system', *Proceedings of the Tenth Symposium on Operating System Principles*, December 1985.
17. Y. Artsy and R. Finkel, 'Designing a process migration facility: The Charlotte experience', *Computer*, **22**(9), 47–56 (1989).
18. M. L. Powell and B. P. Miller, 'Process migration in DEMOS/MP', *Proceedings of the 9th Symposium on Operating System Principle* (1983).
19. F. Douglis and J. Ousterhout, 'Transparent process migration: Design alternatives and the Sprite implementation', *Software—Practice and Experience*, **21**(8), 757–785 (1991).
20. A. Bricker, M. Litzkow and M. Livny, 'Condor technical summary', *Technical report*, Computer Sciences Department, University of Wisconsin-Madison, January 1992.
21. E. R. Zayas, 'Attacking the process migration bottleneck', *Symposium on Operating System Principles*, Austin, TX, November 1987, pp. 13–22.
22. M. Nuttall, 'A brief survey of systems providing process of object migration facilities', *Operating Systems Review*, **64** (1994).
23. D. Milojicic, F. Douglis, Y. Paindaveine, R. Wheeler and S. Zhou, 'Process migration', *http://www.opengroup.org/ dejan/papers/indx.htm.*
24. A. Black, N. Hutchinson, E. Jul, H. Levy and L. Carter, 'Distribution and abstract types in Emerald', *IEEE Transactions on Software Engineering*, **13**(1), 65–76 (1987).
25. R. K. Raj, E. Tempero, H. M. Levy, A. P. Black, N. C. Hutchinson and E. Jul, 'Emerald: A general-purpose programming language', *Software—Practice and Experience*, **21**(1), 91–118 (1991).
26. B. Achauer, 'The DOWL distributed object oriented language', *Communications of the ACM*, **36**(9), 48 (1993).
27. A. B. Schill and M. U. Mock, 'DCE++: Distributed object-oriented system support on top of OSF DCE', *Technical report*, Institute of Telematics. University of Karlsruhe, Germany.
28. H. Moons and P. Verbaeten, 'Object migration in a heterogeneous world–a multi-dimensional affair', *Proceedings of the Third International Workshop on Object Orientation in Operating Systems*, Asheville, NC, December 1993, pp. 62–72.
29. R. Lea, C. Jacquemot and E. Pillevesse, 'COOL: System support for distributed programming', *Communications of the ACM*, **36**(9), 37–46 (1993).

30. Charles M. Shub, 'Native code process-originated migration in a heterogeneous environment', *ACM Conference on Computer Science*, ACM, New York, 1990, pp. 266–270.
31. F. Brent Dubach, R. M. Rutherford and C. M. Shub, 'Process-originated migration in a heterogeneous environment', *ACM Conference on Computer Science*, ACM, New York, 1989.
32. D. G. Von Bank, C. M. Shub and R. W. Sebesta, 'A unified modelor pointwise equivalence of procedural computations', *ACM Transactions on Programming Languages and Systems*, **16**(6), 1842–1874 (1994).
33. M. M. Theimer and B. Hayes, 'Heterogeneous process migration by recompilation', *11th International Conference on Distributed Computing Systems*, May 1991.
34. V. Strumpen and B. Ramkumar, 'Portable checkpointing and recovery in heterogeneous environments', *Technical Report ECE-96-6-1*, Department of Electrical and Computer Engineering, University of Iowa, July 1996.
35. M. Bishop, M. Valence and L. F. Wisniewski, 'Process migration for heterogeneous distributed systems', *Technical Report PCS-TR95-264*, Department of Computer Science, Dartmouth College, August 1995.
36. J. S. Plank, M. Beck and G. Kingsley, 'Libckpt: Transparent checkpointing under Unix', *USENIX Technical Conference*, 1995.
37. M. Beck, J. S. Plank and G. Kingsley, 'Compiler-assisted checkpointing', *Technical Report CS-94-269*, University of Tennessee, Knoxville, December 1994.
38. S. Pope, 'Application migration for mobile computers', *3rd International Workshop on Services in Distributed and Networked Environments (SDNE 96)*, June 1996.
39. M. Copperman, 'Debugging optimized code without being misled', *ACM Transactions on Programming Languages and Systems*, **16**(3), 387–427 (1994).
40. U. Holzle, C. Chambers and D. Ungar, 'Debugging optimized code with dynamic deoptimization', *ACM SIGPLAN 1993 Conference on Programming Language Design and Implementation*, 1992.
41. 'DWARF Debugging Information Format', *Industry Review Draft*, UNIX International, July 1993.
42. Sun Microsystems, *Open Network Computer: RPC Programming*. The official documentation for Sun RPC and XDR.
43. *Information Technology–Abstract Syntax Notation One (ASN.1)–Specification of Basic Notation*, International Organization for Standardization, February 1994.
44. H. Herlihy and B. Liskov, 'A value transmission method for abstract data types', *ACM Transactions on Programming Languages and Systems*, 1982.
45. P. R. Wilson, 'Uniprocessor garbage collection techniques', *ACM Computing Surveys* (submitted 1996).
46. H.-J. Boehm and M. Weiser, 'Garbage collection in an uncooperative environment', *Software—Practice and Experience*, **18**(9), 807–820 (1988).
47. A. Diwan, E. Moss and R. Hudson, 'Compiler support for garbage collection in a statically typed language', *SIGPLAN '92 Conference on Programming Language Design and Implementation*, June 1992, pp. 273–282.
48. D. B. Wortman, S. Zhou and S. Fink, 'Automating data conversion for heterogeneous distributed shared memory', *Software—Practice and Experience*, **24**(1), 111–125 (1994).
49. S. Zhou, M. Stumm, K. Li and D. Wortman, 'Heterogeneous distributed shared memory', *IEEE Transactions on Parallel and Distributed Systems*, 540 (1992).
50. R. L. Sites, A. Chernoff, M. B. Kirk, M. P. Marks and S. G. Robinson, 'Binary translation', *Communications of the ACM*, **36**(2), 69–81 (1993).
51. G. M. Silberman and K. Ebcioglu, 'An architectural framework for supporting heterogeneous instruction-set architectures', *IEEE Computer*, **26**(6), 39–56 (1993).

Chapter 6 Migration Policies

The last two papers in this chapter address one of the primary uses of process migration: load balancing. **Cabrera**'s work demonstrates that long-lived processes are viable candidates for migration-based load balancing schemes [Cabrera, 1986]. Specifically, his measurements show that 40 percent of processes with a lifetime of T units of time will in fact end after running 2T units of time. This observation is the key to predicting which processes are worthwhile candidates for migration: a process is a good candidate when the cost of migration is less than its expected remaining lifetime.

Harchol-Balter and **Downey** present a model based on measurements that demonstrates the utility of preemptive process migration [Harchol-Balter, 1997]. This work demonstrates an additional benefit of migration-based load balancing. In addition to improving the performance of the migrated process, they show that migrating a long-running job away from a busy host can improve the performance of other jobs on that original host, even short-lived jobs, by reducing contention for local resources.

Other relevant studies of policies that use process migration include work by Eager, and others [Eager, 1986a, b; 1988], Krueger and Livny [Krueger, 1988], Leland [Leland, 1986], Shivaratri and Krueger [Shivaratri, 1992], Kunz [Kunz, 1991], Bryant and Finkel [Bryant, 1981], Zhou and Ferrari [Zhou, 1988], and Zhu [Zhu, 1995].

6.1 Cabrera, Luis-Felipe, The Influence of Workload on Load Balancing Strategies

Cabrera, L.-F., "The Influence of Workload on Load Balancing Strategies," *Proceedings of the USENIX Summer Conference*, pp. 446–458, June 1986.

The Influence of Workload on Load Balancing Strategies

Luis-Felipe Cabrera

Computer Science Department
Office System Laboratory
IBM Almaden Research Center [1]
650 Harry Road
San Jose, California 95120-6099
cabrera@ibm.com.ARPA

Abstract

We present an empirical analysis of process lifetimes in several Unix™ installations. We have found consistent processor resource consumption patterns across installations. Moreover, there are process creation patterns present in all the workloads. This analysis shows that several modelling hypotheses used in the load balancing literature do not hold in these installations. We also conclude that for these workloads several proposed load balancing strategies would be detrimental to the user perceived response time. We assert that load balancing strategies in uncontrolled Unix workload environments, such as research and development installations, need to be based on different schedulers than those currently available. We conclude that general purpose load balancing strategies should be based on a process migration mechanism and driven by the detection of long-lived processes. We also conclude that load balancing techniques based on specific services can be implemented at the command level with success.

The tools we have used are available to any Unix user without need of modifying the system.

Section 1. Introduction

All load balancing schemes use the observation that given a set of interconnected processors and a process stream to be executed, there are situations where distributing the processes among processors will be of benefit. Load balancing benefits are sought both in loosely coupled systems (internetworks of computers) as well as in tightly coupled systems (shared memory multiprocessors). The benefit function is usually stated in terms of response time reduction or throughput maximi-

---

1 This work was done while the author was with the Computer Science Division of the Electrical Engineering and Computer Sciences Department of the University of California at Berkeley. All the systems measured are in the Computer Science Division of the University of California at Berkeley.

zation. The appeal of load balancing schemes in communities of computers is clear: it is to the best of the community's interest to fully exploit their computer resources.

Research in load balancing strategies has been approached from the modelling and the practical viewpoints by several authors (Barak, Shiloh; Bershad; Bokhari; Bryant, Finkel; Chow, Kohler; Eager, Lazowska, Zahorjan; Krueger, Finkel; Livny, Melman; Stone; Tantawi, Towsley; Wang, Morris). In all modelling approaches the interrelationships between local scheduling cost, distributed load balancing information maintenance cost, and network access and transport cost, are normally oversimplified. Some of the simplifications are required to make the modelling effort mathematically tractable. However, even though it is known that some of these costs, like the network related ones, are non-negligible and non-constant (Cabrera, Karels, Mosher; Hunter), hypotheses which don't take them into account repeatedly surface in the modelling efforts. Workload hypotheses which we have found not to hold, like assuming that all processes belong to a same resource consumption class, are also often found. The overhead associated with the periodic distribution of workload data is seldom taken into account.

The empirical approaches to load balancing we know about have been of the specific command type. In this mode of operation a given system service identified by a command, like text formatting or compiling, is endowed with a front end which determines where to route the request for "better" service. These schemes have been implemented for single services at a time, and not as part of a more general load balancing scheme.

In this paper we report on an initial study of process characteristics in research and development workloads. We have obtained process lifetime data from several installations on different days, as well as process survival rates and process creation patterns. The data collected have characteristics which conflict with several assumptions made about possible load balancing strategies. They also affect their possible benefits. In particular we have seen that most Unix processes are very short lived, thus making very unattractive load balancing mechanisms which operate at the command interpreter level examining all incoming user commands. We have also quantified the fact that "long lived processes tend to live long". We have concluded that schedulers best for load balancing need to identify long lived processes in a system. Current BSD schedulers (Straathof, Thareja, Agrawala) fail to do this.

The rest of the paper is divided as follows. In Section 2 we discuss our data gathering method. Section 3 analyzes process lifetimes and process creation patterns. Section 4 has process survival rates while Section 5 discusses the relevance of our findings to alternative load balancing strategies.

Section 2. Gathering Data

Given that we were interested in the statistical characteristics of processes's resource consumption rather than in very precise measurements about any individual process, we decided to use standard BSD utilities in spite of their inaccuracy. The current sampling method used by BSD accounting facilities is biased as a consequence of the software clock losing interrupts. These clock interrupts are usually set to occur every 10 milliseconds. CPU consumption accounting is also inaccurate as whole time slices are charged to the process which happens to be executing when a clock interrupt occurs, irrespective of its initiation time within the quantum time slice. Short lived processes, which we have found to be predominant, are the most affected.

From the files updated by the administration utilities of the system we obtained the desired resource consumption statistics. We have concentrated on processor consumption as this is most indicative of possible gains from load balancing. If a process takes little CPU time to complete there is not much improvement in terms of response time and throughput which may come from shipping it to a remote site for execution. Understanding the CPU consumption behavior of processes seems a prerequisite to the adoption of a load balancing strategy.

Our data is of two types. On the one hand we monitored three multi-user systems for a substantial number of days to see, on a day by day basis, the number of processes created in each of these systems. This has given us an indication of how much work such a system can be expected to have. We have also seen how frequent bursts of process creations can be, as we have data showing the number of processes created per 20 minute interval. This data was gathered using a user-level *script*. Our second kind of data is about process birth-time and CPU consumption during its lifetime. This came out of the standard accounting files. We chose five different installations and obtained data about them in different sessions. We repeated our analysis on different days to assess if the data was consistent or not. (It was.) Most of our data gathering sessions were quite long. They typically lasted more than 16 consecutive hours. We also did two short sessions to see if those results matched the others. (They did.) Section 3 and Section 4 have our data and analysis.

For this paper we have labelled "Run 1" up to "Run 10" our data gathering sessions. The five installations considered represent 3 different kinds of processors: VAX 11/750, VAX 11/785, and VAX 8600. Table 1 has the name of the run, the type of processor, an installation identity, and the number of data points in each of the data gathering sessions. All the data depicted in the Figures of this paper comes from VAX 11/785 processors. The installation identities allow comparisons of different data gathering sessions from a given installation.

Label	Installation	Processor	Number of Processes Created
Run 1	c	VAX 11/785	22,732
Run 2	b	VAX 11/785	16,988
Run 3	c	VAX 11/785	21,168
Run 4	c	VAX 11/785	20,490
Run 5	b	VAX 11/785	14,485
Run 6	a	VAX 11/785	26,258
Run 7	e	VAX 8600	1,975
Run 8	d	VAX 11/750	374
Run 9	b	VAX 11/785	11,683
Run 10	a	VAX 11/785	20,471
Total number of processes:			156,624

Table 1: Classification of the data used in this study. Identity of the runs, installations, processors, and number of process creations in each measurement session.

Section 3. Process Lifetimes and Process Creation Patterns

CPU consumption and process creation patterns need to be well understood for load balancing efforts. We shall deal with them in the following two subsections. Indeed process behavior should also be considered when designing or refining the process handling mechanisms of a system.

Section 3.1. Process CPU consumption

Let's define a process lifetime to be the amount, in units of time, of processor cycles it needs to complete.

In BSD environments an approximation of process lifetime is the sum of system time and user time as given by the *time* command. Our measurements show that most processes have a very short lifetime. Indeed faster processors will make this assertion valid for larger fractions of processes. Thus, network latency costs, those incurred when placing an outgoing message on the network transport medium, will have an increasingly important impact in balancing processor loads in an internetwork of fast processors. Say l is the network latency, and t the network transport time between two nodes. Then the (conservative) minimum overhead cost of remote execution is $2l + 2t$, as this is a lower bound on a message round trip time. Given a faster CPU, then, all those processes whose lifetimes are reduced in $2l + 2t$ will automatically not benefit from remote execution.

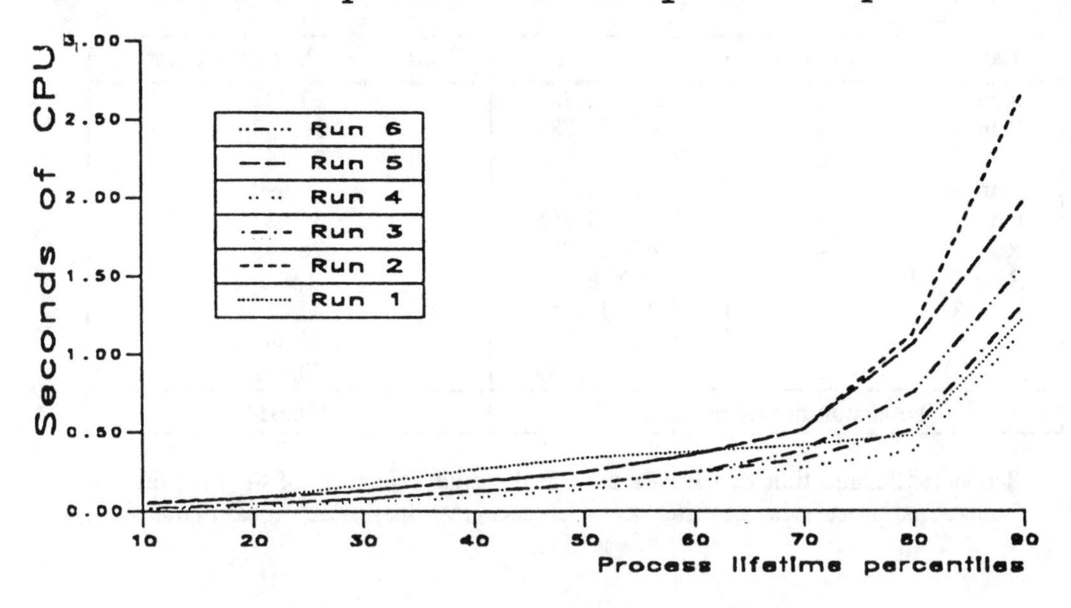

Figure 1: CPU consumption versus process lifetime percentiles.

The high skewness we have observed in process lifetime distributions tells us that the percentage of processes which won't benefit from remote execution increases more than linearly with a linear increase of the CPU power.

Figure 1 shows how the vast majority of processes requires very little CPU. Figure 3 supplements it by displaying longer process lifetimes. The data plotted in these two figures corresponds to more than 122,000 processes, all run on VAX 11/785. Between 70% and 80% of all processes in all the measured installations required less than 0.5 seconds of CPU to execute. (Between 78% and 95% of all processes required less than 1.0 seconds.) As an aside, we can observe from the data that independent of load balancing considerations BSD systems should be tailored for handling short lived processes. This is not the case today. Current process creation and process scheduling mechanisms are not geared for such short lived processes.

Figure 1 also shows that the median lifetime of processes is small. Indeed not only the median lifetime is small but the distributions of process lifetimes are also very skewed. In all systems at least 78% of the processes had lifetimes of less than one second. Figure 2 shows a closer look of processes taking less than one second of CPU, depicting the breakdown in tenths of a second of all processes from three of the runs. Each run was from a different installation. We notice that between 24% and 42% of the 1 second lifetime processes use less that 0.1 seconds of CPU. 73% to 84% require less than 0.4 seconds of CPU. A process based computing environment promotes this behavior.

For short lived processes the soundest processing strategy is to execute them locally. Minimum system overhead should be used when dealing with them. We believe that given today's computing

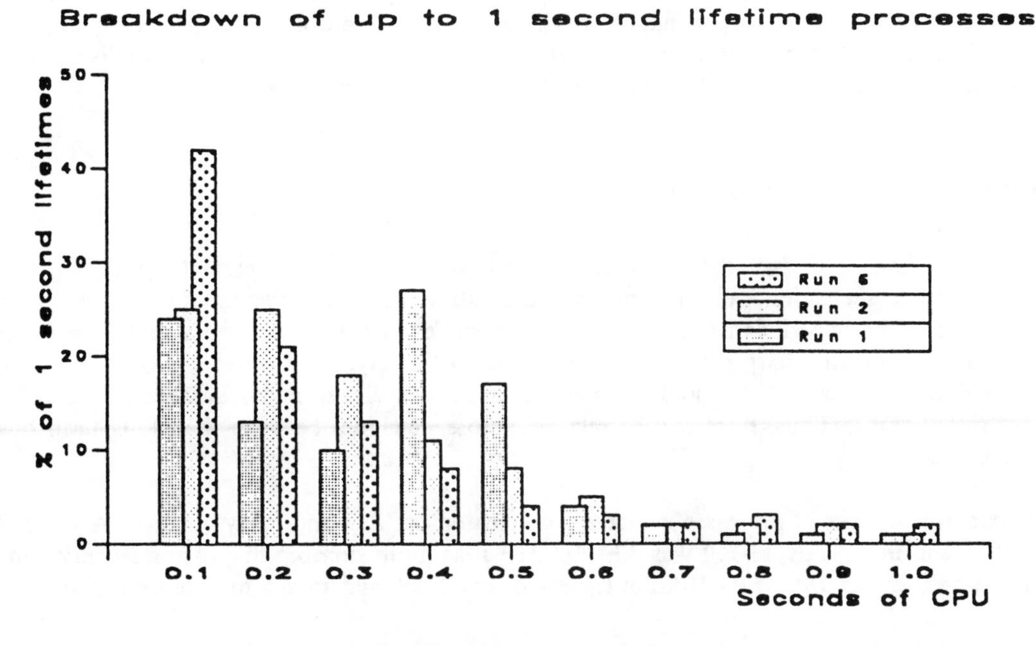

Figure 2: Breakdown of up to 1 second lifetime processes in three different VAX 11/785 installations.

power available even in workstations all load balancing strategies should contemplate local processing of short lived computations. This is certainly compatible with minimization of user perceived response time, and a very good heuristic for throughput optimization of an internetwork of hosts. From the observation that there is a substantial fraction of processes with short lifetimes, it is clear that any load balancing strategy built layered on top of the command interpreter which were to analyze all incoming commands would err on the side of unnecessary overhead. The system should not spend any time deciding to process locally short lived processes Process oriented operating systems should minimize the overhead of process creation, process dispatching, and process rescheduling.

There are services, however, which are known to require nonnegligible amounts of processor time. It is simple to instrument the system to gather data about them. A command driven approach based on such specific services, like text formatting or compilations, would be able to determine a best processing site with a high degree of confidence. We believe this selective view of command driven load balancing can be successful both for throughput maximization as for response time minimization.

A complementary aspect to the above discussion is that of overall process lifetimes. For how long do the longer lived processes stay alive? Figure 3 is an expanded version of Figure 1, in that we begin with lifetimes of 0.5 seconds and include lifetimes of up to 16 seconds. In Figure 3 we see that in all of the installations not more than 4% of the processes had lifetimes of more than 8 seconds, and not more than 2% had lifetimes of more than 16 seconds. In other words 96% of the processes used less than 8 seconds of CPU to complete and 98% used less than 16 seconds.

It should be remarked that even some of these long lived processes may not be amenable to processing on another host. A long and intense editor session, for example, can consume more than 16 seconds of CPU. It is clearly a mistake to ship that session away for processing. The associated network costs to transport data files back and forth will probably outweighs any response time benefit derived from a faster processor in the target installation. In Section 4 we discuss how load balancing techniques should benefit from these long lived processes.

Section 3.2. Process creations

Workloads are known to exhibit cyclic behavior patterns (Cabrera, Rodriguez-Galant). Figure 4 depicts the process creation behavior of one installation on four different days. The graphs correspond to 24 hour periods based on 20 minute samples. Unix connoisseurs should recognize *cron's* doings behind the sharp peak at the early hour of the day. In this installation most of the activity is done between 9:30AM and 4:30PM. This same cyclic behavior has been observed in different installations. Thus, there is a correlation across machines between their long term busy and idle cycles.

The maximum number of processes we have seen created per day on a heavily used VAX 11/785 was 33,800 and in a VAX 11/750 was 13,870. The maximum number of processes created in a 20 minute period was 1819, in a 10 minute period was 1189, and in a 5 minute period was 670.

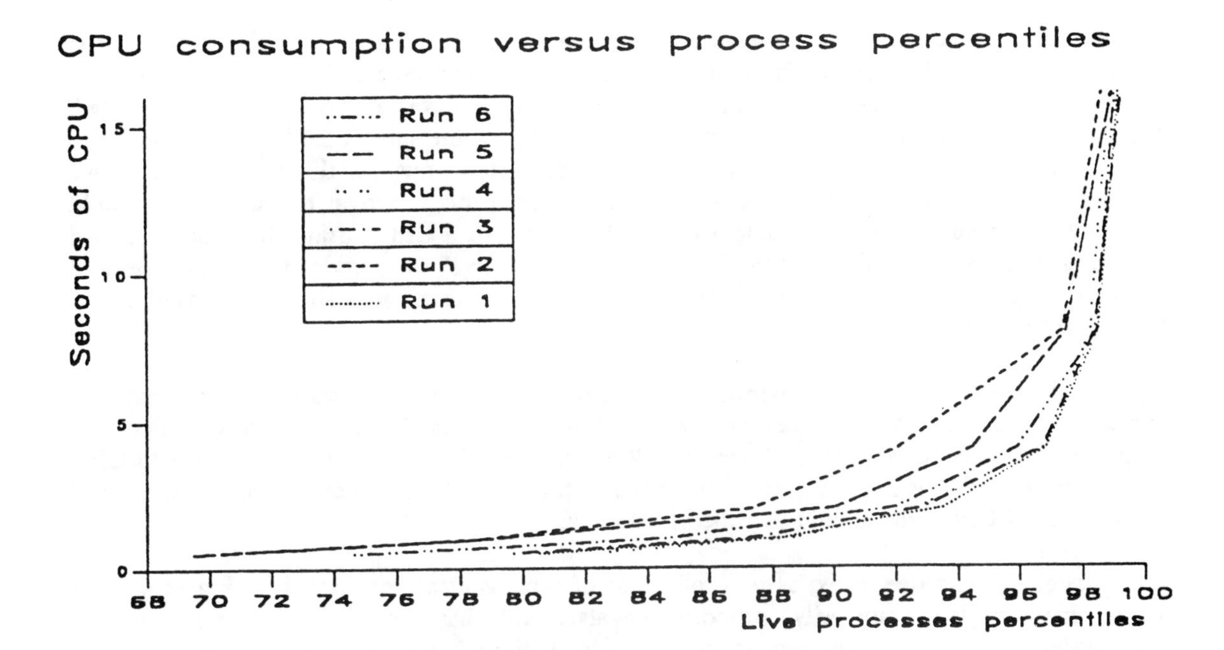

Figure 3: Lifetime percentiles versus total CPU utilization.

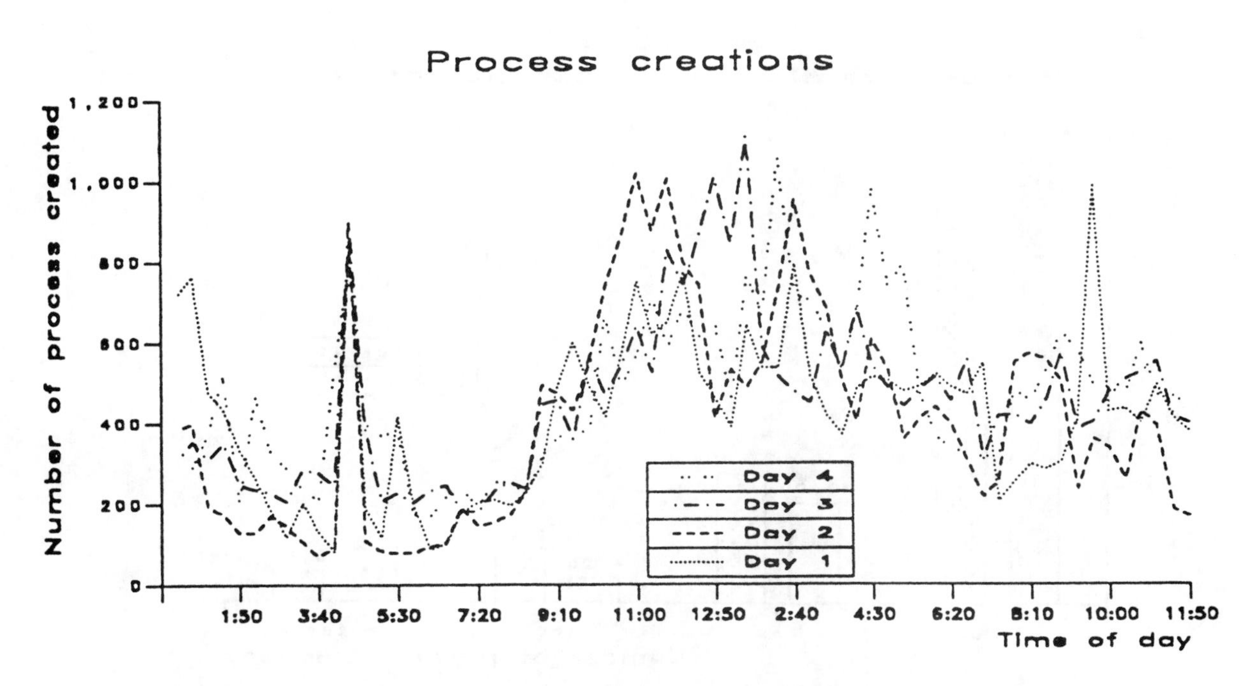

Figure 4: Process creations in 20 minute intervals in four different days. Installation a.

Figures 5 and 6 shed light on process creation frequencies on Run 9 and 10. Figure 5 depicts the frequency of periods in which different amounts of processes were created in one second windows. In Run 9 we see that in 94% of the one second intervals there was at most one process created. This figure was 86% for Run 10. In Section 3.1 we have seen the substantial fraction of processes that requires less than 0.5 seconds of CPU time to complete. Thus, for this time granularity we see that load balancing techniques based exclusively on process arrival considerations should probably do nothing between 86% to 94% of the time. An added consideration for such load balancing strategies is the high cost they would pay to keep "live" the workload data in one second intervals. Experience with high level BSD utilities, such as *rwhod*, have shown this overhead to be substantial.

Figure 6 depicts the frequency of periods in which different amounts of processes were created in five second windows. We now see that for Run 9 in 60% of the five second intervals there was at most one process created. This figure was 40% for Run 10. A ratio of elapsed time to CPU utilization time of 10 is high. Figures 5 and 6 tell us that all of those processes which use at most 0.5 seconds of CPU time should be processed locally.

The possible coincidence across systems of bursts of process creations, hinted by Figure 4 albeit for an inappropriate granularity, supports the statement that load balancing techniques for internetworks of computers with uncontrolled workloads should only consider long lived processes.

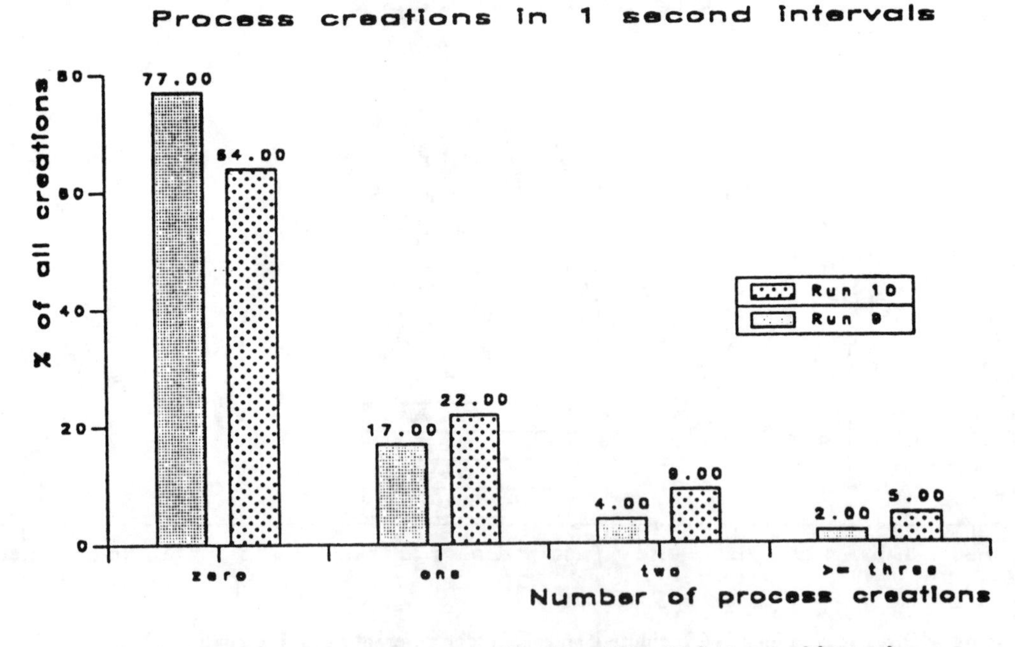

Figure 5: Process creations on Run 9 and Run 10 in 1 second intervals.

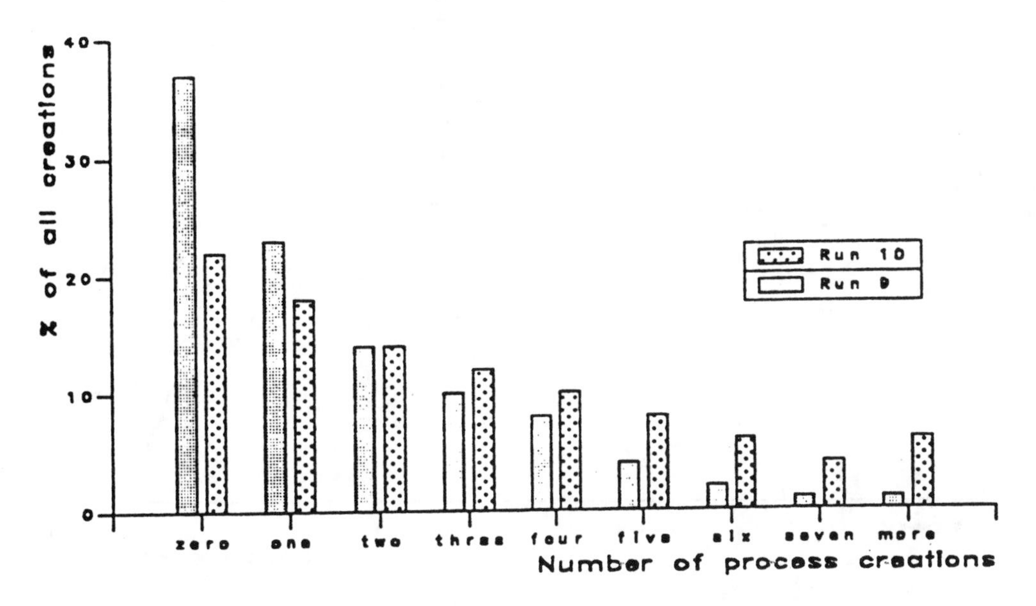

Figure 6: Process creations on Run 9 and Run 10 in 5 second intervals.

Section 4. Process Survival Rates

We have seen that most processes have very short lifetimes. How do the very long lived processes behave? Figure 7 shows this. We have plotted the percentage of survivor processes as a function of process lifetimes in the range of 0.5 seconds to 1024 seconds. The value plotted for the 0.5 second lifetime corresponds to the percentage of processes whose lifetime was at least 0.5 seconds, i. e., the percentage of processes created who took at least 0.5 seconds of CPU to complete. Beginning with processes whose lifetime was 0.5 seconds, we then calculated the percentage of them whose lifetime was larger that 1.0 seconds. The plots show that for all systems at least 60% of those processes whose lifetime was 0.5 seconds had a lifetime of at least 1.0 second. We then looked at the percentage of processes with 1.0 second lifetime who had 2.0 second lifetimes. The plots show that for all systems at least 44% of those processes whose lifetime was 1.0 seconds had a lifetime of at least 2.0 seconds. The rest of the plots proceed analogously looking at the percentages of processes with lifetimes of T seconds whose lifetimes was 2T seconds. The X axis in Figure 7 has a logarithmic scale.

The remarkable result is that for a wide range of lifetimes and systems at least 40% of the processes which have a lifetime of T units of time have in fact a lifetime of 2T units of time. This confirms that general purpose load balancing techniques should be concerned only about very long lived processes, relying on process migration for their redistribution. To achieve this other parts of the system, like the scheduler, should collaborate by detecting the long-lived processes.

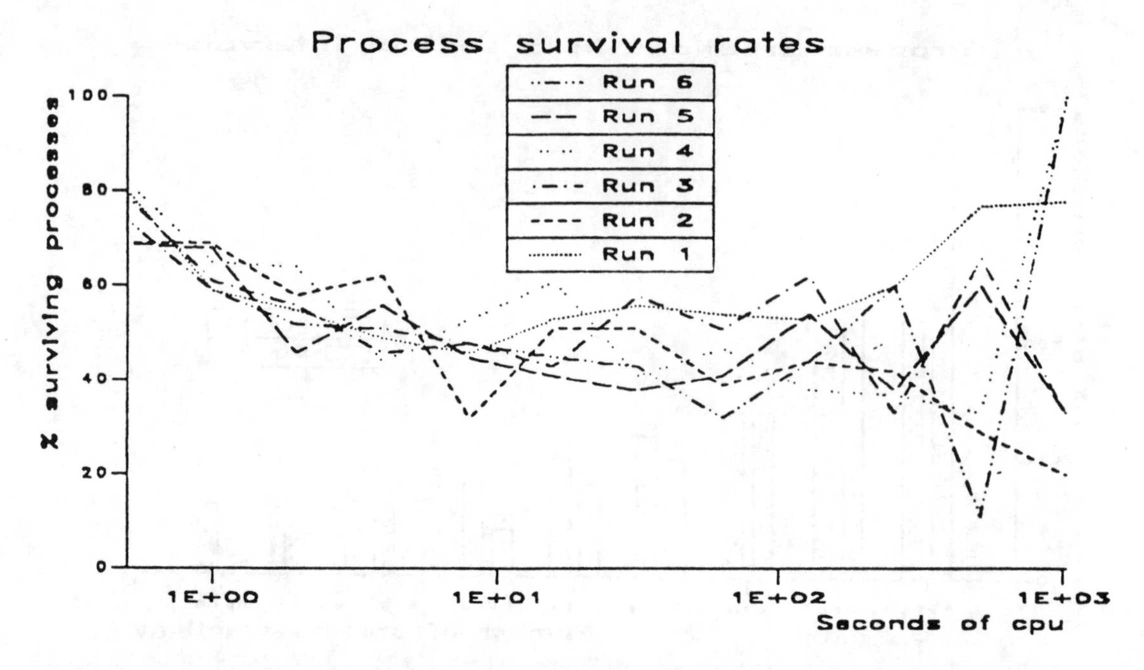

Figure 7: Process survival rates versus total CPU utilization. For each lifetime beginning at 0.5 seconds, the value of Y represents the fraction of processes whose lifetimes exceeded twice the previous lifetime.

Section 5. Conclusions on Load Balancing Strategies

Our previous sections have shown the following assertions for the measured BSD systems:

1. The median lifetime of a process is 0.4 seconds.

2. More than 78% of processes have lifetimes of less than 1 second.

3. 97% of processes have a lifetime of less than 8 seconds.

4. 98% of processes have a lifetime of less than 16 seconds.

5. Installations exhibit cyclic process creation patterns.

6. There are installations where at most one process is created in 94% of all one second intervals of wall clock time.

7. There are installations where at most one process is created in 60% of all five second intervals of wall clock time.

8. For most systems and process lifetimes, for processes that have lived longer than 0.5 seconds the probability that a process whose lifetime is T has a lifetime of 2T units of time is at least 0.4.

The main relevance of (1) and (2) is that those load balancing schemes proposed to act at the command interpreter level analyzing all user commands seem bound to do so in detriment of user perceived response time. The overhead incurred by them in deciding not to execute remotely would produce no benefit. (1) also tells us that in practice, network costs and those of obtaining resource consumption data from distant nodes surpasses the cost of locally servicing most requests. This is especially true if gateways are to be crossed. There should be lean mechanisms to deal with short lifetime processes.

(5) warns us about assumptions of statistical independence among process arrivals in an internetwork of computers, albeit the data is at a granularity three orders of magnitude over the median process lifetime. (6) and (7) tell us that there would be few process creation bursts. Because of (5) we believe some of these bursts would occur within the same time frame in all systems.

(6) and (7) add support to the claim that load balancing strategies should be triggered by the detection of long lived computations. (8) says that the long-lived processes are worthy of attention. Once the system discovers a long-lived process the overhead associated with its possible relocation is well spent.

Looking at the scheduler of most Unix installations, we see that there is a pervasive use of Round-Robin with priority queues. There is not, however, a scheduler understood notion of long-lived processes. This would allow, for example, to suspend and reschedule at a future time in possibly a different machine a long-lived process. In high load conditions when several long-lived processes are competing among themselves for processor cycles the fairness criterion built into the current Unix schedulers prevent suspension of such processes. Schedulers will have to change to make good use of the knowledge that a particular process is long-lived (specially if it is a noninteractive one). A categorization along these lines would allow triggering a load sharing mechanism which could query other systems about their long term resource demands. These long term resource demands can be better predicted if one has an explicit category of long-term jobs.

In light of these measurements, one should also consider possible changes to the mechanisms which manipulate processes. The basic observation is to minimize process creation costs. One can also think of making the system data statistics optional. They should be allowed to be turned off at will. Lastly process dispatching and process scheduling priority (re)computations should have minimum overhead.

In synthesis, the empirical analysis of process lifetimes shows that the workload has an important impact on load balancing strategies. Our data clearly shows that in the measured environments a command interpreter based general purpose load balancing scheme which were to analyze all incoming user commands would be detrimental to both response time and throughput. We have also seen that the scheduler will need to play an important role in successful load balancing strategies. The observation that between 78% and 95% of all processes have a lifetime of less than 1.0 seconds points us to the long lived processes as those worthy of attention for load balancing. The fact that long-lived processes survive longer also calls for this long-lived process based approach. Our data suggests that the only viable general purpose load balancing schemes should be based on a process migration mechanism and driven by the detection of long-lived processes. Moreover, there should also be a mechanism to differentiate long lived interactive processes from batch oriented intense computations. The former should not be migrated at all while the latter are the prime candidate for load balancing. We also conclude that load balancing techniques based on specific services can be implemented at the command level with success.

We conjecture that other workloads of systems which favor the process model of computing will also exhibit the behaviors described in this paper. For such workloads all of our considerations apply.

Acknowledgements. Thanks to Ricardo Gusella for providing me with some data. Also thanks to Roger Haskin, John Palmer, and Jim Wyllie for their helpful comments and careful reading of earlier versions of this manuscript.

Section 6. References

1. A. Barak, and A. Shiloh, A Distributed Load Balancing Policy for a Multicomputer, Department of Computer Science, The Hebrew University of Jerusalem, 1984.

2. B. Bershad, The Garcon-Maitre d' System. Computer Science Division, Department of Electrical Engineering and Computer Sciences, University of California, Berkeley.

3. R. Bryant, and R. A. Finkel, A Stable Distributed Scheduling Algorithm, Proceedings of the 2nd International Conference on Distributed Computing Systems, April 1981.

4. L.-F. Cabrera, and G. Rodriguez-Galant, Predicting Performance in Unix Systems From Portable Workload Estimators Based on the Terminal Probe Method, Computer Science Division, Department of Electrical Engineering and Computer Sciences, University of California, Berkeley, Research Report UCB/CSD 84/194, August 1984.

5. L.-F. Cabrera, E. Hunter, M. Karels, and D. Mosher, A User-Process Oriented Performance Study of Ethernet Networking Under Berkeley Unix 4.2BSD. Computer Science Division, Department of Electrical Engineering and Computer Sciences, University of California, Berkeley, Research Report UCB/CSD 84/217, December 1984.

6. L.-F. Cabrera, M. Karels, and D. Mosher, The Impact of Buffer Management on Networking Software Performance in Berkeley Unix 4.2BSD: A Case Study. Proceedings of the 1985 Summer Usenix Conference, Portland, Oregon, pp. 507-517. June 1985.

7. D. L. Eager, E. D. Lazowska, and J. Zahorjan, Adaptive Load Sharing in Homogeneous Distributed Systems. IEEE Transactions on Software Engineering, December 1985.

8. D. L. Eager, E. D. Lazowska, and J. Zahorjan, A Comparison of Receiver-Initiated and Sender-Initiated Adaptive Load Sharing. To appear in Performance Evaluation, Spring 1986.

9. E. Hunter, A Performance Study of the Ethernet Under Berkeley Unix 4.2BSD. Proceedings of CMG XV, pp. 373-382, December 1984.

10. P. Krueger, and R. A. Finkel, An Adaptive Load Balancing Algorithm for a Multicomputer. Technical Report 539, Department of Computer Science, University of Wisconsin, April 1984.

11. M. Livny, and M. Melman, Load Balancing in Homogeneous Broadcast Distributed Systems. Proceedings of ACM Computer Network Performance Symposium, 1982.

12. H. S. Stone, Multiprocessor Scheduling with the Aid of Network Flow Algorithms. IEEE Transactions on Software Engineering, January 1977.

13. H. S. Stone, Critical Load Factors in Two Processors Distributed Systems. IEEE Transactions on Software Engineering, May 1978.

14. J. H. Straathof, A. K. Thareja, and A. K. Agrawala, Unix Scheduling for Large Systems, Proceedings of the 1986 Winter Usenix Technical Conference, Denver, Colorado, pp. 111-139.

15. A. N. Tantawi, and D. Towsley, Optimal Static Load Balancing in Distributed Computer Systems. JACM 32, No. 2, April 1985.

16. K. Thompson, Unix Implementation, Bell System Technical Journal, Vol. 57, No. 6, Part 2, pp. 1931-1946, July/August 1978.

17. Y.-T. Wang, and R. J. T. Morris, Load Sharing in Distributed Systems. IEEE Transactions on Computers, C-34, No. 3, March 1985.

6.2 Harchol-Balter, M. and Downey, A.B., Exploiting Process Lifetime Distributions for Dynamic Load Balancing

Harchol-Balter, M. and Downey, A.B., "Exploiting Process Lifetime Distributions for Dynamic Load Balancing," *ACM Transactions on Computer Systems,* 15(3):253–285, August 1997.

Exploiting Process Lifetime Distributions for Dynamic Load Balancing

Mor Harchol-Balter
and
Allen B. Downey
University of California, Berkeley

We consider policies for CPU load balancing in networks of workstations. We address the question whether preemptive migration (migrating active processes) is necessary, or whether remote execution (migrating processes only at the time of birth) is sufficient for load balancing. We show that resolving this isssue is strongly tied to understanding the process lifetime distribution. Our measurements indicate that the distribution of lifetimes for UNIX process is Pareto (heavy-tailed), with a consistent functional form over a variety of workloads. We show how to apply this distribution to derive a preemptive migration policy that requires no hand-tuned parameters. We use a trace-driven simulation to show that our preemptive migration strategy is far more effective than remote execution, even when the memory transfer cost is high.

Categories and Subject Descriptors: unknown [**unknown**]: unknown—***unknown***

General Terms: unknown

Additional Key Words and Phrases: Load balancing, load sharing, migration, remote execution, workload modeling, trace-driven simulation, network of workstations, heavy-tailed, Pareto distribution

1. INTRODUCTION

Most systems that perform load balancing use remote execution (i.e. non-preemptive migration) based on a priori knowledge of process behavior, often in the form of a list of process names eligible for migration. Although some systems are capable of migrating active processes, most do so only for reasons other than load balancing, such as preserving autonomy. A previous analytic study by Eager et al. discourages

Mor Harchol-Balter supported by National Physical Science Consortium (NPSC) Fellowship and NSF grant number CCR-9201092. Allen Downey partially supported by NSF (DARA) grant DMW-8919074.
An earlier version of this paper appeared in the *Proceedings of the ACM Sigmetrics Conference on Measurement and Modeling of Computer Systems* (May 23-26,1996) pp. 13-24.
Address: {harchol,downey}@cs.berkeley.edu, University of California, Berkeley, CA 94720.

implementing preemptive migration for load balancing, showing that the additional performance benefit of preemptive migration is small compared with the benefit of simple non-preemptive migration schemes [Eager et al. 1988]. But simulation studies, which can use more realistic workload descriptions, and implemented systems have shown greater benefits for preemptive migration [Krueger and Livny 1988] [Barak et al. 1993]. This paper uses a measured distribution of process lifetimes and a trace-driven simulation to investigate these conflicting results.

1.1 Load balancing taxonomy

On a network of shared processors, *CPU load balancing* is the idea of migrating processes across the network from hosts with high loads to hosts with lower loads. The motivation for load balancing is to reduce the average completion time of processes and improve the utilization of the processors. Analytic models and simulation studies have demonstrated the performance benefits of load balancing, and these results have been confirmed in existing distributed systems (see Section 1.4).

An important part of the load balancing strategy is the ***migration policy***, which determines when migrations occur and which processes are migrated. This is the question we address in this paper. The other half of a load balancing strategy is the location policy — the selection a new host for the migrated process. Previous work has suggested that simply choosing the target host with the shortest CPU run queue is both simple and effective [Zhou 1987] [Kunz 1991]. Our work confirms the relative unimportance of location policy.

Process migration for purposes of load balancing comes in two forms: ***remote execution***, also called ***non-preemptive*** migration, in which some new processes are (possibly automatically) executed on remote hosts, and ***preemptive*** migration, in which running processes may be suspended, moved to a remote host, and restarted. In non-preemptive migration only newborn processes are migrated.

Load balancing may be done explicitly (by the user) or implicitly (by the system). Implicit migration policies may or may not use a priori information about the function of processes, how long they will run, etc.

Since the cost of remote execution is usually significant relative to the average lifetime of processes, implicit non-preemptive policies require some a priori information about job lifetimes. This information is often implemented as an eligibility list that specifies by process name which processes are worth migrating [Svensson 1990] [Zhou et al. 1993].

In contrast, most preemptive migration policies do not use a priori information, since this is often difficult to maintain and preemptive strategies can perform well without it. These systems use only system-visible data like the current age of each process or its memory size.

This paper examines the performance benefits of **preemptive, implicit** load balancing strategies that assume **no a priori information** about processes.

We answer the following two questions:

(1) Is preemptive migration worthwhile, given the additional cost (CPU and latency) associated with migrating an active process?

(2) Which active processes, if any, are worth migrating?

1.2 Process model

In our model, processes use two resources: CPU and memory (we do not consider I/O). Thus, we use "age" to mean CPU age (the CPU time a process has used thus far) and "lifetime" to mean CPU lifetime (the total CPU time from start to completion). We assume that processors implement time-sharing with round-robin scheduling; in Section 7 we discuss the effect of other local scheduling policies. Since processes may be delayed while on the run queue or while migrating, the slowdown imposed on a process is

$$\text{Slowdown of process } p = \frac{\textit{wall-time}(p)}{\textit{CPU-time}(p)}$$

where wall time is the total time a process spends running, waiting in queue, or migrating.

1.3 Outline

The effectiveness of load balancing — either by remote execution or preemptive migration — depends strongly on the nature of the workload, including the distribution of process lifetimes and the arrival process. This paper presents empirical observations about the workload on a network of UNIX workstations, and uses a trace-driven simulation to evaluate the impact of this workload on proposed load balancing strategies.

Section 2 presents a study of the distribution of process lifetimes for a variety of workloads in an academic environment, including instructional machines, research machines, and machines used for system administration. We find that the distribution is predictable with goodness of fit greater than 99% and consistent across a variety of machines and workloads. As a rule of thumb, the probability that a process with CPU age of one second uses more than T seconds of total CPU time is $1/T$ (see Figure 1).

Our measurements are consistent with those of Leland and Ott [Leland and Ott 1986], but this prior work has been incorporated in few subsequent analytic and simulator studies of load balancing. This omission is unfortunate, since the results of these are sensitive to the lifetime model (see Section 2.2).

Our observations of lifetime distributions have the following consequences for load balancing:

—They suggest that it is preferable to migrate older processes because these processes have a higher probability of living long enough (eventually using enough CPU) to amortize their migration cost.

—A functional model of the distribution provides an analytic tool for deriving the eligibility of a process for migration as a function of its current age, migration cost, and the loads at its source and target host.

In Section 3 we derive a migration eligibility criterion that guarantees that the slowdown imposed on a migrant process is lower in expectation than it would be without migration. According to this criterion, a process is eligible for migration only if its

$$\text{CPU age} > \frac{1}{n-m} \cdot \text{migration cost}$$

where n (respectively m) is the number of processes at the source (target) host.

In Section 5 we use a trace-driven simulation to compare our preemptive migration policy with a non-preemptive policy based on name-lists. The simulator uses start times and durations from traces of a real system, and migration costs chosen from a measured distribution.

We use the simulator to run three experiments: first we evaluate the effect of migration cost on the relative performance of the two strategies. Not surprisingly, we find that as the cost of preemptive migration increases, it becomes less effective. Nevertheless, preemptive migration performs better than non-preemptive migration even with surprisingly large migration costs, despite several conservative assumptions that give non-preemptive migration an unfair advantage.

Next we choose a specific model of preemptive and non-preemptive migration costs based on real systems (see Section 4), and use this model to compare the two migration strategies in more detail. We find that preemptive migration reduces the mean delay (queueing and migration) by 35–50%, compared to non-preemptive migration. We also propose several alternative metrics intended to measure users' perception of system performance. By these metrics, the additional benefits of preemptive migration compared to non-preemptive migration appear even more significant.

In Section 5.4 we discuss in detail why a simple preemptive migration policy is more effective than even a well-tuned non-preemptive migration policy. In Section 5.5 we use the simulator to compare our preemptive migration strategy with previously proposed preemptive strategies.

We finish with a criticism of our model in Section 6, a discussion of future work in Section 7 and conclusions in Section 8.

1.4 Related work

1.4.1 *Systems.* Although several systems have the mechanism to migrate active jobs, few have implemented implicit load balancing ***policies***. Most systems only allow for ***explicit*** load balancing. That is, there is no load balancing policy; the user decides which processes to migrate, and when. Examples include Accent [Zayas 1987], Locus [Thiel 1991], Utopia [Zhou et al. 1993], DEMOS/MP [Powell and Miller 1983], V [Theimer et al. 1985], NEST [Agrawal and Ezzet 1987], RHODOS [De Paoli and Goscinski 1995], and MIST [Casas et al. 1995].

A few systems have ***implicit*** load balancing policies, however they are strictly non-preemptive policies (active processes are only migrated for purposes other than load balancing, such as preserving workstation autonomy). Examples include Amoeba [Tanenbaum et al. 1990], Charlotte [Artsy and Finkel 1989], Sprite [Douglis and Ousterhout 1991], Condor [Litzkow et al. 1988], and Mach [Milojicic 1993]. In general, non-preemptive load balancing policies depend on ***a priori*** information about processes; e.g., explicit knowledge about the runtimes of processes or user-provided lists of migratable processes [Agrawal and Ezzet 1987] [Litzkow and Livny 1990] [Douglis and Ousterhout 1991] [Zhou et al. 1993].

One existing system that has implemented implemented automated preemptive load balancing is MOSIX [Barak et al. 1993]. Our results support the MOSIX claim that their scheme is effective and robust.

1.4.2 *Studies.* Although few systems incorporate migration policies, there have been many simulation and analytical studies of various migration policies. Most of these studies have focused on load balancing by remote execution [Livny and Melman 1982] [Wang and Morris 1985] [Casavant and Kuhl 1987] [Zhou 1987] [Pulidas et al. 1988] [Kunz 1991] [Bonomi and Kumar 1990] [Evans and Butt 1993] [Lin and Raghavendra 1993] [Mirchandaney et al. 1990] [Zhang et al. 1995] [Zhou and Ferrari 1987] [Hać and Jin 1990] [Eager et al. 1986].

Only a few studies address preemptive migration policies [Leland and Ott 1986] [Krueger and Livny 1988]. The Leland and Ott migration policy is also age based, but doesn't take migration cost into account.

Eager et. al.,[Eager et al. 1988], conclude that the additional performance benefit of preemptive migration is too small compared with the benefit of non-preemptive migration to make preemptive migration worthwhile. This result has been widely cited, and in several cases used to justify the decision not to implement preemptive migration, as in the Utopia system, [Zhou et al. 1993]. Our work differs from [Eager et al. 1988] in both system model and workload description. [Eager et al. 1988] model a server farm in which incoming jobs have no affinity for a particular processor, and thus the cost of initial placement (remote execution) is free. This is different from our model, a network of workstations, in which incoming jobs arrive at a particular host and the cost of moving them away, even by remote execution, is significant compared to most process lifetimes. Also, [Eager et al. 1988] use a degenerate hyperexponential distribution of lifetimes that includes few jobs with non-zero lifetimes. When the coefficient of variation of this distribution matches the distributions we observed, fewer than 4% of the simulated processes have non-zero lifetimes. With so few jobs (and balanced initial placement) there is seldom any load imbalance in the system, and thus little benefit for preemptive migration. Furthermore, the [Eager et al. 1988] process lifetime distribution is exponential for jobs with non-zero lifetimes, the consequences of which we discuss in Section 2.2. For a more detailed explanation of this distribution and its effect on the study, see [Downey and Harchol-Balter 1995].

Krueger and Livny investigate the benefits of supplementing non-preemptive migration with preemptive migration and find that preemptive migration is worthwhile. They use a hyperexponential lifetime distribution that approximates closely the distribution we observed; as a result, their findings are largely in accord with ours. One difference between their work and ours is that they used a synthetic workload with Poisson arrivals. The workload we observed, and used in our trace-driven simulations, exhibits serial correlation; i.e. it is more bursty than a Poisson process. Another difference is that their migration policy requires several hand-tuned parameters. In Section 3.1 we show how to use the distribution of lifetimes to eliminate these parameters.

Like us, Bryant and Finkel discuss the distribution of process lifetimes and its effect on preemptive migration policy, but their hypothetical distributions are not based on system measurements [Bryant and Finkel 1981]. Also like us, they choose migrant processes on the basis of expected slowdown on the source and target hosts, but their estimation of those slowdowns is very different from ours. In particular, they use the distribution of process lifetimes to predict a host's future load as a function of its current load and the ages of the processes running there. We have examined this issue and found (1) that this model fails to predict future loads because it ignores future arrivals, and (2) that current load is the best predictor of future load (see Section 3.1). Thus, in our estimates of slowdown, we assume that the future load on a host is equal to the current load.

2. DISTRIBUTION OF LIFETIMES

The general shape of the distribution of process lifetimes in an academic environment has been known for a long time [Rosin 1965]: there are many short jobs and a few long jobs, and the variance of the distribution is greater than that of an exponential distribution.

In 1986, Cabrera measured UNIX processes and found that over 40% doubled their current age [Cabrera 1986]. That same year, Leland and Ott proposed a functional form for the process lifetime distribution, based on measurements of the lifetimes of 9.5 million UNIX processes between 1984 and 1985 [Leland and Ott 1986]. They conclude that process lifetimes have a UBNE (used-better-than-new-in-expectation) type of distribution. That is, the greater the current CPU age of a process, the greater its expected remaining CPU lifetime. Specifically, they find that for $T > 3$ seconds, the probability of a process's lifetime exceeding T seconds is rT^k, where $-1.25 < k < -1.05$ and r normalizes the distribution.

In contrast, Rommel [Rommel 1991] claims that his measurements show that "long processes have exponential service times." Many subsequent studies assume an exponential lifetime distribution.

Because of the importance of the process lifetime distribution for load balancing policies, we performed an independent study of this distribution, and found that the functional form proposed by Leland and Ott fits the observed distributions well, for processes with lifetimes greater than 1 second. This functional form is consistent across a variety of machines and workloads, and although the parameter, k, varies from -1.3 to -.8, it is generally near -1. Thus, as a *rule of thumb*,

—The probability that a process with age 1 second uses at least T seconds of total CPU time is about $1/T$.

—The probability that a process with age T seconds uses at least an additional T seconds of CPU time is about $1/2$. Thus, the median remaining lifetime of a process is equal to its current age.

Section 2.1 describes our measurements and the distribution of lifetimes we observed. Section 2.2 discusses other models for the distribution of lifetimes, and argues that the particular shape of this distribution is critical for evaluating migration policies.

2.1 Lifetime distribution when lifetime $> 1s$

To determine the distribution of lifetimes for UNIX processes, we measured the lifetimes of over one million processes, generated from a variety of academic workloads, including instructional machines, research machines, and machines used for system administration. We obtained our data using the UNIX command `lastcomm`, which outputs the CPU time used by each completed process.

Figure 1 shows the distribution of lifetimes from one of the machines. The plot shows only processes whose lifetimes exceed one second. The dotted (heavy) line

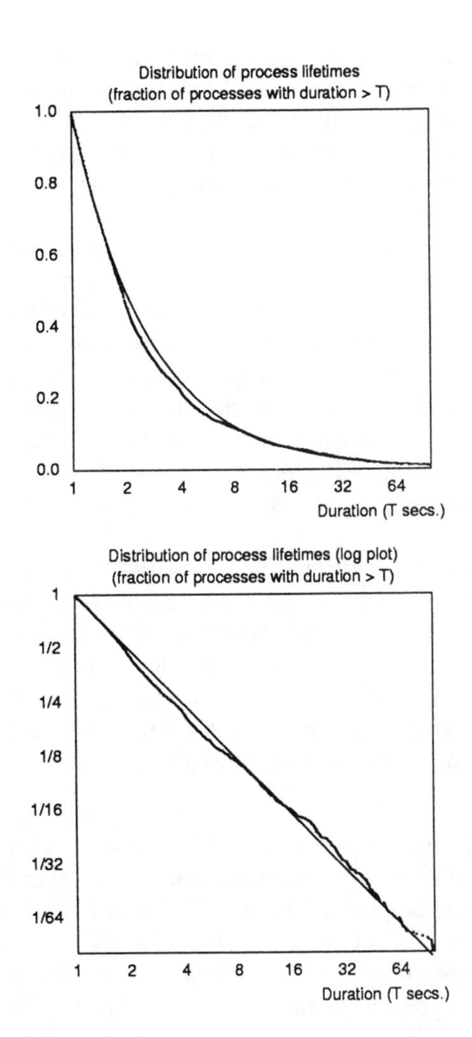

Fig. 1. (a) Distribution of lifetimes for processes with lifetimes greater than 1 second, observed on machine po mid-semester. The dotted (thicker) line shows the measured distribution; the solid (thinner) line shows the least squares curve fit. (b) The same distribution on a log-log scale. The straight line in log-log space indicates that the distribution can be modeled by T^k, where k is the slope of the line.

shows the measured distribution; the solid (thinner) line shows the least-squares fit to the data using the proposed functional form $Pr\{Lifetime > T\} = T^k$.

By visual inspection, it is clear that the proposed model fits the observed data well. In contrast, Figure 2 shows that it is impossible to find an exponential curve that fits the distribution of lifetimes we observed.

For all the machines we studied, the distribution of process lifetimes fits a curve of the form T^k, with k varying from -1.3 to -0.8 for different machines. Table 1 shows the value of the parameter for each machine we studied, estimated by an iteratively weighted least-squares fit (with no intercept, in accordance with the functional model). We calculated these estimates with the BLSS command `robust` [Abrahams and Rizzardi 1988].

The standard error associated with each estimate gives a confidence interval for that parameter (all of these parameters are statistically significant at a high degree of certainty). The R^2 value indicates the goodness of fit of the model — the values shown here indicate that the fitted curve accounts for greater than 99% of the variation of the observed values. Thus, the goodness of fit of these models is high (for an explanation of R^2 values, see [Larsen and Marx 1986]).

Table 2 shows the cumulative distribution function, probability density function, and conditional distribution function for process lifetimes. The second column shows these functions when $k = -1$, which we will assume for our analysis in Section 3.

The functional form we are proposing (the fitted distribution) has the property that its moments (mean, variance, etc.) are infinite. Of course, since the observed distributions have finite sample size, they have finite mean (0.4 seconds) and coefficient of variation (5–7). One must be cautious when summarizing long-tailed distributions, though, because calculated moments tend to be dominated by a few outliers. In our analyses we use more robust summary statistics (order statistics like the median, or the estimated parameter k) to summarize distributions, rather than moments.

2.1.1 *Process with lifetime* < 1 *second.* For processes with lifetimes less than 1 second, we did not find a consistent functional form; however, for the machines we studied these processes had an even lower hazard rate than those of age > 1 second. That is, the probability that a process of age $T < 1$ seconds lives another T seconds is always greater than 1/2. Thus for jobs with lifetimes less than 1 second, the median remaining lifetime is greater than the current age.

2.2 Why the distribution is critical

Many prior studies of process migration assume an exponential distribution of process lifetimes, both in analytical papers [Lin and Raghavendra 1993] [Mirchandaney et al. 1990] [Eager et al. 1986] [Ahmad et al. 1991] and in simulation studies [Kunz 1991] [Pulidas et al. 1988] [Wang and Morris 1985] [Evans and Butt 1993] [Livny and Melman 1982] [Zhang et al. 1995] [Chowdhury 1990]. The reasons for this assumption include: (1) analytic tractability, and (2) the belief that even if the actual lifetime distribution is in fact not exponential, assuming an exponential distribution will not affect the results of load balancing studies.

Regarding the first point, although the functional form that we and Leland and

Name of Host	Number of Procs	Number Procs > 1 sec	Estimated Distrib.	Std Error	R^2
po1	77440	4107	$T^{-0.97}$	.016	0.997
po2	154368	11468	$T^{-1.22}$	.012	0.999
po3	111997	7524	$T^{-1.27}$	.021	0.997
cory	182523	14253	$T^{-0.88}$	.030	0.982
pors	141950	10402	$T^{-0.94}$	.015	0.997
bugs	83600	4940	$T^{-0.82}$	.007	0.999
faith	76507	3328	$T^{-0.78}$	.045	0.964

Table 1. The estimated lifetime distribution for each machine measured, and the associated goodness of fit statistics. Description of machines: po is a heavily-used DECserver5000/240, used primarily for undergraduate coursework. Po1, po2, and po3 refer to measurements made on po mid-semester, late-semester, and end-semester. Cory is a heavily-used machine, used for coursework and research. Porsche is a less frequently-used machine, used primarily for research on scientific computing. Bugs is a heavily-used machine, used primarily for multimedia research. Faith is an infrequently-used machine, used both for video applications and system administration.

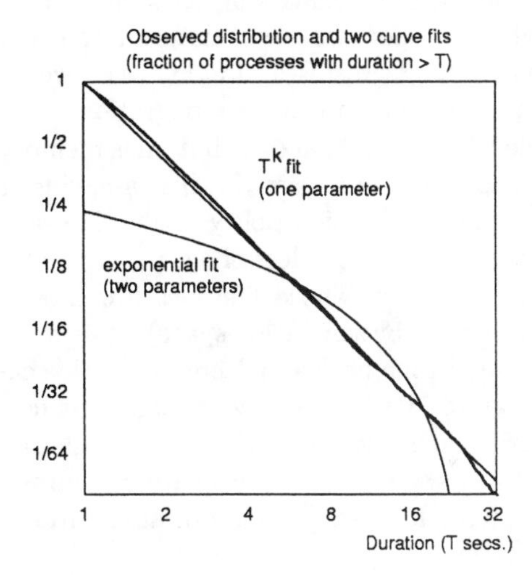

Fig. 2. In log-log space, this plot shows the distribution of lifetimes for the 13000 processes from our traces with lifetimes greater than second, and two attempts to fit a curve to this data. One of the fits is based on the model proposed in this paper, T^k. The other fit is an exponential curve, $c \cdot e^{-\lambda T}$. Although the exponential curve is given the benefit of an extra free parameter, it fails to model the observed data. The proposed model fits well. Both fits were performed by iteratively-weighted least squares.

Distribution of lifetimes for processes > 1 sec	When $k = -1$
$\mathbf{Pr}\{L > T \text{ sec}\} = T^k$	$1/T$
$\mathbf{Pr}\{T < L < T + dT \text{ sec}\} = -kT^{k-1}dT$	$1/T^2 \cdot dT$
$\mathbf{Pr}\{L > b \text{ sec} \mid \text{age} = a\} = \left(\frac{b}{a}\right)^k$	a/b

Table 2. The cumulative distribution function, probability density function, and conditional distribution function for the process lifetime L. The second column shows the functional form of each for the typical value $k = 1$.

Ott propose cannot be used in queueing models as easily as an exponential distribution, it nevertheless lends itself to some forms of analysis, as we show in Section 3.1.

Regarding the second point, we argue that the particular shape of the lifetime distribution affects the performance of migration policies, and therefore that it is important to model this distribution accurately. Specifically, the choice of a migration policy depends on how the expected remaining lifetime of a job varies with age. In our observations we found a distribution with the UBNE property — the expected remaining lifetime of a job increases linearly with age. As a result, we chose a migration policy that migrates only old jobs.

But different distributions yield in different relationships between the age of a process and its remaining lifetime. For example, a uniform distribution has the NBUE property — the expected remaining lifetime *decreases* linearly with age. Thus if the distribution of lifetimes were uniform, the migration policy should choose to migrate only young processes. In this case, we expect non-preemptive migration to perform better than preemptive migration.

As another example, the exponential distribution is memoryless — the remaining lifetime of a job is independent of its age. In this case, since all processes have the same expected lifetimes, the migration policy might choose to migrate the process with the lowest migration cost, regardless of age.

As a final example, processes whose lifetimes are chosen from a uniform log distribution (a uniform distribution in log-space) have a remaining lifetime that increases up to a point and then begins to decrease. In this case, the best migration policy might be to migrate jobs that are old enough, but not too old.

Thus different distributions, even with the same mean and variance, can lead to different migration policies. In order to evaluate a proposed policy, it is critical to choose a distribution model with the appropriate relationship between expected remaining lifetime and age.

Some studies have used hyperexponential distributions to model the distribution of lifetimes. These distributions may or may not have the right behavior, depending on how accurately they fit observed distributions. [Krueger and Livny 1988] use a three-stage hyperexponential with parameters estimated to fit observed values. This distribution has the appropriate UBNE property. But the two-stage hyperexponential distribution [Eager et al. 1988] use is memoryless; the remaining lifetime of a job is independent of its age (for jobs with nonzero lifetimes). According to this distribution, migration policy is irrelevant; all processes are equally good candidates for migration. This result is clearly in conflict with our observations.

Assuming the wrong lifetime distribution may also underestimate the benefits of preemptive migration. The heavy tail of our measured lifetime distribution implies that a tiny fraction of the jobs require more CPU than all the other jobs combined. As we'll discuss in Section 5.4, part of the power of preemptive migration is its ability to identify those few hogs. In a lifetime distribution without such a heavy tail, preemptive migration might not be as effective.

3. MIGRATION POLICY

A migration policy is based on two decisions: when to migrate processes and which processes to migrate. The first question concerns how often or at what times the system checks for eligible migrants. We address this issue briefly in Section 5.1.1. The focus of this paper is the second question, also known as the ***selection policy***:

> Given that the load at a host is too high, how do we choose ***which*** process to migrate?

Our heuristic is to migrate processes that are expected to have long remaining lifetimes. The motivation for this heuristic is twofold. From the process's perspective, migration time has a large impact on response time. A process would choose to migrate only if the migration overhead could be amortized over a longer lifetime. From the perspective of the source host, it takes a significant amount of work to package a process for migration. The host would only choose to migrate processes that are likely to be more expensive to run than to migrate.

Many existing migration policies only migrate newborn processes (no preemption), because these processes have no allocated memory and thus their migration cost is low. The idea of migrating newborn processes might also stem from the fallacy that process lifetimes have an exponential distribution, implying that all processes have equal expected remaining lifetimes regardless of their age, so one should migrate the cheapest processes. The problem with only migrating newborn processes is that, according to the process lifetime distribution, newborn processes are unlikely to live long enough to justify the cost of remote execution. In fact, our measurements show that over 70% of processes have lifetimes smaller than the smallest non-preemptive migration cost (see Table 3).

Thus a newborn migration policy is only justified if the system has prior knowledge about processes and can selectively migrate processes likely to be CPU hogs. We have found, though, that the ability of the system to predict process lifetimes by name is limited (Section 5.4).

Can we do better? The distribution of lifetimes implies that we expect and old process to run longer than a young process; thus, it is preferable to migrate old processes.

There are two potential problems with this approach. First, since the vast majority of processes are short, there might not be enough old processes to have a significant load balancing effect. In fact, although there are few long-lived processes, they account for a large part of the total CPU load. According to our measurements, typically fewer than 4% of processes live longer than 2 seconds, yet these processes make up more than 60% of the total CPU load. This is due to the long tail of the process lifetime distribution. Furthermore, we will see that the ability to migrate even a few large jobs can have a large effect on system performance, since a single long process on a busy host imposes slowdowns on many short processes.

A second problem with migrating old processes is that the migration cost for an active process is much greater than the cost of remote execution. If preemptive migration is done carelessly, this additional cost might overwhelm the benefit of migrating processes with longer expected lives.

3.1 Our Migration Policy

For this reason, we propose a strategy that guarantees that every migration improves the expected performance of the migrant process and the other processes at the source host. This strategy migrates a process ***only if it improves the expected slowdown of the process***, where slowdown is defined as in Section 1.2. Of course, processes on the target host are slowed by an arriving migrant, but on a moderately-loaded system there are almost always idle hosts; thus the number of processes at the target host is usually zero. In any case, the number of processes at the target is always less than the number at the source.

If there is no process on the host that satisfies the above migration criterion, no migration is done. If migration costs are high, few processes will be eligible for migration; in the extreme there will be no migration at all. But in no case is the performance of the system worse (in expectation) than the performance without migration.

Using the distribution of process lifetimes, we calculate the expected slowdown imposed on a migrant process, and use this result to derive a minimum age for migration based on the cost of migration. Denoting the age of the migrant process by a; the cost of migration by c; the (eventual total) lifetime of the migrant by L, the number of processes at the source host by n; and the number of processes at the target host (including the migrant) by m, we have:

$$
\begin{aligned}
&\mathbf{E}\,\{\text{slowdown of migrant}\} \\
&= \int_{t=a}^{\infty} \mathbf{Pr}\left\{\begin{array}{c}\text{Lifetime of}\\ \text{migrant is t}\end{array}\right\} \cdot \left\{\begin{array}{c}\text{Slowdown given}\\ \text{lifetime is t}\end{array}\right\} dt \\
&= \int_{t=a}^{\infty} \mathbf{Pr}\,\{t \le L < t + dt | L \ge a\} \cdot \frac{na + c + m(t-a)}{t} \\
&= \int_{t=a}^{\infty} \frac{a}{t^2} \cdot \frac{na + c + m(t-a)}{t} dt \\
&= \frac{1}{2}\left(\frac{c}{a} + m + n\right)
\end{aligned}
$$

If there are n processes at a heavily loaded host, then a process should be eligible for migration only if its expected slowdown after migration is less than n (which is the slowdown it expects in the absence of migration).

Thus, we require $\frac{1}{2}(\frac{c}{a} + m + n) < n$, which implies

$$
\boxed{\text{Minimum migration age} = \frac{\text{Migration cost}}{n - m}}
$$

We can extend this analysis to the case of heterogeneous processor speeds by applying a scale factor to n or m.

This analysis assumes that current load predicts future load; that is, that the load at the source and target hosts will be constant during the migration. In an attempt to evaluate this assumption, and possibly improve it, we considered a number of alternative load predictors, including (1) taking a load average (over an interval of time), (2) summing the ages of the processes running on the host, and a (3) calculating a prediction of survivors and future arrivals based on the distribution model proposed here. We found that current (instantaneous) load is the best single predictor, and that using several predictive variables in combination did not greatly improve the accuracy of prediction. These results are in accord with Zhou [Zhou 1987] and Kunz [Kunz 1991].

3.2 Prior Preemptive Policies

Only a few preemptive strategies have been implemented in real systems or proposed in prior studies. The three that we have found are, like ours, based on the principle that a process should be migrated if it is old enough.

In many cases, the definition of *old enough* depends on a "voodoo" constant[1]: a free parameter whose value is chosen without explanation, and that would need to be re-tuned for a different system or another workload.

Under Leland and Ott's policy, a process p is eligible for migration if

$$\text{age}(p) \; > \; \text{ages of } k \text{ younger jobs at host}$$

where k is a free parameter called MINCRIT [Leland and Ott 1986]. Krueger and Livny's policy, like ours, takes the job's migration cost into account. A process p is eligible for migration if

$$\text{age}(p) \; > 0.1 * \text{migration cost}(p)$$

but they do not explain how they chose the value 0.1 [Krueger and Livny 1988]. The MOSIX policy is similar [Barak et al. 1993]; a process is eligible for migration if

$$\text{age}(p) \; > 1.0 * \text{migration cost}(p).$$

The choice of the constant (1.0) in the MOSIX policy ensures that the slowdown of a migrant process is never more than 2, since in the worst case the migrant completes immediately upon arrival at the target.

Despite this justification, the choice of the maximum slowdown (2) is arbitrary. We expect the MOSIX policy to be too restrictive, for two reasons. First, it ignores the slowdown that would be imposed at the source host in the absence of migration (presumably there is more than one process there, or the system would not be attempting to migrate processes away). Second, it is based on the worst-case slowdown rather than the expected slowdown. In Section 5.5, we show that the best choice for this parameter, for our workload, is usually near 0.4, but it depends on load.

[1]This term was coined by Professor John Ousterhout at U.C. Berkeley.

4. MODEL OF MIGRATION COSTS

Migration cost has such a large effect on the performance of preemptive load balancing; this section presents the model of migration costs we use in our simulation studies.

We model the cost of migrating an active process as the sum of a ***fixed migration cost*** for migrating the process's system state and a ***memory transfer cost*** that is proportional to the amount of the process's memory that must be transferred.

We model ***remote execution cost*** as a fixed cost; it is the same for all processes. The cost of remote execution includes sending the command and arguments to the remote host, logging in or otherwise authenticating the process, and creating a new shell and environment on the remote host.

Throughout this paper, we use the following notation:

—r: the cost of remote execution, in seconds
—f: the fixed cost of preemptive migration, in seconds
—b: the memory transfer bandwidth, in MB per second
—m: the memory size of migrant processes, in MB

and thus:

$$\text{cost of remote execution} = r$$
$$\text{cost of preemptive migration} = f + m/b$$

where the quotient m/b is the memory transfer cost.

4.1 Memory transfer costs

The amount of a process's memory that must be transferred during preemptive migration depends on properties of the distributed system. Douglis and Ousterhout [Douglis and Ousterhout 1991] have an excellent discussion of this issue, and we borrow from them here.

At the most, it might be necessary to transfer a process's entire memory. On a system like Sprite, which integrates virtual memory with a distributed file system, it is only necessary to write dirty pages to the file system before migration. When the process is restarted at the target host, it will retrieve these pages. In this case the cost of migration is proportional to the size of the resident set rather than the size of memory.

In systems that use precopying, such as V [Theimer et al. 1985], pages are transferred while the program continues to run at the source host. When the job stops execution at the source, it will have to transfer again any pages that have become dirty during the precopy. Although the number of pages transferred might be increased, the delay imposed on the migrant process is greatly decreased. Additional techniques can reduce the cost of transferring memory even more [Zayas 1987].

4.2 Migration costs in real systems

The specific parameters of migration cost depend not only on the nature of the system (as discussed above) but also on the speed of the network. Tables 3 and 4 show reported costs from a variety of real systems. Later we will use a trace-driven simulator to evaluate the effect of these parameters on system performance. We

System	Hardware	Cost of rexec, r
Sprite [Douglis and Ousterhout 1991]	SPARCstation1 10Mb/sec Ethernet	0.33 sec
GLUNIX [Vahdat et al. 1994] [Vahdat 1995]	HP workstations ATM network	0.25 to 0.5 sec
MIST [Prouty 1996]	HP9000/720 10Mb/sec Ethernet	0.33 sec
Utopia [Zhou et al. 1993]	DEC 3100 and SPARC IPC Ethernet	0.1 sec

Table 3. Cost of non-preemptive migration in various systems. Some of these numbers were obtained from personal communication with the authors.

System	Hardware	Fixed Cost, f	Inverse Bandwidth, $1/b$
Sprite [Douglis and Ousterhout 1991]	SPARCstation1 10Mb/sec Ethernet	0.33 sec	2.00 sec/MB
MOSIX [Barak et al. 1993] [Braverman 1995]	Intel Pentium 90MHz and 486 66MHz	0.006 sec	0.44 sec/MB
MIST [Prouty 1996]	HP9000/720s 10Mb/sec Ethernet	0.24 sec	0.99 sec/MB

Table 4. Values for preemptive migration costs from various systems. Many of these numbers were obtained from personal communication with the authors. The memory transfer cost is the product of the inverse bandwidth, $1/b$, and the amount of memory that must be transferred, m.

will make the pessimistic simplification that a migrant's entire memory must be transferred, although, as pointed out above, this is not necessarily the case.

5. TRACE-DRIVEN SIMULATION

In this section we present the results of a trace-driven simulation of process migration. We compare two migration strategies: our proposed age-based preemptive migration strategy (Section 3.1) and a non-preemptive strategy that migrates newborn processes according to the process name (similar to strategies proposed by [Wang et al. 1993] and [Svensson 1990]). With the intention of finding a conservative estimate of the benefit of preemptive migration, we give the name-based strategy the benefit of several unrealistic advantages; for example, the name-lists are derived from the same trace data used by the simulator.

Section 5.1 describes the simulator and the two strategies in more detail. We use the simulator to run three experiments. First, in Section 5.2, we evaluate the sensitivity of each strategy to the migration costs r, f, b, and m discussed in Section 4. Next, in Section 5.3, we choose values for these parameters that are representative of current systems and compare the performance of the two strategies in detail. In Section 5.4 we discuss why the preemptive policy outperforms the non-preemptive policy. Lastly, in Section 5.5, we evaluate the analytic criterion for migration age proposed in Section 3.1, compared to criteria used in previous studies.

5.1 The simulator

We have implemented a trace-driven simulation of a network of six identical workstations.[2] We selected six daytime intervals from the traces on machine po (see Section 2.1), each from 9:00 a.m. to 5:00 p.m. From the six traces we extracted the start times and CPU durations of the processes. We then simulated a network where each of six hosts executes (concurrently with the others) the process arrivals from one of the daytime traces.

Although the workloads on the six hosts are homogeneous in terms of the job mix and distribution of lifetimes, there is considerable variation in the level of activity during the eight-hour trace. For most of the traces, every arriving process finds at least one idle host in the system, but in the two busiest traces, a small fraction of processes (0.1%) arrive to find all hosts busy. In order to evaluate the effect of changes in system load, we divided the eight-hour trace into eight one-hour intervals. We refer to these as *runs* 0 through 7, where the runs are sorted from lowest to highest load. Run 0 has a total of 15000 processes submitted to the six simulated hosts; Run 7 has 30000 processes. The average duration of processes (for all runs) is 0.4 seconds. Thus the total utilization of the system, ρ, is between 0.27 and 0.54.

The birth process of jobs at our hosts is burstier than a Poisson process. For a given run and a given host, the serial correlation in interarrival times is typically between .08 and .24, which is significantly higher than one would expect from a Poisson process (uncorrelated interarrival times yield a serial correlation of 0.0; perfect correlation is 1.0).

Although the start times and durations of the processes come from trace data, the memory size of each process, which determines its migration cost, is chosen randomly from a measured distribution (see Section 5.2). This simplification obliterates any correlations between memory size and other process characteristics, but it allows us to control the mean memory size as a parameter and examine its effect on system performance.

In our system model, we assume that processes are always ready to run; i.e. they are never blocked on I/O. During a given time interval, we divide CPU time equally among the processes on the host (processor sharing).

In real systems, part of the migration time is spent on the source host packaging the transferred pages, part in transit in the network, and part on the target host unpacking the data. The size of these parts and whether they can be overlapped depend on details of the system. In our simulation we charge the entire cost of migration to the source host. This simplification is conservative in the sense that it makes preemptive migration less effective.

5.1.1 *Strategies.* We compare the preemptive migration strategy proposed in Section 3.1 with a non-preemptive migration strategy, where the non-preemptive strategy is given unfair advantages. For purposes of comparison, we have tried to make the policies as simple and as similar as possible. For both types of migration, we consider performing a migration only when a new process is born, even though a

[2]The trace-driven simulator and the trace data are available at http://http.cs.berkeley.edu/~harchol/loadbalancing.html.

preemptive strategy might benefit by initiating migrations at other times. Also, for both strategies, a host is considered heavily-loaded any time it contains more than one process; in other words, any time it would be sensible to consider migration. Finally, we use the same location policy in both cases: the host with the lowest instantaneous load is chosen as the target host (ties are broken by random selection).

Thus the only difference between the two migration policies is which processes are considered eligible for migration:

Name-based non-preemptive migration. A process is eligible for migration only if its name is on a list of processes that tend to be long-lived. If an eligible process is born at a heavily-loaded host, the process is executed remotely on the selected host. Processes cannot be migrated once they have begun execution.

The performance of this strategy depends on the list of eligible process names. We derived this list by sorting the processes from the traces according to name and duration and selecting the 15 common names with the longest *mean durations.* We chose a threshold on mean duration that is empirically optimal (for this set of runs). Adding more names to the list detracts from the performance of the system, as it allows more short-lived processes to be migrated. Removing names from the list detracts from performance as it becomes impossible to migrate enough processes to balance the load effectively. Since we used the trace data itself to construct the list, our results may overestimate the performance benefits of this strategy.

Age-based preemptive migration. A process is eligible for migration only if it has aged for some fraction of its migration cost. Based on the derivation in Section 3.1, this fraction is $\frac{1}{n-m}$, where n (respectively m) is the number of processes at the source (target) host. When a new process is born at a heavily-loaded host, all processes that satisfy the migration criterion are migrated away.

This strategy understates the performance benefits of preemptive migration, because it does not allow the system to initiate migrations except when a new process arrives.

As described in Section 3.1, we also modeled other location policies based on more complicated predictors of future loads, but none of these predictors yielded significantly better performance than the instantaneous load we use here.

We also considered the effect of allowing preemptive migration at times other than when a new process arrives. Ideally, one would like to initiate a migration whenever a process becomes eligible (since the eligibility criterion guarantees that the performance of the migrant will improve in expectation). One of the strategies we considered performs periodic checks of each process on a heavily-loaded host to see if any satisfy the criterion. The performance of this strategy is significantly better than that of the simpler policy (migrating only at process arrival times).

5.1.2 *Metrics.* We evaluate the effectiveness of each strategy according to the following performance metrics:

Mean slowdown. Slowdown is the ratio of wall-clock execution time to CPU time (thus, it is always greater than one). The average slowdown of all jobs is a common metric of system performance. When we compute the ratio of mean slowdowns (as from different strategies) we will use normalized slowdown, which is the ratio of

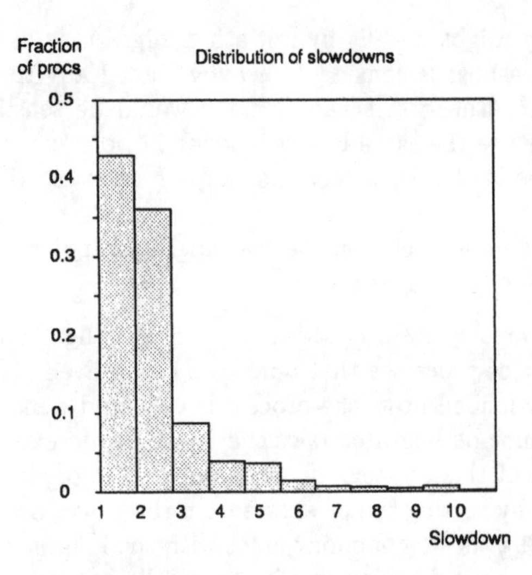

Fig. 3. Distribution of process slowdowns for run 0 (with no migration). Most processes suffer small slowdowns, but the processes in the tail of the distribution are more noticeable and annoying to users.

inactive time (the *excess* slowdown caused by queueing and migration delays) to CPU time. For example, if the (unnormalized) mean slowdown drops from 2.0 to 1.5, the ratio of normalized mean slowdowns is 0.5/1.0 = 0.5: a 50% reduction in delay.

Mean slowdown alone is not a sufficient measure of the difference in performance of the two strategies; it understates the advantages of the preemptive strategy for these two reasons:

—Skewed distribution of slowdowns: Even in the absence of migration, the majority of processes suffer small slowdowns (typically 80% are less than 3.0. See Figure 3). The value of the mean slowdown is dominated by this majority.

—User perception: From the user's point of view, the important processes are the ones in the tail of the distribution, because although they are the minority, they cause the most noticeable and annoying delays. Eliminating these delays might have a small effect on the mean slowdown, but a large effect on a user's perception of performance.

Therefore, we will also consider the following three metrics:

Variance of slowdown. This metric is often cited as a measure of the unpredictability of response time [Silberschatz et al. 1994], which is a nuisance for users trying to schedule tasks. In light of the distribution of slowdowns, however, it may be more meaningful to interpret this metric as a measure of the length of the tail of the distribution; i.e. the number of jobs that experience long delays. (See Figure 5b).

Number of severely slowed processes. In order to quantify the number of noticeable delays explicitly, we consider the number (or percentage) of processes that are severely impacted by queueing and migration penalties. (See Figures 5c and 5d).

Mean slowdown of long jobs. Delays in longer jobs (those with lifetimes greater than .5 seconds) are more perceivable to users than delays in short jobs. (See Figure 6).

5.2 Sensitivity to migration costs

In this section we compare the performance of the non-preemptive and preemptive strategies over a range of values of r, f, b and m (the migration cost parameters defined in Section 4).

For the following experiments, we chose the remote execution cost $r = .3$ seconds. We considered a range for the fixed migration cost of $.1 < f < 10$ seconds.

The memory transfer cost is the quotient of m (the memory size of the migrant process) and b (the bandwidth of the network). We chose the memory transfer cost from a distribution with the same shape as the distribution of process lifetimes, setting the mean memory transfer cost (MMTC) to a range of values from 1 to 64 seconds. The shape of this distribution is based on an informal study of memory-use patterns on the same machines from which we collected trace data. The important feature of this distribution is that there are many jobs with small memory demands and a few jobs with very large memory demands. Empirically, the exact form of this distribution does not affect the performance of either migration strategy strongly, but of course the mean (MMTC) does have a strong effect.

Figures 4a and 4b are contour plots of the ratio of the performance of the two migration strategies using normalized slowdown. Specifically, for each of the eight one-hour runs we calculate the mean (respectively standard deviation) of the slowdown imposed on all processes that complete during the hour. For each run, we then take the ratio of the means (standard deviations) of the two strategies. Lastly we take the geometric mean of the eight ratios (for discussion of the geometric mean, see [Hennessy and Patterson 1990]).

The two axes in Figure 4 represent the two components of the cost of preemptive migration, namely the fixed cost, f, and the MMTC, m/b. The cost of non-preemptive migration, r, is fixed at 0.3 seconds. As expected, increasing either the fixed cost of migration or the MMTC hurts the performance of preemptive migration. The contour line marked 1.0 indicates the crossover where the performance of preemptive and non-preemptive migration is equal (the ratio is 1.0). For smaller values of the cost parameters, preemptive migration performs better; for example, if the fixed migration cost is 0.3 seconds and the MMTC is 2 seconds, the normalized mean slowdown with preemptive migration is almost 40% lower than with non-preemptive migration. When the fixed cost of migration or the MMTC are very high, almost all processes are ineligible for preemptive migration; thus, the preemptive strategy does almost no migrations. The non-preemptive strategy is unaffected by these costs so the non-preemptive strategy can be more effective.

Figure 4b shows the effect of migration costs on the standard deviation of slowdowns. The crossover point — where non-preemptive migration surpasses preemptive migration — is considerably higher here than in Figure 4a. Thus there is a region where preemptive migration yields a higher mean slowdown than non-preemptive migration, but a lower standard deviation. The reason for this is that non-preemptive migration occasionally chooses a process for remote execution that turns out to be short-lived. These processes suffer large delays (relative to their

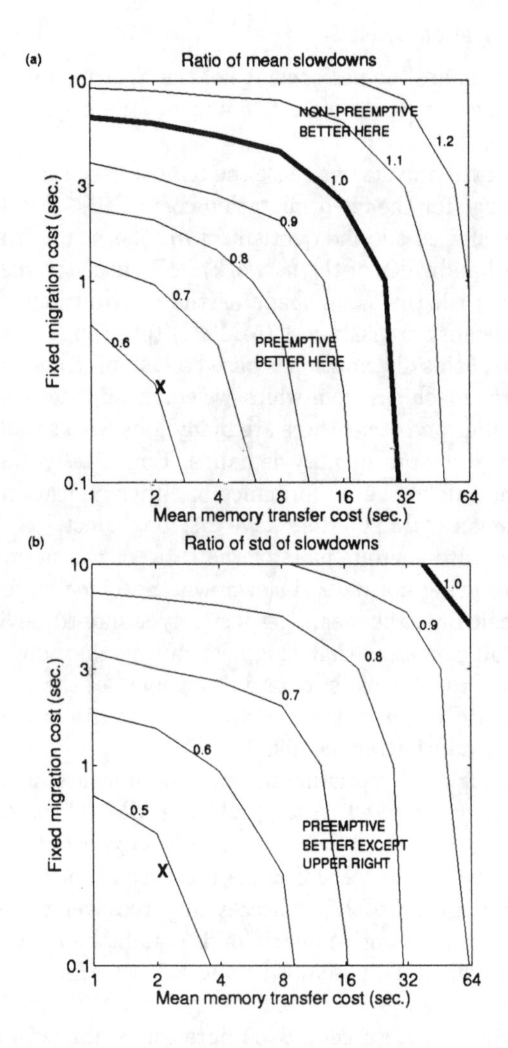

Fig. 4. (a) The performance of preemptive migration relative to non-preemptive migration deteriorates as the cost of preemptive migration increases. The two axes are the two components of the preemptive migration cost. The cost of non-preemptive migration is held fixed. The X marks the particular set of parameters we will consider in the next section. (b) The standard deviation of slowdown may give a better indication of a user's perception of system performance than mean slowdown. By this metric, the benefit of preemptive migration is even more significant.

run times) and add to the tail of the distribution of slowdowns. In the next section, we show cases in which the standard deviation of slowdowns is actually worse with non-preemptive migration than with no migration at all (three of the eight runs).

5.3 Comparison of preemptive and non-preemptive strategies

In this section we choose migration cost parameters representative of current systems (see Section 4.2) and use them to examine more closely the performance of the two migration strategies. The values we chose are:

—r: the cost of remote execution, 0.3 seconds

—f: the fixed cost of preemptive migration, 0.3 seconds

—b: the memory transfer bandwidth, 0.5 MB per second

—m: the mean memory size of migrant processes, 1 MB

In Figures 4a and 4b, the point corresponding to these parameter values is marked with an X. Figure 5 shows the performance of the two migration strategies at this point (compared to the case of no migration).

Non-preemptive migration reduces the normalized mean slowdown (Figure 5a) by less than 20% for most runs (and 40% for the two runs with the highest loads). Preemptive migration reduces the normalized mean slowdown by 50% for most runs (and more than 60% for two of the runs). The performance improvement of preemptive migration over non-preemptive migration is typically between 35% and 50%.

As discussed above, we feel that the mean slowdown (normalized or not) understates the performance benefits of preemptive migration. We have proposed other metrics to try to quantify these benefits. Figure 5b shows the standard deviation of slowdowns, which reflects the number of severely impacted processes. Figures 5c and 5d explicitly measure the number of severely impacted processes, according to two different thresholds of acceptable slowdown. By these metrics, the benefits of migration in general appear greater, and the discrepancy between preemptive and non-preemptive migration appears much greater. For example in Figure 5d, in the absence of migration, 7 – 18% of processes are slowed by a factor of 5 or more. Non-preemptive migration is able to eliminate 42–62% of these, which is a significant benefit, but preemptive migration consistently eliminates nearly all (86–97%) severe delays.

An important observation from Figure 5b is that for several runs, non-preemptive migration actually makes the performance of the system worse than if there were no migration at all. For the preemptive migration strategy, this outcome is nearly impossible, since migrations are only performed if they improve the slowdowns of all processes involved (in expectation). In the worst case, then, the preemptive strategy will do no worse than the case of no migration (in expectation).

Another benefit of preemptive migration is graceful degradation of system performance as load increases (as shown in Figure 5). In the presence of preemptive migration, both the mean and standard deviation of slowdown are nearly constant, regardless of the overall load on the system.

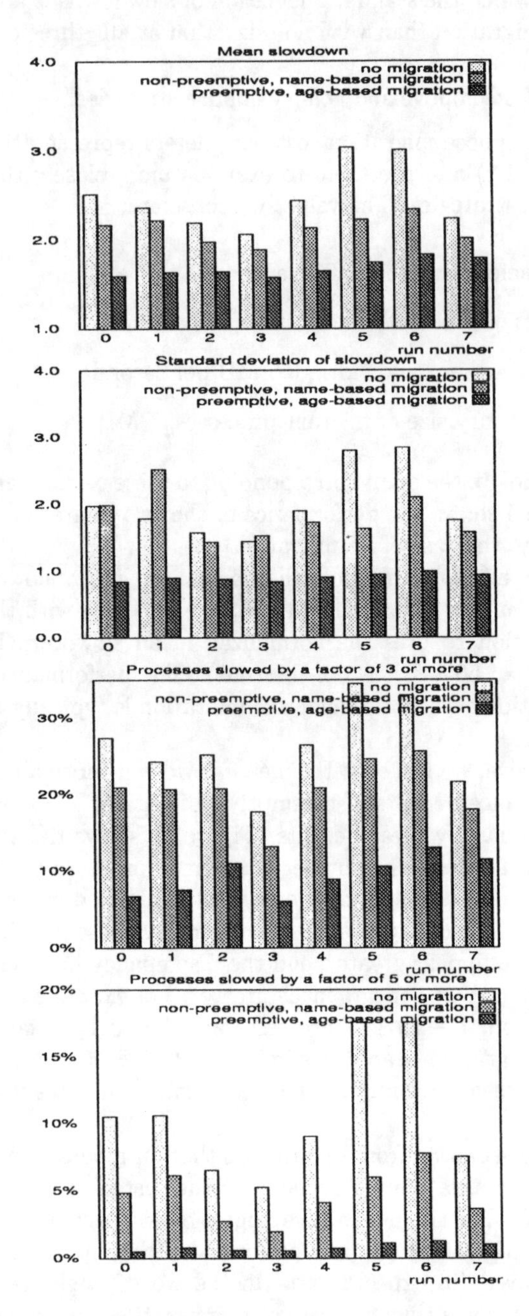

Fig. 5. (a) Mean slowdown. (b) Standard deviation of slowdown. (c) Percentage of processes slowed by a factor of 3 or more. (d) Percentage of processes slowed by a factor of 5 or more.

5.4 Why preemptive migration outperforms non-preemptive migration

The alternate metrics discussed above shed some light on the reasons for the performance difference between preemptive and non-preemptive migration. We consider two kinds of mistakes a migration system might make:

Failing to migrate long-lived jobs. This type of error imposes moderate slowdowns on a potential migrant, and, more importantly, inflicts delays on short jobs that are forced to share a processor with a CPU hog. Under non-preemptive migration, this error occurs whenever a long-lived process is not on the name-list, possibly because it is an unknown program or an unusually long execution of a typically short-lived program. Preemptive migration can correct these errors by migrating long jobs later in their lives.

Migrating short-lived jobs. This type of error imposes large slowdowns on the migrated process, wastes network resources, and fails to effect significant load balancing. Under non-preemptive migration, this error occurs when a process whose name is on the eligible list turns out to be short-lived. Our preemptive migration strategy all but eliminates this type of error by guaranteeing that the performance of a migrant improves in expectation.

Even occasional mistakes of the first kind can have a large impact on performance, because one long job on a busy machine will impede many small jobs. This effect is aggravated by the serial correlation between arrival times (see Section 5.1), which suggests that a busy host is likely to receive many future arrivals.

Thus, an important feature of a migration policy is its ability to identify long-lived jobs for migration. To evaluate this ability, we consider the average lifetime of the processes chosen for migration under each policy. Under non-preemptive migration, the average lifetime of migrant processes was 2.0 seconds (the mean lifetime for all processes is 0.4 seconds), and the median lifetime of migrants was 0.9 seconds. The non-preemptive policy migrated about 1% of all jobs, which accounted for 5.7% of the total CPU.

The preemptive migration policy was better able to identify long jobs; the average lifetime of migrant processes under preemptive migration was 4.9 seconds; the median lifetime of migrants was 2.0 seconds. The preemptive policy migrated 4% of all jobs, but since these migrants were long-lived, they accounted for 55% of the total CPU.

Thus the primary reason for the success of preemptive migration is its ability to identify long jobs accurately and to migrate those jobs away from busy hosts.

The second type of error did not have as great an impact on the mean slowdown for all processes, but it did impose large slowdowns on some small processes. These outliers are reflected in the standard deviation of slowdowns — because the non-preemptive policy sometimes migrates very short jobs, it can make the standard deviation of slowdowns worse than with no migration (see Figure 5b). The age-based preemptive migration criterion eliminates most errors of this type by guaranteeing that the performance of the migrant will improve in expectation.

There is, however, one type of migration error that is more problematic for preemptive migration than for non-preemptive migration: stale load information. A target host may have a low load when a migration is initiated, but its load may have increased by the time the migrant arrives. This is more likely for a preemptive migration because the migration time is longer. In our simulations, we found that these errors do occur, although infrequently enough that they do not have a severe impact on performance.

Specifically, we counted the number of migrant processes that arrived at a target host and found that the load was higher than it had been at the source host when migration began. For most runs, this occurred less than 0.5% of the time (for two runs with high loads it was 0.7%). Somewhat more often, 3% of the time, a migrant process arrived at a target host and found that the load at the target was greater than the *current* load at the source. These results suggest that the performance of a preemptive migration strategy might be improved by a reservation system as in MOSIX.

One other potential problem with preemptive migration is the volume of network traffic that results from large memory transfers. In our simulations, we did not model network congestion, on the assumption that the traffic generated by migration would not be excessive. This assumption seems to be reasonable: under our preemptive migration strategy fewer than 4% of processes are migrated once and fewer than .25% of processes are migrated more than once. Furthermore, there is seldom more than one migration in progress at a time.

In summary, the advantage of preemptive migration — its ability to identify long jobs and move them away from busy hosts — overcomes its disadvantages (longer migration times and stale load information).

5.4.1 *Effect of migration on short and long jobs.* We have claimed that identifying long jobs and migrating them away from busy hosts helps not only the long jobs (which run on more lightly-loaded hosts) but also the short jobs that run on the source host. To test this claim, we divided the processes into three lifetime groups and measured the performance benefit for each group due to migration. The number of jobs in short group is roughly ten times the number in the medium group, which in turn is roughly ten times the number in the long group. Figure 6 shows that migration reduces the mean slowdown of all three groups: for non-preemptive migration the improvement is the same for all groups; under preemptive migration the long jobs enjoy a slightly greater benefit.

This breakdown by lifetime group is useful for evaluating various metrics of system performance. The metric we are using here, slowdown, gives equal weight to all jobs; as a result, the mean slowdown metric is dominated by the most populous group, short jobs. Another common metric, residence time, effectively weights jobs according to their lifetimes. Thus the mean residence time metric reflects, primarily, the performance benefit for long jobs. Under the mean residence time metric, then, preemptive migration appears even more effective.

5.5 Evaluation of analytic migration criterion

As derived in Section 3.1, the minimum age for a migrant process according to our *analytic criterion* is

$$\begin{array}{c}\text{Minimum}\\\text{migration age}\end{array} = \frac{\text{Migration cost}}{n - m}$$

where n is the load at the source host and m is the load at the target host (including

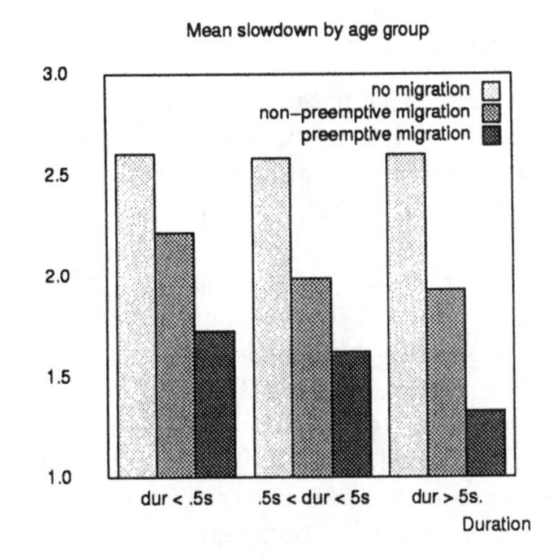

Fig. 6. Mean slowdown broken down by lifetime group.

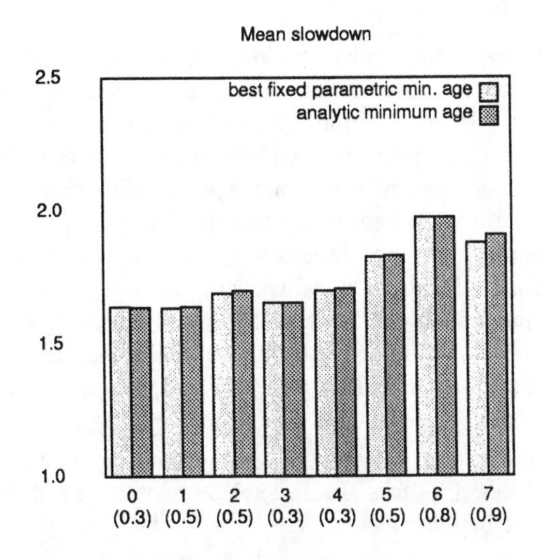

Fig. 7. The mean slowdown for eight runs, using the two criteria for minimum migration age. The value of the best fixed parameter α is shown in parentheses for each run.

the potential migrant).

We compare the analytic criterion with the ***fixed parameter criterion***:

$$\text{Minimum migration age} = \alpha * \text{Migration cost}$$

where α is a free parameter. This parameter is meant to model preemptive migration strategies in the literature, as discussed in Section 3.2. For comparison, we will use the ***best fixed parameter***, which is, for each run, the value that yields the smallest mean slowdown. Of course, this gives the fixed parameter criterion a considerable advantage.

Figure 7 compares the performance of the analytic minimum age criterion with the best fixed parameter. The best fixed parameter varies considerably from run to run, and appears to be roughly correlated with the average load during the run (the runs are sorted in increasing order of total load).

The performance of the analytic criterion is always within a few percent of the performance of the best fixed value criterion. The advantage of the analytic criterion is that it is parameterless, and therefore more robust across a variety of workloads. We feel that the elimination of one free parameter is a useful result in an area with so many (usually hand-tuned) parameters.

This result also suggests that the parameter used by Krueger and Livny ($\alpha = 0.1$) is too low, and the parameter used in MOSIX ($\alpha = 1.0$) is too high, at least for this workload (see Section 3.2).

6. WEAKNESSES OF THE MODEL

Our simulation ignores a number of factors that would affect the performance of migration in real systems:

CPU-bound jobs only. Our model considers all jobs CPU-bound; thus, their response time necessarily improves if they run on a host with a lighter load. For I/O bound jobs, however, CPU contention has little effect on response time. These jobs would benefit less from migration. To see how large a role this plays in our results, we noted the names of the processes that appear most frequently in our traces (with CPU time greater than 1 second, since these are the processes most likely to be migrated). The most common names were cc1plus and cc1, both of which are CPU bound. Next most frequent were: trn, cpp, ld, jove (a version of emacs), and ps. So although some jobs in our traces are in reality interactive, our simple model is reasonable for many of the most common jobs. In Section 7 we discuss further implications of a workload including interactive, I/O-bound, and non-migratable jobs.

Environment. Our migration strategy takes advantage of the used-better-than-new property of process lifetimes. In an environment with a different distribution, this strategy will not be effective.

Local scheduling. We assume that local scheduling on the hosts is similar to round-robin. Other policies, like feedback scheduling, can reduce the impact of long jobs on the performance of short jobs, and thereby reduce the need for load balancing. We explore this issue in more detail in Section 7 and find that preemptive migration is still beneficial under feedback scheduling.

Memory size. One weakness of our model is that we chose memory sizes from a measured distribution and therefore our model ignores any correlation between memory size and other process characteristics. To justify this simplification, we conducted an informal study of processes in our department, and found no correlation between memory size and process CPU usage. Krueger and Livny report a similar observation [Krueger and Livny 1988]. Thus, this may be a reasonable simplification.

Network contention. Our model does not consider the effect of increased network traffic as a result of process migration. We observe, however, that for the load levels we simulated, migrations are occasional (one every few seconds), and that there is seldom more than one migration in progress at a time.

7. FUTURE WORK

In our workload model we have assumed that all processes are CPU-bound. Of primary interest in future work is including interactive, I/O-bound, and non-migratable jobs into our workload.

In a workload that includes interactive jobs, I/O-bound jobs, and daemons, there will be some jobs that should not or cannot be migrated. An I/O-bound job, for example, will not necessarily run faster on a more lightly-loaded host, and might run slower if it is migrated away from the disk or other I/O device it uses. A migrated interactive job might benefit by running on a more lightly-loaded host if it uses significant CPU time, but will suffer performance penalties for all future interactions. Finally, some jobs (e.g. many daemons) cannot be migrated away from their hosts.

The policy we proposed for preemptive migration can be extended to deal appropriately with interactive and I/O bound jobs by including in the definition of migration cost the additional costs that will be imposed on these jobs after migration, including network delays, access to non-local data, etc. The estimates of these costs might be based on the recent behavior of the job; e.g. the number and frequency of I/O requests and interactions. Jobs that are explicitly forbidden to migrate could be assigned an infinite migration cost.

The presence of a set of jobs that are either expensive or impossible to migrate might reduce the ability of the migration policy to move work around the network and balance loads effectively. However, we observe that the majority of long-lived jobs are, in fact, CPU-bound, and it is these long-lived jobs that consume the majority of CPU time. Thus, even if the migration policy were only able to migrate a subset of the jobs in the system, it could still have a significant load-balancing effect.

Another way in which the proposed migration policy should be altered in a more general environment is that n (the number of jobs at the source) and m (the number of jobs at the target host) should distinguish between CPU-bound jobs and other types of jobs, since only CPU-bound jobs affect CPU contention, and therefore are significant in CPU load balancing.

Another important consideration in load balancing is the effect of local scheduling at the hosts. Most prior studies of load balancing have assumed, as we do, that the local scheduling is round-robin (or processor-sharing). A few assume first-come-first-serve (FCFS) scheduling, but fewer still have studied the effect of feedback scheduling, where processes that have used the least CPU time are given priority over older processes. We simulated feedback scheduling and found that it greatly reduced mean slowdown (from approximately 2.5 to between 1.2 and 1.7, depending on load) even without migration. Thus, the potential benefit of either type of migration is greatly reduced.

We evaluated the non-preemptive migration policy from Section 5.1.1 under feedback scheduling, and found that it often makes things worse, increasing the mean slowdown in 5 of the 8 runs, and only decreasing it by 11% in the best case (highest load).

To evaluate our preemptive policy, we had to change the migration criterion to reflect the effect of local scheduling. Under processor sharing, we assume that the slowdown imposed on a process is equal to the number of processes on the host. Under feedback scheduling, the slowdown is closer to the number of ***younger*** processes, since older processes have lower priority. Thus, we modified the migration criterion in Section 3.1 so that n and m are the number of processes at the source and target hosts that are younger than the migrant process. Using this criterion, preemptive migration reduces mean normalized slowdown by 12–32%, and reduces the number of severely slowed processes (slowdown greater than 5) by 30–60%.

An issue that remains unresolved is whether feedback scheduling is as effective in real systems as it was in our simulations. For example, decay-usage scheduling as used in UNIX has some characteristics of both round-robin and feedback policies [Epema 1995]. Young jobs do have some precedence, but old jobs that perform interaction or other I/O are given higher priority, which allows them to interfere with short jobs. In our experiments on a SPARC workstation running SunOS, we found that a long-running job that performs periodic I/O can obtain more than 50% of the CPU time, even if it is sharing a host with much younger processes. The more recent lottery scheduling behaves more like processor-sharing [Waldspurger and Weihl 1994]. To understand the effect of local scheduling on load balancing requires a process model that includes interaction and I/O.

8. CONCLUSIONS

—To evaluate migration strategies, it is important to model the distribution of process lifetimes accurately. Assuming an exponential distribution can underestimate the benefits of preemptive migration, because it ignores the fact that old jobs are expected to be long-lived. Even a lifetime distribution that matches the measured distribution in both mean and variance may be misleading in designing and evaluating load balancing policies.

—Preemptive migration outperforms non-preemptive migration even when memory-transfer costs are high, for the following reason: non-preemptive name-based strategies choose processes for migration that are expected to have long lives. If this prediction is wrong, and a process runs longer than expected, it cannot be migrated away, and many subsequent small processes will be delayed. A preemptive strategy is able to predict lifetimes more accurately (based on age) and, more importantly, if the prediction is wrong, the system can recover by migrating the process later.

—Migrating a long job away from a busy host helps not only the long job, but also the many short jobs that are expected to arrive at the host in the future. A busy host is expected to receive many arrivals because of the serial correlation (burstiness) of the arrival process.

—Using the functional form of the distribution of process lifetimes, we have derived a criterion for the minimum time a process must age before being migrated. This criterion is parameterless and robust across a range of loads.

—Exclusive use of mean slowdown as a metric of system performance understates the benefits of load balancing as perceived by users, and especially understates the benefits of preemptive load balancing. One performance metric which is more related to user perception is the number of severely slowed processes. While non-preemptive migration eliminates half of these noticeable delays, preemptive migration reduces them by a factor of ten.

—Although preemptive migration is difficult to implement, several systems have chosen to implement it for reasons other than load balancing. Our results suggest these systems would benefit from incorporating a preemptive load balancing policy.

ACKNOWLEDGMENTS

We would like to thank Tom Anderson and the members of the NOW group for comments and suggestions on our experimental setup. We are also greatly indebted to the anonymous reviewers from SIGMETRICS and SOSP, and to John Zahorjan, whose comments greatly improved the quality of this paper.

REFERENCES

ABRAHAMS, D. M. AND RIZZARDI, F. 1988. *The Berkeley Interactive Statistical System.* W. W. Norton and Co.

AGRAWAL, R. AND EZZET, A. 1987. Location independent remote execution in NEST. *IEEE Transactions on Software Engineering 13*, 8 (August), 905–912.

AHMAD, I., GHAFOOR, A., AND MEHROTRA, K. 1991. Performance prediction of distributed load balancing on multicomputer systems. In *Supercomputing* (1991), pp. 830–839.

ARTSY, Y. AND FINKEL, R. 1989. Designing a process migration facility: The Charlotte experience. *IEEE Computer*, 47–56.

BARAK, A., SHAI, G., AND WHEELER, R. G. 1993. *The MOSIX Distributed Operating System: Load Balancing for UNIX.* Springer Verlag, Berlin.

BONOMI, F. AND KUMAR, A. 1990. Adaptive optimal load balancing in a nonhomogeneous multiserver system with a central job scheduler. *IEEE Transactions on Computers 39*, 10 (October), 1232–1250.

BRAVERMAN, A. 1995. Personal Communication.

BRYANT, R. M. AND FINKEL, R. A. 1981. A stable distributed scheduling algorithm. In *2nd International Conference on Distributed Computing Systems* (1981), pp. 314–323.

CABRERA, F. 1986. The influence of workload on load balancing strategies. In *Proceedings of the Usenix Summer Conference* (June 1986), pp. 446–458.

CASAS, J., CLARK, D. L., KONURU, R., OTTO, S. W., PROUTY, R. M., AND WALPOLE, J. 1995. Mpvm: A migration transparent version of pvm. *Computing Systems 8*, 2 (Spring), 171–216.

CASAVANT, T. L. AND KUHL, J. G. 1987. Analysis of three dynamic distributed load-balancing strategies with varying global information requirements. In *7th International Conference on Distributed Computing Systems* (September 1987), pp. 185–192.

CHOWDHURY, S. 1990. The greedy load sharing algorithm. *Journal of Parallel and Distributed Computing 9*, 93–99.

DE PAOLI, D. AND GOSCINSKI, A. 1995. The rhodos migration facility. *Journal of Systems and Software*. Submitted. See also http://www.cm.deakin.edu.au/rhodos/.

DOUGLIS, F. AND OUSTERHOUT, J. 1991. Transparent process migration: Design alternatives and the sprite implementation. *Software - Practice and Experience 21*, 8 (August), 757–785.

DOWNEY, A. B. AND HARCHOL-BALTER, M. 1995. A note on "The limited performance benefits of migrating active processes for load sharing". Technical Report UCB/CSD-95-888 (November), University of California, Berkeley.

EAGER, D. L., LAZOWSKA, E. D., AND ZAHORJAN, J. 1986. Adaptive load sharing in homogeneous distributed systems. *IEEE Transactions on Software Engineering 12*, 5 (May), 662–675.

EAGER, D. L., LAZOWSKA, E. D., AND ZAHORJAN, J. 1988. The limited performance benefits of migrating active processes for load sharing. In *ACM Sigmetrics Conference on Measuring and Modeling of Computer Systems* (May 1988), pp. 662–675.

EPEMA, D. 1995. An analysis of decay-usage scheduling in multiprocessors. In *ACM Sigmetrics Conference on Measurement and Modeling of Computer Systems* (1995), pp. 74–85.

EVANS, D. J. AND BUTT, W. U. N. 1993. Dynamic load balancing using task-transfer probabilites. *Parallel Computing 19*, 897–916.

HAĆ, A. AND JIN, X. 1990. Dynamic load balancing in a distributed system using a sender-initiated algorithm. *Journal of Systems Software 11*, 79–94.

HENNESSY, J. L. AND PATTERSON, D. A. 1990. *Computer Architecture A Quantitative Approach.* Morgan Kaufmann Publishers, San Mateo, CA.

KRUEGER, P. AND LIVNY, M. 1988. A comparison of preemptive and non-preemptive load distributing. In *8th International Conference on Distributed Computing Systems* (June 1988), pp. 123–130.

KUNZ, T. 1991. The influence of different workload descriptions on a heuristic load balancing scheme. *IEEE Transactions on Software Engineering 17*, 7 (July), 725–730.

LARSEN, R. J. AND MARX, M. L. 1986. *An introduction to mathematical statistics and its applications* (2nd ed.). Prentice Hall, Englewood Cliffs, N.J.

LELAND, W. E. AND OTT, T. J. 1986. Load-balancing heuristics and process behavior. In *Proceedings of Performance '86 and ACM Sigmetrics*, Volume 14 (1986), pp. 54–69.

LIN, H.-C. AND RAGHAVENDRA, C. 1993. A state-aggregation method for analyzing dynamic load-balancing policies. In *IEEE 13th International Conference on Distributed Computing Systems* (May 1993), pp. 482–489.

LITZKOW, M. AND LIVNY, M. 1990. Experience with the Condor distributed batch system. In *IEEE Workshop on Experimental Distributed Systems* (1990), pp. 97–101.

LITZKOW, M., LIVNY, M., AND MUTKA, M. 1988. Condor - a hunter of idle workstations. In *8th International Conference on Distributed Computing Systems* (June 1988).

LIVNY, M. AND MELMAN, M. 1982. Load balancing in homogeneous broadcast distributed systems. In *ACM Computer Network Performance Symposium* (April 1982), pp. 47–55.

MILOJICIC, D. S. 1993. *Load Distribution: Implementation for the Mach Microkernel.* PhD Dissertation, University of Kaiserslautern.

MIRCHANDANEY, R., TOWSLEY, D., AND STANKOVIC, J. A. 1990. Adaptive load sharing in heterogeneous distributed systems. *Journal of Parallel and Distributed Computing 9*, 331–346.

POWELL, M. AND MILLER, B. 1983. Process migrations in DEMOS/MP. In *6th ACM Symposium on Operating Systems Principles* (November 1983), pp. 110–119.

PROUTY, R. 1996. Personal Communication.

PULIDAS, S., TOWSLEY, D., AND STANKOVIC, J. A. 1988. Imbedding gradient estimators in load balancing algorithms. In *8th International Conference on Distributed Computing Systems* (June 1988), pp. 482–490.

ROMMEL, C. G. 1991. The probability of load balancing success in a homogeneous network. *IEEE Transactions on Software Engineering 17*, 922–933.

ROSIN, R. F. 1965. Determining a computing center environment. *Communications of the ACM 8*, 7.

SILBERSCHATZ, A., PETERSON, J., AND GALVIN, P. 1994. *Operating System Concepts, 4th Edition.* Addison-Wesley, Reading, MA.

SVENSSON, A. 1990. History, an intelligent load sharing filter. In *IEEE 10th International Conference on Distributed Computing Systems* (1990), pp. 546–553.

TANENBAUM, A., VAN RENESSE, R., VAN STAVEREN, H., AND SHARP, G. 1990. Experiences with the Amoeba distributed operating system. *Communications of the ACM*, 336–346.

THEIMER, M. M., LANTZ, K. A., AND CHERITON, D. R. 1985. Preemptable remote execution facilities for the V-System. In *10th ACM Symposium on Operating Systems Principles* (December 1985), pp. 2–12.

THIEL, G. 1991. Locus operating system, a transparent system. *Computer Communications 14*, 6, 336–346.

VAHDAT, A. 1995. Personal Communication.

VAHDAT, A. M., GHORMLEY, D. P., AND ANDERSON, T. E. 1994. Efficient, portable, and robust extension of operating system functionality. Technical Report UCB//CSD-94-842, University of California, Berkeley.

WALDSPURGER, C. A. AND WEIHL, W. E. 1994. Lottery scheduling: Flexible proportional-share resource management. In *Proceedings of the First Symposium on Operating System Design and Implementation* (November 1994), pp. 1–11.

WANG, J., ZHOU, S., K.AHMED, AND LONG, W. 1993. LSBATCH: A distributed load sharing batch system. Technical Report CSRI-286 (April), Computer Systems Research Institute, University of Toronto.

WANG, Y.-T. AND MORRIS, R. J. 1985. Load sharing in distributed systems. *IEEE Transactions on Computers c-34*, 3 (March), 204–217.

ZAYAS, E. R. 1987. Attacking the process migration bottleneck. In *11th ACM Symposium on Operating Systems Principles* (1987), pp. 13–24.

ZHANG, Y., HAKOZAKI, K., KAMEDA, H., AND SHIMIZU, K. 1995. A performance comparison of adaptive and static load balancing in heterogeneous distributed systems. In *Proceedings of the 28th Annual Simulation Symposium* (April 1995), pp. 332–340.

ZHOU, S. 1987. *Performance studies for dynamic load balancing in distributed systems.* Ph.D. Dissertation, University of California, Berkeley.

ZHOU, S. AND FERRARI, D. 1987. A measurement study of load balancing performance. In *IEEE 7th International Conference on Distributed Computing Systems* (October 1987), pp. 490–497.

ZHOU, S., WANG, J., ZHENG, X., AND DELISLE, P. 1993. Utopia: a load-sharing facility for large heterogeneous distributed computing systems. *Software – Practice and Expeience 23*, 2 (December), 1305–1336.

Chapter 7 Other Sources of Information

Process migration is a well researched area. In this section we offer some pointers to books, conferences, journals, and useful Web sites related to process migration.

Books. There are a few books specifically about process migration or systems that support migration:

Barak, A., Guday, S., and Wheeler, R. G. (1993). *The MOSIX Distributed Operating System.* Springer Verlag.

Milojičić, D. (1994). *Load Distribution,* Implementation for the Mach Microkernel. Vieweg, Wiesbaden.

Popek, G. and Walker, B. (1985). *The Locus Distributed System Architecture.* MIT Press, Cambridge, Massachusetts.

Conferences and Workshops. Most work in process migration is published in the various USENIX technical conferences, conferences on distributed systems (for example, the *International Conference on Distributed Computing Systems*), or other operating system events like the SOSP conferences.

Useful sites. The following sites provide pointers to ongoing work-in-process migration and actual implementations that support migration:

- **MOSIX:** http://www.mosix.cs.huji.ac.il.
- **Platform Computing:** http://www.platform.com/Products/TheLSFSuite.
- **Tui:** http://www.cs.ubc.ca/spider/psmith/tui.html.

Migration and other interesting research topics are also discussed in several Internet news groups, most notably comp.os.research.

Part III

Physical Mobility: Mobile Computing

Chapter 8 Mobile Computing

The advent of widespread portable computers, in the 1980s and beyond, has led to a wide variety of interesting devices, and interesting hardware and software issues. Laptops have long been popular, especially among traveling businessmen. The vast majority of laptop computers run in a stand-alone mode, at least when away from one's home or office. (This mode is sometimes termed "nomadic computing.") A smaller number, though increasing, have wireless network connectivity within a building or even a metropolitan area (permitting true "mobile computing"). Although some devices have been rather transient, such as the combined pen-based computer and cellular phone in the AT&T EO Communicator, others, such as the 3Com PalmPilot, have become cultural icons.

Though this belief is not entirely uncontroversial, we postulate that small wireless computers will be as ubiquitous in a few years as cellular phones are today. In fact, many of today's cellular phones are equipped with textual displays that provide Web access, e-mail, stock quotes, and other functionality. Furthermore, the cellular telephony industry is driving the Cellular Digital Packet Data (CDPD) standard [Taylor, 1997], and several telecom companies are working together on wireless networking applications and protocols [Wireless, 1998].

The purpose of this chapter is to explore the impact of physical mobility on computers and communication. Mobility affects the initial ability to communicate, the reliability and performance of communication, and the characteristics of the individual devices used for communication.

8.1 Benefits and Challenges of Mobile Computing

Benefits

There are many obvious benefits to being able to move a computer around, with or without the additional ability to communicate during or after movement. In addition, the same system support that enables such movement can help to survive network or server failures. Finally, when wireless LANs and small computers can link to fully connected computers wherever one goes, new applications and services become available.

Untethered access. Mobile computing gives one the opportunity to work with computer resources from almost anywhere. With a wireless network, it has the appearance of a traditional connected environment (though usually with poorer performance). In-building wireless LANs allow users to read e-mail, browse the Web, and perform other operations during meetings—something that the organizers of those meetings might well wish was not possible. Wide-area networks will even let one check on-line driving directions while stopped at a traffic light—something the *police* probably wish was not possible.

Enhanced reliability. Systems that are designed to operate only when connected with certain other systems, such as file servers, may not function well when a file server is unavailable. In contrast, systems that are designed to expect no connectivity may be able to ignore the outage—when a host cannot tell the difference between a nonfunctional network and a nonfunctional computer connected to it—and recover later when the server returns to active status.

Ubiquitous access. What are the ramifications of having numerous small portable computing devices in one's everyday environment, so small and numerous as to be unnoticeable? The vision of Xerox PARC's *Ubiquitous Computing* project was for computers to fade into the background. Small wireless computers could tell one where to find a book on the bookshelf, or get one's attention while walking down a hallway. However, integrating computers seamlessly into everyday life has numerous challenges as well, including issues of privacy, security, and performance.

Challenges

Connectivity is perhaps the greatest problem associated with mobile computing: how can one work when completely or intermittently disconnected; how can one work when connected over a channel that may have low bandwidth and/or high error rates; and how can one keep in communication over the Internet while moving among multiple cells in a wireless network. The restrictive user interfaces of small mobile devices present an additional challenge.

Functioning without connectivity: nomadic computing. Users often run in complete isolation, either out of necessity or by choice. For instance, even though one might connect a laptop on an airplane to a server at one's office via an air-to-ground wireless phone link, the cost might be prohibitive. Support in file systems and other applications for *disconnected operation* permits users to take a portable computer away, work with it, and integrate it later into the networked environment. Both the *correctness* and the *cost-performance* of the reintegration are important.

Functioning with limited connectivity: mobile computing. Wireless networks permit computers to communicate, but they often offer substantially degraded performance (and at much higher cost) by comparison to today's wired LANs. For instance, the NCR Wavelan 2-Mbps wireless network offers one-fifth the bandwidth of the 10-Mbps Ethernets that have been commonplace for nearly two decades, and orders of magnitude less bandwidth than today's fastest networks. Wide-area wireless networks like the Metricom Ricochet [Metricom Corp.] are still slower, more like modems of a few years ago, while they are shared with possibly many simultaneous users. Therefore, some systems attempt to filter the amount of data that goes over the wireless link. Additionally, coverage of the "cells" through which a wireless computer travels may be limited, causing physical movement to result in intermittent connectivity. Cell movements can result in dropped packets, which can be incorrectly interpreted as congestion in the network.

Interprocess communication. Before the advent of mobile computing, the Internet Protocol (IP) assumed that an address mapped to a static location: the first few octets of an IP address specified a network, and the last part of the address specified a host on that network. Since a mobile computer can change which physical network it is connected to at any given time, keeping a static IP address requires another way to route communication to and from a mobile computer. *Mobile IP* has evolved as a protocol to support the physical migration of a computer among different networks.

User interfaces. Some mobile devices, such as laptop computers, are as functional as their desktop counterparts. Others, like palmtops, provide a restricted viewing area, poor resolution, and limited user interactions. Permitting these restricted interfaces to interoperate with the rest of the world, such as the Web, requires special techniques to make the interface manageable.

8.2 Applications

Unlike the other two areas of emphasis in this book, process migration and mobile agents, the applications for mobile computing require little elaboration. This is because the goal of mobile computing is often to duplicate our familiar desktop environment. Still, there are some differences.

Web access. The Web is one of the dominant applications of computing, accounting for the vast majority of Internet traffic. Just because one is mobile does not mean that one need not access resources on the Internet. With Metricom and other wireless wide-area networks, one can access the Web as easily from a mobile computer as from a wired computer, just more slowly. However, caching plays a more important role as the overhead of using the wireless network increases.

E-mail. A well-connected user can synchronously read and send mail over the Internet in "real time." A disconnected user can download mail, read and compose replies while disconnected, and transmit the replies upon reconnection.

Personal information management. Applications on a device like a PalmPilot include a datebook, address book, "to do" list, and notepad. While most of these are personal, and are synchronized with a desktop computer more as a backup than anything else, the datebook application is evolving into a shared resource. If multiple users synchronize their calendars at a single point, one can schedule an event for all of them and download the appointment to each one automatically.

Personalized use of computer resources. In a mobile environment, the choice of which resources to use may vary over time. For instance, one might print on a nearby printer, or see a reminder on a nearby monitor that is selected based on one's physical location.

8.3 Myths and Facts

Mobile computing is becoming increasingly commonplace, and is relatively well understood. Here we discuss a few issues that are perhaps not as clearly understood.

Myth: *Bandwidth or bust.* **Fact:** As we have already discussed, compared to today's best wireline networks, wireless networks are low-bandwidth. But compared to yesterday's networks, wireless is competitive, and it is improving steadily. Multimedia applications have yet to become mainstream in a wireless environment, despite projects such as the InfoPad project at University of California at Berkeley [Le, 1994], but Web access and e-mail over today's wireless networks are generally acceptable. In the near future, broadband wireless networking may even supplant wired telephony and cable modems in some environments.

Myth: *Wireless provides too little of a shared medium.* **Fact:** Bandwidth is very much application-dependent. Existing wireless networks, particularly indoors, provide ample bandwidth for many applications (such as e-mail). Sharing the medium has potential benefits as well. Since the wireless network is inherently a broadcast medium, one can broadcast popular data to many users for the same cost (in bandwidth) as sending it to just one. Several research projects have investigated the use of broadcast networks to access databases [Acharya, 1995; 1996] and also the Web [Almeroth, 1998].

Myth: *Who needs wireless?* **Fact:** Most users of laptops work standalone, and connect via standard phone lines when they wish to connect to the rest of the world. For them, the existing infrastructure is sufficient, and "mobile computing" is an uninteresting topic. On the other hand, Metricom's wide-area wireless networks have been successful enough to expand into multiple metropolitan areas, and in some office environments wireless LANs are commonplace. The "nomads" who currently work without connectivity should welcome the ability to get more general and continuous access to resources within the office environment and from the Internet as a whole.

Myth: *Mobile computing and wireless computing are the same thing.* **Fact:** Though we referred to wireless computing as "true mobile computing" earlier in this chapter, the two are certainly not interchangeable. For instance, AT&T has touted "fixed wireless" as a method to get direct access to households without the cost of running a wired connection to every home. Wireless networking raises issues that are separate from mobile computing, and vice versa, even though in many cases the two are tightly coupled. Mobile computing in the absence of wireless networking—termed nomadic computing—has a separate set of problems.

8.4 Overview

Mobile computing is a broad topic, and indeed there are entire books devoted to it [Goodman, 1997; Imieliński, 1996; Perkins, 1998a; Solomon, 1998; Taylor, 1997]. Here we focus on physical mobility, primarily wireless networking. We do not consider a range of other topics relating to mobile computing, particularly the limitations that arise from portability itself: limited battery capacities, limited storage, and so forth. Some of these other topics are touched on in the following articles, particularly Forman and Zahorjan's **overview** [Forman, 1994]. That overview, which appeared in *Computer* in 1994, is largely applicable today. (The primary area in which it dates itself applies to the tables of hardware specifications of the devices that were available at that time.) Since the overview touches on many of the subjects above, we recommend that the reader peruse this article and the **Afterword** Forman provided for this collection, then proceed with the subsequent chapters, each of which is introduced separately.

8.5 Forman, G.H. and Zahorjan, J., The Challenges of Mobile Computing

Forman, G.H., and Zahorjan, J., "The Challenges of Mobile Computing," *Computer*, 27(4):38–47, April 1994.

The Challenges of Mobile Computing

George H. Forman

John Zahorjan

Computer Science & Engineering
University of Washington

March 9, 1994

Abstract

Advances in wireless networking technology have engendered a new paradigm of computing, called *mobile computing*, in which users carrying portable devices have access to a shared infrastructure independent of their physical location. This provides flexible communication between people and continuous access to networked services. Mobile computing is expected to revolutionize the way computers are used.

This paper is a survey of the fundamental software design pressures particular to mobile computing. The issues discussed arise from three essential requirements: the use of wireless networking, the ability to change locations, and the need for unencumbered portability. Promising approaches to address these challenges are identified, along with their shortcomings.

Keywords: mobile computing, hand-held computers, PDAs, surveys, wireless communication, networks, disconnection, low bandwidth, data security, mobility, location dependence, portability, low power, small user interfaces

Available as UW CSE Tech Report # 93-11-03 from ftp.cs.washington.edu.

Contents

1 Introduction

Recent advances in technology enable portable computers to be equipped with wireless interfaces, allowing networked communication even while mobile. Whereas today's notebook computers and personal digital assistants (PDAs) are self-contained, tomorrow's networked *mobile computers*[1] are part of a greater computing infrastructure. *Mobile computing* constitutes a new paradigm of computing that is expected to revolutionize the way computers are used.

Wireless networking greatly enhances the utility of carrying a computing device. It provides mobile users with versatile communication to other people and expedient notification of important events, yet with much more flexibility than cellular phones or pagers. It also permits continuous access to the services and resources of the land-based network. The combination of networking and mobility will engender new applications and services, such as collaborative software to support impromptu meetings, electronic bulletin boards that adapt their contents according to the people present, self-adjusting lighting and heating, and navigation software to guide users in unfamiliar places and on tours[14].

[1] We use the term *mobile computer* to denote a portable computer that is capable of wireless networking.

The technical challenges to establishing this paradigm of computing are non-trivial, however. In this paper we survey the principal challenges faced by the software design of a mobile computing system, as distinguished from the design of today's stationary networked systems. We discuss those issues pertinent to the software designer, without delving into the lower level details of the hardware realization of the mobile computers themselves. Where appropriate, we identify promising approaches that researchers have applied, as well as their limitations.

The issues described herein divide cleanly into three sections, each stemming from an essential property of mobile computing. Section 2 considers the implications of using *wireless communication*, for example, susceptibility to disconnection, low bandwidth availability, and highly variable network conditions. Section 3 discusses the consequences of *mobility*, including dynamically changing network addresses, location-dependent answers to user queries and system configuration, and communication locality that deteriorates as mobile users move away from their servers. Section 4 investigates the pressures that *portability* places on the design of a mobile system, such as low power, risk of data loss, and small surface area available for the user interface.

In order to expose a greater assortment of issues, the target in mind is large scale, hand-held mobile computing. Of course, special purpose systems may avoid some design pressures by doing without certain desirable properties. For instance, mobile computers installed in the dashboards of cars would be less concerned with the portability pressures than would hand-held mobile computers.

Within the notion of mobile computing, there is considerable latitude regarding the role of the portable device. Is it a terminal or an independent, stand-alone computer? How many purposes shall the device serve? Should it incorporate a telephone (as does the AT&T EO)? Should it provide the work environment of a general purpose workstation, or something more restrictive, such as the Apple Newton MessagePad? These design choices greatly affect the severity of the issues in the following sections. For example, a portable terminal, such as the PARC Tab[14], is more dependent on the network but less prone to loss of storage media than a stand-alone computer. It is important to consider such questions in relation to the issues presented below.

2 Wireless Communication

Mobile computers require wireless network access, although sometimes they may physically attach to the network for a better or cheaper connection when they remain stationary, such as during meetings or while at a desk.

Wireless communication is much more difficult to achieve than wired communication because the surrounding environment interacts with the signal, blocking signal paths and introducing noise and echoes. As a result, wireless connections are of lower quality than wired connections: lower bandwidths, higher error rates, and more frequent spurious disconnections. These factors can in turn increase communication latency due to retransmissions, retransmission timeout delays, error control protocol processing, and short disconnections.

Wireless connections can be lost or degraded also by mobility. Users may outstep

...ansceivers or enter areas of high interference. Unlike typical ... of devices in a cell varies dynamically, and large concentra... s at conventions and public events, may overload network ...

... the design challenges resulting from the need for wireless ... more frequent disconnections, lower bandwidth, greater variation in ... bandwidth, greater network heterogeneity, and increased security risks.

2.1 Disconnection

Today's computer systems often depend heavily on the network, and may cease to function during network failures. For example, distributed file systems may block waiting for other servers, and application processes may fail altogether if the network stays down too long.

Network failure is of greater concern to mobile computing designs than traditional designs, because wireless communication is so susceptible to disconnection. One can either spend more resources on the network trying to prevent disconnections, or spend those resources enabling systems to cope with disconnections more gracefully and work around them where possible.

The more autonomous a mobile computer, the better it can tolerate network disconnection. For example, some applications can reduce communication by running entirely locally on the mobile unit, rather than splitting the application and the user interface across the network. In environments with frequent disconnections it is more important for the mobile device to operate as a stand-alone computer, as opposed to a portable terminal.

In some cases both round-trip latency and short disconnections can be hidden by operating asynchronously. The X11 Window system uses this technique to achieve good performance. As opposed to the synchronous remote procedure call paradigm where the client waits for a reply after each request, in asynchronous operation a client sends multiple requests before asking for any acknowledgment. Similarly, prefetching and lazy write-back also decouple the act of communication from the actual time a program consumes or produces data, allowing it to make progress during network disconnections. These techniques, therefore, have the potential to mask some network failures.

The Coda file system provides a good example of handling network disconnections, although it is designed for today's notebook computers where disconnections may be less frequent, more predictable and longer lasting than in mobile computing[6]. Information from the user's profile is used to help keep the best selection of files in an on-board cache. It is important to cache whole files rather than blocks so that entire files can be read during a disconnection. When the network reconnects, the cache is automatically reconciled with the replicated master repository. Coda allows files to be modified even during disconnections. More conservative file systems disallow this to prevent multiple users from making inconsistent versions. Coda's optimism is justified by studies showing that only rarely are files actually shared in a distributed system; less than 1% of all writes are followed by a write by a different user[6]. In those cases where strong consistency guarantees are needed, clients can ask for them explicitly. Hence, providing flexible consistency semantics can allow better autonomy.

Of course, not all network disconnections can be masked. In these cases good user interfaces can help by providing feedback to the user about which operations are unavailable due to network disconnections.

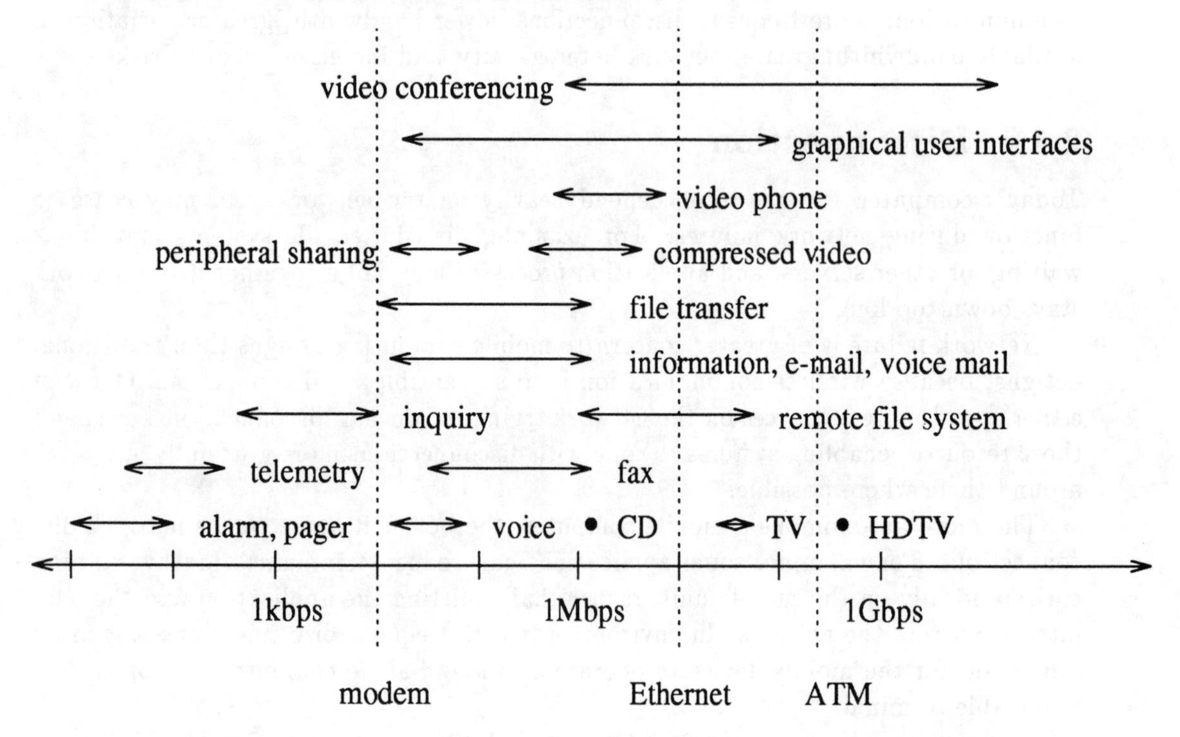

Figure 1:
This figure shows application bandwidth requirements laid out on a horizontal log-scale axis in bits per second (bps). The vertical lines show the bandwidth capability of a few network technologies. This figure clarifies which applications are suitable for a given bandwidth technology. The newest cellular modems are achieving speeds adequate for the everyday informational needs of mobile users, such as electronic mail, and some day may be able to support remote file systems.

2.2 Low Bandwidth

Mobile computing designs need be more concerned about bandwidth consumption and constraints than designs for stationary computing, because wireless networks deliver lower bandwidth than wired networks— cutting-edge products for portable wireless communications achieve only 1 Mbps for infrared communication, 2 Mbps for radio communication, and 9–14 kbps for cellular telephony, while Ethernet provides 10 Mbps, FDDI 100 Mbps, and ATM 155 Mbps. Even non-portable wireless networks, such as the Motorola Altair, barely achieve 5.7 Mbps.

Network bandwidth is divided among the users sharing a cell. The deliverable bandwidth per user, therefore, is a more useful measure of network capacity than raw transmission bandwidth. But because this measure depends on the size and distribution of a user population, Weiser and others promote measuring a wireless network's capacity by its bandwidth per cubic meter[15].

To improve network capacity, one can install more wireless cells to service a user population. There are two ways of doing this: overlap cells on different wavelengths, or reduce transmission ranges so that more cells fit in a given area.

The scalability of the first technique is limited, because the electromagnetic spectrum available for public consumption is scarce. This technique is more flexible, however, because it allows (and requires) software to allocate bandwidth among users.

The second technique is generally preferred. It is arguably simpler, reduces power requirements (see section 4.1), and may decrease corruption of the signal because it may interact with fewer objects in the environment. Also, there is a hardware tradeoff between bandwidth and coverage area— transceivers covering less area can achieve higher bandwidths.

Certain software techniques can also help cope with the low bandwidth of wireless links. Modems typically use compression to increase their effective bandwidth, sometimes almost doubling throughput. Because bulk operations are usually more efficient than many short transfers, logging can improve bandwidth usage by making large requests out of many short ones. Logging in conjunction with compression can further improve throughput because larger blocks compress better.

Certain software techniques for coping with disconnection can also help cope with low bandwidth. Typical network usage occurs in bursts, and disconnections are similar to bursts in that demand temporarily exceeds available bandwidth. For example, lazy write-back and prefetching use the valleys to reduce demand at the peaks. Lazy write-back can even reduce overall communication when the data to be transmitted are further mutated or deleted before they are transmitted. Prefetching involves knowing or guessing which files will be needed soon and downloading them over the network before they are demanded[10]. Bad guesses can waste network bandwidth, however.

System performance can be improved by scheduling communication intelligently. When available bandwidth does not satisfy the demand, priority should be given to those processes for which the user is waiting. Backups should be performed only with "leftover" bandwidth. Mail can be trickle fed onto the mobile computer slowly before the user is notified. Although these techniques do not increase effective bandwidth, they are equally important to improving user satisfaction.

2.3 High Bandwidth Variability

Mobile computing designs also contend with much greater variation in network bandwidth than traditional designs. Bandwidth can shift one to four orders of magnitude between being plugged in versus using wireless access. Fluctuant traffic load seldom causes this much variation in available bandwidth on today's networks.

An application can approach this variability in one of three ways: it can assume high bandwidth connections and operate only while plugged in, it can assume low bandwidth connections and not take advantage of higher bandwidth when it is available, or it can adapt to the currently available resources, providing the user with a variable level of detail or quality. Different choices make sense for different applications.

2.4 Heterogeneous Networks

In contrast to most stationary computers, which stay connected to a single network, mobile computers encounter more heterogeneous network connections. As they leave the range of one network transceiver they switch to another. In different places they may experience different network qualities, for example, a meeting room may have better wireless equipment installed than a hallway. There may be places where they can access multiple transceivers on different frequencies. Even when plugged in, they may concurrently use wireless access.

Also, they may need to switch interfaces when moving from indoors to outdoors. For example, infrared interfaces cannot be used outside because sunlight drowns out the signal. Even if only radio frequency transmission is used, the interface may still need to change access protocols for different networks, for example when switching from cellular coverage in a city to satellite coverage in the country. This heterogeneity makes mobile networking more complex than traditional networking.

2.5 Security Risks

Precisely because it is so easy to connect to a wireless link, the security of wireless communication can be compromised much more easily than wired communication, especially if the transmission range encompasses a large area. This increases pressure on mobile computing designs to include security measures.

Security is further complicated if users are to be allowed to cross security domains, for example, allowing the untrusted mobile computers of hospital patients to use nearby printers while disallowing access to distant printers and resources designated for personnel only.

Secure communication over insecure channels is accomplished by encryption, which can be done in software, or more quickly by specialized hardware, such as the recently proposed CLIPPER chip. The security depends upon a secret encryption key being known only to the authorized parties. Managing these keys securely is difficult, but can be automated by software such as MIT's Kerberos[9].

Kerberos provides secure authentication services, provided the Kerberos server itself is trusted. It authenticates users without exposing their passwords on the network and generates secret encryption keys that can be selectively shared between mutually suspicious parties. It also allows roaming mobile units to authenticate themselves in foreign domains where they are unknown, thus enhancing the scale of mobility. Methods have also been devised to use Kerberos for authorization control and accounting. Its security is limited, however.For example, the current version is susceptible to off-line password guessing attacks and to replay attacks for a limited time window.

3 Mobility

The ability to change locations while connected to the network increases the volatility of some information. Certain data considered static for stationary computing becomes dynamic for mobile computing. For example, although a stationary computer can be configured statically to prefer the nearest server, a mobile computer needs a mechanism to determine which server to use.

As volatility increases, cost-benefit tradeoff points shift, calling for appropriate modifications in the design. For example, greater volatility of a data object reduces its ratio of uses per modification. For lower ratios, it makes less sense to cache the data, or even to store it at all if it can be recomputed from scratch easily enough. As another example, where management of static information is often done by hand, automated methods are required to handle higher rates of change. Even where automated methods exist, many are ill-suited for the dynamicism of mobile computing.

The following three sections discuss the main problems introduced by mobility: the network address of a mobile computer changes dynamically; its current location affects configuration parameters as well as answers to user queries; and as it wanders away from a nearby server, the communication path between the two grows.

3.1 Address Migration

As people move, their mobile computers will use different network access points, or 'addresses.' Today's networking is not designed for dynamically changing addresses. Active network connections usually cannot be moved to a new address. Once an address for a host name is known to a system, it is typically cached with a long expiration time and with no way to invalidate out-of-date entries. In the Internet Protocol (IP), for example, a host IP name is inextricably bound with its network address— moving to a new location means acquiring a new IP name. Human intervention is often required to coordinate the use of addresses.

In order to communicate with a mobile computer, messages must be sent to its most recent address. There are four basic mechanisms for determining the current address of a mobile computer: broadcast[5, 4], central services[8], home bases[12], and forwarding pointers[5]. These are the building blocks of the current proposals for 'mobile-IP' schemes.

Selective Broadcast: With the broadcast method, a message is sent to all network cells, asking the mobile computer sought to reply with its current address. This becomes too expensive for frequent use in a large network, but if the mobile computer is known to be in some small set of cells, selectively broadcasting in those cells alone is workable. Hence, the methods described below can employ selective broadcast to obtain the current address when only approximate location information is known. For example, a slightly out-of-date cell address may suffice if adjacent cells are known.

Central Services: With the central service method, the current address for each mobile computer is maintained in a logically centralized database. Each time a mobile computer changes its address, it sends a message to update the database. Although this database is logically centralized, the common techniques of distribution, replication, and caching can be employed to improve availability and response time.

Home Bases: The home base method is essentially the limiting case of distributing a central service— only a single server knows the current location of a mobile computer. This brings with it the availability problems of aggressive distribution without replication. For example, if a home base is down or inaccessible, the mobile computers it tracks cannot be contacted. Note that if users sometimes change home bases, another instance of the address migration problem arises, albeit with much lower volatility.

Forwarding Pointers: With the forwarding pointer method, each time a mobile computer changes its address, a copy of the new address is deposited at the old location. Each message is forwarded along the chain of pointers until it reaches the mobile computer. To avoid the inefficient routing that can result from long chains, pointers at message sources can be updated to reflect more recent addresses.

Although the forwarding pointer method is among the fastest, it is prone to failures anywhere along the trail of pointers, and in its simplest form, does not allow forwarding pointers to be forgotten. Hence, forwarding pointers are often employed only to speed the common case and another method is used to fall back on for failures and to allow reclamation of old pointers.

Note that the forwarding pointer method requires an active entity at the old address to receive and forward messages. This does not fit standard networking models, where a network address either is a passive entity, such as an Ethernet cable, or is specific to the mobile computer, which cannot remain to forward its own messages. This mismatch introduces subtle difficulties in implementing forwarding efficiently (such as with intra-cell traffic, or when multiple gateways are attached to a network address).

3.2 Location Dependent Information

Because traditional computers do not move, information that depends on location is configured statically, such as the local name server, available printers, and the time zone. A challenge for mobile computing is to factor out this information intelligently and provide mechanisms to obtain configuration data appropriate to the present location.

Besides this dynamic configuration problem, mobile computers need access to more location sensitive information than stationary computers if they are to serve as guides in places unfamiliar to their users, for example, to answer queries like "where is the fiction section (in this library)?" or "where is the nearest open gas station heading north?"

Whereas such queries require static location information about the world, Badrinath and Imielinski are studying a related class of queries that depends on the dynamic locations of other mobile objects, for example, determining where the nearest taxi is[4].

Privacy: Answering these queries requires knowing the location of another mobile user. In some cases this may be sensitive information, more so if given at a fine resolution. Even where it is not particularly sensitive, such information should be protected against misuse, for example, to prevent a burglar from determining when the inhabitants of a house are far away.

Privacy can be ensured by denying users the ability to know another's location. The challenge for mobile computing is to allow more flexible access to this information without violating privacy, for there are many legitimate uses of location information, including contacting colleagues, routing telephone calls, logging meetings in personal diaries, and tailoring the content of electronic announcement displays to the viewers[15].

3.3 Migrating Locality

Mobile computing engenders a new kind of locality that migrates as users move. Even if a mobile computer spends the effort to find the server that is nearest for a given service, over time it may cease to be the nearest due to migration. Because the physical distance between two points does not necessarily reflect the network distance, the communication path can grow disproportionately to actual movement. For example, a small movement can result in a much longer path when crossing network administrative boundaries. A longer network path means communication traverses more intermediaries, resulting in longer latency and greater risk of disconnection. This also consumes more network capacity, even though the bandwidth between the mobile unit and the server may not degrade.

To avoid these disadvantages, service connections may be dynamically transferred to servers that are closer[3]. When many mobile units converge, such as during meetings, load balancing concerns may outweigh the importance of communication locality.

Table 1: Characteristics of Personal Digital Assistant products and the AT&T EO tablet computer. Each has a pen interface and a black & white reflective LCD screen. The portable PC is included for comparison. (These data were gathered from advertisements, company representatives, and product reviews, such as those in PC Magazine, October 1993.)

Product	RAM	MHz	CPU	Batteries (hours,# & type)	Weight (lbs.)	Display (pixels, sq.inches)
Amstrad Pen Pad PDA600	128 KB	20	Z80	40, 3 AAs	0.9	240×320, 10.4
Apple Newton MessagePad	640 KB	20	ARM	6–8, 4 AAAs	0.9	240×336, 11.2
Apple N. MessagePad 110	1 MB	20	ARM	50, 4 AAs	1.25	240×320, 11.8
Casio Z-7000 PDA	1 MB	7.4	8086	100, 3 AAs	1.0	320×256, 12.4
Sharp ExpertPad	640 KB	20	ARM	20, 4 AAAs	0.9	240×336, 11.2
Tandy Z-550 Zoomer PDA	1 MB	8	8086	100, 3 AAs	1.0	320×256, 12.4
AT&T EO 440 Personal Communicator	4–12 MB	20	Hobbit	1-6, NiCad	2.2	640×480, 25.7
Portable PC	4–16 MB	33–66	486	1-6, NiCad	5–10	640×480, 84 (or 1024×768)

4 Portability

Today's desktop computers are not intended to be carried, so their design is liberal in their use of space, power, cabling, and heat dissipation. In contrast, the design of a hand-held mobile computer should strive for the properties of a wristwatch: small, light weight, durable, water-resistant and long battery life. Concessions can be made in each of these areas to enhance functionality, but ultimately the value provided to the user must exceed the trouble of carrying the device. Similarly, any specialized hardware to offload from the CPU tasks such as data compression or encryption should justify its consumption of power and space.

In the sections below we describe the design pressures caused by portability constraints: low power, heightened risk of data loss, small user-interfaces, and limited on-board storage. These pressures are evident in the designs of the recent PDA products listed in Table 1, as will be related below.

Table 2: Power consumption of the components of a portable computer and accessories. (The data for the computer components were derived from the Sharp PC 6785 manual. The data for the accessories were obtained from manufacturers; starred figures are estimates for PCMCIA products that are soon to be released.)

Device	Power (Watts)
base system (2MB, 25 MHz CPU)	3.650
base system (2MB, 10 MHz CPU)	3.150
base system (2MB, 5 MHz CPU)	2.800
screen backlight	1.425
hard drive motor	1.100
math co-processor	.650
floppy drive	.500
external keyboard	.490
LCD screen	.315
hard drive active (head seeks)	.125
IC card slot	.100
additional memory (per MB)	.050
parallel port	.035
serial port	.030
Accessories:	
1.8" PCMCIA hard drive	0.7–3.0
cellular telephone active	5.400
cellular telephone standby	.300
infrared network- 1 Mbps*	.250
PCMCIA modem- 14400 bps	1.365
PCMCIA modem- 9600 bps	.625
PCMCIA modem- 2400 bps	.565
global positioning receiver*	.670

4.1 Low Power

Batteries are the largest single source of weight in a portable computer. While reducing battery weight is important, too small a battery can undermine portability— users may have to recharge frequently, carry spare batteries, or use their mobile computer less. Minimizing power consumption can improve portability by reducing battery weight and lengthening the life of a charge.

Power consumption of dynamic components is proportional to CV^2F, where C is the capacitance of the wires, V is the voltage swing, and F is the clock frequency. This function suggests three ways to save power. (1) Capacitance can be reduced by greater levels of VLSI integration and multichip module technology. (2) Voltage can be reduced by redesigning chips to operate at lower voltages. Historically, chips operate at five volts, but to save power, the Apple MessagePad operates at three volts. Manufacturers are rapidly developing a line of low-power chip sets for 2.5 and 3.3 volt operation. (3) Clock frequency can be reduced, trading off computational speed for power savings. PDA products have adopted this concession, as shown in Table 1. In some notebook computers, the clock frequency can be changed dynamically, providing a flexible tradeoff; for example, the Sharp PC 6785 can save power by dynamically shifting its clock from 25 MHz down to 10 MHz or even 5 MHz, as seen in Table 2. In order to retain more computational power at lower frequencies, processors are being designed that perform more work on each clock cycle[1].

Power can be conserved not only by the design but also by efficient operation. Power management software can power down individual components when they are idle, for example, spinning down the internal disk or turning off screen lighting. Li et al. determined recently that for today's notebook computing it is worthwhile to spin down the internal disk drive after it has been idle for just a few seconds[7]. Applications can conserve power by reducing their appetite for computation, communication and memory, and by performing their periodic operations infrequently to amortize the startup overhead. Since cellular telephone transmission typically requires about ten times as much power as reception, trading talking for more listening can also save power. The potential savings of these techniques can be evaluated using Tables 2 and 3, which show example power budgets for notebook computers. Although screen lighting consumes a large amount of power, it has been found to greatly improve readability, for example, on EO models it enhances contrast from 6:1 to 13:1. Nevertheless, PDA products have elected to omit screen lighting in favor of longer battery life.

4.2 Risks to Data

Making computers portable heightens their risk of physical damage, unauthorized access, loss, and theft. This can lead to breaches of privacy or total loss of data. These risks can be reduced by minimizing the essential data that is kept on board— notably, a mobile device that serves only as a portable terminal is less prone to data loss than a stand-alone computer. This is the approach taken for PARC's Tabs and for Bershad's BNU system[13].

To help prevent unauthorized disclosure of information, data stored on disks and removable memory cards can be encrypted. For this to be effective, users must not

Table 3: Power consumption breakdown by subsystems of a portable computer. These data were obtained from the Compaq LTE 386/s20 manual.

System	% Power
display edge-light	35%
CPU/Memory	31%
hard disk	10%
floppy disk	8%
display	5%
keyboard	1%

leave authenticated sessions (logins) unattended.

To safeguard against data loss, one can keep a copy that does not reside on the portable unit. One solution is to have the user make backup copies, but users often neglect this chore, and data modified between backups is not protected. With the addition of wireless networks to portable computers, newly produced data can be copied immediately to secure, remote media. This can be accomplished with replicated file systems such as Coda and Echo[6].

4.3 Small User Interface

The size constraints on a portable computer require a small user interface. Desktop windowing environments may be sufficient for today's notebook computers, but for smaller, more portable devices current windowing technology is inadequate. On small displays it is impractical to have several windows open at the same time regardless of screen resolution, and it can be difficult to locate windows or icons when stacked atop one another deeply. Also, window title bars and borders either consume significant portions of screen space or become difficult to operate with the pointing device.

Duchamp and Feiner have investigated the use of head-mounted virtual reality displays for portable computers[3]. As the user turns his head, the image displayed in the eye shifts to give the sensation that there is a screen all around. This effectively increases the screen area available for windowing systems. Disadvantages of this approach include the hassle of the head gear, low-resolution (one tenth that of conventional displays), eye fatigue, and the requirement for dim lighting conditions.

Buttons vs. Recognition: The shortage of surface area on a small computer can cause us to trade buttons in favor of recognizing the user's intention from analog input devices: handwriting recognition, gesture recognition, and voice recognition. Although handwriting is about three times slower than typing on average, handwriting recognition allows the keyboard to be eliminated, which reduces size and improves durability. This approach has been adopted by all the PDA products in Table 1.

Handwriting recognition rates for high-end systems are typically 96–98% accurate when trained to a specific user. (Tappert et al. give a thorough survey[11].) With context information, recognition rates can be enhanced effectively to 100%, but context constraints do not help for all kinds of input, such as when entering words that are not in the dictionary. Popular reports indicate that the Apple Newton's handwriting recognition, while among the best of the PDAs, is nevertheless a source of frustration. Finally, recognition of the user's intention in a general setting is inherently hard because the interpretation of pen strokes is ambiguous. For example, by drawing a circle a user may intend to select an object or an area, write a zero, degree sign, or the letter 'o'.

Speech production and recognition seem an ideal user interface for a mobile computer in that they require no surface area and allow hands-free and even eye-free operation. The voice commanded VCR programmer by Voice Powered Technology demonstrates the feasibility of this interface on a hand-held device for a narrow domain. Speaker-independent recognition rates of nearly 96% have been reported for the Sphinx research project; 98% for speaker-trained recognition. However, general purpose speech input and output places substantial storage and processing demands on a mobile device. Also, speech is inappropriate in common situations: it disturbs others in quiet environments, it cannot be recognized clearly in noisy environments, and it can compromise privacy. Finally, speech is ill-suited for skimming data because of its sequential nature.

Pointing Devices: The mouse is the standard pointing device for desktop computers, but does not suit mobile computers. Pens have become the standard input device for PDAs because of their ease of use while mobile, their versatility and their ability to supplant the keyboard.

Switching to pens requires changing both the user interface and the software interface because mice and pens are really quite different[3]. Users with pens can jump to absolute screen positions and enter path information more easily than with mice; it is nearly impossible to write with a mouse. Pen positioning resolution on current tablet computers is several times that of screen resolution, for example, on the EO pen resolution is 0.1mm while screen resolution is 0.23–0.3mm. Parallax between the pen tip and the screen image can mislead pointing; with mice, there is no parallax because the mouse cursor provides feedback in the image plane. Finally, the mouse cursor obscures much less of the screen than is obscured by one's hand when writing with a pen.

4.4 Small Storage Capacity

Storage space on a portable computer is limited by physical size and power requirements. Traditionally, disks provide large amounts of non-volatile storage. To a mobile computer, however, disks are a liability. They consume more power than memory chips, except when off-line, and may not be non-volatile when subject to the indelicate treatment a portable device endures. Hence, none of the PDA products have disks.

Coping with limited storage is not a new problem. Solutions include compressing file systems, accessing remote storage over the network, sharing code libraries, and compressing virtual memory pages[2]. Although today's networked computers have had great success with distributed file systems and remote paging, relying on the network is less appropriate for mobile computers that regularly encounter network disconnections.

A novel approach to reducing program code size is to interpret script languages, instead of executing compiled object codes, which are typically many times the size of the source code. This approach is embodied by General Magic's Telescript and Apple Technology Group's Dylan and NewtonScript. An equally important goal of such languages is to enhance portability by supporting a common programming model across different machines.

5 Conclusion

In this paper we have examined the repercussions of three principal features of mobile computing: wireless communication, mobility, and portability. Wireless communication brings challenging network conditions, making access to remote resources often slow or sometimes temporarily unavailable. Mobility causes greater dynamicism of information. Portability entails limited resources available on board to handle the changeable mobile computing environment. The challenge for mobile computing designers is how to adapt the system designs that have worked well for traditional computing. To date, few prototype mobile computing systems have been built that include the wireless network[3, 15].

6 Acknowledgments

Support for this work was provided in part by the National Science Foundation (Grants CCR-9123308 and CCR-9200832), Tektronix Inc. (a graduate fellowship), the Washington Technology Center, and Digital Equipment Corporation (Systems Research Center and External Research Program). We thank Robert Bedichek, Brian Bershad, Blake Hannaford, Marc Fiuczynski, Brian Pinkerton, and Stefan Savage for helpful pointers and clarifying discussions that significantly improved this paper.

References

[1] Anantha Chandrakasan, Samuel Sheng, and R.W. Brodersen. Design Considerations for a Future Portable Multimedia Terminal. In *Third Generation Wireless Information Networks,* Kluwer Academic Publishers, pages 75–97, 1992.

[2] Fred Douglis. On the Role of Compression in Distributed Systems. In *5th SIGOPS Workshop on Models and Paradigms for Distributed Systems Structuring,* 6 pages, Sept 1992.

[3] Dan Duchamp, Steven K. Feiner, and Gerald Q. Maguire, Jr. Software Technology for Wireless Mobile Computing. *IEEE Network Magazine,* pages 12–18, Nov 1991.

[4] T. Imielinski and B. R. Badrinath. Data Management for Mobile Computing. *SIGMOD Record,* 22(1):34–39, March 1993.

[5] John Ioannisdis, Dan Duchamp, and Gerald Q. Maguire Jr. IP-based Protocols for Mobile Internetworking. In *Proceedings of SIGCOMM '91 Symposium,* pages 235–245, Sept 1991.

[6] James J. Kistler and M. Satyanarayanan. Disconnected Operation in the Coda File System. *ACM Transactions on Computer Systems,* 10(1):3–25, Feb 1992.

[7] Kester Li, Roger Kumpf, Paul Horton, and Thomas Anderson. A Quantitative Analysis of Disk Drive Power Management in Portable Computers. Technical Report, Computer Science Division, University of California at Berkeley, 1993.

[8] Chaoying Ma. On Building Very Large Naming Systems. In *5th SIGOPS Workshop on Models and Paradigms for Distributed Systems Structuring,* 5 pages, Sept 1992.

[9] B. Clifford Neuman. Protection and Security Issues for Future Systems. In *Workshop on Operating Systems of the 90s and Beyond,* Springer-Verlag Lecture Notes in Computer Science #563, pages 184–201, July 1991.

[10] Carl D. Tait and Dan Duchamp. Detection and Exploitation of File Working Sets. In *11th International Conference on Distributed Computing Systems,* pages 2–9, May 1991.

[11] C.C. Tappert, C.Y. Suen, and T. Wakahara. On-line Handwriting Recognition—A Survey. In *9th International Conference on Pattern Recognition,* 2:1123–1132, 1988.

[12] Fumio Teraoka and M. Tokoro. Host migration transparency in IP networks: the VIP approach. *Computer Communication Review,* 23(1):45–65, Jan 1993.

[13] T. Watson and B. N. Bershad. Local Area Mobile Computing on Stock Hardware and Mostly Stock Software. In *USENIX Proceedings of the Mobile and Location-Indepent Computing Symposium,* pages 109–116, Aug 1993.

[14] Mark Weiser. The Computer for the 21st Century. *Scientific American,* 265(3):94–104, Sept 1991.

[15] Mark Weiser. Some Computer Science Issues in Ubiquitous Computing. *Communications of the ACM,* 36(7):75–84, July 1993.

Wanted: Programming Support for Ensuring Responsiveness Despite Resource Variability and Volatility

George H. Forman, Hewlett-Packard Labs

Introduction

Our purpose here is to expose and stress the growing need for programming support to build applications that ensure good response time to the user despite underlying resources that may give variable, and at times, unsatisfactory service [1,2]. We begin by motivating the problem of variability and the challenge of writing software to cope with it. We then highlight known techniques for coping with variability and call for research to facilitate the use of these techniques in building applications with robust responsiveness despite environmental variability.

It can be extremely aggravating when the programs we use exhibit poor response time, especially if our expectations of responsiveness are dumbfounded by high variability. Supposing we download an acclaimed program to our wireless, Java-enabled mobile computer—it is prone to exhibit poor responsiveness in our networked[1] mobile computing environment for a number of reasons (some of which are not specific to mobile computing):

1. network bandwidth variation between wired and wireless access—to illustrate, there are 5 orders of magnitude difference between 155 Mbps ATM and 9600 bps wireless modem,
2. variable network latency due to sporadic setup delay—mobile computers may have to establish wireless networking on demand to save on connection costs or to recover from lost connections,
3. variable network delays due to wireless interference—e.g., due to intermittent interference,
4. location-dependent resource variability as we change locations—e.g., in a conference room we may have 2 Mbps wireless network service with a heavy-duty proxy server available, whereas on the road we may resort to a 9600 bps wireless modem and our outdated home machine as proxy server,
5. variations in processing capability—the application might have been written with the expectation of a high-end desktop computer, whereas mobile computers typically have lower performance due to portability constraints,
6. shared resource variability—shared resources, such as CSMA wireless networks, exhibit varying responsiveness according to current user load, and
7. variable data magnitude—the trend toward sporadic inclusion of multimedia content into some documents can drastically vary their transfer and processing time.

Furthermore, applications running on mobile computers are *likely* to be dependent on the performance of the network and remote services. This is because their use is often as a communicator or information utility, and considering their limited computing resources, it is natural to compensate by employing remote services. This dependence on remote services may be transparent to the application, e.g., remote file systems and remote paging.

Challenge to Programming

Scaling applications to fit resource constraints is an old problem. There is an inherent tradeoff between response time and the quality of results presented to the user (amount of work done). For example, repainting a window as it is dragged with the mouse in real time gives a higher quality experience than just dragging an outline of the window, but at the cost of less responsive movement if insufficient computing resources are available.

Traditionally, programmers have managed this balance by *static sizing*: they scale the amount of processing their application requires based on its response time on a specific platform, typically their own. Such applications running on less endowed platforms will exhibit poor responsiveness, and on more endowed platforms will forgo opportunities for higher quality. With variable resources, both disadvantages may be experienced at different times.

The follow-on is *dynamic sizing*: programmers write their application to size its workload according to the environment. For example, measure network bandwidth and scale the image resolution to be transferred so that it meets a response time target. This lies at the heart of much of the work in quality of service negotiation. Consider what is required: (1) applications capable of multiple service levels, (2) a predictive model of the response time at each service level, (3) a mechanism to sense the current resources of the environment for selecting a service level, (4) optionally, the capability to make resource reservations, which may have to be revoked later, (5) the capability to monitor an ongoing operation to detect early that it will not meet its objective due to either changes in the environmental resources/reservations or errors in the performance model, and (6) a mechanism to cancel and renegotiate an operation at a different service level, preferably not throwing away work that has already been done.

This is a lot to get working correctly. In addition, the paradigm of quality of service negotiation breaks down in environments where resource variability may be highly dynamic. In a volatile environment, the measurement of

[1] Our scope here does not include complete network disconnection, which may occur in some mobile environments.

current resources has little bearing on the actual resources available when the operation is finally performed. At high volatility, the system may be reduced to thrashing as it constantly renegotiates service levels. The conceptual graph to the right distinguishes variable resource environments along two axes: the range of variation and the rate of variation. Dynamic sizing is appropriate for a wider range of variation than static sizing; however, "flexible" techniques are required to expand coverage into the space of high volatility. The remainder of this paper addresses techniques and support for building applications that can operate effectively in all these regions.

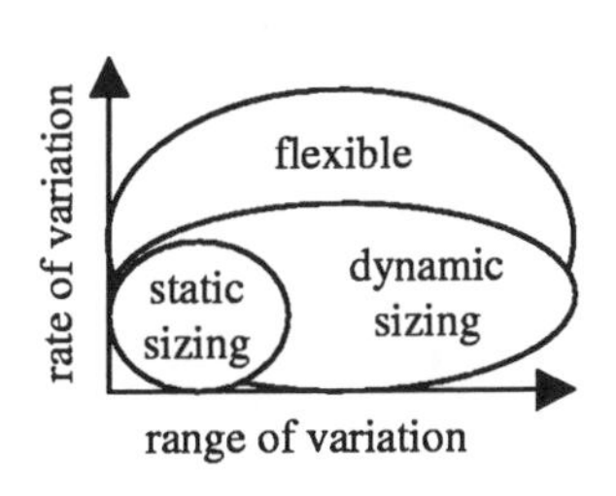

Obtaining Responsive Behavior Despite Variability and Volatility

How can applications behave with good responsiveness towards the user, even though the resources they depend on are unpredictable? We forward four principal techniques:

Incremental Results (a.k.a. multi-resolution encoding, progressive transmission/computation): The notion is to do the most important bits first and improve on quality as resources become available, if at all possible, without taking any more time than it would have taken to do the work in the straightforward order. This technique can be applied to a broad variety of domains: images, movies, audio, object graphics, 3-D models, compound documents—even to the order in which portions of a graphical user-interface are drawn.

Concurrency: In traditional, sequential programming, the time to complete each statement depends on the response time of those before it, even if there is no direct dependency. This is perilous for response time in an environment where some tasks may unpredictably take a long time. By making tasks independent of each other wherever possible, the opportunity for forward progress is maximized. This is especially important in mobile computing, where the dependence on remote services may be implicit and non-obvious to the programmer, making it difficult to predict the portions of their application that need special attention to ensure non-blocking behavior.

Dynamic prioritization: As the user's priorities shift, revealed for example by shifting focus between windows, the outstanding tasks associated with each window may be profitably re-prioritized to allocate most resources to what the user is currently attending to, at the cost of delaying tasks that are not of immediate interest.

Cancellation: In an environment where asynchronous tasks may take a long time to finish, newer user actions can cause outstanding tasks to become obsolete. By detecting these situations and terminating obsolete tasks proactively, we can conserve resources, improving responsiveness for the tasks of current interest.

These responsiveness-enhancing techniques come with a significant cost in programming complexity, and so are used only sparingly by today's programmers. Programming support for these techniques by popular languages and libraries is weak and spotty, at best. Thread interfaces are relatively low level and inconvenient to use for small pieces of concurrent work. We therefore propose an important area for software research: to develop general purpose frameworks that effectively support programmers in building applications that exhibit good responsiveness in volatile, variable-resource environments, e.g., general support for the above techniques [2].

Conclusion

We believe that successful applications in the mobile computing environment must offer the user consistently acceptable response time despite tremendously variable underlying service. Mobility and wireless networking cause much of this variability.

Further, we believe it is both essential and practical to develop programming support for techniques that insulate against service variability. Without this, programmers must expend much more effort to implement such techniques, and so they will be applied only sparingly. This results in software that locks up awkwardly during periods of scarce resources. Which software would you rather have?

Finally, we believe that solutions in this space will find great use beyond mobile computing as well. Users in general are facing increased service variability with the ubiquitous use of wide area networking and resource sharing, and the sporadic inclusion of multimedia content. Interestingly, there is a related need in building troubleshooting tools for IT administrators. Such tools must operate robustly and with robust performance even when the environment is behaving poorly.

Acknowledgements

I am grateful to John Zahorjan for his guidance and clarity of thought in this research.

References

1. D. Duis and J. Johnson. Improving user-interface responsiveness despite performance limitations. In *IEEE COMPCON, Spring '90*, pages 380-386, 1990.
2. G. H. Forman. Obtaining responsiveness in resource-variable environments. Ph.D. dissertation, Computer Science & Engineering Dept., Univ. of Washington, 1996.

Chapter 9 Limits on Connectivity

The 1980s was the decade of networking. Numerous companies, such as Sun Microsystems and Silicon Graphics, made their reputation based on desktop workstations connected via local area networks such as Ethernet™ [Metcalfe, 1976]. Sun's Network File System (NFS) [Sandberg, 1985] protocol was created to permit these computers to share a common file system; it quickly became a *de facto* standard. Numerous other shared file systems were also developed during this time frame, including the Andrew File System (AFS) [Howard, 1988] from Carnegie Mellon University. One distinguishing characteristic of AFS, compared to most other file systems, was that client machines would cache entire files, rather than individual file blocks. This whole-file caching proved to be useful for mobility, as described below.

In a local area network environment, the assumption was that a set of networked computers could virtually always communicate. If a file server was unavailable, or the network was inaccessible, individual workstations would simply be unusable until the problem was rectified. There were some exceptions to this, particularly in networks that were partitionable, where some machines could communicate with each other while others could not.

The advent of laptop computers changed all this. Originally laptops were essentially standalone; any sort of network connection might be an added bonus, but not a requirement. Eventually it was clear that a new mode of operation could become common: one in which a user might be connected to a LAN one hour, a modem another, and nothing at all yet another. In this way, a user would be able to access and edit files regardless of location and connectivity.

Kistler and Satyanarayanan were among the first to recognize the problem of *disconnected operation* and provide a solution [Kistler, 1992; 1993]. They realized there were predominantly two issues:

1. How would a user access files from a file server when the user's machine was unable to communicate with the server?
2. How would updates on the user's machine get reflected on the server once the two machines could again communicate?

They addressed the first issue with the concept of *hoarding*. A file system usually caches data locally in order to avoid having to retrieve it repeatedly over a network. In the case of AFS, upon which their new **Coda** [Kistler, 1992; 1993; Satyanarayanan, 1990; 1993] system was built, all of a file was cached if any of it was. Caching normally was done in some sort of "least recently used" fashion, in that files that had not been accessed recently would be removed from the cache to make room for newly-accessed ones. "Hoarding" is similar to caching, but it attempts to keep files cached not only for performance but also for availability. One can instruct the system to be sure to cache certain files, regardless of when they are actually accessed, in order to prepare for a later disconnection. If one disconnects intentionally, the system could run through its list of files to hoard and make sure it has the most recent copy of each file. If the disconnection is inadvertent, whatever is in the cache suffices.

They addressed the second issue with *logging* and *optimistic concurrency control.* While disconnected a client would keep track of all modifications to the file system. After reconnecting, the client would propagate those changes to their respective file servers. In nearly all cases, the changes could be reintegrated without a problem, because there was very little "write sharing" in the AFS environment, that is, two different users would very rarely modify the same file at the same time. In the rare cases when a problem occurred, the user would be notified.

Weak connectivity means there is an intermediate point between the disconnected operation described above and the strong connectivity of a fast, reliable local area network. Examples of weak connectivity include wireless networks and 28.8Kbps modems. There have been several recent developments in this area.

Mummert, and others, extended **Coda** to support *weak connectivity* [Mummert, 1994; 1995; 1996]. They modified their transport protocol to be more efficient over slow networks. They added volume-level cache consistency, so that rather than revalidating every file upon reconnection, a client could first determine if anything at all on a file system had been modified since disconnection. They allowed clients to slowly propagate updates to file servers asynchronously, something referred to as "trickle reintegration." Finally, they modeled a user's "patience" for waiting for a cache miss, such that a miss in the cache for a large file might be treated as an error, as though the system were completely disconnected. This avoided placing a high load on the slow network for an extended period of time and forcing the user to wait.

Satyanarayanan has provided an **Afterword** to the preceding two articles, in which he summarizes their impact and discusses several follow-on projects. He demonstrates the significant impact Coda has had, both at his own institution (CMU) and at many other organizations. This afterword also includes several additional pointers to recent work.

Terry, and others, have focused on weak connectivity in a different environment. In their **Bayou** system [Demers, 1994; Satyanarayanan, 1990; Terry, 1995], clients can read and write any replica, even if it is possibly inconsistent with other replicas. Hosts interact in a pair-wise fashion to exchange updates, so that updates are eventually propagated to all hosts. Applications are explicitly aware of the possibility of conflict, unlike other systems that have tried to hide replication and the possibility of conflict. Also, application-specific mechanisms handle any conflicts that arise. Example applications in their paper include a meeting room scheduler and a bibliographic database.

In contrast to Bayou, the **Rover** toolkit [Joseph, 1995; 1997a, b] of Joseph, and others, attempts to support both mobile-aware and mobile-transparent applications. It provides two primary abstractions: the relocatable dynamic object (RDO), and the queued remote procedure call (QRPC). An RDO is an interface that can be loaded from a client into a server to reduce communication—or the other way around. QRPC permits applications to make remote procedure calls when disconnected, and have them take place upon reconnection. They demonstrate several sample mobile-aware applications, such as a mail reader, calendar program, address book, and stock market watcher, and proxies to support unmodified (mobile-transparent) applications such as a Web browser. The notion of Web proxies to support application-specific modifications was earlier demonstrated by Brooks and others [Brooks, 1995] and has been used in numerous other systems.

Finally, Barron and Housel's **WebExpress** system [Chang, 1997; Housel, 1996] performs several optimizations specific to Web access in a wireless environment. Like the Rover Web proxy, it intercepts the requests of a normal Web browser. WebExpress performs several optimizations, including caching, "forms differencing" (in which the submission of a CGI request could result in a compact update to generate a new page based on the previous one, a process that has been generalized as "delta-encoding" [Banga, 1997; Mogul, 1997]), protocol reduction, and HTTP header compression. For example, rather than a browser sending the list of accepted data formats on each request, the proxy would omit the list over the wireless link. Then it would be regenerated and sent to the content provider over the remaining network(s).

9.1 Kistler, J.J. and Satyanarayanan, M., Disconnected Operation in the Coda File System

Kistler, J.J. and M. Satyanarayanan. "Disconnected Operation in the Coda File System," *ACM Transactions on Computer Systems,* 10(1):3–25, February 1992. Previous version appeared at the 13th Symposium on Operating System Principles in Pacific Grove, CA in October 1991.

Disconnected Operation in the Coda File System

James J. Kistler and M. Satyanarayanan

School of Computer Science
Carnegie Mellon University
Pittsburgh, PA 15213

Abstract

Disconnected operation is a mode of operation that enables a client to continue accessing critical data during temporary failures of a shared data repository. An important, though not exclusive, application of disconnected operation is in supporting portable computers. In this paper, we show that disconnected operation is feasible, efficient and usable by describing its design and implementation in the Coda File System. The central idea behind our work is that *caching of data,* now widely used for performance, can also be exploited to improve *availability.*

1. Introduction

Every serious user of a distributed system has faced situations where critical work has been impeded by a remote failure. His frustration is particularly acute when his workstation is powerful enough to be used standalone, but has been configured to be dependent on remote resources. An important instance of such dependence is the use of data from a distributed file system.

Placing data in a distributed file system simplifies collaboration between users, and allows them to delegate the administration of that data. The growing popularity of distributed file systems such as NFS [15] and AFS [18] attests to the compelling nature of these considerations. Unfortunately, the users of these systems have to accept the fact that a remote failure at a critical juncture may seriously inconvenience them.

How can we improve this state of affairs? Ideally, we would like to enjoy the benefits of a shared data repository, but be able to continue critical work when that repository is inaccessible. We call the latter mode of operation *disconnected operation,* because it represents a temporary deviation from normal operation as a client of a shared repository.

In this paper we show that disconnected operation in a file system is indeed feasible, efficient and usable. The central idea behind our work is that *caching of data,* now widely used to improve performance, can also be exploited to enhance *availability.* We have implemented disconnected operation in the Coda File System at Carnegie Mellon University.

Our initial experience with Coda confirms the viability of disconnected operation. We have successfully operated disconnected for periods lasting four to five hours. For a disconnection of this duration, the process of reconnecting and propagating changes typically takes about a minute. A local disk of 100MB has been adequate for us during these periods of disconnection. Trace-driven simulations indicate that a disk of about half that size should be adequate for disconnections lasting a typical workday.

2. Design Overview

Coda is designed for an environment consisting of a large collection of untrusted Unix[1] clients and a much smaller number of trusted Unix file servers. The design is optimized for the access and sharing patterns typical of academic and research environments. It is specifically not intended for applications that exhibit highly concurrent, fine granularity data access.

Each Coda client has a local disk and can communicate with the servers over a high bandwidth network. At certain times, a client may be temporarily unable to communicate with some or all of the servers. This may be due to a server or network failure, or due to the detachment of a *portable client* from the network.

This work was supported by the Defense Advanced Research Projects Agency (Avionics Lab, Wright Research and Development Center, Aeronautical Systems Division (AFSC), U.S. Air Force, Wright-Patterson AFB, Ohio, 45433-6543 under Contract F33615-90-C-1465, ARPA Order No. 7597), National Science Foundation (PYI Award and Grant No. ECD 8907068), IBM Corporation (Faculty Development Award, Graduate Fellowship, and Research Initiation Grant), Digital Equipment Corporation (External Research Project Grant), and Bellcore (Information Networking Research Grant).

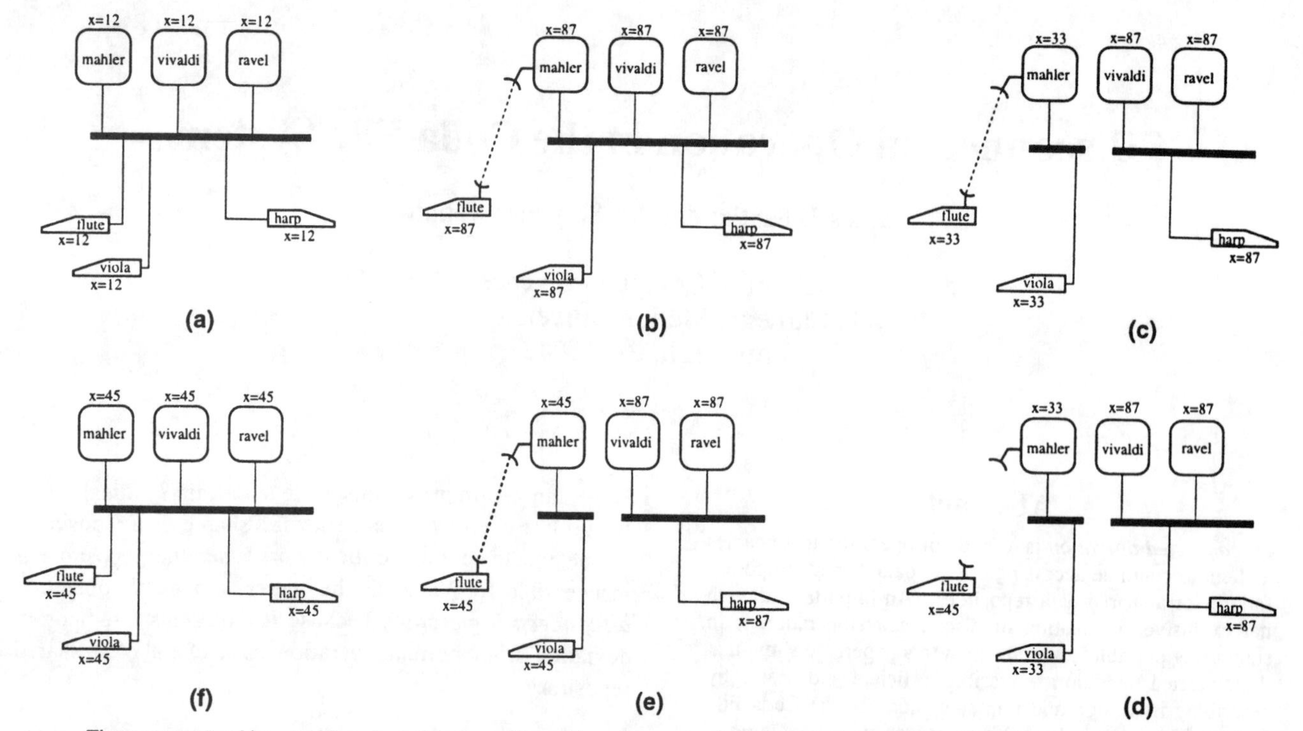

Three servers (*mahler, vivaldi,* and *ravel*) have replicas of the volume containing file *x*. This file is potentially of interest to users at three clients (*flute, viola,* and *harp*). *Flute* is capable of wireless communication (indicated by a dotted line) as well as regular network communication. Proceeding clockwise, the steps above show the value of *x* seen by each node as the connectivity of the system changes. Note that in step (d), *flute* is operating disconnected.

Figure 1: How Disconnected Operation Relates to Server Replication

Clients view Coda as a single, location-transparent shared Unix file system. The Coda namespace is mapped to individual file servers at the granularity of subtrees called *volumes.* At each client, a *cache manager (Venus)* dynamically obtains and caches volume mappings.

Coda uses two distinct, but complementary, mechanisms to achieve high availability. The first mechanism, *server replication,* allows volumes to have read-write replicas at more than one server. The set of replication sites for a volume is its *volume storage group (VSG).* The subset of a VSG that is currently accessible is a client's *accessible VSG (AVSG).* The performance cost of server replication is kept low by caching on disks at clients and through the use of parallel access protocols. Venus uses a cache coherence protocol based on *callbacks* [9] to guarantee that an open of a file yields its latest copy in the AVSG. This guarantee is provided by servers notifying clients when their cached copies are no longer valid, each notification being referred to as a "callback break." Modifications in Coda are propagated in parallel to all AVSG sites, and eventually to missing VSG sites.

Disconnected operation, the second high availability mechanism used by Coda, takes effect when the AVSG becomes empty. While disconnected, Venus services file system requests by relying solely on the contents of its cache. Since cache misses cannot be serviced or masked, they appear as failures to application programs and users. When disconnection ends, Venus propagates modifications and reverts to server replication. Figure 1 depicts a typical scenario involving transitions between server replication and disconnected operation.

Earlier Coda papers [17, 18] have described server replication in depth. In contrast, this paper restricts its attention to disconnected operation. We discuss server replication only in those areas where its presence has significantly influenced our design for disconnected operation.

3. Design Rationale

At a high level, two factors influenced our strategy for high availability. First, we wanted to use conventional, off-the-shelf hardware throughout our system. Second, we wished to preserve *transparency* by seamlessly integrating the high availability mechanisms of Coda into a normal Unix environment.

At a more detailed level, other considerations influenced our design. These include the need to *scale* gracefully, the advent of *portable workstations,* the very different *resource, integrity,* and *security* assumptions made about clients and servers, and the need to strike a balance between *availability* and *consistency*. We examine each of these issues in the following sections.

3.1. Scalability

Successful distributed systems tend to grow in size. Our experience with Coda's ancestor, AFS, had impressed upon us the need to prepare for growth *a priori*, rather than treating it as an afterthought [16]. We brought this experience to bear upon Coda in two ways. First, we adopted certain mechanisms that enhance scalability. Second, we drew upon a set of general principles to guide our design choices.

An example of a mechanism we adopted for scalability is callback-based cache coherence. Another such mechanism, *whole-file caching*, offers the added advantage of a much simpler failure model: a cache miss can only occur on an `open`, never on a `read`, `write`, `seek`, or `close`. This, in turn, substantially simplifies the implementation of disconnected operation. A partial-file caching scheme such as that of AFS-4 [21], Echo [8] or MFS [1] would have complicated our implementation and made disconnected operation less transparent.

A scalability principle that has had considerable influence on our design is the *placing of functionality on clients* rather than servers. Only if integrity or security would have been compromised have we violated this principle. Another scalability principle we have adopted is the *avoidance of system-wide rapid change*. Consequently, we have rejected strategies that require election or agreement by large numbers of nodes. For example, we have avoided algorithms such as that used in Locus [22] that depend on nodes achieving consensus on the current partition state of the network.

3.2. Portable Workstations

Powerful, lightweight and compact laptop computers are commonplace today. It is instructive to observe how a person with data in a shared file system uses such a machine. Typically, he identifies files of interest and downloads them from the shared file system into the local name space for use while isolated. When he returns, he copies modified files back into the shared file system. Such a user is effectively performing manual caching, with write-back upon reconnection!

Early in the design of Coda we realized that disconnected operation could substantially simplify the use of portable clients. Users would not have to use a different name space while isolated, nor would they have to manually propagate changes upon reconnection. Thus portable machines are a champion application for disconnected operation.

The use of portable machines also gave us another insight. The fact that people are able to operate for extended periods in isolation indicates that they are quite good at predicting their future file access needs. This, in turn, suggests that it is reasonable to seek user assistance in augmenting the cache management policy for disconnected operation.

Functionally, *involuntary* disconnections caused by failures are no different from *voluntary* disconnections caused by unplugging portable computers. Hence Coda provides a single mechanism to cope with all disconnections. Of course, there may be qualitative differences: user expectations as well as the extent of user cooperation are likely to be different in the two cases.

3.3. First vs Second Class Replication

If disconnected operation is feasible, why is server replication needed at all? The answer to this question depends critically on the very different assumptions made about clients and servers in Coda.

Clients are like appliances: they can be turned off at will and may be unattended for long periods of time. They have limited disk storage capacity, their software and hardware may be tampered with, and their owners may not be diligent about backing up the local disks. Servers are like public utilities: they have much greater disk capacity, they are physically secure, and they are carefully monitored and administered by professional staff.

It is therefore appropriate to distinguish between *first class replicas* on servers, and *second class replicas* (i.e., cache copies) on clients. First class replicas are of higher quality: they are more persistent, widely known, secure, available, complete and accurate. Second class replicas, in contrast, are inferior along all these dimensions. Only by periodic revalidation with respect to a first class replica can a second class replica be useful.

The function of a cache coherence protocol is to combine the performance and scalability advantages of a second class replica with the quality of a first class replica. When disconnected, the quality of the second class replica may be degraded because the first class replica upon which it is contingent is inaccessible. The longer the duration of disconnection, the greater the potential for degradation. Whereas server replication preserves the quality of data in the face of failures, disconnected operation forsakes quality for availability. Hence server replication is important because it reduces the frequency and duration of disconnected operation, which is properly viewed as a measure of last resort.

Server replication is expensive because it requires additional hardware. Disconnected operation, in contrast, costs little. Whether to use server replication or not is thus a tradeoff between quality and cost. Coda does permit a volume to have a sole server replica. Therefore, an installation can rely exclusively on disconnected operation if it so chooses.

3.4. Optimistic vs Pessimistic Replica Control

By definition, a network partition exists between a disconnected second class replica and all its first class associates. The choice between two families of replica control strategies, *pessimistic* and *optimistic* [5], is therefore central to the design of disconnected operation. A pessimistic strategy avoids conflicting operations by disallowing all partitioned writes or by restricting reads and writes to a single partition. An optimistic strategy provides much higher availability by permitting reads and writes everywhere, and deals with the attendant danger of conflicts by detecting and resolving them after their occurence.

A pessimistic approach towards disconnected operation would require a client to acquire shared or exclusive control of a cached object prior to disconnection, and to retain such control until reconnection. Possession of exclusive control by a disconnected client would preclude reading or writing at all other replicas. Possession of shared control would allow reading at other replicas, but writes would still be forbidden everywhere.

Acquiring control prior to voluntary disconnection is relatively simple. It is more difficult when disconnection is involuntary, because the system may have to arbitrate among multiple requestors. Unfortunately, the information needed to make a wise decision is not readily available. For example, the system cannot predict which requestors will actually use the object, when they will release control, or what the relative costs of denying them access would be.

Retaining control until reconnection is acceptable in the case of brief disconnections. But it is unacceptable in the case of extended disconnections. A disconnected client with shared control of an object would force the rest of the system to defer all updates until it reconnected. With exclusive control, it would even prevent other users from making a copy of the object. Coercing the client to reconnect may not be feasible, since its whereabouts may not be known. Thus, an entire user community could be at the mercy of a single errant client for an unbounded amount of time.

Placing a time bound on exclusive or shared control, as done in the case of *leases* [7], avoids this problem but introduces others. Once a lease expires, a disconnected client loses the ability to access a cached object, even if no else in the system is interested in it. This, in turn, defeats the purpose of disconnected operation which is to provide high availability. Worse, updates already made while disconnected have to be discarded.

An optimistic approach has its own disadvantages. An update made at one disconnected client may conflict with an update at another disconnected or connected client. For optimistic replication to be viable, the system has to be more sophisticated. There needs to be machinery in the system for detecting conflicts, for automating resolution when possible, and for confining damage and preserving evidence for manual repair. Having to repair conflicts manually violates transparency, is an annoyance to users, and reduces the usability of the system.

We chose optimistic replication because we felt that its strengths and weaknesses better matched our design goals. The dominant influence on our choice was the low degree of write-sharing typical of Unix. This implied that an optimistic strategy was likely to lead to relatively few conflicts. An optimistic strategy was also consistent with our overall goal of providing the highest possible availability of data.

In principle, we could have chosen a pessimistic strategy for server replication even after choosing an optimistic strategy for disconnected operation. But that would have reduced transparency, because a user would have faced the anomaly of being able to update data when disconnected, but being unable to do so when connected to a subset of the servers. Further, many of the previous arguments in favor of an optimistic strategy also apply to server replication.

Using an optimistic strategy throughout presents a uniform model of the system from the user's perspective. At any time, he is able to read the latest data in his *accessible universe* and his updates are immediately visible to everyone else in that universe. His accessible universe is usually the entire set of servers and clients. When failures occur, his accessible universe shrinks to the set of servers he can contact, and the set of clients that they, in turn, can contact. In the limit, when he is operating disconnected, his accessible universe consists of just his machine. Upon reconnection, his updates become visible throughout his now-enlarged accessible universe.

4. Detailed Design and Implementation

In describing our implementation of disconnected operation, we focus on the client since this is where much of the complexity lies. Section 4.1 describes the physical structure of a client, Section 4.2 introduces the major states of Venus, and Sections 4.3 to 4.5 discuss these states in detail. A description of the server support needed for disconnected operation is contained in Section 4.5.

different volumes, depending on failure conditions in the system.

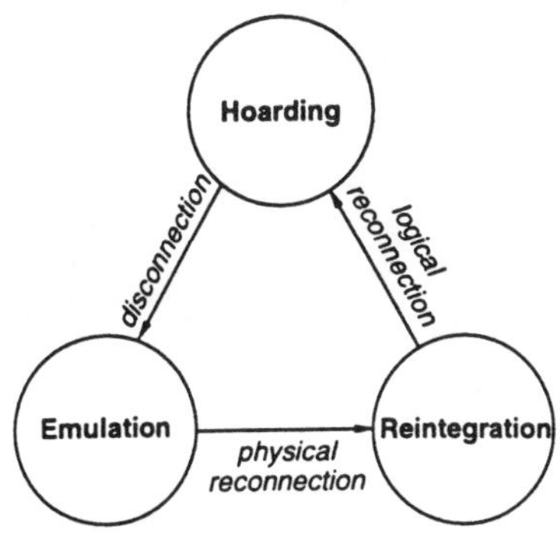

When disconnected, Venus is in the emulation state. It transits to reintegration upon successful reconnection to an AVSG member, and thence to hoarding, where it resumes connected operation.

Figure 3: Venus States and Transitions

4.1. Client Structure

Because of the complexity of Venus, we made it a user level process rather than part of the kernel. The latter approach may have yielded better performance, but would have been less portable and considerably more difficult to debug. Figure 2 illustrates the high-level structure of a Coda client.

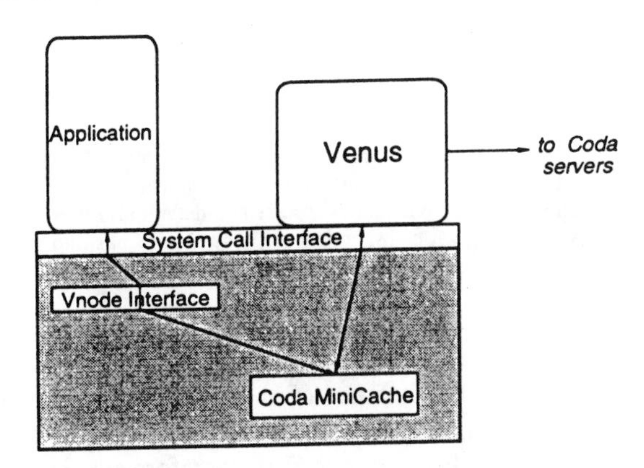

Figure 2: Structure of a Coda Client

Venus intercepts Unix file system calls via the widely-used Sun Vnode interface [10]. Since this interface imposes a heavy performance overhead on user-level cache managers, we use a tiny in-kernel *MiniCache* to filter out many kernel-Venus interactions. The MiniCache contains no support for remote access, disconnected operation or server replication; these functions are handled entirely by Venus.

A system call on a Coda object is forwarded by the Vnode interface to the MiniCache. If possible, the call is serviced by the MiniCache and control is returned to the application. Otherwise, the MiniCache contacts Venus to service the call. This, in turn, may involve contacting Coda servers. Control returns from Venus via the MiniCache to the application program, updating MiniCache state as a side effect. MiniCache state changes may also be initiated by Venus on events such as callback breaks from Coda servers. Measurements from our implementation confirm that the MiniCache is critical for good performance [20].

4.2. Venus States

Logically, Venus operates in one of three states: *hoarding, emulation,* and *reintegration.* Figure 3 depicts these states and the transitions between them. Venus is normally in the hoarding state, relying on server replication but always on the alert for possible disconnection. Upon disconnection, it enters the emulation state and remains there for the duration of disconnection. Upon reconnection, Venus enters the reintegration state, resynchronizes its cache with its AVSG, and then reverts to the hoarding state. Since all volumes may not be replicated across the same set of servers, Venus can be in different states with respect to different volumes, depending on failure conditions in the system.

4.3. Hoarding

The hoarding state is so named because a key responsibility of Venus in this state is to hoard useful data in anticipation of disconnection. However, this is not its only responsibility. Rather, Venus must manage its cache in a manner that balances the needs of connected and disconnected operation. For instance, a user may have indicated that a certain set of files is critical but may currently be using other files. To provide good performance, Venus must cache the latter files. But to be prepared for disconnection, it must also cache the former set of files.

Many factors complicate the implementation of hoarding:

- File reference behavior, especially in the distant future, cannot be predicted with certainty.
- Disconnections and reconnections are often unpredictable.
- The true cost of a cache miss while disconnected is highly variable and hard to quantify.
- Activity at other clients must be accounted for, so that the latest version of an object is in the cache at disconnection.
- Since cache space is finite, the availability of less critical objects may have to be sacrificed in favor of more critical objects.

To address these concerns, we manage the cache using a *prioritized algorithm,* and periodically reevaluate which objects merit retention in the cache via a process known as *hoard walking.*

```
# Personal files
a /coda/usr/jjk d+
a /coda/usr/jjk/papers 100:d+
a /coda/usr/jjk/papers/sosp 1000:d+

# System files
a /usr/bin 100:d+
a /usr/etc 100:d+
a /usr/include 100:d+
a /usr/lib 100:d+
a /usr/local/gnu d+
a /usr/local/rcs d+
a /usr/ucb d+
```

(a)

```
# X11 files
# (from X11 maintainer)
a /usr/X11/bin/X
a /usr/X11/bin/Xvga
a /usr/X11/bin/mwm
a /usr/X11/bin/startx
a /usr/X11/bin/xclock
a /usr/X11/bin/xinit
a /usr/X11/bin/xterm
a /usr/X11/include/X11/bitmaps c+
a /usr/X11/lib/app-defaults d+
a /usr/X11/lib/fonts/misc c+
a /usr/X11/lib/system.mwmrc
```

(b)

```
# Venus source files
# (shared among Coda developers)
a /coda/project/coda/src/venus 100:c+
a /coda/project/coda/include 100:c+
a /coda/project/coda/lib c+
```

(c)

These are typical hoard profiles provided by a Coda user, an application maintainer, and a group of project developers. Each profile is interpreted separately by the HDB front-end program. The 'a' at the beginning of a line indicates an add-entry command. Other commands are delete an entry, clear all entries, and list entries. The modifiers following some pathnames specify non-default priorities (the default is 10) and/or meta-expansion for the entry. Note that the pathnames beginning with '/usr' are actually symbolic links into '/coda'.

Figure 4: Sample Hoard Profiles

4.3.1. Prioritized Cache Management

Venus combines implicit and explicit sources of information in its priority-based cache management algorithm. The implicit information consists of recent reference history, as in traditional caching algorithms. Explicit information takes the form of a per-workstation *hoard database (HDB),* whose entries are pathnames identifying objects of interest to the user at that workstation.

A simple front-end program allows a user to update the HDB using command scripts called *hoard profiles,* such as those shown in Figure 4. Since hoard profiles are just files, it is simple for an application maintainer to provide a common profile for his users, or for users collaborating on a project to maintain a common profile. A user can customize his HDB by specifying different combinations of profiles or by executing front-end commands interactively. To facilitate construction of hoard profiles, Venus can record all file references observed between a pair of start and stop events indicated by a user.

To reduce the verbosity of hoard profiles and the effort needed to maintain them, Venus supports *meta-expansion* of HDB entries. As shown in Figure 4, if the letter 'c' (or 'd') follows a pathname, the command also applies to immediate children (or all descendants). A '+' following the 'c' or 'd' indicates that the command applies to all future as well as present children or descendents. A hoard entry may optionally indicate a *hoard priority,* with higher priorities indicating more critical objects.

The current priority of a cached object is a function of its hoard priority as well as a metric representing recent usage. The latter is updated continuously in response to new references, and serves to age the priority of objects no longer in the working set. Objects of the lowest priority are chosen as victims when cache space has to be reclaimed.

To resolve the pathname of a cached object while disconnected, it is imperative that all the ancestors of the object also be cached. Venus must therefore ensure that a cached directory is not purged before any of its descendants. This *hierarchical cache management* is not needed in traditional file caching schemes because cache misses during name translation can be serviced, albeit at a performance cost. Venus performs hierarchical cache management by assigning infinite priority to directories with cached children. This automatically forces replacement to occur bottom-up.

4.3.2. Hoard Walking

We say that a cache is in *equilibrium,* signifying that it meets user expectations about availability, when no uncached object has a higher priority than a cached object. Equilibrium may be disturbed as a result of normal activity. For example, suppose an object, *A*, is brought into the cache on demand, replacing an object, *B*. Further suppose that *B* is mentioned in the HDB, but *A* is not. Some time after activity on *A* ceases, its priority will decay below the hoard priority of *B*. The cache is no longer in equilibrium, since the cached object *A* has lower priority than the uncached object *B*.

Venus periodically restores equilibrium by performing an operation known as a *hoard walk.* A hoard walk occurs every 10 minutes in our current implementation, but one may be explicitly requested by a user prior to voluntary disconnection. The walk occurs in two phases. First, the *name bindings* of HDB entries are reevaluated to reflect update activity by other Coda clients. For example, new children may have been created in a directory whose pathname is specified with the '+' option in the HDB. Second, the priorities of all entries in the cache and HDB are reevaluated, and objects fetched or evicted as needed to restore equilibrium.

Hoard walks also address a problem arising from callback breaks. In traditional callback-based caching, data is refetched only on demand after a callback break. But in Coda, such a strategy may result in a critical object being unavailable should a disconnection occur before the next reference to it. Refetching immediately upon callback break avoids this problem, but ignores a key characteristic of Unix environments: once an object is modified, it is likely to be modified many more times by the same user within a short interval [14, 6]. An immediate refetch policy would increase client-server traffic considerably, thereby reducing scalability.

Our strategy is a compromise that balances availability, consistency, and scalability. For files and symbolic links, Venus purges the object on callback break, and refetches it on demand or during the next hoard walk, whichever occurs earlier. If a disconnection were to occur before refetching, the object would be unavailable. For directories, Venus does not purge on callback break, but marks the cache entry suspicious. A stale cache entry is thus available should a disconnection occur before the next hoard walk or reference. The acceptability of stale directory data follows from its particular callback semantics. A callback break on a directory typically means that an entry has been added to or deleted from the directory. It is often the case that other directory entries and the objects they name are unchanged. Therefore, saving the stale copy and using it in the event of untimely disconnection causes consistency to suffer only a little, but increases availability considerably.

4.4. Emulation

In the emulation state, Venus performs many actions normally handled by servers. For example, Venus now assumes full responsibility for access and semantic checks. It is also responsible for generating temporary *file identifiers* (fids) for new objects, pending the assignment of permanent fids at reintegration. But although Venus is functioning as a *pseudo-server,* updates accepted by it have to be revalidated with respect to integrity and protection by real servers. This follows from the Coda policy of trusting only servers, not clients. To minimize unpleasant delayed surprises for a disconnected user, it behooves Venus to be as faithful as possible in its emulation.

Cache management during emulation is done with the same priority algorithm used during hoarding. Mutating operations directly update the cache entries of the objects involved. Cache entries of deleted objects are freed immediately, but those of other modified objects assume infinite priority so that they are not purged before reintegration. On a cache miss, the default behavior of Venus is to return an error code. A user may optionally request Venus to block his processes until cache misses can be serviced.

4.4.1. Logging

During emulation, Venus records sufficient information to replay update activity when it reintegrates. It maintains this information in a per-volume log of mutating operations called a *replay log.* Each log entry contains a copy of the corresponding system call arguments as well as the version state of all objects referenced by the call.

Venus uses a number of optimizations to reduce the length of the replay log, resulting in a log size that is typically a few percent of cache size. A small log conserves disk space, a critical resource during periods of disconnection. It also improves reintegration performance by reducing latency and server load.

One important optimization to reduce log length pertains to `write` operations on files. Since Coda uses whole-file caching, the `close` after an `open` of a file for modification installs a completely new copy of the file. Rather than logging the `open`, `close`, and intervening `write` operations individually, Venus logs a single `store` record during the handling of a `close`.

Another optimization consists of Venus discarding a previous `store` record for a file when a new one is appended to the log. This follows from the fact that a `store` renders all previous versions of a file superfluous. The `store` record does not contain a copy of the file's contents, but merely points to the copy in the cache.

We are currently implementing two further optimizations to reduce the length of the replay log. The first generalizes the optimization described in the previous paragraph such that any operation which *overwrites* the effect of earlier operations may cancel the corresponding log records. An example would be the cancelling of a `store` by a subsequent `unlink` or `truncate`. The second optimization exploits knowledge of *inverse* operations to cancel both the inverting and inverted log records. For example, a `rmdir` may cancel its own log record as well as that of the corresponding `mkdir`.

4.4.2. Persistence

A disconnected user must be able to restart his machine after a shutdown and continue where he left off. In case of a crash, the amount of data lost should be no greater than if the same failure occurred during connected operation. To provide these guarantees, Venus must keep its cache and related data structures in non-volatile storage.

Meta-data, consisting of cached directory and symbolic link contents, status blocks for cached objects of all types, replay logs, and the HDB, is mapped into Venus' address space as *recoverable virtual memory* (RVM). Transactional access to this memory is supported by the RVM library [12] linked into Venus. The actual contents of cached files are not in RVM, but are stored as local Unix files.

The use of transactions to manipulate meta-data simplifies Venus' job enormously. To maintain its invariants Venus need only ensure that each transaction takes meta-data from one consistent state to another. It need not be concerned with crash recovery, since RVM handles this transparently. If we had chosen the obvious alternative of placing meta-data in local Unix files, we would have had to follow a strict discipline of carefully timed synchronous writes and an ad-hoc recovery algorithm.

RVM supports local, non-nested transactions and allows independent control over the basic transactional properties of atomicity, permanence, and serializability. An application can reduce commit latency by labelling the commit as *no-flush,* thereby avoiding a synchronous write to disk. To ensure persistence of no-flush transactions, the application must explicitly flush RVM's write-ahead log from time to time. When used in this manner, RVM provides *bounded persistence*, where the bound is the period between log flushes.

Venus exploits the capabilities of RVM to provide good performance at a constant level of persistence. When hoarding, Venus initiates log flushes infrequently, since a copy of the data is available on servers. Since servers are not accessible when emulating, Venus is more conservative and flushes the log more frequently. This lowers performance, but keeps the amount of data lost by a client crash within acceptable limits.

4.4.3. Resource Exhaustion

It is possible for Venus to exhaust its non-volatile storage during emulation. The two significant instances of this are the file cache becoming filled with modified files, and the RVM space allocated to replay logs becoming full.

Our current implementation is not very graceful in handling these situations. When the file cache is full, space can be freed by truncating or deleting modified files. When log space is full, no further mutations are allowed until reintegration has been performed. Of course, non-mutating operations are always allowed.

We plan to explore at least three alternatives to free up disk space while emulating. One possibility is to compress file cache and RVM contents. Compression trades off computation time for space, and recent work [2] has shown it to be a promising tool for cache management. A second possibility is to allow users to selectively back out updates made while disconnected. A third approach is to allow portions of the file cache and RVM to be written out to removable media such as floppy disks.

4.5. Reintegration

Reintegration is a transitory state through which Venus passes in changing roles from pseudo-server to cache manager. In this state, Venus propagates changes made during emulation, and updates its cache to reflect current server state. Reintegration is performed a volume at a time, with all update activity in the volume suspended until completion.

4.5.1. Replay Algorithm

The propagation of changes from client to AVSG is accomplished in two steps. In the first step, Venus obtains permanent fids for new objects and uses them to replace temporary fids in the replay log. This step is avoided in many cases, since Venus obtains a small supply of permanent fids in advance of need, while in the hoarding state. In the second step, the replay log is shipped in parallel to the AVSG, and executed independently at each member. Each server performs the replay within a single transaction, which is aborted if any error is detected.

The replay algorithm consists of four phases. In phase one the log is parsed, a transaction is begun, and all objects referenced in the log are locked. In phase two, each operation in the log is validated and then executed. The validation consists of conflict detection as well as integrity, protection, and disk space checks. Except in the case of `store` operations, execution during replay is identical to execution in connected mode. For a `store`, an empty *shadow file* is created and meta-data is updated to reference it, but the data transfer is deferred. Phase three consists exclusively of performing these data transfers, a process known as *back-fetching*. The final phase commits the transaction and releases all locks.

If reintegrations succeeds, Venus frees the replay log and resets the priority of cached objects referenced by the log. If reintegration fails, Venus writes out the replay log to a local *replay file* in a superset of the Unix `tar` format. The log and all corresponding cache entries are then purged, so that subsequent references will cause refetch of the current contents at the AVSG. A tool is provided which allows the user to inspect the contents of a replay file, compare it to the state at the AVSG, and replay it selectively or in its entirety.

Reintegration at finer granularity than a volume would reduce the latency perceived by clients, improve concurrency and load balancing at servers, and reduce user effort during manual replay. To this end, we are revising our implementation to reintegrate at the granularity of subsequences of dependent operations within a volume. Dependent subsequences can be identified using the *precedence graph* approach of Davidson [4]. In the revised implementation Venus will maintain precedence graphs during emulation, and pass them to servers along with the replay log.

4.5.2. Conflict Handling

Our use of optimistic replica control means that the disconnected operations of one client may conflict with activity at servers or other disconnected clients. The only class of conflicts we are concerned with are *write/write* conflicts. *Read/write* conflicts are not relevant to the Unix file system model, since it has no notion of atomicity beyond the boundary of a single system call.

The check for conflicts relies on the fact that each replica of an object is tagged with a *storeid* that uniquely identifies the last update to it. During phase two of replay, a server compares the storeid of every object mentioned in a log entry with the storeid of its own replica of the object. If the comparison indicates equality for all objects, the operation is performed and the mutated objects are tagged with a new storeid specified in the log entry.

If a storeid comparison fails, the action taken depends on the operation being validated. In the case of a `store` of a file, the entire reintegration is aborted. But for directories, a conflict is declared only if a newly created name collides with an existing name, if an object updated at the client or the server has been deleted by the other, or if directory attributes have been modified at the server and the client. This strategy of *resolving* partitioned directory updates is consistent with our strategy in server replication [11], and was originally suggested by Locus [22].

Our original design for disconnected operation called for preservation of replay files at servers rather than clients. This approach would also allow damage to be confined by marking conflicting replicas inconsistent and forcing manual repair, as is currently done in the case of server replication. We are awaiting more usage experience to determine whether this is indeed the correct approach for disconnected operation.

5. Status and Evaluation

Today, Coda runs on IBM RTs, Decstation 3100s and 5000s, and 386-based laptops such as the Toshiba 5200. A small user community has been using Coda on a daily basis as its primary data repository since April 1990. All development work on Coda is done in Coda itself. As of July 1991 there were nearly 350MB of triply-replicated data in Coda, with plans to expand to 2GB in the next few months.

A version of disconnected operation with minimal functionality was demonstrated in October 1990. A more complete version was functional in January 1991, and is now in regular use. We have successfully operated disconnected for periods lasting four to five hours. Our experience with the system has been quite positive, and we are confident that the refinements under development will result in an even more usable system.

In the following sections we provide qualitative and quantitative answers to three important questions pertaining to disconnected operation. These are:

1. How long does reintegration take?
2. How large a local disk does one need?
3. How likely are conflicts?

5.1. Duration of Reintegration

In our experience, typical disconnected sessions of editing and program development lasting a few hours require about a minute for reintegration. To characterize reintegration speed more precisely, we measured the reintegration times after disconnected execution of two well-defined tasks. The first task is the Andrew benchmark [9], now widely used as a basis for comparing file system performance. The second task is the compiling and linking of the current version of Venus. Table 1 presents the reintegration times for these tasks.

The time for reintegration consists of three components: the time to allocate permanent fids, the time for the replay at the servers, and the time for the second phase of the update protocol used for server replication. The first component will be zero for many disconnections, due to the preallocation of fids during hoarding. We expect the time for the second component to fall, considerably in many cases, as we incorporate the last of the replay log optimizations described in Section 4.4.1. The third component can be avoided only if server replication is not used.

One can make some interesting secondary observations from Table 1. First, the total time for reintegration is roughly the same for the two tasks even though the Andrew benchmark has a much smaller elapsed time. This is because the Andrew benchmark uses the file system more intensively. Second, reintegration for the Venus make takes longer, even though the number of entries in the replay log is smaller. This is because much more file data is back-fetched in the third phase of the replay. Finally, neither task involves any think time. As a result, their reintegration times are comparable to that after a much longer, but more typical, disconnected session in our environment.

5.2. Cache Size

A local disk capacity of 100MB on our clients has proved adequate for our initial sessions of disconnected operation. To obtain a better understanding of the cache size requirements for disconnected operation, we analyzed file reference traces from our environment. The traces were obtained by instrumenting workstations to record information on every file system operation, regardless of whether the file was in Coda, AFS, or the local file system.

	Elapsed Time (seconds)	Reintegration Time (seconds)				Size of Replay Log		Data Back-Fetched (Bytes)
		Total	AllocFid	Replay	COP2	Records	Bytes	
Andrew Benchmark	288 (3)	43 (2)	4 (2)	29 (1)	10 (1)	223	65,010	1,141,315
Venus Make	3,271 (28)	52 (4)	1 (0)	40 (1)	10 (3)	193	65,919	2,990,120

This data was obtained with a Toshiba T5200/100 client (12MB memory, 100MB disk) reintegrating over an Ethernet with an IBM RT-APC server (12MB memory, 400MB disk). The values shown above are the means of three trials. Figures in parentheses are standard deviations.

Table 1: Time for Reintegration

Our analysis is based on simulations driven by these traces. Writing and validating a simulator that precisely models the complex caching behavior of Venus would be quite difficult. To avoid this difficulty, we have modified Venus to act as its own simulator. When running as a simulator, Venus is driven by traces rather than requests from the kernel. Code to communicate with the servers, as well code to perform physical I/O on the local file system are stubbed out during simulation.

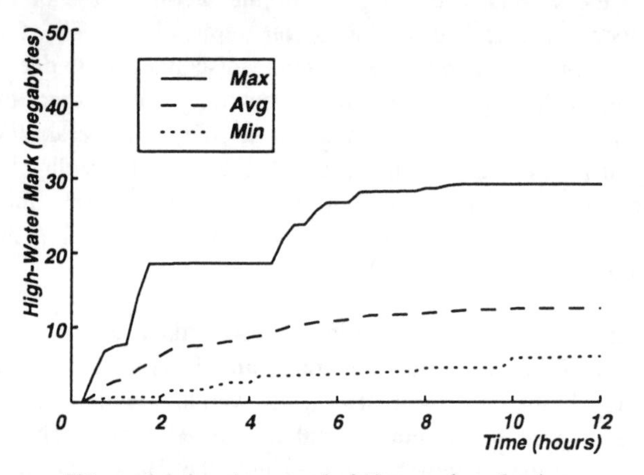

This graph is based on a total of 10 traces from 5 active Coda workstations. The curve labelled "Avg" corresponds to the values obtained by averaging the high-water marks of all workstations. The curves labelled "Max" and "Min" plot the highest and lowest values of the high-water marks across all workstations. Note that the high-water mark does not include space needed for paging, the HDB or replay logs.

Figure 5: High-Water Mark of Cache Usage

Figure 5 shows the *high-water mark* of cache usage as a function of time. The actual disk size needed for disconnected operation has to be larger, since both the explicit and implicit sources of hoarding information are imperfect. From our data it appears that a disk of 50-60MB should be adequate for operating disconnected for a typical workday. Of course, user activity that is drastically different from what was recorded in our traces could produce significantly different results.

We plan to extend our work on trace-driven simulations in three ways. First, we will investigate cache size requirements for much longer periods of disconnection. Second, we will be sampling a broader range of user activity by obtaining traces from many more machines in our environment. Third, we will evaluate the effect of hoarding by simulating traces together with hoard profiles that have been specified *ex ante* by users.

5.3. Likelihood of Conflicts

In our use of optimistic server replication in Coda for nearly a year, we have seen virtually no conflicts due to multiple users updating an object in different network partitions. While gratifying to us, this observation is subject to at least three criticisms. First, it is possible that our users are being cautious, knowing that they are dealing with an experimental system. Second, perhaps conflicts will become a problem only as the Coda user community grows larger. Third, perhaps extended voluntary disconnections will lead to many more conflicts.

To obtain data on the likelihood of conflicts at larger scale, we instrumented the AFS servers in our environment. These servers are used by over 400 computer science faculty, staff and graduate students for research, program development, and education. Their usage profile includes a significant amount of collaborative activity. Since Coda is descended from AFS and makes the same kind of usage assumptions, we can use this data to estimate how frequent conflicts would be if Coda were to replace AFS in our environment.

Every time a user modifies an AFS file or directory, we compare his identity with that of the user who made the previous mutation. We also note the time interval between mutations. For a file, only the `close` after an `open` for update is counted as a mutation; individual `write` operations are not counted. For directories, all operations that modify a directory are counted as mutations.

Type of Volume	Number of Volumes	Type of Object	Total Mutations	Same User	Different User					
					Total	< 1min	< 10 min	< 1hr	< 1 day	< 1 wk
User	529	Files	3,287,135	99.87 %	0.13 %	0.04 %	0.05 %	0.06 %	0.09 %	0.09 %
		Directories	4,132,066	99.80 %	0.20 %	0.04 %	0.07 %	0.10 %	0.15 %	0.16 %
Project	108	Files	4,437,311	99.66 %	0.34 %	0.17 %	0.25 %	0.26 %	0.28 %	0.30 %
		Directories	5,391,224	99.63 %	0.37 %	0.00 %	0.01 %	0.03 %	0.09 %	0.15 %
System	398	Files	5,526,700	99.17 %	0.83 %	0.06 %	0.18 %	0.42 %	0.72 %	0.78 %
		Directories	4,338,507	99.54 %	0.46 %	0.02 %	0.05 %	0.08 %	0.27 %	0.34 %

This data was obtained between June 1990 and May 1991 from the AFS servers in the `cs.cmu.edu` cell. The servers stored a total of about 12GB of data. The column entitled "Same User" gives the percentage of mutations in which the user performing the mutation was the same as the one performing the immediately preceding mutation on the same file or directory. The remaining mutations contribute to the column entitled "Different User".

Table 2: Sequential Write-Sharing in AFS

Table 2 presents our observations over a period of twelve months. The data is classified by volume type: *user* volumes containing private user data, *project* volumes used for collaborative work, and *system* volumes containing program binaries, libraries, header files and other similar data. On average, a project volume has about 2600 files and 280 directories, and a system volume has about 1600 files and 130 directories. User volumes tend to be smaller, averaging about 200 files and 18 directories, because users often place much of their data in their project volumes.

Table 2 shows that over 99% of all modifications were by the previous writer, and that the chances of two different users modifying the same object less than a day apart is at most 0.75%. We had expected to see the highest degree of write-sharing on project files or directories, and were surprised to see that it actually occurs on system files. We conjecture that a significant fraction of this sharing arises from modifications to system files by operators, who change shift periodically. If system files are excluded, the absence of write-sharing is even more striking: more than 99.5% of all mutations are by the previous writer, and the chances of two different users modifying the same object within a week are less than 0.4%! This data is highly encouraging from the point of view of optimistic replication. It suggests that conflicts would not be a serious problem if AFS were replaced by Coda in our environment.

6. Related Work

Coda is unique in that it exploits caching for both performance and high availability while preserving a high degree of transparency. We are aware of no other system, published or unpublished, that duplicates this key aspect of Coda.

By providing tools to link local and remote name spaces, the Cedar file system [19] provided rudimentary support for disconnected operation. But since this was not its primary goal, Cedar did not provide support for hoarding, transparent reintegration or conflict detection. Files were versioned and immutable, and a Cedar cache manager could substitute a cached version of a file on reference to an unqualified remote file whose server was inaccessible. However, the implementors of Cedar observe that this capability was not often exploited since remote files were normally referenced by specific version number.

Birrell and Schroeder pointed out the possibility of "stashing" data for availability in an early discussion of the Echo file system [13]. However, a more recent description of Echo [8] indicates that it uses stashing only for the highest levels of the naming hierarchy.

The FACE file system [3] uses stashing but does not integrate it with caching. The lack of integration has at least three negative consequences. First, it reduces transparency because users and applications deal with two different name spaces, with different consistency properties. Second, utilization of local disk space is likely to be much worse. Third, recent usage information from cache management is not available to manage the stash. The available literature on FACE does not report on how much the lack of integration detracted from the usability of the system.

An application-specific form of disconnected operation was implemented in the PCMAIL system at MIT [Lambert88]. PCMAIL allowed clients to disconnect, manipulate existing mail messages and generate new ones, and re-synchronize with a central repository at reconnection. Besides relying heavily on the semantics of mail, PCMAIL was less transparent than Coda since it required manual re-synchronization as well as pre-registration of clients with servers.

The use of optimistic replication in distributed file systems was pioneered by Locus [22]. Since Locus used a peer-to-peer model rather than a client-server model, availability was achieved solely through server replication. There was no notion of caching, and hence of disconnected operation.

Coda has benefited in a general sense from the large body of work on transparency and performance in distributed file systems. In particular, Coda owes much to AFS [18], from which it inherits its model of trust and integrity, as well as its mechanisms and design philosophy for scalability.

7. Future Work

Disconnected operation in Coda is a facility under active development. In earlier sections of this paper we described work in progress in the areas of log optimization, granularity of reintegration, and evaluation of hoarding. Much additional work is also being done at lower levels of the system. In this section we consider two ways in which the scope of our work may be broadened.

An excellent opportunity exists in Coda for adding *transactional support to Unix.* Explicit transactions become more desirable as systems scale to hundreds or thousands of nodes, and the informal concurrency control of Unix becomes less effective. Many of the mechanisms supporting disconnected operation, such as operation logging, precedence graph maintenance, and conflict checking would transfer directly to a transactional system using optimistic concurrency control. Although transactional file systems are not a new idea, no such system with the scalability, availability, and performance properties of Coda has been proposed or built.

A different opportunity exists in extending Coda to support *weakly-connected operation,* in environments where connectivity is intermittent or of low bandwidth. Such conditions are found in networks that rely on voice-grade lines, or that use wireless technologies such as packet radio. The ability to mask failures, as provided by disconnected operation, is of value even with weak connectivity. But techniques which exploit and adapt to the communication opportunities at hand are also needed. Such techniques may include more aggressive write-back policies, compressed network transmission, partial file transfer, and caching at intermediate levels.

8. Conclusion

Disconnected operation is a tantalizingly simple idea. All one has to do is to pre-load one's cache with critical data, continue normal operation until disconnection, log all changes made while disconnected, and replay them upon reconnection.

Implementing disconnected operation is not so simple. It involves major modifications and careful attention to detail in many aspects of cache management. While hoarding, a surprisingly large volume and variety of interrelated state has to be maintained. When emulating, the persistence and integrity of client data structures become critical. During reintegration, there are dynamic choices to be made about the granularity of reintegration.

Only in hindsight do we realize the extent to which implementations of traditional caching schemes have been simplified by the guaranteed presence of a lifeline to a first-class replica. Purging and refetching on demand, a strategy often used to handle pathological situations in those implementations, is not viable when supporting disconnected operation. However, the obstacles to realizing disconnected operation are not insurmountable. Rather, the central message of this paper is that disconnected operation is indeed feasible, efficient and usable.

One way to view our work is to regard it as an extension of the idea of *write-back caching.* Whereas write-back caching has hitherto been used for performance, we have shown that it can be extended to mask temporary failures too. A broader view is that disconnected operation allows graceful transitions between states of *autonomy* and *interdependence* in a distributed system. Under favorable conditions, our approach provides all the benefits of remote data access; under unfavorable conditions, it provides continued access to critical data. We are certain that disconnected operation will become increasingly important as distributed systems grow in scale, diversity and vulnerability.

Acknowledgments

We wish to thank Lily Mummert for her invaluable assistance in collecting and postprocessing the file reference traces used in Section 5.2, and Dimitris Varotsis, who helped instrument the AFS servers which yielded the measurements of Section 5.3. We also wish to express our appreciation to past and present contributors to the Coda project, especially Puneet Kumar, Hank Mashburn, Maria Okasaki, and David Steere.

References

[1] Burrows, M.
Efficient Data Sharing.
PhD thesis, University of Cambridge, Computer Laboratory, December, 1988.

[2] Cate, V., Gross, T.
Combining the Concepts of Compression and Caching for a Two-Level File System.
In *Proceedings of the 4th ACM Symposium on Architectural Support for Programming Languages and Operating Systems*. April, 1991.

[3] Cova, L.L.
Resource Management in Federated Computing Environments.
PhD thesis, Department of Computer Science, Princeton University, October, 1990.

[4] Davidson, S.B.
Optimism and Consistency in Partitioned Distributed Database Systems.
ACM Transactions on Database Systems 9(3), September, 1984.

[5] Davidson, S.B., Garcia-Molina, H., Skeen, D.
Consistency in Partitioned Networks.
ACM Computing Surveys 17(3), September, 1985.

[6] Floyd, R.A.
Transparency in Distributed File Systems.
Technical Report TR 272, Department of Computer Science, University of Rochester, 1989.

[7] Gray, C.G., Cheriton, D.R.
Leases: An Efficient Fault-Tolerant Mechanism for Distributed File Cache Consistency.
In *Proceedings of the 12th ACM Symposium on Operating System Principles*. December, 1989.

[8] Hisgen, A., Birrell, A., Mann, T., Schroeder, M., Swart, G.
Availability and Consistency Tradeoffs in the Echo Distributed File System.
In *Proceedings of the Second Workshop on Workstation Operating Systems*. September, 1989.

[9] Howard, J.H., Kazar, M.L., Menees, S.G., Nichols, D.A., Satyanarayanan, M., Sidebotham, R.N., West, M.J.
Scale and Performance in a Distributed File System.
ACM Transactions on Computer Systems 6(1), February, 1988.

[10] Kleiman, S.R.
Vnodes: An Architecture for Multiple File System Types in Sun UNIX.
In *Summer Usenix Conference Proceedings*. 1986.

[11] Kumar, P., Satyanarayanan, M.
Log-Based Directory Resolution in the Coda File System.
Technical Report CMU-CS-91-164, School of Computer Science, Carnegie Mellon University, 1991.

[12] Mashburn, H., Satyanarayanan, M.
RVM: Recoverable Virtual Memory User Manual
School of Computer Science, Carnegie Mellon University, 1991.

[13] Needham, R.M., Herbert, A.J.
Report on the Third European SIGOPS Workshop: "Autonomy or Interdependence in Distributed Systems".
SIGOPS Review 23(2), April, 1989.

[14] Ousterhout, J., Da Costa, H., Harrison, D., Kunze, J., Kupfer, M., Thompson, J.
A Trace-Driven Analysis of the 4.2BSD File System.
In *Proceedings of the 10th ACM Symposium on Operating System Principles*. December, 1985.

[15] Sandberg, R., Goldberg, D., Kleiman, S., Walsh, D., Lyon, B.
Design and Implementation of the Sun Network Filesystem.
In *Summer Usenix Conference Proceedings*. 1985.

[16] Satyanarayanan, M.
On the Influence of Scale in a Distributed System.
In *Proceedings of the 10th International Conference on Software Engineering*. April, 1988.

[17] Satyanarayanan, M., Kistler, J.J., Kumar, P., Okasaki, M.E., Siegel, E.H., Steere, D.C.
Coda: A Highly Available File System for a Distributed Workstation Environment.
IEEE Transactions on Computers 39(4), April, 1990.

[18] Satyanarayanan, M.
Scalable, Secure, and Highly Available Distributed File Access.
IEEE Computer 23(5), May, 1990.

[19] Schroeder, M.D., Gifford, D.K., Needham, R.M.
A Caching File System for a Programmer's Workstation.
In *Proceedings of the 10th ACM Symposium on Operating System Principles*. December, 1985.

[20] Steere, D.C., Kistler, J.J., Satyanarayanan, M.
Efficient User-Level Cache File Management on the Sun Vnode Interface.
In *Summer Usenix Conference Proceedings*. June, 1990.

[21] *Decorum File System*
Transarc Corporation, 1990.

[22] Walker, B., Popek, G., English, R., Kline, C., Thiel, G.
The LOCUS Distributed Operating System.
In *Proceedings of the 9th ACM Symposium on Operating System Principles*. October, 1983.

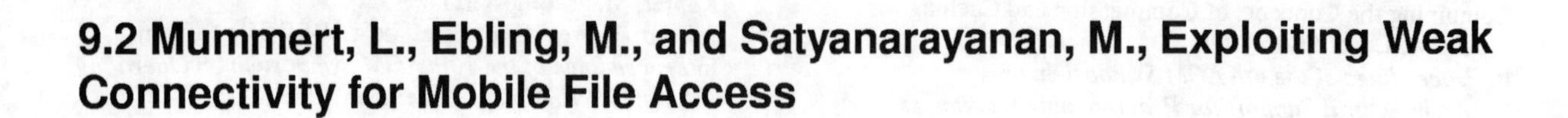

9.2 Mummert, L., Ebling, M., and Satyanarayanan, M., Exploiting Weak Connectivity for Mobile File Access

Mummert, L., Ebling, M., and Satyanarayanan, M., "Exploiting Weak Connectivity for Mobile File Access," *Proceedings of the 15th ACM Symposium on Operating System Principles,* pp. 143–155, December 1995.

Exploiting Weak Connectivity for Mobile File Access

Lily B. Mummert, Maria R. Ebling, M. Satyanarayanan
School of Computer Science
Carnegie Mellon University

Abstract

Weak connectivity, in the form of intermittent, low-bandwidth, or expensive networks is a fact of life in mobile computing. In this paper, we describe how the Coda File System has evolved to exploit such networks. The underlying theme of this evolution has been the systematic introduction of *adaptivity* to eliminate hidden assumptions about strong connectivity. Many aspects of the system, including communication, cache validation, update propagation and cache miss handling have been modified. As a result, Coda is able to provide good performance even when network bandwidth varies over four orders of magnitude — from modem speeds to LAN speeds.

1. Introduction

For the forseeable future, mobile clients will encounter a wide range of network characteristics in the course of their journeys. Cheap, reliable, high-performance connectivity via wired or wireless media will be limited to a few oases in a vast desert of poor connectivity. Mobile clients must therefore be able to use networks with rather unpleasant characteristics: intermittence, low bandwidth, high latency, or high expense. We refer to connectivity with one or more of these properties as *weak connectivity.* In contrast, typical LAN environments have none of these shortcomings and thus offer *strong connectivity.*

In this paper, we report on our work toward exploiting weak connectivity in the Coda File System. Our mechanisms preserve usability even at network speeds as low as 1.2 Kb/s. At a typical modem speed of 9.6 Kb/s, performance on a family of benchmarks is only about 2% slower than at 10 Mb/s. When a client reconnects to a network, synchronization of state with a server typically takes only about 25% longer at 9.6 Kb/s than at 10 Mb/s. To make better use of a network, Coda may solicit advice from the user. But it preserves usability by limiting the frequency of such interactions.

Since *disconnected operation* [13] represents an initial step toward supporting mobility, we begin by reviewing its strengths and weaknesses. We then describe a set of *adaptive* mechanisms that overcome these weaknesses by exploiting weak connectivity. Next, we evaluate these mechanisms through controlled experiments and empirical observations. Finally, we discuss related work and close with a summary of the main ideas.

This research was supported by the Air Force Materiel Command (AFMC) and ARPA under contract number F196828-93-C-0193. Additional support was provided by the IBM Corp., Digital Equipment Corp., Intel Corp., Xerox Corp., and AT&T Corp. The views and conclusions contained here are those of the authors and should not be interpreted as necessarily representing the official policies or endorsements, either express or implied, of AFMC, ARPA, IBM, DEC, Intel, Xerox, AT&T, CMU, or the U.S. Government.

2. Starting Point: Disconnected Operation

2.1. Benefits and Limitations

Disconnected operation is a mode of operation in which a client continues to use data in its cache during temporary network or server failures. It can be viewed as the extreme case of weakly-connected operation — the mobile client is effectively using a network of zero bandwidth and infinite latency.

The ability to operate disconnected can be useful even when connectivity is available. For example, disconnected operation can extend battery life by avoiding wireless transmission and reception. It can reduce network charges, an important feature when rates are high. It allows radio silence to be maintained, a vital capability in military applications. And, of course, it is a viable fallback position when network characteristics degrade beyond usability.

But disconnected operation is not a panacea. A disconnected client suffers from many limitations:

- *Updates are not visible* to other clients.
- *Cache misses* may impede progress.
- *Updates are at risk* due to theft, loss or damage.
- *Update conflicts* become more likely.
- *Exhaustion of cache space* is a concern.

Our goal is to alleviate these limitations by exploiting weak connectivity. How successful we are depends on the quality of the network. With a very weak connection, a user is little better off than when disconnected; as network quality improves, the limitations decrease in severity and eventually vanish.

To attain this goal, we have implemented a series of modifications to Coda. Since Coda has been extensively described in the literature [13, 25, 26], we provide only a brief review here.

2.2. Implementation in Coda

Coda preserves the model of security, scalability, and Unix compatibility of AFS [5], and achieves high availability through the use of two complementary mechanisms. One mechanism is disconnected operation. The other mechanism is *server replication,* which we do not discuss further in this paper because it is incidental to our focus on mobility.

A small collection of trusted Coda servers exports a location-transparent Unix file name space to a much larger collection of untrusted clients. These clients are assumed to be general-purpose computers rather than limited-function devices such as InfoPads [28] and ParcTabs [29]. Files are grouped into *volumes,* each forming a partial subtree of the name space and typically containing the files of one user or project. On each client, a user-level process, *Venus,* manages a file cache on the local disk. It is Venus that bears the brunt of disconnected operation.

As described by Kistler [13], Venus operates in one of three states: *hoarding, emulating,* and *reintegrating.* It is normally in the hoarding state, preserving cache coherence via *callbacks* [5]. Upon disconnection, Venus enters the emulating state and begins logging updates in a *client modify log (CML).* In this state, Venus performs *log optimizations* to improve performance and reduce resource usage. Upon reconnection, Venus enters the reintegrating state, synchronizes its cache with servers, propagates updates from the CML, and returns to the hoarding state. Since consistency is based on optimistic replica control, update conflicts may occur upon reintegration. The system ensures their detection and confinement, and provides mechanisms to help users recover from them [14].

In anticipation of disconnection, users may *hoard* data in the cache by providing a prioritized list of files in a per-client *hoard database (HDB).* Venus combines HDB information with LRU information to implement a cache management policy addressing both performance and availability concerns. Periodically, Venus *walks* the cache to ensure that the highest priority items are present, and consistent with the servers. A user may also explicitly request a hoard walk at any time.

3. Design Rationale and Overview

3.1. Strategy

We chose an incremental approach to extending Coda, relying on usage experience and measurements at each stage. The underlying theme of this evolution was the identification of hidden assumptions about strong connectivity, and their systematic elimination through the introduction of adaptivity.

Our design is based on four guiding principles:

- *Don't punish strongly-connected clients.*
  It is unacceptable to degrade the performance of strongly-connected clients on account of weakly-connected clients. This precludes use of a broad range of cache write-back schemes in which a weakly-connected client must be contacted for token revocation or data propagation before other clients can proceed.

- *Don't make life worse than when disconnected.*
  While a minor performance penalty may be an acceptable price for the benefits of weakly-connected operation, a user is unlikely to tolerate substantial performance degradation.

- *Do it in the background if you can.*
  Network delays in the foreground affect a user more acutely than those in the background. As bandwidth decreases, network usage should be moved into the background whenever possible. The effect of this strategy is to replace intolerable performance delays by a degradation of availability or consistency — lesser evils in many situations.

- *When in doubt, seek user advice.*
  As connectivity weakens, the higher performance penalty for suboptimal decisions increases the value of user advice. Users also make mistakes, of course, but they tend to be more forgiving if they perceive themselves responsible. The system should perform better if the user gives good advice, but should be able to function unaided.

More generally, we were strongly influenced by two classic principles of system design: favoring *simplicity* over unwarranted generality [15], and respecting the *end-to-end argument* when layering functionality [24].

3.2. Evolution

We began by modifying Coda's RPC and bulk transfer protocols to function over a serial-line IP (SLIP) connection [23]. These modifications were necessary because the protocols had been originally designed for good LAN performance. Once they functioned robustly down to 1.2 Kb/s, we had a reliable means of reintegrating and servicing critical cache misses from any location with a phone connection. Performance was atrocious because Venus used the SLIP connection like a LAN. But users were grateful for even this limited functionality, because the alternative would have been a significant commute to connect to a high-speed network.

Phone reintegration turned out to be much slower than even the most pessimistic of our estimates. The culprit was the validation of cache state on the first hoard walk after reconnection. Our solution raises the granularity at which cache coherence is maintained. In most cases, this renders the time for validation imperceptible even at modem speeds.

Next, we reduced update propagation delays by allowing a user to be logically disconnected while remaining physically connected. In this mode of use, Venus logged updates in the CML but continued to service cache misses. It was the user's responsibility to periodically initiate reintegration. Cache misses hurt performance, but there were few of them if the user had done a good job of hoarding.

Our next step was to eliminate manual triggering of reintegration when weakly connected. Since this removed user control over an important component of network usage, we had to be confident that a completely automated strategy could perform well even on very slow networks. This indeed proved possible, using a technique called *trickle reintegration.*

The last phase of our work was to substantially improve the handling of cache misses when weakly connected. Examination of misses showed that they varied widely in importance and cause. We did not see a way of automating the handling of all misses while preserving usability. So we decided to handle a subset of the misses transparently, and to provide users with a means of influencing the handling of the rest.

As a result of this evolution, Coda is now able to effectively exploit networks of low bandwidth and intermittent connectivity. Venus and the transport protocols transparently adapt to variations in network bandwidth spanning nearly four orders of magnitude — from a few Kb/s to 10 Mb/s. From a performance perspective, the user is well insulated from this variation. Network quality manifests itself mainly in the promptness with which updates are propagated, and in the degree of transparency with which cache misses are handled.

4. Detailed Design and Implementation

We provide more detail on four aspects of our system:

- Transport protocol refinements.
- Rapid cache validation.
- Trickle reintegration.
- User-assisted miss handling.

Although rapid cache validation has been described in detail an earlier paper [18], we provide a brief summary here for completeness. We also augment our earlier evaluation with measurements from the deployed system. The other three aspects of Coda are described here for the first time.

4.1. Transport Protocol Refinements

Coda uses the *RPC2* remote procedure call mechanism [27], which performs efficient transfer of file contents through a specialized streaming protocol called *SFTP*. Both RPC2 and SFTP are implemented on top of UDP. We made two major changes to them for slow networks.

One change addressed the isolation between RPC2 and SFTP. While this isolation made for clean code separation, it generated duplicate keepalive traffic. In addition, Venus generated its own higher-level keepalive traffic. Our fix was to share keepalive information between RPC2 and SFTP, and to export this information to Venus.

The other change was to modify RPC2 and SFTP to monitor network speed by estimating round trip times (*RTT*) using an adaptation of the timestamp echoing technique proposed by Jacobson [10]. The RTT estimates are used to dynamically adapt the retransmission parameters of RPC2 and SFTP. Our strategy is broadly consistent with Jacobson's recommendations for TCP [8].

With these changes, RPC2 and SFTP perform well over a wide range of network speeds. Figure 1 compares the performance of SFTP and TCP over three different networks: an Ethernet, a WaveLan wireless network, and a modem over a phone line. In almost all cases, SFTP's performance exceeds that of TCP.

Opportunities abound for further improvement to the transport protocols. For example, we could perform header compression as in TCP [9], and enhance the SLIP driver to prioritize traffic as described by Huston and Honeyman [7]. We could also enhance SFTP to ship file differences rather than full contents. But we have deliberately tried to minimize efforts at the transport level.

Protocol	Network	Nominal Speed	Receive (Kb/s)	Send (Kb/s)
TCP	Ethernet	10 Mb/s	1824 (64)	2400 (224)
	WaveLan	2 Mb/s	568 (136)	760 (80)
	Modem	9.6 Kb/s	6.8 (0.06)	6.4 (0.04)
SFTP	Ethernet	10 Mb/s	1952 (104)	2744 (96)
	WaveLan	2 Mb/s	1152 (64)	1168 (48)
	Modem	9.6 Kb/s	6.6 (0.02)	6.9 (0.02)

This table compares the observed throughputs of TCP and SFTP. The data was obtained by timing the disk-to-disk transfer of a 1MB file between a DECpc 425SL laptop client and a DEC 5000/200 server on an isolated network. Both client and server were running Mach 2.6. Each result is the mean of five trials. Numbers in parentheses are standard deviations.

Figure 1: Transport Protocol Performance

As the rest of this paper shows, mechanisms at higher levels of the system offer major benefits for weakly-connected operation. Additional transport level improvements may enhance those mechanisms, but cannot replace them.

4.2. Rapid Cache Validation

Coda's original technique for cache coherence while connected was based on *callbacks* [5, 25]. In this technique, a server remembers that a client has cached an object[1], and promises to notify it when the object is updated by another client. This promise is a *callback*, and the invalidation message is a *callback break*. When a callback break is received, the client discards the cached copy and refetches it on demand or at the next hoard walk.

When a client is disconnected, it can no longer rely on callbacks. Upon reconnection, it must validate all cached objects before use to detect updates at the server.

4.2.1. Raising the Granularity of Cache Coherence

Our solution preserves the correctness of the original callback scheme, while dramatically reducing reconnection latency. It is based upon the observation that, in most cases, the vast majority of cached objects are still valid upon reconnection. The essence of our solution is for clients to track server state at multiple levels of *granularity*. Our current implementation uses only two levels: entire volume and individual object. We have not yet found the need to support additional levels.

A server now maintains version stamps for each of its volumes, in addition to stamps on individual objects. When an object is updated, the server increments the version stamp of the object and that of its containing volume. A client caches volume version stamps at the end of a hoard walk. Since all cached objects are known to be valid at this point, mutual consistency of volume and object state is achieved at minimal cost.

When connectivity is restored, the client presents these volume stamps for validation. If a volume stamp is still valid, so is every object cached from that volume. In this case, validation of all

[1] For brevity, we use "object" to mean a file, directory, or symbolic link.

those objects has been achieved with a single RPC. We batch multiple volume validation requests in a single RPC for even faster validation. If a volume stamp is not valid, nothing can be assumed; each cached object from that volume must be validated individually. But even in this case, performance is no worse than in the original scheme.

4.2.2. Volume Callbacks

When a client obtains (or validates) a volume version stamp, a server establishes a *volume callback* as a side effect. This is in addition to (or instead of) callbacks on individual objects. The server must break a client's volume callback when another client updates any object in that volume. Once broken, a volume callback is reacquired only on the next hoard walk. In the interim, the client must rely on object callbacks, if present, or obtain them on demand.

Thus, volume callbacks improve speed of validation at the cost of precision of invalidation. This is an excellent performance tradeoff for typical Unix workloads [2, 19, 22]. Performance may be poorer with other workloads, but Coda's original cache coherence guarantees are still preserved.

4.3. Trickle Reintegration

Trickle reintegration is a mechanism that propagates updates to servers asynchronously, while minimally impacting foreground activity. Its purpose is to relieve users of the need to perform manual reintegration. The challenge is to meet this goal while remaining unobtrusive.

4.3.1. Relationship to Write-Back Caching

Trickle reintegration is conceptually similar to *write-back caching,* as used in systems such as Sprite [20] and Echo [16]. Both techniques strive to improve client performance by deferring the propagation of updates to servers. But they are sufficiently different in their details that it is appropriate to view them as distinct mechanisms.

First, write-back caching preserves strict Unix write-sharing semantics, since it is typically intended for use in strongly-connected environments. In contrast, trickle reintegration has the opportunity to trade off consistency for performance because its users have already accepted the lower consistency offered by optimistic replication.

Second, the focus of write-back caching is minimizing file system latency; reducing network traffic is only an incidental concern. In contrast, reducing traffic is a prime concern of trickle reintegration because network bandwidth is precious.

Third, write-back caching schemes maintain their caches in volatile memory. Their need to bound damage due to a software crash typically limits the maximum delay before update propagation to some tens of seconds or a few minutes. In contrast, local persistence of updates on a Coda client is assured by the CML. Trickle reintegration can therefore defer propagation for many minutes or hours, bounded only by concerns of theft, loss, or disk damage.

4.3.2. Structural Modifications

Supporting trickle reintegration required major modifications to the structure of Venus. Reintegration was originally a transient state through which Venus passed *en route* to the hoarding state. Since reintegration is now an ongoing background process, the transient state has been replaced by a stable one called the *write disconnected* state. Figure 2 shows the new states of Venus and the main transitions between them.

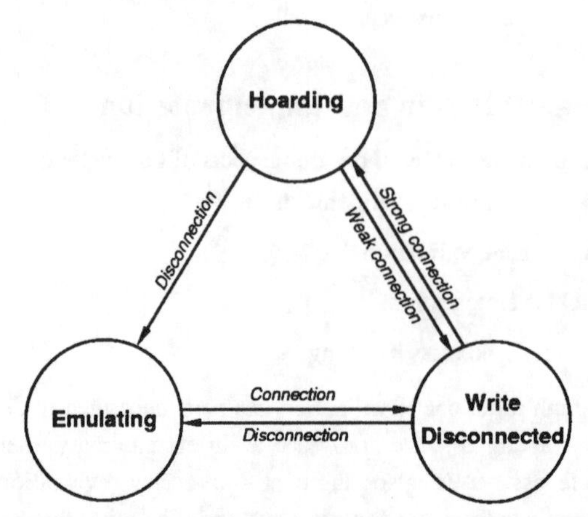

This figure shows the states of Venus, as modified to handle weak connectivity. The state labelled "Write Disconnected" replaces the reintegrating state in our original design. In this state, Venus relies on trickle reintegration to propagate changes to servers. The transition from the emulating to the write disconnected state occurs on any connection, regardless of strength. All outstanding updates are reintegrated before the transition to the hoarding state occurs.

Figure 2: Venus States and Transitions

As in our original design, Venus is in the hoarding state when strongly connected, and in the emulating state when disconnected. When weakly connected, it is in the write disconnected state. In this state, Venus' behavior is a blend of its connected and disconnected mode behaviors. Updates are logged, as when disconnected; they are propagated to servers via trickle reintegration. Cache misses are serviced, as when connected; but some misses may require user intervention. Cache coherence is maintained as explained earlier in Section 4.2.2.

A user can force a full reintegration at any time that she is in the write disconnected state. This might be valuable, for example, if she wishes to terminate a long distance phone call or realizes that she is about to move out of range of wireless communication. It is also valuable if she wishes to ensure that recent updates have been propagated to a server before notifying a collaborator via telephone, e-mail, or other out-of-band mechanism.

Our desire to avoid penalizing strongly-connected clients implies that a weakly-connected client cannot prevent them from updating an object awaiting reintegration. This situation results in a callback break for that object on the weakly-connected client. Consistent with our optimistic philosophy, we ignore the callback break and proceed as usual. When reintegration of the object is eventually attempted, it may be resolved successfully or may fail. In the latter case, the conflict becomes visible to the user just as if it had occured after a disconnected session. The existing Coda mechanisms for conflict resolution [14] are then applied.

4.3.3. Preserving the Effectiveness of Log Optimizations

Early trace-driven simulations of Coda indicated that log optimizations were the key to reducing the volume of reintegration data [26]. Measurements of Coda in actual use confirm this prediction [21].

Applying log optimizations is conceptually simple. Except for `store` records, a CML record contains all the information needed to replay the corresponding update at the server. For a `store` record, the file data resides in the local file system. Before a log record is appended to the CML, Venus checks if it cancels or overrides the effect of earlier records. For example, consider the `create` of a file, followed by a `store`. If they are followed by an `unlink`, all three CML records and the data associated with the `store` can be eliminated.

Trickle reintegration reduces the effectiveness of log optimizations, because records are propagated to the server earlier than when disconnected. Thus they have less opportunity to be eliminated at the client. A good design must balance two factors. On the one hand, records should spend enough time in the CML for optimizations to be effective. On the other hand, updates should be propagated to servers with reasonable promptness. At very low bandwidths, the first concern is dominant since reduction of data volume is paramount. As bandwidth increases, the concerns become comparable in importance. When strongly connected, prompt propagation is the dominant concern.

Our solution, illustrated in Figure 3, uses a simple technique based on *aging*. A record is not eligible for reintegration until it has spent a minimal amount of time in the CML. This amount of time, called the *aging window*, (A), establishes a limit on the effectiveness of log optimizations.

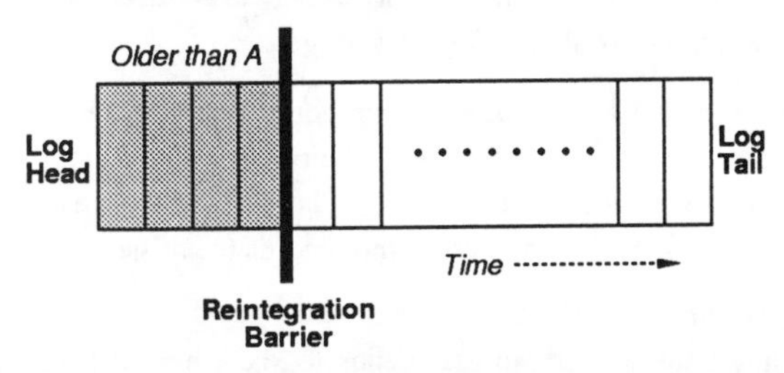

This figure depicts a typical CML scenario while weakly connected. A is the aging window. The shaded records in this figure are being reintegrated. They are protected from concurrent activity at the client by the reintegration barrier. For `store` records, the corresponding file data is locked; if contention occurs later, a shadow copy is created and the lock released.

Figure 3: CML During Trickle Reintegration

Since the CML is maintained in temporal order, the aging window partitions log records into two groups: those older than A, and those younger than A. Only the former group is eligible for reintegration. At the beginning of reintegration, a logical divider called the *reintegration barrier* is placed in the CML. During reintegration, which may take a while on a slow network, the portion of the CML to the left of the reintegration barrier is frozen. Only records to the right are examined for optimization.

If reintegration is successful, the barrier and all records to its left are removed. If a network or server failure causes reintegration to be aborted, the barrier as well as any records rendered superfluous by new updates are removed. Our implementation of reintegration is atomic, ensuring that a failure leaves behind no server state that would hinder a future retry. Until the next reintegration attempt, all records in the CML are again eligible for optimization.

Discovery of records old enough for reintegration is done by a periodic daemon. Once the daemon finds CML records ripe for reintegration, it notifies a separate thread to do the actual work.

4.3.4. Selecting an Aging Window

What should the value of A be? To answer this question, we conducted a study using file reference traces gathered from workstations in our environment [19]. We chose five week-long traces (used in an earlier analysis [26]) in which there were extended periods of sustained high levels of user activity.

The traces were used as input to a Venus simulator. This simulator is the actual Venus code, modified to accept requests from a trace instead of the operating system. The output of the simulator includes the state of the CML at the end of the trace, and data on cancelled CML records. Our analysis includes all references from the traces, whether to the local file system, AFS, Coda, or NFS.

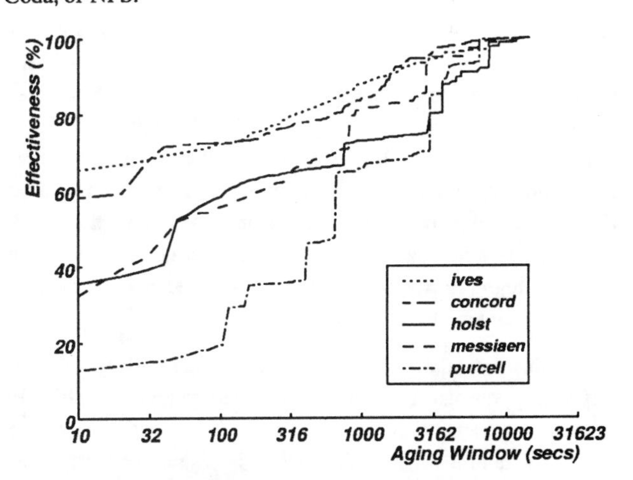

The X axis of this graph shows the aging window (A) on a logarithmic scale. Only CML records of age A or less are subject to optimization. Each curve corresponds to a different trace, and a point on a curve is the ratio of two quantities. The numerator is the amount of data saved by optimizations for the value of A at that point. The denominator is the savings when A is four hours (14,400 seconds). The value of the denominator is 84 MB for *ives*, 817 MB for *concord*, 40 MB for *holst*, 152 MB for *messiaen*, and 44 MB for *purcell*.

Figure 4: Effect of Aging on Optimizations

Figure 4 presents the results of our analysis. For each trace, this graph shows the impact of the aging window on the effectiveness of log optimizations. The results have been normalized with respect to a maximum aging window of four hours. We chose this period because it represents half a typical working day, and is a reasonable upper bound on the amount of work loss a user might be willing to tolerate.

The graph shows that there is considerable variation across traces. Values of A below 300 seconds barely yield an effectiveness of 30% on some traces, but they yield nearly 80% on others. For effectiveness above 80% on all traces, A must be nearly one hour. Since 600 seconds yields nearly 50% effectiveness on all traces, we have chosen it as the default value of A. This value can easily be changed by the user.

4.3.5. Reducing the Impact of Reintegration

Reintegrating all records older than A in one *chunk* could saturate a slow network for an extended period. The performance of a concurrent high priority network event, such as the servicing of a cache miss, could then be severely degraded. To avoid this problem, we have made reintegration chunk size adaptive.

The choice of a chunk size, (C), must strike a balance between two factors affecting performance. A large chunk size is more appropriate at high bandwidths because it amortizes the fixed costs of reintegration (such as transaction commitment at the server) over many log records. A small chunk size is better at low bandwidths because it reduces the maximum time of network contention. We have chosen a default value of 30 seconds for this time. This corresponds to C being 36 KB at 9.6 Kb/s, 240 KB at 64 Kb/s, and 7.7 MB at 2 Mb/s.

Before initiating reintegration, we estimate C for the current bandwidth. We then select a maximal prefix of CML records whose age is greater than A and whose sizes sum to C or less. Most records are small, except for `store` records, whose sizes include that of the corresponding file data. In the limit, we select at least one record even if its size is larger than C. This prefix is the chunk for reintegration. The reintegration barrier is placed after it, and reintegration proceeds as described in Section 4.3.3. This procedure is repeated a chunk at a time, deferring between chunks to high priority network use, until all records older than A have been reintegrated.

With this procedure, the size of a chunk can be larger than C only when it consists of a single `store` record for a large file. In this case, we transfer the file as a series of *fragments* of size C or less. If a failure occurs, file transfer is resumed after the last successful fragment. Atomicity is preserved in spite of fragmentation because the server does not logically attempt reintegration until it has received the entire file. Note that this is the reverse of the procedure at strong connectivity, where the server verifies the logical soundness of updates before fetching file contents. The change in order reflects a change in the more likely cause of reintegration failure in the two scenarios.

We are considering a refinement that would allow a user to force immediate reintegration of updates to a specific directory or subtree, without waiting for propagation of other updates. Implementing this would require computing the precedence relationships between records, and ensuring that a record is not reintegrated before its antecedents. This computation is not necessary at present because the CML and every possible chunk are already in temporal order, which implies precedence order. We are awaiting usage experience to decide whether the benefits of this refinement merit its implementation cost.

4.4. Seeking User Advice

When weakly connected, the performance impact of cache misses is often too large to ignore. For example, a cache miss on a 1 MB file at 10 Mb/s can usually be serviced in a few seconds. At 9.6 Kb/s, the same miss causes a delay of nearly 20 minutes!

From a user's perspective, this lack of performance transparency can overshadow the functional transparency of caching. The problem is especially annoying because cache miss handling, unlike trickle reintegration, is a foreground activity. In most cases, a user would rather be told that a large file is missing than be forced to wait for it to be fetched over a weak connection.

But there are also situations where a file is so critical that a user is willing to suffer considerable delay. We refer to the maximum time that a user is willing to wait for a particular file as her *patience threshold* for that file. The need for user input arises because Venus has to find out how critical a missing object is.

Since the hoarding mechanism already provided a means of factoring user estimates of importance into cache management, it was the natural focal point of our efforts. Our extensions of this mechanism for weak connectivity are in two parts: an interactive facility to help augment the hoard database (HDB), and another to control the amount of data fetched during hoard walks. Together these changes have the effect of moving many cache miss delays into the background.

4.4.1. Handling Misses

When a miss occurs, Venus estimates its service time from the current network bandwidth and the object's size (as given by its status information). If the object's status information is not already cached, Venus obtains it from the server. The delay for this is acceptable even on slow networks because status information is only about 100 bytes long.

The estimated service time is then compared with the patience threshold. If the service time is below the threshold, Venus transparently services the miss. If the threshold is exceeded, Venus returns a cache miss error and records the miss.

4.4.2. Augmenting the Hoard Database

At any time, a user can ask Venus to show her all the misses that have occured since the previous such request. Venus displays each miss along with contextual information, as shown in Figure 5. The user can then select objects to be added to the HDB. This action does not immediately fetch the object; that is deferred until a future hoard walk. Hoard walks occur once every 10 minutes, or by explicit user request.

4.4.3. Controlling Hoard Walks

A hoard walk is executed in two phases. In the first phase, called the *status walk*, Venus obtains status information for missing objects and determines which objects, if any, should be fetched. Because of volume callbacks, the status walk usually involves little network traffic. During the second phase, called the *data walk*, Venus fetches the contents of objects selected by the status walk. Even if there are only a few large objects to be fetched, this phase can be a substantial source of network traffic.

File/Directory	Program	HDB?
/coda/usr/hqb/papers/s15/s15.bib	emacs	■
/coda/misc/tex/i386_mach/lib/macros/art10.sty	virtex	■
/coda/misc/emacs/i386_mach/lisp/tex-mode.elc	emacs	■
/coda/misc/X11-others/i386_mach/bin/xloadimage	csh	□
/coda/misc/others/lib/weather/latest	csh	□
/coda/project/coda/alpha/src/venus/fso0.c	more	■

Cancel Done

This screen shows the name of each missing object and the program that referenced it. To add an object to the HDB, the user clicks the button to its right. A pop-up form (not shown here) allows the user to specify the hoard priority of the object and other related information.

Figure 5: Augmenting the Hoard Database

By introducing an interactive phase between the status and data walks, we allow users to limit the volume of data fetched in the data walk. Each object whose estimated service time is below the user's patience threshold is pre-approved for fetching. The fetching of other objects must be explicitly approved by the user.

Cache Files: Allocated = 6250 Occupied = 389 Available = 5861

Cache Space (KB): Allocated = 50000 Occupied = 8244 Available = 41756

Speed of Network Connection (b/s) = 9600

Number of Objects Preapproved for Fetch = 3

Object Name	Priority	Cost (s)	Fetch?	Stop Asking?
/coda/project/coda/alpha/src/venus/fso0.c	20	48	■	□
/coda/misc/emacs/i386_mach/bin/emacs	600	611	□	■

Total Expected Fetch Time (s) = 65

Total Number of Objects to be Fetched = 4

Cache Space (KB) After Walk: Alloc'd = 50000 Occ'd = 8301 Avail = 41699

Cancel Done

This screen enables the user to suppress fetching of objects selectively during a hoard walk. The priority and estimated service time of each object are shown. The user approves the fetch of an object by clicking on its "Fetch" button. By clicking on its "Stop Asking" button, she can prevent the prompt and fetch for that object until strongly connected. The cache state that would result from the data walk is shown at the bottom of the screen. This information is updated as the user clicks on "Fetch" buttons.

Figure 6: Controlling the Data Walk

Figure 6 shows an example of the screen displayed by Venus between the status and data walks. If no input is provided by the user within a certain time, the screen disappears and all the listed objects are fetched. This handles the case where the client is running unattended.

4.4.4. Modelling User Patience

Our goal in modelling user patience is to improve usability by reducing the frequency of user interaction. In those cases where we can predict a user's response with reasonable confidence, we can avoid the corresponding interactions. As mentioned earlier, a user's patience threshold, (τ), depends on how important she perceives an object to be: for a very important object, she is probably willing to wait many minutes.

Since user perception of importance is the notion captured by the hoard priority, (P), of an object, we posit that τ should be a function of P. At present, we are not aware of any data that could be the scientific basis for establishing the form of this relationship. Hence we use a function based solely on intuition, but have structured the implementation to make it easy to substitute a better alternative.

We conjecture that patience is similar to other human processes such as vision, whose sensitivity is logarithmic [3]. This suggests a relationship of the form $\tau = \alpha + \beta e^{\gamma P}$, where β and γ are scaling parameters and α represents a lower bound on patience. Even if an object is unimportant, the user prefers to tolerate a delay of α rather than dealing with a cache miss. We chose parameter settings based on their ability to yield plausible patience values for files commonly found in the hoard profiles of Coda users. The values we chose were $\alpha = 2$ seconds, $\beta = 1$, $\gamma = 0.01$.

Figure 7 illustrates the resulting model of user patience. Rather than expressing τ in terms of seconds, we have converted it into the size of the largest file that can be fetched in that time at a given bandwidth. For example, 60 seconds at a bandwidth of 64 Kb/s yields a maximum file size of 480KB. Each curve in Figure 7 shows τ as a function of P for a given bandwidth. In the region below this curve, cache misses are transparently handled and pre-approval is granted during hoard walks.

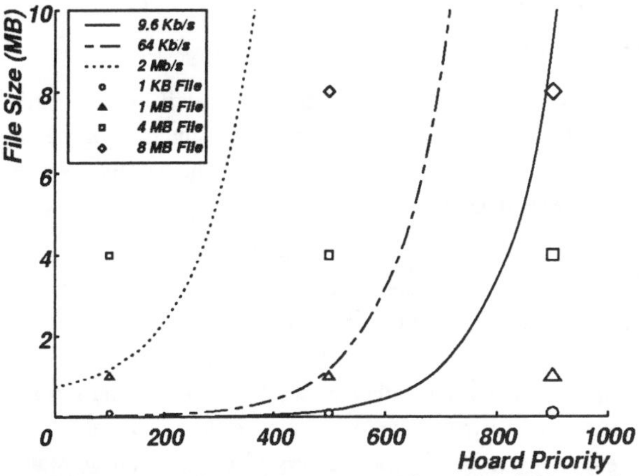

Each curve in this graph expresses patience threshold, (τ), in terms of file size. Superimposed on these curves are points representing files of various sizes hoarded at priorities 100, 500, and 900. At 9.6 Kb/s, only the files at priority 900 and the 1KB file at priority 500 are below τ. At 64 Kb/s, the 1MB file at priority 500 is also below τ. At 2Mb/s, all files except the 4MB and 8MB files at priority 100 are below τ.

Figure 7: Patience Threshold versus Hoard Priority

The user patience model is the source of adaptivity in cache miss handling. It maintains usability at all bandwidths by balancing two factors that intrude upon transparency. At very low bandwidths, the delays in fetching large files annoy users more than the need for interaction. As bandwidth rises, delays shrink and interaction becomes more annoying. To preserve usability, we handle more cases transparently. In the limit, at strong connectivity, cache misses are fully transparent.

5. Deployment Status

The mechanisms described in this paper are being deployed to a user community of Coda developers and other computer science researchers. We have over 40 user accounts, of which about 25 are used regularly. Many users run Coda on both their desktop workstations and their laptops. We have a total of about 35 Coda clients, evenly divided between workstations and laptops. These clients access almost 4.0 GB of data stored on Coda servers.

The evolution described in Section 3.2 has spanned over two years. Early usage experience with each mechanism was invaluable in guiding further development. The transport protocol extensions were implemented in early 1993, and incorporated into the deployed system later that year. The rapid cache validation mechanism was implemented in late 1993, and has been deployed since early 1994. The trickle reintegration and user advice mechanisms were implemented between 1994 and early 1995, and have been released for general use.

6. Evaluation

6.1. Rapid Cache Validation

Two questions best characterize our evaluation of Coda's rapid cache validation mechanism:

- Under ideal conditions, how much do volume callbacks improve cache validation time?
- In practice, how close are conditions to ideal?

The first question was discussed in detail in an earlier paper [18]. Hence, we only present a brief summary of the key results. More recently, we have addressed the second question, and present the detailed results here.

6.1.1. Performance Under Ideal Conditions

For a given set of cached objects, the time for validation is minimal when two conditions hold. First, at disconnection, volume callbacks must exist for all cached objects. Second, while disconnected, the volumes containing these objects must not be updated at the server. Then, upon reconnection, communication is needed only to verify volume version stamps. Fresh volume callbacks are acquired as a side effect, at no additional cost.

Under these conditions, the primary determinants of performance are network bandwidth and the composition of cache contents. We conducted experiments to measure validation time as a function of these two variables. To study variation due to cache composition, we used the hoard profiles of five typical Coda users. To vary bandwidth, we used a network emulator.

Figure 8 shows that for all users, and at all bandwidths, volume callbacks reduce cache validation time. The reduction is modest at high bandwidths, but becomes substantial as bandwidth decreases. At 9.6 Kb/s, the improvement is dramatic, typically taking only about 25% longer than at 10 Mb/s.

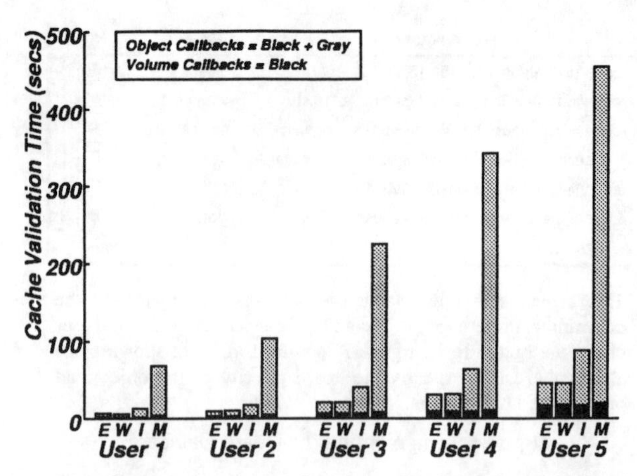

This figure compares the time for validation using object and volume callbacks. Cache contents were determined by the hoard profiles of five Coda users. The network speeds correspond to the nominal speeds of Ethernet (E, 10 Mb/s), WaveLan (W, 2 Mb/s), ISDN (I, 64 Kb/s), and Modem (M, 9.6 Kb/s). The client and server were DECstation 5000/200s running Mach 2.6. Bandwidth was varied using an emulator on Ethernet.

Figure 8: Validation Time Under Ideal Conditions

6.1.2. Conditions Observed in Practice

There are two ways in which a Coda client in actual use may find conditions less than ideal. First, a client may not possess volume stamps for some objects at disconnection. If frequent, this event would indicate that our strategy of waiting for a hoard walk to acquire volume callbacks is not aggressive enough. Second, a volume stamp may prove to be stale when presented for validation. This would mean that the volume was updated on the server while the client was disconnected. If frequent, this event would indicate that acquiring volume stamps is futile, because it rarely speeds up validation. It could also be symptomatic of a volume being too large a granularity for cache coherence, for reasons analogous to false sharing in virtual memory systems with too large a page size.

To understand how serious these concerns are, we instrumented Coda clients to record cache validation statistics. Figure 9 presents data gathered from 26 clients. The data shows that our fears were baseless. On average, clients found themselves without a volume stamp only in 3% of the cases. The data on successful validations is even more reassuring. Most success rates were over 97%, and each successful validation saved roughly 53 individual validations.

6.2. Trickle Reintegration

How much is a typical user's update activity slowed when weakly connected? This is the question most germane to trickle reintegration, because the answer will reveal how effectively foreground activity is insulated from update propagation over slow networks.

The simplest way to answer this question would be to run a standard file system benchmark on a write-disconnected client over a wide range of network speeds. The obvious candidate is the *Andrew benchmark* [5] since it is compact, portable, and widely used. Unfortunately, this benchmark is of limited value in evaluating trickle reintegration.

Client	Missing Stamp	Validation Attempts	Fraction Successful	Objs per Success
bach	2%	970	99%	89
berlioz	8%	1178	97%	48
brahms	0%	542	99%	5
chopin	4%	1674	97%	102
copland	3%	1387	94%	171
dvorak	2%	5536	98%	75
gershwin	11%	467	95%	32
gs125	0%	897	99%	22
holst	0%	474	99%	29
ives	0%	1532	98%	56
mahler	1%	566	97%	6
messiaen	0%	827	98%	31
mozart	1%	1633	98%	126
varicose	0%	568	98%	32
verdi	6%	2370	98%	64
vivaldi	7%	344	89%	28
Mean	**3%**	**1310**	**97%**	**57**

(a) Desktops

Client	Missing Stamp	Validation Attempts	Fraction Successful	Objs per Success
caractacus	2%	650	97%	40
deidamia	2%	2257	98%	112
finlandia	13%	541	99%	32
gloriana	2%	1457	97%	29
guntram	0%	2977	99%	26
nabucco	1%	1301	96%	28
prometheus	6%	1617	97%	74
serse	8%	1790	98%	32
tosca	1%	652	99%	60
valkyrie	4%	759	96%	32
Mean	**4%**	**1400**	**98%**	**47**

(b) Laptops

These tables present data collected for approximately four weeks in July and August 1995 from 16 desktops and 10 laptops. The first column indicates how often validation could not be attempted because of a missing volume stamp. The last column gives a per-client average of object validations saved by a successful volume validation.

Figure 9: Observed Volume Validation Statistics

First, the running time of the benchmark on current hardware is very small, typically less than three minutes. This implies that no updates would be propagated to the server during an entire run of the benchmark for any reasonable aging window. Increasing the total time by using multiple iterations is not satisfactory because the benchmark is not idempotent. Second, although the benchmark captures many aspects of typical user activity, it does not exhibit overwrite cancellations. Hence, its file references are only marginally affected by log optimizations. Third, the benchmark involves no user think time, which we believe to be atypical of mobile computing applications.

For these reasons, our evaluation of trickle reintegration is based on *trace replay,* which is likely to be a much better indicator of performance in real use.

6.2.1. Trace Replay: Experiment Design

The ultimate in realism would be to measure trickle reintegration in actual use by mobile users. But this approach has serious shortcomings. First, a human subject cannot be made to repeat her behavior precisely enough for multiple runs of an experiment. Second, many confounding factors make timing results from actual use difficult to interpret. Third, such experiments cannot be replicated at other sites or in the future.

To overcome these limitations, we have developed an experimental methodology in which trace replay is used in lieu of human subjects. Realism is preserved since the trace was generated in actual use. Timing measurements are much less ambiguous, since experimental control and replicability are easier to achieve. The traces and the replay software can be exported.

Note that a trace replay experiment differs from a trace-driven simulation in that traces are replayed on a live system. Our replay software [19] generates Unix system calls that are serviced by Venus and the servers just as if they had been generated by a human user. The only difference is that a single process performs the replay, whereas the trace may have been generated by multiple processes. It would be fairly simple to extend our replay software to exactly emulate the original process structure.

How does one incorporate the effect of human think time in a trace replay experiment? Since a trace is often used many months or years after it was collected, the system on which it is replayed may be much faster than the original. But a faster system will not speed up those delays in the trace that were caused by human think time. Unfortunately, it is difficult to reliably distinguish think time delays from system-limited delays in a trace.

Our solution is to perform sensitivity analysis for think time, using a parameter called *think threshold,* (λ). This parameter defines the smallest delay in the input trace that will be preserved in the replay. When λ is 0, all delays are preserved; when it is infinity, the trace is replayed as fast as possible.

We rejected both extremities as parameter values for our experiments. At $\lambda = 0$, there is so much opportunity for overlapping data transmission with think time that experiments would be biased too much in favor of trickle reintegration. At $\lambda =$ infinity, the absence of think time makes the experiment as unrealistic as the Andrew benchmark. In the light of these considerations, we chose values of λ equal to 1 second and 10 seconds for our experiments. These are plausible values for typical think times during periods of high activity, and they are not biased too far against or in favor of trickle reintegration.

Since log optimizations play such a critical role in trickle reintegration, we also conducted a sensitivity analysis for this factor. We divided the traces mentioned in Section 4.3.4 into 45-minute segments, selected segments with the highest activity levels, and analyzed their susceptibility to log optimizations. A segment longer than 45 minutes would have made the duration of each experiment excessive, allowing us to explore only a few parameter combinations.

We define the *compressibility* of a trace segment as the ratio of two quantities obtained when the segment is run through the Venus simulator. The numerator is the amount of data optimized out; the denominator is the length of the unoptimized CML. Figure 10 shows the observed distribution of compressibility in those trace segments with a final CML of 1MB or greater.

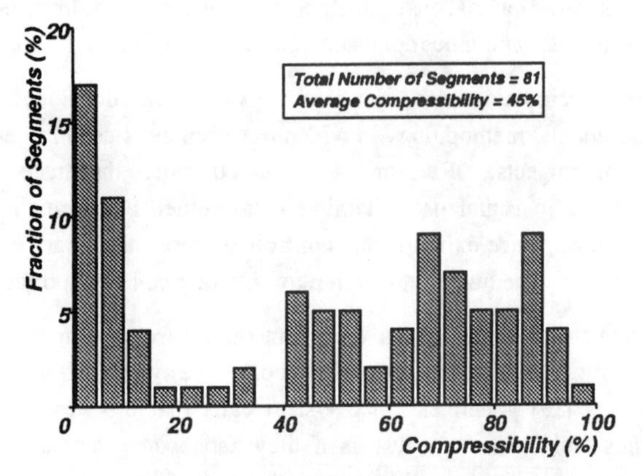

Figure 10: Compressibility of Trace Segments

The data shows that the compressibilities of roughly a third of the segments are below 20%, while those of the remaining two-thirds range from 40% to 100%. For our experiments, we chose one segment from each quartile of compressibility. The characteristics of these segments are shown in Figure 11.

Trace Segment	No. of References	No. of Updates	Unopt. CML (KB)	Opt. CML (KB)	Compress-ibility
Purcell	51681	519	2864	2625	8%
Holst	61019	596	3402	2302	32%
Messiaen	38342	188	6996	2184	69%
Concord	160397	1273	34704	2247	94%

Each of these segments is 45 minutes long. Since Coda uses the `open-close` session semantics of AFS, individual `read` and `write` operations are not included. Hence "Updates" in this table only refers to operations such as `close` after write, and `mkdir`. "References" includes, in addition, operations such as `close` after read, `stat`, and `lookup`.

Figure 11: Segments Used in Trace Replay Experiments

6.2.2. Trace Replay: Results

Figure 12 presents the results of our trace replay experiments. The same data is graphically illustrated in Figure 13. To ensure a fair comparison, we forced Venus to remain write disconnected at all bandwidths. We also deferred the beginning of measurements until 10 minutes into each run, thus warming the CML for trickle reintegration. The choice of 10 minutes corresponds to the largest value of *A* used in our experiments.

Figures 12 and 13 cover 64 combinations of experimental parameters: two aging windows (A = 300 and 600 seconds), two think thresholds (λ = 1 and 10 seconds), four trace compressibilities (8, 32, 69, and 94%), and four bandwidths (10 Mb/s, 2 Mb/s, 64 Kb/s, and 9.6 Kb/s).

These measurements confirm the effectiveness of trickle reintegration over the entire experimental range. Bandwidth varies over three orders of magnitude, yet elapsed time remains almost unchanged. On average, performance is only about 2% slower at 9.6 Kb/s than at 10 Mb/s. Even the worst case, corresponding to the Ethernet and ISDN numbers for Concord in Figure 12(d), is only 11% slower.

Trickle reintegration achieves insulation from network bandwidth by decoupling updates from their propagation to servers. Figure 14 illustrates this decoupling for one combination of λ and A. As bandwidth decreases, so does the amount of data shipped. For example, in Figure 14(b), the data shipped decreases from 2254 KB for Ethernet to 1536 KB for Modem. Since data spends more time in the CML, there is greater opportunity for optimization: 1067 KB versus 1081 KB. At the end of the experiment, more data remains in the CML at lower bandwidths: 70KB versus 2289 KB.

7. Related Work

Effective use of low bandwidth networks has been widely recognized as a vital capability for mobile computing [4, 11], but only a few systems currently provide this functionality. Of these, Little Work [6] is most closely related to our system.

Like Coda, Little Work provides transparent Unix file access to disconnected and weakly-connected clients, and makes use of log optimizations. But, for reasons of upward compatibility, it makes no changes to the AFS client-server interface. This constraint hurts its ability to cope with intermittent connectivity. First, it renders the use of large-granularity cache coherence infeasible. Second, it weakens fault tolerance because transactional support for reintegration cannot be added to the server.

Little Work supports *partially connected operation* [7], which is analogous to Coda's write disconnected state. But there are important differences. First, users cannot influence the servicing of cache misses in Little Work. Second, update propagation is less adaptive than trickle reintegration in Coda. Third, much of Little Work's efforts to reduce and prioritize network traffic occur in the SLIP driver. This is in contrast to Coda's emphasis on the higher levels of the system.

AirAccess 2.0 is a recent product that provides access to Novell and other DOS file servers over low-bandwidth networks [1]. Its implementation focuses on the lower levels, using techniques such as data compression and differential file transfer. Like Little Work, it preserves upward compatibility with existing servers and therefore suffers from the same limitations. AirAccess has no analog of trickle reintegration, nor does it allow users to influence the handling of cache misses.

From a broader perspective, application packages such as Lotus Notes [12] and cc:Mail [17] allow use of low-bandwidth networks. These systems differ from Coda in that support for mobility is entirely the responsibilty of the application. By providing this support at the file system level, Coda obviates the need to modify individual applications. Further, by mediating the resource demands of concurrent applications, Coda can better manage resources such as network bandwidth and cache space.

Trace Segment	Ethernet 10 Mb/s	WaveLan 2 Mb/s	ISDN 64 Kb/s	Modem 9.6 Kb/s
Purcell	2025 (16)	1999 (15)	2002 (20)	2096 (32)
Holst	1960 (3)	1961 (5)	1964 (5)	1983 (5)
Messiaen	1950 (2)	1970 (9)	1959 (3)	1995 (6)
Concord	1897 (9)	1952 (20)	1954 (43)	2002 (13)

(a) λ = 1 second, A = 300 seconds

Trace Segment	Ethernet 10 Mb/s	WaveLan 2 Mb/s	ISDN 64 Kb/s	Modem 9.6 Kb/s
Purcell	2086 (28)	2064 (6)	2026 (20)	2031 (4)
Holst	2004 (13)	1984 (11)	1970 (11)	2009 (17)
Messiaen	1949 (2)	1974 (8)	1969 (16)	1986 (3)
Concord	2078 (49)	2051 (38)	2017 (39)	2079 (10)

(b) λ = 1 second, A = 600 seconds

Trace Segment	Ethernet 10 Mb/s	WaveLan 2 Mb/s	ISDN 64 Kb/s	Modem 9.6 Kb/s
Purcell	1747 (20)	1622 (12)	1624 (5)	1744 (8)
Holst	1026 (6)	1000 (3)	1005 (10)	1047 (2)
Messiaen	1234 (2)	1241 (2)	1238 (5)	1278 (9)
Concord	1254 (7)	1323 (16)	1312 (17)	1362 (18)

(c) λ = 10 seconds, A = 300 seconds

Trace Segment	Ethernet 10 Mb/s	WaveLan 2 Mb/s	ISDN 64 Kb/s	Modem 9.6 Kb/s
Purcell	1704 (9)	1658 (14)	1664 (23)	1683 (16)
Holst	1060 (10)	1027 (8)	1021 (8)	998 (3)
Messiaen	1234 (3)	1265 (13)	1263 (11)	1279 (7)
Concord	1258 (7)	1383 (27)	1402 (30)	1340 (16)

(d) λ = 10 seconds, A = 600 seconds

This table presents the elapsed time, in seconds, of the trace replay experiments described in Section 6.2.1. The think threshold is λ, and the aging window is *A*. Each data point is the mean of five trials; figures in parentheses are standard deviations. The experiments were conducted using a DEC pc425SL laptop client and a DECstation 5000/200 server, both with 32 MB of memory, and running Mach 2.6. The client and server were isolated on a separate network. The Ethernet, WaveLan and Modem experiments used actual networks of the corresponding type. The ISDN experiments were conducted on an Ethernet using a network emulator. Measurements began after a 10 minute warming period.

Figure 12: Performance of Trickle Reintegration on Trace Replay

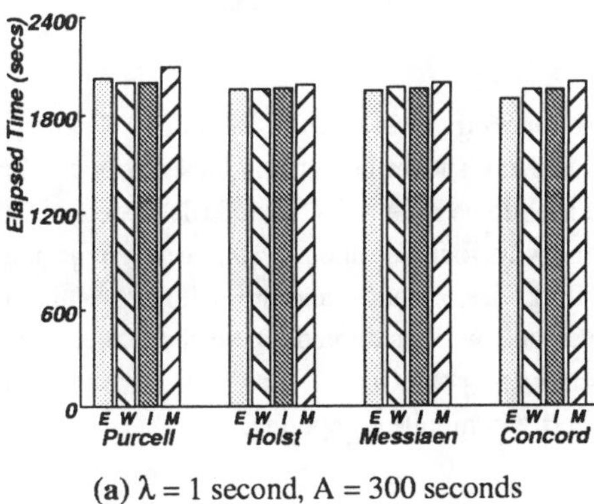

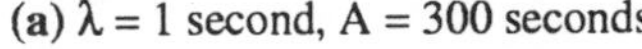
(a) λ = 1 second, A = 300 seconds

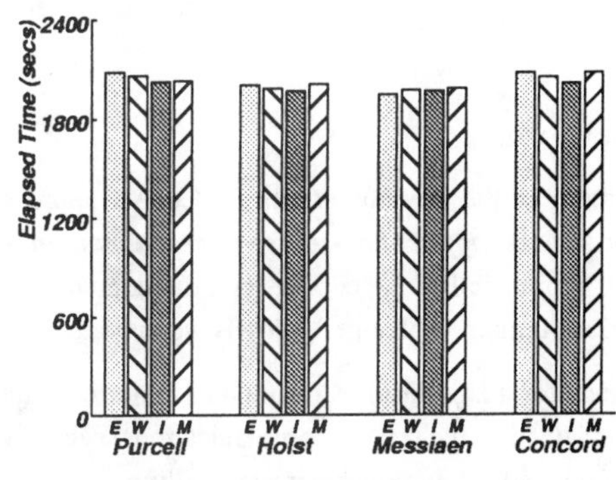

(b) λ = 1 second, A = 600 seconds

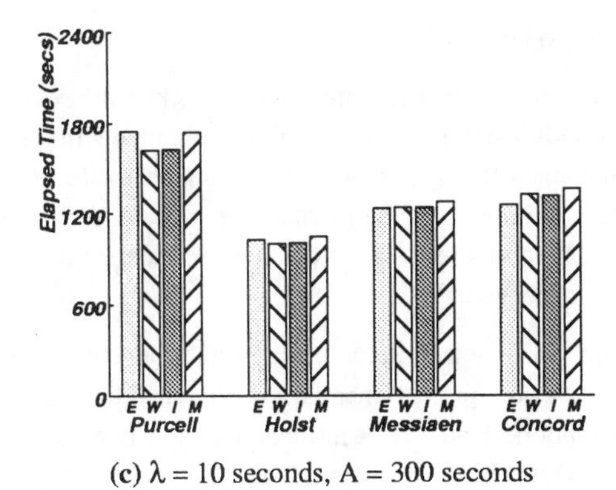

(c) λ = 10 seconds, A = 300 seconds

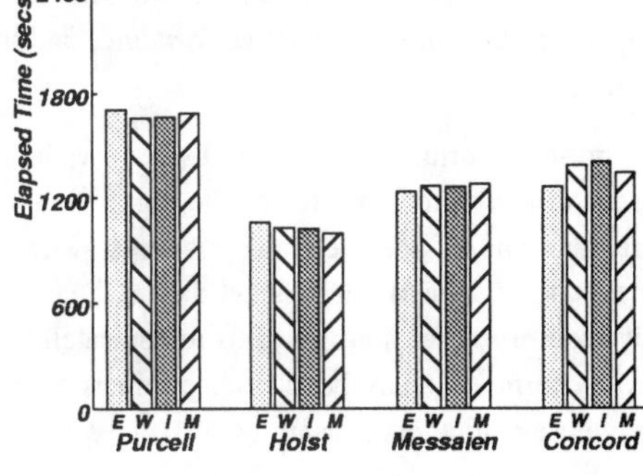

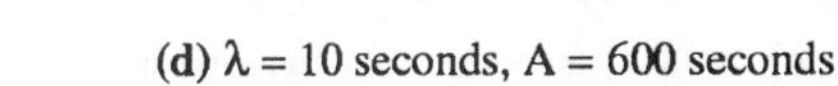
(d) λ = 10 seconds, A = 600 seconds

These graphs illustrate the data in Figure 12. Network speed is indicated by **E** (Ethernet), **W** (WaveLan), **I** (ISDN), or **M** (Modem).

Figure 13: Comparison of Trace Replay Times

Network Type	Begin CML (KB)	End CML (KB)	Shipped (KB)	Optimized (KB)
Ethernet	0 (0)	59 (0)	2618 (0)	238 (0)
WaveLan	0 (0)	59 (0)	2618 (0)	238 (0)
ISDN	0 (0)	2128 (0)	800 (105)	238 (0)
Modem	0 (0)	2538 (0)	114 (19)	238 (0)

(a) Trace Segment = Purcell

Network Type	Begin CML (KB)	End CML (KB)	Shipped (KB)	Optimized (KB)
Ethernet	2133 (0)	70 (0)	2254 (0)	1067 (0)
WaveLan	2133 (0)	70 (0)	2254 (0)	1067 (0)
ISDN	2133 (0)	70 (0)	2252 (0)	1069 (0)
Modem	2133 (0)	2289 (0)	1536 (68)	1081 (0)

(b) Trace Segment = Holst

Network Type	Begin CML (KB)	End CML (KB)	Shipped (KB)	Optimized (KB)
Ethernet	896 (0)	0 (0)	2270 (0)	3022 (0)
WaveLan	896 (0)	0 (0)	2270 (0)	3022 (0)
ISDN	896 (0)	0 (0)	2270 (0)	3022 (0)
Modem	896 (0)	1060 (0)	1309 (16)	3103 (0)

(c) Trace Segment = Messiaen

Network Type	Begin CML (KB)	End CML (KB)	Shipped (KB)	Optimized (KB)
Ethernet	63 (0)	2103 (0)	2496 (0)	30209 (0)
WaveLan	63 (0)	2103 (0)	2496 (0)	30209 (0)
ISDN	63 (0)	2103 (0)	2407 (0)	30291 (0)
Modem	63 (0)	2180 (0)	1142 (46)	32322 (0)

(d) Trace Segment = Concord

This table shows components of the data generated in the experiments of Figure 12(b). Results for other combinations of λ and A are comparable. The columns labelled "Begin CML" and "End CML" give the amount of data in the CML at the beginning and end of the measurement period. This corresponds to the amount of data waiting to be propagated to the servers at those times. The column labelled "Shipped" gives the amount of data actually transferred over the network; "Optimized" gives the amount of data saved by optimizations.

It may appear at first glance that the sum of the "End CML", "Shipped", and "Optimized" columns should equal the "Unopt. CML" column of Figure 11. But this need not be true for the following reasons. First, optimizations that occur prior to the measurement period are not included in "Optimized". Second, if an experiment ends while a large file is being transferred as a series of fragments, the fragments already transferred are counted both in the "End CML" and "Shipped" columns. Third, log records are larger when shipped than in the CML.

Figure 14: Data Generated During Trace Replay (λ = 1 second, A = 600 seconds)

8. Conclusion

Adaptation is the key to mobility. Coda's approach is best characterized as *application-transparent adaptation* — Venus bears full responsibility for coping with the demands of mobility. Applications remain unchanged, preserving upward compatibility.

The quest for adaptivity has resulted in major changes to many aspects of Coda, including communication, cache validation, update propagation, and cache miss handling. In making these changes, our preference has been to place functionality at higher levels of Venus, with only the bare minimum at the lowest levels. Consistent with the end-to-end argument, we believe that this is the best approach to achieving good performance and usability in mobile computing.

In its present form, Coda can use a wide range of communication media relevant to mobile computing. Examples include regular phone lines, cellular modems, wireless LANs, ISDN lines, and cellular digital packet data (CDPD) links. But Coda may require further modifications to use satellite networks, which have enormous delay-bandwidth products, and cable TV networks, whose bandwidth is asymmetric.

Our work so far has assumed that performance is the only metric of cost. In practice, many networks used in mobile computing cost real money. We therefore plan to explore techniques by which Venus can electronically inquire about network cost, and base its adaptation on both cost and quality. Of course, full-scale deployment of this capability will require the cooperation of network providers and regulatory agencies.

Weak connectivity is a fact of life in mobile computing. In this paper, we have shown how such connectivity can be exploited to benefit mobile users of a distributed file system. Our mechanisms allow users to focus on their work, largely ignoring the vagaries of network performance and reliability. While many further improvements will undoubtedly be made, Coda in its present form is already a potent and usable tool for exploiting weak connectivity in mobile computing.

Acknowledgements

Brian Noble, Puneet Kumar, David Eckhardt, Wayne Sawdon, and Randy Dean provided insightful comments that substantially strengthened this paper. Our SOSP shepherd, Mary Baker, was helpful in improving the presentation. Brent Welch helped with the TCL mud-wrestling involved in implementing user-assisted cache management.

This work builds upon the contributions of many past and present Coda project members. Perhaps the most important contribution of all has been made by the Coda user community, through its bold willingness to use and help improve an experimental system.

References

[1] *AirSoft AirAccess 2.0: Mobile Networking Software*
AirSoft, Inc., Cupertino, CA, 1994.

[2] Baker, M.G., Hartmann, J.H., Kupfer, M.D., Shirriff, K.W., Ousterhout, J.K.
Measurements of a Distributed File System.
In *Proceedings of the Thirteenth ACM Symposium on Operating Systems Principles*. Pacific Grove, CA, October, 1991.

[3] Cornsweet, T.N.
Visual Perception.
Academic Press, 1971.

[4] Forman, G.H., Zahorjan, J.
The Challenges of Mobile Computing.
IEEE Computer 27(4), April, 1994.

[5] Howard, J.H., Kazar, M.L., Menees, S.G., Nichols, D.A., Satyanarayanan, M., Sidebotham, R.N., West, M.J.
Scale and Performance in a Distributed File System.
ACM Transactions on Computer Systems 6(1), February, 1988.

[6] Huston, L., Honeyman, P.
Disconnected Operation for AFS.
In *Proceedings of the 1993 USENIX Symposium on Mobile and Location-Independent Computing*. Cambridge, MA, August, 1993.

[7] Huston, L., Honeyman, P.
Partially Connected Operation.
In *Proceedings of the Second USENIX Symposium on Mobile and Location-Independent Computing*. Ann Arbor, MI, April, 1995.

[8] Jacobson, V.
Congestion Avoidance and Control.
In *Proceedings of SIGCOMM88*. Stanford, CA, August, 1988.

[9] Jacobson, V.
RFC 1144: Compressing TCP/IP Headers for Low-Speed Serial Links.
February, 1990.

[10] Jacobson, V., Braden, R., Borman, D.
RFC 1323: TCP Extensions for High Performance.
May, 1992.

[11] Katz, R.H.
Adaptation and Mobility in Wireless Information Systems.
IEEE Personal Communications 1(1), 1994.

[12] Kawell Jr., L., Beckhardt, S., Halvorsen, T., Ozzie, R., Greif, I.
Replicated Document Management in a Group Communication System.
In Marca, D., Bock, G. (editors), *Groupware: Software for Computer-Supported Cooperative Work*, pages 226-235. IEEE Computer Society Press, Los Alamitos, CA, 1992.

[13] Kistler, J.J., Satyanarayanan, M.
Disconnected Operation in the Coda File System.
ACM Transactions on Computer Systems 10(1), February, 1992.

[14] Kumar, P.
Mitigating the Effects of Optimistic Replication in a Distributed File System.
PhD thesis, School of Computer Science, Carnegie Mellon University, December, 1994.

[15] Lampson, B. W.
Hints for Computer System Design.
In *Proceedings of the Ninth ACM Symposium on Operating Systems Principles*. Bretton Woods, NH, October, 1983.

[16] Mann, T., Birrell, A., Hisgen, A., Jerian, C., Swart, G.
A Coherent Distributed File Cache with Directory Write-Behind.
ACM Transactions on Computer Systems 12(2), May, 1994.

[17] Moeller, M.
Lotus Opens cc:Mail to Pagers.
PC Week 11(35):39, September, 1994.

[18] Mummert, L.B., Satyanarayanan, M.
Large Granularity Cache Coherence for Intermittent Connectivity.
In *Proceedings of the 1994 Summer USENIX Conference*. Boston, MA, June, 1994.

[19] Mummert, L.B., Satyanarayanan, M.
Long-Term Distributed File Reference Tracing: Implementation and Experience.
Technical Report CMU-CS-94-213, School of Computer Science, Carnegie Mellon University, November, 1994.

[20] Nelson, M.N., Welch, B.B., Ousterhout, J.K.
Caching in the Sprite Network File System.
ACM Transactions on Computer Systems 6(1), February, 1988.

[21] Noble, B., Satyanarayanan, M.
An Empirical Study of a Highly-Available File System.
In *Proceedings of the 1994 ACM Sigmetrics Conference.* Nashville, TN, May, 1994.

[22] Ousterhout, J., Da Costa, H., Harrison, D., Kunze, J., Kupfer, M., Thompson, J.
A Trace-Driven Analysis of the 4.2BSD File System.
In *Proceedings of the Tenth ACM Symposium on Operating System Principles*. Orcas Island, WA, December, 1985.

[23] Romkey, J.
RFC 1055: A Nonstandard for Transmission of IP Datagrams Over Serial Lines: SLIP.
June, 1988.

[24] Saltzer, J.H., Reed, D.P., Clark, D.D.
End-to-End Arguments in System Design.
ACM Transactions on Computer Systems 2(4), November, 1984.

[25] Satyanarayanan, M., Kistler, J.J., Kumar, P., Okasaki, M.E., Siegel, E.H., Steere, D.C.
Coda: A Highly Available File System for a Distributed Workstation Environment.
IEEE Transactions on Computers 39(4), April, 1990.

[26] Satyanarayanan, M., Kistler, J.J., Mummert, L.B., Ebling, M.R., Kumar, P., Lu, Q.
Experience with Disconnected Operation in a Mobile Computing Environment.
In *Proceedings of the 1993 USENIX Symposium on Mobile and Location-Independent Computing*. Cambridge, MA, August, 1993.

[27] Satyanarayanan, M. (Editor).
RPC2 User Guide and Reference Manual
Department of Computer Science, Carnegie Mellon University, 1995 (Last revised).

[28] Sheng, S., Chandrakasan, A., Brodersen, R.W.
A Portable Multimedia Terminal.
IEEE Communications Magazine 30(12), December, 1992.

[29] Weiser, M.
The Computer for the Twenty-First Century.
Scientific American 265(3), September, 1991.

Afterword

M. Satyanarayanan
School of Computer Science
Carnegie Mellon University

In retrospect, the impact of our work on disconnected operation has proved to be considerably greater than any of us expected or hoped for in October 1991. With the benefit of hindsight, it is easy to see that this is in large measure due to fortuitous timing — we were at the right place and time with the right ideas.

The concepts and terminology introduced by our work addressed problems whose importance was beginning to be recognized by the distributed systems community. Fortunately, there was no established or competing viewpoint or system design that our ideas had to displace. Replication, the classic approach to high availability, was not rejected by Coda. Rather, it was embraced in the form of server replication; disconnected operation was offered as an adjunct, not a substitute. Indeed, our characterization of these two alternatives as distinct flavors (first and second class replication) of the same core concept provided a unification I had not anticipated at the outset of our work. Personally, the most valuable insight I gained from the writing of the 1991 Coda paper was in providing this rationale — it clarified in my mind the true relationship between caching and replication from the viewpoint of availability.

The conceptual simplicity of our ideas also helped in their acceptance. That caching, one of the oldest ideas in computer science, could be exploited for a purpose hitherto unsuspected was a pleasant surprise. It is rare for a single mechanism to simultaneously contribute toward two very different goals: performance and availability in this case. More commonly, a mechanism is a compromise betweeen competing considerations. Of course, conceptual simplicity need not translate to simplicity of implementation. Bridging this gap between a simple idea and its effective realization continues to animate the dialog on disconnected operation to this day.

Another factor contributing to the impact of our work was the emergence of portable computers with enough resources to be credible Coda clients. Again, timing played a big part: the appearance of our 1991 paper coincided with the earliest availability of such machines in the 6 to 8 lb. weight class — the heaviest that users are typically willing to carry with them. Mobile computing along the lines advocated by Coda thus became an immediate reality, not just a distant possibility. Today, Coda and disconnected operation are most commonly associated with mobile computing rather than the broader issues of data availability in distributed systems that motivated them originally. Thus, mobile computing has served as a "champion application" for Coda: that is, an application domain where the merits of a new technology have such overwhelming benefits that they overshadow the cost and inconvenience of adopting an untried technology, discovering its shortcomings and fixing them, and living with these shortcomings until they are overcome.

An important metric of impact is the extent of follow-on work spawned by a piece of research. Unlike citations, which are almost cost-free, follow-on work involves investment of time and energy (and often money, especially in cases of commercialization). It thus implicitly endorses the importance of the original work, even if the goal of the follow-on work is to debunk the original. The endorsement is particularly strong if the follow-on work is conducted by people without close ties to the original investigators.

By this metric, the research reported in our 1991 paper has indeed had considerable impact. The most obvious follow-on pieces of work are the extension of Coda to exploit weak connectivity (as described in the second of the two papers included here, and elaborated in Mummert's PhD dissertation [11]), and its extension to render caching translucent for improved usability, as elaborated in Ebling's PhD dissertation [2]. The use of volume callbacks in Coda has led to a generalization called *predicate callback* that is applicable to databases and other repositories, and which offers an interesting approach to maintaining data consistency in the presence of high network latency [15].

Less obvious is the impact that Coda has had on Odyssey, a platform for mobile computing currently being developed by our research group. As described in our 1997 SOSP paper [12] and elaborated in Noble's PhD dissertation [13], Odyssey offers an application-aware approach to mobile computing in contrast to Coda's application-transparent paradigm. Coda's conceptual contribution to Odyssey is two-fold: first, it helped us formulate a taxonomy of approaches to mobile information access and to understand the corresponding trade-offs; second, its design helped us realize that consistency represented a particular dimension of the broader concept of fidelity of data, a critical insight in the design of Odyssey.

There has also been substantial follow-on activity outside Carnegie Mellon. For example, there have been a number of efforts to implement variants of the ideas in our 1991 paper on non-Mach or non-Unix platforms. Examples include the work by Huston [5, 6], Huizinga [4], and Mazer [10]. An implementation of disconnected operation for the OS/2 operating system called Mobile FileSync was exhibited by IBM in 1994 [14], though not subsequently released as a product. Currently, Clement Yui-Wah Lee of the Chinese University of Hong Kong is working on a PhD dissertation that is exploring recomputation on strongly-connected surrogate Coda clients to reduce the volume of reintegrated file data over slow links.

The concept of *hoarding,* first identified in Coda, has been the subject of at least two PhD dissertations outside Carnegie Mellon: Tait at Columbia [16, 17, 18] and Kuenning at UCLA [7, 8]. The goal of these dissertations is the development of automated techniques to reduce the amount of explicit cache advice provided by a user.

From a broader perspective, there have been efforts to extend the concept of disconnected operation to domains other than distributed file systems. For example, in the domain of electronic mail, Tso at Xerox PARC [19] extended a mail client to support disconnected operation; more fundamentally, the IMAP4 protocol [1] specifies disconnected operation as part of its core. Other examples include the CaubWeb system for disconnected Web browsing by LoVerso and Mazer [9] and the extensions to the Thor objected-oriented database reported by Gruber et al [3].

In the commercial world, Microsoft has adopted the ideas of disconnected operation and conflict resolution from Coda as key architectural features of the IntelliMirror component of the forthcoming Windows NT 5.0 file system. In a most satisfying way, this confirms that the classic model of technology transfer from academia to industry through the medium of refereed publication in highly selective research forums is alive and well.

References

[1] Crispin, M.
Internet Message Access Protocol - Version 4rev1.
Technical Report RFC 2060, University of Washington, December, 1996.

[2] Ebling, M.R.
Translucent Cache Management for Mobile Computing.
PhD thesis, Carnegie Mellon University, 1998.

[3] Gruber, R., Kaashoek, F., Liskov, B., Shrira, L.
Disconnected Operation in the Thor Object-Oriented Database System.
In *Proceedings of the Workshop on Mobile Computing Systems and Applications.* Santa Cruz, CA, December, 1994.

[4] Huizinga, D.M., Heflinger, K.A.
Experience with Connected and Disconnected Operation of Portable Notebook Computers in Distributed Systems.
In *Proceedings of the Workshop on Mobile Computing Systems and Applications.* Santa Cruz, CA, December, 1994.

[5] Huston, L., Honeyman, P.
Disconnected Operation for AFS.
In *Proceedings of the 1993 USENIX Symposium on Mobile and Location-Independent Computing.* Cambridge, MA, August, 1993.

[6] Huston, L., Honeyman, P.
Partially Connected Operation.
In *Proceedings of the Second USENIX Symposium on Mobile and Location-Independent Computing.* Ann Arbor, MI, April, 1995.

[7] Kuenning, G. H., Popek, G.J.
Automated Hoarding for Mobile Computers.
In *Proceedings of the Sixteenth ACM Symposium on Operating Systems Principles.* Saint-Malo, France, October, 1997.

[8] Kuenning, G.H.
Seer: Predictive File Hoarding for Disconnected Mobile Operation.
PhD thesis, University of California, Los Angeles, 1997.

[9] LoVerso, J.R., Mazer, M.S.
Caubweb: Detaching the Web with Tcl.
In *Proceedings of the Fifth Annual Tcl/Tk Workshop.* Boston, MA, July, 1997.

[10] Mazer, M.S., Tardo, J.J.
A Client-Side-Only Approach to Disconnected File Access.
In *Proceedings of the Workshop on Mobile Computing Systems and Applications.* Santa Cruz, CA, December, 1994.

[11] Mummert, L.B.
Exploiting Weak Connectivity in a Distributed File System.
PhD thesis, Carnegie Mellon University, 1996.

[12] Noble, B. D., Satyanarayanan, M., Narayanan, D., Tilton, J.E., Flinn, J., Walker, K.R.
Agile, Application-Aware Adaptation for Mobility.
In *Proceedings of the Sixteenth ACM Symposium on Operating Systems Principles.* Saint-Malo, France, October, 1997.

[13] Noble, B.D.
Mobile Data Access.
PhD thesis, Carnegie Mellon University, 1998.

[14] Satyanarayanan, M.
Workshop Digest.
In *Proceedings of the Workshop on Mobile Computing Systems and Applications.* Santa Cruz, CA, December, 1994.

[15] Satyanarayanan, M.
Fundamental Challenges in Mobile Computing.
In *Proceedings of the Fifteenth Annual ACM Symposium on Principles of Distributed Computing.* Philadelphia, PA, May, 1996.

[16] Tait, C., Duchamp, D.
Discovery and Exploitation of File Working Sets.
In *Proceedings of the Eleventh International Conference on Distributed Computing Systems.* May, 1991.

[17] Tait, C.D.
A File System for Mobile Computing.
PhD thesis, Columbia University, 1993.

[18] Tait, C., Lei, H., Acharya, S., Chang, H.
Intelligent File Hoarding for Mobile Computers.
In *Proceedings of the First Annual International Conference on Mobile Computing and Networking.* Berkeley, CA, November, 1995.

[19] Tso, M.
Using Property Specifications to Achieve Graceful Disconnected Operation in an Intermittent Mobile Computing Environment.
Technical Report CSL-93-8, XEROX Palo Alto Research Center, 1993.

9.3 Terry, Douglas B., Theimer, Marvin M., Petersen, Karin, Demers, Alan J., Spreitzer, Mike J., and Hauser, Carl H., Managing Update Conflicts in Bayou, A Weakly Connected Replicated Storage System

Terry, D., Theimer, M., Petersen, K., Demers, A., Spreitzer, M., and Hauser, C., "Managing Update Conflicts in a Weakly Connected Replicated Storage System," *Proceedings of the 15th ACM Symposium on Operating System Principles*, pp. 173–183, December 1995.

Managing Update Conflicts in Bayou, a Weakly Connected Replicated Storage System

Douglas B. Terry, Marvin M. Theimer, Karin Petersen, Alan J. Demers, Mike J. Spreitzer and Carl H. Hauser

Computer Science Laboratory
Xerox Palo Alto Research Center
Palo Alto, California 94304 U.S.A.

Abstract

Bayou is a replicated, weakly consistent storage system designed for a mobile computing environment that includes portable machines with less than ideal network connectivity. To maximize availability, users can read and write any accessible replica. Bayou's design has focused on supporting application-specific mechanisms to detect and resolve the update conflicts that naturally arise in such a system, ensuring that replicas move towards eventual consistency, and defining a protocol by which the resolution of update conflicts stabilizes. It includes novel methods for conflict detection, called dependency checks, and per-write conflict resolution based on client-provided merge procedures. To guarantee eventual consistency, Bayou servers must be able to rollback the effects of previously executed writes and redo them according to a global serialization order. Furthermore, Bayou permits clients to observe the results of all writes received by a server, including tentative writes whose conflicts have not been ultimately resolved. This paper presents the motivation for and design of these mechanisms and describes the experiences gained with an initial implementation of the system.

1. Introduction

The Bayou storage system provides an infrastructure for collaborative applications that manages the conflicts introduced by concurrent activity while relying only on the weak connectivity available for mobile computing. The advent of mobile computers, in the form of laptops and personal digital assistants (PDAs) enables the use of computational facilities away from the usual work setting of users. However, mobile computers do not enjoy the connectivity afforded by local area networks or the telephone system. Even wireless media, such as cellular telephony, will not permit continuous connectivity until per-minute costs decline enough to justify lengthy connections. Thus, the Bayou design requires only occasional, pair-wise communication between computers. This model takes into consideration characteristics of mobile computing such as expensive connection time, frequent or occasional disconnections, and that collaborating computers may never be all connected simultaneously [1, 13, 16].

The Bayou architecture does not include the notion of a "disconnected" mode of operation because, in fact, various degrees of "connectedness" are possible. Groups of computers may be partitioned away from the rest of the system yet remain connected to each other. Supporting disconnected workgroups is a central goal of the Bayou system. By relying only on pair-wise communication in the normal mode of operation, the Bayou design copes with arbitrary network connectivity.

A weak connectivity networking model can be accommodated only with weakly consistent, replicated data. Replication is required since a single storage site may not be reachable from mobile clients or within disconnected workgroups. Weak consistency is desired since any replication scheme providing one copy serializability [6], such as requiring clients to access a quorum of replicas or to acquire exclusive locks on data that they wish to update, yields unacceptably low write availability in partitioned networks [5]. For these reasons, Bayou adopts a model in which clients can read and write to any replica without the need for explicit coordination with other replicas. Every computer eventually receives updates from every other, either directly or indirectly, through a chain of pair-wise interactions.

Unlike many previous systems [12, 27], our goal in designing the Bayou system was *not* to provide transparent replicated data support for existing file system and database applications. We believe that applications must be aware that they may read weakly consistent data and also that their write operations may conflict with those of other users and applications. Moreover, applications must be involved in the detection and resolution of conflicts since these naturally depend on the semantics of the application.

To this end, Bayou provides system support for application-specific conflict detection and resolution. Previous systems, such as Locus [30] and Coda [17], have proven the value of semantic conflict detection and resolution for file directories, and several systems are exploring conflict resolution for file and database contents [8, 18, 26]. Bayou's mechanisms extend this work by letting applications exploit domain-specific knowledge to achieve automatic conflict resolution at the granularity of individual update operations without compromising security or eventual consistency.

Automatic conflict resolution is highly desirable because it enables a Bayou replica to remain available. In a replicated system with the weak connectivity model adopted by Bayou, conflicts may be detected arbitrarily far from the users who introduced the conflicts. Moreover, conflicts may be detected when no user is present. Bayou does not take the approach of systems that mark conflicting data as unavailable until a person resolves the conflict. Instead, clients can read data at all times, including data whose conflicts have not been fully resolved either because human intervention is needed or because other conflicting updates may be propagating through the system. Bayou provides interfaces that make the state of a replica's data apparent to the application.

The contributions presented in this paper are as follows: we introduce per-update dependency checks and merge procedures as a general mechanism for application-specific conflict detection and resolution; we define two states of an update, committed and tentative, which relate to whether or not the conflicts potentially introduced by the update have been ultimately resolved; we present mechanisms for managing these two states of an update both from the perspective of the clients and the storage management requirements of the replicas; we describe how replicas move towards eventual consistency; and, finally, we discuss how security is provided in a system like Bayou.

2. Bayou Applications

The Bayou replicated storage system was designed to support a variety of non-real-time collaborative applications, such as shared calendars, mail and bibliographic databases, program development, and document editing for disconnected workgroups, as well as applications that might be used by individuals at different hosts at different times. To serve as a backdrop for the discussion in following sections, this section presents a quick overview of two applications that have been implemented thus far, a meeting room scheduler and a bibliographic database.

2.1 Meeting room scheduler

Our meeting room scheduling application enables users to reserve meeting rooms. At most one person (or group) can reserve the room for any given period of time. This meeting room scheduling program is intended for use after a group of people have already decided that they want to meet in a certain room and have determined a set of acceptable times for the meeting. It does not help them to determine a mutually agreeable place and time for the meeting, it only allows them to reserve the room. Thus, it is a much simpler application than one of general meeting scheduling.

Users interact with a graphical interface for the schedule of a room that indicates which times are already reserved, much like the display of a typical calendar manager. The meeting room scheduling program periodically re-reads the room schedule and refreshes the user's display. This refresh process enables the user to observe new entries added by other users. The user's display might be out-of-date with respect to the confirmed reservations of the room, for example when it is showing a local copy of the room schedule on a disconnected laptop.

Users reserve a time slot simply by selecting a free time period and filling in a form describing the meeting that is being scheduled. Because the user's display might be out-of-date, there is a chance that the user could try to schedule a meeting at a time that was already reserved by someone else. To account for this possibility, users can select several acceptable meeting times rather than just one. At most one of the requested times will eventually be reserved.

A user's reservation, rather than being immediately confirmed (or rejected), may remain "tentative" for awhile. While tentative, a meeting may be rescheduled as other interfering reservations become known. Tentative reservations are indicated as such on the display (by showing them grayed). The "outdatedness" of a calendar does not prevent it from being useful, but simply increases the likelihood that tentative room reservations will be rescheduled and finally "committed" to less preferred meeting times.

A group of users, although disconnected from the rest of the system, can immediately see each other's tentative room reservations if they are all connected to the same copy of the meeting room schedule. If, instead, users are maintaining private copies on their laptop computers, local communication between the machines will eventually synchronize all copies within the group.

2.2 Bibliographic database

Our second application allows users to cooperatively manage databases of bibliographic entries. Users can add entries to a database as they find papers in the library, in reference lists, via word of mouth, or by other means. A user can freely read and write any copy of the database, such as one that resides on his laptop. For the most part, the database is append-only, though users occasionally update entries to fix mistakes or add personal annotations.

As is common in bibliographic databases, each entry has a unique, human-sensible key that is constructed by appending the year in which the paper was published to the first author's last name and adding a character if necessary to distinguish between multiple papers by the same author in the same year. Thus, the first paper by Jones *et al.* in 1995 might be identified as "Jones95" and subsequent papers as "Jones95b", "Jones95c", and so on.

An entry's key is tentatively assigned when the entry is added. A user must be aware that the assigned keys are only tentative and may change when the entry is "committed." In other words, a user must be aware that other concurrent updaters could be trying to assign the same key to different entries. Only one entry can have the key; the others will be assigned alternative keys by the system. Thus, for example, if the user employs the tentatively assigned key in some fashion, such as embedding it as a citation in a document, then he must also remember later to check that the key assigned when the entry was committed is in fact the expected one.

Because users can access inconsistent database copies, the same bibliographic entry may be concurrently added by different users with different keys. To the extent possible, the system detects duplicates and merges their contents into a single entry with a single key.

Interestingly, this is an application where a user may choose to operate in disconnected mode even if constant connectivity were possible. Consider the case where a user is in a university library looking up some papers. He occasionally types bibliographic references into his laptop or PDA. He may spend hours in the library but only enter a handful of references. He is not likely to want to keep a cellular phone connection open for the duration of his visit. Nor will he want to connect to the university's local wireless network and subject himself to student hackers. He will more likely be content to have his bibliographic entries integrated into his database stored by Bayou upon returning to his home or office.

3. Bayou's Basic System Model

In the Bayou system, each *data collection* is replicated in full at a number of *servers*. Applications running as *clients* interact with the servers through the Bayou application programming interface (API), which is implemented as a client stub bound with the application. This API, as well as the underlying client-server RPC protocol, supports two basic operations: *Read* and *Write*. Read operations permit queries over a data collection, while Write operations can insert, modify, and delete a number of data items in a collection. Figure 1 illustrates these components of the Bayou architecture. Note that a client and a server may be co-resident on a host, as would be typical of a laptop or PDA running in isolation.

Access to one server is sufficient for a client to perform useful work. The client can read the data held by that server and submit Writes to the server. Once a Write is accepted by a server, the client has no further responsibility for that Write. In particular, the client does not wait for the Write to propagate to other servers. In other words, Bayou presents a weakly consistent replication model with a *read-any/write-any* style of access. Weakly consistent replication has been used previously for availability, simplicity and scalability in a variety of systems [3, 7, 10, 12, 15, 19].

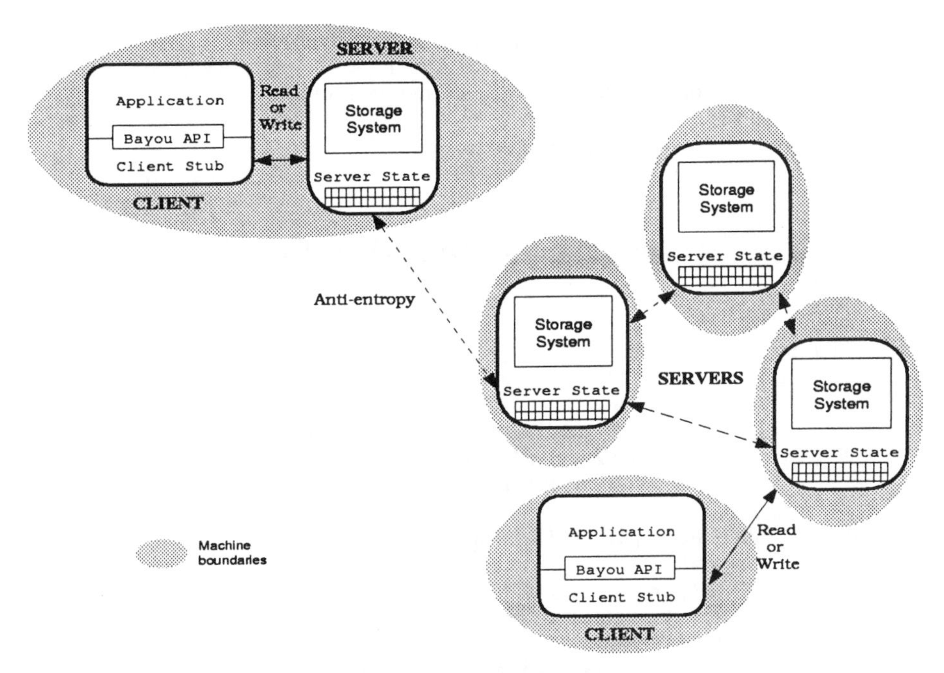

Figure 1. Bayou System Model

While individual Read and Write operations are performed at a single server, clients need not confine themselves to interacting with a single server. Indeed, in a mobile computing environment, switching between servers is often desirable, and Bayou provides *session guarantees* to reduce client-observed inconsistencies when accessing different servers. The description of session guarantees has been presented elsewhere [29].

To support application-specific conflict detection and resolution, Bayou Writes must contain more than a typical file system write or database update. Along with the desired updates, a Bayou Write carries information that lets each server receiving the Write decide if there is a conflict and if so, how to fix it. Each Bayou Write also contains a globally unique *WriteID* assigned by the server that first accepted the Write.

The storage system at each Bayou server conceptually consists of an ordered log of the Writes described above plus the data resulting from the execution of these Writes. Each server performs each Write locally with conflicts detected and resolved as they are encountered during the execution. A server immediately makes the effects of all known Writes available for reading.

In keeping with the goal of requiring as little of the network as possible, Bayou servers propagate Writes among themselves during pair-wise contacts, called *anti-entropy* sessions [7]. The two servers involved in a session exchange Write operations so that when they are finished they agree on the set of Bayou Writes they have seen and the order in which to perform them.

The theory of epidemic algorithms assures that as long as the set of servers is not permanently partitioned each Write will eventually reach all servers [7]. This holds even for communication patterns in which at most one pair of servers is ever connected at once. In the absence of new Writes from clients, all servers will eventually hold the same data. The rate at which servers reach convergence depends on a number of factors including network connectivity, the frequency of anti-entropy, and the policies by which servers select anti-entropy partners. These policies may vary according to the characteristics of the network, the data, and its servers. Developing optimal anti-entropy policies is a research topic in its own right and not further discussed in this paper.

4. Conflict Detection and Resolution

4.1 Accommodating application semantics

Supporting application-specific conflict detection and resolution is a major emphasis in the Bayou design. A basic tenet of our work is that storage systems must provide means for an application to specify its notion of a conflict along with its policy for resolving conflicts. In return, the system implements the mechanisms for reliably detecting conflicts, as specified by the application, and for automatically resolving them when possible. This design goal follows from the observation that different applications have different notions of what it means for two updates to conflict, and that such conflicts cannot always be identified by simply observing conventional reads and writes submitted by the applications.

As an example of application-specific conflicts, consider the meeting room scheduling application discussed in Section 2.1. Observing updates at a coarse granularity, such as the whole-file level, the storage system might detect that two users have concurrently updated different replicas of the meeting room calendar and conclude that their updates conflict. Observing updates at a fine granularity, such as the record level, the system might detect that the two users have added independent records and thereby conclude that their updates do not conflict. Neither of these conclusions are warranted. In fact, for this application, a conflict occurs when two meetings scheduled for the same room overlap in time.

Bibliographic databases provide another example of application-specific conflicts. In this application, two bibliographic entries conflict when either they describe different publications but have been assigned the same key by their submitters or else they describe the same publication and have been assigned distinct keys. Again, this definition of conflicting updates is specific to this application.

The steps taken to resolve conflicting updates once they have been detected may also vary according to the semantics of the application. In the case of the meeting room scheduling application, one or more of a set of conflicting meetings may need to be

```
Bayou_Write (update, dependency_check, mergeproc) {
    IF (DB_Eval (dependency_check.query) <> dependency_check.expected_result)
        resolved_update = Interpret (mergeproc);
    ELSE
        resolved_update = update;
    DB_Apply (resolved_update);
}
```

Figure 2. Processing a Bayou Write Operation

```
Bayou_Write(
    update = {insert, Meetings, 12/18/95, 1:30pm, 60min, "Budget Meeting"},
    dependency_check = {
        query = "SELECT key FROM Meetings WHERE day = 12/18/95
            AND start < 2:30pm AND end > 1:30pm",
        expected_result = EMPTY},
    mergeproc = {
        alternates = {{12/18/95, 3:00pm}, {12/19/95, 9:30am}};
        newupdate = {};
        FOREACH a IN alternates {
            # check if there would be a conflict
            IF (NOT EMPTY (
                SELECT key FROM Meetings WHERE day = a.date
                 AND start < a.time + 60min AND end > a.time))
                    CONTINUE;
            # no conflict, can schedule meeting at that time
            newupdate = {insert, Meetings, a.date, a.time, 60min, "Budget Meeting"};
            BREAK;
        }
        IF (newupdate = {})  # no alternate is acceptable
            newupdate = {insert, ErrorLog, 12/18/95, 1:30pm, 60min, "Budget Meeting"};
        RETURN newupdate;}
)
```

Figure 3. A Bayou Write Operation

moved to a different room or different time. In the bibliographic application, an entry may need to be assigned a different unique key or two entries for the same publication may need to be merged into one.

The Bayou system includes two mechanisms for automatic conflict detection and resolution that are intended to support arbitrary applications: *dependency checks* and *merge procedures*. These mechanisms permit clients to indicate, for each individual Write operation, how the system should detect conflicts involving the Write and what steps should be taken to resolve any detected conflicts based on the semantics of the application. They were designed to be flexible since we expect that applications will differ appreciably in both the procedures used to handle conflicts, and, more generally, in their ability to deal with conflicts.

Techniques for semantic-based conflict detection and resolution have previously been incorporated into some systems to handle special cases such as file directory updates. For example, the Locus [30], Ficus [12], and Coda [17] distributed file systems all include mechanisms for automatically resolving certain classes of conflicting directory operations. More recently, some of these systems have also incorporated support for "resolver" programs that reduce the need for human intervention when resolving other types of file conflicts [18, 26]. Oracle's symmetric replication product also includes the notion of application-selected resolvers for relational databases [8]. Other systems, like Lotus Notes [15], do not provide application-specific mechanisms to handle conflicts, but rather create multiple versions of a document, file, or data object when conflicts arise. As will become apparent from the next couple of sections, Bayou's dependency checks and merge procedures are more general than these previous techniques.

4.2 Dependency checks

Application-specific conflict detection is accomplished in the Bayou system through the use of *dependency checks*. Each Write operation includes a dependency check consisting of an application-supplied query and its expected result. A conflict is detected if the query, when run at a server against its current copy of the data, does not return the expected result. This dependency check is a precondition for performing the update that is included in the Write operation. If the check fails, then the requested update is not performed and the server invokes a procedure to resolve the detected conflict as outlined in Figure 2 and discussed below.

As an example of application-defined conflicts, Figure 3 presents a sample Bayou Write operation that might be submitted by the meeting room scheduling application. This Write attempts to reserve an hour-long time slot. It includes a dependency check with a single query, written in an SQL-like language, that returns information about any previously reserved meetings that overlap with this time slot. It expects the query to return an empty set.

Bayou's dependency checks, like the version vectors and timestamps traditionally used in distributed systems [12, 19, 25, 27], can be used to detect Write-Write conflicts. That is, they can be used to detect when two users update the same data item without one of them first observing the other's update. Such conflicts can be detected by having the dependency check query the current values of any data items being updated and ensure that they have not changed from the values they had at the time the Write was submitted, as is done in Oracle's replicated database [8].

Bayou's dependency checking mechanism is more powerful than the traditional use of version vectors since it can also be used to detect Read-Write conflicts. Specifically, each Write operation can explicitly specify the expected values of any data items on which the update depends, including data items that have been read but are not being updated. Thus, Bayou clients can emulate the optimistic style of concurrency control employed in some distributed database systems [4, 6]. For example, a Write operation that installs a new program binary file might only include a dependency check of the sources, including version stamps, from which it was derived. Since the binary does not depend on its previous value, this need not be included.

Moreover, because dependency queries can read any data in the server's replica, dependency checks can enforce arbitrary, multi-item integrity constraints on the data. For example, suppose a Write transfers $100 from account A to account B. The application, before issuing the Write, reads the balance of account A and discovers that it currently has $150. Traditional optimistic concurrency control would check that account A still had $150 before performing the requested Write operation. The real requirement, however, is that the account have at least $100, and this can easily be specified in the Write's dependency check. Thus, only if concurrent updates cause the balance in account A to drop below $100 will a conflict be detected.

4.3 Merge procedures

Once a conflict is detected, a *merge procedure* is run by the Bayou server in an attempt to resolve the conflict. Merge procedures, included with each Write operation, are general programs written in a high-level, interpreted language. They can have embedded data, such as application-specific knowledge related to the update that was being attempted, and can perform arbitrary Reads on the current state of the server's replica. The merge procedure associated with a Write is responsible for resolving any conflicts detected by its dependency check and for producing a revised update to apply. The complete process of detecting a conflict, running a merge procedure, and applying the revised update, shown in Figure 2, is performed atomically at each server as part of executing a Write.

In principle, the algorithm in Figure 2 could be imbedded in each merge procedure, thereby eliminating any special mechanisms for dependency checking. This approach would require servers to create a new merge procedure interpreter to execute each Write, which would be overly expensive. Supporting dependency checks separately allows servers to avoid running the merge procedure in the expected case where the Write does not introduce a conflict.

The meeting room scheduling application provides good examples of conflict resolution procedures that are specific not only to a particular application but also to a particular Write operation. In this application, users, well aware that their reservations may be invalidated by other concurrent users, can specify alternate scheduling choices as part of their original scheduling updates. These alternates are encoded in a merge procedure that attempts to reserve one of the alternate meeting times if the original time is found to be in conflict with some other previously scheduled meeting. An example of such a merge procedure is illustrated in Figure 3. A different merge procedure altogether could search for the next available time slot to schedule the meeting, which is an option a user might choose if any time would be satisfactory.

In practice, Bayou merge procedures are written by application programmers in the form of templates that are instantiated with the appropriate details filled in for each Write. The users of applications do not have to know about merge procedures, and therefore about the internal workings of the applications they use, except when automatic conflict resolution cannot be done.

In the case where automatic resolution is not possible, the merge procedure will still run to completion, but is expected to produce a revised update that logs the detected conflict in some fashion that will enable a person to resolve the conflict later. To enable manual resolution, perhaps using an interactive merge tool [22], the conflicting updates must be presented to a user in a manner that allows him to understand what has happened. By convention, most Bayou data collections include an error log for unresolvable conflicts. Such conventions, however, are outside the domain of the Bayou storage system and may vary according to the application.

In contrast to systems like Coda [18] or Ficus [26] that lock individual files or complete file volumes when conflicts have been detected but not yet resolved, Bayou allows replicas to always remain accessible. This permits clients to continue to Read previously written data and to continue to issue new Writes. In the meeting room scheduling application, for example, a user who only cares about Monday meetings need not concern himself with scheduling conflicts on Wednesday. Of course, the potential drawback of this approach is that newly issued Writes may depend on data that is in conflict and may lead to cascaded conflict resolution.

Bayou's merge procedures resemble the previously mentioned resolver programs, for which support has been added to a number of replicated file systems [18, 26]. In these systems, a file-type-specific resolver program is run when a version vector mismatch is detected for a file. This program is presented with both the current and proposed file contents and it can do whatever it wishes in order to resolve the detected conflict. An example is a resolver program for a binary file that checks to see if it can find a specification for how to derive the file from its sources, such as a Unix makefile, and then recompiles the program in order to obtain a new, "resolved" value for the file. Merge procedures are more general since they can vary for individual Write operations rather than being associated with the type of the updated data, as illustrated above for the meeting room scheduling application.

5. Replica Consistency

While the replicas held by two servers at any time may vary in their contents because they have received and processed different Writes, a fundamental property of the Bayou design is that all servers move towards *eventual consistency*. That is, the Bayou system guarantees that all servers *eventually* receive all Writes via the pair-wise anti-entropy process and that two servers holding the same set of Writes will have the *same* data contents. However, it cannot enforce strict bounds on Write propagation delays since these depend on network connectivity factors that are outside of Bayou's control.

Two important features of the Bayou system design allows servers to achieve eventual consistency. First, Writes are performed in the same, well-defined order at all servers. Second, the conflict detection and merge procedures are deterministic so that servers resolve the same conflicts in the same manner.

In theory, the execution history at individual servers could vary as long as their execution was *equivalent to* some global Write ordering. For example, Writes known to be commutative could be performed in any order. In practice, because Bayou's Write operations include arbitrary merge procedures, it is effectively impossible either to determine whether two Writes commute or to transform two Writes so they can be reordered as has been suggested for some systems [9].

When a Write is accepted by a Bayou server from a client, it is initially deemed *tentative*. Tentative Writes are ordered according to timestamps assigned to them by their accepting servers. Eventually, each Write is *committed*, by a process described in the next section. Committed Writes are ordered according to the times at which they commit and before any tentative Writes.

The only requirement placed on timestamps for tentative Writes is that they be monotonically increasing at each server so that the pair <timestamp, ID of server that assigned it> produce a total order on Write operations. There is no requirement that servers have synchronized clocks, which is crucial since trying to ensure clock synchronization across portable computers is problematic. However, keeping servers' clocks reasonably close is desirable so that the induced Write order is consistent with a user's perception of the order in which Writes are submitted. Bayou servers maintain logical clocks [20] to timestamp new Writes. A server's logical clock is generally synchronized with its real-time system clock, but, to preserve the causal ordering of Write operations, the server may need to advance its logical clock when Writes are received during anti-entropy.

Enforcing a global order on tentative, as well as committed, Writes ensures that an isolated cluster of servers will come to agreement on the tentative resolution of any conflicts that they encounter. While this is not strictly necessary since clients must be prepared to deal with temporarily inconsistent servers in any case, we believe it desirable to provide as much internal consistency as possible. Moreover, clients can expect that the tentative resolution of conflicts within their cluster will correspond to their eventual permanent resolution, provided that no further conflicts are introduced outside the cluster.

Because servers may receive Writes from clients and from other servers in an order that differs from the required execution order, and because servers immediately apply all known Writes to their replicas, servers must be able to undo the effects of some previous tentative execution of a Write operation and reapply it in a different order. Interestingly, the number of times that a given Write operation is re-executed depends only on the order in which Writes arrive via anti-entropy and not on the likelihood of conflicts involving the Write.

Conceptually, each server maintains a log of all Write operations that it has received, sorted by their committed or tentative timestamps, with committed Writes at the head of the log. The server's current data contents are generated by executing all of the Writes in the given order. Techniques for pruning a server's Write log and for efficiently maintaining the corresponding data contents by undoing and redoing Write operations are given in Section 7.

Bayou guarantees that merge procedures, which execute independently at each server, produce consistent updates by restricting them to depend only on the server's current data contents and on any data supplied by the merge procedure itself. In particular, a merge procedure cannot access time-varying or server-specific "environment" information such as the current system clock or server's name. Moreover, merge procedures that fail due to exceeding their limits on resource usage must fail deterministically. This means that all servers must place uniform bounds on the CPU and memory resources allocated to a merge procedure and must consistently enforce these bounds during execution. Once these conditions are met, two servers that start with identical replicas will end up with identical replicas after executing a Write.

6. Write Stability and Commitment

A Write is said to be *stable* at a server when it has been executed for the last time by that server. Recall that as servers learn of new updates by performing anti-entropy with other servers, the effects of previously executed Write operations may need to be undone and the Writes re-executed. Thus, a given Write operation may be executed several times at a server and may produce different results depending on the execution history of the server. A Write operation becomes stable when the set of Writes that precede it in the server's Write log is fixed. This means that the server has already received and executed any Writes that could possibly be ordered before the given Write. Bayou's notion of stability is similar to that in ordered multicast protocols, such as those provided in the ISIS toolkit [2].

In many cases, an application can be designed with a notion of "confirmation" or "commitment" that corresponds to the Bayou notion of stability. As an example, in the Bayou meeting room scheduling application, two users may try to schedule separate meetings for the same time in the same room. Only when one of the users discovers that his Write has become stable and his schedule still shows that he has reserved the room for the desired time, can he be sure that his tentative reservation has been confirmed.

Since clients may want to know when a Write has stabilized, the Bayou API provides means for inquiring about the stability of a specific Write. Given a Write's unique identifier, a client can ask a server whether the given Write is stable at the server. The answer may vary, of course, depending on which server is contacted. Bayou also provides support for clients that may choose to access only stable data.

How does a server determine whether a Write is stable? One approach would be to have each server include in the information passed during anti-entropy not only any Writes that have been accepted by this server but also the current value of the clock that it uses to timestamp new Writes. With suitable assumptions about the propagation order of Writes, a server could then determine that a Write is stable when it has a lower timestamp than all servers' clocks. The main drawback of this approach is that a server that remains disconnected can prevent Writes from stabilizing, which could cause a large number of Writes to be rolled back when the server reconnects.

To speed up the rate at which updates stabilize in an environment where communication with some servers may not be possible for extended periods of time, the Bayou system uses a *commit* procedure. That is, a Write becomes stable when it is explicitly committed, and, in fact, we generally use the terms "stable" and "committed" interchangeably in the Bayou system. Committed Writes, in commit order, are placed ahead of any tentative Writes in each server's Write log. This, along with Bayou's anti-entropy protocol ensuring that servers learn of committed Writes in the order that they were committed, provides stability.

In the Bayou system, we use a *primary commit* scheme [28]. That is, one server designated as the *primary* takes responsibility for committing updates. Knowledge of which Writes have committed and in which order they were committed then propagates to other servers during anti-entropy. In all other respects, the primary behaves exactly like any other server. Each replicated data collection can have a different server designated as its primary.

Any commit protocol that prevents different groups of servers from committing updates in different orders would meet Bayou's needs. In our anticipated weak connectivity environment, using a primary to commit data is more attractive than the standard two-phase commit protocol since it alleviates the need to gather a majority quorum of servers. Consider the case of data that is replicated among laptops that are mostly disconnected. Requiring a majority of these laptops to be in communication with each other

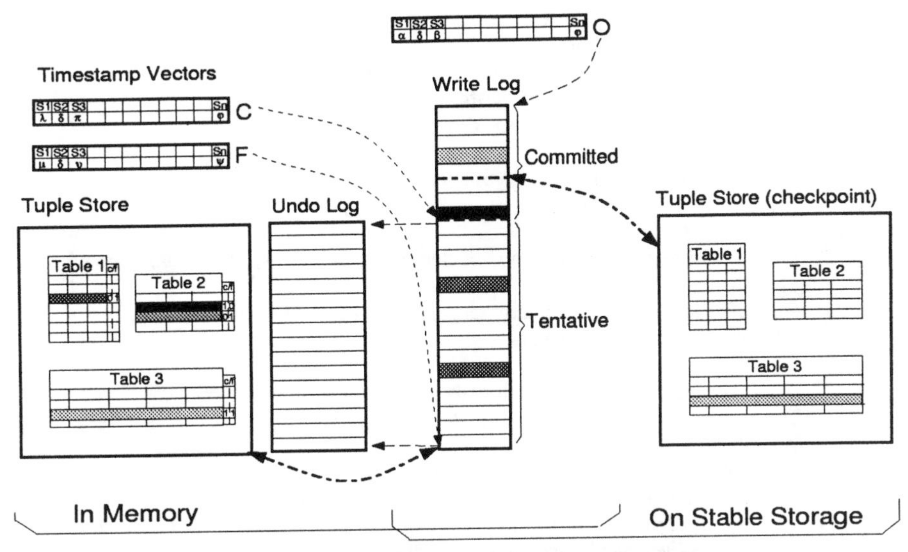

Figure 4. Bayou Database Organization

at the same time in order to commit updates would be unreasonable.

The primary commit approach also enables updates to commit on a disconnected laptop that acts as the primary server. For example, suppose a user keeps the primary copy of his calendar with him on his laptop and allows others, such as a spouse or secretary, to keep secondary, mostly read-only copies. In this case, the user's updates to his own calendar commit immediately. This example illustrates how one might choose the primary to coincide with the locus of update activity, thereby maximizing the rate at which Writes get committed.

Unlike other distributed storage systems in which the ability to commit data is of primary importance, the Bayou design readily accommodates the temporary unavailability of the primary. The inability of a client to communicate with the primary server, for instance if the primary crashes or is disconnected, does not prevent it from performing useful Read and Write operations. Writes accepted by other servers simply remain tentative until they eventually reach the primary.

Bayou tries to arrange, but cannot ensure, that the order in which Writes are committed is consistent with the tentative order indicated by their timestamps. Writes from a given server are committed in timestamp order. Writes from different servers, however, may commit in a different order based on when the servers perform anti-entropy with the primary and with each other. Writes held on a disconnected non-primary server, for instance, will commit only after the server reconnects to the rest of the system and could be committed after Writes with later timestamps.

7. Storage System Implementation Issues

The Bayou design places several demands on the underlying storage system used by each server including the need for space-efficient Write logging, efficient undo/redo of Write operations, separate views of committed and tentative data, and support for server-to-server anti-entropy. We implemented a storage system tailored to these special needs.

Our implementation is factored into three main components as shown in Figure 4: the *Write Log*, the *Tuple Store*, and the *Undo Log*. The Write Log contains Writes that have been received by a Bayou server, sorted by their global committed or tentative order. The server's Tuple Store is a database that is obtained by executing the Writes in order and is used to process Read requests. The Undo Log facilitates rolling back tentative Writes that have been applied to the Tuple Store so that they can be re-executed in a different order, such as when a newly received Write gets inserted into the middle of the Write Log or when existing Writes get reordered through the commit process.

The Write Log conceptually contains all Writes ever received by the server, as discussed in Section 5. In practice, a server can discard a Write from the Write Log once it becomes stable, since by definition the server will never need to rollback and re-execute a stable Write. Bayou servers do, in fact, hold onto a few recently committed Writes to facilitate incremental anti-entropy, the details of which are beyond the scope of this paper. Thus, the Write Log is actually an ordered set of Writes containing a tail of the committed Writes and all tentative Writes known to the server.

Each server must keep track of which Writes it has received but are no longer explicitly held in its Write Log. This is to ensure that the server does not re-accept the same Writes from another server during anti-entropy. Each server maintains a timestamp vector, called the "O vector", to indicate in a compact way the "omitted" prefix of committed Writes. This O vector records, for each server, the timestamp of the latest Write from the given server that has been discarded. A single timestamp vector can precisely characterize the set of discarded Writes because: (1) servers discard a prefix of their Write Log, and (2) Writes that originate from any given server propagate and get committed in timestamp order.

The Tuple Store we implemented is an in-memory relational database, providing query processing in a subset of SQL, local transaction support, and some integrity constraints. Requiring a database to fit in virtual memory is, admittedly, a practical limitation in our current implementation, but is not intrinsic to the overall Bayou design. The Tuple Store, and its associated language for specifying queries and updates, is the principal place in the Bayou architecture where the issue of data model arises. We chose the relational model for our initial prototype because of its power and flexibility. It naturally supports fine-grain access to specific fields of tuples as well as queries and updates to all tuples in the database.

A unique aspect of the Tuple Store is that it must support the two distinct views of a Bayou database that are of interest to clients: *committed* and *full*. When a Write is tentative, its effect appears in the full view but not in the committed view. Once the Write has been committed, its effect appears in both views. A ten-

```
Receive_Writes (writeset, received_from) {
    IF (received_from = CLIENT) {
        # Received one write from the client, insert at end of WriteLog
        # first increment the server's timestamp
        logicalclock = MAX(systemclock, logicalclock + 1);
        write = First(writeset);
        write.WID = {logicalclock, myServerID};
        write.state = TENTATIVE;
        WriteLog_Append(write);
        Bayou_Write(write.update, write.dependency_check, write.mergeproc);
    } ELSE {
        # Set of writes received from another server during anti-entropy,
        # therefore writeset is ordered
        write = First(writeset);
        insertionPoint = WriteLog_IdentifyInsertionPoint(write.WID);
        TupleStore_RollbackTo(insertionPoint);
        WriteLog_Insert(writeset);
        # Now roll forward
        FOREACH write IN WriteLog AFTER insertionPoint DO
            Bayou_Write(write.update, write.dependency_check, write.mergeproc);
        # Maintain the logical clocks of servers close
        write = Last(writeset);
        logicalclock = MAX(logicalclock, write.WID.timestamp);
    }
}
```

Figure 5. Applying Sets of Bayou Writes to the Database

tative deletion may result in a tuple that appears in the committed view but not in the full view. For many servers, certainly those that communicate regularly with the primary, the committed and full views will be nearly identical. However, neither view is a subset of the other.

Our Tuple Store maintains the union of the two views. Each tuple is tagged with a 2-bit characteristic vector identifying the set of views that contain it. The bits of all tuples affected by a Write get set when the Write is applied to the Tuple Store. Therefore, re-executing a Write when it gets committed is necessary so that all corresponding committed bits get set appropriately. Our query processor respects and propagates these bits, so that in the result of a query each tuple is tagged with the views for which that tuple would be produced if the identical query were run conventionally. Propagating these bits through a relational algebra query is straightforward. Assuming the tentative and committed views are nearly identical, this technique reduces the space occupied by the Tuple Store, compared to maintaining two separate full and committed databases, by nearly a factor of two without substantially increasing the query processing cost. In addition, our query processor can easily guarantee that identical tuples occurring in the two views of a query result will always be merged and delivered as a single tuple with both bits in the characteristic vector set. This makes it convenient for clients to base decisions on the *difference* between the two views without having to merge the results of independent queries.

To support anti-entropy efficiently, the running state of each server also includes two timestamp vectors that represent the committed and full views. The "C vector" characterizes the state of the Tuple Store after executing the last committed Write in the Write Log while the "F vector" characterizes the state after executing the last tentative Write in the Write Log, that is, the current Tuple Store. These timestamp vectors are *not* used for conflict detection; they simply enable server pairs to identify precisely the sets of Writes that need to be exchanged during anti-entropy.

The Undo Log permits a server to undo any effects on the Tuple Store of Writes performed after a given position in the Write Log. As each new Write is received via anti-entropy, a server inserts it into its Write Log. Newly committed Writes are inserted immediately following the current set of committed Writes known to the server, which may in turn require that some of these Writes be removed from their previous tentative positions in the Write Log. After all the Writes have been received, the server uses its Undo Log to roll back its Tuple Store to a state corresponding to the position where the first newly received Write was inserted. It then enumerates and executes all following Writes from the Write Log, bringing its Tuple Store and Undo Log up-to-date. This procedure is illustrated in Figure 5.

For crash recovery purposes, both the full Write Log and a checkpoint of the Tuple Store are maintained in stable storage, while for performance the Write Log and the current Tuple Store are maintained in memory as shown in Figure 4. The Undo Log is maintained only in memory. The stable checkpoint of the Tuple Store reflects only a prefix of the committed Writes. This checkpoint must contain the effects of any Writes that have been truncated from the Write Log. At all times, a valid Tuple Store can be recovered by reading this checkpoint and applying a suffix of the Write Log to it. Thus, to make the database recoverable, Bayou stably records the unique identifier of the last Write reflected in the Tuple Store checkpoint (making it possible to identify the correct suffix of the Write Log) and makes the Write Log itself recoverable using conventional techniques for logging high-level changes to the Write Log.

8. Access Control

Providing access control and authentication in Bayou posed interesting challenges because of our minimal connectivity assumptions. In particular, the design cannot rely on an online, trusted authentication server [23] to mediate the establishment of secure channels between a client and server or between two Bayou servers. As an example, suppose two users holding Bayou replicas on their portable computers are in a meeting together. Before performing anti-entropy, each of the two mutually suspicious servers must verify that the other is authorized to manage the data. Similarly, if one machine simply wants to act as a client for the data stored on the other, it will want to make sure that the server is legitimate and then must prove that it is authorized to access the data.

The access control model currently implemented in the Bayou system provides authorization at the granularity of a whole data collection, which is the unit of replication. A user may be granted Read and Write privileges to a data collection. A user may also be granted "Server" privileges to maintain a replica of the data on his workstation or portable computer, that is, to run a server for the data collection. Enabling servers to run on mobile platforms radically departs from the notion of physically protected servers.

Mutual authentication and access control in Bayou is based on public-key cryptography. Every user possesses a public/private key pair and a set of digitally signed *access control certificates* granting him access to various data collections. Client applications and Bayou servers operate on behalf of users and obtain the key pair and access control certificates from the corresponding user at start-up time. Currently, we use a single trusted signing authority with a well-known public key to sign all access-granting certificates, though moving to a hierarchy or web of signing authorities would not be difficult.

Bayou uses three types of certificates to grant, delegate and revoke access to a data collection:

- AC[PU, P, D] - certificate that grants privilege P (one of Read, Write, or Server) on data collection D to the user whose public key is PU. AC certificates are signed by the well-known signing authority.
- D[PU, C, PY] - certificate signed by the user whose public key is PY to delegate his privileges encoded in certificate C to another user whose public key is PU.
- R[C, PY] - certificate signed by the user whose public key is PY to revoke some user's privileges encoded in certificate C; the user whose public key is PY must also have originally signed certificate C.

Revocation certificates are stored by Writes to, and hence propagated with, the data collections to which they apply. Certificates that revoke server privileges may also be kept by client users to ensure protection against malicious servers. Users maintain a *certificate purse*, which applications running under their identity can both read and append to.

For a server to determine whether a client has some privilege for the server's data, the server first authenticates the client's identity using a challenge/response protocol. The client also hands the server a certificate that asserts the privilege in question. The server must verify that the certificate is legitimate, that the certificate and any enclosed certificates for a delegation have not been revoked to the server's knowledge, and that it grants the necessary access rights. Server-to-client and server-to-server authentication and access control checking is done in a similar fashion. The establishment of mutual trust between a server and a client is performed at the beginning of a *secure session* and covers all Read or Write operations performed as part of that session. A server will preempt the session if it is notified of a revocation that affects a certificate associated with the session.

For Write operations, the submitter's access rights are checked once by the accepting server, and then again at the primary when the Write is committed. Servers, other than the primary, when receiving a Write during anti-entropy trust that the accepting server has correctly checked the user's privileges and rejected Writes with unsuitable access rights. This level of trust is reasonable since a server ensures that any server with which it performs anti-entropy is authorized to hold a replica of the data collection.

Having access controls checked for a second time at the primary server ensures that revocations of Write privileges can be applied at the primary and guarantees that any "bad" Write attempting to commit after such a revocation will be rejected. In particular, revocation of Write access for a malicious user can be enforced without having to ensure that every server to which such a user could connect has been notified of the revocation.

Even though a Write's merge procedure may perform different Read operations on the data and perform different updates when it is executed at different times, checking access control once is sufficient because of the whole-data-collection access control model. More fine-grained access control would require careful design modifications.

9. Status and Experience

The implementation of the Bayou system has two distinguishable components: the client stub and the server. The client stub is a runtime library linked into applications that use Bayou for storage management. It provides mechanisms for server location, session guarantees, secure sessions, Read and Write operations, and miscellaneous utilities. The server implements the Bayou storage management including the mechanisms for conflict detection and resolution, server to server communication, and persistent database management. Bayou's implementation is Posix compliant and developed in ANSI C so that the same sources run on Intel-based laptops with Linux and on our regular development platform of Sun SPARCstations with SunOS.

In the current implementation, ILU [14], a language-independent RPC package developed at Xerox PARC, is used for communication between Bayou clients and servers, as well as between servers. Server location, by both clients and other servers, uses a simple decentralized registration and lookup service for key-value pairs that are made visible across a network via multicast. Bayou merge procedures are Tcl programs [24] that are run in a Tcl interpreter modified to enforce the limits described in Section 5. We foresee that these components may change as the system evolves.

The two running applications have demonstrated how to use Bayou's conflict detection and resolution mechanisms effectively. Interestingly, one of the lessons we learned immediately from these applications was that the Bayou server had to supply a per-database library mechanism for Tcl code invoked by the merge procedures. Otherwise, Writes are bloated by the large amount of repeated code in their merge procedures. For both the meeting room scheduler and the shared bibliographic database manager only two of roughly 100 lines of Tcl in the original merge procedures changed from one Write to another.

The performance of Bayou depends on several factors, such as the schema of the data being stored, the amount of data stored at a server, the location of clients and servers, and the platforms on which the components are running. This section shows how Bayou performs for a particular instance of the system: a server and client for the bibliographic database described in Section 2. The database is composed of a single table containing 1550 tuples, obtained from a bibtex source [21]. Each tuple was inserted into the database with a single Bayou Write operation. Results are presented for five different configurations of the database characterized by the number of Writes that are tentative. For each configuration we measured storage requirements and the execution times for three operations in the system: undoing and redoing the effect of all tentative Writes, executing a client Read operation against the database, and adding a new Write to the database.

Table 1: Size of Bayou Storage System for the Bibliographic Database with 1550 Entries
(sizes in Kilobytes)

Number of Tentative Writes	0 (none)	50	100	500	1550 (all)
Write Log	9	129	259	1302	4028
Tuple Store Ckpt	396	384	371	269	1
Total	**405**	**513**	**630**	**1571**	**4029**
Factor to 368K bibtex source	1.1	1.39	1.71	4.27	10.95

Table 2: Performance of the Bayou Storage System for Operations on Tentative Writes in the Write Log
(times in milliseconds with standard deviations in parentheses)

Tentative Writes	0	50		100		500		1550	
	Server running on a Sun SPARC/20 with Sunos								
Undo all	0	31	(6)	70	(20)	330	(155)	866	(195)
(avg. per Write)		.62		.7		.66		.56	
Redo all	0	237	(85)	611	(302)	2796	(830)	7838	(1094)
(avg. per Write)		4.74		6.11		5.59		5.05	
	Server running on a Gateway Liberty Laptop with Linux								
Undo all	0	47	(3)	104	(7)	482	(15)	1288	(62)
(avg. per Write)		.94		1.04		.96		.83	
Redo all	0	302	(91)	705	(134)	3504	(264)	9920	(294)
(avg. per Write)		6.04		7.05		7.01		6.4	

Table 3: Performance of the Bayou Client Operations
(times in milliseconds with standard deviations in parentheses)

Server Client	Sun SPARC/20 same as server		Gateway Liberty same as server		Sun SPARC/20 Gateway Liberty	
Read: 1 tuple	27	(19)	38	(5)	23	(4)
100 tuples	206	(20)	358	(28)	244	(10)
Write: no conflict	159	(32)	212	(29)	177	(22)
with conflict	207	(37)	372	(17)	223	(40)

Table 1 shows that the size of a tentative Write for this database is about 10 times that of a committed Write. Over half of a tentative Write's size is taken by the access control certificate required for security. The server's storage requirements decrease significantly as data gets committed. When most of the Writes in the database are committed, its size is almost identical to that of the bibtex file from which the data was obtained.

Table 2 illustrates the execution times for a Bayou server to undo and then redo all Writes that are tentative in each configuration of the bibliographic database. Each result corresponds to the average over 100 executions of the undo/redo operations. The cost incurred by the server is a function of the number of Writes being undone and redone. While in general the size of the Tuple Store may affect the performance of executing a Write, the cost of redoing each tentative Write for this database is close to constant because dependency checks are selections on the database's primary key index, and are therefore independent of the Tuple Store size. The standard deviations on the Sun tend to be higher than those for the laptop since the Sun workstation was running a much higher workload of other applications than the laptop.

Table 3 shows the performance of client/server interactions for the bibliographic database. Measurements were taken for three computing platform combinations: a Bayou server and bibliographic database client running on the same Sun SPARCstation/20, both server and client running on the same Gateway Liberty laptop, and, finally, the server running on the Sun SPARCstation/20 and the client running on the Gateway Liberty. The numbers for both Reads and Writes include the costs of session guarantee management, the RPC proper, and a database query, which for Writes is part of the dependency check. Additionally, Writes require two file system synchronization operations and, in case of conflict, the execution of the merge procedure, which runs another database query. For Writes involving conflicts, the bibliographic entry key presented in each Write is not unique and, hence, must be reassigned in the merge procedure as discussed in Section 2.2; the key presented is changed after each set of five Write operations. Because Reads operate on the in-memory Tuple Store, running selections on the primary key, and Writes are appended to the Write Log, execution times across the different database configurations vary little. Hence we present the combined average of the 500 executions of each operation over all configurations.

10. Conclusions

Bayou is a storage infrastructure for mobile applications that relies only on weak connectivity assumptions. To cope with arbitrary network partitions, the system is built around pair-wise client-server and server-server communications. To provide high availability, Bayou employs weakly consistent replication where clients are able to connect to any available server to perform Reads and Writes. Support for automatic conflict detection and resolution enables applications to deal with concurrent updates effectively. The system guarantees eventual consistency by ensuring that all updates eventually propagate to all servers, that servers perform updates in a global order, and that any update conflicts are resolved in a consistent manner at all servers.

Bayou's management of update conflicts differs significantly from previous replicated systems, including file systems like Coda [18] and Ficus [26] as well as Oracle's recent commercial database offering [8], in the following main areas:

Non-transparency. Previous systems have tried to support existing file and database applications by detecting and resolving conflicts without the applications' knowledge. In contrast, Bayou adopts the philosophy that applications must be aware of and integrally involved in conflict detection and resolution. Bayou applications can take advantage of the semantics of their data to minimize false conflict detections and maximize the ability to resolve detected conflicts automatically.

Application-specific conflict detection. File systems that rely on version vectors and database systems that employ optimistic concurrency control detect update conflicts by observing clients' Reads and Writes. Using application-provided rules for detecting conflicts, called dependency checks, Bayou can detect a wider class of conflicts, particularly those that depend on application semantics.

Per-write conflict resolvers. Whereas Coda, Ficus, and Oracle all permit clients to write custom procedures to resolve conflicts, these resolvers are stored within the system and invoked based on the type of the file or data in conflict. In Bayou, each Write operation includes, in addition to the desired update and dependency check, a merge procedure that gets executed if the Write is determined to have caused a conflict. The power of this approach is demonstrated in applications, such as Bayou's meeting room scheduler, where the merge procedure varies for each Write.

Partial and multi-object updates. Bayou's Write operations can atomically perform insertions, partial modifications, and deletions to one or more data objects. This means that, unlike systems with a whole-file update model where storing the most recent data contents is sufficient, Bayou servers must apply every Write operation. Techniques have been devised in Bayou for propagating, ordering, and undoing/redoing Write operations to ensure eventual consistency for arbitrary updates and conflict resolution procedures. These techniques are needed not only for relational database models, as in the current Bayou system, but also for file systems supporting record-level updates and multi-file atomic transactions.

Tentative and stable resolutions. The Bayou system is novel in maintaining both full and committed views of the data while permitting clients to read either. Both clients and servers in Bayou want to know when any conflicts involving a Write have been fully resolved. Committing a Write ensures that its outcome is stable, including the resolution of conflicts involving the Write. The rate at which Writes stabilize is independent of the probability of conflict. In keeping with the goal of minimal connectivity requirements, Bayou commits Writes using a primary server.

Security. Bayou executes each merge procedure within a secure environment in which the only allowable external actions are reading and writing data using the access credentials of the user who submitted the conflicting Write. Public-key, digitally-signed certificates permit authentication and access control outside the presence of an authentication server or user.

Applications that can best utilize Bayou's replication scheme are those for which reading weakly consistent, tentative data is acceptable and for which the chance of update conflicts is low or the success of automatic resolution is high. Provided that the penalty for conflict is not excessive, humans would rather deal with the occasional unresolvable conflict than incur the adverse impact on availability inherent in systems that avoid conflicts altogether, such as those based on pessimistic locking. A number of shared databases, such as phone books and bulletin boards, meet these characteristics, as do many asynchronous collaborative applications [22].

We have built an initial version of the Bayou system and our measurements indicate that its performance and overhead are acceptable. In particular, running Bayou servers and applications on today's laptop computers is reasonable. Our measurements also confirm that much of the extra overhead imposed by Bayou's heavier-weight Write operations is present only so long as a Write operation is tentative. Committed data is no more expensive than in other, simpler storage systems. We are also building a number of applications on top of Bayou and experimenting with them to gain better insights into their needs.

Issues we are planning to explore further in the context of Bayou include partial replication, policies for choosing servers for anti-entropy, building servers with conventional database managers, alternate data models, and finer grain access control. Our current focus is on supporting partial replicas that contain subsets of a data collection, which is important for some laptop-based applications and raises a number of difficult problems ranging from characterizing a partial replica to resolving conflicts in a consistent manner across partial replicas. The next steps in the implementation will include the development of other applications, such as an e-mail reader, porting refdbms [11], a widely used shared bibliographic database manager, to run on the Bayou storage system, and experimenting with wireless connectivity for servers and clients running on a laptop.

11. Acknowledgments

The Bayou design has benefitted from discussions with a number of colleagues, including our fellow PARC researchers and Tom Anderson, Mary Baker, Brian Bershad, Hector Garcia-Molina, and Terri Watson. We especially thank Brent Welch for his technical contributions in the early stages of the Bayou project. Atul Adya and Xinhua Zhou helped implement the first Bayou applications, from which we learned a tremendous amount. Surendar Chandra contributed significantly to making the network environment on our laptops work. Sue Owicki helped guide the final revisions to this paper. Mark Weiser and Craig Mudge, as managers of the Computer Science Lab, have been supportive throughout.

12. References

[1] R. Alonso and H. F. Korth. Database system issues in nomadic computing. *Proceedings ACM SIGMOD International Conference on Management of Data*, Washington, D.C., May 1993, pages 388-392.

[2] K. Birman, A. Schiper, and P. Stephenson. Lightweight, causal and atomic group multicast. *ACM Transactions on Computer Systems* 9(3):272-314, August 1991.

[3] A. Birrell, R. Levin, R. M. Needham, and M. D. Schroeder. Grapevine: An exercise in distributed computing. *Communications of the ACM* 25(4):260-274, April 1982.

[4] M. J. Carey and M. Livny. Conflict detection tradeoffs for replicated data. *ACM Transactions on Database Systems* 16(4):703-746, December 1991.

[5] B. A. Coan, B. M. Oki, and E. K. Kolodner. Limitations on database availability when networks partition. *Proceedings Fifth ACM Symposium on Principles of Distributed Computing*, Calgary, Alberta, Canada, August 1986, pages 187-194.

[6] S. Davidson, H. Garcia-Molina, and D. Skeen. Consistency in a partitioned network: A survey. *ACM Computing Surveys* 17(3):341-370, September 1985.

[7] A. Demers, D. Greene, C. Hauser, W. Irish, J. Larson, S. Shenker, H. Sturgis, D. Swinehart, and D. Terry. Epidemic algorithms for replicated database maintenance. *Proceedings Sixth Symposium on Principles of Distributed Computing*, Vancouver, B.C., Canada, August 1987, pages 1-12.

[8] A. Downing. Conflict resolution in symmetric replication. *Proceedings European Oracle User Group Conference*, Florence, Italy, April 1995, pages 167-175.

[9] C. Ellis and S. Gibbs. Concurrency control in groupware systems. *Proceedings ACM SIGMOD International Conference on Management of Data*, Portland, Oregon, June 1989, pages 399-407.

[10] R. A. Golding. A weak-consistency architecture for distributed information services. *Computing Systems* 5(4):379-405, Fall 1992.

[11] R. Golding, D. Long, and J. Wilkes. The refdbms distributed bibliographic database system. *Proceedings Winter USENIX Conference,* San Francisco, California, January 1994, pages 47-62.

[12] R.G. Guy, J.S. Heidemann, W. Mak, T.W. Page, Jr., G.J. Popek, and D. Rothmeier. Implementation of the Ficus replicated file system. *Proceedings Summer USENIX Conference*, June 1990, pages 63-71.

[13] T. Imielinski and B. R. Badrinath. Mobile wireless computing: Challenges in data management. *Communications of the ACM* 37(10):18-28, October 1994.

[14] B. Janssen and M. Spreitzer. *Inter-Language Unification - ILU.* ftp://ftp.parc.xerox.com/pub/ilu/ilu.html.

[15] L. Kalwell Jr., S. Beckhardt, T. Halvorsen, R. Ozzie, and I. Greif. Replicated document management in a group communication system. In *Groupware: Software for Computer-Supported Cooperative Work*, edited by D. Marca and G. Bock, IEEE Computer Society Press, 1992, pages 226-235.

[16] J. J. Kistler and M. Satyanarayanan. Disconnected operation in the Coda file system. *ACM Transactions on Computer Systems* 10(1): 3-25, February 1992.

[17] P. Kumar and M. Satyanarayanan. Log-based directory resolution in the Coda file system. *Proceedings Second International Conference on Parallel and Distributed Information Systems*, San Diego, California, January 1993.

[18] P. Kumar and M. Satyanarayanan. Flexible and safe resolution of file conflicts. *Proceedings USENIX Technical Conference*, New Orleans, Louisiana, January 1995, pages 95-106.

[19] R. Ladin, B. Liskov, L. Shrira, and S. Ghemawat. Providing high availability using lazy replication. *ACM Transactions on Computer Systems* 10(4):360-391, November 1992.

[20] L. Lamport. Time, clocks, and the ordering of events in a distributed system. *Communications of the ACM* 21(7):558-565, July 1978.

[21] L. Lamport. *LaTeX - a document preparation system.* Addison-Wesley Publishing Company, 1986.

[22] J. P. Munson and P. Dewan. A flexible object merging framework. *Proceedings ACM Conference on Computer Supported Cooperative Work (CSCW)*, Chapel Hill, North Carolina, October 1994, pages 231-242.

[23] R. M. Needham and M. D. Schroeder. Using encryption for authentication in large networks of computers. *Communications of the ACM* 21(12): 993-999, December 1978.

[24] J. Ousterhout. *Tcl and the Tk Toolkit*. Addison-Wesley Publishing Company, 1994.

[25] D. S. Parker, G. J. Popek, G. Rudisin, A. Stoughton, B. J. Walker, E. Walton, J. M. Chow, D. Edwards, S. Kiser, and C. Kline. Detection of mutual inconsistency in distributed systems. *IEEE Transactions on Software Engineering* SE-9(3):240-246, May 1983.

[26] P. Reiher, J. Heidemann, D. Ratner, G. Skinner, and G. Popek. Resolving file conflicts in the Ficus file system. *Proceedings Summer USENIX Conference*, June 1994, pages 183-195.

[27] M. Satyanarayanan, J.J. Kistler, P. Kumar, M.E. Okasaki, E.H. Siegel, and D.C. Steere. Coda: a highly available file system for a distributed workstation environment. *IEEE Transactions on Computers* 39(4):447-459, April 1990.

[28] M. Stonebraker. Concurrency control and consistency of multiple copies of data in distributed INGRES. *IEEE Transactions on Software Engineering* SE-5(3):188-194, May 1979.

[29] D. B. Terry, A. J. Demers, K. Petersen, M. J. Spreitzer, M. M. Theimer and B. B. Welch. Session guarantees for weakly consistent replicated data. *Proceedings Third International Conference on Parallel and Distributed Information Systems*, Austin, Texas, September 1994, pages 140-149.

[30] B. Walker, G. Popek, R. English, C. Kline, and G. Thiel. The LOCUS distributed operating system. Proceedings Ninth Symposium on Operating Systems Principles, Bretton Woods, New Hampshire, October 1983, pages 49-70.

9.4 Joseph, A.D., Tauber, J.A., and Kaashoek, M.F., Mobile Computing with the Rover Toolkit

This work was supported in part by the Advanced Research Projects Agency under contract DABT63-95-C-005, by an NSF National Young Investigator Award, by an Intel Graduate Fellowship, and by grants from AT&T, IBM and Intel.

Joseph, A.D., Tauber, J.A., and Kaashoek. M.F., "Mobile Computing with the Rover Toolkit," *IEEE Transactions on Computers,* 46(3):337–352, March 1997.

Mobile Computing with the Rover Toolkit

Anthony D. Joseph, Joshua A. Tauber, and M. Frans Kaashoek
M.I.T. Laboratory for Computer Science
Cambridge, MA 02139, U.S.A.
{adj, josh, kaashoek}@lcs.mit.edu

Abstract

Rover is a software toolkit that supports the construction of both *mobile-transparent* and *mobile-aware* applications. The mobile-transparent approach aims to enable existing applications to run in a mobile environment without alteration. This transparency is achieved by developing proxies for system services that hide the mobile characteristics of the environment from applications. However, to excel, applications operating in the harsh conditions of a mobile environment must often be aware of and actively adapt to those conditions. Using the programming and communication abstractions present in the Rover toolkit, applications obtain increased availability, concurrency, resource allocation efficiency, fault tolerance, consistency, and adaptation. Experimental evaluation of a suite of mobile applications demonstrates that use of the toolkit requires relatively little programming overhead, allows correct operation, substantially increases interactive performance, and dramatically reduces network utilization.

1 Introduction

Mobile computing environments present application designers with a unique set of communication and data integrity constraints that are absent in traditional distributed computing settings. For example, although mobile communication infrastructures are becoming more common, network bandwidth in mobile environments is often severely limited, unavailable, or expensive. Therefore, system facilities in mobile computing environments should minimize dependence upon continuous network connectivity, provide tools to optimize the utilization of available network bandwidth, minimize dependence on data stored on remote servers, and allow dynamic division of work between clients and servers. In this paper, we describe the *Rover toolkit*, a set of software tools that supports applications that operate obliviously to the underlying environment, while also enabling the construction of applications that use awareness of the mobile environment to adapt to its limitations. We illustrate the effectiveness of the toolkit using a number of distributed applications, each of which runs well over networks that differ by three orders of magnitude in bandwidth and latency.

This work was supported in part by the Defense Advanced Research Projects Agency under contract DABT63-95-C-005, by an NSF National Young Investigator Award, by an Intel Graduate Fellowship, and by grants from AT&T, IBM, and Intel.

Appears in IEEE Transactions on Computers: Special issue on Mobile Computing, 46(3). March 1997. p. 337-352.

1.1 Mobile versus Stationary Environment

Designers of applications for mobile environments must address several differences between the mobile environment and the stationary environment. Issues that represent minor inconveniences in stationary distributed systems are significant problems for mobile computers. This distinction requires a rethinking of the classical distributed systems techniques normally used in stationary environments.

Computers in a stationary environment are usually very reliable. Relative to their stationary counterparts, mobile computers are quite fragile: a mobile computer may run out of battery power, be damaged in a fall, be lost, or be stolen. As pointed out by the designers of Coda, given these threats, primary ownership of data should reside with stationary computers rather than with mobile computers [1]. Furthermore, application designers should take special precautions to enhance the resilience of the data stored on mobile computers.

Relative to most stationary computers, a mobile computer often has fewer computational resources available. In addition, the available resources may change dynamically (*e.g.*, a "docked" mobile computer has access to a larger display, auxiliary graphics or math coprocessors, additional stable storage, unlimited power, etc.).

A stationary environment can distribute an application's components and rely upon the use of high-bandwidth, low-latency networks to provide good interactive application performance. Mobile computers operate primarily in a limited bandwidth, high-latency, and intermittently-connected environment; nevertheless, users want the same degree of responsiveness and performance as in a fully-connected environment.

Network partitions are an infrequent occurrence in stationary networks, and most applications consider them to be failures that are exposed to users. In the mobile environment, applications will face frequent, lengthy network partitions. Some of these partitions will be involuntary (*e.g.*, due to a lack of network coverage), while others will be voluntary (*e.g.*, due to a high dollar cost). Mobile applications should handle such partitions gracefully and as transparently as possible. In addition, users should be able, as far as possible, to continue working as if the network was still available. In particular, users should be able to modify local copies of global data.

When users modify local copies of global data, consistency becomes an issue. In a mobile environment, optimistic concurrency control [2] is useful because pessimistic methods are inappropriate (a disconnected user cannot acquire or release locks), as pointed out by the designers of Coda [1]. However, using an optimistic approach does involve some difficulties. In particular, long duration partitions will cause a greater incidence of apparent write-write conflicts than in stationary environments. It is therefore important to use application-specific semantic information to detect when such conflicts are false positives and can be resolved.

1.2 The Argument for Mobile-Aware Computing

The attributes of the typical stationary environment have guided the development of classical distributed computing techniques for building client-server applications. These applications are usually unaware of the actual state of the environment; therefore, they make certain implicit assumptions about the location and availability of resources.

Such *mobile-transparent* applications can be used unmodified in mobile environments by having the system shield or hide the differences between the stationary and mobile environments from applications. Coda [1] and Little Work [3] use this approach in providing a file system interface to applications. These systems consist of a local *proxy* for some service (the file system) running on the mobile host and providing the standard service interface to the application, while attempting to mitigate any adverse effects of the mobile environment. The proxy on the mobile host cooperates with a remote server running on a well-connected, stationary host.

This mobile-transparent approach simplifies mobile applications, but sacrifices functionality and performance. Although the system hides mobility issues from the application, it usually requires manual intervention by the user (*i.e.,* having the user indicate which data to prefetch onto the user's computer). Similarly, conflict resolution is complicated because the interface between the application and its data was designed for a stationary environment. Consider an application writing records into a file shared among stationary and mobile hosts. While disconnected, the application on the mobile host inserts a new record. The local file system proxy records the write in a log. Meanwhile, an application on a stationary host alters another record in the same file. Upon reconnection, the file system can detect that conflicting updates have occurred: one file contains a record that does not appear in the other and one file contains a record that differs from the same record in the other file. Depending upon how records are named, it may not be possible to distinguish between the instance where the same record is modified in one file and not the other and the instance where a record is deleted and a new one is added. However, the file system alone cannot resolve the conflict.

The cause of this ambiguity is that these systems change the contract between the application and the file system in order to hide the condition of the underlying network. The read/write interface no longer applies to a single file, but to possibly inconsistent replicas of the file. Therefore, any applications that depend on the semantics of the standard read/write interface for synchronization and ordering may fail.

Coda recognizes this limitation in conflict resolution and provides support for *application-specific resolvers* (ASRs) [4], programs that are invoked based on the file name extension to resolve conflicts during synchronization. However, ASRs alone are insufficient, because the only information provided to an ASR is the two conflicting files, not the two operations and contexts that led to the differing results. An application has no way of knowing that the file system contract has been changed by Coda, and therefore no way to save the additional information the ASR would need to resolve the conflict. Thus, in the above example, there is no way for an ASR to use the file system interface to determine whether the mobile host inserted a new record or the stationary host deleted an old one.

So, although the mobile-transparent approach is appealing in that it offers to run existing applications without alteration, it is fundamentally limited in that it hides the information needed to allow applications to remain correct and to perform well in an intermittently-connected environment. The alternative to hiding environmental information from applications is to expose information and involve applications and users in decision-making. This alternative yields the class of *mobile-aware* applications.

In the above case, a mobile-aware application can store not only the *value* of a write, but also the *operation* associated with the write. That operation can include any relevant context. Storing the operation allows the application to use application-specific semantic and contextual information; for example, it allows for "on-the-fly" dynamic construction of conflict avoidance and conflict resolution procedures.

The mobile-aware argument can be viewed as applying the end-to-end argument ("Communication functionality can be implemented only with the knowledge and help of the application standing at the endpoints of the communications system." [5]) to mobile applications. The file system example described above illustrates that some applications need to be aware of intermittent network connectivity to achieve consistency. Similar examples exist for performance, reliability, low-power operation, etc.

The mobile-aware argument does not require that every application use its own, *ad-hoc* approach to mobile computing. Rather, it allows the underlying communication and programming systems to define an application programming interface that optimizes common cases and supports the transfer of appropriate information between the layers. Since mobile-aware applications share common design goals, they can share design features and techniques. The Rover toolkit provides such a mobile-aware application programming interface that, unlike previous systems, is designed to support both mobile-aware and mobile-transparent approaches.

1.3 Rover: The Toolkit Approach

The Rover toolkit offers applications a distributed object system based on a client/server architecture with client caching and optimistic concurrency control [6] (see Fig. 1). Clients are Rover applications that typically run on mobile hosts, but could run on stationary hosts as well. Servers, which may be replicated, typically run on stationary hosts and hold the long term state of the system.

The toolkit provides mobile communication support based on two ideas: *relocatable dynamic objects* (RDOs) and *queued remote procedure call* (QRPC). A relocatable dynamic object is an object (code and data) with a well-defined interface that can be dynamically loaded into a client computer from a server computer, or *vice versa,* to reduce client-server communication requirements. Queued remote procedure call is a communication system that permits applications to continue to make non-blocking remote procedure calls [7] even when a host is disconnected — requests and responses are exchanged upon network reconnection.

Rover applications employ an optimistic, primary-copy, tentative-update, distributed object model of data sharing: they call into the Rover library to *import* RDOs and to *export* logs of operations that mutate RDOs. Client-side applications *invoke* operations directly on locally cached RDOs. Server-side applications are responsible for resolving conflicts and notifying clients of resolutions. A network scheduler (operating above the operating system level) drains the stable QRPC log, which contains the RPCs that must still be performed at the server. Where possible, in addition to opening and closing network connections, the network scheduler also interacts with the operating system to initiate or terminate connection-based transport links (*e.g.*, by dial-up) or to power-up/power-down wireless transmitters.

Servers can run either at stationary hosts or at mobile hosts. Running servers at mobile hosts is useful when forming "mobile workgroups," where a group of clients isolated from other hosts can interact with local servers. However, for two reasons, we mostly use stationary hosts: first, an operation cannot be committed (*i.e.*, it is tentative) until it reaches and is accepted by a server; second, a server's copy of an object is the canonical version of the object. Thus, clients can ascertain the latest value for an object only by contacting the server that owns the object. Both commit and import operations require well-connected servers. If servers are on poorly connected mobile hosts, clients outside of the workgroup may encounter delays in committing operations or importing objects.

The key task of the programmer when building a mobile-aware application with Rover is to define RDOs for the data types manipulated by the application and for data transported between client and server. The programmer then divides the program into portions that run on the client and portions that run on the server; these parts communicate by means of QRPC. The programmer then defines methods that update objects, including code for conflict detection, avoidance, and resolution.

To use the Rover toolkit, a programmer links the modules that compose the client and server portions of an application with the Rover toolkit. The application can then actively cooperate with the runtime system to *import* objects onto the local machine, *invoke* well-defined methods on those objects, *export* logs of method invocations on those objects to servers, and *reconcile* the client's copies of the objects with the server's.

1.4 Main results

This paper extends previous work on the Rover architecture [6, 8] by providing a more detailed discussion of the design and implementation of queued RPC and relocatable dynamic objects, including compressed and batched QRPCs; presenting in greater depth the argument for making applications mobile-aware; and explaining how applications use that awareness and the Rover toolkit to mitigate the effects of intermittent communication on application performance. We draw four main conclusions from our experimental data and experience developing Rover:

1. QRPC is well suited to intermittently connected environments. Queuing enables RPCs to be scheduled, batched, and compressed for increased network performance. QRPC performance is acceptable in the target environment even if every RPC is stored in stable logs at clients and servers. For lower-bandwidth networks, the overhead of writing the logs is dwarfed by the underlying communication costs.

2. Use of RDOs allows mobile-aware applications to migrate functionality dynamically to either side of a slow network connection and thereby minimize reliance on network connectivity. Caching RDOs reduces latency and bandwidth consumption. User interface functionality can run at full speed on a mobile host, while large data manipulations can be performed on the well-connected server.

3. Porting applications to Rover and making them mobile-aware generally requires relatively little change to the original application. In addition, building Rover proxies is easy and has allowed the use of applications (*e.g.*, Netscape and USENET browsers) without modification. We have implemented several mobile-aware applications (Rover Exmh, Webcal) by modifying conventional instances of the same applications. We have built others applications from scratch (Irolo, Stock Market Watcher). Most were made mobile-aware with a few simple changes (to approximately 10% of the original code and with as little as three weeks of work). Only one of the applications required more than a month of work. In part, the extra time was due to additional evolution and redesign of the toolkit.

4. Measurements of end-to-end mobile application performance show that, by using Rover, mobile-transparent and mobile-aware applications perform significantly better on slow networks than their original versions. For example, when performing routine Web browsing using Netscape with the mobile-transparent Rover HTTP proxy, we observe performance improvements of up to 17% over Netscape alone. When using mobile-aware applications running over slow networks to perform routine tasks, we observe performance improvements of up to a factor of 7.5 over the original versions.

5. We also observe significant improvements in the performance of the user interfaces of client/server-based mobile-aware applications in comparison to those of conventional applications used remotely across slow networks. Without Rover, scrolling and refreshing operations are extremely slow and pressing buttons and selecting text are unpleasant because of the lag between mouse clicks and display updates. With Rover, the user sees excellent GUI performance across a range of networks that varies by three orders of magnitude in both bandwidth and latency.

2 Related Work

The need for mobile-aware applications and complementary system services to expose mobility to applications was identified concurrently by several groups. Katz noted the need for adaptation of mobile systems to a variety of networking environments [9]. Davies *et al.* cited the need for protocols to provide feedback about the network to applications in a vertically integrated application environment [10].

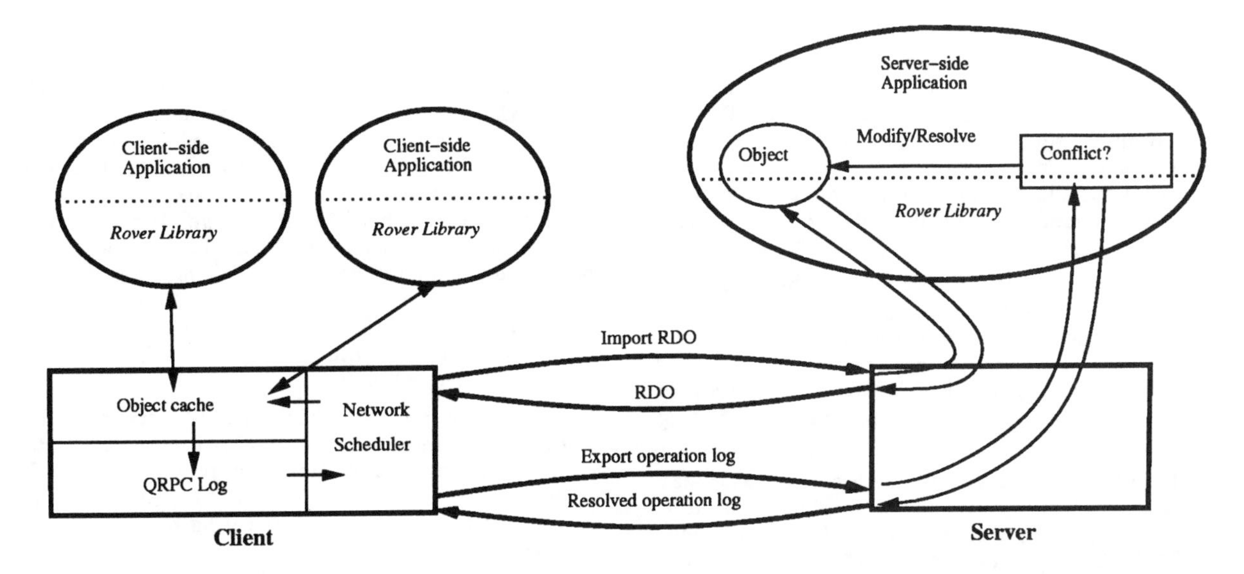

Figure 1: Rover offers applications a client/server distributed object system with client caching and optimistic concurrency control. This figure shows the control flow within the toolkit.

Kaashoek *et al.* created a Web browser that exposed the mobile environment to code implementing mobile-aware Web pages [11]. The Bayou project proposed and implemented an architecture for mobile-aware databases [12]. Baker identified the dichotomy between mobile-awareness and mobile-transparency in general application and system design [13]. However, Rover is the first implemented general application-architecture to support both mobile-transparent system service proxies and mobile-aware applications.

Several previous projects have studied building mobile-transparent services for mobile clients. The Coda project pioneered distributed services for mobile clients. In particular, it investigated how to build a mobile-transparent file system proxy for mobile computers by using optimistic concurrency control and prefetching [1]. Coda logs all updates to the file system during disconnection and replays the log on reconnection; automatic conflict resolution mechanisms are provided for directories and files, using Unix file naming semantics to invoke ASRs at the file system level [4]. A manual repair tool is provided for conflicts that cannot be resolved automatically. A newer version of Coda supports low-bandwidth networks as well as intermittent communication [14]. Odyssey is a mobile-aware follow-on project that focuses on system support to enable mobile-aware applications to use "data fidelity" to control resource utilization. Data fidelity is defined as the degree to which a copy of data matches the original [15].

The Ficus file system is a mobile-transparent, user-level file system supporting disconnected operation and peer-to-peer data sharing [16]. The Little Work project caches files to smooth disconnection from an AFS file system. Conflicts are detected and reported to the user. Little Work is also able to support partial connection over low-bandwidth networks [3].

The BNU project implements an RPC-driven mobile-transparent application framework on mobile computers. It enables code shipping by downloading Scheme functions for interpretation [17]. The BNU environment includes mobile-transparent proxies on stationary hosts for hiding the mobility of the system. BNU applications do not dynamically adjust to the environment, nor do they have a concept of tentative or stale data. No additional support for disconnected operation, such as Rover's queued RPC is included in BNU. A follow-up project, Wit, addresses some of these shortcomings and shares many of the goals of Rover, but employs different solutions [18].

A number of proposals have been made for various degrees of mobile-awareness in operating system services and applications. The Bayou project [12] defines a mobile-aware database architecture for sharing data among mobile users. Bayou supports tentative operation logs and tentative data values combined with session guarantees for weakly-consistent replicated data [19]. Each database write is associated with a dependency check and a deterministic, application-specific merge procedure. Each Bayou host replicates the entire database and may receive incremental updates from any other host. To illustrate these concepts, the authors built a calendar tool and a bibliographic database. Rover shares the notions of tentative operations and data, session guarantees, and the calendar tool example with the Bayou project. Rover's RDO method invocation via QRPC is more general than Bayou database writes in that it supports computation relocation and communication scheduling for dealing with intermittent communication, limited bandwidth, and resource-poor clients.

The DATAMAN project provided a framework for mobile-aware applications around an object-oriented, language-level event delivery architecture [20]. Applications interested in events in the environment declare EventObjects, which support both polling and triggers (callbacks). The DATAMAN application framework does not include support for data replication and consistency management.

The InfoPad [21], Daedalus [22], GloMop [23], and W4 [24] projects focus on mobile-aware wireless information access. The InfoPad project employs a dumb terminal and offloads all functionality from the client to the server with the specific goal of reducing power consumption. Daedalus and GloMop use dynamic "transcoding" or "distillation" to reduce the bandwidth consumed by data transmitted to a mobile host. Transcoding technology is completely compatible with Rover's architecture. Applications on the mobile host cooperate with mobile-aware proxies on a stationary host to define the characteristics of the desired network connections. Similarly, to enable Web browsing, W4 divides application functionality between a small PDA and a more powerful, stationary host. Rover is designed for more flexible, dynamic divisions. Depending on the power of the mobile host and available bandwidth, Rover allows mobile-aware browsers to move functionality between the client and the server dynamically.

The DeckScape WWW browser [25] modifies the user interface of a Web browser, turning it into a "click-ahead" browser. DeckScape was developed simultaneously with the Rover HTTP proxy which transparently provides click-ahead for existing browsers. Since they implemented a browser from scratch, DeckScape is not compatible with existing browsers.

The BARWAN project address the problems associated with "overlay networks," networks of varying bandwidth, latency, coverage, and application-level performance visibility [26]. BARWAN supports mobile, "data type aware" applications by using proxies on both sides of the client–server link to monitor the performance attributes of the network. The monitoring, in conjunction with mobile code and a dynamically extensible type system, enables mobile-aware applications to trade processing cycles for network traffic. Similarly, Davies' Adaptive Services [10] uses a protocol-centric approach to expose information about the mobile environment to applications. In contrast, Rover is designed to focus on dynamic adaptation of program functionality and data types.

A number of successful commercial mobile-aware applications have been developed for mobile hosts and limited-bandwidth channels. For example, Qualcomm's Eudora is a mail browser that allows efficient remote access over low-bandwidth links. Lotus Notes [27] is a groupware application allowing users to share data in a weakly-connected environment. Notes supports two forms of update operations: append and time-stamped. Conflicts are referred to the user. TimeVision and Meeting Maker are group calendar tools that allow a mobile user to download portions of a calendar for off-line use. The Rover toolkit and its applications provide functionality that is similar to these proprietary approaches, but in an application-independent manner. Using the Rover toolkit, standard workstation applications such as *Exmh* and *Ical* can easily be turned into mobile-aware applications.

Gray *et al.* perform a thorough theoretical analysis of the options for database replication in a mobile environment and conclude that primary-copy replication with tentative updates is the most appropriate approach for mobile environments [28].

3 Design of the Rover Toolkit

The Rover toolkit is designed to support the construction of mobile-aware applications and proxies for mobile-transparent applications. In this section, we describe the design of the key components of the Rover toolkit. In the following section, we discuss implementation details for each component of the toolkit.

3.1 Object Design and QRPC

The central structures in Rover are relocatable dynamic objects (RDOs). All Rover design decisions are based upon RDOs; thus, they provide the key point of control in Rover applications. RDOs have four components: mobile code, encapsulated data, a well-defined interface, and the ability to make outcalls to other RDOs. The mobile code and encapsulated data components make an RDO relocatable from one machine to another. Each RDO provides methods for marshaling and unmarshaling itself so that it can be relocated to another machine and stored in a client's cache. An RDO can invoke the methods of another RDO by using that RDO's well-defined interface.

RDOs may execute at either clients or servers. Each RDO has a "home" server that maintains its primary, canonical copy. Clients use library functions provided by the Rover toolkit to import secondary copies of RDOs into their local caches and to export logs of mutating method invocations back to servers.

All application code and all application-touched data are contained within RDOs. RDOs may vary in complexity from simple data items with a small set of operations (*e.g.*, calendar entries) to modules that encapsulate a significant part of an application (*e.g.*, the graphical user interface for an e-mail browser). The toolkit provides functions that enable complex RDOs optionally to create threads of control when they are imported. The toolkit ensures the safe execution of RDOs by using authentication and by executing RDOs in a controlled environment. For our research prototype, we assume that RDOs may be faulty, but not malicious. Thus, the toolkit executes RDOs in separate address spaces, but does not rely on code inspection or other techniques to detect malicious code. These safety measures are appropriate for the sharing of objects between mobile hosts and servers in the framework of specific applications. However, there are several safety issues relating to the general use of mobile code that are not addressed by our current implementation. These safety issues represent an area of active research that is beyond the scope of this paper.

At the level of RDO design, application builders have semantic knowledge that is useful in attaining the goals of mobile computing. By tightly coupling data with program code, applications can manage resource utilization more carefully than is possible with a replication system that handles only generic data. For example, an RDO can include compression and decompression methods along with data in order to obtain application-specific and situation-specific compression, reducing both network and storage utilization.

A QRPC may contain an RDO, a request for an RDO, or arbitrary application data. Rover clients use QRPC to fetch RDOs from servers lazily (see Fig. 1). When an application issues a QRPC, Rover stores the QRPC in a local stable log and immediately returns control to the application. If the application has registered a *callback* routine, it will be invoked when the requested QRPC completes. Applications may poll the state of the QRPC or simply block to wait for critical data. However, this choice is undesirable, especially when the mobile host is disconnected. Upon reconnection, the Rover network scheduler drains the log in the background, forwarding any queued QRPCs to the server.

When a Rover application modifies a locally cached RDO, the cached copy is marked *tentative*. The RDO is only marked *committed* after the client knows that the Rover server has applied the mutating operations to the canonical RDO. These updates and resolutions are lazily propagated between the client and server using QRPC. In the meantime, the application may choose to use tentative RDOs. This choice allows the application to continue execution even if the mobile host is disconnected. Cached copies of the RDO at other clients are updated either by client polling or server callbacks.

3.2 Communication Scheduling

The Rover network scheduler may deliver QRPCs out of order (*i.e.*, non-FIFO) based on consistency requirements and application-specified operation priorities. Reordering is important for usability in an environment with intermittent connectivity, as it allows the user, through applications, to identify important operations. For example, a user may choose to send urgent updates as soon as possible while delaying other sends until inexpensive communication is available.

Multiple QRPCs for the same server may be batched together reducing communication overhead. If the operating system provides the appropriate interface, the network scheduler can use batching to reduce the amount of time that a wireless transmitter is powered up and idle.

QRPC supports split-phase operation; if a mobile host is disconnected between sending the request and receiving the reply, a Rover server will periodically attempt to contact the mobile host and deliver the reply. The split-phase communication model enables Rover to use different communication channels for the request and the response and to close channels during the intervening period. Several wired and wireless technologies offer asymmetric communication options, such as cable television modems, direct broadcast satellite, receive-only pagers, and PCS phones that can initiate calls, but cannot receive them. By splitting the request and response pair, communication can be directed over the most efficient, available channel. Closing the channel while waiting is particularly useful when the waiting period is long and the client must pay for connection time.

The combination of the split-phase and stable nature of QRPCs allows a mobile host to be completely powered-down while waiting for a pending operation to complete. When the mobile host resumes normal operation, the results of the RDO invocation will be relayed reliably from the server and any unsent messages at the mobile host will be delivered to servers. Thus, long-lived computation (*e.g.*, a database search) can occur at the server while the mobile host conserves power.

3.3 Computation Relocation

Rover gives applications control over the location where computation will be performed and where data will be stored. In other words, Rover allows both function-shipping and data-shipping. In an intermittently-connected environment, the network often separates an application from the data upon which it is dependent. By moving RDOs across the network, applications can move data and/or computation from client to server and vice-versa. Computation relocation is useful when a large body of data can be distilled down to a small amount of data or code that actually traverses the network and when remote functionality is needed during periods of disconnection.

For example, migrating a Graphical User Interface (GUI) to the client serves both these purposes. The code to implement a GUI is small compared to the graphical display updates it generates. At the same time, the GUI together with the application's RDOs can process user actions locally, avoiding additional network traffic and enabling disconnected operation.

Clients can also use RDOs to export computation to servers. Such RDOs are particularly useful for two operations: performing filtering actions against a dynamic data stream and performing complex actions against a large amount of data. With RDOs, the desired processing can be performed at the server, with only the processed results returned to the client.

3.4 Notification

Since the mobile environment is dynamic, it is important to present the user and the application with information about the current environment. The Rover toolkit provides applications with environmental information for use in dynamic decision making or for presentation to the user. Applications may use either polling or callback models to determine the state of the mobile environment.

Applications can forward notifications to users or use them for silent policy changes. For example, in our calendar application (see Section 5), appointments that have been modified but not propagated to the server are displayed in a distinctive color, a technique that was borrowed from the Bayou room scheduling tool [12]. This color scheme informs users that the appointment might be canceled due to a conflict.

3.5 Object Caching and Consistency

An essential component to accomplishing useful work while disconnected is having the necessary information locally available [1]. RDO replica caching is the chief technique provided by Rover to achieve high availability, concurrency, and reliability. In this section, we discuss strategies for selecting objects to cache and for reducing the costs related to maintaining consistency.

3.5.1 Caching

During periods of network connectivity, Rover fills the mobile host's cache with useful RDOs. Rover leaves it up to applications to decide which objects to prefetch. We believe that the usability of applications will be critically dependent upon simple user interface metaphors for indicating collections of objects to be prefetched. Requiring users to list the names of objects that they wish to prefetch is inherently confusing and error-prone. Instead, Rover applications can provide prioritized prefetch lists based upon high-level user actions. For example, the Rover e-mail browser automatically generates prefetch operations for the user's inbox folder and recently received messages, as well as folders the user visits or selects.

The Rover toolkit provides clients with cache access and manipulation functions, including checking an RDO's consistency information, flushing RDOs from the cache, and verifying that an RDO is up-to-date. The toolkit also provides server portions of applications with functions for automatically handling RDO validation and cache tags for managing the consistency of client caches. The cache tags provide support for: verifying, if possible, an RDO before use; "leasing" an RDO for a fixed period of time; relying on server callbacks for notification of a stale RDO; and indicating that an RDO is immutable.

While caching can bring great benefits, application designers must be careful to avoid unnecessary communication, increased latencies, and dead-lock. Applications should avoid caching more data than is necessary to provide good availability, since unnecessary prefetching consumes valuable network resources. In addition, applications should strive to keep update messages small.

3.5.2 Consistency

Applications generally require consistency control when clients are allowed to perform concurrent updates on shared RDOs. The Rover toolkit provides significant flexibility in the choice of mechanism, ranging from application-level locking to application-specific algorithms for resolving uncoordinated updates to a single RDO. Certain applications will be structured as a collection of independent atomic actions [29], where the importing action uses application-level locks, version vectors, or dependency-set checks to implement fully-serializable transactions within Rover method calls. Others may impose no ordering requirement, allowing all operations, no matter when they reach the server.

Since no single scheme is appropriate for all applications, Rover leaves the selection of consistency scheme to the application. However, only a limited number of schemes lend themselves naturally to mobile environments. For example, using pessimistic concurrency control may cause long blocking periods in a mobile environment; optimistic concurrency control schemes, on the other hand, allow immediate updates by any host on any local data. We therefore expect that optimistic schemes will be widely used, and Rover provides substantial, but not exclusive, support for primary-copy, tentative-update optimistic concurrency control. Specifically, the Rover library supports logging, rollback, and replay of operations; log manipulation; and automatically-maintained RDO consistency vectors. All Rover applications built to date use primary-copy consistency control.

The server is responsible for maintaining the consistent view of the system. Update conflicts are detected and resolved by the server, and the results of reconciliation are treated by clients as overriding tentative state stored at the client. Thus, to reconcile its state and to assure that any updates are durable a client needs to communicate only with the server.

Rover provides clients with facilities for automatically logging the operations performed by mutating method invocations. The operations consist of conflict avoidance and detection checks, the mutating operation, and the data for the operation. Logging operations, rather than only new data values, increases application flexibility in avoiding or resolving conflicts by allowing applications to use type-specific concurrency control [30]. For example, a financial account object with debit, credit, and balance methods provides a great deal more semantic information to the application than a simple account file containing only the balance. Debit and credit operations from multiple clients could be arbitrarily interleaved as long as the balance never becomes negative. In contrast, consistently updating a balance value by overwriting the old value would require the use of an exclusive lock on the global balance.

When the QRPC containing the mutating operation arrives at a server, the server invokes the operation on the the primary copy of the object. Typically, the operation first checks whether the object has changed since it was imported by a mobile host. The operation may also check other objects to determine if they have changed since they were imported (so that potential read-write conflicts can be detected). The definition of conflicting modifications is strongly application- and data-specific. Therefore, Rover does not try to detect conflicts directly, although it maintains version vectors for each RDO to aid conflict detection.

Rover uses type-specific concurrency control [30] to enable applications to avoid conflicts. Using Rover, conflict avoidance may depend not only on the application, but on the data or even the operation involved. When a write-write conflict is detected, the application must determine whether the preconditions for the mutating operation are actually violated (yielding a genuine conflict). Since the submitted operation was originally performed at the client on potentially stale data, the result of performing the operation at the server may not be exactly what the client expected. However, the server can use application-specific semantics to create an acceptable value.

4 Implementation of the Rover Toolkit

As shown in Fig. 2, the Rover toolkit consists of four key components: the access manager, the object cache (client-side only), the operation log, and the network scheduler. These pieces make up the the minimal "kernel" of Rover. Further functionality may be imported on demand. This feature is particularly important for mobile hosts with limited resources: small memory or small screen versions of applications may be loaded by default. However, if the application finds more hardware and network resources available (*e.g.*, if the mobile host is docked) further RDOs may be loaded to handle these cases [11]. We now discuss each of the four components in turn.

Each machine has a local Rover *access manager*, which is responsible for handling all interactions between client-side and server-side applications and among client-side applications. The access manager services requests for objects (RDOs), mediates network access, logs modifications to objects, and manages the client object cache. Client-side applications communicate with the access manager to import objects from servers, cache them locally, and dynamically link them into existing computation. Server-side applications are invoked by the access manager to handle requests from client-side applications. Applications invoke the methods provided by the objects and, using the access manager, make changes globally visible by exporting them back to the servers.

In the current implementation, RDOs are implemented using the Tcl and Tk languages [31]. However, since the toolkit interface is designed to be language-independent, it will be easy to explore the use of other interpreted or byte-compiled languages (*e.g.*, Java [32]).

Within the access manager, RDOs are imported into the object cache, while QRPCs are exported to the operation log. The access manager routes invocations and responses

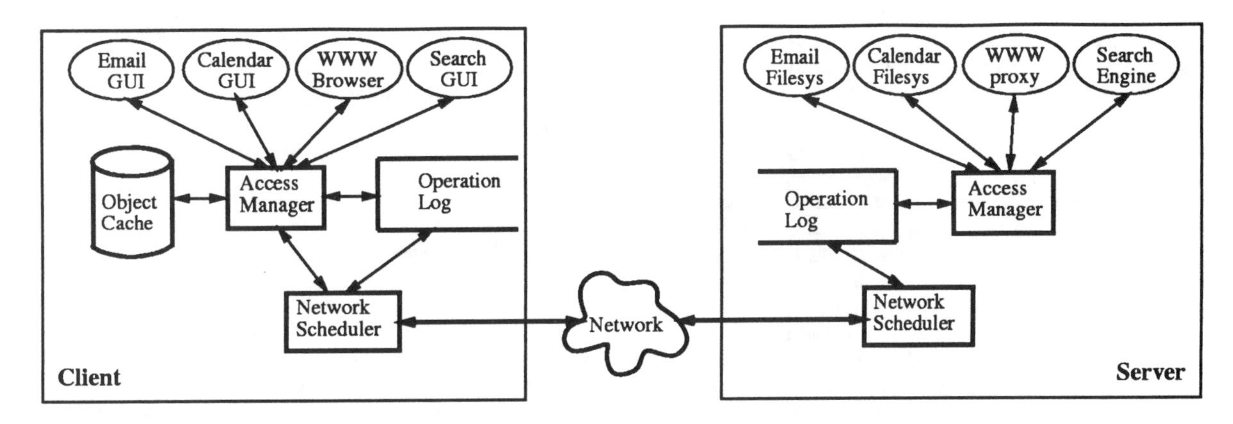

Figure 2: The Rover architecture consists of four key components: access managers, operation logs, network schedulers, and object caches. The first three of these exist on both the client and the server.

between applications, the cache, and the operation log. The log is drained by the network scheduler, which mediates between the rest of the toolkit and the various communication protocols and network interfaces.

The access manager also handles failure recovery. This task is eased somewhat by our use of both a persistent cache and an operation log. After a failure, the access manager requeues any incomplete QRPCs for redelivery. At-most-once delivery semantics are provided by unique identifiers and the persistent log. One issue that remains an open question is how to handle error responses from resent QRPCs for client-side applications that no longer are running. Our implementation currently ignores such responses, although the same reliability techniques we applied to servers [8] could also be applied to client applications.

The *object cache* provides stable storage for local copies of imported objects. The object cache consists of a local private cache located within the application's address space and a global shared cache located within the access manager's address space. Client-side applications do not usually interact directly with the object cache. When a client-side application issues an *import* or *export* operation, the toolkit satisfies the request based upon whether the object is found in the local cache and the cache consistency option specified for the object (see Section 3.5.1).

The *operation log* provides stable storage for QRPCs. Once an object has been imported into the client-side application's local address space, method invocations without side effects are serviced locally by the object. At the application's discretion, method invocations with side effects may also be processed locally, inserting tentative data into the object cache. Operations with side effects also insert a QRPC into the operation log located at the client. Each insertion is synchronous. Optionally, insertions can be logged asynchronously. This option allows the cost of logging to be amortized across multiple QRPCs, but opens a small (several second) window of vulnerability, where a client failure would cause loss of data.

The stable log is implemented as an ordinary file. Rover performs both a flush and a synchronize operation to force new QRPCs to the log. Thus, the log update is on the critical path for message sending. Support for intermittent-network connectivity is accomplished by allowing the log to be incrementally flushed back to the server. Thus, as network connectivity comes and goes, the client will make progress towards reaching a consistent state.

One issue Rover addresses with an application-specific approach is operation log growth during disconnected operation. Operation logging may lead to an operation log that grows in size at a rate exceeding that of a simple value log. The traditional approach is log compaction [1]. In addition to providing limited automatic log compaction, Rover also directly involves applications in log compaction by enabling them to download procedures into the access manager to manipulate their log records. For example, an application can filter multiple object verification QRPCs to leave only a single QRPC. In addition, applications can apply their own notion of "overwriting" to the operations in the log.

The original *network scheduler* examines each unsent operation to determine the destination and selects the appropriate transport protocol and medium over which to send it. Rover is capable of using a variety of network transports. Rover supports both connection-based protocols (*e.g.*, HTTP over TCP/IP networks) and connectionless protocols (*e.g.*, SMTP over IP or non-IP networks). Different protocols have different strengths. For example, while SMTP has extremely high latency, it is fundamentally a queued background process; it is more appropriate than more interactive protocols for fetching extremely large documents, such as stored video, which require large amounts of time regardless of the protocol. Another advantage is that the IP networks required for HTTP or TCP are not always available, whereas SMTP often reaches even the most obscure locations.

We have experimented with two network schedulers. The original network scheduler sent a request as soon as it was received from a client application. The new improved scheduler uses the following heuristic to batch requests that are destined for the same server:

- When a request is received from one client application, it uses the access manager to check all the client applications (including the one that sent the original request) to see if any are in the process of sending a request.
- If there are additional requests pending from any client applications, the scheduler delays sending the original request. Upon receipt of the next request, the scheduler repeats the pending request check.
- When there are no pending requests from client applications (or the first request has been delayed for a

preset maximum time), the scheduler batches the requests and sends them on the same connection; the results are also received on the same connection.

There are two situations in which we expect batching to occur: when the mobile host is disconnected and when a client issues a series of requests almost simultaneously (*e.g.*, prefetching or importing several objects, exporting changes to multiple objects, or verifying the consistency of multiple objects. This heuristic imposes a small delay on requests: the time for the access manager to check each client application for pending requests and to receive the requests). However, since the check allows the scheduler to automatically batch requests, it is a small penalty to pay relative to that caused by a high roundtrip time. Thus, an application that issues several requests in a series will have the requests automatically batched and sent to the server using a single connection. The first few requests will incur increased latency; however, throughput and average latency will be substantially improved. We are also investigating alternatives that rely upon applications to specify the set of requests that should be batched together.

The network scheduler leverages the queuing of QRPCs performed by the log to gain transmission efficiency. The result is a potentially significant reduction in per-operation transmission overhead and an increase in connection efficiency through amortization of connection setup and teardown across multiple requests and responses. This amortization is especially important when connection setup is expensive, either in terms of added latency or dollar cost. For example, the latency for a null RPC over a 9.6 Kbit/s Cellular CSLIP link is 2.23 seconds; batching offers a substantial performance benefit.

The network scheduler also applies compression to the headers associated with requests and, in the absence of application-specified compression, applies compression to application data. This compression offers significant performance advantages, especially when combined with batching. Typical compression ratios for the applications we have studied range from 1.5:1 to 9.7:1. The combination of batching and compression over low-bandwidth networks yields (on average) a two- to four-fold reduction in execution times over uncompressed, single requests.

5 Mobile Computing Using Rover

In this section, we discuss the steps involved in implementing mobile-aware applications, porting existing applications to a mobile-aware environment, the programming interface provided by the Rover toolkit, and the set of sample applications that we constructed using the toolkit. Many of the same steps are involved in building a proxy to support a mobile-transparent service. Note that Rover provides a consistent framework for developing mobile applications rather than providing any mechanical tools for building applications.

5.1 Using Objects Instead of Files

There are several steps involved in porting an existing application to Rover or creating a new Rover-based application. Each step requires the application developer to make one of several implementation decisions. The choices we used in developing the initial set of Rover applications are presented in Table 1.

We illustrate the development process with an e-mail browser that we have constructed, Rover Exmh. The e-mail browser is a port of Brent Welch's Exmh Tcl/Tk-based e-mail browser.

5.1.1 Object Design

The important first step is to split the application into components and carefully identify which components and functions should be present on each side of the network link. The division will be mostly static, as most of the file system components will remain on the server and most of the GUI components will remain on the client. However, those components that are dependent upon the computing environment (network or computational resources) or are infrequently used may be dynamically generated. For example, the search operation performed by an Exmh client is dynamically customized to the current link attributes: over a low-latency link, more work is done at the client and less at the server, and vice versa for a high-latency link.

In migrating to the mobile environment, reading files is replaced by importing objects, and writing files is replaced by exporting changes to those objects. The file-system interface still exists in the server-side portion of the application; however, inserted between the server-side file-system interface and the client-side graphical user interface is an object layer.

5.1.2 Caching

Once the application has been split into components, the next step is to appropriately encapsulate the application's state within objects that can be replicated and sent to multiple clients. For example, a user's electronic mail consists of messages and folders. In a traditional distributed file system, one method of encapsulation is to store each message in an individual file and use directories to group the messages into folders. Information about the size or modification date of a message is determined by using file system status operations. With Rover, the corresponding encapsulation stores messages as objects and folders as objects containing references to message objects. Each object encapsulates both the message or folder data and the appropriate metadata.

5.1.3 Computation Migration

One of the primary purposes of the object layer is to provide a means of reducing the number of network messages that must be sent between the client and server; this reduction is done by migrating computation. Consider the e-mail folder scan operation, which returns a list of messages and information about the messages in a folder. Using a file system-based approach means scanning the directory for the folder, opening each message, and extracting the relevant information. All of this information would flow across the link between the client and server; thus, this function is an appropriate operation for a well-connected host, but would be very expensive and time-consuming over a high-latency link. Using Rover's object-based approach, the server-side portion of the application constructs a folder object containing the metadata for the messages contained in the folder. The client-side portion

Issue	Design Decision
Object Design	Use RDOs that encapsulate sufficient state to effectively service local requests, but are small enough to easily prefetch
Caching	Use RDOs to cache information
Computation Migration	Use RDOs to migrate computation that requires high bandwidth access
Object Prefetching	Tradeoff of RDO size versus easier prefetching, but have to avoid overly aggressive prefetching
Notification	Use colors and text to notify users of tentative information
Consistency	Use logs of operations to detect conflicts and either avoid or help resolve them

Table 1: Implementation choices for the initial application set built using the Rover toolkit.

of the application then imports the folder object in a single roundtrip request, avoiding multiple roundtrip requests. The multiple requests are replaced by local computation — querying the folder object about the messages it contains. Batching the metadata together trades an increase in the latency to view the metadata for the first message for reduced network communication and fast interactive use of the rest of the metadata.

5.1.4 Object Prefetching

Prefetching decisions may be made at design time or run time. The former relies on the designer's knowledge of the application structure. For example, the main portion of Rover Exmh's help information is prefetched by a client, but less frequently referenced portions are loaded on demand. These run time decisions add contextual information to the designer's knowledge. For example, the e-mail application automatically prefetches the contents of the most recently visited folders and the messages in the user's inbox.

5.1.5 Notification

The next step is to add support for interacting with the environment. For example, in the e-mail application, one of the important pieces of message metadata that a folder object contains is the message's size and the size of any attachments. This information is used by the application and conveyed to users allowing them to make decisions about which mail messages to import. These sizes could be further used in combination with network performance metrics to support application prefetching decisions.

The application developer also must decide which mechanisms to use for notifying users of the cache status of displayed data. In the e-mail application, color is used to distinguish operations that have not been propagated to a server.

5.1.6 Consistency

The final important step is the addition of application-specific conflict avoidance and resolution. Whereas conflicts are infrequent in most stationary environments, in mobile environments, they will be more common. Fortunately, application developers can leverage the additional semantic information that is available with Rover's operation-based (instead of value-based) approach to object updating.

Conflict avoidance has two aspects: flexible precondition tests for operations and operations that are, where possible, commutative. Commutative operations have the advantage that the server may arbitrarily interleave operations presented by several clients without altering the client operations. The precondition test determines whether the operation can still be executed. Consider, for example, a banking application. Given an operation that subtracts $10 from an account, the precondition should test not for an expected balance, but that the balance be equal to or greater than $10. Likewise, the operation should not simply write an expected new balance, rather it should read the current balance, subtract $10, and write the new balance.

Even with application-specific information, it may not be possible to automatically resolve a conflict. In such cases, the application will have to involve the user. The advantage with Rover is that the application can present the user with the current value for an object and the particular operation that was being performed, instead of simply providing the user with the new value and the value the application tried to write.

5.2 Toolkit Programming Interface

The programming interface between Rover and its client applications contains four primary functions: *create session*, *import*, *invoke*, and *export*. Client applications call *create session* once with authentication information to set up a connection with the local access manager and receive a session identifier. The authentication information is used by the access manager to authenticate client requests sent to Rover servers.

Rover objects are referenced using *Uniform Resource Locators*. Each URL is composed of a server name and a unique (to that server) identifier. To import an object, an application calls *import* and provides the object's URL, the session identifier, a callback, and arguments. In addition, the application specifies a priority that is used by the network scheduler to reorder QRPCs. The *import* function immediately returns a promise [33] to the application. The application can then wait on this promise or continue execution. Rover transparently queues QRPCs for each *import* operation in the stable log. When the requested object is received by the access manager, the access manager updates the promise with the returned information. In addition, if a callback was specified, the access manager invokes it. Once an object is imported, an application can *invoke* methods on it to read

Rover Application	Base code	New Rover client code	New Rover server code
Exmh	24,000 Tcl/Tk	1,700 Tcl/Tk 220 C	140 Tcl/Tk 2,700 C
Webcal	26,000 C++ and Tcl/Tk	2,600 C++ and Tcl/Tk	1,300 C++ and Tcl/Tk
HTTP Proxy	none	250 Tcl/Tk 1,500 C	740 C
Irolo	470 Tcl/Tk	400 Tcl/Tk 220 C	280 Tcl/Tk
NNTP Proxy	none	525 Tcl/Tk 476 C	350 C
Stock Watcher	none	84 Tcl/Tk 220 C	260 Perl 65 Tcl/Tk

Table 2: Lines of code changed or added in porting *Exmh* and *Webcal* to Rover and using Rover to implement the HTTP and NTTP proxies, *Irolo*, and the *Stock Watcher* applications.

and change it. Applications export each local change to an object back to servers by calling the *export* operation and providing the object's unique identifier, the session identifier, a callback, and arguments. Like *import*, *export* immediately returns a promise. When the access manager receives responses to exports, it updates the affected promises and invokes any application-specified callbacks.

5.3 Rover Application Suite

Section 3 discusses several important issues in designing mobile-aware applications; this section provides examples of how those issues are addressed in several mobile-transparent and mobile-aware applications that have been developed using the Rover toolkit (Table 1 lists the major implementation issues). The two mobile-transparent applications are: *Rover NNTP proxy*, a USENET reader proxy; and *Rover HTTP proxy*, a proxy for Web browsers. The mobile-aware applications are: *Rover Exmh*, an e-mail browser; *Rover Webcal*, a distributed calendar tool; *Rover Irolo*, a graphical Rolodex tool; *Rover Stock Market Watcher*, a tool that obtains stock quotes.

Two of the mobile-aware applications are based upon existing UNIX applications: Rover Exmh is a port of Brent Welch's Exmh Tcl/Tk-based e-mail browser, and Rover Webcal is a port of Ical, a Tcl/Tk and C++ based distributed calendar and scheduling program written by Sanjay Ghemawat. Rover Irolo and the Rover Stock Market Watcher were built from scratch. The former was written first using a standard file system interface and then ported to Rover.

This application suite was chosen to test several hypotheses about the ability to reasonably meet users' expectations in a mobile, intermittently-connected environment. Our application suite represents a set of applications that mobile users are in fact currenly using. However, current applications are not well integrated with the mobile environment. Because RDOs affect the structure of applications, it is important to qualitatively test the ideas contained in the Rover toolkit with complete applications in addition to using standard quantitative techniques.

As can been seen in Table 2, porting these file system-based workstation applications to a mobile-aware Rover applications requires varying amounts of work. Some applications were written or ported in a few weeks. Only Webcal required more than a month of work. In part, the extra time was due to additional evolution and redesign of the conflict resolution semantics of the toolkit. Porting *Exmh* and *Ical* to Rover required simple changes to appoximately 10% of the lines of code. Most of these changes consisted of replacing file system calls with object invocations; these modifications in *Rover Exmh* and *Rover Webcal* were made largely independently of the rest of the code.

The Rover HTTP and NNTP proxies demonstrate how Rover mobile-aware proxies support existing applications (*e.g.*, Netscape and USENET news browsers) without modification. Creating proxies for these services is far easier than modifying all the applications that use these services.

5.3.1 Mobile-Transparent Applications

Rover NNTP proxy. Using the Rover NNTP proxy (located on the client), users can read USENET news with standard news readers while disconnected and receive news updates even over slow links. Whereas most NNTP servers download and store all available news, the Rover proxy cache is filled on a demand-driven basis. When a user begins reading a newsgroup, the NNTP proxy loads the headers for that newsgroup as a single RDO while the articles are prefetched in the background. As the user's news reader requests the header of each article, the NNTP proxy provides them by using the local newsgroup RDO. As new articles arrive at the server, the server-side portion of the proxy constructs operations to update the newsgroup-header object. Thus, when a news reader performs the common operation of rereading the headers in a newsgroup, the NNTP proxy can service the request with minimal communication over the slow link.

Rover HTTP proxy. This application is unique in that it interoperates with most of the popular Web browsers. It allows users of existing Web browsers to "click ahead" of the arrived data by requesting multiple new documents before earlier requests have been satisfied. The proxy intercepts all HTTP requests and, if the requested item is not locally cached, returns a null response to the browser and enqueues the request in the operation log. When a connection becomes available, the page is automatically requested. In the meantime, the user can continue to browse already available pages and issue additional requests for pages without waiting. The granularity of RDOs is individual pages and images.

The client and server cooperate in prefetching. The client specifies the depth of prefetching for pages, while the server automatically prefetches in-lined images.

The proxy uses a separate non-browser window to display the status of a page (loaded or pending). If an uncached file is requested and the network is unavailable, an entry is added to the window. As pages arrive, the window is updated to reflect the changes. This window exposes the object cache and operations log directly to the user and allows the user limited control over them.

5.3.2 Mobile-Aware Applications

The mobile-aware applications all use the same technique for handling conflict detection and avoidance: a log of changes to RDOs. This log allows the server to detect and either avoid or resolve a conflict. The user is notified of any unresolvable conflicts and may then retry the original operation with or without modifying it or abort it. Also, most of the applications use color coding to indicate that an operation is tentative.

Rover Exmh. *Rover Exmh* uses three types of RDOs for manipulating application data, *mail messages*, *mail folders*, and *lists of mail folders*. By using this level of granularity, many user requests can be handled locally without any network traffic. Upon startup, Rover Exmh prefetches the list of mail folders, the mail folders the user has recently visited, and the messages in the user's inbox folder. Alternatively, using a finer level of granularity (*e.g.*, header and message body) would allow for more prefetching, but could delay servicing of user requests, especially during periods of disconnection. In the other direction, using a larger granularity (*e.g.*, entire folders) would seriously affect usability and response times for slow links.

Rover Exmh gives the user some control over the migration of computation to servers. For example, instead of performing a search of mail folders locally at the client (and thus having to import the index across a potentially low-bandwidth link), the user can instruct the client to construct a query request RDO and send it to the server.

Rover Webcal. This distributed calendar tool uses two types of data RDOs: *items* (appointments, daily todo lists, and daily reminders) and *calendars* (lists of items). At this level of granularity, the client fetches calendars and then prefetches the items they contain using a variety of user-specified strategies (*e.g.*, plus or minus one week, a month at a time, etc.).

Rover Irolo. This graphical Rolodex application uses two types of data RDOs, *entries* and *indices* (lists of entries). The GUI displays the last time an entry was updated and indicates whether the item is committed or tentative. For comparison purposes, we also implemented a file system-based version of this application.

Rover Stock Market Watcher. This is a simple financial stock tracking application. The client portion of this application constructs and sends an RDO to the server. The RDO executes at the server and watches the attributes of a user-specified stock for changes that exceed a user-specified threshold (*e.g.*, price, volume, or changes). When the threshold is exceeded, the server notifies the client. The server portion of the application uses fault-tolerant techniques to store the real-time information retrieved from stock ticker services [8].

6 Experiments

Rover runs on several platforms: IBM ThinkPad 701C (25 / 75Mhz i80486DX4) laptops running Linux 1.2.8; Intel Advanced/EV (120 Mhz Pentium) workstations running Linux 1.3.74; DECstation 5000 workstations running Ultrix 4.3; and SPARCstation 5 and 10 workstations running SunOS 4.1.3_U1. The primary mode of operation is to use the laptops as clients of the workstations. However, workstations can also be used as clients of other workstations.

The Rover server executes either as a Common Gateway Interface (CGI) plugin to NCSA's *httpd* 1.5a server (running on Ultrix and SunOS in the non-forking, pool of servers mode), or as a standalone TCP/IP server. The standalone server yields significant performance advantages over the CGI version, as it avoids the fork and exec overheads incurred on each invocation of the CGI version. In addition, because a new copy of the CGI server is started to satisfy each incoming request, any persistent state across connections must be stored in the file system and re-read for each connection, a penalty that the standalone server avoids.

Network options that we have experimented with include 10 Mbit/s switched Ethernet, 2 Mbit/s wireless AT&T WaveLAN, 128 Kbit/s and 64 Kbit/s Integrated Digital Services Network (ISDN) links, and Serial Line IP with Van Jacobson TCP/IP header compression (CSLIP) over 19.2 Kbit/s *V.32terbo* wired and 9.6 Kbit/s *Enhanced Throughput Cellular* (ETC) cellular dial-up links. The configuration used for the cellular experiments was the one suggested by our cellular provider and the cellular modem manufacturer: 9.6 Kbit/s ETC. The client connected to our laboratory's terminal server modem pool through the cellular service provider's pool of ETC cellular modems. This imposes a substantial added latency of approximately 600 ms, but also yields significantly better resilience to errors. Other choices are 14.4 Kbit/s ETC and directly connecting to the terminal server modem pool using 14.4 Kbit/s V.32bis. However, both choices suffer from significantly higher error rates, especially when the mobile host is in motion. Also, V.32bis is significantly less tolerant of the communication interruptions introduced by the in-band signaling used by cellular phones for cell switching and power level change requests.

The test environment consisted of a single server and multiple clients. The server machine was an Intel Advanced/EV workstation running the standalone TCP/IP server. The clients were IBM ThinkPad 701C laptops. All of the machines were otherwise idle during the tests.

To minimize the effects of unrelated network traffic on the experiments, the switched Ethernet was configured such that the server, the ThinkPad Ethernet adapter, and the WaveLAN base station were the only machines on the Ethernet segment and were all on the same switch port. However, network traffic over the wired, cellular, and ISDN links used shared public resources and traversed shared links; thus, there is increased variability in the experimental results for those network transports. To reduce the effects of the variations on the experiments, each experiment was executed multiple times and the results averaged. The standard deviations for our measurements were within 10% of the mean values. It is important to note that ordinary TCP/IP was used on the wireless networks. While Rover applications might benefit from the use of a specialized TCP/IP implementation, it is not necessary. Since a Rover application sends less data than an unmodified application, it is less sensitive to errors on wireless links.

The following experiments are designed to explore the performance characteristics of the Rover toolkit. In particular, the experiments test the following hypotheses:

1. The overhead imposed by logging QRPCs in stable logs at both the client and server is negligible for the target environments (*i.e.*, dial-up links).
2. The batching and compression of multiple requests that is enabled by using QRPC instead of RPC significantly improves the average request latency and throughput.
3. The Rover toolkit substantially improves the performance and usability of mobile-transparent applications.

Transport	TCP		QRPC Latency		
	Throughput 1 Mbyte	Latency null RPC	No Logging	Flash RAM Logging	Disk Logging
Ethernet	4.45	8	22	77	97
WaveLAN	1.09	20	34	79	116
128 ISDN	0.57	74	116	168	189
64 ISDN	0.32	87	117	184	191
19.2 Wired CSLIP	0.027	430	738	769	789
9.6 Cellular CSLIP	0.008	2230	3540	3670	3800

Table 3: The Rover experimental environment. Latencies are in milliseconds, throughput is in Mbit/s.

4. Using mobile-aware versions of applications offers significant performance and usability advantages over unmodified versions.

6.1 QRPC Performance

To establish the baseline performance for QRPC, we repeated the latency and bandwidth measurement experiments from [6], but extended them to include several additional network technologies and the use of Flash RAM for the client and server stable logs. The results are summarized in Table 3. Null RPC latency is a ping-pong over TCP sockets; TCP throughput is the time to send 1 Mbyte of compressible (14.4:1 using GNU's *gzip -6*) ASCII data similar to Rover Tcl-based RDOs; and QRPC latency is the time to perform a null 200-byte QRPC (200 bytes is the average size of a QRPC from our experience with Rover applications). The ISDN, Wired CSLIP, and Cellular CSLIP links perform hardware compression. The cellular times reflect the overhead of the ETC protocol and a non-error-free wireless link.

The cost of a QRPC has several primary components: the transport cost (the base null TCP cost from Table 3 plus the per-byte network transmission cost); the stable client and server logging costs; and the execution cost of the QRPC itself. By using stable logging at clients, Rover can guarantee the delivery of requests from clients to servers. The use of server-side stable logging allows Rover to avoid having to retransmit a request from a client (which might be disconnected) after a server failure [8]. The results show that the relative impact of logging is a function of the transport media. Since we expect that Rover users will often be connected via slower links (*e.g.*, wired or cellular dialup), the cost of stable logging will be a minor component of overall performance (*e.g.*, less than 1% for cellular links when using Flash RAM). Thus, we believe it is acceptable to pay the additional cost for client and server logging of QRPCs.

To understand the effects of batching and compression, we used a synthetic workload and measured the performance of QRPC with asynchronous logging. Using asynchronous logging allows the cost of logging to be amortized across multiple QRPCs. The synthetic work load consisted of a issuing a series of 60 200-byte null QRPCs. For the batching and multiple outstanding requests experiments, there were no interarrival pauses.

The network scheduler's batching algorithm yields the following behavior:

1. When the network scheduler receives each of the first few requests, the queue of pending requests from the test client is empty; so each request is sent in a separate message. No delay is imposed on these messages.

2. As the access manager starts spending additional time processing outgoing messages and incoming responses, the queue of pending requests from the test client grows in size.

3. At this point, the network scheduler constructs large batches of requests. Early messages in each batch incur some delay, while later messages in the same batch incur significantly less delay.

Fig. 3 shows the effects of batching and compression (using the heuristic from Section 4) on the per-request cost using the synthetic workload. In each set of bars, the leftmost bar (compression and batching) shows the performance when both compression and batching are applied. For this test, the compression ratio was approximately twelve to one and the batch size was an average of seven requests per message with individual batches ranging in size from one to 55 requests, depending upon the speed of the network link.

The second bar (compression with a single outstanding request) shows the performance with compression and only a single request outstanding. The compression ratio was 2.5 to one. The third bar (batching and no compression) shows the performance with batching and no compression. Batches contained an average of seven requests per message with individual batches ranging in size from one to 55 requests, depending upon the speed of the network link.

The fourth bar (no compression and multiple outstanding requests) shows the performance without compression or batching, but with multiple outstanding requests. The rightmost bar (No compression with a single outstanding request) shows the performance without compression or batching and with only a single request outstanding.

The results show that together compression and batching offer performance gains for all networks with the largest gains occurring for the slowest networks. The main reason for the batching performance gain is the elimination of multiple roundtrip latencies. Compression offers a significant benefit only when used with batching because it is able to compress multiple QRPC headers within a batch.

6.2 Mobile-Transparent Application Performance

We compared the performance of Netscape using a mobile-transparent Rover HTTP proxy against the same application executing independently. We measured the time to fetch and display 10 WWW pages using a variety of networks

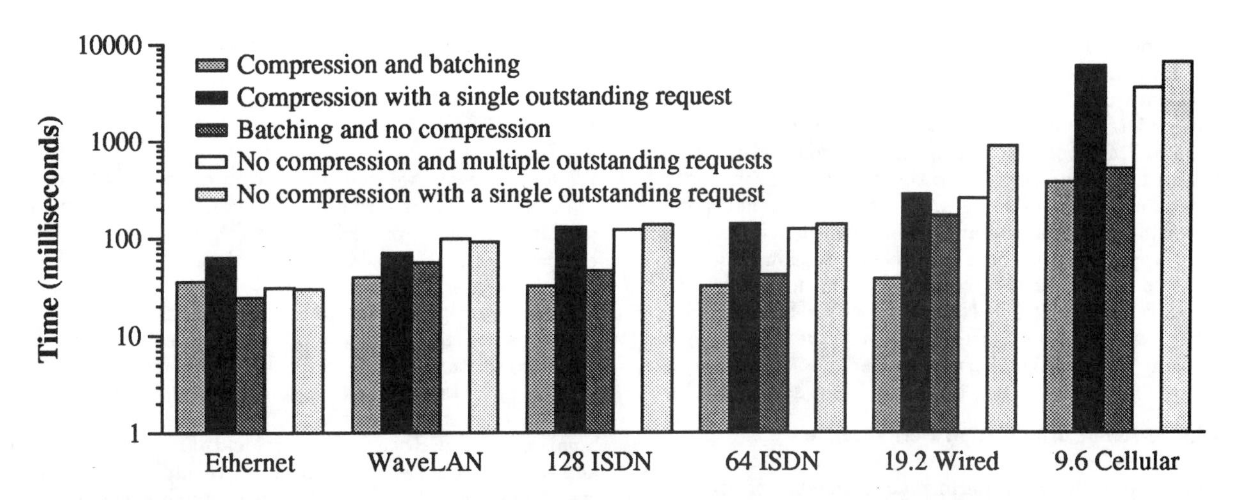

Figure 3: Average time in milliseconds for one null QRPC when using compression and batching of a synthetic workload consisting of null QRPCs with asynchronous log record flushing. The y-axis uses a logarithmic scale.

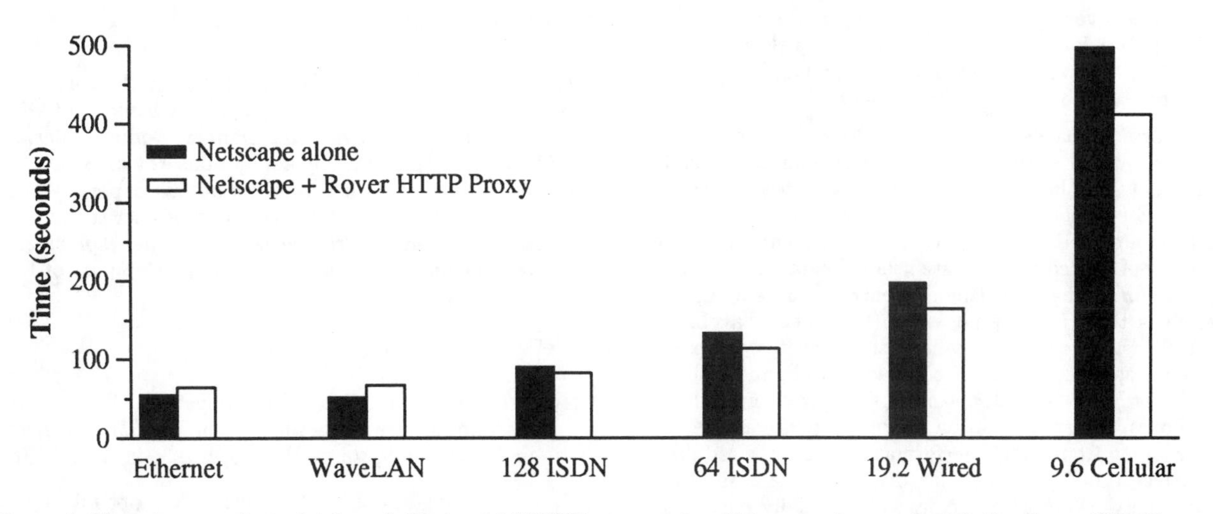

Figure 4: Time in seconds to fetch and display 10 WWW pages using Netscape alone and with the Rover HTTP proxy.

Fig. 4 provides the results of the experiment and shows that performance when using the Rover HTTP proxy is comparable for faster networks and up to 17% faster for the slower networks. The performance gain is a result of the elimination of multiple roundtrip latencies to open connections for inlined objects and the compression of the data. The total data transmitted to the client was 239 Kbytes of compressed data equivalent to 286 Kbytes of uncompressed data. The HTML portion of the pages accounted for 44.5 Kbytes and had a compression ratio of 2.6:1. The majority of the data consisted of images, which were far less compressible using the default compression. We plan to explore the use of application-specific image compression [23]. It is important to note that the experiments do not reflect the "click-ahead" nature of the Netscape+Rover HTTP proxy application, which allows the user to browse the loaded pages while waiting for additional pages to load.

6.3 Mobile-Aware Application Performance

This section quantifies the performance benefits of caching RDOs and presents a comparison between mobile-transparent applications and mobile-aware applications running on both high-bandwidth, low-latency and low-bandwidth, high-latency networks.

To measure the performance benefits of the complete Rover system for mobile-aware applications, we compare the performance of Rover Webcal, Rover Exmh, and Rover Irolo to their unmodified X11-based counterparts, Ical, Exmh, and Irolo. For each application, we designed a workload representative of a typical user's actions and measured the time to perform the complete task. To keep the measurements representative, we did not measure the cost of starting the application and loading the data required for the task. This operation is typical of how most people use mobile computers, where the application is started and frequently used data is loaded over a fast network (*e.g.*, at the office) and then the application is used repeatedly over a slow network or without any network connectivity (*e.g.*, on a plane). Each task was repeated on each of the six network options.

Fig. 5 presents the speedup of the Rover version of each application over the original X11-based application while performing a sample set of tasks. The tasks were: reading eight MIME e-mail messages, viewing one week's appointments from a calendar with 50 items in it (the approximate median size of shared and private user calendars on our systems), and browsing fifty Rolodex entries. In general, the results show that, for fast networks (Ethernet, WaveLAN, and ISDN), the performance when using Rover is comparable (a slight speedup for Irolo, equal for Exmh, and a slight slowdown for Ical). Over slower networks (wired and cellular dial-up links), Rover application performance is consistently better (ranging from a 33% performance gain on wired dial-up to a factor of 7.5 on cellular dial-up). The results for these two networking technologies are especially encouraging, since they represent the target environment for Rover.

When no network is present, it is not possible to use the original X11-based applications. The Rover applications, however, show no change in performance as long as the application data are locally cached.

What the numbers fail to convey is the extreme sluggishness of the user interface when using slower (*e.g.*, cellular) links without Rover. Scrolling and refreshing operations are extremely slow. Pressing buttons and selecting text are very difficult operations to perform because of the lag between mouse clicks and display updates. With Rover, the GUI runs locally and the user sees the same excellent GUI performance across a range of networks that varies by three orders of magnitude in both bandwidth and latency.

7 Conclusions

We have shown that the integration of relocatable dynamic objects and queued remote procedure calls in the Rover toolkit provides a powerful basis for building mobile-transparent and mobile-aware applications. We have found it quite easy to adapt applications to use these Rover facilities, resulting in applications that are far less dependent on high-performance communication connectivity. For example, one might conjecture that it would be difficult to build a mobile version of Netscape that provides useful service in the absence of network connectivity. In practice, we find the combination of the Rover cache, relocatable dynamic objects, and queued remote procedure calls results in a surprisingly useful system. Furthermore, it was much easier to build a proxy using the toolkit than with an *ad-hoc* approach.

RDOs and QRPCs allow application developers to decouple many user-observable delays from network latencies. The result is excellent graphical user interface performance over network technologies that vary by three orders of magnitude in bandwidth and latency. In addition, measurements of end-to-end mobile application performance shows that mobile-transparent and mobile-aware applications perform significantly better than their stationary counterparts. For example, for the mobile-transparent Netscape application, we observe a performance improvement of 17%. For mobile-aware applications, we observe performance improvements of up to a factor of 7.5 over slow networks.

8 Acknowledgments

We thank David Gifford for his efforts in the early design stages of the Rover toolkit, in particular his idea of QRPC. We also thank Atul Adya, Greg Ganger, Eddie Kohler, Massimiliano Poletto, and the anonymous reviewers for their careful readings of earlier versions of this paper. We would also like to thank the rest of the Rover project team: George Candea, Constantine Cristakos, Alan F. deLespinasse, and Michael Shurpik for helping with the development of Rover.

References

[1] J. J. Kistler and M. Satyanarayanan, "Disconnected operation in the Coda file system," *ACM Transactions on Computer Systems*, vol. 10, pp. 3–25, Feb. 1992.

[2] H. T. Kung and J. T. Robinson, "On optimistic methods for concurrency control," *ACM Transactions on Database Systems*, vol. 6, no. 2, pp. 213–226, June 1981.

[3] L. Huston and P. Honeyman, "Partially connected operation," In USENIXMLIC2 [34], pp. 91–97.

[4] P. Kumar, *Mitigating the Effects of Optimistic Replication in a Distributed File System*, Ph.D. thesis, School of Computer Science, Carnegie Mellon University, Dec. 1994.

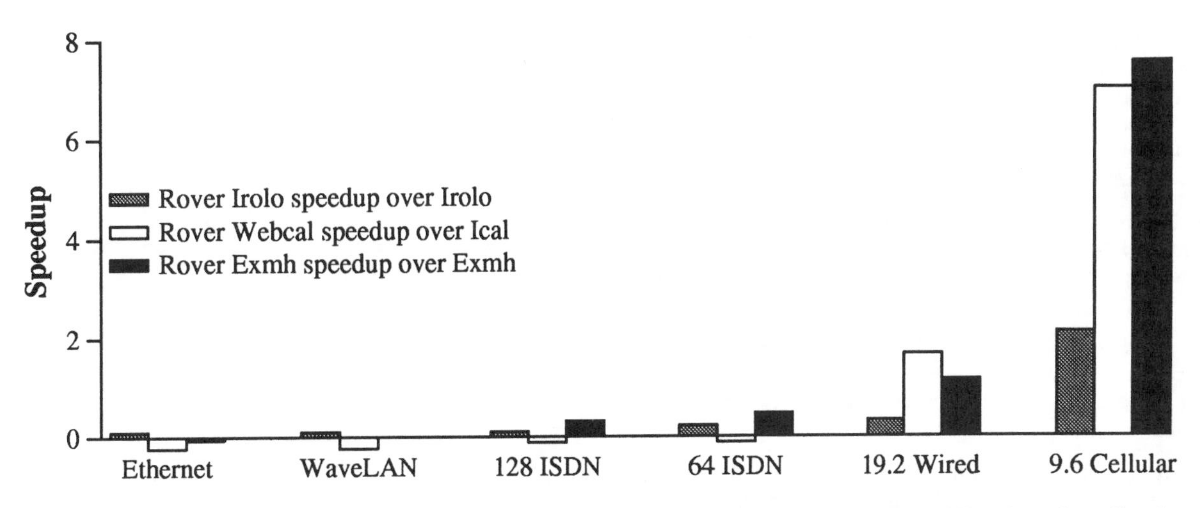

Figure 5: Speedup (or slowdown) of Rover mobile-aware versions of applications over the original X11-based applications while performing a sample set of tasks.

[5] J. H. Saltzer, D. P. Reed, and D. D. Clark, "End-to-end arguments in system design," *ACM Transactions on Computer Systems*, vol. 2, no. 4, pp. 277–28, Nov. 1984.

[6] A. Joseph, A. F. deLespinasse, J. A. Tauber, D. K. Gifford, and F. Kaashoek, "Rover: A toolkit for mobile information access," in *Proc. of the Fifteenth Symposium on Operating Systems Principles (SOSP)*, Copper Mountain Resort, CO, Dec. 1995, pp. 156–171.

[7] A.D. Birrell and B.J. Nelson, "Implementing remote procedure calls," *ACM Transactions on Computer Systems*, vol. 2, no. 1, pp. 39–59, Feb. 1984.

[8] A. Joseph and F. Kaashoek, "Building reliable mobile-aware applications using the Rover toolkit," in *Proc. of the Second International Conference on Mobile Computing and Networking (MOBICOM '96)*, Rye, NY, Nov. 1996, pp. 117–129.

[9] R. H. Katz, "Adaptation and mobility in wireless information systems," *IEEE Personal Communications*, vol. 1, no. 1, pp. 6–17, First Quarter 1994.

[10] N. Davies, G. Blair, K. Cheverst, and A. Friday, "Supporting adaptive services in a heterogeneous mobile environment," in *Proc. of the Workshop on Mobile Computing Systems and Applications*, Santa Cruz, CA, Dec. 1994, pp. 153–157.

[11] F. Kaashoek, T. Pinckney, and J. A. Tauber, "Dynamic documents: Mobile wireless access to the WWW," in *Proc. of the Workshop on Mobile Computing Systems and Applications*, Santa Cruz, CA, Dec. 1994, pp. 179–184.

[12] D. B. Terry, M. M. Theimer, K. Petersen, A. J. Demers, M. J. Spreitzer, and C. H. Hauser, "Managing update conflicts in a weakly connected replicated storage system," in *Proc. of the Fifteenth Symposium on Operating Systems Principles (SOSP)*, Copper Mountain Resort, CO, Dec. 1995, pp. 172–183.

[13] M.G. Baker, "Changing communication environments in MosquitoNet," in *Proc. of the Workshop on Mobile Computing Systems and Applications*, Santa Cruz, CA, Dec. 1994, pp. 64–68.

[14] L. B. Mummert, M. R. Ebling, and M. Satyanarayanan, "Exploiting weak connectivity for mobile file access," in *Proc. of the Fifteenth Symposium on Operating Systems Principles (SOSP)*, Copper Mountain Resort, CO, Dec. 1995, pp. 143–155.

[15] B. D. Noble, M. Price, and M. Satyanarayanan, "A programming interface for application-aware adaptation in mobile computing," In USENIXMLIC2 [34], pp. 57–66.

[16] P. Reiher, J. Heidemann, D. Ratner, G. Skinner, and G. J. Popek, "Resolving file conflicts in the Ficus file system," in *Proc. of the USENIX Summer 1994 Technical Conference*, Boston, MA, 1994, pp. 183–195.

[17] T. Watson and B. Bershad, "Local area mobile computing on stock hardware and mostly stock software," in *Proc. of the First USENIX Symposium on Mobile & Location-Independent Computing*, Cambridge, MA, Aug. 1993, pp. 109–116.

[18] T. Watson, "Application design for wireless computing," in *Proc. of the Workshop on Mobile Computing Systems and Applications*, Santa Cruz, CA, Dec. 1994, pp. 91–94.

[19] D. B. Terry, A. J. Demers, K. Petersen, M. J. Spreitzer, M. M. Theimer, and B. B. Welch, "Session guarantees for weakly consistent replicated data," in *Proc. of the 1994 Symposium on Parallel and Distributed Information Systems*, Sept. 1994, pp. 140–149.

[20] B. R. Badrinath and G. Welling, "Event delivery abstractions for mobile computing," Tech. Rep. LCSR–TR–242, Department of Computer Science, Rutgers University, New Brunswick, NJ 08903, 1996.

[21] M.T. Le, F. Burghardt, S. Seshan, and J. Rabaey, "InfoNet: the networking infrastructure of InfoPad," in *Proc. of the Spring COMPCON Conference*, San Francisco, Mar. 1995, pp. 163–168.

[22] S. Narayanaswamy, S. Seshan, E. Amir, E. Brewer, E. Brodersen, F. Bughardt, A. Burstein, Y. Chang, A. Fox, J. Gilbert, R. Han, R. Katz, A. Long, D. Messerschmitt, and J. Rabaey, "Application and network support for InfoPad," *IEEE Personal Communications*, vol. 3, no. 2, pp. 4–17, Apr. 1996.

[23] A. Fox, S. D. Gribble, E. Brewer, and E. Amir, "Adapting to network and client variability via on-demand dynamic distillation," in *Proc. of the Seventh International Conference on Architectural Support for Programming Languages and Operating Systems (ASPLOS)*, Cambridge, MA, Oct. 1996, pp. 160–173.

[24] J. Bartlett, "W4—the Wireless World-Wide Web," in *Proc. of the Workshop on Mobile Computing Systems and Applications*, Santa Cruz, CA, Dec. 1994, pp. 176–178.

[25] M. H. Brown and R. A. Schillner, "DeckScape: An experimental web browser," Tech. Rep. 135a, Digital Equipment Corporation Systems Research Center, Mar. 1995.

[26] R. Katz, E. A. Brewer, E. Amir, H. Balakrishnan, A. Fox, S. Gribble, T. Hodes, D. Jiang, G. Thanh Nguyen, V. Padmanabhan, and M. Stemm, "The Bay Area Research Wireless Access Network (BARWAN)," in *Proc. of the Spring COMPCON Conference*, Santa Clara, California, Feb. 1996.

[27] L. Kawell Jr., S. Beckhardt, T. Halvorsen, R. Ozzie, and I. Greif, "Replicated document management in a group communication system," Presented at the *Second Conference on Computer-Supported Cooperative Work*, Portland, OR, Sept. 1988.

[28] J. Gray, P. Helland, P. O'Neil, and D. Shasha, "The dangers of replication and a solution," in *Proc. of the 1996 SIGMOD Conference*, Montreal, Quebec, Canada, June 1996, pp. 173–182.

[29] D. K. Gifford and J. E. Donahue, "Coordinating independent atomic actions," in *Proc. of the Spring COMPCON Conference*, San Francisco, CA, Feb. 1985, pp. 92–92.

[30] W. Weihl and B. Liskov, "Implementation of resilient, atomic data types," *ACM Transactions on Programming Languages and Systems*, vol. 7, no. 2, pp. 244–269, Apr. 1985.

[31] J.K. Ousterhout, *Tcl and the Tk Toolkit*, Addison-Wesley, Reading, MA, 1994.

[32] K. Arnold and J. Gosling, *The Java Programming Language*, Addison-Wesley Publishing Co., Reading, Massachusetts, 1996.

[33] B. Liskov and L. Shrira, "Promises: Linguistic support for efficient asynchronous procedure calls," in *Proc. of the SIGPLAN Conference on Programming Language Design and Implementation*, Atlanta, GA, June 1988, pp. 260–267.

[34] *Proc. of the Second USENIX Symposium on Mobile & Location-Independent Computing*, Ann Arbor, MI, Apr. 1995.

9.5 Housel, B.C. and Lindquist, D.B., WebExpress: A System for Optimizing Web Browsing in a Wireless Environment

Housel, B.C. and Lindquist, D.B., "WebExpress: A System for Optimizing Web Browsing in a Wireless Environment," *Proceedings of the Second Annual International Conference on Mobile Computing and Networking*, pp. 108–116, November 1996.

Web*Express*: A System for Optimizing Web Browsing in a Wireless Environment

Barron C. Housel Ph.d
IBM Corporation (BRQA/502)
P.O. Box 12195
RTP, NC 27709
housel@vnet.ibm.com
(919)254-6337

David B. Lindquist
IBM Corporation (BRQA/502)
P.O. Box 12195
RTP, NC 27709
lindqui@raleigh.ibm.com
(919)254-7862

Abstract: This paper describes software technology that makes it possible to run Internet World Wide Web applications in wide area wireless networks. Web technology in conjunction with today's mobile devices (e.g., laptops, notebooks, personal digital assistants) and the emerging wireless technologies (e.g., packet radio, CDPD) offer the potential for unprecedented access to data and applications by mobile workers. Yet, the limited bandwidth, high latency, high cost, and poor reliability of today's wireless wide-area networks greatly inhibits (even to the point of being unfeasible) supporting Web applications over wireless networks. WebExpress significantly reduces user cost and response time of wireless communications by intercepting the HTTP data stream and performing various optimizations including: file caching, forms differencing, protocol reduction, and the elimination of redundant HTTP header transmission. This paper describes these optimizations and presents some experiment results.

Introduction

This paper describes WebExpress, a software system that significantly reduces data volume and latency of wireless communications in the context of World Wide Web (simply "Web") client/server applications. WebExpress provides this traffic reduction using an interception technology that is transparent to Web clients and servers.

Web technology [1] is rapidly being accepted as a universal interface for network access to information. The Web technology is based on the Hyper Text Markup Language (HTML) [2] and the Hyper Text Transport Protocol (HTTP) [3]. HTML provides a common representation for information, and HTTP defines the common protocol for transporting information between Web clients and Web servers. The Web browser serves as the end user interface; it is responsible for sending user requests to the appropriate Web server (normally via a Web proxy gateway) and formatting and displaying HTML data streams returned to the client device.

In addition to information retrieval and browsing, Web technology is being used as the user interface for many forms processing applications. While the Web technology currently lacks certain features useful for forms processing (e.g., field editing and flexible display formatting), the widespread availability of Web browsers, the availability of the Internet to offer worldwide any-to-any connectivity, the ease with which users can create forms using HTML, and the ability to easily write Web application servers (using the HTTP defined Common Gateway Interface (CGI)) make the Web very attractive (and inexpensive) for a large class of forms-based applications. WebExpress is aimed at routine repetitive commercial applications (in contrast to random, ad hoc Web browsing) such as: visiting nurse or medical personnel; mobile salespersons that need to query product data, place orders, and check credit; service workers involved with equipment repair, checking warehouses for parts, etc.

The growth in mobile computing devices [1] and the emergence of wide area wireless technologies [2] [4-8] paves the way for rapid growth in mobile wireless communications. The popularity of the Web technology suggests that Web browsers offer a compelling end user interface for many mobile wireless applications.

The objective of WebExpress is to facilitate the use of Web technology to run typical commercial transaction processing applications over wireless networks. The predictability of this type of usage makes it possible to employ optimizations that make wireless Web access practical from both a usability and cost perspective. The successful deployment of WebExpress extends Web technology to a new usage domain.

Next, we briefly summarize the HTTP protocol and describe the inhibitors to Web browsing over wireless networks.

Hyper Text Transport Protocol (HTTP) Summary

Normally, (see Figure 1) a Web browser communicates directly with a Web server (or proxy server) over a TCP

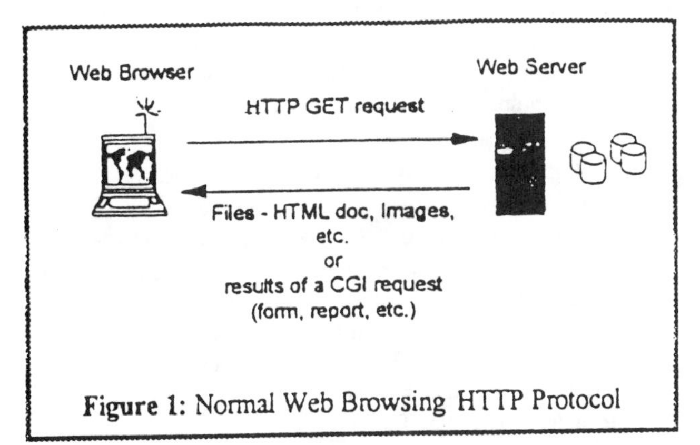

Figure 1: Normal Web Browsing HTTP Protocol

connection using the HTTP protocol. The user specifies a *Universal Resource Locator* (URL) to address the object requested. This object may be a stored (HTML) text document or an HTML data stream that is generated by a program. In the latter case, the Web server invokes the program via the *Common Gateway Interface* (CGI) that is defined as part of the Web technology. The HTML object returned to the Web browser client may contain hyper-links to other HTML objects and directly embed graphic (GIF, JPG...) objects. It is the browser's responsibility to issue additional requests for the embedded objects (on behalf or the end user) until the entire document is complete.

Inhibitors to Web Browsing over WAN Wireless Networks

Wide area wireless communications has four major negative characteristics:

- **High cost:** The cost per byte transmitted is orders of magnitude greater than traditional wire line (LAN/WAN) networks. Sending a 10K HTML page costs from one to two dollars over current wireless networks [3].
- **High latency:** The response time for wireless links in WANs is very slow compared to their wire line or wireless LAN counterparts. WebExpress depends on communications support (e.g., CDPD, ARTour[4] [8]) that provides TCP/IP over wireless links. A typical TCP/IP request takes around 15 seconds. This response time was also reported by Oracle [9].

- **Low bandwidth:** The capacity of WAN wireless links is very limited compared to most wire line links. Depending on the technology, the channel bit rate of wireless links ranges from 4800 bps (e.g., Ardis MDC4800) to 19200 bps (CDPD, RD-LAP) with the effective data rate being substantially lower due to latency, shared media, retransmission. A wireless link is ***shared*** among all the mobile terminals within range of the wireless base station. Thus, the data rate realized by any given wireless device is the link speed divided by the number of devices simultaneously using the link (i.e., similar to a multi-drop link).
- **Very unreliable:** Wireless WAN connections are significantly less reliable than wire line connections. Wireless connections may be intermittently or permanently disrupted for various reasons: the mobile device goes out-of-range; the mobile device goes behind a barrier that blocks the signal; or higher layer communications or applications protocols time-out because of delayed responses. Disruptions may be shielded from the wireless application by error recovery procedures in the link-level and higher layer communications protocols. However, this often results in excessive retransmissions (and extra cost), not to mention user frustration that may result in request resubmissions.

In addition to the limitations of wireless communications, the **Web HTTP protocol** presents the following inefficiencies:

- **Connection overhead:** Each request for an HTML page or graphic object (i.e., GIF or JPEG file) requires the browser to open a TCP/IP socket. This operation adds to data overhead and greatly increases the latency. We have observed connection times in the neighborhood of 15 seconds in a WAN wireless environment.
- **Redundant transmission of capabilities:** The HTTP protocol is stateless; this requires that the browser must (re)send its capabilities, a list of 200–400 bytes in length, with each request. These capabilities are normally the same for any given browser (i.e., client device).
- **Verbose protocol:** HTTP control information is coded in standard ASCII and employs human-friendly keywords, which increases the number of bytes transmitted per request.

While the above overhead can be tolerated [5] in wire line networks, it renders Web access in a wireless environment infeasible because of unacceptable response times and occasional time-outs.

The remainder of the paper describes the WebExpress Intercept model and the variety of data reduction techniques that it employs: caching, differencing, protocol reduction, and header reduction.

---

[3] We accessed a popular quote server over a packet radio network and incurred a charge of $27; the second quote was less than $1. With WebExpress both quotes were about $0.10 a piece .

[4] ARTour is a registered trademark of the IBM Corp.

[5] Even in wire line networks the ability of HTTP to scale is a problem as evidenced by performance improvements in HTTP 1.1 and active research (see references) to improve HTTP.

WebExpress Intercept Model

An important objective of WebExpress is to be able to run with *any Web browser* (e.g., Netscape, Mosaic, ...) and *any Web server* without imposing any changes to either. To accomplish this we use an intercept technique that enables WebExpress to intercept and control communications over the wireless link for the purposes of reducing traffic volume and optimizing the communications protocol to reduce latency. The WebExpress components that implement this intercept technique are shown in Figure 2. We insert two components into the data path between the Web client and the Web server[6]: the *Client Side Intercept (CSI)* process that runs in the end user client mobile device and the *Server Side Intercept (SSI)* process that runs within the wire line network. The CSI intercepts HTTP requests and, together with the SSI, performs optimizations to reduce data transmission over the wireless link.

From the viewpoint of the browser, the CSI appears as a local Web proxy that is co-resident with the Web browser. The CSI communicates with the Web browser over a local TCP connection (using the TCP/IP "loopback" feature) via the HTTP protocol. Therefore, no external communication occurs over the TCP/IP connection between the browser and the CSI. No changes to the browser are required other than specifying the (local) IP address of the CSI as the browser's proxy address. The actual proxy (or *socket server)* address is specified as part of the SSI configuration. The CSI communicates with an SSI process over a TCP connection using a *reduce*d version of HTTP. (see HTTP Protocol Reduction).
The SSI reconstitutes the HTML data stream and forwards it to the designated Web proxy server. Likewise, for responses returned by Web servers (or proxies), the CSI reconstitutes an HTML data stream received from the SSI and sends it to the Web browser over the local TCP connection as though it came directly from the Web server.

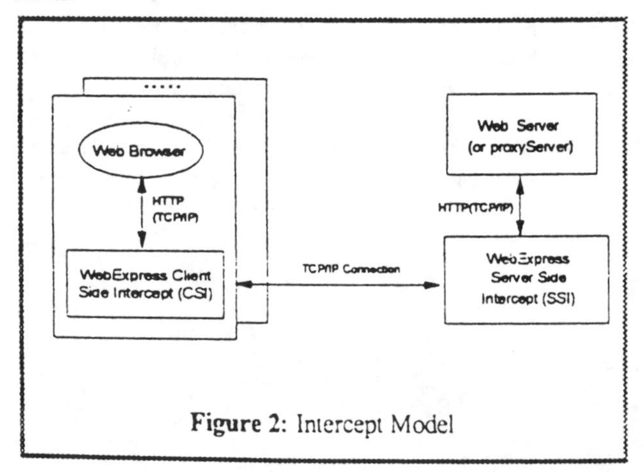

Figure 2: Intercept Model

The intercept model implemented in WebExpress offers a number of advantages: It is transparent to both Web browsers and Web (proxy) servers and, therefore, can be employed with any Web browser. It is largely insensitive to the development of the rapidly maturing HTML/HTTP technology. Thus, WebExpress does not have to be upgraded to run with new (or different) versions of Web browsers that are available in the market place. The CSI/SSI protocols facilitate highly effective data reduction and protocol optimization without limiting any of the Web browser functionality or interoperability. The remainder of this paper describes the WebExpress optimization methods summarized below:

- **Caching:** Both the SSI and CSI cache graphic and HTML objects. If the URL specifies an object in the CSI's cache, it is returned immediately as the browser response. The caching functions guarantee cache integrity within a client-specified time interval. The SSI cache is populated by responses from the requested Web servers. If a requested URL received from a CSI is cached in the SSI, it is returned as the response to the request. As we shall see later the cache plays an important role in the differencing function.
- **Differencing:** CGI requests result in responses that normally vary for multiple requests to the same URL (e.g., a stock quote server). The concept of differencing is to cache a common *base* object on both the CSI and SSI. When a response is received, the SSI computes the difference between the base object and the response and then sends the difference to the CSI. The CSI then merges the difference with its base form to create the browser response. This same technique is used to determine the difference between HTML documents.
- **Protocol reduction:** Each CSI connects to its SSI with a single TCP/IP connection. All requests are routed over this connection to avoid the costly connection establishment overhead. Requests and responses are multiplexed over the connection.
- **Header reduction:** The HTTP protocol is stateless, requiring that each request contain the browser's capabilities[7]. For a given browser, this information is the same for all requests. When the CSI establishes a connection with its SSI, it sends its capabilities only on the first request. This information is maintained by the SSI for the duration of the connection. The SSI includes the capabilities as part of the HTTP request that it forwards to the target server (in the wire line network).

Caching

The WebExpress cache methods are designed to address the browsing of stored documents or files that change relatively infrequently. Information that does change frequently is handled by the differencing methods described in the next section.

Today's Web browsers offer a variety of cache technologies and cache management options. In general, their caches are designed to meet the needs of *wired* users where ad hoc browsing is common and cache methods will trade *inexpensive* network bandwidth for reduced storage consumption. These cache methods are designed around a browser session, where we define a browser session to be the start and close of a browser application. These methods either purge the cache at the end of a session or let the cache objects persist across session with updates occurring once on first reference

[6]Henceforth, the term *Web server* can be a proxy server, a socket server, or the target Web server.

[7]Called *access lists* in HTTP.

per session. Advanced user options are often available to cause cached objects to be updated on every reference or to never be updated. Our experience with wireless systems indicates that cache methods designed for wired networks are not well suited for wireless users of Web application. We found that cross-session persistence of cached objects is critical, that update methods should be optimistic techniques that look for changes to stored objects based on elapsed time, and that changed objects should be updated through differencing versus a complete object update. WebExpress attempts to maximize the value of its client cache by blending these requirements with some additional user controls such that the cache management technique is adaptable to the unique characteristics of each application.

As illustrated in Figure 3, WebExpress supports client and server caching. The algorithm used to manage the client cache is basically Least Recently Used (LRU), with a user option to specify indefinite persistence of specific objects. It is our intention to maximize client cache efficiencies through cache methods that allow a user to *declare* frequently accessed or critical information for object persistence and to enable the system to adapt to the browsing patterns of the individual user. The server cache, which is also LRU managed, is designed to adapt to the browsing patterns of a set of users. Information retrieved by one user may be reused by others, avoiding delays associated with retrieving information from the Web server.

Objects loaded into either the WebExpress client or server caches persist across browser sessions. This decision increased the cache hit ratios but presented us with a cache coherency problem. We needed a mechanism to identify when objects change and the ability to update changed objects, without over-utilizing the wireless link. To satisfy this requirement we designed the cache coherency methods to be based on the age of information. This approach allows WebExpress users to adjust the frequency of update requests to the dynamics of each Web application. To provide this cache coherency model, WebExpress associates a *digital signature*, computed as a CRC, and a *coherency interval* (CI) with each cached object. The CI specifies when the object is to be checked for changes. The CI, measured in minutes, is set by

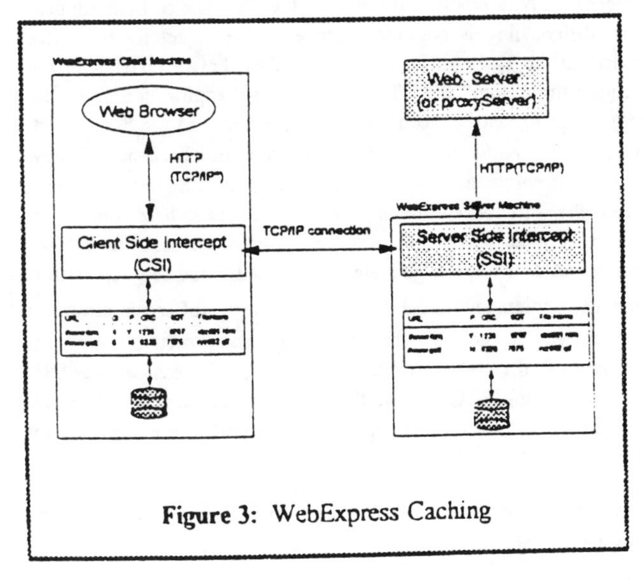

Figure 3: WebExpress Caching

each user or an administrator as a default for all cache objects. The user may override the default CI for any given set of cached objects. When a cached object is referenced, the CSI checks to see if the coherency interval has been exceeded. If it has not, the cached page is used. If the coherency interval *has* been exceeded, the CSI and SSI execute a protocol to determine if a fresh copy of the page has to be fetched. Essentially, the CSI requests that the SSI verify that the object in question has not changed; the CSI provides the SSI the object's URL, a digital signature of the object, and the CI associated with the object. The SSI attempts to satisfy the coherency request with the contents of its cache, depending upon the age of the object within its cache. If the object within the SSI is too old based on the CI or the object isn't in the SSI cache, the SSI will obtain a new copy of the object from the Web server and enter a Store Date Time (SDT) and digital signature for the object into its cache directory. Now, if the digital signatures match between the CSI's and SSI's copy, SSI indicates to the CSI that the object in question is up to date. CSI then updates the store date time (SDT) of the object to reflect the SSI's SDT of the same object; an age scheme is used to avoid clock synchronization problems. If the object has changed, based on the digital signature compare, the SSI indicates to the CSI that the object is out of date and sends the updated object to CSI. CSI then updates its cache and directory appropriately.

The use of a coherency interval allows users to specify how frequently to update cached Web objects, trading off wireless traffic for instantaneous cache coherency. Initiating coherency checks only on object references limits wireless traffic to only those objects actually being referenced. An alternative approach, which we may investigate in a future version of WebExpress, would be to initiate a single asynchronous coherency check on session start (CSI start) for all cached objects that are older than the coherency interval. This *batch* check has the potential to further reduce the per object wireless latency associated with maintaining cache coherency.

The WebExpress cache methods reduce the volume of application data transmitted over the wireless link, through cross browser session persistence, age based coherency algorithms, digital signature based modification verifications, and the user options (CI, and object persistence). When updates to cached objects are required, the differencing methods are invoked to further the reduction of application traffic.

Differencing

The expectation of browsing stored documents or files is that they change relatively infrequently. This fact is fundamental to the caching techniques discussed previously. However, the Web technology can be used to perform transaction processing. The HTML definition allows the specification of HTML streams that enable users to enter data and then *submit* the data (or form) for processing by some executable program located somewhere in the Web. The program is identified by a URL just like non-executable files, and a command (e.g., GET or POST) is sent from the browser to the server specified by the URL. This command may be coded explicitly or generated implicitly as a result of entering data on a displayed form. Input data (if any) follow the URL as part of the HTTP data stream. The rules for invoking programs, enabling them to read parameter data and generate replies, are called the *Common Gateway Interface* (CGI). The term *CGI Processing* refers the to the process of executing programs from Web browsers.

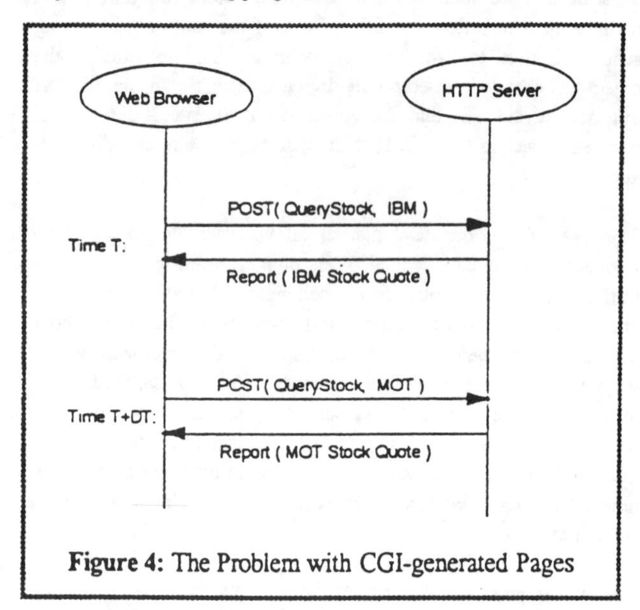

Figure 4: The Problem with CGI-generated Pages

The caching techniques described earlier do not help in CGI processing because no two replies to requests to the same URL are likely to be the same. This is not surprising because users enter different data for different requests and expect to receive different results. Figure 4 shows two different queries to a stock-quote server. Naturally, the reports for IBM and Motorola stocks are different. To minimize responses from CGI programs, we use a different form of caching and differencing technology. This approach is based on the observation that *different replies from the same program (application server) are usually very similar*. For example, the replies to the IBM and Motorola stock query vary only in numbers (e.g., price) and symbols (e.g., MOT or IBM); i.e., HTML byte streams representing query responses often contains lots of unchanging formatting data (including graphics). To illustrate the algorithm, we consider two queries to program (Form) X at time T and T+DT as shown in Figures 5 and 6.

The CSI determines that the HTTP request is a CGI request if the method is *Post* or if the URL is followed by a name/value

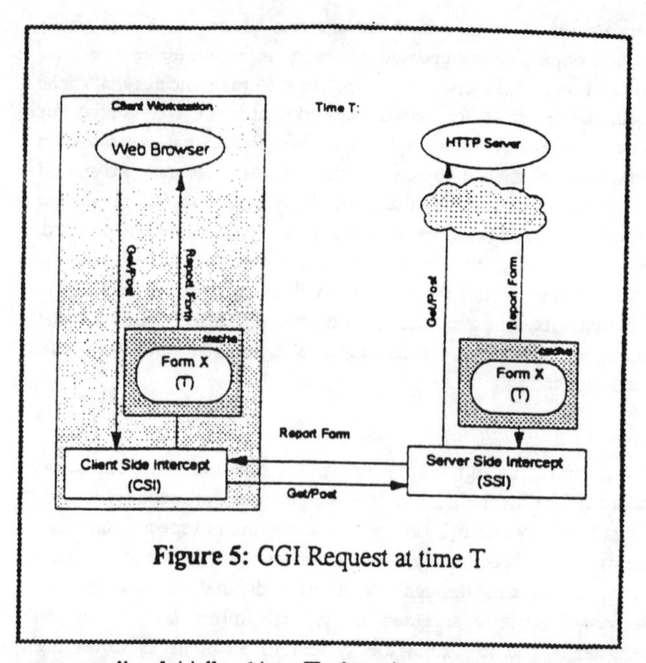

Figure 5: CGI Request at time T

parameter list. Initially, (time T) there is no record of a cached response for the URL at the CSI, and the request is sent to the SSI and forwarded to the server as normal. When the response is received by the SSI, it is cached (and it's CRC is computed) before forwarding it to the CSI. Likewise, the form is cached at the CSI before it is sent to the browser. At this point a *base object* has been established for the CGI URL. The cache coherency methods described previously are not used for base objects[8].

Now, consider the flow when another request is issued to program X using the same URL at time T+DT (see Figure 6). When a request for CGI processing is detected, the CSI checks to see if the URL is cached. In this example, at time T+DT a cached version is found. Now, the CSI forwards the request (i.e., URL plus parameters) to the SSI along with the CRC value of the base object (i.e., the report received for the request at time T). This CRC is maintained as part of the request state. The HTTP data stream is forwarded to the HTTP server to execute the request.

Subsequently, a report is received at the SSI. The SSI determines that differencing is possible because a base object for the URL exists in the cache and its CRC matches that received with the request from the CSI. The differencing engine computes the *difference stream* between the received report and the base object of the URL. A *difference stream*, consisting of a sequence of copy and insert commands, is sent to the CSI. The CSI update engine uses the difference stream and the request's base object to reconstruct the new report. The copy commands tell which byte sequences of the base object are to be copied to the new report and the insert commands cause data received in the difference stream to be copied to the new report during reconstruction. We are guaranteed that the reconstruction is correct because the SSI verified (using CRCs) that the URL's base objects are equal. Finally, the CSI sends the reconstructed report to the browser for display.

[8]By definition, CGI responses are *never* coherent because they change with every request.

The differencing engine uses well known differencing technology

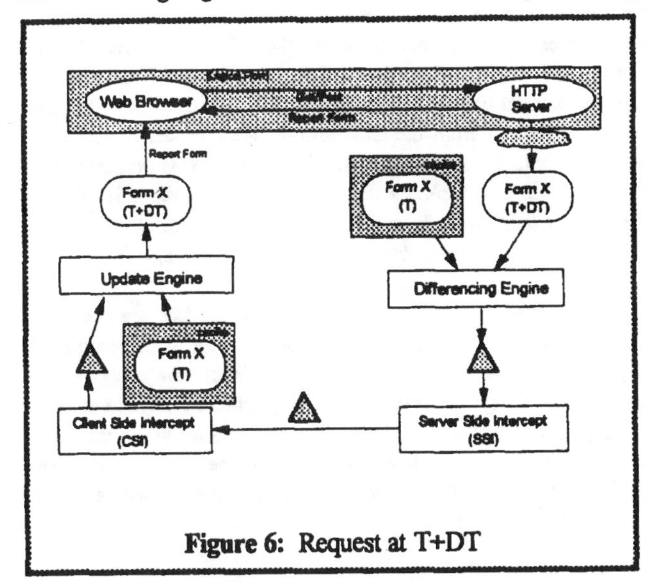

Figure 6: Request at T+DT

[10,11] that has typically been used in code library maintenance. However, it is important that the differencing engine work well on binary files because the concept of a *line* does not exist in HTML; indeed, in our experience, programs that generate HTML streams often do not generate carriage-returns or line feeds. Also, the difference processing should be efficient for small files, since most CGI responses are 10K or less.

Basing clients: An SSI may serve many CSIs (clients). To avoid the SSI maintaining a separate base object (for a given URL) for each client, we need to return the same base object for each client the first time[9] it requests a given URL. This is easily accomplished with a slight variation of the logic described in the previous example. When the SSI receives a response from the CGI server and the CRC received from CSI **does not match** the CRC of the corresponding base object at the SSI, the SSI computes the difference stream and returns **both** the base object and the difference stream to the CSI. The CSI caches the new base object before constructing the browser response. The significant point is that the CSI is prepared to receive two objects from the SSI: the difference stream and the new base object. This also handles the problem of *rebasing* the client which is necessary when the base objects in the CSI and SSI get *out of sync* as described below.

Rebasing: *Rebasing* the client is periodically necessary because, for a variety of reasons, it is possible that the base objects in the CSI and SSI become different. An object in either the CSI or SSI may be flushed from the cache as a result of the LRU policy. Alternatively, the base object in the SSI may be updated because the SSI detects that the difference stream has grown beyond a certain threshold, which often indicates that the response CGI data stream has changed for reasons other than different request parameters (i.e., the applications changes the format for aesthetics or adds information such as a copyright notice). When the SSI updates its base object, the CSI is rebased (as previously described) the next time the URL is requested.

Differencing for non-CGI responses: Although the above discussion has focused on using differencing to accommodate CGI processing, the technique can be generally useful. Presently, differencing is not applied to graphics objects (GIF, JPEG files). However, differencing may produce dramatic reduction for normal HTML files that have incurred minor updates.

Protocol Reduction

The use of caching and differencing significantly reduces the volume of application data (i.e., HTML and graphic objects) transmitted over the wireless link that connects the client work station to its wire line backbone network. However, these provisions do not address the overhead of repeated TCP/IP connections and redundant header transmissions. The WebExpress system for optimizing Web browsing in a wireless environment employs techniques to reduce the overhead of both of these categories.

Reduction of TCP/IP Connection Overhead

The normal HTTP protocol is depicted in Figure 7. The browser establishes a connection with the server and sends a single request to the server. The server sends a response document and closes its end of the connection. The browser receives the document and closes its end of the connection. For HTML documents this scenario is repeated for each image referenced in the document. It is also repeated every time the user clicks a *hyper-link* on the displayed page. The TCP/IP overhead consists of the transmission of extra control information and a severe increase in response time (latency).

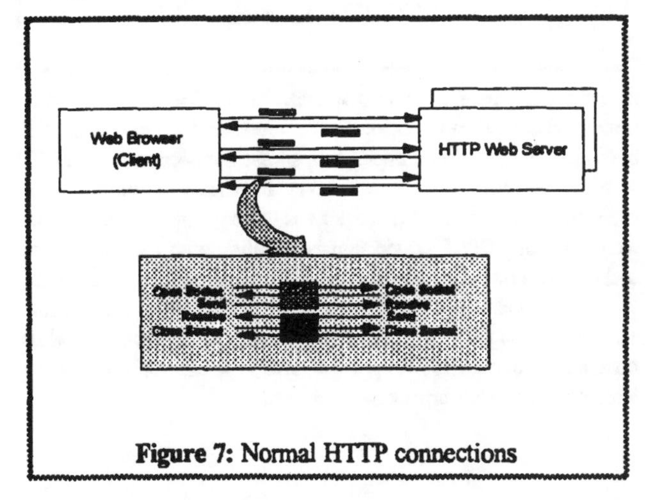

Figure 7: Normal HTTP connections

Virtual sockets: As Figure 8 illustrates, the WebExpress system eliminates most of the overhead of opening and closing connections across the wireless link by establishing a single TCP/IP connection between the CSI and the SSI. The CSI intercepts connection requests and document requests from the browser and sends the document requests over the single TCP/IP

[9]Recalling that objects in the cache may persist across many activations of the client browser and CSI, the *first time* is defined as the initial instantiation of the base object in the cache.

connection to the SSI. For each request received from the CSI, the SSI establishes a connection with the destination server and forwards the request. When the SSI receives the response from the server, it closes the connection with the server, sends the document to the CSI via the single TCP/IP connection but does not close this single TCP/IP connection. The CSI then forwards the document to the browser and closes its TCP/IP connection with the browser. The connection setup and takedown overhead is incurred between the browser and the CSI and the SSI and the Web server but not over the wireless link between the CSI and SSI.

WebExpress uses a mechanism called *virtual sockets* to provide this multiplexing support. Virtual sockets enables a CSI to establish a single TCP/IP connection with an SSI and use the connection for many HTTP requests. Data sent for a given request is prefixed by a small header that contains a virtual socket

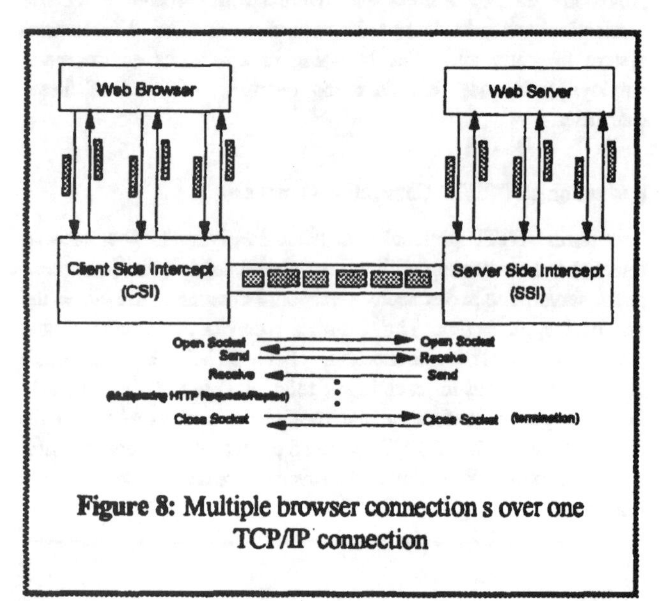

Figure 8: Multiple browser connection s over one TCP/IP connection

id, a command byte, and a length field. At the CSI, the virtual id is associated with a (real) socket to the browser; likewise, at the SSI the virtual socket id is mapped to a socket connection to a HTTP server. A suite of virtual socket interfaces is defined that corresponds to the TCP/IP sockets calls (e.g., open, close, select, etc.). A real TCP/IP connection is establish with the first virtual socket open and closed when a preset time interval expires after the last virtual socket is closed. In summary, this mechanism permits efficient transport of HTTP requests and response while maintaining correct HTTP protocol and WebExpress transparency with respect to Web browsers and servers.

Reduction of HTTP Headers

HTTP requests and responses are prefixed with headers. HTTP *request headers* contain a list of MIME content-types that tell the server the various document formats the browser can handle. This list can be several hundred bytes in length. Since it is usually constant for the browser, it is unnecessary to send it across the wireless link in every request. Instead, the CSI allows this information to flow in the first request that flows after CSI-to-SSI connection has been established. Both the CSI and the SSI save this list as part of the connection state information. For each request received from the browser, the CSI compares the list received with its saved version; if they match, the list is deleted from the request before it is forwarded to the SSI. When the SSI receives a request from the CSI with no access lists, it inserts its saved copy into the request header. If an access list is present in the received request, it replaces the saved version at the SSI if one exists. In either event, the correct access list is sent to the server as though there were a direct browser-server connection.

HTTP *response headers*, unlike the request headers are normally different each time. Normally, however (as with CGI responses), only a few bytes (e.g., date-time) vary from one response to another. Encoding the constant data (e.g., content-type) can reduce the response to just a few bytes. This reduction, while normally inconsequential in wire line networks, can be worthwhile when multiplied by all the mobile wireless units sharing a wireless link (i.e., every little bit helps!).

Usage Scenarios and Results

To demonstrate the effectiveness of WebExpress, we ran a series of transactions against two applications on active Web sites in the Internet: a DB2 World Wide Web Connection demo application and a popular quote server application. The results are summarized in Table 1.

Table 1: Results of running Web applications with and without WebExpress

	Data Volume Comparison (bytes)			Response Time Comparison (seconds)		
Test	Base	WebExpress	%Reduction	Base	WebExpress	%Reduction
1	56779	2302	96	1260	166	87
2	9635	2643	73	343	. 96	72
3	4853	1272	74	114	59	48
4	137649	·456	99.7	1079	30	97
5	2234	533	76	59	37	37
6	1326	515	61	19	11	42

The bytes transferred and the elapsed response time were recorded for each transaction. Each test case was evaluated with and without WebExpress. The **Base** column represents the measurements made without the use of WebExpress. To achieve maximum benefit and usability of WebExpress, the contents of the cache were established in the mobile device (using wire line communications) before the wireless access was attempted. The test environment was not a controlled environment, these tests were performed on a production Mobitex network connected to an enterprise network which was connected to the Internet. Consequently, the utilization of the various network and server components varied during our measurements, typical of a production environment.
Test cases 1, 2 and 3 correspond to the DB2 application. The DB2 application consisted of: querying product information, querying

and updating a customer profile and querying a parts order. This resulted in 10 Web pages, 7 documents and 3 forms, totaling about 30,000 bytes (including images). With all three test cases we exercised the complete application, all 10 pages. Test case 1 was the first pass through the application during peak hours, between 10am and 4pm. Test cases 2 and 3 were subsequent passes through the application with test case 2 occurring during peak hours and test case 3 occurring after 10pm.

With the first test case, the initial pass through the application, we evaluated a worst case scenario for the browser since the browser's cache did not contain any of the application data. This test although a worst case is a reality for a mobile worker when the browser does not support a persistent cache or refreshes pages on first time reference within a session. Without WebExpress each Web page and its corresponding images were fetched by the browser, generating 56KB of traffic and taking in excess of 20 minutes to complete. With WebExpress, the persistent cache methods satisfied the browser requests for the application documents and images. The differencing and communication methods were able to update the query reports with 2KB of network traffic. The complete elapsed time of the job was reduced from 20 minutes to under 3 minutes during peak network utilization.

In test cases 2 and 3, the browser cache contained all the documents and images in its memory cache. Without WebExpress the three queries (CGI processing) generated 9,600 bytes of network traffic during peak hours and 4,800 bytes during off hours. We believe that the difference between the measurements was due to the amount of packet retransmission during the peak hours. With WebExpress, the network traffic did not vary much between test cases 1 and 2 as we expected since the queries still needed to be resolved. And with test case 3 we again measured a significant drop in network traffic during the off hours.

Test cases 4-6 correspond to the quote server application. The quote server application consisted of: a home page, an input form and a report. This resulted in 3 Web pages, totaling about 50,000 bytes (including images). Like the DB2 tests, we exercised the complete application for each test case. Test case 1 was the first pass through the application during peak hours, between 10am and 4pm. Test cases 5 and 6 were subsequent passes through the application with test case 5 occurring during peak hours and test case 6 occurring after 10pm. In the 4th test case, without WebExpress each Web page and its corresponding images were fetched by the browser, generating 137,000 bytes of traffic and taking in excess of 17 minutes to complete. With WebExpress, the persistent cache methods satisfied the browser requests for the application documents and images. The differencing and communication methods were able to update the stock quote requests with about 500 bytes of network traffic. The complete elapsed time of the job was reduced from 17 minutes to 30 seconds during peak network utilization times.

In test cases 5 and 6, the browser cache contained all the documents and images in its memory cache. Without WebExpress, the stock quote requests generated 2.2KB of network traffic during peak hours and 1.3KB during off hours. With WebExpress, the network traffic did not vary much between any of the three test cases 4, 5, or 6. The quote requests were satisfed with about 500 bytes of network traffic.

In summary, our experiments with the WebExpress prototype indicate significant data reductions and response time improvements. As illustrated in Table 1, we measured 61% to 99% reductions in wireless network traffic and 37% to 97% improvements in application response time.

Related Work

A number of studies [12-15] have shown promising results regarding performance improvements of wireless Web browsing. Kaashoek et al. [12] reduced the latency of slow links by modifying the Mosaic client with scripts to perform caching and prefetching. Rover [13] used relocatable dynamic objects and queued RPC to develop a non-blocking operation for most browsers to allow a user to *click ahead*. GloMop [14] uses *lossy* compression while preserving semantic information for documents and images. Liljeberg et al. [15] provides an optimized sockets library and transport for wireless cellular links. Additionally, they implement an agent-proxy model to facilitate an HTTP *batch get* and disconnected operations. Like WebExpress, most of these approaches employ varations of agent-proxy and cache technologies. The novelty of WebExpress lies in its use of differencing technology with forms (i.e., transaction) processing, its distributed cache model with its coherency algorithms, and the virtual socket design and implementation that minimizes the number of network connections needed for HTTP requests. In addition, the WebExpress system combines these technologies via the Intercept Model transparently support today's browsers, servers, and transport stacks.

Conclusions

We have shown that WebExpress makes it feasible, from both a usability and cost viewpoint, to run commercial Web applications over wide area wireless networks. An important key to this success is the repetitive and predictable nature of transaction processing. This predictability enables the WebExpress caches to be preloaded using wire line access before wireless access is attempted, thereby, enabling the caching and differencing functions to work with minimal data transfer. Due to the serious limitation of wireless communications, it is necessary to employ a variety of optimization techniques to achieve a usable system. WebExpress demonstrates one set of optimizations that has proven successful. In particular, the *differencing* and *virtual socket* functions are critical, since they effectively extend caching to work with continually updated objects without requiring that the entire object (response) be transferred.

Acknowledgments

We acknowledge Michael L. Fraenkel and Reed R. Bittinger for their extensive contributions to the design and development of WebExpress.

References

1. T. Berners-Lee et al., *The World-Wide Web*, CACM 37(8):76-82., August 1994.
2. T. Berners-Lee, D. Connolly, *Hypertext Markup Language Specification - 2.0*, Internet-Draft, Internet Engineering Task Force (IETF), HTML Working Group, June 1995. Available at <http://www.ics.uci.edu/pub/ietf/html/html2spec.ps.gz>.
3. T. Berners-Lee, R. Fielding, H. Frystyk, *Hypertext Transfer Protocol - HTTP/1.0 Specification*, IETF, Internet-Draft, August 13, 1995. Available at <http://www.ics.uci.edu/pub/ietf/http/draft-fielding-http-spec-01.ps.Z>.
4. *An Introduction to Wireless Technology*, IBM International Technical Support Center SG24-4465-01, October 1995.
5. George Calhoun, *Wireless Access and the Local Telephone Network*, Boston: Artech House Inc., 1992.
6. *ARDIS Network Connectivity Guide*, Illinois: ARDIS, March 1992.
7. *RAM Mobile Data System Overview*, RAM Mobile Data Limited Partnership USA RMDUS 031-RMDSO-RM Release 5.2, October 1994.
8. *ARTour Technical Overview Release 1*, IBM Corp. SB14-0110-0, March 1995.
9. *Oracle Mobile Agents Technical Product Summary*, Oracle White Paper, Oracle Corp., March 1995.
10. Kris Coppieters, *A Cross-Platform Binary Diff*, Dr. Dobb's Journal, May 1995.
11. D. M. Ludlow, *Compare Process for Quick Determination of Text Changes*, IBM Technical Disclosure Bulletin, Vol. 22 No. 8A, January 1980.
12. M. Frans Kaashoek, et al., *Dynamic Document: Mobile Wireless Access to the WWW*, Proc. IEEE Workshop on Mobile Computing and Applications, Santa Cruz, CA, December 1995.
13. Anthony D. Joseph, et al., *Rover: A Toolkit for Mobile Information Access*, Proc. of the Fifteenth Symposium on Operating Systems Principles, December 1995.
14. *GloMop: Global Mobile Computing By Proxy*, GloMop Group, March 13, 1995 (glomop@full-sail.cs.berkeley.edu).
15. Mika Liljeberg, et al, *Optimizing World-Wide Web for Weakly Connected Mobile Workstations: An Indirect Approach*, Proc. of SDNE '95, Whistler, Canada, June 5-6, 1995 (IEEE 0-8186-7092-4/95).
16. Alan Demers, et al., *The Bayou Architecture: Support for Data Sharing among Mobile Users*, Proc. of Workshop on Mobile Computing Systems and Applications, pp.2-7, Santa Cruz, CA, 1994 (contact: terry@parc.xerox.com).
17. Geoffrey M. Voelker and Brian N. Bershad, *Mobisaic: An Information System for a Mobile Wireless Computing Environment*, Dept. of Computer Science and Engineering, University of Washington, September 19, 1994.

Chapter 10 Mobile IP

The Internet Protocol associates a numeric identifier with each computer (more accurately, with each interface, since computers can have multiple IP addresses). Each IP address is associated with a single physical network. When a "sending" host needs to communicate with a "receiving" host, given its IP address, it has three choices:

- If the receiving host is on the same physical network as the sender, it can direct a packet directly to the receiving host.
- If the network containing the receiving host is found in a routing table and specifies a particular host to direct the packet to, it can send the packet to the specified router.
- Finally, the sending host has a default router to which it sends everything else. The router, which is connected to at least one other network, performs the same operation to forward the packet closer to its destination.

Fixing an IP address to both a host and a physical network suggests that once a host is on a network, using a particular IP address, it cannot move. Without special support for mobility, that would be the case. However, support for mobility within the Internet Protocol has been made possible through a long series of research initiatives, beginning with work by Ioannidis, and others, in the early 1990s [Ioannidis, 1991].

With Mobile IP, a computer need not remain in a single location. Instead, in the simplest case, another host on the mobile computer's logical network—the one corresponding to the IP address it uses in communication with other hosts—acts as a router for the mobile host. Packets destined for the mobile host are sent to this router and then *tunneled* to a temporary IP address that the mobile host uses while in its present location. (Tunneling refers to encapsulating the original IP packet inside another one). The mobile host replies in the same way, by tunneling packets to the router and letting them be sent onward as though they had originated on the expected physical network.

The above approach is simple, but inefficient. In fact, much research has followed this initial design in order to improve upon its performance without sacrificing security. One improvement is to have the mobile host send packets directly to other hosts on the Internet claiming to be from the home IP address of the mobile host. This eliminates an indirection in one direction, but is subject to problems if routers reject packets with IP addresses that could not have originated from the direction whence they come—a common security feature. Another improvement is to modify IP directly so that correspondent hosts can receive an update message that informs them to communicate directly with a mobile host in its current location. This requires special support in those hosts, and also is vulnerable to abuse if these redirection messages are not guaranteed to be secure.

We begin this chapter with one of the original Mobile IP papers, by **Ioannidis and Maguire** [Ioannidis, 1993]. They describe the simple (in retrospect) solution described above and some of the problems with the approach. Some of their other work is described in greater detail in [Ioannidis, 1991; 1993].

We then present a recent **tutorial** on Mobile IP by one of the authors of the IETF specification, Charles **Perkins** [Perkins, 1998b]. That provides substantial additional detail on the latest modifications to the protocol, as well as additional research issues.

The remaining two papers in this chapter draw comparisons among different proposed mechanisms for improving the performance of Mobile IP. **Cheshire and Baker** [Cheshire, 1996] consider the cross-product of two sets of alternatives that depend on whether packets are sent directly or tunneled, or whether the correspondent hosts are aware of mobility and are able to communicate directly with a mobile host or must go through its home proxy, and so on.

Balakrishnan, and others, examine the interactions of TCP and wireless networks [Balakrishnan, 1995; 1997a, b]. Compared to wireline (wired) networks, wireless networks have two special properties that affect TCP. First, they are far more likely to encounter errors that will corrupt data, thereby causing TCP to need to retransmit one or more packets. Second, mobile computers can move between *cells*, which are the areas covered by individual base stations that serve as gateways between the wireline and wireless networks. When a mobile computer leaves one cell and enters another, packets that were destined for it may go to the wrong base station and either be delayed or be dropped altogether. Traditional TCP treats all such dropped packets as *congestion* and significantly reduces the transmission rate [Cáceres, 1995]. Balakrishnan and others consider three approaches:

1. Loss recovery is performed by the sender, as with traditional TCP, but the sender is aware that losses do not necessarily indicate congestion.
2. Recovery is done on a per-link basis, with the base station retransmitting lost packets without involving the sender.
3. The TCP connection is split at the base station. In this approach, the correspondent host communicates with the base station, which acknowledges data, then passes it on to the mobile host. Failures after that acknowledgment can result in the correspondent host mistakenly believing the mobile host has received packets.

They find that link-level recovery is useful, since it can shield the sender from transient losses that would otherwise require duplicate acknowledgments and other unnecessary traffic. They also find that while splitting the TCP connection has good performance, the second approach can outperform it while still providing end-to-end acknowledgment guarantees.

10.1 Ioannidis, J. and Maguire, G.Q. Jr., The Design and Implementation of a Mobile Internetworking Architecture

Ioannidis, J. and Maguire, G.Q. Jr., "The Design and Implementation of a Mobile Internetworking Architecture," *Proceedings of the USENIX Winter 1993 Technical Conference,* pp. 491–502, January 1993.

The Design and Implementation of a Mobile Internetworking Architecture

John Ioannidis, Gerald Q. Maguire Jr.
Department of Computer Science
Columbia University
New York, NY 10027

{ji,maguire}@cs.columbia.edu

Abstract

We present the design, implementation, and evaluation of **Mobile*IP**, a set of IP-based protocols and mechanisms to support host mobility throughout the Internet. The design requires changes only in the mobile hosts and their special routers; leaves transport and higher protocols unaffected, and requires no changes in the device drivers for individual interfaces. No modifications whatsoever are needed in non-mobile hosts and routers, the system scales well, and has no single points of failure. We have implemented Mobile*IP under Mach 2.6, and the code is readily portable to any version of Unix that uses Berkeley networking code.

1 Introduction

1.1 Motivation

The continuing drop in prices and increase in functionality of personal, portable computers, the increasing availability of wireless networking options as well as wide-area research and commercial networking offerings, and an increased desire of users to carry these systems and connections with them while they travel, suggests a marketplace and a user base ripe for introducing transparent network mobility to existing networking architectures. The proliferation of terms such as *Nomadic Computing*, *Personal Communications Networks*, [Ste90] [Cox90] [Gin91] [Ros91] [Lyn91] [SI91], as well as the rapid expansion of more established technologies such as the Cellular Phone System [Lee89], pagers, the new cordless phones [Smi91], etc., suggest that the mobile/wireless industry is moving fast. However, there is no clear sense of what services should be supported or what infrastructure is required.

We expect that in a few years, wireless/mobile support will be as widespread in educational, research, and business environments as Ethernet or other LAN connections are today. Rather than trying to define and predict specific applications, we looked at the general problem of mobile data communications and designed a solution that works across a wide variety of technologies and applications, and interoperates with the Internet. Our design was first described in [IDM91] and fully specified in [IDMD92], and a reference implementation is freely available.

This paper documents our implementation, providing a reasonable amount of detail of the software structure, and evaluates its performance. It is also intended to serve as an guide, in conjunction with [IDMD92], to other people wishing to implement Mobile*IP on their platforms. In the remainder of this section, we outline our design goals, and give a summary overview of how our system works. The next Section discusses the rationale behind the particular addressing and routing mechanisms that we chose. Sections 3 and 4 describe the software design, implementation details, and performance, in the Mobile Support Routers and Mobile Hosts, respectively. Section 5 describes the additional software components needed for Popups (that is, Mobile Hosts migrating outside their home networks), and Section 6 completes the paper.

1.2 Design Goals

The concept of routing for mobile hosts is not a new one [SP80][LNT87][JT87][Int87]; these previous designs, however, are impractical in today's Internet, with its vast number of applications, hosts, and networking infrastructure. We have developed a new approach, optimized for localized mobility, driven by the following design goals:

- Work within the TCP/IP protocol suite [Pos81a] [Pos81b].
- Provide Internet-wide mobility.
- A mobile host always keeps its IP address, called its "Home Address".
- Optimize local-area mobility without sacrificing performance or functionality of the general case.

- Transport-layer and higher protocols should be left untouched.
- No applications should change in order to run on or be used from mobile hosts.
- The infrastructure, that is, non-mobile hosts, routers, routing protocols, etc. should be left untouched.
- Mobility should be handled at the network layer.
- Minimize points of failure.
- Be responsive and scale well.

The rationale behind those choices is presented in the aforementioned papers. In summary, it is crucially important for mobile hosts to maintain their IP address as they move, since their IP address is also used as the Endpoint Identifier (EID) [Per92] for connections, and trying to notify transport protocols and applications that the host address has changed was deemed unrealistic. In addition, we did not expect our approach to be adopted if it required ordinary hosts and routers to be modified in order to talk to mobile machines. Lastly, we wanted to achieve mobility with the least amount of impact to the rest of the network. These goals were met by decoupling routing to/from the mobile hosts from routing to/from the rest of the network; by requiring the Mobile Hosts (MHs) to have IP addresses from a reserved subnet or subnets; and having a special class of routers, called the Mobile Support Routers (MSRs), route packets between the MHs and the rest of the network.

1.3 Overview of Operations

Figure 1 shows a small portion of a campus network that supports mobile hosts. Each MSR defines one or more cells, one for each network interface. The cells may be wireless (RF or IR), wired (Ethernet, token-ring, FDDI) or just groups of point-to-point links where mobiles can attach. In the example shown in Figure 1, there is only one cell per MSR, and all mobiles have RF interfaces. Three MSRs, one MH, and one non-mobile host are shown, as well as the routers (R and Rw) linking the three segments together and to the outside world.

In each cell, the MSR broadcasts a periodic Beacon. MHs entering the cell receive the Beacon, send a Greeting to the MSR, which answers with an Greeting Acknowledgement (as shown by the three thin solid arrows between the mobile and MSR-A. If the mobile was previously in MSR-B's cell, it will also send a Forwarding Pointer (FWDPTR) notifying it of its migration, which should be acknowledged with a Forwarding Acknowledgement (FWDACK). Once in MSR-A's cell, the MH sets it as its default router to the world, and MSR-A marks the MH as "local". IP datagrams from an MH to another MH in the same cell are processed locally. IP datagrams from an MH to a non-mobile host anywhere else

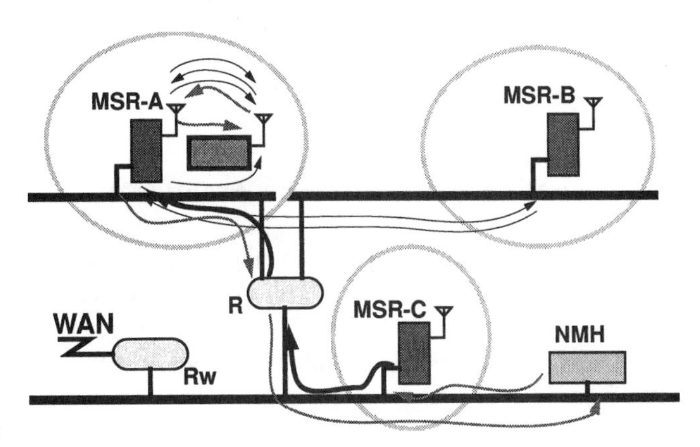

Figure 1: Sample Campus.

on the Internet are routed in the usual manner. IP datagrams from a non-mobile host are sent to the "nearest" MSR; if the target MH is served by that MSR, the datagram is simply forwarded to the MH. Otherwise, if the MSR knows which MSR is currently handling the MH, it encapsulates the datagram in another datagram (shown by the change of gray density in the arrow going from NMH through MSR-C, and on to MSR-A via the router R, and sends it to that other MSR, which decapsulates it and delivers it to the MH. If the original MSR (in the example, MSR-C) does not know where the MH is, it queries the other MSRs (it can also query a name-location server), and the one handling the MH will respond; the response is cached as long as there is traffic to the target MH, and future datagrams do not cause a new query. Finally if an MH sends a datagram to an MH in a different cell, the first MH's MSR will receive it and tunnel the datagram to the MSR handling the other MH.

Mobiles that wish to connect from outside their home network and still appear to be hosts belonging to that network, acquire a temporary address, called the Nonce Address, in the foreign subnet, then use it to first handshake with one of its home MSRs, and then tunnel packets to and from an MSR from their home network.

2 Network Considerations

The main problem of trying to add mobility to IP, is that the IP address of a host is, at once, an *Endpoint Identifier (EID)*, that is, a (unique) name used to identify the connections to and from the host, and also an *address*, that is, an indication of *where* the host is, and thus *how* to reach it (route packets to it). The address has structure (`<network, subnet, [subnet...], host>`) to make routing scale. This structure, however, implies that the IP address of a host is determined by the subnet in which it is connected; ordinarily, if it were to move to another subnet, it would have to change its IP address. Since such changes were deemed unacceptable, we came up with a design that avoids this problem. In this section, we discuss the design alternatives and the reason behind our choice.

2.1 Routing and Addressing Architecture

In order to allow hosts to move within an administrative domain, (e.g., a business network or a university campus) without changing their IP address, some aspect of the nature of their IP address would have to be changed. The options were:

- Discard the 'unique-identifier' aspect. That is, when a host migrates, it is assigned a new IP address. Transport-level and higher protocols, including applications, would need to deal with such changes, and therefore this is not an acceptable option.
- Flatten the structure, at least for the mobile hosts, which would imply distributing *per-host* routes in all routers in the campus. As we shall see in Section 2.2, this is prohibitively expensive for large networks
- Extend the notion of 'subnet' to convey not just a connected network, but also a Virtual subnet which consists of a set of partitioned physical networks (which we shall call *cells*) linked with tunnels. The mobile hosts belong to this virtual network.

In the third option, all Mobile Hosts have addresses in the same logical IP subnet, the Mobile Subnet. This is the so-called 'embedded network' approach [CPR91]. Routes to the Mobile Subnet are via the Mobile Support Routers; thus, to route to a mobile, it is necessary to first route to the nearest MSR, which will tunnel the datagram to the mobile's MSR, which will subsequently deliver the datagram to the target mobile. This way, only the MSRs have to know where each mobile is (as opposed to *every* router), thus reducing the amount of routing updates necessary. Tunneling is necessary because the datagrams may have to traverse multiple routers, and eventually an MSR, in order to reach their destination, and these intermediate routers do not know how to route to the particular mobile; all they know is how to route to the mobile subnet.

MSRs advertise routes to the Mobile Subnet using ordinary Internal Routing Protocols, such as RIP[Hed88], Hello[Mil83], IGRP[Hed91], etc. In selecting the routing protocol and the redistribution parameters of the route(s) to the Mobile Subnet, care must be taken to ensure that the nearest active MSR is always used as an entry router to the Mobile Subnet; poor configuration may result in all the campus routers routing traffic to exactly one MSR, usually as a result of having the internal routers trust each other more than the MSRs as far as routing updates are concerned.

Each MSR 'supports' one or more cells. Since any mobile may be in any cell, all cells have the same subnet number. The result of this architecture is that the mobile subnet is really comprised of many unconnected physical network segments, the cells. In order to make this partitioned network appear as a single subnet, the MSRs exchange information about which mobiles are where, and *tunnel* datagrams between them when they are required to route a datagram destined for a mobile in another MSR's cell.

Two protocols were defined for the purposes of supporting Internet Mobility: The IP-inside-IP Encapsulation protocol (IPIP), and the Mobile Internetworking Control Protocol (MICP). Their numbers, assigned by the Internet Assigned Numbers Authority [RP92] are as follows:

```
#define IPPROTO_IPIP    94
#define IPPROTO_MICP    95
```

Encapsulation is a common technique for delivering data to a remote endpoint through routers that do not know how to route to that endpoint. In the case of Mobile*IP, encapsulation is used between MSRs to deliver IP datagrams whose destination address is a mobile served by the target MSR. The alternative to encapsulation is to use source routing to first route to the MSR handling the target MH, and from there deliver the packet to its destination. Source routing has, however, several problems; it is an IP option, and as such, it needs extra processing in every router the packet goes through; it interferes with a potentially already present SSRR, LSRR, or RR option; the transmitted IP packet may not have enough space left in the header to handle an additional source route; and finally, unless it is specifically stripped at the last-hop router, the option is sent over the (potentially slow) link between the MSR and the MH, thus increasing the traffic in a possibly slow link. It is also impossible to do nested tunneling with source routing, whereas it is possible to encapsulate an already encapsulated packet (a practice not recommended, but still feasible). The overhead in terms of additional network traffic due to larger packets, is higher for encapsulation than it is for source routing (20 octets in the case of IPIP encapsulation, versus 12 octets (option header plus twice four octets for the IP addresses of the source and target MSR, plus padding) in the case of source-routing), but the benefits of using encapsulation outweigh the additional eight octets per packet.

MICP is the protocol used to acquire and distribute information about MHs. The various packet types, as referred to later in the document, are PING, BEACON, GREETING, GRACK, GRNACK, FWDPTR, FWDACK, WHOHAS, IHAVE, OTHERHAS, and POPUP. The exact contents are defined in [IDMD92].

Separate IP protocols were necessary (rather than using UDP) both to keep the datagram size down and because all of IPIP and some of MICP processing is done in the kernel.

The exact formats are documented in [IDMD92], and their implementation is described later in this paper.

In summary, using MICP for discovery and IPIP for tunneling, the MSRs 'conspire' to 'heal' the partitioned network, and make a collection of heterogeneous network fragments appear to the rest of the network as a seamless subnet.

2.2 Route Dissemination

Assume that an organization network has N_R routers, of which N_M are MSRs, and N_m mobile hosts. Let μ be the mobility of MHs, expressed in number of cell switches per MH per unit time. Also, let ν be the average number of hosts, mobile or not, that an MH is communicating with (i.e., receives file service from, has virtual circuit connections to, etc.). Observe that μ is a function of the users' mobility habits, but also of the size of the network, and the size of its cells; a denser higher-speed, network, will tend to have smaller, more confined cells (to support larger numbers of mobiles at higher data rates). The value of μ can vary widely, from 10 switches per second (e.g., a car driving through a densely populated area with very small cells, 'microcells'), to 10^{-4} (switches per second), or less than once an hour (e.g., students moving between classrooms, managers moving between meetings, etc.). The quantity $N_m \times \mu$ gives the number of cell switches per unit time in the network. ν, on the other hand, will tend to be fairly constant and small, probably below 10. Naturally, such numbers are very rough estimates, and are derived by considering what the mobility profile of an average user would be, and how many servers/services that users would be accessing. More experience with mobile data networking is needed before more accurate estimates can be obtained. In the remainder of this section, we shall use $\mu = .001$ (roughly three times an hour), and $\nu = 5$.

Let us now justify the decision to use an embedded network approach, as opposed to simply distributing host routes among the organization's network's routers. If host-routes were to be distributed every time a host moves, as would be the case in a non-embedded-network solution, the load imposed on the network would be proportional to the number of routers involved, the number of mobile hosts, and their mobility. The routing traffic would thus be $T_{hostroutes} = N_R \times N_m \times \mu$, expressed in number of (routing update) messages per unit time. Note that this quantity essentially increases quadratically with the size of the network, as both the number of routers and the number of mobile hosts are a measure of the size of the network.

If host routes are not used, but rather an embedded-network approach is adopted, the problem of how to inform MSRs of where each MH is still remains. There are two extreme solutions: whenever an MH migrates inform all the MSRs, or whenever an MSR needs to know the location of an MH, in queries the other MSRs. Let us examine the traffic imposed on the network by such arrangements. In both cases, tunneling would be necessary to get the packets through, thus increasing the *data* traffic on the network; however, this increase is small (if the average packet size is 500 octets, it is 4%), but more important, it is *predictable* and the resulting traffic increase is smoother and varies linearly with the size of the network (i.e., the number of MHs).

The first case is similar to the host-routes case, except that now only MSRs are involved; presumably, $N_R >> N_M$, (notice that N_R includes N_M) and hence the routing traffic generated, $T_{msrroutes} = N_M \times N_m \times \mu$, is already smaller than $T_{hostroutes}$. This is at the expense of having to encapsulate each message in IPIP.

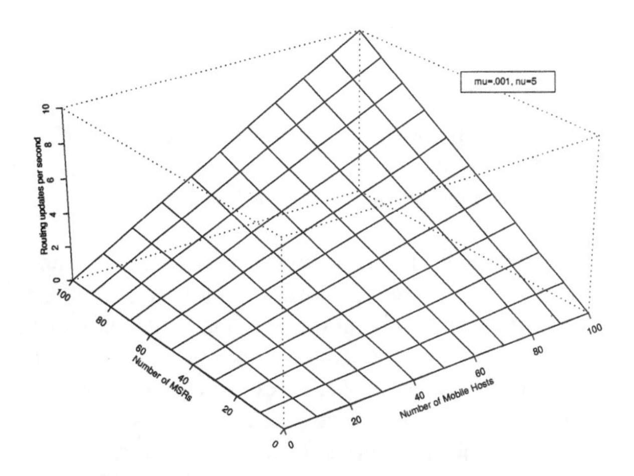

Figure 2: Traffic from distributing routes among MSRs.

The second case is similar to the way ARP[Plu82] works. In ARP, when a host needs to map an IP address to a hardware address, it broadcasts an ARP request on the local network, and the host with the requested IP address responds. Here, an MSR needs the IP address of another MSR handling an MH, and it multicasts a request, asking for the handler. The important points are:

1. The requests only happen when there has not been traffic between the mobile and the host it is trying to communicate with.
2. The results of the queries are cached to minimize requests.
3. The resulting traffic is proportional to the number of other hosts a mobile is communicating with.

The first point implies that no routing traffic (other than forwarding pointers) is generated when a mobile moves, if it generates or receives no network traffic. This is especially significant when the mobile is idle and is just moving with its owner.

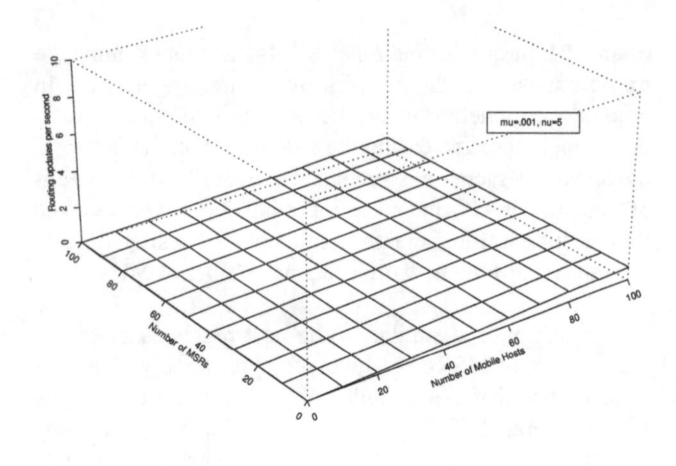

Figure 3: Traffic from requesting routes.

The second point implies that while there is traffic to and from the mobile, forwarding pointers and redirects are sufficient to keep the participating MSRs informed.

The third point is the most important; the number of hosts a mobile is communicating with at any one time, its 'working set' of hosts, that is, is bound to be a small number; judging from modern office workstations, it is a number around 10. This number stays the same **regardless of the network size.** Provided that we are in a steady-state, the number of routing updates as mobile hosts move, $T_{demandroutes} = N_m \times \nu \times \mu$, is thus **proportional only to the number of mobile hosts,** and not proportional to the number of mobile hosts and the number of MSRs (or routers). If traffic to a mobile is so infrequent that the caches of intermediate MSRs get cleared, the extra processing needed to establish the location of the MH from scratch is negligible. This is enough of a reason to choose cached, on-demand route acquisition, as opposed to gratuitously distributing MSR location information to all MSRs, or, even worse, host routes to all hosts.

Figures 2 and 3 show, in the same scale, the traffic, in packets per second, generated by having MSRs advertise MH location information as soon as an MH moves in their cell, and the traffic generated by having MSRs request routes for MH they are interested in. In both cases, $\mu = .001$ and $\nu = 5$. Observe how the second case is not affected by the size of the network (reflected in the number of MSRs).

In summary, the routing and addressing architecture chosen was that of an embedded network, the *Mobile Subnet*, where all mobile hosts belong to. It appears as a single subnet to the rest of the routing architecture, and thus preserves the scaling properties of IP addressing. MH location (i.e., routing information distribution) is done on a demand basis, thus keeping the amount of routing traffic generated low, and proportional only to the number of mobile hosts.

Let us now examine how such a mobile networking architecture is implemented.

3 Mobile Support Router Software

A Mobile Support Router performs four functions:

1. Advertises routes to the Mobile Subnet.
2. Handshakes with the MHs in its cell(s).
3. Exchanges information with other MSRs about the location of MHs in needs to serve.
4. Routes packets to and from its cell(s), tunneling them to other MSRs if necessary.

In the reference implementation, the first two operations are performed by a privileged, user-level process, `msrmicp`.

The routing and tunneling are done in the kernel, with WHOHAS requests also originating from the kernel. `Gated` is used to advertise routes to the Mobile Subnet using HELLO. HELLO was preferred over RIP because its distance metric has much finer granularity than RIP. Our campus uses Cisco routers which use a proprietary routing protocol (IGRP), and which is not supported by `gated`. In addition, the routers are configured to redistribute routes to the Mobile Subnet using HELLO rather than IGRP so that the nearest MSR will be used (IGRP would otherwise decide on the 'best' MSR to use and the routers would route packets only to that MSR, ignoring the fact that another MSR might be closer).

3.1 Data Structures

Each MSR keeps three kinds of information:

Global configuration information: Its "main" IP address, the list of other MSRs (or the multicast address for `all-msrs-on-campus`), the address ranges of mobile subnets, and the rate at which it expires information about MHs.

Per-interface information: For each interface that controls a cell, the MSR keeps the IP address of the cell interface, its broadcast address, the beacon rate, and an epoch number that changes whenever the MSR wants its MHs to re-handshake.

MH information: An array of `struct mhinfo` (see Figure 4, indexed by the offset of the IP address of the corresponding MH from the first mobile IP address. If non-contiguous ranges of addresses are used, more than one such arrays are used, each appropriately labeled. Each structure contains flags indicating whether the MH is local, is a popup, or is handled by another MSR, the IP address of the MSR handling the MH (in the case of local MHs, the IP address of the interface talking to the MH), a timestamp that the MH provided in its greeting, and the remaining time before this entry is expired.

```
struct mhinfo {
        u_short mi_flags;
        u_short mi_timer;
        u_long  mi_tstamp;
        struct in_addr mi_msr;
};

#define MIF_LOCAL        0x0001
#define MIF_DONTEXPIRE   0x0002
#define MIF_POPUP        0x0010
#define MIF_NOREDIRS     0x0020
#define MIF_REMOTE       0x0100
#define MIF_KNOWN
    (MIF_LOCAL | MIF_POPUP | MIF_REMOTE)
```

Figure 4: *Mhinfo* structure and related constants.

3.2 Algorithms and Implementation

3.2.1 User-level processing

`Msrmicp` is the process that handles all the user-level processing. It uses an `IPPROTO_MICP` raw socket to talk to the MHs and the other MSRs, and thus needs to run as `root`. The `msrmicp` process:

- Listens for and responds to MICP messages, both from the MHs in its cell(s) and from the other MSRs.
- Periodically broadcasts a Beacon in each of its cells.
- Periodically scans the `mhinfo` arrays and decrements the timeout counters. If any of them reach zero, it clears the entire entry.

When an MICP_GREET arrives from an MH (or an MICP_POPUP arrives from a popup), and assuming it meets the necessary security criteria, the corresponding `struct mhinfo` is updated: the `MIF_LOCAL` (or `MIF_POPUP`) flag is set; unless otherwise specified by the MICP_POPUP, the `MIF_NOREDIRECTS` flag is also set for a popup (this means that when the MSR receives a tunneled packet for that popup, it will re-tunnel it to the popup rather than send a redirect to the originating MSR); the timestamp from the packet is saved; the IP address of the interface the packet was received on is saved, or, in the case of a popup, the source IP address of the packet is saved instead; the timeout counter is set to its maximum value (configuration dependent); and an MICP_GRACK is sent to the MH informing it of the timeout period. If the MH moved in from another cell, the greeting will contain a list of one or more "pending MSRs", MSRs in whose cells it has been and which have not yet acknowledged its move. For each MSR in this list, the MSR sends it an MICP_FWDPTR packet with the IP address and timestamp of the MH in question. When an MSR receives an MICP_FWDPTR with a more recent timestamp that it has in its corresponding `mhinfo` entry, it changes the flag to `MIF_REMOTE`, saves the address of the source of the Fwdptr, updates the timestamp, resets the timeout value, and replies with an MICP_FWDACK, including the IP address of the MH and its timestamp. When the original MSR receives the fwdack, it passes it along to the MH using an MICP_GRACK packet. Finally, if the MSR wants to deny service to an MH at any time, it just sends it an MICP_GRNACK packet.

MSRs receive MICP packets from other MSRs as well. When an MICP_WHOHAS is received, and the MH in the request packet is marked `MIF_LOCAL` or `(MIF_POPUP | MIF_NOREDIRECTS)`, the MSR replies with an MICP_IHAVE, including the IP address of the MH and its timestamp in the packet. If only the `MIF_POPUP` flag is set, the reply is an MICP_OTHERHAS, giving the same information as before, but also the IP address of the MSR handling the target MH. When an MICP_IHAVE, an MICP_OTHERHAS, or an MICP_REDIRECT is received, the processing is identical to that of MICP_FWDPTR, except that an acknowledgement packet is not sent back to the source. When an MICP_EXPIRED packet is received, and the `MIF_REMOTE` flag is set, the `mhinfo` entry is cleared.

Finally, an MSR keeps a `struct mhinfo` entry for each of its interfaces, and for the broadcast address, marked `MIF_LOCAL` and `MIF_DONTEXPIRE`, so that the expiration mechanism will not touch them. This is done so that the MSR will respond to requests from outside its cell to its interfaces controlling cells.

3.2.2 Kernel Processing

An MSR is essentially a router, with `ip_output()` (in `netinet/ip_output.c`) modified to handle mobility:

> **if** *target is not an MH* **then**
>   deliver/route using the regular IP code;
> **else if** *target is a local MH* **then**
>   deliver locally using the appropriate interface;
> **else if** *target is remote and known* **then**
>   encapsulate in an IPIP packet and
>   deliver to the corresponding MSR;
> **else**
>   attempt to locate MSR and drop packet;

If the `mhinfo` entry for the target MH has the `MIF_REMOTE` or the `MIF_POPUP` flags set, the packet will be tunneled to the MSR handling it. First the expiration timeout is reset to its original value, indicating that there is traffic to that mobile. Then, an IP header is prepended, with the `ip_p` field set to `IPPROTO_IPIP`, the `ip_src` and `ip_dst` fields set to the current and the target MSR's IP addresses, the `ip_ttl` field set to `MAXTTL`, the `ip_len` field set to the length of the encapsulated datagram, and the entire packet is passed to `ip_output()` again, only now it will be sent to the main (non-mobile) address of an MSR, so the procedure will not repeat.

If the target MH is not known, the kernel will send MICP_WHOHAS requests to all the MSRs in its network. This should be preferably done by multicasting, or by sending individual messages to each MSR if multicasting is not supported.

To handle encapsulated packets, an entry for IPIP is set up in `struct protosw inetsw[]`. The input routine, `ipip_input()` is what receives the encapsulated datagram. It strips the encapsulation header (the extra IP header) and saves the address of the originating MSR. The resulting datagram should be destined to an MH. If that MH is unknown, an MICP_EXPIRED message is sent to the originating MSR, and the packet is dropped. If the MH is marked `MIF_REMOTE`, an MICP_REDIRECT with the MH's address and timestamp is sent to the originating MSR, and the packet is passed to `ip_output()` again. Finally, if the MH is local, it is passed to `ip_output()` to be delivered locally.

Finally, to accommodate communication between MHs in different cells, and also hosts which might ARP (on hardware that relies on ARP) instead of routing directly to the MSR, the ARP algorithm (`in_arpinput()`, in `netinet/if_ether.c`) needs some minor changes:

> **if** *target is not an MH* **then**
>   use the regular ARP code;
> **else if** *source is a remote or unknown MH* **or**
>   *both source and target are local* **then**
>   drop the packet;
> **else if** *target is unknown* **then**
>   attempt to locate it;
> **else**
>   proxy-ARP for target;

The MSR then proceeds to route or tunnel any packets it gets because it replied to the ARP.

3.3 Evaluation and Performance

Each MSR only keeps information about MHs in its cell, and also MHs in other cells whose MSRs it has to tunnel packets to. This co-location of location information and routing has two desired effects: (a) information is kept in a distributed fashion, and only acquired on demand, hence lowering the routing-information traffic, and (b) there is no single failure point; if an MSR fails, only the MHs in its cell are affected in terms of discovery and routing, what is also termed "fate sharing". From a global point of view, **the system has no single point of failure**.

3.3.1 Memory Requirements

The memory requirements are proportional to the size of the mobile subnet; 12 bytes per host. Even if an entire Class B network is reserved for mobile hosts, less than one megabyte of main memory is required, an amount readily available in modern machines. The memory requirements to keep the additional required configuration and state information are negligible.

The MSR code is approximately 1200 lines of C code for `msrmicp`, and another 1500 lines for the kernel code (including header files).

3.3.2 Values of Configuration Parameters

In our network, we send Beacons once a second, expire local and popup MHs after 60 seconds of inactivity, and remote MHs (handled by other MSRs) after 30 seconds of inactivity. The once-a-second value for the beacon was derived by our desire to have MHs greet within one second from the time they enter a cell, and keep the delay low in case one or two consecutive beacons are lost. A higher rate would hardly make any difference in the responsiveness of migrating MHs, and a lower rate would make recovering from lost beacons a lengthier process.

The value for the local and popup expiration parameters was chosen to be smaller than the time it takes any of our Mobile Hosts to reboot. The reason for this (and also the reason for expiring these entries) is that an MH that reboots may lose the knowledge that it was in a particular MSR's cell, and when it comes up again, it may greet a different MSR. If the original MSR did not expire the mobile, it would be left thinking that it owns it, and thus respond to WHOHAS requests and attempt to route packets to it. Although such responses would not affect the behavior of other MSRs, since replies would also arrive from the MSR which is really handling the MH, and those replies would have a more recent timestamp, this still remains a problem for hosts directly routing traffic to the errant MSR. By expiring a mobile that has not re-greeted before the timeout period passes, we guarantee that no such routing 'black holes' will occur. The expiration period for remote MHs should be set to a value that reflects what is considered an idle workstation. If the value is too low, entries will be expired too soon, only to be followed by a WHOHAS/IHAVE sequence, wasting bandwidth. If they are too high, the MSR may end up tunneling to other MSRs which no longer handle the target MH, resulting in EXPIRED messages, which restart the WHOHAS/IHAVE cycle. Empirically, we found that 30 seconds is a good value, if somewhat low.

3.3.3 Performance

The MSRs used are generic 386 and 486 machines. The values quoted below are for a 486/33MHz machine running Mach 2.6.

The `msrmicp` process consumes approximately 2% of the CPU time in a mostly idle network. This time is spent sending beacons and scanning the `mhinfo` arrays to expire entries that have timed out. This part is only a function of the beacon rate, the scanning rate, and the expiration timeouts. Over 85% of the processing time was spent just scanning the `mhinfo` entries, according to the profiling data. The remaining time is spent responding to greetings and requests from other MSRs and processing replies that it solicited. Given that the complexity of the code that handles the various requests is roughly the same as the code that handles beacons, the code can handle ten requests a second (reflecting the load from a thousand mobiles moving every minute, according to Figure 3) and still only well under 10% of its CPU.

If multicasting is not used, locating an MH is directly proportional to the number of MSRs, but that only needs to happen the first time the MSR needs to locate an MH. So long as there are data flowing to the MH, the timer keeps getting reset to its maximum value. The actual time to locate an MH depends on how many hops away its MSR is, as each hop increases the delay. Figure 5 shows a box-plot of the response time of 100 WHOHAS requests to an MSR on the other side of a Cisco MGS router. The mean is 5.45ms. For comparison purposes, the ping times to the same machine are also shown; their mean is 4.05ms. Assuming negligible processing time for the ICMP echo, the mean processing time for a WHOHAS request is 1.4ms (figures rounded to the nearest .05 ms). The profiled code reports that `mf_whohas()`, the routine handling the requests, takes 0.06ms to run; the difference between the two values includes two context switches, since `msrmicp` is running as a user-level program, and a small amount of additional code that checks the legality of the packet and selects the routine to call.

Figure 6 shows the time spent encapsulating (in `ipip_send()`), and then sending (in `ip_output()`), datagrams ranging in size from 64 to 1000 octets. The average time to add the encapsulation header, which involves requesting an `mbuf` if no space exists in the header, and filling the fields, is practically constant, and averages 45 microseconds. The time to transmit a packet increases linearly with the size of the packet, from slightly under 300 microseconds to almost 1.2 milliseconds, between 6 and 25 times the amount of time spent encapsulating. Most of the time in `ip_output()` is spent copying data to the Ethernet controller. Assuming that `ip_input()` and the decapsulation part of the IPIP handler take similar amounts of time, the machines we are using can easily handle the routing and encapsulation, even at full network utilization, as their mobile side has only 2Mbps interfaces (NCR WaveLan cards).

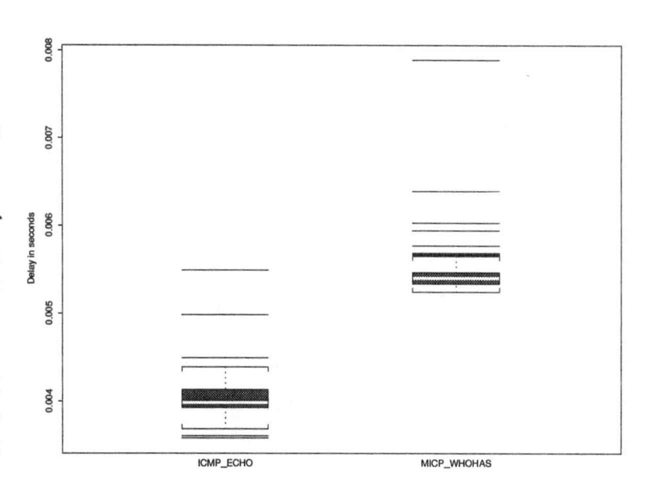

Figure 5: ICMP Echo (ping) and MICP WHOHAS processing.

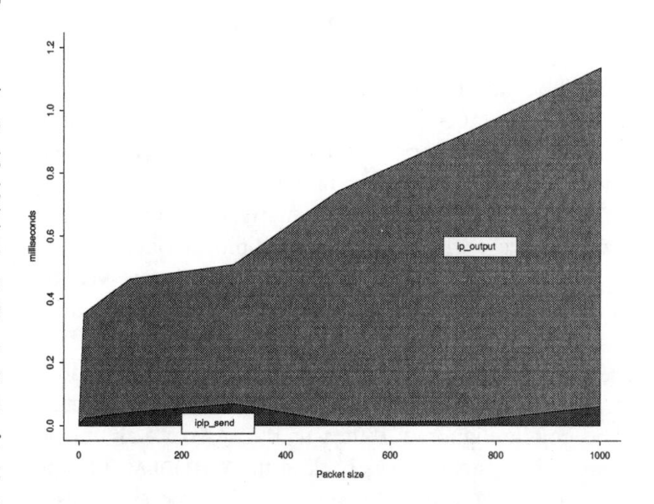

Figure 6: Time spent encapsulating and sending datagrams.

4 Mobile Host Software

An MH that only moves around its own campus needs no kernel modifications, as it always relies on an MSR for service. It needs, however, to handshake with the MSRs and reset the default route whenever it changes cells. The program handling these operations is `mhmicp`; it uses an `IPPROTO_MICP` raw socket to talk to its MSR(s), and thus needs to run as `root`.

4.1 Data Structures

An MH keeps two kinds of information:

Configuration information: Its IP address and netmask, the waiting-for-beacon and waiting-for-ack timeouts, and the the number of beacons to be lost before it starts hunting for a new MSR. Also, the name of a script (`huntscript`) to run when it needs to hunt for a new beacon.

State: The following variables define the state of the `mhmicp` process.

- `currentstate`; one of WTG4BEACON, BCNRXED, INCELL, and INCELLPENDINGPING.
- `timestamp`; with a finer temporal resolution than the time needed for cell-to-cell moves.
- `currentmsr`; IP address of the MSR as advertised in the beacon.
- `currentrouter`; source IP address of the beacon, to be used as the default route to the world.
- the pinging timeouts.
- the list of previous MSRs; MSRs whose cells the MH has been in but has not received an ack for.

4.2 Algorithms and Implementation

Initially, the `mhmicp` state variable `currentstate` is set to WTG4BEACON. If a beacon is not received within `waiting_for_beacon` seconds, `huntscript` is run to change external parameters of interfaces (e.g., spreading code for a spread-spectrum RF interface), or change interfaces in the case where a unit has multiple interfaces, and the process restarts. If an MICP_BEACON packet is received, the state changes to BCNRXED, the source address of the packet is set as the default router and also placed in the previous-MSRs list, and an MICP_GREET packet is sent to the MSR. In addition, the ARP cache of the MH is flushed. While in this state, all beacons are ignored. If the `waiting_for_ack` timeout expires, the system moves back to the WTG4BEACON state. If an MICP_GRACK packet is received, the process moves to the INCELL state, removes the current MSR from the pending list, and schedules a ping timeout to go off in half the expiration interval supplied with the MICP_GRACK packet. When that timeout goes off, the MH will move to the INCELLPENDINGPING state, send an MICP_GREET to the MSR (at which point the MSR will reset its corresponding `mhinfo` entry to the maximum value again. The MH halves the remaining time and schedules another interrupt, and so on, until it gets an `micp_grack` from the MSR, at which point it resets the timer to its original value, and moves back to the INCELL state.

While in all but the WTG4BEACON state, beacons from other MSRs are simply ignored. If the epoch number in the current MSR's beacon changes, the MH must regreet, as this indicates that the MSR lost its internal state. Also, with each beacon received, the timeout watching for lost beacons is reset. If several consecutive beacons are lost (5 in our implementation), the timeout will go off, add the current MSR to the pending MSRs list, and move the process to the WTG4BEACON STATE. The same holds true if at any point an MICP_GRNACK from the MSR is received. Finally, as long as the previous-MSRs list is non-empty, the MH will keep sending it to the MSR, with a linear backoff bounded by the expiration timeouts of the individual MSRs. This is because, if an MSR is unreachable for more than that period, it will expire the corresponding entry anyway, so there is no reason to keep asking the MSR to send the MICP_FWDPTRS to them.

4.3 Performance

The three timeouts are all set to five seconds in our implementation, thus the MH never has to wait more than five seconds if it loses a beacon. The MSRs beacon every second, and advertise an expiration timeout of one minute. This means that the MHs have to process beacons every second, and ping every thirty seconds. On the average, the `mhmicp` process takes about 0.2% of the CPU time on a 20MHz i386 machine. The `mhmicp` and `pumicp` programs share about 2000 lines of C code, and each one has an extra 600 lines.

While signalling overhead is not a problem with MHs, subtle interactions between periods when the MH is off the network (e.g., in the process of changing cells), or when it first tries to establish a connection, and timeouts in transport- and higher layer protocols (such as TCP or NFS) may be felt by the users. Figure 7 shows the time needed to open a TCP connection to a mobile address; for comparison purposes, a stationary host, two different MSRs (known by their mobile addresses), a local MH and a Popup, are used. The box plot summarizes 100 attempts to connect to the `discard` port (TCP port 9) for each host. Connections to the stationary host complete immediately; connections to mobile addresses take almost 6 seconds; this is because the first SYN packet causes the nearest MSR to send out a WHOHAS request. The originating host waits six seconds before it will send a second SYN, which can now get routed to the mobile address and complete the connection. Observe how a small number (4) of packets to the Mobile must have gotten lost, and it took 12 seconds to complete the connection. This indicates the need for caching a packet that causes a WHOHAS to be transmitted. This, however, would add to the complexity of the MSR code.

Another concern[1], is what happens when a mobile has an open TCP connection to another host, which is sending it a continuous stream of data, and the mobile gets temporarily disconnected from the network; the TCP window will fill on the stationary host's side, and as acknowledgements will not be arriving, TCP will back off, and retransmit less and less frequently. When the mobile moves back into the coverage area, it will have to wait for the next retransmission from the stationary host to occur before it can continue receiving its data, and this can take up to a minute. Such behavior can

[1] This problem was pointed out to us by Andrew Myles (School of Mathematics, Physics, Computing, and Electronics, Macquarie University, Sydney, Australia; *andrewm@mpce.mq.edu.au*), during his visit to Columbia University in August 1992.

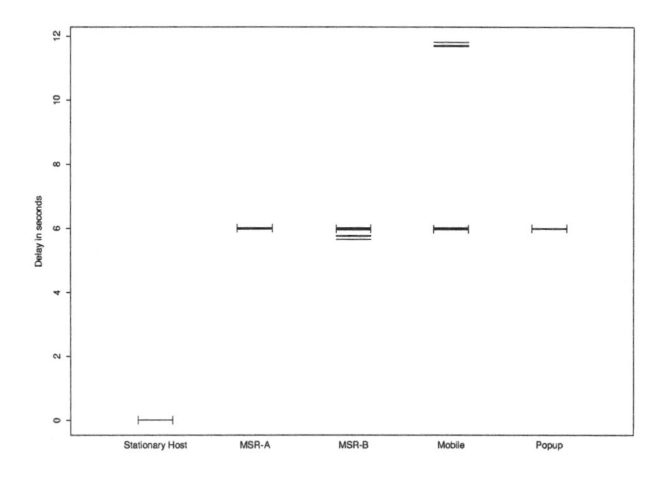

Figure 7: Startup times for TCP.

be fixed by 'kick'-ing TCP whenever network connectivity is reestablished. This, however, is bound to be a problem with a lot of high-level protocols that depend on timeouts, and shows that some adjustments may be needed to transport and higher protocols in order for them to work well in the presence of frequent network outages.

5 Popup Software

A Popup is a mobile that has wandered away from its home campus and wants to communicate back. To achieve that while maintaining its home address, it has to acquire a temporary address, called its Nonce Address, from the network it is visiting. This is because the mobile is now in a different administrative domain, whose MSRs, even if they exist, do not communicate with the MSRs in its home campus. Using the Nonce Address, the mobile sends an MICP_POPUP message to one of its home MSRs. The handshake is similar to that of a local mobile, including pings, except that there is no beacon. The MSR uses the Nonce Address as the remote endpoint to tunnel datagrams back to the popup.

5.1 Algorithms and Implementation

How the nonce address is acquired is not very important; we supply it manually, but a protocol such as DHCP [Dro92] should be used.

The main problem is how to accommodate two addresses (the home address and the nonce address) on a machine with just one interface, and also do tunneling to and from its home campus (tunneling *from* the popup *to* anywhere else is not strictly necessary, but smart routers may see packets coming in from the wrong interface, and drop them. The easiest solution was to define a Virtual Interface (vif), described next. When a popup shows up on a foreign network, it acquires a nonce address which it assigns to its real interface, then assigns its home address to the virtual interface, and then routes all packets not explicitly destined for the home MSR through the virtual interface.

5.2 Virtual Interface

The virtual interface (`vif`), much like the loopback interface, has no input queue. Packets sent to it by `ip_output()` are looped back if their destination address is the same as `vif`'s; otherwise, they are encapsulated in IPIP and tunneled to the home MSR using the real interface. All packets are routed through the virtual interface, by doing the equivalent of:

`/etc/route add default` *home-address* `0`

that is, adding a default route that is the interface itself. Thus (see Figure 8), outgoing traffic is routed through the virtual interface tunneled to the home campus, and routed from there on. A packet *for* the popup is first routed to its home network, where an MSR picks it up, tunnels it to the MSR handling the popup, if necessary, which in turn tunnels it again back to the popup. The packet received has protocol type IPIP, so it is sent to the input routine of IPIP, which strips the header and feeds the packet back into the IP output queue. But now this is a packet for the mobile address of the popup, and it is simply looped back and delivered to the appropriate transport protocol.

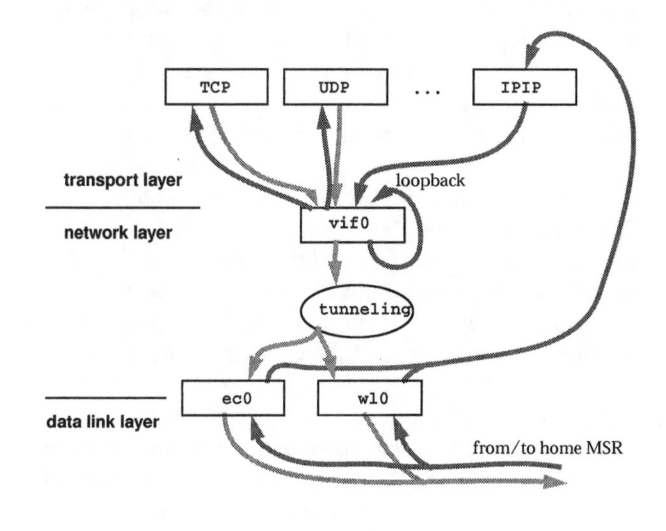

Figure 8: VIF processing.

5.3 Performance

Since there is no constant beacon, the control process, `pumicp` takes even less CPU time, about 0.1% of the CPU on a 20MHz 386 machine. However, the popup has to do its own tunneling, which adds 100 μs to each outgoing packet (this is a slower machine than the 486/33Mhz showing the 45μs encapsulation time). Since, however, network interfaces on mobiles are not expected to run at the full bandwidth, the encapsulation delay merely adds to the latency, which is probably insignificant compared to the latency of traversing a wide area network.

The code for the vif driver (`netinet/if_vif.c` and `netinet/if_vif.h`) is under 300 lines of C code. The interface is `if_attach()`ed when a special device file (`/dev/vif`) is opened, which also allows the use of `ioctl()`s to set the address of the home MSR for use by the tunneling code.

6 Summary

We have presented an infrastructure which enables mobile machines to keep their network connections while they move in a networked environment. A lot of fine detail, such as packet formats, has not been covered here, as it can be found in the official specification [IDMD92]. The emphasis was on describing and justifying the major design decisions, and describing the reference implementation and its performance.

To summarize, our design for mobility follows the embedded network model, whereby Mobile Hosts keep their IP address as they migrate, and the addresses of all MHs in an organization's network belong to the same logical subnet, the Mobile Subnet. Special routers, the Mobile Support Routers, keep track of the MHs' location, in order to be able to route (tunnel) packets to and from them. By allowing Popup MHs to acquire secondary addresses, and use tunneling to communicate to their home MSRs, we extend the ability to be mobile to outside the limits of an organization's network, or even to parts of that network without MSRs. The ability to use MSRs the way we are using them should be viewed as an optimization of the general case. In order to keep the routing updates traffic low and scalable, we use on-demand discovery of MH locations, rather than gratuitously propagate their location to all the MSRs in a network.

Finally, we present the performance of the reference implementation; the figures show that a medium-power machine, such as a 486/33Mhz "clone" can more than adequately perform all the tasks of an MSR: MH registration and tracking, exchange of routing information with the other MSRs as well as the regular routers, and route/tunnel packets between MHs in its cell(s) and the hosts they communicate with. As far as the impact of mobility on the MHs' performance is concerned, the signalling necessary is negligible, although the mechanics of on-demand acquisition of routes adds delays in the setup of connections.

This work was supported in part by National Science Foundation grant CDA-9022123, IBM Corporation, and the Center for Telecommunications Research, an NSF Engineering Research Center funded by grant ECD-88-11111. It has benefitted from discussions, comments, and suggestions by a lot of people, including Dan Duchamp, Steve Deering, Mike O'Dell, Phil Karn, and the Mobile Hosts Working group of the IETF.

John Ioannidis is a last-year graduate student in the Computer Science department at Columbia University, and by the time you read this he should have received his PhD and should be working at Bellcore. In addition to his thesis work on Mobile Internetworking, JI has interests in all aspects of System Design, including networks, operating systems, and security.

Gerald (Chip) Q. Maguire Jr. is associate professor of Computer Science at Columbia University in the City of New York. His research interests include distributed computing, wireless networking, building portable software systems and special-purpose processors, picture archiving and communication systems (PACS), and image processing/computer graphics with an emphasis on medical images.

References

[Cox90] D. C. Cox. Personal Communications — A Viewpoint. *IEEE Communications Magazine*, pages 8–20,92, November 1990.

[CPR91] Danny Cohen, Jonathan B. Postel, and Raphael Rom. IP Addressing and Routing in a Local Wireless Network, July 1991.

[Dro92] R. Droms. Dynamic Host Configuration Protocol. Internet-Draft, available as `draft-ietf-dhc-protocol-04.txt`, August 1992.

[Gin91] Sam Ginn. Personal Communication Services: Expanding the Freedom to Communicate. *IEEE Communications Magazine*, pages 30–39, February 1991.

[Hed88] C. L. Hedrick. Routing Information Protocol. RFC 1058, June 1988.

[Hed91] Charles L. Hedrick. An Introduction to IGRP, August 1991. Available with anonymous ftp from `ftp.cisco.com`, file `igrp.doc`.

[IDM91] John Ioannidis, Dan Duchamp, and Gerald Q. Maguire Jr. IP-Based protocols for mobile internetworking. In *Proceedings of SIGCOMM'91*, pages 235–245. ACM, September 1991.

[IDMD92] John Ioannidis, Dan Duchamp, Gerald Q. Maguire Jr., and Steve Deering. Protocols for Supporting IP Hosts. Internet-Draft, June 1992.

[Int87] SRI International. Network reconstitution protocol. Technical Report RADC-TR-87-38, Rome Air Development Center, June 1987.

[JT87] John Jubin and Janet D. Tornow. The DARPA Packet Radio Network Protocols. *Proceedings of the IEEE*, 75(1):21–32, January 1987.

[Lee89] William C. Y. Lee. *Mobile Cellular Telecommunications Systems*. McGraw-Hill, 1989.

[LNT87] Barry M. Leiner, Donald L. Nielson, and Fouad A. Tobagi. Issues in Packet Radio Network Design. *Proceedings of the IEEE*, 75(1):6–20, January 1987.

[Lyn91] Richard J. Lynch. PCN: Son of Cellular? The Challenges of Providing PCN Service. *IEEE Communications Magazine*, pages 56–57, February 1991.

[Mil83] D. L. Mills. DCN Local-Network Protocols. RFC 891, December 1983.

[Per92] Radia Perlman. *Interconnections: Bridges and Routers*. Addison-Wesley, 1992.

[Plu82] D. C. Plummer. Ethernet Address Resolution Protocol: Or converting network protocol addresses to 48.bit Ethernet address for transmission on Ethernet hardware. RFC 826, November 1982.

[Pos81a] Jon Postel. Internet Protocol. RFC 791, September 1981.

[Pos81b] Jon Postel. Transmission Control Protocol. RFC 793, September 1981.

[Ros91] Ian M. Ross. Wireless Network Directions. *IEEE Communications Magazine*, pages 40–42, February 1991.

[RP92] Joyce Reynolds and Jon Postel. Assigned Numbers. RFC 1340, July 1992.

[SI91] Richard M. Singer and David A. Irwin. Personal Communications Services: The Next Technological Revolution. *IEEE Communications Magazine*, pages 62–66, February 1991.

[Smi91] Douglas G. Smith. Spread Spectrum for Wireless Phone Systems: The Subtle Interplay between Technology and Regulation. *IEEE Communications Magazine*, pages 44–46, February 1991.

[SP80] C. Sunshine and J. Postel. Addressing Mobile Hosts in the ARPA Internet Environment. IEN 135, March 1980.

[Ste90] Raymond Steele. Depoloying Personal Communication Networks. *IEEE Communications Magazine*, pages 12–15, September 1990.

10.2 Perkins, C. E., Mobile Networking with Mobile IP

Perkins, C.E., "Mobile Networking with Mobile IP," *IEEE Internet Computing*, 2(1):58–69, January/February 1998.

Mobile Networking with Mobile IP

Charles E. Perkins

Abstract— **As sales of laptop computers continue to grow, protocols are being designed or modified to assist the users who need connections to the Internet wherever they go. Mobile networking affects every layer of the protocol stack, but the effects are most evident at the lower layers. A general discussion about mobile networking, including social and technical driving forces, is followed in this article by a focussed look at Mobile IP. Mobile IP is a Proposed Standard protocol designed within the IETF to make mobility transparent, as much as is possible or desirable, to applications and higher-level protocols (such as TCP). Basic Mobile IP is described, with details given on the three major components: *Discovery*, *Registration*, and *Tunneling*. IPv6 mobility is discussed in some detail. Some topics of current interest are described.**

Keywords— **Mobile IP, mobility, nomadicity, mobile networking, IETF, IPv6, wireless**

I. Introduction

Mobile IP can help to usher in a new era of Internet computing and interactivity with global data resources. The modern Internet is known around the world as a major new avenue for human expression and interpersonal communication, as well as the path towards realizing age-old dreams of making the libraries of and wealth of human knowledge available to everyone. Today, anyone with a modem and a personal computer, even with a near-minimal configuration, can participate in this fantastic new human enterprise. Doors are opening for citizen participation on an unprecedented scale. Internet users regularly trade recipes for cooking as well as hints on how to make best use of their software. People spend hours in chat rooms, finding new friends and probably making new enemies.

The constant feature in all of this is the fact that today most people don't expect to be hooked up to the Internet until they arrive at some familiar access point. It can be home where their computer is tied into the phone system. It can be at work, where employers are ever more cognizant of the natural place for personal electronic mail and web access in the workpace. But, connectivity to the Internet is typically thought of as originating from one or two particular points of attachment. It is the opinion of this author that such restrictions will soon enough be considered somewhat old-fashioned.

To see the contrast between current Internet connectivity and the future possibilities, consider how people now view the use of telephones. It took many years before people learned to rely on the telephone for interpersonal communications, but once a part of our culture it seems to have become indispensable. For many people it is unusual to be completely away from any telephone for a whole day. Yet, even with this internalized dependence on the telephone for basic communications, people occasionally have to frantically search for a telephone. Cliche movie plots rest on the unavailability of a telephone to ward off disaster, or the frustration evident among people waiting in line for a chance to use a public telephone booth.

Since the introduction of mobile phones, first in cars and more recently in handheld cellular telephones, this acceptance of inconvenience is melting away. People who depend in time-critical ways upon the telephone for their interpersonal communications, and who can afford the increased charges, carry their telephones with them. Cellular telephones are, for many people, no longer a luxury but instead a necessity upon which businesses and livelihoods depend. Even hikers find a new sense of freedom and security knowing that they have communications resources in the unlikely event of emergency.

The analogous transition in the domain of Internet computing, from dependence upon fixed points of attachment to the flexibility afforded by mobility, has just begun. Confident access to the Internet anytime, anywhere, will help provide the right frame of reference to free us from the ties binding us to our desktops. Knowing that we can have the Internet available to us as we move will give us the tools and the creativity to build new computing environments wherever we go, helping to make us all citizens of the world. This is analogous to the way that the mobile telephone has made people more mobile, going beyond merely offering convenience to people who were already mobile. Mobile networking could go so far as to change our existing work ethic. Knowledge workers may visit the office much less often, instead preferring to shape their ideas by interacting with world in other ways; computing power would be available to them almost equally effectively in many different environments.

There are major differences between the evolution of mobile telephony and the likely evolution of mobile computing. For one thing, the endpoints of a telephone connection are typically human, whereas there is likely to be in the future a great deal of emphasis on noninteractive computer applications. The obvious examples are mobile computing devices on airplanes or ships, and in automobiles. Perhaps equally important are the various devices that we may wish to carry with us, most of which have not been invented yet, but will depend for their operation on position-finding devices such as GPS, in tandem with wireless access to the unimaginable future Internet.

Another clear evolutionary distinction is the possible difference in rates of adoption between mobile telephony and mobile computing. It has taken 20 years for mobile phones to become cheap enough and light enough to really offer the required user convenience. Today's mobile computers are not terribly convenient, but other wireless mobile computing devices such as PDAs and pocket organizers are already

C. E. Perkins is with Sun Microsystems, Mountain View, CA. 94303
E-mail: cperkins@eng.sun.com

acceptable. The modern headlong rush of technology directly enables many improvements of convenience. The end result is that the general spread of mobile computing, once begun, may occur at a greater rate than that of mobile telephony.

II. Nomadicity

Mobile IP is a large part, but by no means the only part, of the story of mobile computing and mobile networking. To see Mobile IP in its true place requires a careful understanding of the relationships between the various layers of network protocols. Each layer should present a clear model of operation to the architect, and once the model is identified, the effects of mobility can be understood in relationship to that model. Nomadicity is the name given by a XIWT [15] working team to the organized understanding of the effects of mobility on the entire network computing environment.

Protocol layer two is responsible for link establishment and maintenance. Thus, physical effects resulting from mobility are likely to require changes in the layer-two protocols. Changes in position affect the signal to interference ratio (SIR). Link layers that adapt forward error correction to SIR can exhibit variable bandwidth but far fewer lost packets. Wireless media, which as described above are an irresistible invitation to mobility, typically introduce many other design requirements at layer two. In particular, the desire for confidentiality leads one to incorporate encryption techniques especially for wireless links. Often, lower bandwidth (compared to wired media) suggests the use of compression techniques. And, typically, transmitting a signal causes the local receiver to lose detection of any other signal because of the great difference in effective power levels between local and remote transmitters. This leads to the unavailability of collision detection techniques, such as used with Ethernet, which then must be replaced by less reliable collision avoidance measures and careful etiquette.

Other distinguishing characteristics of wireless communications media include the difficulty of establishing a precise range (i.e., cell size) for connectivity to the medium, and the ability for separate stations to use the media without interference. This latter property of re-use depends upon avoiding interference between neighboring transmitters, and there is a great engineering discipline built up for understanding the optimal placement of such wireless equipment as base stations. To achieve maximal re-use of the physical wireless medium, the cell size should be as small as possible. This means that as demand for wireless communications increases, cell sizes will decrease, and the frequency with which mobile computers will switch cells (changing their point of attachment to the Internet) will grow correspondingly.

IP, the Internet Protocol, is best thought of as a layer-three protocol that selects routes (determines paths) through a loosely confederated association of independent network links. IP offers routing from one network to another, in addition to some minor services such as fragmentation and reassembly, and checksumming. Moving from one place to another can be modeled as changing the network node's point of attachment to the Internet. Supporting mobility at layer 3 is therefore naturally modeled as changing the routing of datagrams destined for the mobile node so that they arrive at the new point of attachment. This turns out to be a very convenient choice, and forms the main focus of discussion later within this article.

TCP [40] and other transport protocols attempt to offer a more convenient abstraction for data services than the characteristically chaotic stream of data emanating from IP. The vagaries and time dependencies of routers and Internet congestion often cause datagrams to be delivered out of order, duplicated, or even dropped entirely before reaching the destination. TCP attempts to solve those problems, but offers little or no help in supplying a steady (constant bandwidth) stream of data, or in making any effort to get the data delivered within specified time bounds. Over time, TCP has been modified to treat dropped packets as an indication of network congestion, and therefore to throttle transmissions [46] as soon as a lost packet was detected (by managing sequence numbers). This is the wrong strategy when packets are corrupted by transmission over a noisy wireless channel, because for such packets immediate retransmission is much better than delayed retransmission. Ways to change this behavior are still a matter of current discussion.

Above the transport layer, and depending to the transport model employed, application protocols are largely freed from much of the drudgery of error correction, retransmission, flow control, etc. However, mobility creates new needs at the application layer which require additional protocol support:

- automatic configuration
- service discovery
- link awareness
- environment awareness

These protocol support mechanisms will form a set of middleware services. For example, a mobile computer might need to be reconfigured differently at each different point of attachment. A new DNS server, new IP address, new link MTU, and new default router may be required, among other things. These configuration items are usually thought of as being worked out at setup time for desktop systems, but for mobile computers no single answer can be sufficient. Recent deployment of DHCP [12], [1] goes some way towards resolving configuration difficulties, but is not the whole answer. Discovering services can be modeled as a requirement for automatic configuration, but is more naturally useful when services are located upon demand and according to the needs of application protocols. This need is just now being met by the Service Location Protocol [48].

One of the more challenging aspects of architecting the middleware lies in offering applications the opportunity to detect the state of the physical link, which changes dynamically and can easily affect the desired operation of the application. The simplest example of this is the need for Web applications to adjust their presentation of graphic data depending upon the available end-to-end bandwidth.

Today that bandwidth is largely constrained by the link conditions at the endpoints, and the congestion status of infrastructure connectivity. Mobile computers introduce more variability into this mix, and thus exacerbate the growing need for multimedia applications to detect and act on dynamic connection parameters such as link bandwidth, error rate, round-trip times (RTT), etc. Other logical parameters such as cost and security may eventually exhibit similar dynamic behavior, and complicate application response to connection status information.

Lastly, a word should be said about the granularity of protocol response to node movement. Today's typical user must be content with portable computing, which requires reinitialization and re-establishment of connections at each new point of attachment to the Internet. Since nothing better has been widely available, people have learned to accept this mode of operation. However, this acceptance would evaporate if the reinitialization process had to be performed a lot more frequently; left unchecked, the expected decreases in cell sizes would require exactly that in the future. The existing methods typify portable network computing, which means establishing the availability of network computing when one arrives at a new point of attachment but being unable to continue previous computing activities.

Portable computing is to be contrasted to truly mobile computing, wherein computing activities are not disrupted by changing the point of attachment of the mobile computer to the network. It is possible to have all the needed reconnection occur automatically and noninteractively; that's the point of Mobile IP, DHCP, and Service Location Protocol, among others. Even if one does not wish to work on any computing applications while moving, it will still be much more convenient to be able to resume one's previous applications upon reconnection. This is especially true in an office wireless LAN environment, where the boundaries between attachment points are not sharp and often not visible to the user. And, as noted above, cell sizes are likely to shrink over time. In view of these circumstances, portable network operation cannot fulfill the promise of mobile wireless computing, and the convenience of mobile networking protocols will be seen instead as a necessity.

III. Overview of Mobile IP

Mobile IP [38] builds on major characteristics of the Internet Protocol (IP) [41] as a network layer protocol in order to offer application transparency. IP is responsible for connecting together the networks of today's Internet. IP routes packets from a source endpoint to a destination, by allowing routers to forward packets from incoming network interfaces to outbound interfaces according to routing tables. The routing tables typically maintain the next-hop (outbound interface) information for each destination IP address, according to the number of network to which that IP address is connected. The network number is derived from the IP address by masking off some of the low order bits. Thus, today the IP address typically carries with it information specifying the point of attachment of the IP node.

In order to maintain existing transport-layer connections as the mobile node moves from place to place, the mobile node is required to keep its IP address the same. In the case of TCP (which accounts for the overwhelming majority of connections in today's Internet) connections are indexed by the quadruplet containing the IP addresses and port numbers of both endpoints of the connection. Changing any of those four numbers will cause the connection to be disrupted and lost. On the other hand, correct delivery of packets to the mobile node's current point of attachment depends on the network number contained within the mobile node's IP address, which changes at new points of attachments. To change the routing requires a new IP address associated with the new point of attachment.

Mobile IP has been designed by a working group within the Internet Engineering Task Force (IETF) to solve this dilemma, by allowing the mobile node to use two IP addresses. One is called the *home address*, which is static (does not change) and which is used, for instance, to identify TCP connections. The other is called the *care-of address*, which changes at each new point of attachment, and can be thought of as the mobile node's *topologically significant* address since it reflects the network topology at the mobile node's point of attachment. The home address makes it appear that the mobile node is continually able to receive data on its *home network*, and on the home network Mobile IP requires the existence of a network node known as the *home agent*. Whenever the mobile node is not attached to its home network, the home agent gets all the packets that are supposed to received by the mobile node and makes arrangements for further delivery of those packets to the mobile node's current point of attachment.

Whenever the mobile node moves, it *registers* its new care-of address with its home agent. To get a packet to a mobile node from its home network, the home agent delivers the packet from the home network to the care-of address. The delivery requires that the packet be modified so that the care-of address appears as the destination IP address. This modification can be understood as a packet transformation (or, more specifically, a *redirection*). Upon arrival at the care-of address, the reverse transformation is applied so that the packet once again appears to have the mobile node's home address (i.e, the static address) as the destination IP address. When the packet arrives at the mobile node, addressed to the home address, it will be processed properly by TCP or whatever higher-level protocol logically receives it from the mobile node's IP (i.e., layer-3) processing layer. More on this abstract modeling of Mobile IP as a way to perform layer-3 redirection on packets delivered to the home agent can be found in [2].

In Mobile IP, redirection of packets from the home network to the care-of address of the mobile node is accomplished by constructing a new IP header containing the mobile node's care-of address as the destination IP address (see section VI). This new header then "shields" or encapsulates the original packet, causing the mobile node's home address to have no effect on the routing of the encapsulated packet until it arrives at the care-of address. Such encapsulation is also called *tunneling*, suggesting a model of helping the packet to burrow through the Internet, bypassing the usual effects of IP routing.

Mobile IP, then, is best understood as the cooperation of three separable mechanisms:

- Discovering the Care-of Address
- Registering the Care-of Address
- Tunneling to the Care-of address

IV. Discovering the Care-of Address

Mobile IP provides mechanisms to help a mobile node discover a care-of address when one is needed at its new point of attachment to the Internet. The discovery process used by Mobile IP relies on advertisements from mobility agents (foreign agents and home agents), and provides a way for a mobile node to solicit an advertisement if one is needed but has not been detected.

The Mobile IP discovery process has been built on top of an existing standard protocol, Router Advertisement, specified in RFC 1256 [11]. The original fields of the existing router advertisements have not been modified; instead, extensions have been defined to associate mobility functions with the original advertisements. Thus, a router advertisement can carry information about default routers, just as before, but in addition carry information about one or more care-of addresses. Home agents also use router advertisements to make themselves known, even if they do not offer any care-of addresses. In contrast to the default routers, however, it is not possible to associate preferences to various care-of addresses in the router advertisements. The IETF working group was concerned that dynamic preference values might destabilize the operation of Mobile IP. Since no one could defend static preference assignments except for the case of backup mobility agents, which do not help distribute the load, the preference assignments were eventually dropped from use with the care-of address list in the router advertisements.

A Router Advertisement, augmented by Mobile IP's Mobility Agent Extension, is then called an Agent Advertisement, and performs the following functions:

- Allow for the detection of mobility agents
- List one or more available care-of addresses
- Inform the mobile node about special features provided by foreign agents – for instance, alternate encapsulation techniques
- Allow mobile nodes to determine the status of their link to the Internet
- Let the mobile node know whether the agent is a home agent, a foreign agent, or both

RFC 1256 also defines Router Solicitations. They may be used without change by mobile nodes wishing to detect any change in the set of mobility agents available at the current point of attachment. If advertisements are no longer detectable from a foreign agent which previously had offered a care-of address to the mobile node, the mobile node should presume that the foreign agent is no longer within range of the mobile node's network interface. In this situation, the mobile node should begin to hunt for a new care-of address, or possibly use a known care-of address known from advertisements which are still being received. The mobile node may choose whether to wait for another advertisement if it has not received any recently advertised care-of addresses, or to send a solicitation.

V. Registration

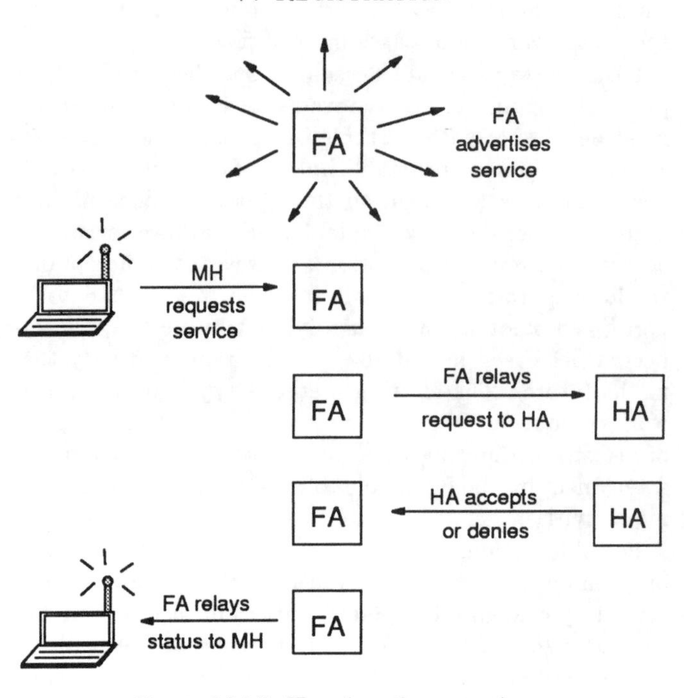

Fig. 1. Mobile IP registration operations

Once a mobile node has a care-of address, its home agent has to find out about it. Mobile IP defines a *registration* process for this purpose (illustrated in figure 1), whereby the mobile node, possibly with the assistance of a foreign agent, sends a Registration Request with the care-of address information. Upon receipt of the request, a home agent will (typically) add the necessary information to its routing table, approve the request, and send a Registration Reply back to the mobile node. Although it is not required that the home agent handle registration requests by updating entries in its routing table, doing so offers a natural implementation strategy, and all implementations known to the author take such an approach.

Registration Requests contain parameters and flags that characterize the tunnel through which the home agent will deliver packets to the care-of address. Tunnels can be constructed in various ways [33], [34], discussed briefly in section VI. When a home agent accepts the request, it begins to associate the home address of the mobile node with the care-of address, and maintain this association until the *registration lifetime* expires. The triplet containing the home address, care-of address, and registration lifetime is called a *binding* for the mobile node. Using that terminology, a registration request can also be considered a *binding update* sent by the mobile node.

A binding update is also an example of a kind of *remote*

redirect, send remotely to the home agent to affect the home agent's routing table. Thinking of the registration in this way should make the need for authentication clear [49]. The home agent has to be certain that the registration was originated by the mobile node, and not by some other malicious node pretending to be the mobile node. Otherwise, if the home agent were to alter its routing table with erroneous care-of address information, the mobile node would be unreachable to all incoming communications from the Internet.

This need for authenticating registration information has had a major role in determining the acceptable design parameters for Mobile IP. Each mobile node and home agent have to share a security association, and be able to make use of Message Digest 5 (MD5) [42] with 128-bit keys to create unforgeable digital signatures for the registration requests. The signature is computed by performing MD5's one-way hash algorithm over all the data within the registration message header and the extensions which precede the signature.

One important requirement for securing the registration request is that each separate registration request be made to have unique data, so that two different registrations will practically never have the same MD5 hash. Otherwise, the protocol would be susceptible to *replay attacks*, whereby a malicious node could record valid registrations for later replay, effectively disrupting the ability of the home agent to tunnel to the current care-of address of the mobile node at that later time. This is accomplished by including within the registration message a special *identification* field, which changes with every new registration. The exact semantics of the identification field depend upon several details which are described completely in the protocol specification [38], but in general terms there are two main ways to make the identification field unique.

One is to use a timestamp. Then, each new registration will have a later timestamp, and differ from previous registrations. The other method is to cause the identification to be pseudo-random number. With enough bits of randomness, it is extremely unlikely that two independently chosen values for the identification field will be the same. When random values are used for this purpose, Mobile IP defines a method by which both registration request and reply are protected from replay, and each message then has 32 bits of randomness in the identification, which ought to be plenty. If the mobile node and the home agent get too far out of synchronization for timestamps, or if they lose track of the expected random numbers, the home agent will reject the registration request and within the reply include information to allow resynchronization.

The identification is also used by the foreign agent to match up pending registration requests to registration replies when they arrive at the home agent, and subsequently be able to relay the reply to the mobile node. Other information which is stored by the foreign agent for pending registrations include the mobile node's home address, the mobile node's MAC address, the source port number for the registration request from the mobile node, the registration lifetime proposed by the mobile node, and the home agent's address. The foreign agent is allowed to limit registration lifetimes to a configurable value which it puts into its agent advertisements. This registration lifetime can be reduced by the home agent, but never increased; the home agent includes the registration lifetime as part of the registration reply.

In Mobile IP, foreign agents are mostly passive agents, as indicated in Figure 1. They decapsulate traffic from the home agent and forward it to the mobile node. They relay registration requests and replies back and forth between the home agent and the mobile node, doing mostly as they are told to do. They do not have to authenticate themselves to the mobile node or home agent. A bogus foreign agent could impersonate a real foreign agent, just by following protocol, and offering agent advertisements to the mobile node. The bogus agent could, for instance, then refuse to forward decapsulated packets to the mobile node when they were received. This is no worse than the result of any node being tricked into using the wrong default router, which is possible using unauthenticated Router Advertisements as specified in RFC 1256.

In cases where the mobile node cannot contact its home agent, there is a mechanism provided for the mobile node to try to register with another unknown home agent on its home network. This method of *automatic home agent discovery* works by using a broadcast IP address instead of the home agent's IP address as the target for the registration request. When the broadcast packet gets to the home network, other home agents on the network will send a rejection to the mobile node; but, their rejection notice will contain their address for the mobile node to use in a freshly attempted registration message. Note that the broadcast is not an Internet-wide broadcast, but instead a *directed* broadcast which reaches only IP nodes on the home network.

VI. Tunneling

Tunneling for Mobile IP is illustrated in figure 2. In the figure, the source is node X and the destination is the mobile node; the tunnel source is the home agent, and the tunnel destination is the care-of address. The default encapsulation mechanism, which must be supported by all mobility agents using Mobile IP, is IP-within-IP [33]. Using IP-within-IP, the home agent inserts a new IP header in front of the IP header of any datagram addressed to the mobile node's home address. The new IP header uses the mobile node's care-of address as the destination IP address (known as the tunnel destination). The tunnel source IP address is the home agent, and the tunnel header uses 4 as the higher-level protocol number, indicating that the next protocol header is again an IP header (the original IP header as sent to the mobile node). When IP-within-IP is used, the entire original IP header is preserved as the first part of the payload of the tunnel IP header, and to recover the original packet, the foreign agent merely has to eliminate the tunnel IP header and deliver the rest to the mobile node.

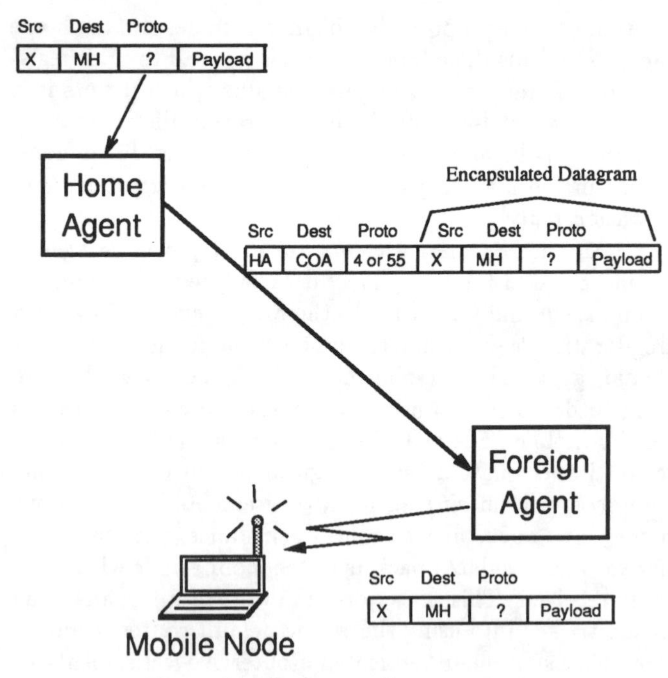

Fig. 2. Mobile IP tunneling operations

The figure also indicates that sometimes the tunnel IP header uses protocol number 55 as the inner header. This happens when the home agent uses *minimal encapsulation* [34] instead of IP-within-IP. Processing for the minimal encapsulation header is slightly more complicated than that for IP-within-IP, because some of the information from the tunnel header is combined with the information in the inner minimal encapsulation header to reconstitute the original IP header with the mobile node's IP address as the destination IP address.

VII. IPv6

IP version 6 (IPv6) [10] [17] is the product of a major effort within the IETF to engineer an eventual replacement for the current version of IP [41]. IPv6 includes many features missing in the current version of IP (henceforth called IPv4), including Stateless Address Autoconfiguration [47] and Neighbor Discovery [28], which are useful in streamlining mobility support. IPv6 also attempts to drastically simplify the process of *renumbering*, which could be critical to the future routability of the Internet [5]. Since the future population of the Internet is likely to be largely composed of mobile computers, efficient support for mobility will make a decisive difference in the Internet's future performance. This effect, along with the growing importance of the Internet and the World Wide Web, indicates the need for paying attention to supporting mobility, as indicated in [3].

IPv6 mobility support [19], in its major outline, follows the design for Mobile IPv4, retaining the ideas of a home network, home agent, and use of encapsulation to deliver packets from the home network to the mobile node's current point of attachment to the Internet. Discovery of a care-of address is also needed, but the means by which a mobile node can configure its care-of address are already specified as part of Stateless Address Autoconfiguration and Neighbor Discovery. Thus, foreign agents are not required to support mobility in IPv6. IPv6-within-IPv6 tunneling is also already specified [9]. IPv6 mobility also borrows heavily from the route optimization ideas specified for IPv4 [37], in particular the idea of delivering *binding updates* directly to correspondent nodes. When it has knowledge of the mobile node's current care-of address, a correspondent node can deliver packets directly to the mobile node's home address without any assistance from the home agent. Route optimization is expected to dramatically improve performance for IPv6 mobile nodes. It is realistic to expect this extra functionality required in all IPv6 nodes for two reasons:

- It is still possible to place requirements on IPv6 nodes, since the IPv6 standards documents are at an early stage of standardization, and
- Processing binding updates can be implemented as a fairly simple modification to IPv6's use of the *destination cache* [28].

One of the biggest differences between IPv6 and IPv4, is that all IPv6 nodes are expected to implement cryptographic and strong authentication and encryption features [24] [25] to improve the security of tomorrow's Internet. This affords a major simplification for Mobile IPv6, since all the authentication procedures can be assumed to exist when needed, and do not have to appear in the Mobile IPv6 protocol specification. Even so, Mobile IPv6 specifies the use of authentication procedures as infrequently as is reasonably possible. For one thing, good authentication comes at the cost of performance, so should only be required occasionally. For another thing, questions about the availability of Internet-wide key management are far from resolved at this time.

In contrast to route optimization for IPv4, correspondent nodes do not tunnel packets to mobile nodes. Instead, correspondent nodes use IPv6 *routing headers*, which implement a variation on IPv4's *source routing* option. In fact, a number of early proposals for supporting mobility in IPv4 specified a similar use of source routing options (for example [32], [20]), but the realities of today's Internet precludes their use. Two existing conditions are mainly responsible for this:

- IPv4 source routing options require that the receiver of source routed packets follow the reversed path to the sender back along the indicated intermediate nodes.
- Existing routers exhibit terrible performance when handling source routes.

The first condition enables malicious nodes using source routes from remote locations within the Internet to impersonate other nodes, exacerbated by the lack of authentication protocols. The second condition has resulted in drastically unfavorable results for the attempted deployment of other protocols using source routes.

IPv6's more careful specification eliminates the need for source route reversal, and allows routers to ignore options that do not need their attention. Thus, the objections to the use of source routes do not apply to IPv6. Consequently, correspondent nodes can use routing headers without penalty. With routing headers, a mobile node can then easily determine when a correspondent node does not have the right care-of address. Packets delivered by encapsulation instead of by source routes in a routing header must have been sent by correspondent nodes that need to receive a binding update from the mobile node. Binding updates to correspondent nodes are delivered from the mobile node instead of from the home agent as is specified by route optimization for IPv4. It seems more likely in IPv6 that key management between the mobile node and correspondent node will be deployed and available.

Other features supported by IPv6 mobility include:

- coexistence with Internet *ingress filtering* [14]
- *smooth handoffs*, which in IPv4 is specified for foreign agents as part of route optimization
- renumbering home networks
- automatic home agent discovery

VIII. Current/Future Work

A great deal of attention is currently focussed on making Mobile IP coexist with the security features now coming into use within the Internet. In particular, *firewalls* [7] cause difficulty for Mobile IP. They block all classes of incoming packets that do not meet specified criteria. Of particular importance to Mobile IP, enterprise firewalls are typically configured to disallow packets to enter the enterprise from the Internet that emanate from internal computers. While this is useful in allowing internal Internet nodes to be managed without such great attention to security, it certainly presents difficulties for mobile nodes wishing to communicate with other nodes within their home enterprise networks. Such communications, originating from the mobile node, carry the mobile node's home address, and would thus be blocked by the firewall.

Complications are presented by *ingress filtering* [14] operations. Many border routers discard packets coming from within the enterprise, if those packets do not contain a source IP address configured for one of the enterprise's internal networks. Since mobile nodes would otherwise use their home address as the source IP address of the packets they transmit, this presents difficulty. Solutions to this problem for IPv4 typically involve tunneling outgoing packets from the care-of address, but then the difficulty becomes that of finding a suitable target for the tunneled packet from the mobile node. The only universally agreed on possibility is the home agent, but that target introduces yet another serious routing anomaly for communications between the mobile node and the rest of the Internet. IPv6 offers a much better solution, the *home address* destination option [19].

Mobile IP forms the basis of many current research efforts. At the University of California at Berkeley [23], Mobile IP is used to construct *vertical handoffs* between dissimilar media (e.g., infrared, radio LANs, wide-are cellular, satellite) depending upon error rates and bandwidth availability. Other factors such as cost and predictive service might also be taken into account. CMU's Monarch project [22] has been the focus of investigation into campus wireless networks, Mobile IP, Mobile IPv6, and ad-hoc networking [21]. Other academic efforts have been proceeding at University of Portland, University of Alabama, University of Texas, UCLA, Macquarie University, SUNY Binghampton, University of Singapore, Swedish Royal Institute of Technology, and many others. Two books about Mobile IP have recently been published [36], [45].

Part of the IETF standardization process requires that interoperability between various independent implementations be shown before the protocol can advance. FTP Software (Andover, MA) has hosted two interoperability testing sessions, and many vendors have taken advantage of the opportunity. The test results have given added confidence that the protocol specification is sound, implementable, and of diverse interest throughout the Internet community. Experience gained a result of the interoperability tests is being used to try to advance the protocol specification from Proposed Standard [38] to Draft Standard [39]. Only a few minor revisions have been needed to ensure that the specification can be interpreted in only one way by the members of its intended audience, which are network protocol engineers and programmers.

Mobile IP can be viewed as a secure tunnel establishment protocol. Efforts along these lines are being made at BBN as part of the MOIPS [13] project to extend Mobile IP operation across firewalls, even when there are multiple security domains involved. Mobile IP is also used as part of the MNCRS initiative, and the tunnel establishment aspect has been borrowed for use with Tunnel Establishment Protocol (TEP) [4]. In a related effort, the ability to advertise multiple foreign agents is being put to use to arrange *hierarchies* of mobility agents, in an attempt to cut down the number of registrations that have to transit the global Internet between the home and foreign networks [31]. Investigations have been made about the suitability of using DHCP to provide care-of addresses to mobile nodes [35], and a new extension to PPP [43] will enable dial-up users to more efficiently employ their dynamic IP addresses as care-of addresses [44].

IX. Counterindications

Circumstances remain under which there are unresolved questions about the deployment of Mobile IP. The most pressing need is for solutions to the abovementioned security problems [6]. A firewall traversal solution is offered in [16]. The use of reverse tunnels to the home agent to counter the restriction imposed by ingress filtering is specified in [27].

The design of Mobile IP is based on the premise that connections based on TCP should survive cell changes. But, opinion is not unanimous on the need for this feature. Many people believe that computer communications to laptop computers are sufficiently bursty that there is no need to increase the reliability of the connections supporting the communications. The analogy is made to fetching Web pages by selecting the appropriate URLs. If a transfer fails, people are used to trying it again. This is tantamount to making the user responsible for the retransmission protocol, and depends for its acceptability on a widespread perception that computers and the Internet cannot be trusted to do things right the first time. Naturally, such assumptions are strongly distasteful to many Internet protocol engineers, this author included. Nevertheless, the economic impact of this perception held by many users cannot be denied. Better product engineering can hopefully counter or erase perception once Mobile IP is fully deployed.

Mobile IP creates the perception that the mobile node is always attached to its home network. This forms the basis for the reachability of the mobile node at an IP address which can be conventionally associated with its *fully-qualified domain name*(FQDN) [26]. If the FQDN is associated with one or more other IP addresses, perhaps dynamically, then those alternative IP addresses may deserve equal standing with the mobile node's home address. Moreover, it is possible that such an alternative IP address would offer shorter a routing path if, for instance, the address were apparently located on a physical link nearer to the care-of address of the mobile node, or if the alternative address were the care-of address itself. Finally, many communications are short-lived and depend neither uponupon the actual identity of the mobile node, nor its FQDN, and thus do not take advantage of the simplicity afforded by use of the mobile node's home address. The issues surrounding the selection by the mobile node of an appropriate long-term (or not so long-term) address for use in establishing connections are complex and far from being resolved.

Mobile IP has been engineered as a solution for wireless LAN location management and communications, but the wireless LAN market has been slow to develop. It is difficult to make general statements about the reasons for this slow development, but with the recent ratification of the IEEE802.11 MAC protocol [18], there is additional hope that wireless LANs will become more popular. There is also the fact that bandwidth for wireless devices has been constantly improving, so that radio and infrared devices on the market today offer multi-Mbyte/sec data rates. Faster wireless access over standardized MAC layers could be a major catalyst for growth of this market.

Mobile IP may face competition from alternative tunneling protocols such as PPTP [29] and L2TP [30]. These other protocols, based on PPP, offer at least portability to the mobile computers. Arguments advanced earlier in this article indicate that portable operation is not a long term solution, but in the short term, and in the absence of full Mobile IP deployment, these alternative access methods can be quite attractive. If they are made widely available, it is interesting to think about whether the use of Mobile IP will be displaced, or instead made more immediately desirable, as people gain more experience with the convenience of mobile computing. Mobile IP could also in the future specify use of such alternative tunneling protocols to capitalize on their deployment on platforms not otherwise supporting IP-within-IP encapsulation.

X. Summary

A quick search for Mobile IP related items on the Web turned up over 30,000 hits, which is impressive even given the notorious lack of selectivity for such procedures. Apparently, there is a high degree of interest in mobile computing and in Mobile IP as a way to provide for it. Numerous products are being engineered based on Mobile IP either explicitly or as part of other products. As one example, the Cellular Digital Packet Data (CDPD) [8] has created a widely deployed communications infrastructure which was based on a previous draft specification of Mobile IP. Other products may be expected within a year. The outlook is far from clear, but once the security solutions are solid, nomadic users may finally begin to enjoy the convenience of untethered seamless roaming and effective application transparency that is the promise of Mobile IP.

It is possible that the pace of deployment Mobile IP will track that of IPv6 deployment, or that the requirement for supporting mobility in IPv6 nodes will give additional impetus to the deployment of both IPv6 and mobile networking. The increased user convenience and the reduced need for application awareness of mobility can be a major driving force for adoption. Since both IPv6 and Mobile IP have little direct effect on the operating systems of mobile computers, outside of the network layer of the protocol stack, applications designers will find this to be an acceptable programming environment. Of course, this depends heavily on the willingness of platform and router vendors to implement Mobile IP and/or IPv6, but indications are strong that most major vendors already have implementations either finished or underway.

We hope this brief introduction to mobile networking will highlight the great potential for Mobile IP, and illuminate its place in the highly mobile world of the future global Internet. The security needs are getting active attention, and will benefit from the deployment efforts underway now. Interested readers should join the mobile IP mailing list, by sending mail to majordomo@Smallworks.COM, including the line "subscribe mobile-ip" in the body of the message. The IETF manages many other interesting areas of protocol engineering. Access to the mobile-IP working group's charter and mail archives, as well as that of all other working groups, can be found by selecting the appropriate links on the IETF Web page `http://www.ietf.org`. The author will also gladly answer electronic mail sent to `charles.perkins@sun.com`.

References

[1] S. Alexander and R. Droms. DHCP Options and BOOTP Vendor Extensions. RFC 2132, March 1997.

[2] Pravin Bhagwat, Charles Perkins, and Satish K. Tripathi. Network Layer Mobility: an Architecture and Survey. *IEEE Personal Communications Magazine*, 3(3):54–64, June 1996.

[3] S. Bradner and A. Mankin. The Recommendation for the IP Next Generation Protocol. RFC 1752, January 1995.

[4] P. Calhoun and C. Perkins. Tunnel Establishment Protocol (TEP). draft-calhoun-tep-00.txt, August 1997. (work in progress).

[5] I. Castineyra, J. Chiappa, and M. Steenstrup. RFC 1992: The Nimrod Routing Architecture, August 1996.

[6] S. Cheshire and M. Baker. Internet mobility 4x4. In *Proceedings of the ACM SIGCOMM Conference on Applications, Technologies, Architectures, and Protocols for Computer Communications*, volume 26,4 of *ACM SIGCOMM Computer Communication Review*, pages 318–329, New York, August 26–30 1996. ACM Press.

[7] William R. Cheswick and Steven Bellovin. *Firewalls and Internet Security*. Addison-Wesley, Reading, Massachusetts, 1994. (ISBN: 0-201-63357-4).

[8] CDPD consortium. *Cellular Digital Packet Data Specification*. P.O. Box 809320, Chicago, Illinois, July 1993.

[9] A. Conta and S. Deering. Generic Packet Tunneling in IPv6. draft-ietf-ipngwg-ipv6-tunnel-07.txt, December 1996. (work in progress).

[10] S. Deering and R. Hinden. Internet Protocol, Version 6 (IPv6) Specification. RFC 1883, December 1995.

[11] Stephen E. Deering, Editor. ICMP Router Discovery Messages. RFC 1256, September 1991.

[12] R. Droms. Dynamic Host Configuration Protocol. RFC 2131, March 1997.

[13] John Zao et.al. A Public-Key Based Secure Mobile IP. In *ACM Mobicom 97*, October 1997.

[14] Paul Ferguson and Daniel Senie. Ingress Filtering in the Internet. draft-ferguson-ingress-filtering-03.txt, October 1997. (work in progress).

[15] Corporation for National Research Initiatives (CNRI). XIWT: Cross Industry Working Team, 1994. http://www.cnri.reston.va.us:3000/XIWT/public.html.

[16] V. Gupta and S. Glass. Firewall Traversal for Mobile IP: Guidelines for Firewalls and Mobile IP entities. draft-ietf-mobileip-firewall-trav-00.txt, March 1997. (work in progress).

[17] R. Hinden and S. Deering. IP Version 6 Addressing Architecture. RFC 1884, December 1995.

[18] Wireless LAN MAC and PHY Specifications. IEEE Document P802.11/D6.1.97/5, Jun 1997. AlphaGraphics #35, 10201 N.35th Avenue, Phoenix AZ 85051.

[19] D. Johnson and C. Perkins. Mobility Support in IPv6. draft-ietf-mobileip-ipv6-03.txt, July 1997. (work in progress).

[20] David B. Johnson. Mobile Host Internetworking Using IP Loose Source Routing. Technical Report CMU-CS-93-128, February 1993. (Carnegie Mellon University).

[21] David B. Johnson and David A. Maltz. *Dynamic Source Routing in Ad Hoc Wireless Networks*, chapter 5, pages 153–181. Kluwer Academic Publishers, 1996. (in Mobile Computing, edited by Tomasz Imielinski and Hank Korth).

[22] David B. Johnson and David A. Maltz. Protocols for Adaptive Wireless and Mobile Networking. *IEEE Personal Communications Magazine*, 3(1):34–42, February 1996.

[23] R.H. Katz. Adaptation and mobility in wireless information systems. *IEEE Personal Communications Magazine*, 1(1):6–17, 1994.

[24] Stephen Kent and Randall Atkinson. IP Authentication Header. draft-ietf-ipsec-auth-header-01.txt, July 1997. (work in progress).

[25] Stephen Kent and Randall Atkinson. IP Encapsulating Security Payload (ESP). draft-ietf-ipsec-esp-v2-00.txt, July 1997. (work in progress).

[26] P. Mockapetris. Domain Names - Concepts and Facilities. STD 13, November 1987.

[27] G. Montenegro. Reverse Tunneling for Mobile IP. draft-ietf-mobileip-tunnel-reverse-04.txt, August 1997. (work in progress).

[28] T. Narten, E. Nordmark, and W. Simpson. Neighbor Discovery for IP version 6 (IPv6). RFC 1970, August 1996.

[29] G. Pall, K. Hamzeh, W. Verthein, J. Taarud, and W. Little. Point-to-Point Tunneling Protocol–PPTP. draft-ietf-pppext-pptp-02.txt, July 1997. (work in progress).

[30] William Palter, T. Kolar, G. Pall, M. Littlewood, A. Valencia, K. Hamzeh, W. Verthein, J. Taarud, and W. Mark Townsley. Layer Two Tunneling Protocol 'L2TP'. draft-ietf-pppext-l2tp-08.txt, November 1997. (work in progress).

[31] C. Perkins. Mobile-IP Local Registration with Hierarchical Foreign Agents. draft-perkins-mobileip-hierfa-00.txt, February 1996. (work in progress).

[32] C. Perkins and P. Bhagwat. A Mobile Networking System based on Internet Protocol(IP). In *USENIX Symposium on Mobile and Location-Independent Computing*, August 1993.

[33] Charles Perkins. IP Encapsulation within IP. RFC 2003, May 1996.

[34] Charles Perkins. Minimal Encapsulation within IP. RFC 2004, May 1996.

[35] Charles Perkins and Jagannadh Tangirala. DHCP for Mobile Networking with TCP/IP. In *Proceedings of IEEE International Symposium on Systems and Communications*, pages 255–261, June 1995.

[36] Charles E. Perkins. *Mobile IP: Design Principles and Practice*. Addison-Wesley Longman, Reading, Massachusetts, 1998.

[37] Charles E. Perkins and David B. Johnson. Route Optimization in Mobile-IP. draft-ietf-mobileip-optim-07.txt, November 1997. (work in progress).

[38] C. Perkins, Editor. IP Mobility Support. RFC 2002, October 1996.

[39] C. Perkins, Editor. IP Mobility Support version 2. draft-ietf-mobileip-v2-00.txt, November 1997. (work in progress).

[40] J. B. Postel, editor. Transmission Control Protocol. RFC 793, September 1981.

[41] J. B. Postel, Editor. Internet Protocol. RFC 791, September 1981.

[42] Ronald L. Rivest. The MD5 Message-Digest Algorithm. RFC 1321, April 1992.

[43] William Allen Simpson, editor. The Point-to-Point Protocol (PPP). RFC 1661, July 1994.

[44] J. Solomon and S. Glass. Mobile-IPv4 Configuration Option for PPP IPCP. draft-ietf-pppext-ipcp-mip-02.txt, July 1997. (work in progress).

[45] James Solomon. *Mobile IP: The Internet Unplugged*. Prentice Hall, Englewood Cliffs, 1998.

[46] W. Stevens. TCP Slow Start, Congestion Avoidance, Fast Retransmit, and Fast Recovery Algorithms. RFC 2001, January 1997.

[47] S. Thomson and T. Narten. IPv6 Stateless Address Autoconfiguration. RFC 1971, August 1996.

[48] J. Veizades, E. Guttman, C. Perkins, and S. Kaplan. Service Location Protocol. RFC 2165, July 1997.

[49] V.L. Voydock and S.T. Kent. Security Mechanisms in High-level Networks. *ACM Computer Surveys*, 15(2), June 1983.

10.3 Cheshire, S. and Baker, M., Internet Mobility 4x4

Cheshire, S. and Baker, M., "Internet Mobility 4x4," *Proceedings of the ACM SIGCOMM'96 Conference on Applications, Technologies, Architectures, and Protocols for Computer Communications*, pp. 318–329, August 1996.

Internet Mobility 4x4

Stuart Cheshire and Mary Baker
Computer Science Department, Stanford University
Stanford, California 94305, USA
{cheshire,mgbaker}@cs.stanford.edu

Abstract

Mobile IP protocols allow mobile hosts to send and receive packets addressed with their home network IP address, regardless of the IP address of their current point of attachment in the Internet.

While some recent work in Mobile IP focuses on a couple of specific routing optimizations for sending packets to and from mobile hosts [Joh96] [Mon96], we show that a variety of different optimizations are appropriate in different circumstances. The best choice, which may vary on a connection-by-connection or even on a packet-by-packet basis, depends on three factors: the characteristics the protocol should optimize, the permissiveness of the networks over which the packets travel, and the level of mobile-awareness of the hosts with which the mobile host corresponds.

Of the sixteen possible routing choices that we identify, we describe the seven that are most useful and discuss their benefits and limitations. These optimizations range from the most costly, which provides completely transparent mobility in all networks, to the most economical, which does not attempt to conceal location information. In particular, hosts should retain the option to communicate conventionally without using Mobile IP whenever appropriate.

Further, we show that all optimizations can be described using a 4x4 grid of packet characteristics. This makes it easier for a mobile host, through a series of tests, to determine which of the currently available optimizations is the best to use for any given correspondent host.

1. Introduction

The increasing number of portable computers, combined with the growth of wireless services, makes supporting Internet mobility important. Mobile hosts need to switch between networks in different administrative domains as they move around the network, and they need to switch between different types of networks (cellular telephone, packet radio, Ethernet, etc.) to achieve the best possible connectivity wherever they are located.

SIGCOMM '96 - Stanford, California, USA, August 1996

We believe that IP is the correct layer at which to implement basic mobility support, rather than lower or higher layers. The problem with link layer solutions is that they are limited to a single medium. For example, one of the current ways of providing mobile Internet access is to use a cellular telephone and modem to dial into a central PPP [RFC1661] server. While this method provides connectivity (albeit at a cost of about 40¢ per minute), it is limited to a single technology — cellular telephony. Since its inception in the early 1970s, one of the most important guiding principles of the Internet has been to seek general-purpose solutions that work for all network technologies, not special-purpose hardware-specific solutions [Cer74]. Similarly, mobility solutions that require widespread changes at layers above IP would be highly impractical.

The challenge for supporting mobility at the IP layer is handling address changes. Even if the IP address of a mobile host's network interface changes when it moves from one network to another, the mobile host should be able to continue corresponding with other machines in the Internet. Most mobile IP protocols [Bla94] [Gup96] [Hag93] [Ioa93] [Myl93] [Per94] [Per96a] [Tas94] [Ter94] address this problem by allowing a mobile host to use its home network IP address no matter where it currently resides. Of these protocols, the protocol specified by the Internet Engineering Task Force (IETF) [Per96a] has become the most popular. It is simple, compatible with existing applications and hosts, and places no special burdens on normal IP routers in the Internet.

When trying to implement the IETF protocol, however, we ran into several problems that led us to conclude that one size does not fit all. There are at least two areas warranting further investigation. Firstly, the basic protocol does not work in some security-conscious networks. Secondly, the route used for packets from other hosts to the mobile host is not the most efficient. We believe that different optimizations are appropriate in different circumstances. For instance, a mobile host corresponding with a host that is physically connected to the same Ethernet segment should not require every packet to travel via its home agent. Choosing the best way to send packets depends upon what characteristics should be optimized, the permissiveness of the networks through which the packets must travel, and whether or not the correspondent hosts have any awareness of mobility. The best choice may even vary on a packet-by-packet basis.

In the next section we briefly outline our Mobile IP protocol, which is based strongly on the Draft IETF Mobile IP proposal (soon to become an official RFC). In Section 3 we list our optimization goals and give examples of some of the difficulties we faced in trying to achieve them.

We then show how to implement the desired optimizations and describe the circumstances in which each is suitable. Section 4 describes the four ways that a mobile host can send packets to a correspondent host, and Section 5 describes the four ways that a correspondent host can send packets to a mobile host. We then describe in Section 6 how these two different choices influence each other. The choices are not completely independent, and the capabilities of the correspondent host constrain which combinations are applicable for the mobile host to use. In any particular situation, a mobile host should choose the best alternative that is available to it.

In Section 7 we give the current status of our implementation and propose some future work.

Although this paper assumes that a mobile host is communicating with a conventional non-mobile correspondent host, the same techniques and optimizations apply equally well if both hosts are mobile.

2. Basic Mobile IP

The goal of Mobile IP is to allow a mobile host to send and receive packets addressed with its home IP address regardless of its current point of attachment to the Internet, and to maintain communication associations (such as TCP connections) even if the point of attachment changes during their lifetime. Users should not have to restart their applications whenever they change location, especially applications such as remote logins that build up significant state at the remote endpoint. It is important to note that computers do not need to be turned on continuously in order to maintain quiescent TCP connections; putting a laptop computer to sleep while moving it from place to place does not necessarily break connections. On our laptop computers running Linux we frequently have idle telnet connections that are preserved for hours, and sometimes even for days or weeks, while the laptop computer is sitting unused in 'sleep' mode.

In order to meet these goals of location transparency and connection durability, each mobile host has a permanent home IP address that does not change. This unchanging address enables conventional Internet hosts, which are unaware of mobility issues, to communicate with the mobile host. If the address changed each time the mobile host changed its point of connection to the Internet, then our requirement of connection durability would not be met, because TCP connections to other Internet hosts would break every time the mobile host moved.

When the mobile host is at home, it sends and receives packets using its home IP address and functions like a normal non-mobile Internet host.

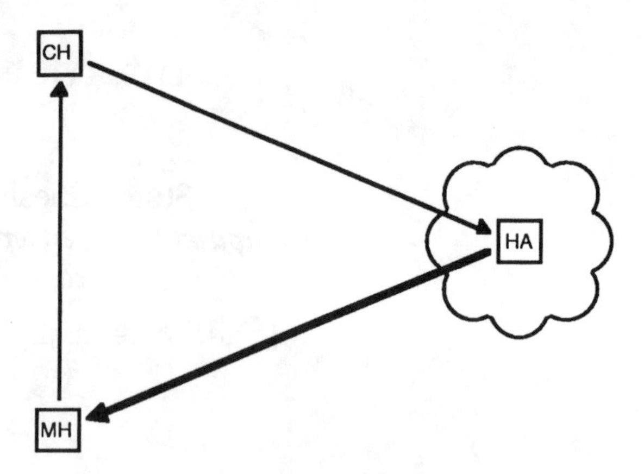

Figure 1. Basic Mobile IP.
This figure shows communication involving a mobile host (MH) away from its home network, its home agent (HA) and a correspondent host (CH). In these figures, squares represent Internet hosts, clouds represent administrative network domains, thin arrows represent normal IP packets being sent from one host to another, and thick arrows represent IP packets that carry an encapsulated IP packet. Here, the administrative domain represented by the cloud is the mobile host's home domain.

When it is away from home, the mobile host obtains a temporary 'guest' connection to the Internet at the site it is visiting. This connection may be obtained by connecting to an Ethernet segment and asking a friendly network administrator to assign an IP address to the visiting host; it may be obtained by connecting to an Ethernet segment and having an address assigned automatically by DHCP [RFC1541], or the connection may be obtained via communication with an IETF 'foreign agent' that has been placed on the network expressly for the purpose of supporting visiting mobile hosts.

After the mobile host has connected to the visited network (directly, or via a foreign agent), it registers its new location with its home agent. The home agent is a machine on the mobile host's home network that acts as a proxy on behalf of the mobile host for the duration of its absence. The home agent uses gratuitous proxy ARP [RFC1027] to capture all IP packets addressed to the mobile host. When packets addressed to the mobile host arrive on its home network, the home agent intercepts them and uses encapsulation (often called 'tunneling') to forward them to the mobile host's current location, as shown in Figure 1.

If the mobile host moves again to a different point of attachment on the Internet, then it must again inform its home agent of its new location. During this transition period it may be possible to lose packets, but higher-level Internet protocols are already responsible for mechanisms to ensure reliable packet delivery where required, so it is not necessary to duplicate this functionality within Mobile IP.

As shown in Figure 1, delivering packets in the opposite direction, from the mobile host to the correspondent host, is simpler. The mobile host transmits its packets into the Internet, addressed directly to the correspondent host.

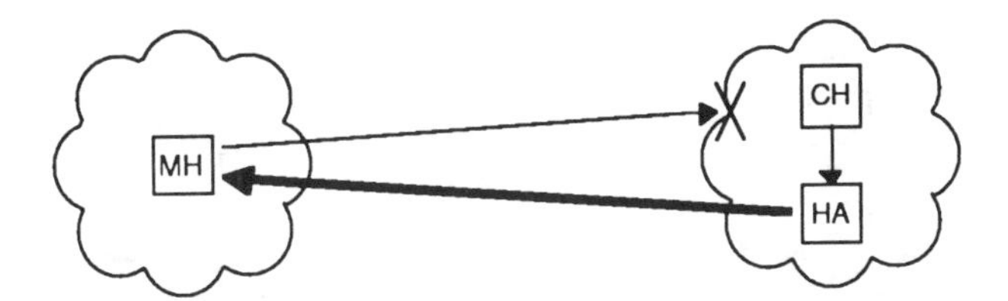

Figure 2. Problem with Source Address Filtering
The home agent encapsulates the correspondent host's packets and correctly forwards them to the mobile host. Unfortunately, the mobile host's replies are discarded by a security-conscious boundary router and never reach the correspondent host to which they are addressed.

Figure 3. Bi-directional Tunneling
By tunneling all of its packets via the home agent, the mobile host avoids their being discarded by the routers at the boundary of its home domain.

Because the source address in the packets is the mobile host's permanent home source address, when the correspondent host replies to those packets, the replies will find the mobile host even if it moves; all replies will travel indirectly via the home agent. The latency and available bandwidth over the two different paths may be significantly different, but this is not unusual for IP. The IP specification makes no promises about the path that packets will take, and much of the current Internet backbone already routes packets going in different directions over different paths [Par96].

The IETF Mobile IP proposal says that mobile hosts may connect directly to the visited network or indirectly via a "foreign agent". When connecting via a foreign agent, the home agent tunnels packets to this foreign agent, which decapsulates them and delivers the enclosed packet to the mobile host.

Our implementation of the protocol emphasizes self-sufficiency for mobile hosts. They connect directly to the Internet and operate independently without requiring a foreign agent. It is impractical for mobile hosts to assume that foreign agent services will be available everywhere. Fortunately it is not difficult for mobile hosts to provide their own mobility support in the absence of foreign agents [Bak96]. Foreign agents may be able to provide useful services to mobile hosts, but they also restrict the freedom of the mobile host to choose from the full range of possible optimizations.

The most important of these optimizations, which foreign agents prevent, is the freedom to forgo the services of Mobile IP for communications that do not need them. A lot of work has been done to make protocols client-originated wherever possible. The trend towards using POP [RFC1725] to retrieve electronic mail is one such example. Mobile IP makes it possible to send packets addressed to a mobile host without knowing its current location and to be able to set up durable connections. These are important facilities, but they should not be provided at the expense of the ability to operate with the efficiency of a normal Internet host.

Some researchers [Tas94] have proposed a different approach to the mobility problem, by assigning a new unique permanent identifier to every host on the Internet. They propose rewriting transport protocols like TCP to identify connection endpoints using this new identifier instead of the location-dependent IP address. While this may appear to result in a more 'elegant' solution, it offers no real benefits over the more pragmatic solution proposed by the IETF. The information contained in this hypothetical enlarged TCP header with its unique endpoint identifiers is semantically identical to the information contained in the IETF's encapsulated IP and TCP headers taken together. Although adding an encapsulated IP header to the packet consumes slightly more space than a redesigned TCP header might, this overhead can be minimized by use of Generic Routing Encapsulation [RFC1702] or Minimal Encapsulation [Per95]. In addition, there is no need to invent a new namespace when the existing IP address namespace is already well understood, and we already have established mechanisms for allocating, assigning, and managing unique identifiers in that namespace.

3. Project Goals

In this section we describe the areas for optimization and improvement over the basic Mobile IP protocol, and the constraints to which we must adhere. The purpose of the optimizations is to achieve efficient delivery of packets, in terms of their size and the path they take through the network. These factors affect the delivery latency and the load on the shared resources in the Internet. There are two constraints. The first is that we may not assume any special support from routers, except for normal IP routing. This constraint is motivated by the end-to-end argument [Sal84], which states that we should not burden the network with functions that can be performed equally well, or better, at the endpoints. The second constraint is that packets be correctly deliverable to their destination: the choice of feasible optimizations is constrained by the permissiveness of the networks over which the packets travel and the level of mobility awareness of the correspondent host. (The term "correctly deliverable" is used in the normal context of a "best-effort" datagram network, meaning that with high probability packets are successfully delivered. Existing causes of packet loss still exist, even when using Mobile IP.)

3.1 Ensure Deliverability

An important goal is that all Mobile IP systems be able to work correctly in the current Internet and interoperate correctly with current hosts, but there are situations where even the IETF Mobile IP solution fails. For security or policy reasons, many networks will not deliver packets which are sent the way the IETF specification describes.

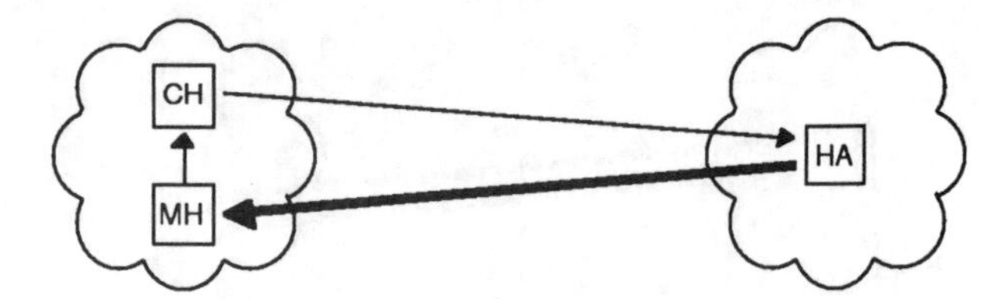

Figure 4. Behavior when CH is Close to MH
The correspondent host addresses packets to the mobile host's permanent home address. The Internet naturally delivers the packets to the mobile host's home network, where the home agent encapsulates them and sends them to the mobile host. Outgoing packets from the mobile host are delivered in the normal fashion, directly to the correspondent host.

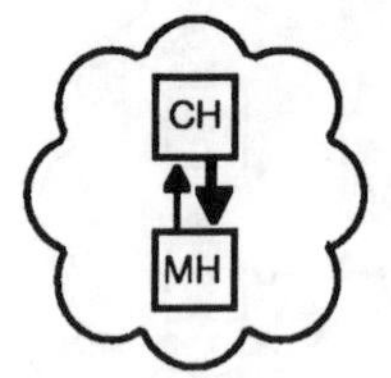

Figure 5. A Smart Correspondent Host.
The correspondent host knows the mobile host's temporary care-of address, so it encapsulates the packet itself and sends it directly.

In most networks, the packets from the mobile host will never reach the correspondent host, for the reason illustrated in Figure 2. When the packet arrives at the boundary of the home institution, the boundary router will see a packet coming from outside the home network, with a source address claiming that the packet originates from a machine inside the home network. Most network administrators, concerned about security, will configure boundary routers to drop such packets. Many network services, including the majority of NFS servers, determine whether or not they can safely trust the host sending the packet solely based on the source address of the packet. If we allow machines *outside* our network to send in packets with source addresses claiming to originate from trusted machines *within* our network, we effectively allow any machine on the Internet to impersonate any machine in our organization. Although in most cases replies will not get back to the machine originating the attack, many kinds of attack can be performed without needing to see any replies.

Another reason that packets sent by the mobile host might be discarded is that most end-user networks have a policy forbidding *transit traffic*. (Transit traffic is traffic passing through an intermediate network on the way to its final destination.) Some network administrators enforce this policy by configuring routers to discard packets with source addresses that appear to be invalid. Most traffic on the Internet backbone is transit traffic, but tail circuits, such as a 100Base-T connection to a desktop computer, are typically not expected to carry transit traffic. Packets appearing on such a tail circuit with source addresses belonging to a foreign network normally indicate some inappropriate use of the network, and would be discarded by the router.

The solution to these problems, which has also been described in [Mon96], is shown in Figure 3. By having the mobile host encapsulate outgoing packets and send them via the home agent, the inner packets are protected from scrutiny by routers. The boundary router only sees packets coming from a machine at some other institution (the mobile host using a temporary care-of address belonging to that institution) going to a local machine (the home agent). These packets are able to travel through the router, and the home agent can send the enclosed packets on the local network on behalf of the mobile host. This lengthens the distance that the packets travel but meets the deliverability requirement.

Note that it is not just 'firewall' routers that will drop these packets. Even the most forgiving of boundary routers would be expected to perform the rudimentary source address checks described above. Firewall routers usually impose much stricter restrictions. In situations where a mobile user is communicating with home services protected by a firewall, we anticipate that the firewall itself would be set up to act as the mobile user's home agent, sitting as it does on the boundary between the untrusted outside world and the trusted world inside.

3.2 Minimize Latency

Subject to the constraint that the packet must be successfully deliverable to the destination, our next goal is to minimize the distance that it travels through the Internet. Packets delivered via the home agent typically travel further through the Internet than they would if they were delivered by the optimal unicast route. As well as increasing the round-trip delay observed by the communicating parties, this also affects other users by increasing the overall load on the shared resources of the Internet.

In Figures 2 and 3 the extra distance added by indirect delivery is small compared to the distance that the packets would travel anyway. Even if the mobile host had been communicating directly with the correspondent host, the packets would still have had to make the long journey across the Internet between the two sites.

Unfortunately in Figure 4 the extra distance is not small. When they travel indirectly via the home agent, packets sent by the correspondent host travel significantly further than is necessary. It would be more efficient if a correspondent host could discover that the mobile host is nearby, and send the packets directly to it. A correspondent host that is aware of mobility issues should be allowed to do this.

We and others [Joh96] have approached this problem by developing an optional routing optimization mechanism to avoid the overhead of indirect delivery via the home agent, as shown in Figure 5. A correspondent host with enhanced networking software can learn the mobile host's temporary care-of address, and then perform the encapsulation itself, sending the packet directly to the mobile host. This avoids the overhead of indirect delivery.

There are several ways that a smart correspondent host can learn that a host is mobile and learn its current temporary care-of address. We are implementing two mechanisms. The first is that when the home agent forwards a packet to the mobile host, it may also send an ICMP message back to the packet's source, informing it of the mobile host's current temporary care-of address. The second is an extension to the Domain Name Service [RFC1034], similar to the current MX records which provide alternative addresses for mail delivery [RFC974]. A mobile host that is away from home, but not currently changing location frequently, could register its care-of address with the extended DNS service. When a smart correspondent looks up a host name and sees that it has a temporary address record in addition to the normal permanent address record, it then knows that it has the option to send packets directly to that temporary address.

It has also been proposed [Per96b] that support for route optimization should be included in the base IPv6 specification [RFC1883] for all IPv6 hosts.

3.3 Minimize Size

In addition to the overhead that indirect delivery adds, encapsulation also adds overhead by increasing the size of the packets. Encapsulation typically adds 20 bytes to the size of the packet in IPv4, and more in IPv6. If the addition of the extra 20 bytes makes the packet exceed the IP maximum transmission unit (MTU) for a particular link, then the packet will be fragmented, doubling the packet count. To avoid this overhead, we should avoid encapsulation when possible.

4. Outgoing Packets

In this section we look at how to achieve our previously described optimization goals for outgoing packets from the mobile host. Although we could use loose source routing, this achieves little that can't be done equally well using an encapsulating header [Per96c]. Current IP routers typically handle packets with options much more slowly than they handle normal unadorned IP packets. In IPv6, source routing is performed exclusively using routing headers, which is equivalent to encapsulation.

Other than loose source routing the only way to influence the path that the packets take through the Internet is by the choice of source and destination addresses in the IP header. If we choose to encapsulate the packet, then we also have the freedom to choose the source and destination addresses in the encapsulating IP packet.

If we choose not to encapsulate IP packets, then the mobile host sends out normal IP packets exactly as a conventional non-mobile host does. The source address of such a packet identifies the entity with which the correspondent host is communicating. If the mobile host uses its home IP address as the source address, then its mobility remains transparent to the correspondent host. As described in Section 3.1, some networks will discard such packets if they are sent directly to the correspondent host. If the mobile host instead uses its temporary care-of address, then the packets will not be discarded by the network, but transparent mobility is lost and TCP connections will be unceremoniously broken when the mobile host moves. Both of these addressing techniques are appropriate in some situations and not in others.

To achieve transparent mobility *and* successful delivery in security conscious networks, we use encapsulation. Encapsulation increases the size of the packet, but it has advantages. It allows us to use different source addresses in the inner and outer headers. We use the home IP address as the source address of the inner packet, to preserve location transparency. We use the temporary care-of address as the source address of the outer packet, so that security conscious routers will not discard it.

When using encapsulation, we have to choose which host will perform the decapsulation. The most conservative choice is to send the packets back to the home agent for decapsulation and subsequent delivery to the correspondent host. We know that we can rely on our own home agent to perform decapsulation for us, but the packets may travel significantly further through the network than is necessary. The most aggressive choice is to send the packets directly to the correspondent host. This avoids indirect delivery via the home agent but can only be used if the correspondent host is able to process encapsulated packets. As before, both of these techniques have situations where they are appropriate, and situations where they are not.

Figures 6 and 7 show the choices the mobile host must make. For packets sent unencapsulated, it has a choice of two possible source addresses for the packet: the permanent home address or the temporary care-of address. For packets sent encapsulated, it must decide whether to send the encapsulating packet to its home agent or directly to the correspondent host. Below we summarize these four delivery choices available to the mobile host and give examples of situations where each is useful. In our notation 'S' denotes the source address, and 'D' denotes the destination address. In the cases where the packet is encapsulated, 's' and 'd' denote the source and destination addresses in the encapsulating (outer) header.

Out-IE: Outgoing, Indirect, Encapsulated (Conservative mode)

s = From temporary care-of address
d = To home agent
S = From permanent home address
D = To correspondent host

Advantages: 1. Avoids the risk of an intervening router discarding the packet because it appears to have an invalid source address for the network from which it originates. 2. The correspondent host is unaware that the packet originated on a mobile host, and needs no special software to receive the packet.

Disadvantages: 1. Indirect delivery. 2. Encapsulation overhead.

Motivation: All mobile hosts must support tunneling through the home agent, since this is the only method that can be relied upon to work in all situations. For example, a

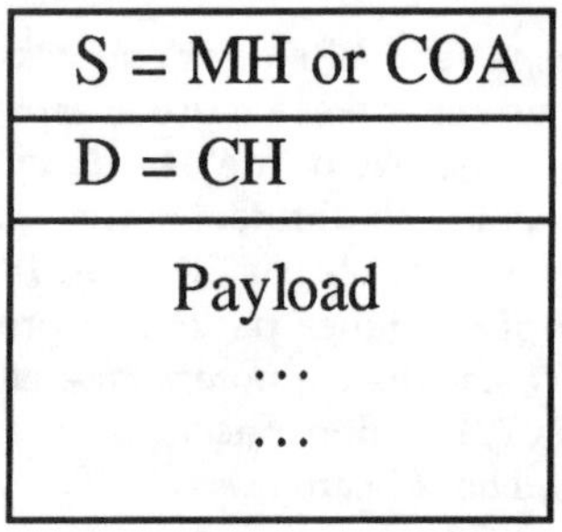

S = MH or COA
D = CH
Payload ... ...

Figure 6. Outgoing Packet Sent Unencapsulated
The source address (shaded) may be either the mobile host's permanent home address (MH) or its temporary care-of address (COA). The destination address (D) is the address of the correspondent host (CH).

mobile host in a network with source address filtering, communicating with a correspondent host that is not mobile-aware, has no choice but to use Out-IE.

Privacy concerns provide another motivation for tunneling through the home agent. In some situations, mobile users may not wish to reveal their current location to the correspondent host. In these cases, sending all outgoing packets indirectly via the home agent may be the method the user wants, even when other more efficient alternatives are also available.

Out-DE: Outgoing, Direct, Encapsulated
(Decapsulation-capable correspondent host)

s = From temporary care-of address
d = To correspondent host
S = From permanent home address
D = To correspondent host

Advantages: 1. Direct delivery. 2. Avoids the risk of an intervening router discarding the packet because it appears to have an invalid source address for the network from which it originates.

Disadvantages: 1. Encapsulation overhead. 2. The correspondent host must have the capability of decapsulating encapsulated IP packets.

Motivation: Out-DE is the best choice for a mobile host in a network with source address filtering, communicating with a correspondent host that is able to process encapsulated packets.

Out-DH: Outgoing, Direct, Home Address
(No source address filtering)

S = From permanent home address
D = To correspondent host

Advantages: 1. Direct delivery. 2. No encapsulation overhead. 3. The correspondent host is unaware that the packet

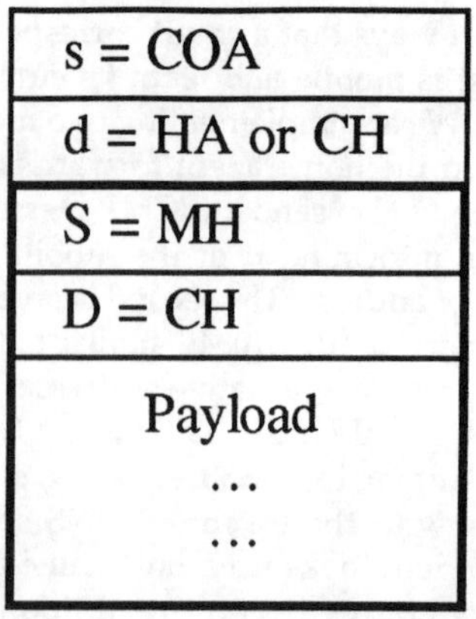

s = COA
d = HA or CH
S = MH
D = CH
Payload ... ...

Figure 7. Outgoing Packet Sent Encapsulated
For the inner (encapsulated) packet, the source address (S) is the mobile host's permanent home address (MH), and the destination address (D) is the address of the correspondent host (CH). For the outer packet, the source address (s) is always the mobile host's temporary care-of address (COA), but the destination address (shaded) is either the address of the home agent (HA), or the address of the correspondent host (CH), depending on which host is to decapsulate the packet.

originated on a mobile host, and it needs no special software to receive the packet.

Disadvantages: 1. An intervening router may discard the packet because it appears to have an invalid source address for the network from which it originates.

Motivation: Out-DH is the best choice when none of the routers on the path from the mobile host to the correspondent host performs source address filtering.

Out-DT: Outgoing, Direct, Temporary Address
(No Mobile IP)

S = From temporary care-of address
D = To correspondent host

Advantages: 1. Direct delivery. 2. No encapsulation overhead. 3. Avoids the risk of an intervening router discarding the packet because it appears to have an invalid source address for the network from which it originates.

Disadvantages: 1. The permanent home address is not used at all. Hence, packets sent this way forgo the benefits of Mobile IP — if the mobile host moves to a new location then reply packets addressed to the old temporary address will be lost.

Motivation: Out-DT is the best choice when transparent mobility is not required. For example, HTTP connections are frequently very short lived, and if the host does move during the brief life of the connection, causing it to break, the user has the option of clicking the Web browser's 'reload' button. In many cases the user may prefer the small risk of an occasional incomplete image, rather than the large cost of slowing down all Web browsing with the overhead of using Mobile IP for every connection. Connectionless datagram transactions, such as DNS name lookups, may also be usefully performed this way.

S = CH

D = MH or COA

Payload

...

...

Figure 8. Incoming Packet Sent Unencapsulated
The source address (S) is the address of the correspondent host (CH). The destination address (D) may be either the mobile host's permanent home address (MH) or its temporary care-of address (COA).

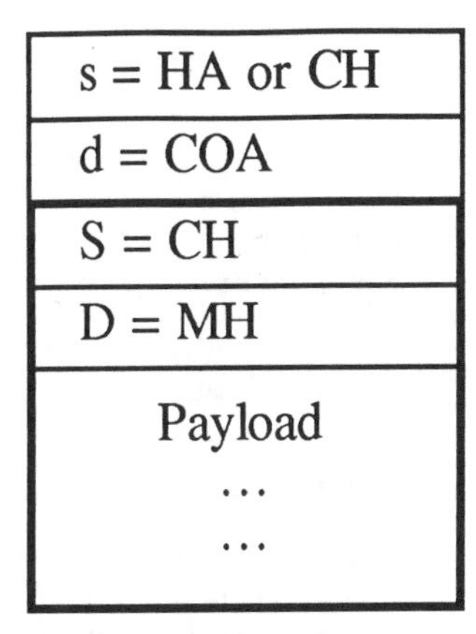

Figure 9. Incoming Packet Sent Encapsulated
For the inner (encapsulated) packet, the source address (S) is the address of the correspondent host (CH), and the destination address (D) is the mobile host's permanent home address (MH). For the outer packet, the destination address (d) is the mobile host's temporary care-of address (COA). When the mobile host receives the packet, the source address (shaded) may be either the home agent's address (HA), or the address of the correspondent host (CH), depending on who performed the encapsulation.

5. Incoming Packets

In this section we look at how to achieve our previously described optimization goals for incoming packets sent to the mobile host from a correspondent host. Normally, packets sent by a correspondent host will be straightforward IP packets addressed to the mobile host's permanent home address, since today's correspondent hosts run conventional IP networking software that is unaware of mobility issues. However, over time, hosts will be enhanced so that their networking software is aware of mobility issues. As with the mobile host, the only way a correspondent host can influence the path that the packets take through the Internet is through its choice of source and destination addresses in the IP header. If it chooses to encapsulate the packet, then it also has freedom to choose the source and destination addresses in the encapsulating IP packet.

The destination address identifies the entity with which the correspondent host is communicating. If the correspondent host uses the mobile host's home IP address as the destination address, then the packets will be successfully deliverable regardless of where in the Internet the mobile host is connected, but they may not travel by the most efficient route. If the correspondent host instead uses the mobile host's current temporary care-of address, then the packets will be delivered efficiently, but transparent mobility is lost and TCP connections will be broken without warning when the mobile host moves. Both of these addressing techniques have situations where they are appropriate and situations where they are not.

To achieve transparent mobility *and* efficient delivery, we use encapsulation. Encapsulation increases the size of the packet, but it allows us to use different destination addresses in the inner and outer headers. We use the home IP address as the destination address of the inner packet, to preserve location transparency, and we use the temporary care-of address as the destination address of the outer packet, so that it will be routed directly to the destination.

When both the mobile host and the correspondent host are physically connected to the same link-layer network segment, there is a better alternative than any of the three choices listed above. In this situation, the IP packet need not pass through any Internet routers at all. It can be delivered directly to the mobile host in a link-layer packet, without invoking IP-layer routing mechanisms. In this case, the IP packet that the correspondent host sends looks exactly the same as the packet that a host with no mobility awareness would send. The only difference is in the link-layer destination to which the packet is addressed.

Figures 8 and 9 show the possible kinds of packets that can arrive at the mobile host. For packets sent unencapsulated, directly from the correspondent host, the destination address will either be the temporary care-of address or, in the special case of two hosts on the same link-layer segment, it may be the home address. For packets sent encapsulated, the source address will depend on whether the packet was encapsulated at the correspondent host or whether it was first sent to the home network and encapsulated there by the home agent.

Below we summarize the four ways that a correspondent host can send packets to a mobile host, and give examples of situations where each is useful. Note that these four ways are *not* the same as the four options for outgoing packets, although when the choices for incoming and outgoing packets are compared, some symmetry does emerge.

In-IE: Incoming, Indirect, Encapsulated
(Correspondent unaware that host is mobile)

S = From Correspondent host
D = To Permanent home address

On arrival at the home agent, the packet is encapsulated to make:

s = From home agent
d = To temporary care-of address
S = From correspondent host
D = To permanent home address

Advantages: 1. The correspondent host is unaware of the special status of the mobile host, and needs no special software to send the packet.

Disadvantages: 1. Indirect delivery. 2. Encapsulation overhead.

Motivation: All Mobile IP systems must support tunneling through the home agent, since this is the only method that can be relied upon to work in all situations. For example, current Internet hosts will simply address packets to the mobile-host's home IP address, so the home agent must be present and able to forward those packets to the mobile host.

In-DE: Incoming, Direct, Encapsulated (Mobile-aware correspondent host)

s = From correspondent host
d = To temporary care-of address
S = From correspondent host
D = To permanent home address

Advantages: 1. Direct delivery.

Disadvantages: 1. Encapsulation overhead. 2. The correspondent host needs to be aware of the special status of the mobile host, and needs special software to look up the temporary address and perform the encapsulation.

Motivation: In any situation where the correspondent host is mobile-aware and knows the mobile host's current care-of address, sending the packets directly is preferable to sending them via the home agent.

In-DH: Incoming, Direct, Home Address (Same physical network segment)

S = From correspondent host
D = To permanent home address

Advantages: 1. Direct delivery. 2. No encapsulation overhead.

Disadvantages: 1. Only applicable when the correspondent host and the mobile host are connected to the same network segment.

Motivation: In-DH is the best choice when visiting another institution and connecting to their network to access data or services on that network. As well as being a fairly common case, the benefit of avoiding communicating through the home agent can be significant, especially if the visited institution is in Japan and the home agent is at MIT.

In another context, this delivery technique is already used when a mobile host operates using a separate foreign agent. The foreign agent uses this delivery technique to deliver the packet over the final hop to the mobile host.

In-DT: Incoming, Direct, Temporary Address (No Mobile IP)

S = From correspondent host
D = To temporary care-of address

Advantages: 1. Direct delivery. 2. No encapsulation overhead.

Disadvantages: 1. The permanent home address is not used at all. Hence, packets sent this way forgo the benefits of Mobile IP — if the mobile host moves to a new location then packets addressed to the old temporary address will be lost.

Motivation: As described in Section 4, this may be useful for short-lived connections and short connectionless datagram exchanges. Also, when a mobile host chooses to initiate a direct communication using its temporary care-of address, replies from the correspondent host will implicitly be sent back to that temporary address without it ever being aware that any mobility issues are involved.

6. 4x4 Choices

The choices presented in Sections 4 and 5 are not independent. For some communication mechanisms, such as dissemination of information via unreliable multicast streams, one-way packet delivery may be sufficient. However, the majority of protocols require two-way communication in order to operate. An NFS request requires a response, and a TCP data segment requires an acknowledgement. This means that for any conversation between two hosts, two decisions must be made: How packets *from* the mobile host are to be sent, and how packets *to* the mobile host are to be sent. Because these decisions are not independent, not all of the sixteen possible combinations are useful.

Figure 10 shows the possible combinations. Below we describe the useful combinations starting with the first row, which is the most conservative and most location transparent, and ending with the last row, which is the least transparent. We then describe why the darkly shaded options in the fourth row and fourth column are not useful.

6.1.1 Row A (Communication with Conventional Correspondent Host)

The first row of the chart shows combinations that are useful when the mobile host is communicating with a conventional Internet host that is not aware of mobility issues. All packets the correspondent host sends addressed to the mobile host's permanent address will be routed naïvely to the home agent, hence all incoming packets will be delivered by the indirect, encapsulated method.

However, the mobile host still has a choice about how it sends outgoing packets back to the correspondent host:

In-IE/Out-IE. The mobile host can send outgoing packets, encapsulated, to the home agent. This is the most conservative approach to Mobile IP. All that is required is for the mobile host to be able to send and receive packets from a single other host on the Internet — its home agent. If it cannot do even this, then it can be reasonably claimed that

		Out-IE **Outgoing** Indirect, Encapsulated	Out-DE **Outgoing** Direct, Encapsulated	Out-DH **Outgoing** Direct, Home Address	Out-DT **Outgoing** Direct, Temp. Address
Conventional Correspondent Host	Row A: In-IE **Incoming** Indirect, Encapsulated	Most conservative: most reliable, least efficient	Requires only decapsulation capability of the correspondent host	Requires there to be no security-conscious routers on the path	
Mobile-Aware Correspondent Host	Row B: In-DE **Incoming** Direct, Encapsulated		Requires fully mobile-aware correspondent host	Requires there to be no security-conscious routers on the path	
Both Hosts on Same Network Segment	Row C: In-DH **Incoming** Direct, Home Address			Requires both hosts to be on same network segment	
Forgoing Mobility Support	Row D: In-DT **Incoming** Direct, Temp. Address				Most efficient, but forgoes benefits of Mobile IP

Figure 10. Internet Mobility 4x4

This figure shows useful ways that a mobile host can communicate with a correspondent host. The different rows show routing options for incoming packets to the mobile host, and the different columns show options for outgoing packets from the mobile host. Each box lists key attributes of that particular communication mode. Lightly shaded boxes indicate combinations that would work correctly with current protocols such as TCP, but for other reasons would not normally be used. Darkly shaded boxes indicate combinations that would not work correctly with current protocols such as TCP. Note that a single host may have many different conversations in progress at the same time, choosing for each of them the communication mode that is most appropriate.

the mobile host is not in any meaningful sense connected to the Internet at all.

In-IE/Out-DE. The mobile host can send outgoing packets directly to the correspondent host, while still encapsulating them to shield the home source address from examination by routers along the path. For this method to work, the correspondent host does not need to be fully mobile-aware, but it does need to be able to decapsulate encapsulated IP packets. Some operating systems, such as recent versions of Linux, have this capability built-in. However, automatic decapsulation is a feature that should be used with caution. Hosts that perform automatic decapsulation lose some degree of firewall protection — automatic decapsulation makes it easy to spoof packet source addresses — so automatic decapsulation should only be done on hosts that use strong authentication mechanisms instead of simply trusting the packet addresses.

In-IE/Out-DH. The most efficient choice for the mobile host is to send outgoing packets directly to the correspondent host, unencapsulated. This does not require any special capabilities on the part of the correspondent host, but some routers may discard such packets.

6.1.2 Row B (Mobile-Aware Correspondent Host)

If the correspondent host is mobile-aware and knows the mobile host's current temporary care-of address, then choices from row B become available. The correspondent host can bypass the step of sending the packet to the home agent for encapsulation, and instead it can encapsulate the packet itself and send it directly to the mobile host.

The mobile host may choose to reply directly with an encapsulated packet (In-DE/Out-DE), or to avoid encapsulation overhead, it may choose to reply directly with an unencapsulated packet (In-DE/Out-DH). The first category (In-DE/Out-IE) is also valid, but is unlikely to be used. If the correspondent host is able to send packets directly to the mobile host, then the mobile host should also send its replies directly.

6.1.3 Row C (Both Hosts on Same Network Segment)

When the correspondent host and the mobile host are connected to the same network segment, routers need not be involved with the communication at all. The correspondent host can simply generate the IP packet, and then send it directly to the mobile host, even though naïve examination of the destination IP address would suggest that it does not belong on this network segment.

When a mobile host receives a packet this way, it should reply the same way, using (In-DH/Out-DH). The first two categories, (In-DH/Out-IE) and (In-DH/Out-DE), are also valid, but are unlikely to be used. If the correspondent host is able to send packets directly to the mobile host in a single link-layer hop, then the mobile host should reply the same way.

6.1.4 Row D (Forgoing Mobility Support)

We believe that the most important option, and the least emphasized in the current Mobile IP literature, is that mobile-aware applications should be able to specify when they *do not* want the services of Mobile IP. They should be allowed to send and receive normal, non-mobile IP packets. The last element of the table (In-DT/Out-DT) represents this option. In effect this is the way that most people currently connect their portable computers to the network when they visit some other institution. In the absence of Mobile IP, they have no other choice.

However, even when a host is capable of using Mobile IP, there are many cases where it might choose not to. Communicating directly incurs no Mobile IP overhead and can be used beneficially in some situations without requiring any mobile-awareness on the part of the correspondent host. Some applications may not need to have connections maintained when the mobile host moves, especially if connections are short, moves are rare, and the application has its own higher-level recovery mechanism. A simple example is viewing Web pages. HTTP connections are typically very short-lived, and if the connection is broken then the Web browser handles it by displaying a broken icon in place of the missing picture. The user can choose to either accept the broken icon, or to click the 'Reload' button to try again.

Our Mobile IP support software itself communicates using the temporary address when registering with the home agent. It has no choice, since until it has registered with the home agent the other Mobile IP delivery services are not available.

Another case where applications should be given the option to bypass Mobile IP services is when using IP multicast [RFC1112]. One of the goals of IP multicast is to reduce unnecessary replication of network traffic. Tunneling multicast packets from the home network to the visited network is therefore a little self-defeating. It would be better if the multicast application were able to join the multicast group through its real physical interface on the current local network, rather than through its virtual interface on its distant home network.

6.1.5 Inapplicable Combinations

The other entries in the fourth row and fourth column of the 4x4 table (shaded dark grey) are not especially useful. The choices in the fourth column denote cases where the mobile host sends packets using the temporary care-of address, not simply as the source address of an outer encapsulating header, but as the sole means of determining whence the packet originates. If the mobile host sends packets using only its temporary care-of address to identify their source, then the correspondent host would almost certainly reply to those packets using that same address. The networking software on the correspondent host would not be expected to have any way of even knowing that the host it's communicating with has other addresses.

The choices in the fourth row denote cases where the correspondent host sends packets addressed to the mobile host's temporary care-of address. If the mobile host receives packets addressed to its temporary care-of address, it ought to reply using that as its source address, or the correspondent host will have no way to associate the reply with the packet that caused it. For these reasons, the use of the temporary care-of address for communication in one direction effectively mandates the use of the same address for the corresponding return communication. Except in contrived circumstances, trying to mix temporary care-of addresses with permanent addresses as communication endpoints is not of any use.

7. Implementation

We have implemented our Mobile IP protocol in Linux. We override the IP route lookup routine and replace it with a routine that consults a mobility policy table before the usual route table. This allows us to control, on a packet by packet basis, whether a packet should use Mobile IP, and if so which interface to use. For the unencapsulated options, the interface is a physical interface. If the packet is to be encapsulated, then the routine directs IP to send the packet to our virtual interface, which encapsulates the packet and resubmits it to IP. This framework allows us to use all of the alternatives that we have described.

The choice of source IP address, and whether or not to encapsulate, needs to be made not only when sending a packet, but also at certain other times. For instance, this decision must also be made when TCP decides what address to use as the endpoint identifier for a TCP connection. Overriding the IP route lookup routine (instead of modifying the IP send packet routine) allows us to capture all of these crucial decision points automatically, without any extra special-case work.

Having provided the framework that allows us to control how packets are sent, we are now experimenting with various ways to make the actual decision about which method to use in each case. Below we describe the choices to be made by the mobile host, and the choices to be made by the correspondent host. We will be making our software freely available at http://mosquitonet.stanford.edu.

7.1 Mobile Host Choices

For the mobile host, there are two decisions to make. The first is whether to use the home address or the temporary address. If using the home address, then the second decision is which of the three home address methods to use.

7.1.1 Temporary Address or Home Address?

There are two ways to make the decision of whether to use the home address or the temporary address. One way is for a mobile-aware application to make the decision explicitly, and the other is for the host's networking software to make the decision based on heuristics.

In our Linux implementation, mobile-aware applications indicate their preferences to the networking software by binding their sockets to specific addresses. If the application binds its socket to the source address of (any of) the machine's physical interface(s), then the packets sent through that socket are sent directly through that interface using Out-DT, honoring the application's desired source address. If a socket is not bound to a particular address, or is bound to the host's permanent home address, then that is taken as an indication that the application is not mobile-aware, and our Mobile IP software should use its heuristics to decide which kind of source address to use. One of the heuristics we are experimenting with for TCP is to make the decision based on port numbers. For example, connections to port 80 are likely to be HTTP requests and can safely use Out-DT. Similarly, UDP packets addressed to UDP port 53 are likely to be DNS requests and can also safely use Out-DT.

7.1.2 Which Home Address Method to use

If the mobile host has decided to use its permanent home address, then it must decide which of the three home address methods to use. The mobile host keeps a cache of the currently selected delivery method associated with each target IP address. This saves it from having to make the decision afresh for every packet and allows it to build up a history, for each correspondent host, of which communication methods have proven to be successful and which have not.

One way the mobile host can choose which home address delivery method to use is to start with the most conservative (Out-IE), and then over the lifetime of the conversation tentatively try each of the more aggressive options (Out-DE and Out-DH), at each stage being prepared to return to the conservative method if the more aggressive method fails [Fox96]. Unfortunately, this can be wasteful, because in many cases either one or both of Out-DH and Out-DE will work fine, and having every conversation start out overly conservative is wasteful.

Another way for the mobile host to choose which home address delivery method to use is to start with the most aggressive (Out-DH). If this fails it can then try the more conservative options (Out-DE and then Out-IE) until one succeeds. Unfortunately, this can also be wasteful because in some easily identifiable circumstances, such as connecting to resources behind a protective gateway at the home institution, Out-DH is known to fail every time.

One solution to the question of which delivery method to start with is to allow the user, as part of the configuration of a Mobile IP machine, to specify rules stating which addresses Mobile IP should begin using in an optimistic mode and which addresses it should begin using in a pessimistic mode. These rules could be specified similarly to the way routing table entries are currently specified, as an address and a mask value. This would allow a single rule to identify, for example, the entire home network as a region where Out-IE should always be used.

In the discussion above, we tacitly assume that the IP layer has some way to tell whether delivery is 'succeeding' or 'failing', but in current operating systems this information is not readily available. This is not a new problem. The Ethernet Address Resolution Protocol (ARP) Specification [RFC826], written fourteen years ago, mentions the problem of stale ARP cache information. It suggests that when transport-level protocol software suspects that packets are not being delivered correctly it should indicate this to the lower layer software, but it also says that "implementation of these is outside the scope of this protocol." We propose that the required behavior could be obtained by a simple addition to the IP programming interface: all IP clients (e.g. TCP) could indicate, for every IP packet they send and receive, whether the packet is an 'original' packet or a retransmission. If the IP layer sees repeated retransmissions to a particular address, then this suggests that the currently selected delivery method may not be working. Similarly, if the IP layer sees repeated retransmissions *from* a particular address, then that suggests that acknowledgements are not getting through, which also indicates that the currently selected delivery method is not working. We have not yet implemented this.

7.2 Correspondent Host Choices

For the correspondent host, the choices are relatively simple. If the correspondent host is not mobile-aware then it will simply send normal IP packets, which means it is using the In-IE method. The same is true of a correspondent host that is mobile-aware, but is not yet aware that the host with which it is communicating is a mobile host.

If a mobile-aware correspondent host knows that the host with which it is communicating is a mobile host, and it knows the current care-of address, then it can encapsulate the packets and send them directly to that address. In this case it is using the In-DE method.

If the correspondent host knows that the mobile host is on the same Ethernet segment then it should also reply directly, using the In-DH method.

Finally, if the mobile host has chosen to initiate communication using its temporary care-of address, then the correspondent host, whether or not it is mobile-aware, will necessarily reply using that address, which means it is using the In-DT method.

8. Conclusions

One size does not fit all. Different situations call for different solutions, and our Mobile IP protocol gives mobile hosts the freedom to use the best solution for each situation. We are able to optimize for latency, packet size and Internet resource utilization. The best choice for each individual packet or conversation depends on what characteristics the protocol should optimize, the permissiveness of the networks over which the packets must travel, and the level of mobile-awareness of the hosts with which the mobile host corresponds.

Most communication does not need to use Mobile IP. We believe that all hosts should retain the ability to communicate using normal IP when that is appropriate. Mobile IP provides useful services, but these facilities should not be provided at the expense of losing the ability to operate as a normal Internet host.

Nevertheless, with the growing use of mobile computers and wireless networking, it is increasingly important that IP evolve to support mobile connections. Even though telnet connections may generate much less traffic than Web browsing, they are still important, and in a future world of ubiquitous mobile computing it is vital that long-lived connections be supported as well as short-lived communications.

9. Acknowledgements

We are grateful for the very helpful comments on this paper from Armando Fox, Hugh Holbrook, Nick McKeown, Venkat Padmanabhan, Craig Partridge, Charles Perkins, Elliot Poger, Xinhua Zhao, the anonymous SIGCOMM reviewers, and the anonymous student reviewers in Craig Partridge's Stanford CS341 networking class.

We are also very grateful to Jonathan Stone for his contributions to the ideas in this paper.

This work was supported by an NSF Faculty Career Development Award, a Robert N. Noyce Family Junior Faculty Chair, and the Center for Telecommunications at Stanford University.

10. References

[Bak96] Mary Baker, Xinhua Zhao, Stuart Cheshire & Jonathan Stone. Supporting Mobility in MosquitoNet. 1995 Winter USENIX, January 1996.

[Bla94] Trevor Blackwell et al. Secure Short-Cut Routing for Mobile IP. 1994 Summer USENIX, June 1994.

[Cer74] V.G. Cerf and R.E. Kahn. A Protocol for Packet Network Interconnection. IEEE Trans. on Communications, Vol 22, No. 5, May 1974, pp. 637-648.

[Gup96] Vipul Gupta and Abhijit Dixit. The Design and Deployment of a Mobility Supporting Network. To appear in the International Symposium on Parallel Architectures, Algorithms, and Networks, June 1996.

[Fox96] Armando Fox, Personal communication, April 1996.

[Hag93] R. Hager, A. Klemets, G. Maguire, M. Smith, F. Reichert. MINT - A Mobile Internet Router. 43rd IEEE Vehicular Technology Conference, New Jersey, USA, May 93.

[Ioa93] John Ioannidis and Gerald Q. Maguire Jr. The Design and Implementation of a Mobile Internetworking Architecture. 1993 Winter USENIX, January 1993.

[Joh96] David B. Johnson and Charles E. Perkins. Route Optimization in Mobile IP. draft-ietf-mobileip-optim-04.txt— work in progress, February 1996.

[Mon96] G. Montenegro. Bi-directional Tunneling for Mobile IP. draft-montenegro-tunneling-00.txt — work in progress, January 1996.

[Myl93] Andrew Myles and David Skellern. Comparing Four IP Based Mobile Host Protocols. Computer Networks and ISDN Systems, vol. 26, pp. 349-355, 1993. Also Proceedings of 4th Joint European Networking Conference, Trondheim, Norway, pp. 191-196, 10-13 May 1993.

[Par96] Craig Partridge, CS341, Stanford University, Spring 1996.

[Per94] Charles E. Perkins, Andrew Myles, and David B. Johnson. The Internet Mobile Host Protocol (IMHP). Proceedings of INET '94, June 1994.

[Per95] Charles E. Perkins. Minimal Encapsulation within IP. draft-ietf-mobileip-minenc-01.txt — work in progress, 25 October 1995.

[Per96a] Charles E. Perkins. IP Mobility Support. draft-ietf-mobileip-protocol-16.txt — work in progress, 22 April 1996.

[Per96b] Charles E. Perkins and David B. Johnson. Mobility Support in IPv6. draft-ietf-mobileip-ipv6-00.txt — work in progress, 26 January 1996.

[Per96c] Charles E. Perkins. IP Encapsulation within IP. draft-ietf-mobileip-ip4inip4-02.txt — work in progress, May 1996.

[RFC826] David C. Plummer. An Ethernet Address Resolution Protocol. RFC 826, November 1982.

[RFC974] Craig Partridge. Mail Routing and The Domain System. RFC 974, January 1986.

[RFC1027] Smoot Carl-Mitchell. Using ARP to Implement Transparent Subnet Gateways. RFC 1027, October 1987.

[RFC1034] P. Mockapetris. Domain Names — Concepts and Facilities. RFC 1034, November 1987.

[RFC1112] S. Deering. Host Extensions for IP Multicasting. RFC 1112, August 1989

[RFC1541] Ralph Droms. Dynamic Host Configuration Protocol. RFC 1541, October 1993.

[RFC1661] William Simpson. The Point-to-Point Protocol (PPP). RFC 1661, July 1994.

[RFC1702] Stan Hanks, Tony Li, Dino Farinacci and Paul Traina. Generic Routing Encapsulation over IPv4 networks. RFC 1702, October 1994.

[RFC1725] John G. Myers and Marshall T. Rose. Post Office Protocol - Version 3. RFC 1725, November 1994.

[RFC1883] Steve Deering and Bob Hinden. Internet Protocol, Version 6 (IPv6) Specification. RFC 1883, December 1995.

[Sal84] J. H. Saltzer, D. P. Reed and D. D. Clark, End-To-End Arguments in System Design. ACM Transactions on Computer Systems, Vol.2, No.4, November 1984, pp. 277-288.

[Tas94] Mitchell Tasman. Protocols and Caching Strategies in Support of Internetwork Mobility. Ph.D Thesis, University of Wisconsin, October 1994.

[Ter94] Fumio Teraoka, Keisuke Uehara, Hideki Sunahara and Jun Murai. VIP: A Protocol Providing Host Mobility. Communications of the ACM, August 1994.

10.4 Balakrishnan, H., Padmanabhan, V.N., Seshan, S., and Katz, R.H., A Comparison of Mechanisms for Improving TCP Performance over Wireless Links

Balakrishnan, H., Padmanabhan, V.N., Seshan, S., and Katz, R.H., "A Comparison of Mechanisms for Improving TCP Performance over Wireless Links," *IEEE/ACM Transactions on Networking*, 5(6):756–769, December 1997.

A Comparison of Mechanisms for Improving TCP Performance over Wireless Links

Hari Balakrishnan, Venkata N. Padmanabhan, Srinivasan Seshan and Randy H. Katz[1]
{hari,padmanab,ss,randy}@cs.berkeley.edu
Computer Science Division, Department of EECS, University of California at Berkeley

Abstract

Reliable transport protocols such as TCP are tuned to perform well in traditional networks where packet losses occur mostly because of congestion. However, networks with wireless and other lossy links also suffer from significant losses due to bit errors and handoffs. TCP responds to all losses by invoking congestion control and avoidance algorithms, resulting in degraded end-to-end performance in wireless and lossy systems. In this paper, we compare several schemes designed to improve the performance of TCP in such networks. We classify these schemes into three broad categories: end-to-end protocols, where loss recovery is performed by the sender; link-layer protocols, that provide local reliability; and split-connection protocols, that break the end-to-end connection into two parts at the base station. We present the results of several experiments performed in both LAN and WAN environments, using throughput and goodput as the metrics for comparison.

Our results show that a reliable link-layer protocol that is TCP-aware provides very good performance. Furthermore, it is possible to achieve good performance without splitting the end-to-end connection at the base station. We also demonstrate that selective acknowledgments and explicit loss notifications result in significant performance improvements.

1. Introduction

The increasing popularity of wireless networks indicates that wireless links will play an important role in future internetworks. Reliable transport protocols such as TCP [24, 26] have been tuned for traditional networks comprising wired links and stationary hosts. These protocols assume *congestion* in the network to be the primary cause for packet losses and unusual delays. TCP performs well over such networks by adapting to end-to-end delays and congestion losses. The TCP sender uses the cumulative acknowledgments it receives to determine which packets have reached the receiver, and provides reliability by retransmitting lost packets. For this purpose, it maintains a running average of the estimated round-trip delay and the mean linear deviation from it. The sender identifies the loss of a packet either by the arrival of several duplicate cumulative acknowledgments or the absence of an acknowledgment for the packet within a *timeout* interval equal to the sum of the smoothed round-trip delay and four times its mean deviation. TCP reacts to packet losses by dropping its transmission (congestion) window size before retransmitting packets, initiating congestion control or avoidance mechanisms (e.g., slow start [13]) and backing off its retransmission timer (Karn's Algorithm [16]). These measures result in a reduction in the load on the intermediate links, thereby controlling the congestion in the network.

Unfortunately, when packets are lost in networks for reasons other than congestion, these measures result in an unnecessary reduction in end-to-end throughput and hence, sub-optimal performance. Communication over wireless links is often characterized by sporadic high bit-error rates, and intermittent connectivity due to handoffs. TCP performance in such networks suffers from significant throughput degradation and very high interactive delays [8].

Recently, several schemes have been proposed to the alleviate the effects of non-congestion-related losses on TCP performance over networks that have wireless or similar high-loss links [3, 7, 28]. These schemes choose from a variety of mechanisms, such as local retransmissions, split-TCP connections, and forward error correction, to improve end-to-end throughput. However, it is unclear to what extent each of the mechanisms contributes to the improvement in performance. In this paper, we examine and compare the effectiveness of these schemes and their variants, and experimentally analyze the individual mechanisms and the degree of performance improvement due to each.

There are two different approaches to improving TCP performance in such lossy systems. The first approach hides any non-congestion-related losses from the TCP sender and therefore requires no changes to existing sender implementations. The intuition behind this approach is that since the problem is local, it should be solved locally, and that the transport layer need not be aware of the characteristics of the individual links. Protocols that adopt this approach attempt to make the lossy link appear as a higher quality link with a reduced effective bandwidth. As a result, most of the losses seen by the TCP sender are caused by congestion.

1. Web page URL http://daedalus.cs.berkeley.edu.
Srinivasan Seshan is now at IBM T.J. Watson Research Center, Hawthorne, NY (srini@watson.ibm.com).

Examples of this approach include wireless links with reliable link-layer protocols such as AIRMAIL [1], split connection approaches such as Indirect-TCP [3], and TCP-aware link-layer schemes such as the snoop protocol [7]. The second class of techniques attempts to make the sender aware of the existence of wireless hops and realize that some packet losses are not due to congestion. The sender can then avoid invoking congestion control algorithms when non-congestion-related losses occur — we describe some of these techniques in Section 3. Finally, it is possible for a wireless-aware transport protocol to coexist with link-layer schemes to achieve good performance.

We classify the many schemes into three basic groups, based on their fundamental philosophy: end-to-end proposals, split-connection proposals and link-layer proposals. The end-to-end protocols attempt to make the TCP sender handle losses through the use of two techniques. First, they use some form of selective acknowledgments (SACKs) to allow the sender to recover from multiple packet losses in a window without resorting to a coarse timeout. Second, they attempt to have the sender distinguish between congestion and other forms of losses using an Explicit Loss Notification (ELN) mechanism. At the other end of the solution spectrum, split-connection approaches completely hide the wireless link from the sender by terminating the TCP connection at the base station. Such schemes use a separate reliable connection between the base station and the destination host. The second connection can use techniques such as negative or selective acknowledgments, rather than just standard TCP, to perform well over the wireless link. The third class of protocols, link-layer solutions, lie between the other two classes. These protocols attempt to hide link-related losses from the TCP sender by using local retransmissions and perhaps forward error correction [e.g., 18] over the wireless link. The local retransmissions use techniques that are tuned to the characteristics of the wireless link to provide a significant increase in performance. Since the end-to-end TCP connection passes through the lossy link, the TCP sender may not be fully shielded from wireless losses. This can happen either because of timer interactions between the two layers [10], or more likely because of TCP's duplicate acknowledgments causing sender fast retransmissions even for segments that are locally retransmitted. As a result, some proposals to improve TCP performance use mechanisms based on the knowledge of TCP messaging to shield the TCP sender more effectively and avoid competing and redundant retransmissions [7].

In this paper, we evaluate the performance of several end-to-end, split-connection and link-layer protocols using end-to-end throughput and goodput as performance metrics, in both LAN and WAN configurations. In particular, we seek to answer the following specific questions:

1. What combination of mechanisms results in best performance for each of the protocol classes?
2. How important is it for link-layer schemes to be aware of TCP algorithms to achieve high end-to-end throughput?
3. How useful are selective acknowledgments in dealing with lossy links, especially in the presence of burst losses?
4. Is it important for the end-to-end connection to be split in order to effectively shield the sender from wireless losses and obtain the best performance?

We answer these questions by implementing and testing the various protocols in a wireless testbed consisting of Pentium PC base stations and IBM ThinkPad mobile hosts communicating over a 915 MHz AT&T Wavelan, all running BSD/OS 2.0. For each protocol, we measure the end-to-end throughput, and goodputs for the wired and (one-hop) wireless paths. For any path (or link), goodput is defined as the ratio of the actual transfer size to the total number of bytes transmitted over that path. In general, the wired and wireless goodputs differ because of wireless losses, local retransmissions and congestion losses in the wired network. These metrics allow us to determine the end-to-end performance as well as the transmission efficiency across the network. While we used a wireless hop as the lossy link in our experiments, we believe our results are applicable in a wider context to links where significant losses occur for reasons other than congestion. Examples of such links include high-speed modems and cable modems.

We show that a reliable link-layer protocol with some knowledge of TCP results in very good performance. Our experiments indicate that shielding the TCP sender from duplicate acknowledgments caused by wireless losses improves throughput by 10-30%. Furthermore, it is possible to achieve good performance without splitting the end-to-end connection at the base station. We also demonstrate that selective acknowledgments and explicit loss notifications result in significant performance improvements. For instance, the simple ELN scheme we evaluated improved the end-to-end throughput by a factor of more than two compared to TCP Reno, with comparable goodput values.

The rest of this paper is organized as follows. Section 2 briefly describes some proposed solutions to the problem of reliable transport protocols over wireless links. Section 3 describes the implementation details of the different protocols in our wireless testbed, and Section 4 presents the results and analysis of several experiments. Section 5 discusses some miscellaneous issues related to handoffs, ELN implementation and selective acknowledgments. We present our conclusions in Section 6, and mention some future work in Section 7.

2. Related Work

In this section, we summarize some protocols that have been proposed to improve the performance of TCP over wireless links. We also briefly describe some proposed methods to add SACKs to TCP.

- **Link-layer protocols**: There have been several proposals for reliable link-layer protocols. The two main classes of techniques employed by these protocols are: error correction, using techniques such as forward error correction (FEC), and retransmission of lost packets in response to automatic repeat request (ARQ) messages. The link-layer protocols for the digital cellular systems in the U.S. — both CDMA [15] and TDMA [22] — primarily use ARQ techniques. While the TDMA protocol guarantees reliable, in-order delivery of link-layer frames, the CDMA protocol only makes a limited attempt and leaves eventual error recovery to the (reliable) transport layer. Other protocols like the AIRMAIL protocol [1] employ a combination of FEC and ARQ techniques for loss recovery.

  The main advantage of employing a link-layer protocol for loss recovery is that it fits naturally into the layered structure of network protocols. The link-layer protocol operates independently of higher-layer protocols and does not maintain any per-connection state. The main concern about link-layer protocols is the possibility of adverse effect on certain transport-layer protocols such as TCP, as described in Section 1. We investigate this in detail in our experiments.

- **Split connection protocols [3, 28]:** Split connection protocols split each TCP connection between a sender and receiver into two separate connections at the base station — one TCP connection between the sender and the base station, and the other between the base station and the receiver. Over the wireless hop, a specialized protocol tuned to the wireless environment may be used. In [28], the authors propose two protocols — one in which the wireless hop uses TCP, and another in which the wireless hop uses a selective repeat protocol (SRP) on top of UDP. They study the impact of handoffs on performance and conclude that they obtain no significant advantage by using SRP instead of TCP over the wireless connection in their experiments. However, our experiments demonstrate benefits in using a simple selective acknowledgment scheme with TCP over the wireless connection.

  Indirect-TCP [Bakre95] is a split-connection solution that uses standard TCP for its connection over the wireless link. Like other split-connection proposals, it attempts to separate loss recovery over the wireless link from that across the wireline network, thereby shielding the original TCP sender from the wireless link. However, as our experiments indicate, the choice of TCP over the wireless link results in several performance problems. Since TCP is not well-tuned for the lossy link, the TCP sender of the wireless connection often times out, causing the original sender to stall. In addition, every packet incurs the overhead of going through TCP protocol processing twice at the base station (as compared to zero times for a non-split-connection approach), although extra copies are avoided by an efficient kernel implementation. Another disadvantage of split connections is that the end-to-end semantics of TCP acknowledgments is violated, since acknowledgments to packets can now reach the source even before the packets actually reach the mobile host. Also, since split-connection protocols maintain a significant amount of state at the base station per TCP connection, handoff procedures tend to be complicated and slow. Section 5.1 discusses some issues related to cellular handoffs and TCP performance.

- **The Snoop Protocol [7]:** The snoop protocol introduces a module, called the *snoop agent*, at the base station. The agent monitors every packet that passes through the TCP connection in both directions and maintains a cache of TCP segments sent across the link that have not yet been acknowledged by the receiver. A packet loss is detected by the arrival of a small number of duplicate acknowledgments from the receiver or by a local timeout. The snoop agent retransmits the lost packet if it has it cached and suppresses the duplicate acknowledgments. In our classification of the protocols, the snoop protocol is a link-layer protocol that takes advantage of the knowledge of the higher-layer transport protocol (TCP).

  The main advantage of this approach is that it suppresses duplicate acknowledgments for TCP segments lost and retransmitted locally, thereby avoiding unnecessary fast retransmissions and congestion control invocations by the sender. The per-connection state maintained by the snoop agent at the base station is *soft*, and is not essential for correctness. Like other link-layer solutions, the snoop approach could also suffer from not being able to completely shield the sender from wireless losses.

- **Selective Acknowledgments**: Since standard TCP uses a cumulative acknowledgment scheme, it often does not provide the sender with sufficient information to recover quickly from multiple packet losses within a single transmission window. Several studies [e.g., 11] have shown that TCP enhanced with selective acknowledgments performs better than standard TCP in such situations. SACKs were added as an option to TCP by RFC 1072 [14]. However, disagreements over the use of SACKs prevented the specification from being adopted, and the SACK option was removed from later TCP RFCs. Recently, there has been renewed interest in adding SACKs to TCP. Two relevant proposals are the

Name	Category	Special Mechanisms
E2E	end-to-end	standard TCP-Reno
E2E-NEWRENO	end-to-end	TCP-NewReno
E2E-SMART	end-to-end	SMART-based selective acks
E2E-IETF-SACK	end-to-end	IETF selective acks
E2E-ELN	end-to-end	Explicit Loss Notification (ELN)
E2E-ELN-RXMT	end-to-end	ELN with retransmit on first dupack
LL	link-layer	none
LL-TCP-AWARE	link-layer	duplicate ack suppression
LL-SMART	link-layer	SMART-based selective acks
LL-SMART-TCP-AWARE	link-layer	SMART and duplicate ack suppression
SPLIT	split-connection	none
SPLIT-SMART	split-connection	SMART-based wireless connection

Table 1. Summary of protocols studied in this paper.

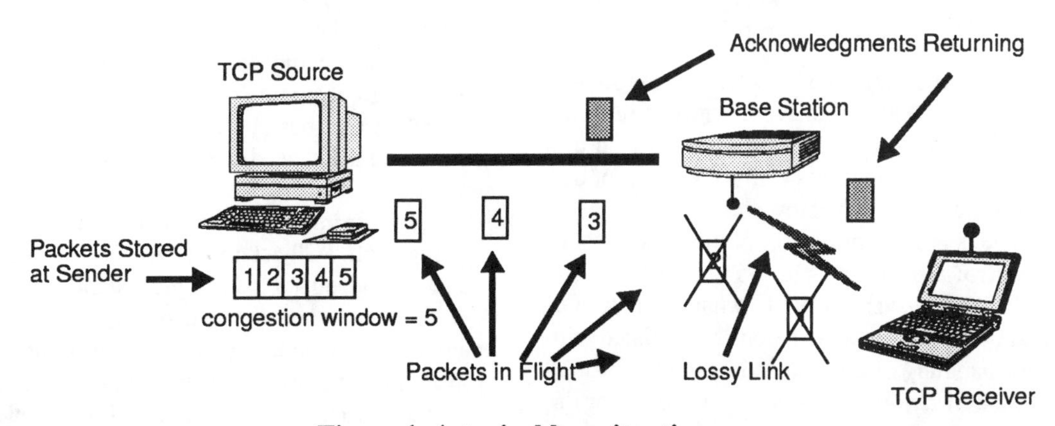

Figure 1. A typical loss situation

recent RFC on TCP SACKs [19] and the SMART scheme [17].

The SACK RFC proposes that each acknowledgment contain information about up to three non-contiguous blocks of data that have been received successfully by the receiver. Each block of data is described by its starting and ending sequence number. Due to the limited number of blocks, it is best to inform the sender about the most recent blocks received. The RFC does not specify the sender behavior, except to require that standard TCP congestion control actions be performed when losses occur.

An alternate proposal, SMART, uses acknowledgments that contain the cumulative acknowledgment and the sequence number of the packet that caused the receiver to generate the acknowledgment (this information is a subset of the three-blocks scheme proposed in the RFC). The sender uses this information to create a bitmask of packets that have been delivered successfully to the receiver. When the sender detects a gap in the bitmask, it immediately assumes that the missing packets have been lost without considering the possibility that they simply may have been reordered. Thus this scheme trades off some resilience to reordering and lost acknowledgments in exchange for a reduction in overhead to generate and transmit acknowledgments.

3. Implementation Details

This section describes the protocols we have implemented and evaluated. Table 1 summarizes the key ideas in each scheme and the main differences between them. Figure 1 shows a typical loss situation over the wireless link. Here, the TCP sender is in the middle of a transfer across a two-hop network to a mobile host. At the depicted time, the sender's congestion window consists of 5 packets. Of the five packets in the network, the first two packets are lost on the wireless link. As described in the rest of this section, each protocol reacts to these losses in different ways and generates messages that result in loss recovery. Although this figure only shows data packets being lost, our experiments have wireless errors in both directions.

3.1 End-To-End Schemes

Although a wide variety of TCP versions are used on the Internet, the current de facto standard for TCP implementations is TCP Reno [26]. We call this the E2E protocol, and use it as the standard basis for performance comparison.

The E2E-NEWRENO protocol improves the performance of TCP-Reno after multiple packet losses in a window by remaining in fast recovery mode if the first new acknowledgment received after a fast retransmission is "partial", i.e, is less than the value of the last byte transmitted when the fast retransmission was done. Such partial acknowledgements are indicative of multiple packet losses within the original window of data. Remaining in fast recovery mode enables the connection to recover from losses at the rate of one segment per round trip time, rather than stall until a coarse timeout as TCP-Reno often would [9, 12].

The E2E-SMART and E2E-IETF-SACK protocols add SMART-based and IETF selective acknowledgments respectively to the standard TCP Reno stack. This allows the sender to handle multiple losses within a window of outstanding data more efficiently. However, the sender still assumes that losses are a result of congestion and invokes congestion control procedures, shrinking its congestion window size. This allows us to identify what percentage of the end-to-end performance degradation is associated with standard TCP's handling of error detection and retransmission. We used the SMART-based scheme [17] only for the LAN experiments. This scheme is well-suited to situations where there is little reordering of packets, which is true for one-hop wireless systems such as ours. Unlike the scheme proposed in [17], we do not use any special techniques to detect the loss of a retransmission. The sender retransmits a packet when it receives a SMART acknowledgment only if the same packet was not retransmitted within the last round-trip time. If no further SMART acknowledgments arrive, the sender falls back to the coarse timeout mechanism to recover from the loss. We used the IETF selective acknowledgement scheme both for the LAN and the WAN experiments. Our implementation is based on the RFC and takes appropriate congestion control actions upon receiving SACK information [4].

The E2E-ELN protocol adds an Explicit Loss Notification (ELN) option to TCP acknowledgments. When a packet is dropped on the wireless link, future cumulative acknowledgments corresponding to the lost packet are marked to identify that a non-congestion related loss has occurred. Upon receiving this information with duplicate acknowledgments, the sender may perform retransmissions without invoking the associated congestion-control procedures. This option allows us to identify what percentage of the end-to-end performance degradation is associated with TCP's incorrect invocation of congestion control algorithms when it does a fast retransmission of a packet lost on the wireless hop. The E2E-ELN-RXMT protocol is an enhancement of the previous one, where the sender retransmits the packet on receiving the first duplicate acknowledgement with the ELN option set (as opposed to the third duplicate acknowledgement in the case of TCP Reno), in addition to not shrinking its window size in response to wireless losses.

In practice, it might be difficult to identify which packets are lost due to errors on a lossy link. However, in our experiments we assume sufficient knowledge at the receiver about wireless losses to generate ELN information. We describe some possible implementation policies and strategies for the ELN mechanism in Section 5.2.

3.2 Link-Layer Schemes

Unlike TCP for the transport layer, there is no de facto standard for link-layer protocols. Existing link-layer protocols choose from techniques such as Stop-and-Wait, Go-Back-N, Selective Repeat and Forward Error Correction to provide reliability. Our base link-layer algorithm, called LL, uses cumulative acknowledgments to determine lost packets that are retransmitted locally from the base station to the mobile host. To minimize overhead, our implementation of LL leverages off TCP acknowledgments instead of generating its own. Timeout-based retransmissions are done by maintaining a smoothed round-trip time estimate, with a minimum timeout granularity of 200 ms to limit the overhead of processing timer events. This still allows the LL scheme to retransmit packets several times before a typical TCP Reno transmitter would time out. LL is equivalent to the snoop agent that does not suppress any duplicate acknowledgments, and does not attempt in-order delivery of packets across the link (unlike protocols proposed in [15], [22]).

While the use of TCP acknowledgments by our LL protocol renders it atypical of traditional ARQ protocols, we believe that it still preserves the key feature of such protocols: the ability to retransmit packets locally, independently of and on a much faster time scale than TCP. Therefore, we expect the qualitative aspects of our results to be applicable to general link-layer protocols.

We also investigated a more sophisticated link-layer protocol (LL-SMART) that uses selective retransmissions to improve performance. The LL-SMART protocol performs this by applying a SMART-based acknowledgment scheme at the link layer. Like the LL protocol, LL-SMART uses TCP acknowledgments instead of generating its own and limits its minimum timeout to 200 ms. LL-SMART is equivalent to the snoop agent performing retransmissions based on selective acknowledgements but not suppressing duplicate acknowledgments at the base station.

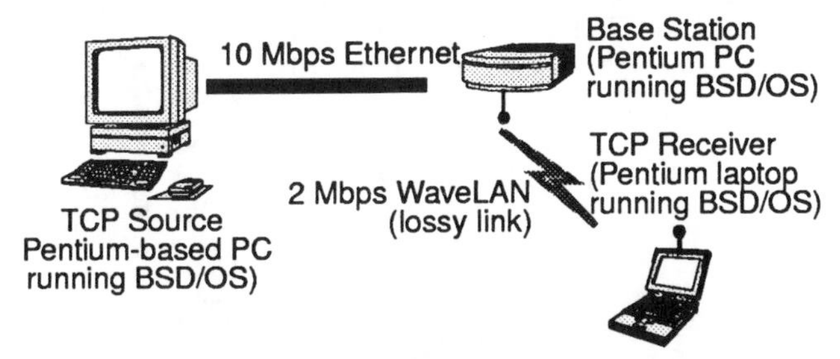

Figure 2. Experimental topology. There were an additional 16 Internet hops between the source and base station during the WAN experiments.

We added TCP awareness to both the LL and LL-SMART protocols, resulting in the LL-TCP-AWARE and LL-SMART-TCP-AWARE schemes. The LL-TCP-AWARE protocol is identical to the snoop protocol, while the LL-SMART-TCP-AWARE protocol uses SMART-based techniques for further optimization using selective repeat. LL-SMART-TCP-AWARE is the best link-layer protocol in our experiments — it performs local retransmissions based on selective acknowledgments and shields the sender from duplicate acknowledgments caused by wireless losses.

3.3 Split-Connection Schemes

Like I-TCP, our SPLIT scheme uses an intermediate host to divide a TCP connection into two separate TCP connections. The implementation avoids data copying in the intermediate host by passing the pointers to the same buffer between the two TCP connections. A variant of the SPLIT approach we investigated, SPLIT-SMART, uses a SMART-based selective acknowledgment scheme on the wireless connection to perform selective retransmissions. There is little chance of reordering of packets over the wireless connection since the intermediate host is only one hop away from the final destination.

4. Experimental Results

In this section, we describe the experiments we performed and the results we obtained, including detailed explanations for observed performance. We start by describing the experimental testbed and methodology. We then describe the performance of the various link-layer, end-to-end and split-connection schemes.

4.1 Experimental Methodology

We performed several experiments to determine the performance and efficiency of each of the protocols. The protocols were implemented as a set of modifications to the BSD/OS TCP/IP (Reno) network stack. To ensure a fair basis for comparison, none of the protocols implementations introduce any additional data copying at intermediate points from sender to receiver.

Our experimental testbed consists of IBM ThinkPad laptops and Pentium-based personal computers running BSD/OS 2.1 from BSDI. The machines are interconnected using a 10 Mbps Ethernet and 915 MHz AT&T WaveLANs [27], a shared-medium wireless LAN with a raw signalling bandwidth of 2 Mbps. The network topology for our experiments is shown in Figure 2. The peak throughput for TCP bulk transfers is 1.5 Mbps in the local area testbed and 1.35 Mbps in the wide area testbed in the absence of congestion or wireless losses. These testbed topologies represent typical scenarios of wireless links and mobile hosts, such as cellular wireless networks. In addition, our experiments focus on data transfer to the mobile host, which is the common case for mobile applications (e.g., Web accesses).

In order to measure the performance of the protocols under controlled conditions, we generate errors on the lossy link using an exponentially distributed bit-error model. The receiving entity on the lossy link generates an exponential distribution for each bit-error rate and changes the TCP checksum of the packet if the error generator determines that the packet should be dropped. Losses are generated in both directions of the wireless channel, so TCP acknowledgments are dropped too. The TCP data packet size in our experiments is 1400 bytes. We first measure and analyze the performance of the various protocols at an average error rate of one every 64 KBytes (this corresponds to a bit-error rate of about 1.9×10^{-6}). Note that since the exponential distribution has a standard deviation equal to its mean, there are several occasions when multiple packets are lost in close succession. We then report the results of some burst error situations, where between two and six packets are dropped in every burst (Section 4.5). Finally, we investigate the performance of many of these protocols across a range of error rates from one every 16 KB to one every 256 KB. The choice of the exponentially distributed error model is motivated by our desire to understand the precise dynamics of each protocol in response to a wireless loss, and is not an attempt to empirically model a wireless channel. While the actual performance numbers will be a function of the exact error model, the relative performance is dependent on how the protocol behaves after one or more losses in a single TCP window. Thus, we expect our overall conclusions to be applicable under other patterns of wireless loss as well. Finally, we believe that though wireless errors are generated artificially in our experiments, the use of a real testbed is still valuable in that it introduces realistic effects such as wireless bandwidth limitation, media access contention, protocol processing delays, etc., which are hard to model realistically in a simulation.

In our experiments, we attempt to ensure that losses are only due to wireless errors (and not congestion). This allows us to focus on the effectiveness of the mechanisms in handling such losses. The WAN experiments are performed across 16

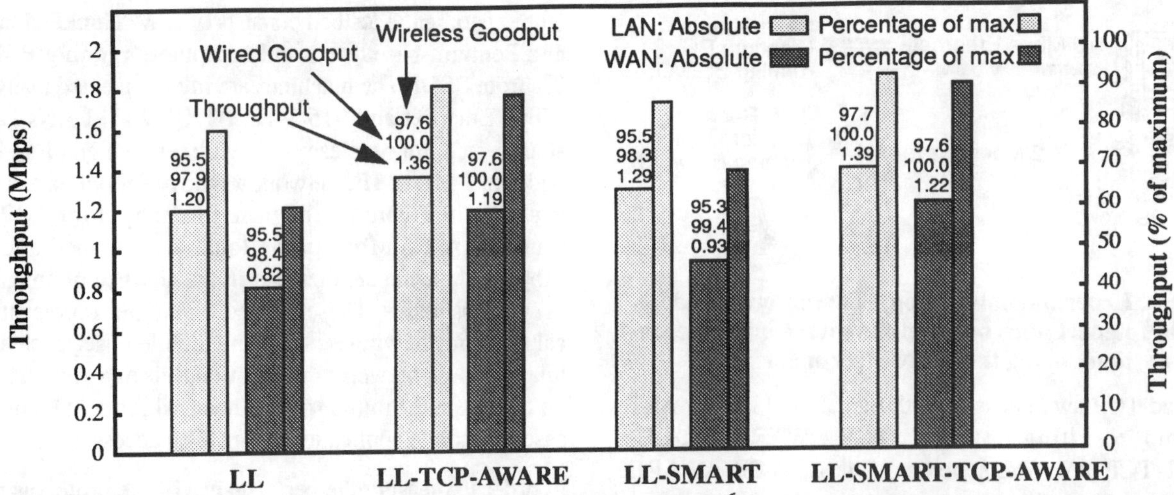

Figure 3. Performance of link-layer protocols: bit-error rate = 1.9×10^{-6} (1 error/65536 bytes), socket buffer size = 32 KB. For each case there are two bars: the thick one corresponds to the scale on the left and denotes the throughput in Mbps; the thin one corresponds to the scale on the right and shows the throughput as a percentage of the maximum, i.e. in the absence of wireless errors (1.5 Mbps in the LAN environment and 1.35 Mbps in the WAN environment).

Internet hops with minimal congestion[2] in order to study the impact of large delay-bandwidth products.

Each run in the experiment consists of an 8 MByte transfer from the source to receiver across the wired net and the WaveLAN link. We chose this rather long transfer size in order to limit the impact of transient behavior at the start of a TCP connection. During each run, we measure the throughput at the receiver in Mbps, and the wired and wireless goodputs as percentages. In addition, all packet transmissions on the Ethernet and WaveLan are recorded for analysis using tcpdump [20], and the sender's TCP code instrumented to record events such as coarse timeouts, retransmission times, duplicate acknowledgment arrivals, congestion window size changes, etc. The rest of this section presents and discusses the results of these experiments.

4.2 Link-Layer Protocols

Traditional link-layer protocols operate independently of the higher-layer protocol, and consequently, do not necessarily shield the sender from the lossy link. In spite of local retransmissions, TCP performance could be poor for two reasons: (i) competing retransmissions caused by an incompatible setting of timers at the two layers, and (ii) unnecessary invocations of the TCP fast retransmission mechanism due to out-of-order delivery of data. In [10], the effects of the first situation are simulated and analyzed for a TCP-like transport protocol (that *closely* tracks the round-trip time to set its retransmission timeout) and a reliable link-layer protocol. The conclusion was that unless the packet loss rate is high (more than about 10%), competing retransmissions by the link and transport layers often lead to significant performance degradation. However, this is not the dominating effect when link layer schemes, such as LL, are used with TCP Reno and its variants. These TCP implementations have coarse retransmission timeout granularities that are typically multiples of 500 ms, while link-layer protocols typically have much finer timeout granularities. The real problem is that when packets are lost, link-layer protocols that do not attempt in-order delivery across the link (e.g., LL) cause packets to reach the TCP receiver out-of-order. This leads to the generation of duplicate acknowledgments by the TCP receiver, which causes the sender to invoke fast retransmission and recovery. This can potentially cause degraded throughput and goodput, especially when the delay-bandwidth product is large.

Our results substantiate this claim, as can be seen by comparing the LL and LL-TCP-AWARE results (Figure 3 and Table 2). For a packet size of 1400 bytes, a bit error rate of 1.9×10^{-6} (1/65536 bytes) translates to a packet error rate of about 2.2 to 2.3%. Therefore, an optimal link-layer protocol that recovers from errors locally and does not compete with TCP retransmissions should have a wireless goodput of 97.7% and a wired goodput of 100% in the absence of congestion. In the LAN experiments, the throughput difference between LL and LL-TCP-AWARE is about 10%. However, the LL wireless goodput is only 95.5%, significantly less than LL-TCP-AWARE's wireless goodput of 97.6%, which is close to the maximum achievable goodput. When a loss occurs, the LL protocol performs a local retransmission relatively quickly. However, enough packets are typically in transit to create more than 3 duplicate acknowledgments. These duplicates eventually propagate to the sender and trigger a fast retransmission and the associated congestion control mechanisms. These fast retransmissions result in

2. WAN experiments across the US were performed between 10 pm and 4 am, PST and we verified that no congestion losses occurred in the runs reported.

	LL	LL-TCP-AWARE	LL-SMART	LL-SMART-TCP-AWARE
LAN (8 KB)	1.20 (95.6%,97.9%)	1.29 (97.6%,100%)	1.29 (96.1%,98.9%)	1.37 (97.6%,100%)
LAN (32 KB)	1.20 (95.5%,97.9%)	1.36 (97.6%,100%)	1.29 (95.5%,98.3%)	1.39 (97.7%,100%)
WAN (32 KB)	0.82 (95.5%,98.4%)	1.19 (97.6%,100%)	0.93 (95.3%,99.4%)	1.22 (97.6%,100%)

Table 2. This table summarizes the results for the link-layer schemes for an average error rate of one every 65536 bytes of data. Each entry is of the form: throughput (wireless goodput, wired goodput). Throughput is measured in Mbps. Goodput is expressed as a percentage.

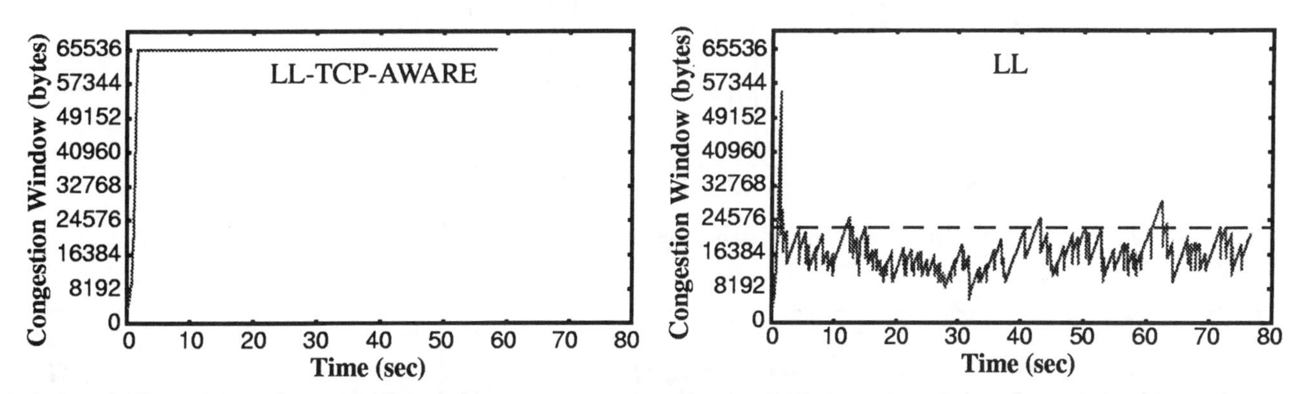

Figure 4. Congestion window size for link-layer protocols in wide area tests. The horizontal dashed line in the LL graph shows the 23000 byte WAN bandwidth-delay product.

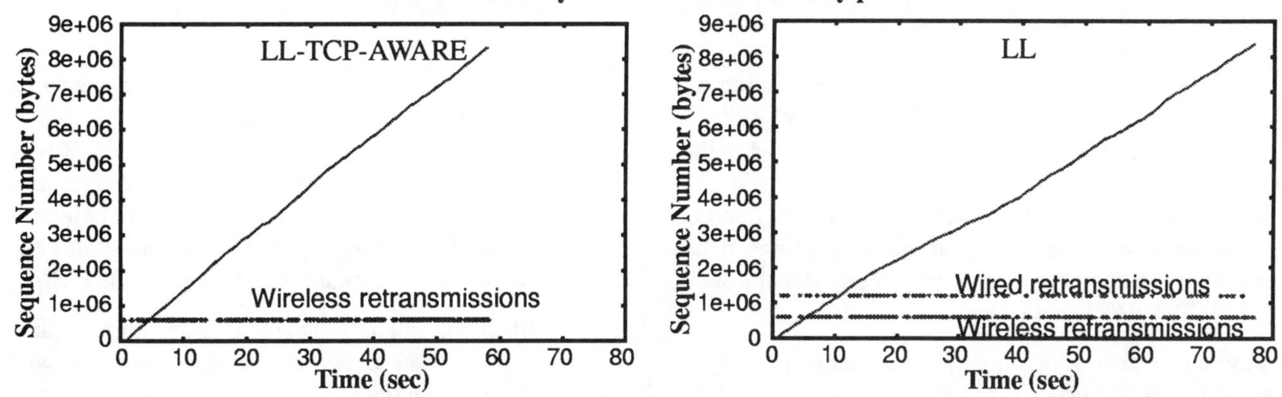

Figure 5. Packet sequence traces for LL-TCP-AWARE and LL. No coarse timeouts occur in either case. For LL-TCP-AWARE, the horizontal row of dots shows the times of wireless link retransmissions. For LL, the top row shows sender fast retransmission times and the bottom row shows both local wireless and sender retransmissions.

reduced goodput; about 90% of the lost packets are retransmitted by both the source and the base station.

The effects of this interaction are much more pronounced in the wide-area experiments — the throughput difference is about 30% in this case. The cause for the more pronounced deterioration in performance is the higher bandwidth-delay product of the wide-area connection. The LL scheme causes the sender to invoke congestion control procedures often due to duplicate acknowledgments and causes the average window size of the transmitter to be lower than for LL-TCP-AWARE. This is shown in Figure 4, which compares the congestion window size of LL and LL-TCP-AWARE as a function of time. Note that the number of outstanding data bytes in the network is the minimum of the congestion window and the receiver advertised window. This is bounded by the receiver's socket buffer size. In the congestion window graphs for each protocol, the receiver socket buffer is 32KB.

In the wide area, the bandwidth-delay product is about 23000 bytes (1.35 Mbps * 135 ms), and the congestion window drops below this value several times during each TCP transfer. On the other hand, the LAN experiments do not suffer from such a large throughput degradation because LL's lower congestion-window size is usually still larger than the connection's delay-bandwidth product of about 1900 bytes (1.5 Mbps * 10 ms). Therefore, the LL scheme can maintain a nearly full "data pipe" between the sender and receiver in the local connection but not in the wide area one. The 10% LAN degradation is almost entirely due to the

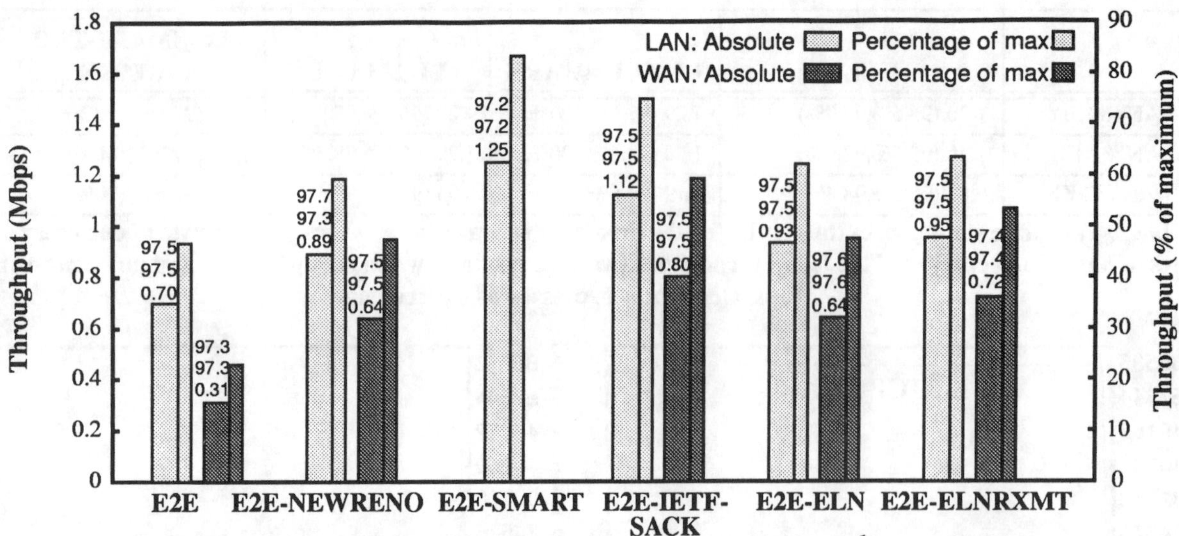

Figure 6. Performance of end-to-end protocols: bit error rate = 1.9×10^{-6} (1 error/65536 bytes).

	E2E	E2E-NEWRENO	E2E-SMART	E2E-IETF-SACK	E2E-ELN	E2E-ELN-RXMT
LAN (8 KB)	0.55 (97.0,96.0)	0.66 (97.3,97.3)	1.12 (97.6,97.6)	0.68 (97.3,97.3)	0.69 (97.3,97.2)	0.86 (97.4,97.3)
LAN (32 KB)	0.70 (97.5,97.5)	0.89 (97.7,97.3)	1.25 (97.2,97.2)	1.12 (97.5,97.5)	0.93 (97.5,97.5)	0.95 (97.5,97.5)
WAN (32 KB)	0.31 (97.3,97.3)	0.64 (97.5,97.5)	N.A.	0.80 (97.5,97.5)	0.64 (97.6,97.6)	0.72 (97.4,97.4)

Table 3. This table summarizes the results for the end-to-end schemes for an average error rate of one every 65536 bytes of data. The numbers in the cells follow the same convention as in Table 2.

excessive retransmissions over the wireless link and to the smaller average congestion window size compared to LL-TCP-AWARE. Another important point to note is that LL successfully prevents coarse timeouts from happening at the source. Figure 5 shows the sequence traces of TCP transfers for LL-TCP-AWARE and LL.

In summary, our results indicate that a simple link-layer retransmission scheme does not entirely avoid the adverse effects of TCP fast retransmissions and the consequent performance degradation. An enhanced link-layer scheme that uses knowledge of TCP semantics to prevent duplicate acknowledgments caused by wireless losses from reaching the sender and locally retransmits packets achieves significantly better performance.

4.3 End-To-End Protocols

The performance of the various end-to-end protocols is summarized in Figure 6 and Table 3. The performance of TCP Reno, the baseline E2E protocol, highlights the problems with TCP over lossy links. At a 2.3% packet loss rate (as explained in Section 4.2), the E2E protocol achieves a throughput of less than 50% of the maximum (i.e., throughput in the absence of wireless losses) in the local-area and less than 25% of the maximum in the wide-area experiments. However, all the end-to-end protocols achieve goodputs close to the optimal value of 97.7%. The primary reason for the low throughput is the large number of timeouts that occur during the transfer (Figure 7). The resulting average window size during the transfer is small, preventing the "data pipe" from being kept full and reducing the effectiveness of the fast retransmission mechanism (Figure 8).

The modified end-to-end protocols improve throughput by retransmitting packets known to have been lost on the wireless hop earlier than they would have been by the baseline E2E protocol, and by reducing the fluctuations in window size. The E2E-NEWRENO, E2E-ELN, E2E-SMART and E2E-IETF-SACK protocols each use new TCP options and more sophisticated acknowledgment processing techniques to improve the speed and accuracy of identifying and retransmitting lost packets, as well as by recovering from multiple losses in a single transmission window without timing out. The remainder of this section discusses the benefits of three techniques — partial acknowledgments, explicit loss notifications, and selective acknowledgments.

Partial acknowledgments: E2E-NEWRENO, which uses partial acknowledgment information to recover from multiple losses in a window at the rate of one packet per round-trip time, performs between 10 and 25% better than E2E over a LAN and about 2 times better than E2E in the WAN experiments. The performance improvement is a function of the socket buffer size — the larger the buffer size, the better the relative performance. This is because in situations that

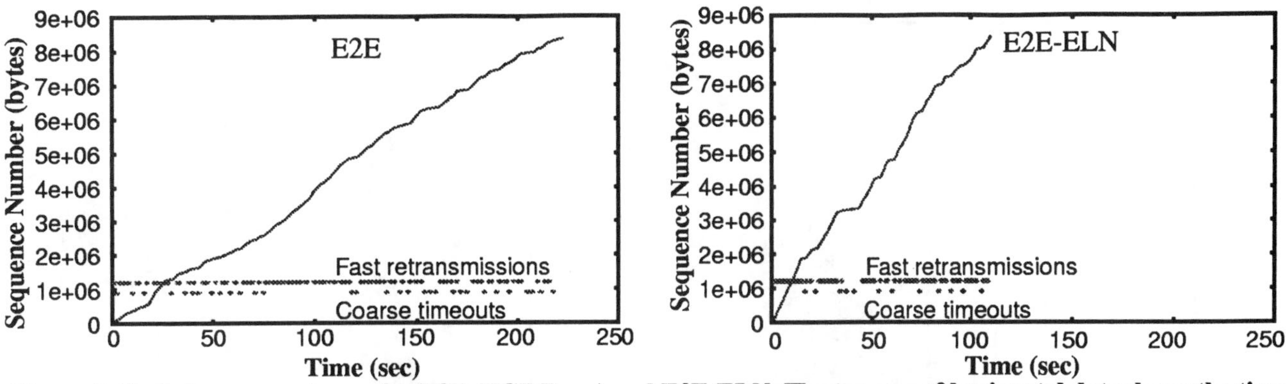

Figure 7. Packet sequence traces for E2E (TCP Reno) and E2E-ELN. The top row of horizontal dots shows the times when fast retransmissions occur; the bottom row shows the coarse timeouts.

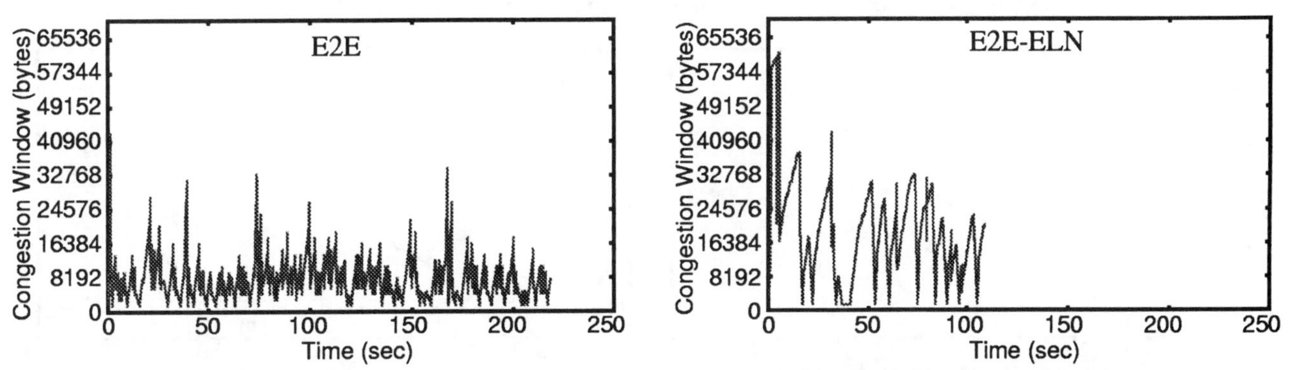

Figure 8. Congestion window size as a function of time for E2E (TCP Reno) and E2E-ELN. This figure clearly shows the utility of ELN in preventing rapid fluctuations, thereby maintaining a larger average congestion window size.

E2E suffers a coarse timeout for a loss, the probability that E2E-NEWRENO does not, increases with the number of outstanding packets in the network.

Explicit Loss Notification: One way of eliminating the long delays caused by coarse timeouts is to maintain as large a window size as possible. E2E-NEWRENO remains in fast recovery if the new acknowledgment is only partial, but reduces the window size to half its original value upon the arrival of the first new acknowledgment. The E2E-ELN and E2E-ELN-RXMT protocols use ELN information (Section 3.1) to prevent the sender from reducing the size of the congestion window in response to a wireless loss. Both these schemes perform better than E2E-NEWRENO, and over two times better than E2E. This is a result of the sender's explicit awareness of the wireless link, which reduces the number of coarse timeouts (Figure 7) and rapid window size fluctuations (Figure 8). The E2E-ELN-RXMT protocol performs only slightly better than E2E-ELN when the socket buffer size is 32 KB. This is because there is usually enough data in the pipe to trigger a fast retransmission for E2E-ELN. The performance benefits of E2E-ELN-RXMT are more pronounced when the socket buffer size is smaller, as the numbers for the 8 KB socket buffer size indicate (Table 3). This is because E2E-ELN-RXMT does not wait for three duplicate acknowledgments before retransmitting a packet, if it has ELN information for it. The maximum socket buffer size of 8 KB limits the number of unacknowledged packets to a small number at any point in time, which reduces the probability of three duplicate acknowledgments arriving after a loss and triggering a fast retransmission.

Despite explicit awareness of wireless losses, timeouts sometimes occur in the ELN-based protocols. This is a result of our implementation of the ELN protocol, which does not convey information about multiple wireless-related losses to the sender. Since it is coupled with only cumulative acknowledgments, the sender is unaware of the occurrence of multiple wireless-related losses in a window; we plan to couple SACKs and ELN together in future work. Section 5.2 discusses some possible implementation strategies and policies for ELN.

Selective acknowledgments: We experimented with two different SACK schemes. In the LAN case, we used a simple SACK scheme based on a subset of the SMART proposal. This protocol was the best of the end-to-end protocols in this situation, achieving a throughput of 1.25 Mbps (in contrast, the best local scheme, LL-SMART-TCP-AWARE, obtained a throughput of 1.39 Mbps).

In the WAN case, we based our SACK implementation [4] on RFC 2018. For the exponentially-distributed loss pattern we used, the throughput was about 0.8 Mbps, significantly

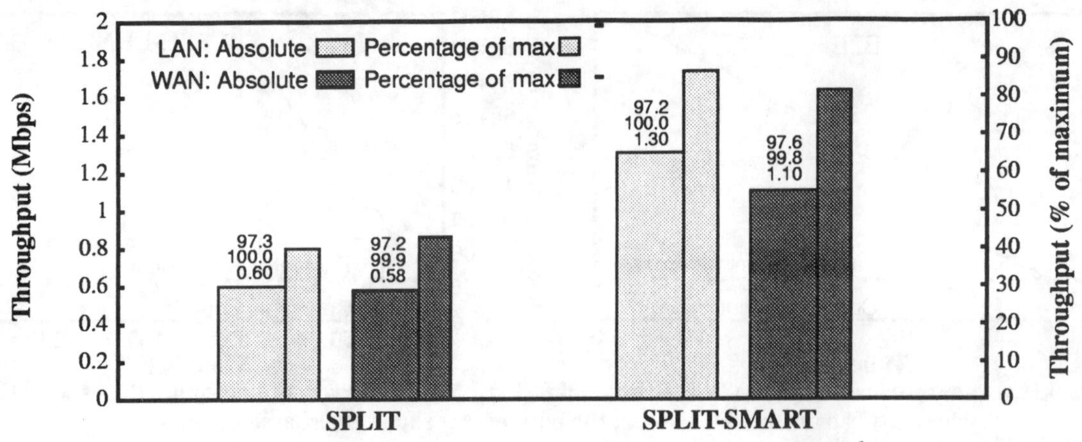

Figure 9. Performance of split-connection protocols: bit error rate = 1.9×10^{-6} (1 error/65536 bytes).

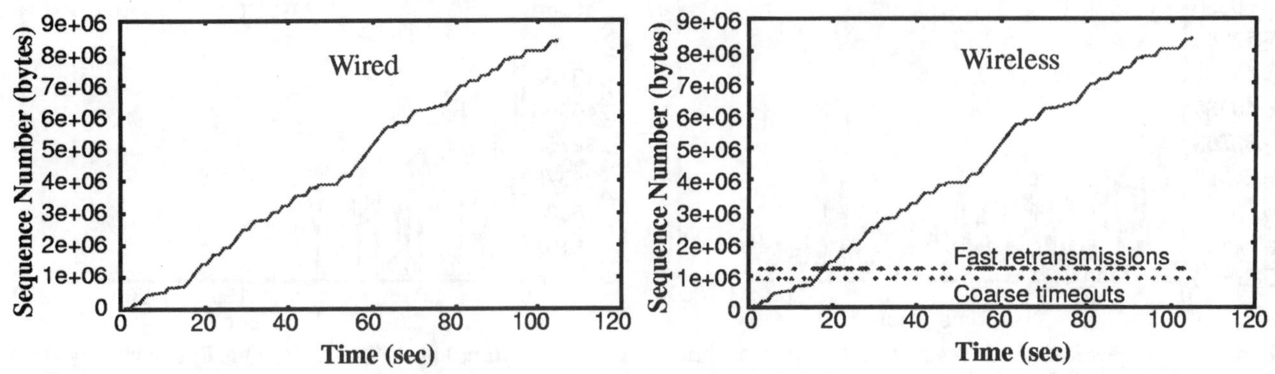

Figure 10. Packet sequence trace for the wired and wireless parts of the SPLIT protocol. The wireless part has two rows of horizontal dots: the top one shows the times of fast retransmissions and the bottom one the times of the timeout-based ones.

higher than the 0.31 Mbps throughput of TCP Reno. However, this is still about 35% worse than LL-OPT. Even though SACKs allow the sender to often recover from multiple losses without timing out, the sender's congestion window decreases every time there is a packet dropped on the wireless link, causing it to remain small.

In summary, E2E-NEWRENO is better than E2E, especially for large socket buffer sizes. Adding ELN to TCP improves throughput significantly by successfully preventing unnecessary fluctuations in the transmission window. Finally, SACKs provide significant improvement over TCP Reno, but perform about 10-15% worse than the best link-layer schemes in the LAN experiments, and about 35% worse in the WAN experiments. These results suggest that an end-to-end protocol that has both ELN and SACKs will result in good performance, and is an area of current work.

4.4 Split-Connection Protocols

The main advantage of the split-connection approaches is that they isolate the TCP source from wireless losses. The TCP sender of the second, wireless connection performs all the retransmissions in response to wireless losses.

Figure 9 and Table 4 show the throughput and goodput for the split connection approach in the LAN and WAN environments. We report the results for two cases: when the wireless connection uses TCP Reno (labeled SPLIT) and when it uses the SMART-based selective acknowledgment scheme described earlier (labeled SPLIT-SMART). We see that the throughput achieved by the SPLIT approach (0.6 Mbps) is quite low, about the same as that for end-to-end TCP Reno (labeled E2E in Figure 6). The reason for this is apparent from Figures 10 and 19, which show the progress of the data transfer and the size of the congestion window for the wired and wireless connections. We see that the wired connection neither has any retransmissions nor any timeouts, resulting in a wired goodput of 100%. However, it (eventually) stalls whenever the sender of the wireless connection experiences a timeout, since the amount of buffer space at the base station (64 KB in our experiments) is bounded[3]. In the WAN case, the throughput of the SPLIT approach is about 0.58 Mbps which is better than the 0.31

	SPLIT	SPLIT-SMART
LAN (8 KB)	0.54 (97.4%,100%)	1.30 (97.6%,100%)
LAN (32 KB)	0.60 (97.3%,100%)	1.30 (97.2%,100%)
WAN (32 KB)	0.58 (97.2%,100%)	1.10 (97.6%,100%)

Table 4. Summary of results for the split-connection schemes at an average error rate of 1 every 64 KB.

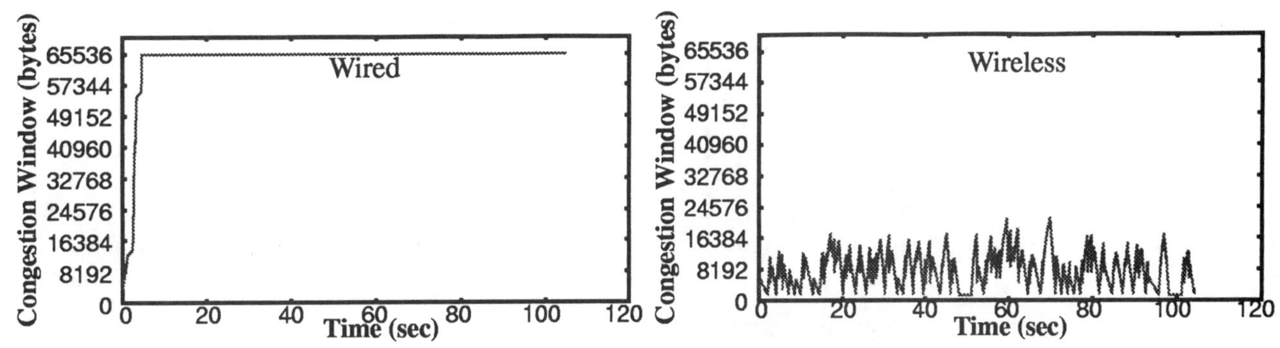

Figure 11. Congestion window sizes as a function of time for the wired and wireless parts of the split TCP connection. The wired sender never sees any losses and maintains a 64 KB congestion window. However, the wireless TCP connection's congestion window fluctuates rapidly.

Mbps that the E2E approach achieves (Figure 6), but not as good as several other protocols described earlier. The large congestion window size of the wired sender in SPLIT enables a higher bandwidth utilization over the wired network, compared to an end-to-end TCP connection where the congestion window size fluctuates rapidly.

As expected, the throughput for the SPLIT-SMART scheme is much higher. It is about 1.3 Mbps in the LAN case and about 1.1 Mbps in the WAN case. The SMART-based selective acknowledgment scheme operating over the wireless link performs very well, especially since no reordering of packets occurs over this hop. However, there are a few times when both the original transmission and the first retransmission of a packet get lost, which sometimes results in a coarse timeout (as described in Section 3.1). This explains the difference in throughput between the SPLIT-SMART scheme and the LL-SMART-TCP-AWARE scheme (Figure 3).

In summary, while the split-connection approach results in good throughput if the wireless connection uses special mechanisms, the performance is worse than that of a well-tuned, TCP-aware link-layer protocol (LL-TCP-AWARE or LL-SMART-TCP-AWARE). Moreover, the link-layer protocol preserves the end-to-end semantics of TCP acknowledgments. This demonstrates that the end-to-end connection need not be split at the base station in order to achieve good performance.

4.5 Reaction to Burst Errors

In this section, we report the results of some experiments that illustrate the benefit of selective acknowledgments in handling burst losses. We consider two of the best performing local protocols: LL-TCP-AWARE (Snoop) and LL-SMART-TCP-AWARE (Snoop with SMART-based selective acknowledgments). LL-TCP-AWARE recovers from a single loss by retransmitting the lost packet when two duplicate acknowledgments arrive for it. It also keeps track of the number of expected duplicate acknowledgments and the next expected new acknowledgment after this local retransmission. If this loss is part of a burst, the first new acknowledgment to arrive after the duplicates will be less than the next expected new one; this causes an immediate retransmission of the lost segment. This is similar to the mechanism used by E2E-NEWRENO (Section 3.1). LL-SMART-TCP-AWARE uses the additional useful information provided by the SMART scheme — the sequence number of the segment that caused the duplicate acknowledgment — to accurately determine losses and recover from them.

Burst Length	LL-TCP-AWARE (Mbps)	LL-SMART-TCP-AWARE (Mbps)
2	1.25	1.28
4	1.02	1.20
6	0.84	1.10

Table 5. Throughputs of LL-TCP-AWARE and LL-SMART-TCP-AWARE at different burst lengths. This illustrates the benefits of SACKs, even for a high-performance, TCP-aware link protocol.

Table 5 shows the performance of the two protocols for bursts of lengths 2, 4, and 6 packets. These errors are generated at an average rate of one every 64 KBytes of data, and 2, 4, or 6 packets are destroyed in each case. Selective acknowledgments improve the performance of LL-SMART-TCP-AWARE over LL-TCP-AWARE by up to 30% in the presence of burst errors. While this is a fairly simplistic burst-error model, it does illustrate the problems caused by the loss of multiple packets in succession. We are in the process of experimenting with a *temporal* burst-loss model based on average lengths of fades and other causes of wireless losses. The parameters of this model are derived from a trace-based modeling and characterization of the WaveLAN network [23].

---

3. A larger buffer at the base station will not necessarily improve performance for two reasons: (1) we measure performance in terms of receiver throughput, which is limited by the small congestion window size of the wireless connection, and (2) a long enough transfer will still fill up the buffer.

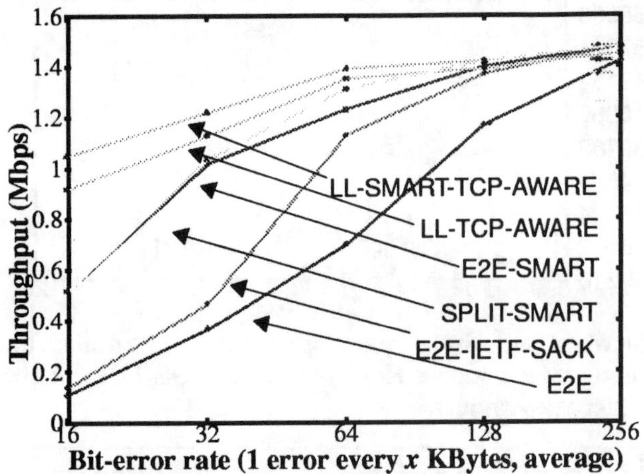

Figure 12. Performance of six protocols (LAN case) across a range of bit-error rates, ranging from 1 error every 16 KB to 1 every 256 KB shown on a log-scale.

4.6 Performance at Different Error Rates

In this section, we present the results of several experiments performed across a range of bit-error rates, for some of the protocols described earlier — E2E (the baseline case), LL-TCP-AWARE, LL-SMART-TCP-AWARE, E2E-SMART, E2E-IETF-SACK, and SPLIT-SMART. We chose the best performing protocols from each category, as well as some other protocols (e.g., E2E-IETF-SACK) to illustrate some interesting effects.

Figure 12 shows the performance of these protocols for an 8 MByte end-to-end transfer in a LAN environment, across exponentially distributed error rates ranging from 1 error every 16 KB to 1 error every 256 KB, in increasing powers of two. We find that the overall qualitative results and conclusions are similar to those presented earlier for the 64 KB error rate. At low error rates (128 KB and 256 KB points in the graph), all the protocols shown perform almost equally well in improving TCP performance. At the 16 KB error rate, the performance of the TCP-aware link-layer schemes is about 1.75-2 times better than E2E-SMART and about 9 times better than TCP Reno.

Another interesting point to note is the relative performance of E2E-IETF-SACK and E2E-SMART, especially at the high error rates. The congestion window does not grow larger than a few packets in the steady state at these error rates where there are multiple losses in many windows. E2E-IETF-SACK does not retransmit any packet using SACK information unless it receives three duplicate acknowledgments (to overcome potential reordering of packets in the network), which implies that no fast retransmissions are triggered if the number of packets in the window is less than four or five[4]. The sender's congestion window is often smaller than this, resulting in timeouts and degraded performance. In contrast, our implementation of E2E-SMART assumes no reordering of packets (which is justified in the LAN case) and retransmits the lost packet when the first duplicate acknowledgment with loss information arrives. This reduces the number of timeouts and results in better end-to-end performance. In Section 5.3, we outline a scheme in which the IETF protocol can be modified to work well even when the sender's congestion window is not large enough to provide enough duplicate acknowledgments.

4. This depends on whether delayed acknowledgments are used.

5. Discussion

In this section, we present a discussion of some miscellaneous issues. We discuss the effects of handoff on TCP performance, some implementation strategies and policies for the ELN mechanism introduced in Section 3.1, and some issues related to SMART-based and IETF selective acknowledgment schemes.

5.1 Wireless Handoffs

Wireless networks are usually organized in a cellular topology where each cell includes a base station that acts as a router between the wireless subnet and a wireline backbone. Mobile hosts typically communicate via the base station in the cell they are currently located in. Examples of networks organized in this fashion include cellular telephone networks and wireless local-area networks.

As a mobile host moves, it may get out of the range of its current base station but still be within the range of other neighboring base stations. To maintain the mobile host's connectivity, a *handoff* procedure is invoked to re-route traffic to and from the mobile host via the new base station. However, depending on the details of the handoff algorithms, this procedure could lead to packet losses and reordering, which in turn could cause significant deterioration in the performance of ongoing TCP transfers [8].

Several proposals have been made for achieving fast handoffs. Two examples include multicast-based handoffs [25] and hierarchical handoffs [9]. In both these schemes, handoffs are made fast by restricting updates to the immediate vicinity of the mobile host. As a result the handoff latency in a WaveLAN-based wireless local-area network is of the order of 10-30 ms.

A small amount of buffering and retransmission from base stations prevents packet loss during the short handoff period. In [9], the buffering happens at the mobile host's old base station, which forwards packets to the new base station at the time of handoff. In [25], one or more base stations in the vicinity join a multicast group corresponding to the mobile host and receive all packets destined to it, in anticipation of a handoff. When the handoff happens, the new base station is readily able to forward the buffered and the newly arriving packets without introducing any reordering, thereby preventing unnecessary invocations of TCP fast retransmissions. Experimental results reported in [25] indicate that such fast handoffs have a minimal adverse effect on TCP performance, even when the handoff frequency is as high as once per second.

In contrast to the above schemes that operate at the network layer, handoffs in a split-connection context, such as in I-TCP [3], involve the transfer of transport-layer state from the old base station to the new one. This results in significantly higher latency; for example, [2] reports I-TCP handoff latencies of the order of hundreds of milliseconds in a WaveLAN-based network.

5.2 Implementation Strategies for ELN

Section 3.1 described the ELN mechanism by which the transport protocol can be made aware of losses unrelated to network congestion and react appropriately to such losses. In this section, we outline possible implementation strategies and policies for this mechanism.

A simple strategy for implementing ELN would be to do so at the receiver, as we did for the results presented in this paper. In this method, the corruption of a packet at the link-layer, indicated by a CRC error, is passed up to the transport layer, which sends an ELN message with the duplicate acknowledgments for the lost packet. In practice, it may be hard to determine the connection that a corrupted packet belongs to, since the header could itself be corrupted: this can be handled by protecting the TCP/IP header using an FEC scheme. However, there are circumstances in which entire packets, including link-level headers, are dropped over a wireless link. In such circumstances, the base station generates ELN messages to the sender (in-band, as part of the acknowledgment stream) when it observes duplicate TCP acknowledgments arriving from the mobile host.

We expect Explicit Loss Notifications to be useful in the context of multi-hop wireless networks, and are exploring this in on-going work. Such networks (e.g., Metricom's Ricochet network [21]) typically use packet radio units to route packets to and from a wired infrastructure. Here, in order to implement ELN, periodic messages are exchanged between adjacent packet radio units about queue lengths and this information is used as a heuristic to distinguish between congestion and packet corruption, especially when entire packets (including headers) are corrupted or dropped over a wireless link. This, coupled with a simple link-level scheme to convey NACK information about missing packets, is sufficient to generate ELN messages to the source.

5.3 Selective Acknowledgment Issues

Our experience with the IETF SACK scheme highlights some weaknesses with it both when sender window sizes are small. This situation can be improved by enhancing the sender's loss recovery algorithm as follows. In general, the arrival of one duplicate acknowledgment at the receiver indicates that one segment has successfully reached the receiver. Rather than wait for three duplicate acknowledgments and perform a fast retransmission, the sender now transmits a *new* segment from beyond the "right edge" of the current window upon the arrival of the first and second duplicate acks. This probes the network for sustained congestion and generates duplicate acknowledgments. Note that we have not violated standard congestion control procedures by doing this: we only send out a segment when one has left the data pipe, following the principle of conservation of packets [13]. This enhancement can coexist with SACKs to further avoid timeouts, since the arrival of an acknowledgment with a SACK block indicating the reception of the newly transmitted segment is a strong indicator that the original segment was lost, independent of whether three duplicate acknowledgments arrive or not. Thus, this mechanism will improve performance when the sender's window is small and losses occur, and is further explored and described in [6].

6. Conclusions

In this paper, we have presented a comparative analysis of several techniques to improve the end-to-end performance of TCP over lossy, wireless hops. We categorize these techniques as end-to-end, link-layer or split-connection based. We use the end-to-end throughput, and the wired and wireless goodputs as metrics for comparison.

Our results lead to the following conclusions:

1. A reliable link-layer protocol that uses knowledge of TCP (LL-TCP-AWARE) to shield the sender from duplicate acknowledgments arising from wireless losses gives a 10-30% higher throughput than one (LL) that operates independently of TCP and does not attempt in-order delivery of packets. Also, the former avoids redundant retransmissions by both the sender and the base station, resulting in a higher goodput. Of the schemes we investigated, the TCP-aware link-layer protocol with selective acknowledgements performs the best.

2. The split-connection approach, with standard TCP used for the wireless hop, shields the sender from wireless losses. However, the sender often stalls due to timeouts on the wireless connection, resulting in poor end-to-end throughput. Using a SMART-based selective acknowledgment mechanism for the wireless hop yields good throughput. However, the throughput is still slightly less than that for a well-tuned link-layer scheme that does not split the connection. This demonstrates that splitting the end-to-end connection is not a requirement for good performance.

3. The SMART-based selective acknowledgment scheme we used is quite effective in dealing with a high packet loss rate when employed over the wireless hop or by a sender in a LAN environment. In the WAN experiments, the SACK scheme based on the IETF Draft resulted in significantly improving end-to-end performance, although its performance was not as good as in the best link schemes. From our results we conclude that selective acknowledgment schemes are very useful in the presence of lossy links, especially when losses occur in bursts.

4. End-to-end schemes, while not as effective as local techniques in handling wireless losses, are promising since significant performance gains can be achieved without any extensive support from intermediate nodes in the network. The explicit loss notification scheme we evaluated resulted in a throughput improvement of more than a factor of two over TCP-Reno, with comparable goodput values.

7. Future Work

Our experiments with various SACK and ELN mechanisms demonstrate the significant benefits of such schemes, as described in Section 5. We are in the process of evaluating protocol enhancements based on these ideas in the presence of both network congestion and wireless losses in different network topologies, especially in networks with multiple wireless hops. In addition, we are evaluating the performance of several of the protocols described in this paper under other patterns of loss derived from traces in [23].

We are investigating the impact of large variations in connection round-trip times and the impact of bandwidth and latency asymmetry on transport performance [5]. Large round-trip variations are common in networks like the Metricom Ricochet wireless network [21], especially in the presence of bidirectional traffic. Bandwidth asymmetry is prevalent in many cable and satellite networks with low-bandwidth return channels.

8. Acknowledgments

We are grateful to Steven McCanne and the anonymous reviewers for ACM SIGCOMM '96 and IEEE/ACM Transactions on Networking for several comments and suggestions that helped improve the quality of this paper. We thank Sally Floyd and Vern Paxson for useful discussions on SACKs and related topics.

This work was supported by DARPA contract DAAB07-95-C-D154, by the State of California under the MICRO program, and by the Hughes Aircraft Corporation, Metricom, Fuji Xerox, Daimler-Benz, Hybrid Networks, and IBM. Hari is partially supported by a research grant from the Okawa Foundation.

9. References

[1] E. Ayanoglu, S. Paul, T. F. LaPorta, K. K. Sabnani, and R. D. Gitlin. AIRMAIL: A Link-Layer Protocol for Wireless Networks. *ACM ACM/Baltzer Wireless Networks Journal*, 1:47–60, February 1995.

[2] A. Bakre and B. R. Badrinath. Handoff and System Support for Indirect TCP/IP. In *Proc. Second Usenix Symp. on Mobile and Location-Independent Computing*, April 1995.

[3] A. Bakre and B. R. Badrinath. I-TCP: Indirect TCP for Mobile Hosts. In *Proc. 15th International Conf. on Distributed Computing Systems (ICDCS)*, May 1995.

[4] H. Balakrishnan. An Implementation of TCP Selective Acknowledgments. ftp://daedalus.cs.berkeley.edu/pub/tcpsack/, 1996.

[5] H. Balakrishnan, V. N. Padmanabhan, and R.H. Katz. The Effects of Asymmetry on TCP Performance. In *Proc. ACM MOBICOM '97*, September 1997.

[6] H. Balakrishnan, V. N. Padmanabhan, S. Seshan, M. Stemm, and R.H. Katz. TCP Behavior of a Busy Web Server: Analysis and Improvements. Technical Report UCB/CSD-97-966, University of California at Berkeley, 1997.

[7] H. Balakrishnan, S. Seshan, and R.H. Katz. Improving Reliable Transport and Handoff Performance in Cellular Wireless Networks. *ACM Wireless Networks*, 1(4), December 1995.

[8] R. Caceres and L. Iftode. Improving the Performance of Reliable Transport Protocols in Mobile Computing Environments. *IEEE Journal on Selected Areas in Communications*, 13(5), June 1995.

[9] R. Caceres and V. N. Padmanabhan. Fast and Scalable Handoffs in Wireless Internetworks. In *Proc. 1st ACM Conf. on Mobile Computing and Networking*, November 1996.

[10] A. DeSimone, M. C. Chuah, and O. C. Yue. Throughput Performance of Transport-Layer Protocols over Wireless LANs. In *Proc. Globecom '93*, December 1993.

[11] K. Fall and S. Floyd. Simulation-based Comparisons of Tahoe, Reno, and Sack TCP. *Computer Communications Review*, 1996.

[12] J. C. Hoe. Start-up Dynamics of TCP's Congestion Control and Avoidance Schemes. Master's thesis, Massachusetts Institute of Technology, 1995.

[13] V. Jacobson. Congestion Avoidance and Control. In *Proc. ACM SIGCOMM 88*, August 1988.

[14] V. Jacobson and R. T. Braden. *TCP Extensions for Long Delay Paths*. RFC, Oct 1988. RFC-1072.

[15] P. Karn. The Qualcomm CDMA Digital Cellular System. In *Proc. 1993 USENIX Symp. on Mobile and Location-Independent Computing*, pages 35–40, August 1993.

[16] P. Karn and C. Partridge. Improving Round-Trip Time Estimates in Reliable Transport Protocols. *ACM Transactions on Computer Systems*, 9(4):364–373, November 1991.

[17] S. Keshav and S. Morgan. SMART Retransmission: Performance with Overload and Random Losses. In *Proc. Infocom '97*, 1997.

[18] S. Lin and D. J. Costello. *Error Control Coding: Fundamentals and Applications*. Prentice-Hall, Inc., 1983.

[19] Mathis, M. and Mahdavi, J. and Floyd, S. and Romanow, A. *TCP Selective Acknowledgment Options*, 1996. RFC-2018.

[20] S. McCanne and V. Jacobson. The BSD Packet Filter: A New Architecture for User-Level Packet Capture. In *Proc. Winter '93 USENIX Conference*, San Diego, CA, January 1993.

[21] Metricom, Inc. http://www.metricom.com, 1997.

[22] S. Nanda, R. Ejzak, and B. T. Doshi. A Retransmission Scheme for Circuit-Mode Data on Wireless Links. *IEEE Journal on Selected Areas in Communications*, 12(8), October 1994.

[23] G. T. Nguyen, R. H. Katz, B. D. Noble, and M. Satyanarayanan. A Trace-based Approach for Modeling Wireless Channel Behavior. In *Proc. Winter Simulation Conference*, Dec 1996.

[24] J. B. Postel. *Transmission Control Protocol*. RFC, Information Sciences Institute, Marina del Rey, CA, September 1981. RFC-793.

[25] S. Seshan, H. Balakrishnan, and R. H. Katz. Handoffs in Cellular Wireless Networks: The Daedalus Implementation and Experience. *Kluwer Journal on Wireless Personal Communications*, January 1997.

[26] W. R. Stevens. *TCP/IP Illustrated, Volume 1*. Addison-Wesley, Reading, MA, Nov 1994.

[27] AT&T WaveLAN: PC/AT Card Installation and Operation. AT&T manual, 1994.

[28] R. Yavatkar and N. Bhagwat. Improving End-to-End Performance of TCP over Mobile Internetworks. In *Mobile 94 Workshop on Mobile Computing Systems and Applications*, December 1994.

Hari Balakrishnan (S '95 / ACM S '95) is a Ph.D. candidate in Computer Science at the University of California at Berkeley. His research interests are in the areas of computer networks, wireless and mobile computing, and distributed computing and communication systems. His current research is in the area of reliable data transport over heterogeneous networking technologies.

Hari received a B. Tech. degree in Computer Science and Engineering from the Indian Institute of Technology, Madras, in 1993 and an M.S. degree in Computer Science from Berkeley in 1995. He received best student paper awards at the Winter Usenix '95 and at the ACM Mobicom '95 conferences, and is the recipient of a research grant from the Okawa Foundation. On the WWW, his URL is http://www.cs.berkeley.edu/~hari and his e-mail address is hari@cs.berkeley.edu.

Venkata N. Padmanabhan (IEEE S '94 / ACM S '94) is a Ph.D. candidate in Computer Science at the University of California at Berkeley. He received his B.Tech. degree from the Indian Institute of Technology, Delhi in 1993 and his M.S. degree from the University of California at Berkeley in 1995, both in Computer Science.

Venkat has done research in the areas of Computer Networking, Mobile Computing and Operating Systems. The focus of his current work is network support for efficient Web access, and data transport over asymmetric networks. He received the best student paper award at the Usenix '95 conference. He may be reached via e-mail at padmanab@cs.berkeley.edu and on the Web at http://www.cs.berkeley.edu/~padmanab.

Srinivasan Seshan (ACM M '92) received a B.S. in Electrical Engineering, an M.S., and a Ph.D. in Computer Science from the University of California at Berkeley in 1990, 1993 and 1995 respectively. Since 1995, he has been a research staff member at the IBM T.J. Watson Research Center. His research interests include computer networks, mobile computing and distributed computing. His e-mail address is srini@watson.ibm.com and his WWW home page is at http://www.research.ibm.com/people/s/srini.

Randy H. Katz (F '96, ACM F '96) is a professor of computer science at the University of California at Berkeley, and is a principal investigator in the Bay Area Research Wireless Access Network (BARWAN) project. He has taught at Berkeley since 1983, with the exception of 1993 and 1994 when he was a program manager and deputy director of the Computing Systems Technology Office at the Defense Department's Advanced Research Projects Agency. He has written over 130 technical publications on CAD, database management, multiprocessor architectures, high performance storage systems, video server architectures, and computer networks.

Dr. Katz received a B.S. degree at Cornell University, and an M.S. and a Ph.D. at the University of California at Berkeley, all in computer science. His e-mail address is randy@cs.berkeley.edu and his WWW home page is http://www.cs.berkeley.edu/~randy.

Chapter 11 Ubiquitous Computing

We end the discussion of papers on mobile computing with an article by Mark Weiser [Weiser, 1993b], the "father of ubiquitous computing," and a recent article by Fox, and others [Fox, 1998], on support for the 3Com PalmPilot as a ubiquitous computing device.

Weiser, who at that time was head of Xerox Palo Alto Research Center's Computer Science Laboratory (PARC CSL), was a proponent of a vision that computers would soon become so cheap and so small that they would blend into the background. **Ubiquitous computing**, or "ubicomp," was a theme of much of the work at PARC CSL from 1988 until recently [Want, 1995; 1996; Weiser, 1991; 1993a, b].

PARC focused initially on three hardware prototypes. The first was a "Liveboard," which was a computer with a large screen such that people could effectively "write" on the board to interact with a computer. The second was a "Tab," which was a hand-held device with buttons. It was intended to be like an electronic Post-It note. The third was a "Pad," which was a notebook-sized portable pen computer. It seems that ultimately a compromise between these two devices was realized in the marketplace by commercial products such as the PalmPilot. Finally, one of the additional tools used in the Ubicomp project was the Active Badge, from Olivetti [Want, 1992a, b]. PARC used the badges to map people's locations, to display messages on computers nearest an individual, and for many other applications.

The Ubicomp project explored a number of interesting computer science issues, as described in Weiser's article. These included questions of power consumption [Weiser, 1994], wireless networking [Tso, 1993], and especially applications. In addition, they investigated a number of issues of privacy that result from the heightened knowledge of personal activities given by active badges [Spreitzer, 1993].

The 3Com PalmPilot has many similarities with the ParcTab, and since its release it has provided numerous research groups and product organizations with a common platform for developing new applications and enhancements. Researchers in the University of California Berkeley Computer Science Division (EECS) built Top Gun Wingman [Fox, 1998], the first graphical Web browser for the PalmPilot, which has subsequently been commercialized by Proxinet. This work leverages off an infrastructure for certain Internet services that they label TACC for *transformation, aggregation, caching,* and *customization* [Fox, 1997]. They use TACC to simplify HTML markup, to convert postscript to PalmPilot-native text rendering, and to shrink down and simplify graphics [Fox, 1996a, b].

Here we include an article that summarizes their proxy support and discusses several applications, including Top Gun Wingman, Top Gun Mediaboard (a shared whiteboard application for the PalmPilot using reliable multicast) [Chawathe, 1998], and Charon (Kerberos-style authentication for low-bandwidth clients) [Fox, 1996b]. The article has some similarities with the Rover [Joseph, 1997b] and WebExpress [Housel, 1996] papers, and could have been included equally well in "Limits on Connectivity."

11.1 Weiser, Mark, Some Computer Science Issues in Ubiquitous Computing

Weiser, M., "Some Computer Science Issues in Ubiquitous Computing," *Communications of the ACM*, 36(7):74–84, July 1993.

SOME COMPUTER SCIENCE ISSUES IN UBIQUITOUS COMPUTING

Mark Weiser

Ubiquitous computing enhances computer use by making many computers available throughout the physical environment, while making them effectively invisible to the user. This article explains what is new and different about the computer science involved in ubiquitous computing. First, it provides a brief overview of ubiquitous computing, then elaborates through a series of examples drawn from various subdisciplines of computer science: hardware components (e.g., chips), network protocols, interaction substrates (e.g., software for screens and pens), applications, privacy, and computational methods. Ubiquitous computing offers a framework for new and exciting research across the spectrum of computer science.

Since we started this work at Xerox Palo Alto Research Center (PARC) in 1988 a few places have begun work on this possible next-generation computing environment in which each person is continually interacting with hundreds of nearby wirelessly interconnected computers. The goal is to achieve the most effective kind of technology, that which is essentially invisible to the user. To bring computers to this point while retaining their power will require radically new kinds of computers of all sizes and shapes to be available to each person. I call this future world "Ubiquitous Computing" (Ubicomp) [27]. The research method for ubiquitous computing is standard experimental computer science: the construction of working prototypes of the necessary infrastructure in sufficient quantity to debug the viability of the systems in everyday use; ourselves and a few colleagues serving as guinea pigs. This is an important step toward ensuring that our infrastructure research is robust and scalable in the face of the details of the real world.

The idea of ubiquitous computing first arose from contemplating the place of today's computer in actual activities of everyday life. In particular, anthropological studies of work life [14, 22] teach us that people primarily work in a world of shared situations and unexamined technological skills. The computer today is isolated from the overall situation, however, and fails to get out of the way of the work. In other words, rather than being a tool through which we work, and thus disappearing from our awareness, the computer too often remains the focus of attention. And this is true throughout the domain of personal computing as currently implemented and discussed for the future, whether one thinks of personal computers, palmtops, or dynabooks. The characterization of the future computer as the "intimate computer" [12], or "rather like a human assistant" [25] makes this attention to the machine itself particularly apparent.

Getting the computer out of the way is not easy. This is not a graphical user interface (GUI) problem, but is a property of the whole context of usage of the machine and the attributes of its physical properties: the keyboard, the weight and desktop position of screens, and so on. The problem is not one of "interface." For the same reason of context, this is not a multimedia problem, resulting from any particular deficiency in the ability to display certain kinds of real-time data or integrate them into applications. (Indeed, multimedia tries to grab attention, the opposite of the ubiquitous computing ideal of invisibility.) The challenge is to create a new kind of relationship of people to computers, one in which the computer would have to take the lead in becoming vastly better at getting out of the way, allowing people to just go about their lives.

In 1988, when I started PARC's work on ubiquitous computing, virtual reality (VR) came the closest to enacting the principles we believed important. In its ultimate envisionment, VR causes the computer to become effectively invisible by taking over the human sensory and affector systems [19]. VR is extremely useful in scientific visualization and entertainment, and will be very significant for those niches. But as a tool for productively changing everyone's relationship to computation, it has two crucial flaws: first, at the present time (1992), and probably for decades, it cannot produce a simulation of significant verisimilitude at reasonable cost (today, at any cost). This means that users will not be fooled and the computer will not be out of the way. Second, and most important, it has the goal of fooling the user—of leaving the everyday physical world behind. This is at odds with the goal of better integrating the computer into human activities, since humans are of and in the everyday world.

Ubiquitous computing is exploring quite different ground from personal digital assistants, or the idea that computers should be autonomous agents that take on our goals. The difference can be characterized as follows: Suppose you want to lift a heavy object. You can call in your strong assistant to lift it for you, or you yourself can be made effortlessly, unconsciously, stronger and just lift it. There are times when both are good. Much of the past and current effort for better computers has been aimed at the former; ubiquitous computing aims at the latter.

The approach I took was to attempt the definition and construction of new computing artifacts for use in everyday life. I took my inspiration from the everyday objects found in offices and homes, in particular those objects whose purpose is to capture or convey information. The most ubiquitous current informational technology embodied in artifacts is the use of written symbols, primarily words, but including also pictographs, clocks, and other sorts of symbolic communication. Rather than attempting to reproduce these objects inside the virtual computer world, leading to another "desktop model" [2], I wanted to put the new kind of computer out in this world of concrete information conveyers. Since these written artifacts occur in many different sizes and shapes, with many different qualities, I wanted the computer embodiments to be of many sizes and shapes, including tiny inexpensive ones that could bring computing to everyone.

The physical world comes in all sizes and shapes. For practical reasons our ubiquitous computing work begins with just three different sizes of devices: enough to give some scope, not enough to deter progress. The first size is the wall-sized interactive surface, analogous to the office whiteboard or the home magnet-covered refrigerator or bulletin board. The second size is the notepad, envisioned not as a personal computer but as analogous to scrap paper to be grabbed and used easily, with many being used by a person at one time. The cluttered office desk or messy front-hall table are real-life examples. Finally, the third size is the tiny computer, analogous to tiny individual notes or Post-it notes, and also similar to the tiny little displays of words found on book spines, light switches, and hallways. Again, I saw this not as a personal computer, but as a pervasive part of everyday life, with many active at all times. I called these three sizes of computers boards, pads, and tabs, and adopted the slogan that, for each person in an office, there should be hundreds of tabs, tens of pads, and one or two boards. Specifications for some prototypes of these three sizes in use at PARC are shown in Figure 1.

This then is Phase I of ubiquitous computing: to construct, deploy, and

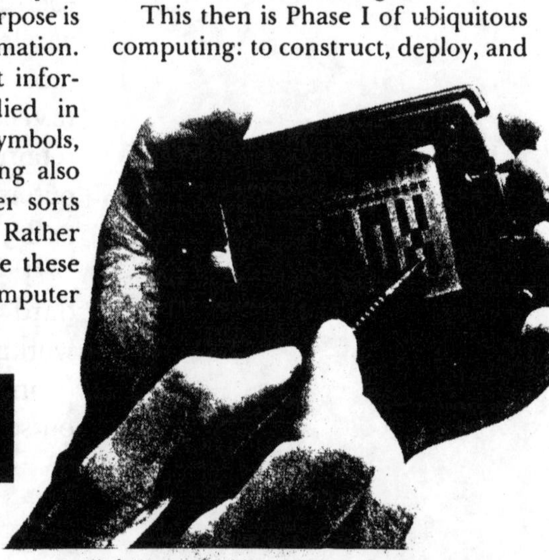

Figure 1. Photographs of Tab, Pad and Board

learn from a computing environment consisting of tabs, pads, and boards. This is only Phase I, because it is unlikely to achieve optimal invisibility. (Later phases are yet to be determined.) But it is a start down the radical direction, for computer science, away from emphasis on the machine and back to the person and his or her life in the world of work, play, and home.

Hardware Prototypes

New hardware systems design for ubiquitous computing has been oriented toward experimental platforms for systems and applications of invisibility. New chips have been less important than combinations of existing components that create experimental opportunities. The first ubiquitous computing technology to be deployed was the Liveboard [6], which is now a Xerox product. Two other important pieces of prototype hardware supporting our research at PARC are the Tab and the Pad.

Tab

The ParcTab is a tiny information doorway. For user interaction it has a pressure-sensitive screen on top of the display, three buttons positioned underneath the natural finger positions, and the ability to sense its position within a building. The display and touchpad it uses are standard commercial units.

The key hardware design problems affecting the pad are physical size and power consumption. With several dozens of these devices sitting around the office, in briefcases, in pockets, one cannot change their batteries every week. The PARC design uses the 8051 chip to control detailed interactions, and includes software that keeps power usage down. The major outboard components are a small analog/digital converter for the pressure-sensitive screen, and analog sense circuitry for the IR receiver. Interestingly, although we have been approached by several chip manufacturers about our possible need for custom chips for the Tab, the Tab is not short of places to put chips. The display size leaves plenty of room, and the display thickness dominates total size. Off-the-shelf components are more than adequate for exploring this design space, even with our severe size, weight, and power constraints.

A key part of our design philosophy is to put devices in everyday use, not just demonstrate them. We can only use techniques suitable for quantity 100 replication. This excludes certain techniques that could make a huge difference, such as the integration of components onto the display surface itself. This technology is being explored at PARC, as well as other research organizations. While it is very promising, it is not yet ready for replication.

The Tab architecture incorporates a careful balance of display size, bandwidth, processing, and memory. For instance, the small display means that even the tiny processor is capable of providing a four-frames-per-second video rate, and the IR bandwidth is capable of delivering this. The bandwidth is also such that the processor can actually time the pulse widths in software timing loops. Our current design has insufficient storage, and we are increasing the amount of nonvolatile RAM in future tabs from 8K to 128K. The tab's goal of casual use similar to that of Post-it notes puts it into a design space generally unexplored in the commercial or research sector.

Pad

The pad is really a family of notebook-sized devices. Our initial pad, the ScratchPad, plugged into a Sun SBus card and provided an X-Window-system-compatible writing and display surface. This same design was used inside our first wall-sized displays, the liveboards, as well. Our later untethered pad devices, the XPad and MPad, continued the system design principles of X-compatibility, ease of construction, and flexibility in software and hardware expansion.

As I write this article, at the end of 1992, commercial portable pen devices have been on the market for two years, although most of the early companies have now gone out of business. Why should a pioneering research lab build its own such device? Each year we ask ourselves the same question, and so far three things always drive us to continue to

design our own pad hardware:

First, we need the correct balance of features—this is the essence of systems design. The commercial devices all aim at particular niches, balancing their design to that niche. For research we need a rather different balance, particularly for ubiquitous computing. For instance, can the device communicate simultaneously along multiple channels? Does the operating system support multiprocessing? What about the potential for high-speed tethering? Is there a high-quality pen? Is there a high-speed expansion port sufficient for video in and out? Is sound in/out and ISDN connectivity available? Optional keyboard? Any one commercial device tends to satisfy some of these needs, ignore others, and choose a balance of the ones it does satisfy, optimizing its niche, rather than ubiquitous computing-style scrap computing. The balance for us emphasizes communication, system memory, multimedia, and expansion ports.

Second, apart from balance of features are the requirements for particular features. Key among these are a pen emphasis, connection to research environments such as Unix, and communication emphasis. A high-speed (>64KB/sec) wireless capability is not built into any commercial devices, and they do not generally have a sufficiently high-speed port to add such a radio. Commercial devices generally come with DOS or Penpoint, and while we have developed in both, they are not our favorite research vehicles because they lack full access and customizability.

The third factor driving our own pad designs is ease of expansion and modification. We need full hardware specifications, complete operating system source code, and the ability to remove and replace both hardware and software components. Naturally these goals are opposed to best price in a niche market, which orients the documentation to the end user, and keeps prices down by integrated rather than modular design.

We have built and used three generations of Pad designs. Six scratchpads were built, three XPads, and 13 MPads, the latest. The MPad uses an FPGA for almost all random logic, giving extreme flexibility. For instance, changing the power control functions and adding high-quality sound were relatively simple FPGA changes. The MPad has both IR (tab compatible) and radio communication built-in and includes sufficient uncommitted space for adding new circuit boards later. It can be used with a tether that provides it with recharging and operating power and an Ethernet connection. The operating system is a standalone version of the public-domain Portable Common Runtime developed at PARC [28].

The Computer Science of Ubicomp

To construct and deploy tabs, pads, and boards at PARC, we found ourselves having to readdress some of the well-worked areas of existing computer science. The fruitfulness of ubiquitous computing for new computer science problems justified our belief in the ubiquitous computing framework.

The following subsections "ascend" the levels of organization of a computer system, from hardware to application. One or two examples of computer science work required by ubiquitous computing are described for each level. Ubicomp is not yet a coherent body of work, but consists of a few scattered communities. The point of this article is to help others understand some of the new research challenges in ubiquitous computing, and inspire them to work on them. This is more akin to a tutorial than a survey, and necessarily selective. The areas included are hardware components (e.g., chips), network protocols, interaction substrates (e.g., software for screens and pens), applications, privacy, and computational methods.

Issues of Hardware Components

In addition to the new systems of tabs, pads, and boards, ubiquitous computing necessitates some new kinds of devices. Examples of three new kinds of hardware devices are very low-power computing, low-power high-bits/cubic-meter communication, and pen devices.

Low Power

In general the need for high performance has dominated the need for low-power consumption in processor design. However, recognizing the new requirements of ubiquitous computing, a number of people have begun work in using additional chip area to reduce power rather than to increase performance [16]. One key approach is to reduce the clocking frequency of their chips by increasing pipelining or parallelism. Then, by running the chips at reduced voltage, the effect is a net reduction in power, because power falls off as the square of the voltage, while only about twice the area is needed to run at half the clock speed.

$\text{Power} = C_L * V_{dd}^2 * f$

where C_L is the gate capacitance,
V_{dd} the supply voltage,
and f the clocking frequency.

This method of reducing power leads to two new areas of chip design: circuits that will run at low power, and architectures that sacrifice area for power over performance. The second requires some additional comment, because one might suppose one would simply design the fastest possible chip, and then run it at reduced clock and voltage. However, as Lyon illustrates, circuits in chips designed for high speed generally fail to work at low voltages. Furthermore, attention to special circuits may permit operation over a much wider range of voltage operation, or achieve power savings via other special techniques, such as adiabatic switching [16].

Wireless

A wireless network capable of accommodating hundreds of high-speed devices for every person is well beyond the commercial wireless systems planned for the next 10 years [20], which are aimed at one low-speed (64kb/sec or voice) device per person. Most wireless work uses a figure of merit of bits/sec × range, and seeks to increase this product. We believe that a better figure of merit is bits/sec/meter^3. This figure of merit causes the optimization of total bandwidth throughout a 3D space, leading to design points of very tiny cellular systems.

Because we felt the commercial world was ignoring the proper figure

To construct and deploy tabs, pads, and boards at PARC, we found ourselves having to readdress some of the well-worked areas of existing computer science.

of merit, we initiated our own small radio program. In 1989 we built spread-spectrum transceivers at 900MHz, but found them difficult to build and adjust, and prone to noise and multipath interference. In 1990 we built direct frequency-shift-keyed transceivers also at 900MHz, using very low power to be license-free. While much simpler, these transceivers had unexpectedly and unpredictably long range, causing mutual interference and multipath problems. In 1991 we designed and built our current radios, which use the near-field of the electromagnetic spectrum. The near-field has an effective fall-off of r^6 in power, instead of the more usual r^2, where r is the distance from the transmitter. At the proper levels this band does not require an FCC license, permits reuse of the same frequency over and over again in a building, has virtually no multipath or blocking effects, and permits transceivers that use extremely low power and low parts count. We have deployed a number of near-field radios within PARC.

Pens

A third new hardware component is the pen for very large displays. We needed pens that would work over a large area (at least 60in × 40in), not require a tether, and work with back projection. These requirements are generated from the particular needs of large displays in ubiquitous computing—casual use, no training, naturalness, simultaneous multiple use. No existing pens or touchpads could come close to these requirements. Therefore members of the Electronics and Imaging lab at PARC devised a new infrared pen. A camera-like device behind the screen senses the pen position, and information about the pen state (e.g., buttons) is modulated along the IR beam. The pens need not touch the screen, but can operate from several feet away. Considerable DSP and analog design work underlies making these pens effective components of the ubiquitous computing system [6].

Network Protocols

Ubicomp changes the emphasis in networking in at least four areas: wireless media access, wide-bandwidth range, real-time capabilities for multimedia over standard networks, and packet routing.

Wireless Media Access

A "media access" protocol provides access to a physical medium. Common media access methods in wired domains are collision detection and token-passing. These do not work unchanged in a wireless domain because not every device is assured of being able to hear every other device (this is called the "hidden terminal" problem). Furthermore, earlier wireless work used assumptions of complete autonomy, or a statically configured network, while ubiquitous computing requires a cellular topology, with mobile devices frequently coming on- and off-line. We have adapted a media access protocol called MACA, first described by Karn [11], with some of our own modifications for fairness and efficiency.

The key idea of MACA is for the two stations desiring to communicate to first do a short handshake of Request-To-Send-N-bytes followed by Clear-To-Send-N-bytes. This exchange allows all other stations to hear that there is going to be traffic, and for how long they should remain quiet. Collisions, which are detected by timeouts, occur only during the short Request-To-Send packet.

Adapting MACA for ubiquitous computing use required considerable attention to fairness and real-time requirements. MACA requires stations whose packets collide to back off a random time and try again. If all stations but one back-off, that one can dominate the bandwidth. By requiring all stations to adapt the back-off parameter of their neighbors, we create a much fairer allocation of bandwidth.

. Some applications need guaranteed bandwidth for voice or video. We added a new packet type, NCTS(n) (Not Clear To Send), to suppress all other transmissions for (n) bytes. This packet is sufficient for a basestation to do effective bandwidth allocation among its mobile units. The solution is robust, in the sense that if the basestation stops allocating bandwidth the system reverts to normal contention.

When a number of mobile units share a single basestation, that basestation may be a bottleneck for communication. For fairness, a basestation with $N > 1$ nonempty output queues needs to contend for bandwidth as though it were N stations. We therefore make the basestation contend just enough more aggressively that it is N times more likely to win a contention for media access.

Other Network Issues

Two other areas of networking research at PARC with ubiquitous computing implications are Gb networks and real-time protocols. Gb-per-second speeds are important because of the increasing number of medium-speed devices anticipated by ubiquitous computing, and the growing importance of real-time (multimedia) data. One hundred 256kb/sec portables per office implies a Gb per group of 40 offices, with all of PARC needing an aggregate of some five Gb/sec. This has led us to do research into local-area ATM switches, in association with other Gb networking projects [15].

Real-time protocols are a new area of focus in packet-switched networks. Although real-time delivery has always been important in telephony, a few hundred milliseconds never mattered in typical packet-switched applications such as telnet and file transfer. With the ubiquitous use of packet-switching, even for telephony using ATM, the need for real-time capable protocols has become urgent if the packet networks are going to support multimedia applications.

Again, in association with other members of the research community, PARC is exploring new protocols for enabling multimedia on the packet-switched Internet [4].

The Internet routing protocol, IP, has been in use for over 10 years. However, neither this protocol nor its OSI equivalent, CLNP, provides sufficient infrastructure for highly mobile devices. Both interpret fields in the network names of devices in order to route packets to the device. For instance, the "13" in IP name 13.2.0.45 is interpreted to mean net 13, and network routers anywhere in the world are expected to know how to get a packet to net 13, and all devices whose name starts with 13 are expected to be on that network. This assumption fails as soon as a user of a net 13 mobile device takes her device on a visit to net 36 (Stanford). Changing the device name dynamically depending on location is no solution: higher-level protocols such as TCP assume that underlying names will not change during the life of a connection, and a name change must be accompanied by informing the entire network of the change so that existing services can find the device.

A number of solutions have been proposed to this problem, among them Virtual IP from Sony [24], and Mobile IP from Columbia University [10]. These solutions permit existing IP networks to interoperate transparently with roaming hosts. The key idea of all approaches is to add a second layer of IP address, the "real" address indicating location, to the existing fixed-device address. Special routing nodes that forward packets to the correct real address, and keep track of where this address is, are required for all approaches.*

Interaction Substrates

Ubicomp has led us to explore new substrates for interaction. Four such substrates are mentioned here, spanning the space from virtual keyboards to protocols for window systems.

Tabs have a very small interaction area—too small for a keyboard, too small even for standard handprinting recognition. Handprinting has the further problem of requiring looking at what is written. Improvements in voice recognition are no panacea, because when other people are present, voice will often be inappropriate. As one possible solution, we developed a method of touch-printing that uses only a tiny area and does not require looking. A drawback of our method is it requires a new printing alphabet to be memorized, and reaches only half the speed of a fast typist [8].

Liveboards have a high interaction area, 400 times that of the tab. Using conventional pull-down or pop-up menus might require walking across the room to the appropriate button, a serious problem. We have developed methods of location-independent interaction by which even complex interactions can be popped up at any location. [13].

The X-Window system, although designed for network use, makes it difficult for windows to move once instantiated at a given X server. This is because the server retains considerable state about individual windows, and does not provide convenient ways to move that state. For instance, context and window IDs are determined solely by the server, and cannot be transferred to a new server, so applications that depend on knowing their value (almost all) will break if a window changes servers. However, in the ubiquitous computing world a user may move frequently from device to device, and want to bring windows along.

Christian Jacobi at PARC has implemented a new X toolkit that facilitates window migration. Applications need not be aware that they have moved from one screen to another; or if desired, the user can be so informed with an upcall. We have written a number of applications on top of this toolkit, all of which can be "whistled up" over the network to follow the user from screen to screen. The author, for instance, frequently keeps a single program development and editing environment open for days at a time, migrating its windows back and forth from home to work and back each day.

A final window system problem is bandwidth. The bandwidth available to devices in ubiquitous computing can vary from Kb/sec to Gb/sec, and with window migration a single application may have to dynamically adjust to bandwidth over time. The X-Window system protocol was primarily developed for Ethernet speeds, and most of the applications written in it were similarly tested at 10Mb/sec. To solve the problem of efficient X-Window use at lower bandwidth, the X consortium is sponsoring a "Low Bandwidth X" (LBX) working group to investigate new methods of lowering bandwidth. [7].

Applications

Applications are of course the whole point of ubiquitous computing. Two examples of applications are locating people and shared drawing.

Ubicomp permits the location of people and objects in an environment. This was first pioneered by Olivetti Research Labs in Cambridge, England, in their Active Badge system [26]. In ubiquitous computing we continued to extend this work, using it for video annotation and updating dynamic maps. For instance, Figure 2 shows a portion of CSL early one morning, and the individual faces are the locations of people. This map is updated every few seconds, permitting quick locating of people, as well as quickly noticing a meeting one might want to go to (or where one can find a fresh pot of coffee).

Xerox PARC, EuroPARC, and the Olivetti Research Center have built several different kinds of location servers. Generally these have two parts: a central database of information about location that can be quickly queried and dumped, and a group of servers that collect information about location and update the database. Information about location can be deduced from logins, or collected directly from an active badge system. The location database may be organized to dynamically notify clients, or simply to facilitate frequent polling.

Some example uses of location information are automatic phone forwarding, locating an individual for a meeting, and watching general activity in a building to feel in touch with

*The Internet community has a working group considering standards for this area (contact deering@xerox.com for more information).

its cycles of activity (important for telecommuting).

Xerox PARC has investigated a number of shared meeting tools over the past decade, starting with the CoLab work [21], and continuing with videodraw and commune [23]. Two new tools were developed for investigating problems in ubiquitous computing. The first is Tivoli [18], the second is Slate,—each tool is based on different implementation paradigms. First their similarities: both emphasize pen-based drawing on a surface; both accept scanned input and can print the results; both can have several users simultaneously operating independently on different or the same pages; both support multiple pages. Tivoli has a sophisticated notion of a stroke as spline, and has a number of features making use of processing the contents and relationships among

Table 1. Some hardware prototypes in use at Xerox PARC

ParcTab	MPad	Liveboard
Dimensions: 10.2- × 7.8- × 2.4cm Weight: 7.2 oz Screen: 6.2- × 4.2cm, 128- × 64 monochrome Touch input: passive pressure sensing Sound: Piezo speaker Wireless interfaces: IR at 850nm, DEC/Olivetti active badge compatible, 19.2k baud PWM baseband modulation, CSMA Processor: Intel 8051-type, 8k (v1) 128k (v2) nvram Ports: I^2C external bus, recharge port Battery: 12 hours continuous or est. 2 weeks normal use, rechargeable	Dimensions: 22.2- × 28- × 3.8cm Weight: 5llbs 4oz Screen: 640- × 480 LCD Display (3 levels of grey) Pen: tethered electromagnetic sensing Sound: Built-in microphone, Speaker, Piezo Beeper Wireless Interfaces: 250Kbps Radio, 19.2Kbps IR Processor: Motorola 68302, 4MB of DRAM, 1/2MB of VRAM, 1/4MB of EPR External Ports: Stylus/microphone, PCMCIA, 1MB Serial, RS232, I^2C bus, Keyboard Internal Ports: Second audio channel, ISDN, Expansion Port Battery: rechargeable, 3 hours	Dimensions: 83in, 52in and 30in Weight: 560lb (250kg) Screen: very bright 45- × 65in, 1024- × 768 monochrome pixels, 640- × 480 pixels color, also NTSC video Pen: IR wireless Stereo sound Networking, processor, and ports determined by choice of embedded workstation, either PC or Sun 12 amps at 115 volts

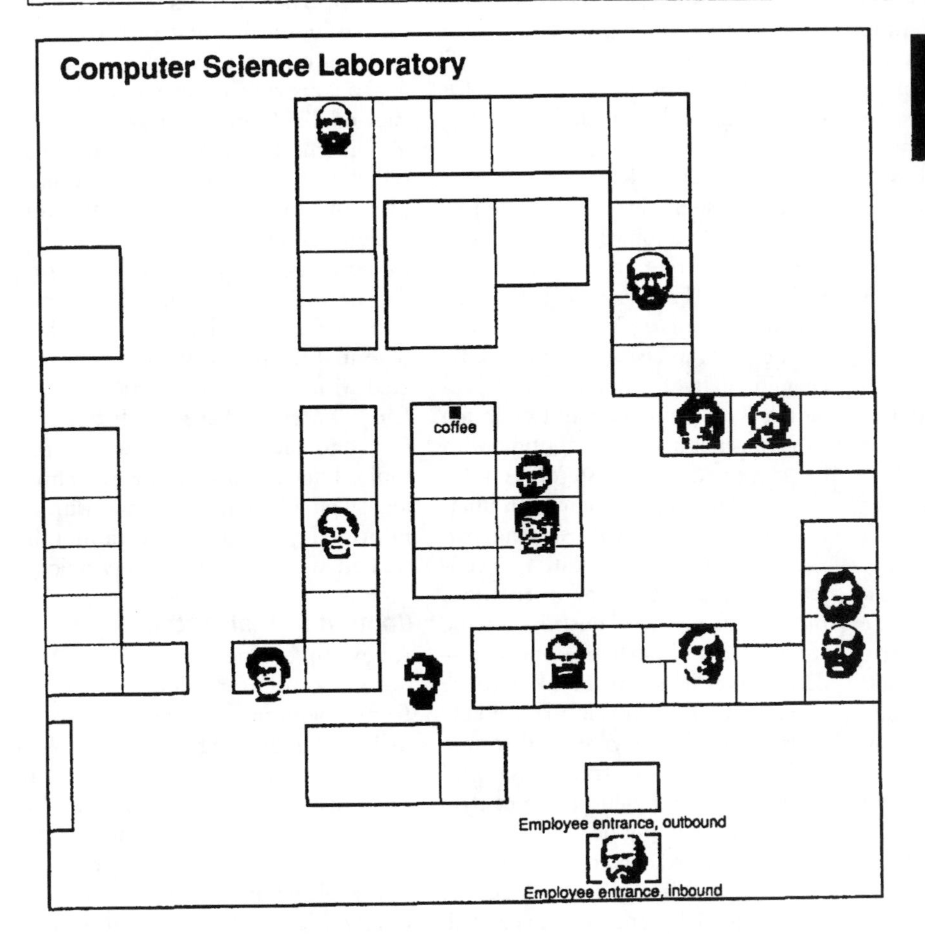

Figure 2. Display of CSL activity from personal locators

strokes. Tivoli also uses gestures as input control to select, move, and change the properties of objects on the screen. When several people use Tivoli, each must be running a separate copy, and connect to the others. On the other hand, Slate is completely pixel-based, simply drawing ink on the screen. Slate manages all the shared windows for all participants, as long as they are running an X-Window server, so its aggregate resource use can be much lower than Tivoli, and it is easier to set up with large numbers of participants. In practice we have used slate from a Sun to support shared drawing with users on Macs and PCs. Both Slate and Tivoli have received regular use at PARC.

Shared drawing tools are a topic of research at many places. For instance, Bellcore has a toolkit for building shared tools [9], and Jacobsen at LBL uses multicast packets to reduce bandwidth during shared tool use. There are some commercial products [3], but these are usually not multipage and so not really suitable for creating documents or interacting over the course of an entire meeting. The optimal shared drawing tool has not been built. For its user interface, there remain issues such as multiple cursors or one, gestures or not, and using an ink or a character recognition model of pen input. For its substrate, is it better to have a single application with multiple windows, or many applications independently connected? Is packet-multicast a good substrate to use? What would it take to support shared drawing among 50 people; 5,000 people? The answers are likely to be both technological and social.

Three new kinds of applications of ubiquitous computing are beginning to be explored at Xerox PARC. One is to take advantage of true invisibility, literally hiding machines in the walls. An example is the Responsive Environment project led by Scott Elrod. This aims to make a building's heat, light, and power more responsive to individually customized needs, saving energy and making a more comfortable environment.

A second new approach is to use so-called "virtual communities" via the technology of MUDs. A MUD, or "Multi-User Dungeon," is a program that accepts network connections from multiple simultaneous users and provides access to a shared database of 'rooms', 'exits', and other objects. MUDs have existed for about 10 years, being used almost exclusively for recreational purposes. However, the simple technology of MUDs should also be useful in other, nonrecreational applications, providing a casual environment integrating virtual and real worlds [5].

A third new approach is the use of collaboration to specify information filtering. Described in the December 1992 issue of *Communications of the ACM*, this work by Doug Terry extends previous notions of information filters by permitting filters to reference other filters, or to depend on the values of multiple messages. For instance, one can select all messages that have been replied to by Smith (these messages do not even mention Smith, of course), or all messages that three other people found interesting. Implementing this required inventing the idea of a "continuous query," which can effectively sample a changing database at all points in time. Called "Tapestry," this system provides new ways for people to invisibly collaborate.

Privacy of Location

Cellular systems inherently need to know the location of devices and their use in order to properly route information. For instance, the traveling pattern of a frequent cellular phone user can be deduced from the roaming data of cellular service providers. This problem could be much worse in ubiquitous computing with its more extensive use of cellular wireless. So a key problem with ubiquitous computing is preserving privacy of location. One solution, a central database of location information, means the privacy controls can be centralized and perhaps done well—on the other hand one break-in there reveals all, and centrality is unlikely to scale worldwide. A second source of insecurity is the transmission of the location information to a central site. This site is the obvious place to try to snoop packets, or even to use traffic analysis on source addresses.

Our initial designs were all central, initially with unrestricted access, gradually moving toward individual users' controlling who can access information about them. Our preferred design avoids a central repository, instead storing information about each person at that person's PC or workstation. Programs that need to know a person's location must query the PC, and proceed through whatever security measures the user has chosen to install. EuroPARC uses a system of this sort.

Accumulating information about individuals over long periods is one of the more useful things to do, but quickly raises hackles. A key problem for location is how to provide occasional location information for clients who need it, while somehow preventing the reliable accumulation of long-term trends about an individual. So far at PARC we have experimented only with short-term accumulation of information to produce automatic daily diaries of activity [17].

It is important to realize there can never be a purely technological solution to privacy, and social issues must be considered in their own right. In the computer science lab we are trying to construct systems that are privacy-enabled, giving power to the individual. But only society can cause the right system to be used. To help prevent future oppressive employers or governments from taking this power away, we are also encouraging the wide dissemination of information about location systems and their potential for harm. We have cooperated with a number of articles in the *San Jose Mercury News*, the *Washington Post*, and the *New York Times* on this topic. The result, we hope, is technological enablement combined with an informed populace that cannot be tricked in the name of technology.

Computational Methods

An example of a new problem in theoretical computer science emerging from ubiquitous computing is optimal cache sharing. This problem originally arose in discussions of optimal disk cache design for portable computer architectures. Bandwidth to the portable machine may be quite low, while its processing power is relatively high. This introduces as a possible design point the compression of pages in a RAM cache, rather than writing them all the way back over a slow link. The question arises of the optimal strategy for partitioning memory between compressed and uncompressed pages.

This problem can be generalized as follows [1]:

The Cache Sharing Problem. A problem instance is given by a sequence of page requests. Pages are of two types, U and C (for uncompressed and compressed), and each page is either IN or OUT. A request is served by changing the requested page to IN if it is currently OUT. Initially all pages are OUT. The cost to change a type-U (type-C) page from OUT to IN is C_U (respectively, C_C). When a requested page is OUT, we say that the algorithm missed. Removing a page from memory is free.

Lower Bound Theorem: No deterministic, on-line algorithm for cache sharing can be c-competitive for

$c < MAX \quad (1 + C_U/(C_U + C_C), \quad 1 + C_C/(C_U + C_C))$

This lower bound for c ranges from 1.5 to 2, and no on-line algorithm can approach closer to the optimum than this factor. Bern et al. [1] also constructed an algorithm that achieves this factor, therefore providing an upper bound as well. They further propose a set of more general symbolic programming tools for solving competitive algorithms of this sort.

Concluding Remarks

As we start to put tabs, pads, and boards into use, the first phase of ubiquitous computing should enter its most productive period. With this substrate in place we can make much more progress both in evaluating our technologies and in choosing our next steps. A key part of this evaluation is using the analyses of psychologists, anthropologists, application writers, artists, marketers, and customers. We believe they will find some features work well; we know they will find some features do not work. Thus we will begin again the cycle of cross-disciplinary fertilization and learning. Ubicomp is likely to provide a framework for interesting and productive work for many more years or decades, but we have much to learn about the details.

Acknowledgments
This work was funded by Xerox PARC. Portions of this work were sponsored under contract #DABT 63-91-0027. Ubiquitous computing is only a small part of the work going on at PARC, and we are grateful for PARC's rich, cooperative, and fertile environment in support of the document company. ◼

References

1. Bern, M., Greene, D., Raghunathan. On-line algorithms for cache sharing. *25th ACM Symposium on Theory of Computing* (San Diego, Calif., 1993).
2. Buxton, W. Smoke and mirrors. *Byte 15*, 7 (July 1990) 205–210.
3. Chatterjee, S. Sun enters computer conferencing market. *Sunworld 5*, 10 (Oct. 1992), Integrated Media, San Francisco, 32–34.
4. Clark, D.D., Shenker, S., Zhang, L. Supporting real-time applications in an integrated services packet network:Architecture and mechanism. *SIGCOMM '92 Conference Proceedings*. Communications Architectures and Protocols. (Baltimore, Md, Aug. 17–20, 1992). *Computer Commun. Rev. 22*, 4 (Oct. 1992) New York, N.Y. pp. 14–26.
5. Curtis, P. MUDDING: Social phenomena in text-based virtual realities. *DIAC—Directions and Implications of Advanced Computing Symposium Proceedings*. Computer Professionals for Social Responsibility (Palo Alto, Calif. May, 1992).
6. Elrod, B., Gold, Goldberg, Halasz, Janssen, Lee, McCall, Pedersen, Pier, Tang and Welch. Liveboard: A large interactive display supporting group meetings, presentations and remote collaboration. *CHI '92 Conference Proceedings*, May 1992. ACM, New York, N.Y. pp. 599–607.
7. Fulton, J. and Kantarjiev, C. An update on low bandwidth X (LBX). In *Proceedings of the Seventh Annual X Technical Conference* (Boston, Mass Jan. 1993), To be published in The X Resource by O'Reilly and Associates.
8. Goldberg, D., Richardson, C. Touch typing with a stylus. To be published in *INTERCHI '93*.
9. Hill, R.D., Brinck, T., Patterson, J.F., Rohall, S.L. and Wilner, W.T. The RENDEZVOUS language and architecture: Tools for constructing multi-user interactive systems. *Commun. ACM, 36*, 1 (Jan. 1993). To be published.
10. Ioannidis, J., Maguire, G.Q., Jr. The design and implementation of a mobile internetworking architecture. *Usenix Conference Proceedings, Usenix '93* (Jan. 1993). To be published.
11. Karn, P. MACA—A new channel access method for packet radio. In *Proceedings of the ARRL 9th Computer Networking Conference* (London Ontario, Canada, Sept. 22, 1990). ISBN 0-87259-337-1.
12. Kay, A. Computers, networks, and education. *Sci. Am.* (Sept. 1991), 138–148.
13. Kurtenbach, G. and Buxton, W. The limits of expert performance using hierarchic marking menus. To be published in *INTERCHI '93*.
14. Lave, J. Situated Learning: Legitimate Peripheral Participation. Cambridge University Press, Cambridge, New York, N.Y. 1991.
15. Lyles, B.J., Swinehart, D.C. The emerging gigabit environment and the role of local ATM. *IEEE Commun.* (Apr. 1992), 52–57.
16. Lyon, R.F. Cost, power, and parallelism in speech signal processing. *Custom Integrated Circuits Conference, IEEE* (San Diego, May 9–12, 1993).
17. Newman, W.M., Eldridge, M.A., Lamming, M.G. PEPYS: Generating autobiographies by automatic tracking. EuroPARC Tech. Rep. Cambridge, England.
18. Pedersen, E., McCall, K., Moran, T.P., Halasz, F., Tivoli, F.G.: An electronic whiteboard for informal workgroup meetings. To be published in *INTERCHI'93*.
19. Rheingold, H. *Virtual Reality*. Summit Books. New York, N.Y., 1991.
20. Rush, C.M. How WARC '92 will affect mobile services. *IEEE Commun.* (Oct. 1992), 90–96.
21. Stefik, M., Foster, G., Bobrow, D.G., Kahn, K., Lanning, S. and Suchman, L. Beyond the chalkboard: Computer support for collaboration and problem solving in meetings. *Commun. ACM 30*, 1 (Jan. 1987), 32–47.
22. Suchman, L.A. Plans and situated actions: The problem of human-machine communication. Xerox PARC Tech. Rep. ISL-6, Feb. 1985.

23. Tang, J.C. and Minneman, S.L. VideoDraw: A video interface for collaborative drawing. *ACM Trans. Off. Inf. Syst. 9,* 2 (Apr. 1991), 170–184.

24. Teraoka, F., Tokote, Y., Tokoro, M. A network architecture providing host migration transparency. In *Proceedings of SIGCOMM'91* (Sept. 1991), pp. 209–220.

25. Tesler, L.G. Networked computing in the 1990's. *Sci. Am.* (Sept. 1991), 86–93.

26. Want, R., Hopper, A., Falcao, V. and Gibbons, J. The active badge location system. *ACM Trans Inf. Syst. 10,* 1 (Jan. 1992), 91–102.

27. Weiser, M. The computer for the twenty-first century. *Sci. Am.* (Sept. 1991), 94–104.

28. Weiser, M., Demers, A. and Hauser, C. The portable common runtime approach to interoperability. In *Proceedings of the ACM Symposium on Operating Systems Principles* (Dec. 1989).

CR Categories and Subject Descriptors: C.3 [Computer Systems Organization]: Special-purpose and application-based systems; H.1.2 [**Information Systems**]: Models and Principles—*User/Machine systems*; H.4.1 [**Information Systems**]: Information Systems Applications—*Office Automation*; J.0 [**Computer Applications**]: General

General Terms: Design, Human Factors

Additional Key Words and Phrases: Ubiquitous computing

About the Author:
MARK WEISER is principal scientist and manager of the Computer Science laboratory at Xerox PARC. Current research interests include new theories of automatic memory reclamation (garbage collection), visualization of operating system internals, ubiquitous computing and embodied virtuality. **Author's Present Address:** Xerox PARC, 3333 Coyote Hill Road, Palo Alto, CA 94304; email: weiser.parc@xerox.com

11.2 Fox, A., Gribble, S.D., Chawathe, Y., and Brewer, E.A., Adapting to Network and Client Variation Using Active Proxies: Lessons and Perspectives

Fox, A., Gribble, S. D., Chawathe, Y., and Brewer, E. A., "Adopting to Network and Client Variation Using Active Proxies: Lessons and Perspectives," *IEEE Personal Communications*, 5(4):10-19, August 1998.

Adapting to Network and Client Variation Using Infrastructural Proxies: Lessons and Perspectives

Armando Fox Steven D. Gribble Yatin Chawathe
Eric A. Brewer
{fox,gribble,yatin,brewer}@cs.berkeley.edu

Today's Internet clients vary widely with respect to both hardware and software properties: screen size, color depth, effective bandwidth, processing power, and the ability to handle different data formats. The order-of-magnitude span of this variation is too large to hide at the network level, making application-level techniques necessary. We show that on-the-fly adaptation by transformational proxies is a widely applicable, cost-effective, and flexible technique for addressing all these types of variation. To support this claim, we describe our experience with datatype-specific distillation (lossy compression) in a variety of applications. We also argue that placing adaptation machinery in the network infrastructure, rather than inserting it into end servers, enables incremental deployment and amortization of operating costs. To this end, we describe a programming model for large-scale interactive Internet services and a scalable cluster-based framework that has been in production use at UC Berkeley since April 1997. We present a detailed examination of TranSend, a scalable transformational Web proxy deployed on our cluster framework, and give descriptions of several handheld-device applications that demonstrate the wide applicability of the proxy-adaptation philosophy.

1 Infrastructural On-the-Fly Adaptation Services

1.1 Heterogeneity: Thin Clients and Slow Networks

The current Internet infrastructure includes an extensive range and number of clients and servers. Clients vary along many axes, including screen size, color depth, effective bandwidth, processing power, and ability to handle specific data encodings, e.g., GIF, PostScript, or MPEG. As shown in tables 1 and 2, each type of variation often spans orders of magnitude. High-volume devices such as smart phones [12] and smart two-way pagers will soon constitute an increasing fraction of Internet clients, making the variation even more pronounced.

These conditions make it difficult for servers to provide a level of service that is appropriate for every client. Application-level adaptation is required to provide a *meaningful* Internet experience across the range of client capabilities. Despite continuing improvements in client computing power and connectivity, we expect the high end to advance roughly in parallel with the low end, effectively maintaining a gap between the two and therefore the need for application-level adaptation.

Platform	SPEC92/ Memory	Screen Size	Bits/ pixel
High-end PC	200/64M	1280x1024	24
Midrange PC	160/32M	1024x768	16
Typ. Laptop	110/16M	800x600	8
Typical PDA	low/2M	320x200	2

Table 1: **Physical variation among clients**

Network	Bandwidth (bits/s)	Round-Trip Time
Local Ethernet	10-100 M	0.5 - 2.0 ms
ISDN	128 K	10-20 ms
Wireline Modem	14.4 - 56 K	350 ms
Cellular/CDPD	9.6 - 19.2 K	0.1 - 0.5 s

Table 2: **Typical Network Variation**

1.2 Approach: Infrastructural Proxy Services

We argue for a *proxy-based approach* to adaptation, in which proxy agents placed between clients and servers perform aggressive computation and storage on behalf of clients. The proxy approach stands in contrast to the *client-based approach*, which attempts to bring all clients up to a least-common-denominator level of functionality (e.g. text-only, HTML-subset compatibility for thin-client Web browsers), and the *server-based approach*, which attempts to insert adaptation machinery at each end server. We believe the proxy approach directly confers three advantages over the client and server approaches.

- **Leveraging the installed infrastructure through incremental deployment.** The enormous installed infrastructure, and its attendant base of existing content, is too valuable to waste; yet some clients cannot handle certain data types effectively. A compelling solution to the problem of client and network heterogeneity should allow interoperability with existing servers, thus enabling incremental deployment while evolving content formats and protocols are tuned and standardized for different target platforms. A proxy-based approach lends itself naturally to transparent incremental deployment, since an application-level proxy appears as a server to existing clients and as a client to existing servers.
- **Rapid prototyping during turbulent standardization cycles.** Software development on "Internet time" does not allow for long deployment cycles. Proxy-based adaptation provides a smooth path for rapid prototyping of new services, formats, and protocols, which can be deployed to servers (or clients) later if the prototypes succeed.
- **Economy of scale.** Basic queueing theory shows that a large central (virtual) server is more efficient in both cost and utilization (though less predictable in per-request performance) than a collection of smaller servers; standalone desktop systems represent the degenerate case of one "server" per user. This supports the argument for Network Computers [28] and suggests that collocating proxy services with infrastructural elements such as Internet points-of-presence (POPs) is one way to achieve effective economies of scale.

Large-scale network services remain difficult to deploy because of three fundamental challenges: scalability, availability and cost effectiveness. By scalability, we mean that when the load offered to the service increases, an incremental and linear increase in hardware can maintain the same per-user level of service. By availability, we mean that the service as a whole must be available 24×7, despite transient partial hardware or software failures. By cost effectiveness, we mean that the service must be economical to administer and expand, even though it potentially comprises many workstation nodes operating as a centralized cluster or "server farm". In section 3 we describe how we have addressed these challenges in our cluster-based proxy application server architecture.

1.3 Contributions and Map of Paper

In section 2 we describe our measurements and experience with *datatype-specific distillation and refinement*, a mechanism that has been central to our proxy-based approach to network and client adaptation. In section 3 we introduce a generalized "building block" programming model for designing and implementing adaptive applications, describe our implemented cluster-based application server that instantiates the model, and present detailed measurements of a particular production application: TranSend, a transformational Web proxy service. In section 4 we present case studies of other services we have built using our programming model, some of which are in daily use by thousands of users, including the Top Gun Wingman graphical Web browser for the 3Com PalmPilot handheld device. We discuss related work in section 5, and attempt to draw some lessons from our experience and guidelines for future research in section 6.

2 Adaptation via Datatype Specific Distillation

We propose three design principles that we believe are fundamental for addressing client variation most effectively.

1. **Adapt to client variation via datatype-specific lossy compression.**

   Datatype-specific lossy compression mechanisms can achieve much better compression than "generic" compressors, because they can make intelligent decisions about what information to throw away based on the semantic type of the data. For example, lossy compression of an image requires discarding color information, high-frequency components, or pixel resolution. Lossy compression of video can additionally include frame rate reduction. Less obviously, lossy compression of formatted text requires discarding some formatting information but preserving the actual prose. In all cases, the goal is to preserve information that has the highest semantic value. We refer to this process generically as *distillation.* A distilled object allows the user to decide whether it is worth asking for a *refinement*: for instance, zooming in on a section of a graphic or video frame, or rendering a particular page containing PostScript text and figures without having to render the preceding pages.

2. **Perform adaptation on the fly.** To reap the maximum benefit from distillation and refinement, a distilled representation must target specific attributes of the client. The measurements reported in section 2.1 show that for typical images and rich-text, distillation time is small in practice, and end-to-end latency is reduced because of the much smaller number of bytes transmitted over low-bandwidth links. *On-demand distillation* provides an easy path for incorporating support for new clients, and also allows distillation aggressiveness to track (e.g.) significant changes in network bandwidth, as might occur in vertical handoffs between different wireless networks [39]. We have successfully implemented useful distillation "workers" that serve clients spanning an order of magnitude in each area of variation, and we have generalized our approach into a common framework, which we discuss in section 3.

3. **Move complexity away from both clients and servers.** Application partitioning arguments have long been used to keep clients simple [40]. However, adaptation through a shared infrastructural proxy enables incremental deployment and legacy client support, as we argued in section 1.2. Therefore, on-demand distillation and refinement should be done at an intermediate proxy that has access to substantial computing resources and is well-connected to the rest of the Internet.

Table 3 lists the "axes" of compression corresponding to three important datatypes: formatted text, images, and video streams. We have found that order-of-magnitude size reductions are often possible without destroying the semantic content of an object (e.g. without rendering an image unrecognizable to the user).

Semantic Type	Specific encodings	Distillation axes
Image	GIF, JPEG, PPM, Postscript	Resolution, color depth, color palette
Text	Plain, HTML, Postscript, PDF	Richness (heavily formatted vs. simple markup vs. plaintext)
Video	NV, H.261, VQ, MPEG	Resolution, frame rate, color depth, progression limit (for progressive encodings)

Table 3: **Three important types and the distillation axes corresponding to each.**

2.1 Performance of Distillation and Refinement On Demand

We now describe and evaluate datatype-specific distillers for images and rich-text.[1] The goal of this section is to support our claim that in the majority of cases, *end-to-end latency is reduced by distillation,* that is, the time to produce a useful distilled object on today's workstation hardware is small enough to be more than compensated by the savings in transmission time for the distilled object relative to the original.

2.1.1 Images

We have implemented an image distiller called *gifmunch*, which implements distillation and refinement for GIF [25] images, and consists largely of source code from the NetPBM Toolkit [35]. Figure 1 shows the result of running gifmunch on a large color GIF image of the Berkeley Computer Science Division's home building, Soda Hall. The image of Figure 1a measures 320×200 pixels—about 1/8 the total area of the original 880×610—and uses 16 grays, making it suitable for display on a typical handheld device.

Due to the degradation of quality, the writing on the building is unreadable, but the user can request a refinement of the subregion containing the writing, which can then be viewed at full resolution.

Image distillation can be used to address all three areas of client variation:

- **Network variation**: The graphs in figure 2

[1]A distiller for real-time network video streams is described separately, in [1].

Left (a) is a distilled image of Soda Hall, and above (b) illustrates refinement. (a) occupies 17 KB at 320x200 pixels in 16 grays, compared with the 492 KB, 880x600 pixel, 249 color original (not shown). The refinement (b) occupies 12 KB. Distillation took 6 seconds on a SPARCstation 20/71, and refinement took less than a second.

Figure 1: Distillation example

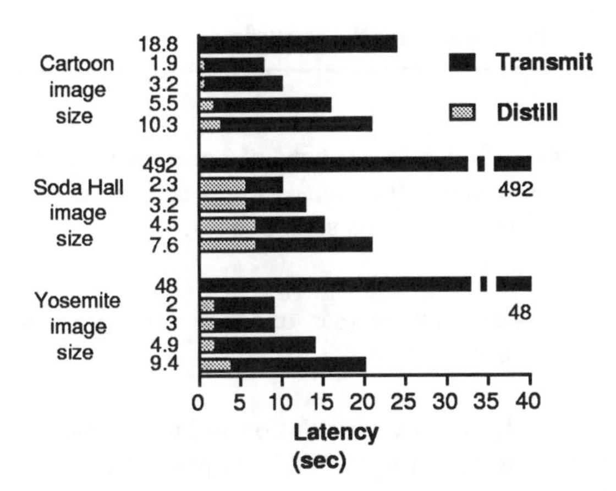

Figure 2: End-to-end latency for images with and without distillation. Each group of bars represents one image with 5 levels of distillation; the top bar represents no distillation at all. The y-axis number is the distilled size in kilobytes (so the top bar gives the original size). Note that two of the undistilled images are off the scale; the Soda Hall image is off by an order of magnitude.

depict end-to-end client latency for retrieving the original and each of four distilled versions of a selection of GIF images: the top set of bars is for a cartoon found on a popular Web page, the middle set corresponds to a large photographic image, and the bottom to a computer rendered image. The images were fetched using a 14.4Kb/s modem with standard compression (V.42bis and MNP-5) through the UC Berkeley PPP gateway, via a process that runs each image through gifmunch. [2] Each bar is segmented to show the distillation latency and transmission latency separately. Clearly, even though distillation adds latency at the proxy, it can result in greatly reduced end-to-end latency. This shows that on the fly distillation is not prohibitively expensive.

- **Hardware variation**: A "map to 16 grays" operation would be appropriate for PDA-class clients with shallow grayscale displays. We can identify this operation as an effective lossy compression technique precisely because we know we are operating on an image, regardless of the particular encoding, and the compression achieved is significantly better than the 2-4 times compression typically achieved by "generic" lossless compression (design principle #1).
- **Software variation**: Handheld devices such as the 3Com PalmPilot frequently have built-in support for proprietary image encodings only. The ability to convert to this format saves code space and decoding latency on the client (design principle #3).

2.1.2 Rich-Text

We have also implemented a rich-text distiller that performs lossy compression of PostScript-encoded

[2]The network and distillation latencies reflect significant overhead due to the naive implementation of *gifmunch* and the latency and slow-start effects of the PPP gateway, respectively. In section 3 we discuss how to overcome some of these problems, but it is worth noting that end-to-end latency is still substantially reduced even in this naive prototype implementation.

Feature	HTML	Rich Text	Post-Script
Different fonts	Y	Y	Y
Bold and Italics	Y	Y	Y
Preserves Font Size	head-ings	Y	Y
Preserves Paragraphs	Y	Y	Y
Preserves Layout	N	Y	Y
Handles Equations	N	some	Y
Preserves Tables	Y	Y	Y
Preserves Graphs	N	N	Y

Table 4: Features for postscript distillation

text using uses a third party postscript-to-text converter [34]. The distiller replaces PostScript formatting information with HTML markup tags or with a custom rich-text format that preserves the position information of the words. PostScript is an excellent target for a distiller because of its complexity and verbosity: both transmission over the network and rendering on the client are resource intensive. Table 4 compares the features available in each format. Figure 3 shows the advantage of rich-text over PostScript for screen viewing. As with image distillation, PostScript distillation yields advantages in all three categories of client variation:

- **Network variation**: Again, distillation reduces the required bandwidth and thus the end-to-end latency. We achieved an average size reduction of a factor of 5 when going from compressed PostScript to gzipped HTML. Second, the pages of a PostScript document are pipelined through the distiller, so that the second page is distilled while the user views the first page. In practice, users only experience the latency of the first page, so the difference in perceived latency is about a factor of 8 for a 28.8K modem. Distillation typically took about 5 seconds for the first page and about 2 seconds for subsequent pages.

- **Hardware variation:** Distillation reduces decoding time by delivering data in an easy-to-parse format, and results in better looking documents on clients with lower quality displays.

- **Software variation:** PostScript distillation allows clients that do not directly support PostScript, such as handhelds, to view these documents in HTML or our rich-text format. The rich-text viewer could be an external viewer similar to *ghostscript*, an applet for a Java-capable browser, or a browser plug-in rendering module.

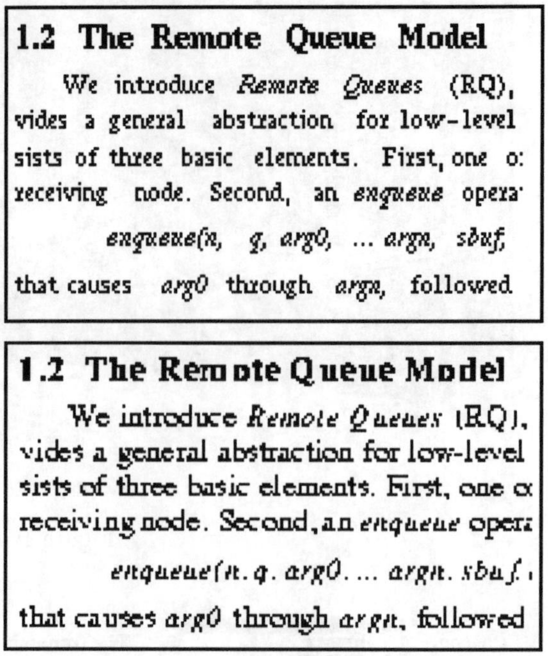

1.2 The Remote Queue Model

We introduce *Remote Queues* (RQ),
vides a general abstraction for low-level
sists of three basic elements. First, one o:
receiving node. Second, an *enqueue* opera·
enqueue(n, q, arg0, ... argn, sbuf,
that causes *arg0* through *argn*, followed

1.2 The Remote Queue Model

We introduce *Remote Queues* (RQ),
vides a general abstraction for low-level
sists of three basic elements. First, one o
receiving node. Second, an *enqueue* opera
enqueue(n, q, arg0, ... argn, sbuf,
that causes *arg0* through *argn*, followed

Figure 3: Screen snapshots of our rich-text (top) versus ghostview (bottom). The rich-text is easier to read because it uses screen fonts.

Overall, rich-text distillation reduces end-to-end latency, results in more readable presentation, and adds new abilities to low-end clients, such as PostScript viewing. The latency for the appearance of the first page was reduced on an average by a factor of 8 using the proxy and PostScript distiller. Both HTML and our rich-text format are significantly easier to read on screen than rendered PostScript, although they sacrifice some layout and graphics accuracy compared to the original PostScript.

2.2 Summary

High client variability is an area of increasing concern that existing servers do not handle well. We have proposed three design principles we believe are fundamental to addressing variation:

- Datatype-specific distillation and refinement achieve better compression than does lossless compression, while retaining useful semantic content and allowing network resources to be managed at the application level.

- When the proxy-to-client bandwidth is substantially smaller than the proxy-to-server bandwidth (as is the case, e.g., in wireless networks or with consumer wireline modems), on-demand distillation and refinement reduce end-to-end latency perceived by the client (sometimes by almost an order of magnitude), are more flexible than reliance on precomputed static representations, and give low-end clients new abilities such as PostScript viewing.
- Performing distillation and refinement in the network infrastructure rather than at the endpoints separates technical as well as economic concerns of clients and servers.

3 Scalable Internet Application Servers

In order to accommodate compute-intensive adaptation techniques by putting resources in the network infrastructure, we must address two important challenges:

1. Infrastructural resources are typically shared, and the sizes of user communities sharing resources such as Internet Points of Presence (POP's) is increasing exponentially. A shared infrastructural service must therefore *scale* gracefully to serve large numbers of users.
2. Infrastructural resources such as the IP routing infrastructure are expected to be reliable, with availability approaching 24 × 7 operation. If we place application-level computing resources such as distillation engines into this infrastructure, we should be prepared to meet comparable expectations.

In this section, we focus on the problem of deploying adaptation-based proxy services to large communities (tens of thousands of users, representative of the subscription size of a medium-sized Internet Service Provider). In particular, we discuss a cluster-friendly programming model for building interactive and adaptive Internet services, and measurements of our implemented prototype of a scalable, cluster-based server that instantiates the model. Our framework reflects the implementation of three real services in use today: TranSend, a scalable transformation proxy for the 25,000 UC Berkeley dialup users (connecting through a bank of 600 modems); Top Gun Wingman, the only graphical Web browser for the 3Com PalmPilot handheld device (commercialized by ProxiNet); and the Inktomi search engine (commercialized as HotBot), which performs over 10 million queries per day against a database of over 100 million web pages. Although HotBot does not demonstrate client adaptation, we use it to validate particular design decisions in the implementation of our server platform, since it pioneered many of the cluster-based scalability techniques generalized in our scalable server prototype.

We focus our detailed discussion and measurements on TranSend, a transformational proxy service that performs on-the-fly lossy image compression. TranSend applies the ideas explored in the preceding section to the World Wide Web.

3.1 TACC: A Programming Model for Internet Services

We focus on a particular subset of Internet services, based on *transformation* (distillation, filtering, format conversion, etc.), *aggregation* (collecting and collating data from various sources, as search engines do), *caching* (both original and transformed content), and *customization* (maintenance of a per-user preferences database that allows workers to tailor their output to the user's needs or device characteristics).

We refer to this model as TACC, from the initials of the four elements above. In the TACC model, applications are built from building blocks interconnected with simple API's. Each building block, or *worker*, specializes in a particular task, for example, scaling/dithering of images in a particular format, conversion between specific data formats, extracting "landmark" information from specific Web pages, etc. Complete applications are built by *composing* workers; roughly speaking, one worker can *chain* to another (similar to processes in a Unix pipeline), or a worker can *call* another as a subroutine or coroutine. This model of composition results in a very general programming model that subsumes transformation proxies [22], proxy filters [43], customized information aggregators, and search engines.

A *TACC server* is a platform that instantiates TACC workers, provides dispatch rules for routing network data traffic to and from them, and provides support for the inter-worker calling and chaining API's. Similar to a Unix shell, a TACC server provides the mechanisms that insulate workers from having to deal directly with low-level concerns such as data routing and exception handling, and gives workers a clean set of API's for communicating with each other, the caches, and the customization database (described below). We describe our prototype implementation of a scalable, commodity-PC cluster-based TACC server later in this section.

3.2 Cluster-Based TACC Server Architecture

We observe that clusters of workstations have some fundamental properties that can be exploited to meet the requirements of a large-scale network services (scalability, high availability, and cost effectiveness). Using commodity PCs as the unit of scaling allows the service to ride the leading edge of the cost/performance curve; the inherent redundancy of clusters can be used to mask transient failures; and "embarrassingly parallel" network service workloads map well onto networks of commodity workstations.

However, developing cluster software and administering a running cluster remain complex. A primary contribution of our work is the design, analysis, and implementation of a layered framework for building adaptive network services that addresses this complexity while realizing the sought-after economies of scale. New services can use this framework as an off-the-shelf solution to scalability, availability, and several other problems, and focus instead on the content of the service being developed.

We now describe our proposed system architecture and service-programming model for building scalable TACC servers using clusters of PC's. The architecture attempts to address the challenges of cluster computing (unwieldy administration, managing partial failures, and the lack of shared state across components) while exploiting the strengths of cluster computing (support for incremental scalability, high availability through redundancy, and the ability to use commodity building blocks). A more detailed discussion of the architecture can be found in [23].

The goal of our architecture is to separate the *content* of network services (i.e., what the services do) from their implementation, by encapsulating the "scalable network service" (SNS) requirements of high availability, scalability, and fault tolerance in a reusable layer with narrow interfaces. Application writers program to the TACC APIs alluded to in the previous section, without regard to the underlying TACC server implementation; the resulting TACC applications automatically receive the benefits of linear scaling, high availability, and failure management when run on our cluster-based TACC server.

The software-component block diagram of a scalable TACC server is shown in figure 4. Each physical workstation in a network of workstations (NOW [2]) supports one or more software components in the figure, but each component in the diagram is confined to one node. In general, the components whose tasks are naturally parallelizable are replicated for scalability, fault tolerance, or both.

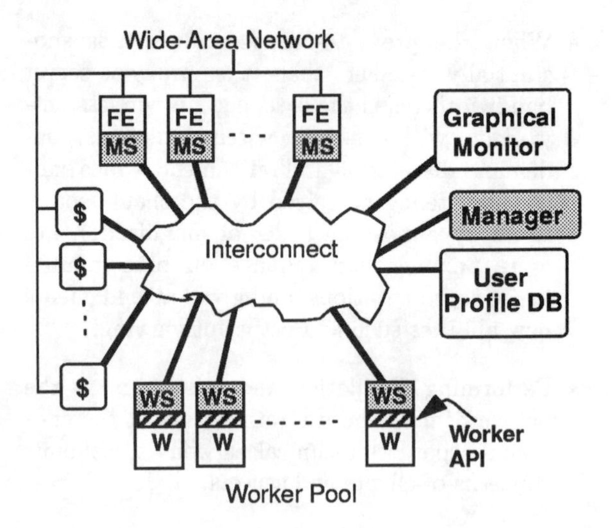

Figure 4: **Architecture of a cluster-based TACC server:** Components include front ends (FE), a pool of TACC workers (W) some of which may be caches ($), a user profile database, a graphical monitor, and a fault-tolerant load manager, whose functionality logically extends into the manager stubs (MS) and worker stubs (WS).

Front Ends provide the interface to the TACC server as seen by the outside world (e.g., HTTP server). They "shepherd" incoming requests by matching them up with the appropriate user profile from the customization database, and queueing them for service by one or more workers. Front ends maximize system throughput by maintaining state for many simultaneous outstanding requests, and can be replicated for both scalability and availability.

The **Worker Pool** consists of caches (currently Harvest [9]) and service-specific modules that implement the actual service (data transformation/filtering, content aggregation, etc.) Each type of module may be instantiated zero or more times, depending on offered load. The TACC API allows all cache workers to be managed as a single virtual cache by providing URL hashing, automatic failover, and dynamic growth of the cache pool.

The **Customization Database** stores user profiles that allow mass customization of request processing. The Manager balances load across workers and spawns additional workers as offered load fluctuates or faults occur. When necessary, it may assign work to machines in the overflow pool, a set of backup machines (perhaps on desktops) that can be harnessed to handle load bursts and provide a smooth transition during incremental growth.

The **Load Balancing/Fault Tolerance manager** keeps track of what workers are running where, autostarts new workers as needed, and balances load across workers. Its detailed operation is described in section 3.3, in the context of the TranSend implementation. Although it is a centralized agent, [23] describes the various mechanisms, including multicast heartbeat and process-peer fault tolerance, that keep this and other system components running and allow the system to survive transient component failures.

The **Graphical Monitor** for system management supports asynchronous error notification via email or pager, temporary disabling of system components for hot upgrades, and visualization of the system's behavior using Tcl/Tk [33]. The benefits of visualization are not just cosmetic: We can immediately detect by looking at the visualization panel what state the system as a whole is in, whether any component is currently causing a bottleneck (such as cache-miss time, distillation queueing delay, interconnect), what resources the system is using, and similar behaviors of interest.

The **Interconnect** provides a low-latency, high-bandwidth, scalable interconnect, such as switched 100-Mb/s Ethernet or Myrinet [32]. Its main goal is to prevent the interconnect from becoming the bottleneck as the system scales.

Components in our TACC server architecture may be replicated for fault tolerance or high availability, but we also use replication to achieve scalability. When the offered load to the system saturates the capacity of some component class, more instances of that component can be launched on incrementally added nodes. The duties of our replicated components are largely independent of each other (because of the nature of the Internet services' workload), which means the amount of additional resources required is a linear function of the increase in offered load.

3.3 Analysis of the TranSend Implementation

TranSend [22], a TACC reimplementation of our earlier prototype called Pythia [19], performs lossy Web image compression on the fly. Each TranSend worker handles compression or markup for a specific MIME type; objects of unsupported types are passed through to the user unaltered.[3]

We took measurements of TranSend using a cluster of 15 Sun SPARC Ultra-1 workstations connected by 100 Mb/s switched Ethernet and isolated from external load or network traffic. For measurements requiring Internet access, the access was via a 10 Mb/s switched Ethernet network connecting our workstation to the outside world. Many of the performance tests are based upon HTTP trace data from the 25,000 UC Berkeley dialup IP users [27], played back using a high-performance playback engine of our own design that can either generate requests at a constant rate or faithfully play back a trace according to the timestamps in the trace file.

In the following subsections we report on experiments that stress TranSend's fault tolerance, responsiveness, and scalability.

3.3.1 Self Tuning and Load Balancing

As mentioned previously, the load balancing and fault tolerance manager is charged with spawning and reaping workers and distributing internal load across them. The mechanisms by which this is accomplished, which include monitoring worker queue lengths and applying some simple hysteresis, are described in [23].

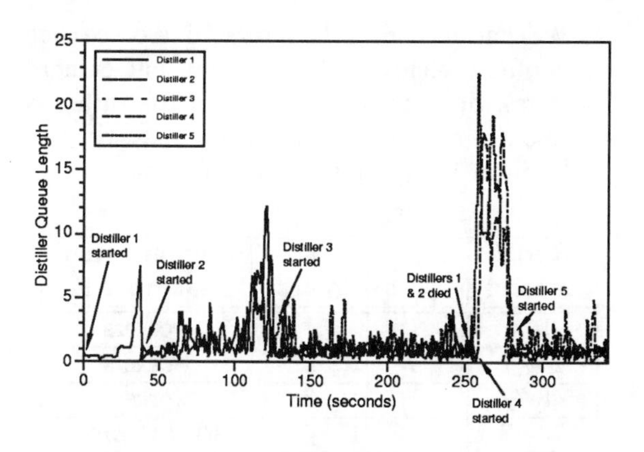

Figure 5: Worker queue lengths observed over time as the load presented to the system fluctuates, and as workers are manually brought down.

Figure 5 shows the variation in worker queue lengths over time. The system was bootstrapped with one front end and the manager, and a single demand-spawned worker. Continuously increasing

[3]The PostScript-to-richtext worker described in section 2 has not yet been added to TranSend.

the load caused the manager to spawn a second and later a third worker. We then manually killed the first two workers; the sudden load increase on the remaining worker caused the manager to spawn one and later another new worker, to stabilize the queue lengths.

3.3.2 Scalability

To demonstrate the scalability of the system, we performed the following experiment:

1. We began with a minimal instance of the system: one front end, one worker, the manager, and a fixed number of cache partitions. (Since for these experiments we repeatedly requested the same subset of images, the cache was effectively not tested.)
2. We increased the offered load until some system component saturated (e.g., worker queues growing too long, front ends no longer accepting additional connections, etc.).
3. We then added more resources to the system to eliminate this saturation (in many cases the system does this automatically, as when it recruits overflow nodes to run more workers), and we recorded the amount of resources added as a function of the increase in offered load, measured in requests per second.
4. We continued until the saturated resource could not be replenished (i.e., we ran out of hardware), or until adding more of the saturated resource no longer resulted in a linear or close-to-linear improvement in performance.

Req./ Second	# FE's	# Wkrs.	Element that saturated
0-24	1	1	workers
25-47	1	2	workers
48-72	1	3	workers
73-87	1	4	FE Ethernet
88-91	2	4	workers
92-112	2	5	workers
113-135	2	6	workers + FE Ethernet
136-159	3	7	workers

Table 5: Results of the scalability experiment. "FE" refers to front end.

Table 5 presents the results of this experiment. At 24 requests per second, as the offered load exceeded the capacity of the single available worker, the manager automatically spawned one additional worker, and then subsequent workers as necessary. (In addition to using faster hardware, the performance-engineering of the cluster-based server has caused a large reduction in the amortized cost of distillation for a typical image, compared to the values suggested by Figure 2.) At 87 requests per second, the Ethernet segment leading into the front end saturated, requiring a new front end to be spawned. We were unable to test the system at rates higher than 159 requests per second, as all of our cluster's machines were hosting workers, front ends, or playback engines. We did observe nearly perfectly linear growth of the system over the scaled range: a worker can handle approximately 23 requests per second, and a 100 Mb/s Ethernet segment into a front-end can handle approximately 70 requests per second. We were unable to saturate the front end, the cache partitions, or fully saturate the interior interconnect during this experiment. We draw two conclusions from this result:

- Even with a commodity 100 Mb/s interconnect, linear scaling is limited primarily by bandwidth into the system rather than bandwidth inside the system.
- Although we originally deployed TranSend on four SPARC 10's, a single Ultra-1 class machine would suffice to serve the entire dialup IP population of UC Berkeley (25,000 users officially, over 8000 of whom surfed during the trace).

4 Other TACC Applications

We now discuss several examples of new services in various stages of deployment, showing how each exploits the TACC model and discussing some of our experiences with the applications. Rather than providing detailed measurements as we did for TranSend in the previous section, the present goal is to demonstrate the flexibility of the TACC framework in accommodating an interesting range of applications, while providing consistent guidelines for approaching application partitioning decisions.

We restrict our discussion here to services that can be implemented using the proxy model, i.e., transparent interposition of computation between clients and servers. (Some of our services do not communicate via HTTP but are conceptually similar.) Also, although we have developed a wider range of applications using the TACC model as part of a graduate seminar [18], we concentrate on those applications that enable adaptation to network and client variation. These services share the following common characteristics, which make them amenable to implementation on our cluster-based framework:

- Compute-intensive transformation or aggregation
- Computation is parallelizable with granularity of a few CPU seconds
- Substantial value added by mass customization

4.1 TranSend as a TACC Application

TranSend is one of the simplest TACC applications we have produced. The dispatch rules simply match the MIME type of the object returned from the origin server to the list of known workers, which (as in all TACC applications) can be updated dynamically. In particular, TranSend does not exploit TACC's ability to compose workers by chaining them into a "pipeline" or having one worker call others as coroutines.

Transformed objects are stored in the cache with "fat URL's" that encode a subset of the transformation parameters, saving the work of re-transforming an original should another user ask for the same degraded version later. Each user can select a desired level of aggressiveness for the lossy compression and choose between HTML and Java-based interfaces for modifying their preferences.

The main difference between TranSend and commercial products based on its ideas (such as Intel's recently-announced QuickWeb [11]) is extensibility: adding support for new datatypes to TranSend is as simple as adding a new worker, and composing workers is as simple as modifying the dispatch rules (or modifying existing workers to hint to the TACC server that they should fall through to new workers). In fact, we have generalized TranSend into a "lazy fixations" system [20] in which users could select from among a variety of available formats for viewing an object; this was implemented by a "graph search" worker that treated all the transformation workers as edges in a directed graph and performed a shortest-paths search to determine what sequence of workers should be run to satisfy a particular request.

One of the goals of TACC is to exploit modularity and composition to make new services easy to prototype by reusing existing building blocks. TranSend's HTML and JPEG workers consist almost entirely of off-the-shelf code, and each took an afternoon to write. A pair of anecdotes illustrates the flexibility of the TACC API's in constructing responsive services. Our original HTML parser was a fast C-language implementation from the W3C. Debugging the pathological cases for this parser was spread out over a period of days—since our prototype TACC server masks transient faults by bypassing original content "around" the faulting worker, we could only deduce the existence of bugs by noticing (on the Graphical Monitor display) that the HTML worker had been restarted several times over a period of hours, although the service as a whole was continuously available.

We later wrote a much slower but more robust parser in Perl to handle proprietary HTML extensions such as inline JavaScript. All HTML pages are initially passed to the slower Perl parser, but if it believes (based on page length and tag density) that processing the page will introduce a delay longer than one or two seconds, it immediately throws an exception and indicates that the C parser should take over. Because the majority of long pages tend to be papers published in HTML rather than complex pages with weird tags, this scheme exploits TACC composition and dispatch to handle common cases well while keeping HTML processing latency barely noticeable.

4.2 Top Gun Wingman

Top Gun Wingman is the only graphical Web browser available for the 3Com PalmPilot, a typical "thin client" device. Based on file downloads, we estimate that 8000 to 10,000 users are using the client software and UC Berkeley's experimental cluster; ProxiNet, Inc. has since commercialized the program and deployed a production cluster to serve additional users. Figure 6 shows a screenshot of the browser.

Previous attempts to provide graphical Web browsing on such small devices have foundered on the severe limitations imposed by small screens, limited computing capability, and austere programming environments, and virtually all have fallen back to simple text-only browsing. Our adaptation approach, combined with the composable-workers model provided by TACC, allows us to approach this problem from a different perspective. The core of Top Gun Wingman consists of three TACC workers: HTML layout, image conversion, and intermediate-form layout to device-specific data format conversion. These three workers address the three areas of variation introduced in section 2:

- Hardware and software adaptation: We have

Figure 6: Screenshot of the Top Gun Wingman browser. This screenshot is taken from the "xcopilot" hardware-level Pilot simulator [15].

built TACC workers that output simplified binary markup and scaled-down images ready to be "spoon fed" to a thin-client device, given knowledge of the client's screen dimensions, image format, and font metrics. This greatly simplifies client-side code since no HTML parsing, layout, or image processing is necessary, and as a side benefit, the smaller and more efficient data representation reduces transmission time to the client. The image worker delivers 2-bit-per-pixel images, since that is what the PalmPilot hardware supports, and the HTML parsing and layout worker ensures that no page description larger than about 32KB is delivered to the client, since that is the approximate heap space limit imposed by the PalmPilot's programming environment. We have also added three "software upgrades" at the proxy since Wingman was first deployed: a worker that delivers data in AportisDoc [4] format (a popular PalmPilot e-book format), a worker that extracts and displays the contents of software archives for download directly to the PalmPilot, and an improved image-processing module contributed by a senior graphics hacker. In terms of code footprint, Wingman weighs in at 40KB of code (compared with 74KB and 109KB for HandWeb and Palmscape 5.0 respectively, neither of which currently support image viewing).

- Network protocol adaptation: In addition to delivering data in a more compact format and exploiting datatype-specific distillation, we have replaced HTTP with a simpler, datagram-oriented protocol based on Application Level Framing [10]. The combined effect of these optimizations makes Wingman two to four times faster than a desktop browser loading the same Web pages over the same bandwidth, and Wingman's performance on text-only pages often exceeds that of HTML/HTTP compliant browsers on the same platform, especially on slow ($<$ 56 Kb/s) links.

4.3 Top Gun Mediaboard

TopGun Mediaboard is an electronic shared whiteboard application for the PalmPilot. This is a derivation of the the desktop *mediaboard* application, which uses SRM (Scalable Reliable Multicast) as the underlying communication protocol. A reliable multicast proxy (RMX) TACC worker participates in the SRM session on behalf of the PDA clients, performing four main types of client adaptation:

- Transport protocol conversion: The PalmPilot's network stack does not support IP multicast. The RMX converts the SRM data into a unicast TCP stream that the client can handle.

- Application protocol adaptation: To keep the client implementation simple, all the complexities of the mediaboard command protocol are handled by the RMX. The protocol adapter transforms the entire sequence of mediaboard commands into a "pseudo-canvas" by executing each command and storing its result in the canvas, transmitting only a sequence of simple draw-ops to the client. The protocol and data format for transmitting the draw-ops is a direct extension of the Top Gun Wingman datagram protocol.

- On-demand distillation: The RMX converts specific data objects according to the client's needs. For example, it transforms the GIF and JPEG images that may be placed on the mediaboard into simpler image representations that the PalmPilot can understand, using the same worker that is part of Wingman. The client application can refine (zoom in on) specific portions of the canvas.

- Intelligent Rate Limiting: Since the proxy has complete knowledge of the client's state, the

RMX can perform intelligent forwarding of data from the mediaboard session to the client. By eliminating redundant draw-ops (for example, *create* followed by *delete* on the same object) before sending data to the client, the RMX reduces the number of bytes that must be sent over the low-bandwidth link. Moreover, although a whiteboard session can consist of a number of distinct pages, the RMX forwards only the data associated with the page currently being viewed on the client.

Top Gun Mediaboard is in prealpha use at UC Berkeley, and performs satisfactorily even over slow links such as the Metricom Ricochet wireless packet radio modem [31].

4.4 Charon: Indirect Authentication for Thin Clients

Although not yet rewritten as a TACC application, Charon [21] illustrates a similar use of adaptation by proxy, for performing indirect authentication. In particular, Charon mediates between thin clients and a Kerberos [38] infrastructure. Charon is necessary because, as we describe in [21], the computing resources required for a direct port of Kerberos to thin clients are forbidding. With Charon, Kerberos can be used both to authenticate clients to the proxy service, and to authenticate the proxied clients to Kerberized servers. Charon relieves the client of a significant amount of Kerberos protocol processing, while limiting the amount of trust that must be placed in the proxy; in particular, if the proxy is compromised, existing user sessions may be hijacked but no new sessions can be initiated, since new sessions require cooperation between the client and proxy. Our Charon prototype client for the Sony MagicLink [14], a once-popular PDA, had a client footprint of only 45KB, including stack and heap usage.

5 Related Work

At the network level, various approaches have been used to shield clients from the effects of poor (especially wireless) networks [5]. At the application level, data transformation by proxy interposition has become particularly popular for HTTP, whose proxy mechanism was originally intended for users behind security firewalls. The mechanism has been harnessed for anonymization [8], Kanji transcoding [36, 41], application-specific stream transformation [7], and personalized "associates" for Web browsing [6, 37]. Some projects provide an integrated solution with both network-level and application-level mechanisms [30, 17, 43], though none propose a uniform application-development model analogous to TACC.

Rover [16], Coda [29], and Wit [40] differ in their respective approaches to partitioning applications between a thin or poorly-connected client and more powerful server. In particular, Rover and Coda provide explicit support for disconnected operation, unlike our TACC work. We find that Rover's application-specific, toolkit-based approach is a particularly good complement to our own; although the TACC model provides a reasonable set of guidelines for thinking about partitioning (leave the client to do what it does well, and move as much as possible of the remaining functionality to the back end), we are working on integrating Rover into TACC to provide a rich abstraction for dealing with disconnection in TACC applications.

SmartClients [42] and SWEB++ [3] have exploited the extensibility of client browsers via Java and JavaScript to enhance scalability of network-based services by dividing labor between the client and server. We note that our system does not preclude, and in fact benefits from, exploiting intelligence and computational resources at the client; we discuss various approaches we have tried in [24].

6 Lessons and Conclusions

We proposed three design principles for adapting to network and client variation and delivering a meaningful Internet experience to impoverished clients: datatype-specific distillation and refinement, adaptation on demand, and moving complexity into the infrastructure. We also offered a high-level description of the TACC programming model (transformation, aggregation, caching, customization) that we have evolved for building adaptive applications, and presented measurements of our scalable, highly-available, cluster-based TACC server architecture, focusing on the TranSend web accelerator application. Finally, we described other applications we have built that are in daily use, including some that push the limits of client adaptation (such as Top Gun Wingman and Top Gun Mediaboard). In this section we try to draw some lessons from what we have learned from building these and similar applications and experimenting with our framework.

Aggressively pushing the adaptation-by-proxy

model to its limits, as we have tried to do with Top Gun Wingman and Top Gun Mediaboard, has helped us validate the proxy-interposition approach for serving thin clients. Our variation on the theme of application partitioning has been to split the application between the client and the proxy, rather than between the client and the server. This has allowed our clients to access existing content with no server modifications. Our guideline for partitioning applications has been to allow the client to perform those tasks it does well in native code, and move as much as possible of the remaining work to the proxy. For example, since most thin clients support some form of toolkit for building graphical interfaces, sending HTML markup is too cumbersome for the client, but sending screen-sized bitmaps is unnecessarily cumbersome for the proxy.

A frequent objection raised against our partitioning approach is that it requires that the proxy service be available at all times, which is more difficult than simply maintaining the reliability of a bank of modems and routers. This observation motivated our work on the cluster-based scalable and highly-available server platform described in section 3, and in fact the TranSend and Wingman proxy services have been running for several months at UC Berkeley with high stability, except for a two-week period in February 1998 when the cluster was affected by an OS upgrade. Other than one part-time undergraduate assistant, the cluster manages itself, yet thousands of users have come to rely on its stability for using Top Gun Wingman, validating the efficacy of our cluster platform. This observation, combined with the current trends toward massive cluster-based applications such as HotBot [13], suggests to us that the adaptive proxy style of adaptation will be of major importance in serving convergent "smart phone"-like devices.

7 Acknowledgments

This project has benefited from the detailed and perceptive comments of countless anonymous reviewers, users, and collaborators. Ken Lutz and Eric Fraser configured and administered the test network on which the TranSend scaling experiments were performed. Cliff Frost of the UC Berkeley Data Communications and Networks Services group allowed us to collect traces on the Berkeley dialup IP network and has worked with us to deploy and promote TranSend within UC Berkeley. Undergraduate researchers Anthony Polito, Benjamin Ling, Andrew Huang, David Lee, and Tim Kimball helped implement various parts of TranSend and Top Gun Wingman. Ian Goldberg and David Wagner helped us debug TranSend, especially through their implementation of the Anonymous Rewebber [26], and Ian implemented major parts of the client side of Top Gun Wingman, especially the 2-bit-per-pixel hacks. Paul Haeberli of Silicon Graphics contributed image processing code for Top Gun Wingman. Murray Mazer at the Open Group Research Institute has provided much useful insight on the structure of Internet applications and future extensions of this work. We also thank the patient students of UCB Computer Science 294–6, *Internet Services*, Fall 97, for being the first real outside developers on our TACC platform and greatly improving the quality of the software and documentation.

We have received much valuable feedback from our UC Berkeley colleagues, especially David Culler, Eric Anderson, Trevor Pering, Hari Balakrishnan, Mark Stemm, and Randy Katz. This research is supported by DARPA contracts #DAAB07-95-C-D154 and #J-FBI-93-153, the California MICRO program, the UC Berkeley Chancellor's Opportunity Fellowship, the NSERC PGS-A fellowship, Hughes Aircraft Corp., and Metricom Inc.

References

[1] Elan Amir, Steve McCanne, and Hui Zhang. An application level video gateway. In *Proceedings ACM Multimedia 1995*, 1995.

[2] Thomas E. Anderson, David E. Culler, and David Patterson. A case for now (networks of workstations). *IEEE Micro*, 12(1):54–64, Feb 1995.

[3] D. Andresen, T. Yang, O. Egecioglu, O. H. Ibarra, and T. R. Smith. Scalability issues for high performance digital libraries on the world wide web. In *Proceedings of IEEE ADL '96, Forum on Research and Technology Advances in Digital Libraries*, Washington D.C., May 1996.

[4] Aportis Inc. AportisDoc Overview, 1998. `http://www.aportis.com/products/AportisDoc/benefits.html`.

[5] Hari Balakrishnan, Venkata N. Padmanabhan, Srinivasan Seshan, and Randy H. Katz. A comparison of mechanisms for improving tcp performance over wireless links. In *Proceedings of the 1996 ACM SIGCOMM Conference*, Stanford, CA, USA, August 1996.

[6] Rob Barrett, Paul P. Maglio, and Daniel C. Kellem. How To Personalize the Web. In *Conference on Human Factors in Computing Systems (CHI 95)*, Denver, CO, May 1995. WBI, developed at IBM

Almaden; see `http://www.raleigh.ibm.com/wbi/wbisoft.htm`.

[7] Charles Brooks, Murray S. Mazer, Scott Meeks, and Jim Miller. Application-Specific Proxy Servers as HTTP Stream Transducers. In *Proceedings of the Fourth International World Wide Web Conference*, Dec 1995.

[8] C2net. Web anonymizer.

[9] A. Chankhunthod, P. B. Danzig, C. Neerdaels, M. F. Schwartz, and K. J. Worrell. A hierarchical internet object cache. In *Proceedings of the 1996 Usenix Annual Technical Conference*, pages 153–163, January 1996.

[10] D.D. Clark and D.L. Tennenhouse. Architectural Considerations for a New Generation of Protocols. *Computer Communication Review*, 20(4):200–208, Sep 1990.

[11] Intel Corp. QuickWeb Web Accelerator.

[12] Nokia Corp. and Geoworks Inc. Nokia 9000 Communicator. `http://www.geoworks.com/htmpages/9000.htm`.

[13] Inktomi Corporation. The hotbot search engine.

[14] Sony Corporation. The Sony MagicLink PDA. `http://www.sel.sony.com/SEL/Magic/`.

[15] Ivan Curtis. xcopilot Pilot simulator, 1998.

[16] Anthony D. Joseph et al. Rover: A toolkit for mobile information access. In *Proceedings of the 15th ACM Symposium on Operating Systems Principles*, Copper Mountain Resort, CO, USA, Dec 1995.

[17] WAP Forum. Wireless application protocol (WAP) forum. http://www.wapforum.org.

[18] Armando Fox and Eric A. Brewer. CS 294-6: Internet services, class proceedings, fall 1997. http://www.cs.berkeley.edu/~fox/cs294.

[19] Armando Fox and Eric A. Brewer. Reducing WWW Latency and Bandwidth Requirements via Real-Time Distillation. In *Proceedings of the Fifth International World Wide Web Conference*, Paris, France, May 1996. World Wide Web Consortium.

[20] Armando Fox and Steven D. Gribble. DOLF: Digital objects with lazy fixations. Unpublished manuscript: CS 294-5 Digital Libraries Seminar, Spring 1996.

[21] Armando Fox and Steven D. Gribble. Security On the Move: Indirect Authentication Using Kerberos. In *Proc. Second International Conference on Wireless Networking and Mobile Computing (MobiCom '96)*, Rye, NY, November 1996.

[22] Armando Fox, Steven D. Gribble, Yatin Chawathe, and Eric Brewer. TranSend Web Accelerator Proxy. Free service deployed by UC Berkeley. See `http://transend.cs.berkeley.edu`, 1997.

[23] Armando Fox, Steven D. Gribble, Yatin Chawathe, Eric A. Brewer, and Paul Gauthier. Cluster-Based Scalable Network Services. In *Proceedings of the 16th ACM Symposium on Operating Systems Principles*, St.-Malo, France, October 1997.

[24] Armando Fox, Steven D. Gribble, Yatin Chawathe, Anthony Polito, Benjamin Ling, Andrew C. Huang, and Eric A. Brewer. Orthogonal Extensions to the WWW User Interface Using Client-Side Technologies. In *User Interface Software and Technology (UIST) 97*, Banff, Canada, October 1997.

[25] Graphics interchange format version 89a (GIF). CompuServe Incorporated, Columbus, Ohio, July 1990.

[26] Ian Goldberg and David Wagner. Taz servers and the rewebber network: Enabling anonymous publishing on the world wide web. Unpublished manuscript available at http://www.cs.berkeley.edu/~daw/cs268/, May 1997.

[27] Steven D. Gribble and Eric A. Brewer. System Design Issues for Internet Middleware Services: Deductions from a Large Client Trace. In *Proceedings of the 1997 USENIX Symposium on Internet Technologies and Systems*, Monterey, California, USA, December 1997.

[28] Tom R. Halfhill. Inside the web pc. *Byte Magazine*, pages 44–56, March 1996.

[29] James J. Kistler and M. Satyanarayanan. Disconnected Operation in the Coda File System. *ACM Transactions on Computer Systems*, 10:3–25, February 1992.

[30] Liljeberg, M., et al. Enhanced Services for World Wide Web in Mobile WAN Environment. Technical Report C-1996-28, University of Helsinki CS Department, April 1996.

[31] Metricom Corp. Ricochet Wireless Modem, 1998. `http://www.ricochet.net`.

[32] Myricom. Myrinet: A Gigabit Per Second Local Area Network. In *IEEE Micro*, February 1995.

[33] John K. Ousterhout. *Tcl and the Tk Toolkit.* Addison-Wesley, 1994.

[34] DEC SRC Personal Communications, Paul MacJones. Postscipt to text converter.

[35] Jef Poskanzer. Netpbm release 7. ftp://wuarchive.wustl.edu/graphics/graphics/packages/NetPBM, 1993.

[36] Y. Sato. DeleGate Server, March 1994. `http://wall.etl.go.jp/delegate/`.

[37] M.A. Schickler, M.S. Mazer, and C. Brooks. Pan-browser support for annotations and other meta-information on the world wide web. In *Proc. Fifth International World Wide Web Conference (WWW-5)*, May 1996.

[38] Jennifer G. Steiner, Clifford Neuman, and Jeffrey I. Schiller. Kerberos: An authentication service for open network systems. In *Proceedings USENIX Winter Conference 1988*, pages 191–202, Dallas, Texas, USA, February 1988.

[39] Mark Stemm and Randy H. Katz. Vertical handoffs in wireless overlay network. *ACM Mobile Networking (MONET), Special Issue on Mobile Networking in the Internet*, Fall 1997.

[40] Terri Watson. Application Design for Wireless Computing. In *Mobile Computing Systems and Applications Workshop*, August 1994.

[41] Ka-Ping Yee. Shoduoka Mediator Service, 1995. `http://www.shoduoka.com`.

[42] C. Yoshikawa, B. Chun, P. Eastham, A. Vahdat, T. Anderson, and D. Culler. Using smart clients to build scalable services. In *Proc. Winter 1997 USENIX Technical Conference*, January 1997.

[43] Bruce Zenel. A Proxy Based Filtering Mechanism for the Mobile Environment. Thesis Proposal, Mar 1996.

Chapter 12 Other Sources of Information

Books. Imieliński and Korth edited a collection of papers on the specific topic of mobile computing [Imieliński, 1996]. It covers much of the material in this part, as well as additional topics such as energy consumption, broadband wireless access, mobile query processing, and wireless multimedia. Perkins [Perkins, 1998a] and Solomon [Solomon, 1998] have authored books specifically on Mobile IP. At a lower level, one can read Goodman's book on personal communications systems [Goodman, 1997] or the book by Taylor, and others, on CDPD [Taylor, 1997].

Web sites. The primary online source of information on mobile computing is the "Virtual Library on Mobile and Wireless Computing" (http://mosquitonet.stanford.edu/mobile/). It contains pointers to numerous projects, conferences, and other sites.

Mailing lists. The Mobile IP Working Group of the IETF maintains a mailing list (mobile-ip@ smallworks.com). The ACM Special Interest Group on Mobile Computing (SIGMOBILE) maintains a mailing list for members (see http://www.acm.org/sigmobile/).

Standards. The Mobile IP Working Group of the IETF (http://www.ietf.cnri.reston.va.us/ html.charters/mobileip-charter.html) is the primary standards body. The IETF charters numerous working groups, others of which may be relevant to this area as well. A full list of working groups is available at http://www.ietf.cnri.reston.va.us/html.charters/wg-dir.html.

The Wireless Application Protocol Forum, Ltd. (http://www.wapforum.org/) is an industry-sponsored consortium to promote standards for applications and services over wireless networks.

Conferences and Workshops. The Annual ACM/IEEE International Conference on Mobile Computing and Networking (Mobicom) is the single most relevant conference in this area. The URL for the conference changes each year, but is available from the SIGMOBILE site (http://www.acm.org/sigmobile/).

There are several other conferences that frequently have a session or more on topics relating to mobile computing:

ACM SIGCOMM and SIGOPS both hold annual or biannual conferences of interest. (http://www.acm.org/sigcomm/ and http://www.acm.org/sigops).

USENIX (http://www.usenix.org/) runs an annual technical conference as well as other events that sometimes cover mobile computing. It ran a conference on Mobile and Location-Independent Computing that was eventually subsumed by Mobicom.

IEEE Comsoc (http://www.comsoc.org/) and IEEE Computer Society (http://www.computer.org/) each hold events in these areas. INFOCOM is roughly analogous to SIGCOMM in scope.

To subscribe, send mail to majordomo@smallworks.com with *subscribe mobile-ip* in the body.

Journals and magazines. SIGMOBILE publishes a newsletter, *Mobile Computing and Communications Review,* and two journals, *Wireless Networks Journal* (WINET), and *Mobile Networks and Applications Journal* (MONET). IEEE publishes *Personal Communications.* Several other journals frequently publish articles or special issues in these areas.

Part IV

Mobility on the Internet: Mobile Agents

Chapter 13 Mobility on the Internet

There are many definitions of agents. For the purpose of this book we define agents by a set of their attributes: active, autonomous, goal-driven, and typically acting on behalf of a user or another agent. Agents are not a new paradigm, as they have been researched in the area of Distributed Artificial Intelligence (DAI) for a number of years. The term **agent oriented programming** has been introduced by Shoham. As a form of object oriented programming, an agent's state consists of beliefs, capabilities, choices, and similar notions, and the computation consists of the interactions, such as informing, offering, accepting, rejecting and competing [Shoham, 1997]. Another predecessor of agents are actors as introduced by Carl Hewitt [Hewitt, 1977] and explored by Gul Agha [Agha, 1986]. However, the widespread use of computers and their connectivity, particularly the Web and the Java programming language, has provided a new influx in the research, development, and deployment of agents. A particular motivation for the use of agents is the huge amount of information available on the Web. Agents have a significant potential in searching for information, filtering it, and extracting it from the source. The ability to represent and act on behalf of the user represents a crucial capability of agents and provides enormous potential for their deployment.

There are at least two communities currently pursuing agent research. The first is the intelligent- and multi-agent systems community. They deal mostly with stationary agents distributed on the network that communicate in order to pursue a common goal. In this book, we address this area to the extent that mobile agents can also be intelligent, or belong to a multi-agent group. The second community deals with mobile agents. It is mainly represented by people doing systems research, with the background in operating systems, distributed systems, and object-oriented research.

Mobile agents can be considered as the end point of the incremental evolution of mobile abstractions, such as mobile code, objects, and processes [White, 1998]. Transferred state of mobile code, such as applets, consists only of code. A mobile object state consists of code and data. Mobile process state in addition includes the thread state. The mobile agent state consists of code, data, thread, and authority of its owner.

If compared according to the movement pattern, mobile agents differ from applets (downloaded from the server to the client), and from servlets (uploaded from client to server) in that they can have multiple hops, and that they can be detached from the client. A mobile agent autonomously visits hosts without the need for continuous interaction with the agent's originating node. The originating node (also called the home node) gets involved only when the agent is sent to the first visiting node, and at the end when the agent returns from the last visited node.

Today's programming languages support mobile agents by means of mobile code, OS independence, object mobility, a remote object model, and language safety.

Mobile Code. Java, TCL/Tk, Python and other modern interpreted or scripting languages support mobile code. This solves problems such as heterogeneity (platform independence), and addresses security and safety issues (see below). Java is particularly suited for mobile agents because of the dynamic class loading, multithreading support, object serialization, and reflection [Lange, 1998]. Another important factor is the wide adoption of Java, and the universal availability of a Java virtual machine. This means that mobile agents can potentially have a large number of docking stations. However scripting languages, such as TCL/Tk [Ousterhout, 1994] and Python [Python], also have a large base of users.

Operating System Independence. Contemporary programming environments, such as Java Virtual Machine (JVM) and TCL/Tk interpreter, run on top of operating systems, and abstract away complex issues that have impacted mobility in the past (for example, file system and the underlying operating system process semantics). In addition, agent applications maintain some of their own execution state, such as the execution point at the moment of migration, and non-transient data. While putting more of a burden on the programmer, for example, a less transparent implementation of the agent application, it results in a much simpler implementation of the underlying environment.

Object Mobility. In order to move objects between heterogeneous environments, an agent system needs to serialize (externalize) objects on the source environment and then deserialize (internalize) them on the destination environment. It is required to save the whole object inheritance graph, and to extract this information at the destination. This simplifies transfer of the state of the agent, which can consist of the arbitrary combination of various objects.

Remote Object Model. There are various packages, such as Java RMI [Wollrath, 1996], CORBA [OMG, 1995], and Voyager [Glass], that support remote method execution by accessing state and executing methods on the remote objects. The remote object model simplifies implementation of the agent system infrastructure, as well as that of the agent application. In many cases, remotely accessing stationary agents represents a plausible alternative to deploying mobile agents.

Language and Run-Time Safety. Mobile code raises the issue of the safety of the languages. Downloading applets to a user's computer requires precautions about their behavior. Safe execution of mobile code is a prerequisite for ensuring the safety of mobile agents. Language and run-time safety, extended by the support for authentication, authorization, and encryption of the mobile code, represents an agent security model.

13.1 Benefits and Challenges of Mobile Agents

Mobile agent systems bear a lot of similarity to process migration; the major difference is in the underlying architecture on which they are used. Whereas process migration is used on clusters where security is less of an issue, mobile agents are typically used on the Internet where security and reliability are much more important. This is reflected in the benefits and challenges of mobile agents, as described in the rest of this section.

Benefits

The benefits of mobile agents consist of overcoming the limitations of a client computer, customization, inherent survivability, representation of a disconnected user, and ease of development.

Overcoming limitations of a client computer, such as communication delays and throughput, memory size, computing power, and small storage, may be achieved if the agent is executed closer to the source of data. For example, sending an agent toward the database that has to be searched according to a custom algorithm can improve performance compared to accessing data remotely. This is a classical argument for locality of reference. In addition, the local computer may not have sufficient data storage to temporarily store large amounts of data, or may have insufficient network bandwidth or computing power.

Customization. Agents can be easily customized. It is hard for the client-server model to adapt to rapid changes. Mobile agents, on the contrary, can be customized to user needs, and sent to the server where customized requests are executed. This is in the spirit of SQL or Postscript, where the requests are expressed in higher level languages and sent to the destination interpreter. Similarly, an agent system represents an "interpreter" that accepts and executes received agents.

Inherent survivability. Because agents transfer both code and state encapsulated within the mobile agent abstraction, they have a higher degree of survivability compared to the client-server model. For example, if a node or a network partially fails, it is possible for an agent to leave this node and execute somewhere else. There are alternative ways to improve survivability of the client-server model, but for large-scale systems this is hard to achieve. Process migration is typically deployed on small-scale systems, whereas agents are designed for use on the Web. Survivability of mobile agents is both a benefit and a challenge. It requires additional support by minimizing the residual state required on the nodes visited by the agent or on server nodes. Security servers, name servers, and a home node are possible exceptions where agent state is required.

Representation of a disconnected user. Many users are mobile today, starting their work in the office, and continuing it on a laptop at a different location. Jobs started from mobile computers frequently need to continue while a user is disconnected. While it is possible to delegate this responsibility to a stationary proxy residing on a network, it is even more convenient to delegate it to a mobile agent that will pursue its owner's tasks even while he or she is disconnected. When the user is back online, the agent can be pulled back from its current location.

Ease of development. For many programmers, it is easier to program when there is an analogy with the real world. For example, traveling salesmen are actually visiting customers that result in a natural representation of an agent who travels around and visits the servers. Another example is a buyer who visits the stores and offers goods. Because of the overwhelming amount of information in the stores, remotely accessing needed data might not be a viable alternative; in other scenarios, for security reasons, local data is not available remotely. Finally, "pipelining" types of applications are suitable for agents in scenarios where large amounts of data are stored at distributed servers and different types of agents successively visit servers to perform required actions.

Challenges

The challenges of mobile agents lie in the lack of applications, security, infrastructure, and standards.

Applications. While there are a lot of potential applications for mobile agents, only a few have emerged in practice. Many agent systems have been developed in industry (for example, Telescript [White, 1996], Aglets [Lange, 1998], Voyager [Glass], Concordia [Wong, 1997], and MOA [Milojičić, 1998b]). However, none of these has been successful in productizing agent systems or widely deploying agent applications. If mobile agents do not begin to be used for commercial applications, they may well experience the fate of AI in the 1980s. The inexistence of widely used commercial applications demonstrates that there is no absolute need for mobile agents, relegating them to a niche. The use of intelligent applications on top of mobile agent infrastructures can expand the base of mobile agents. However, intelligent agents are still being developed, and even in the static form they are in their infancy.

Security is hard to achieve for mobile agents. Security technology in itself is not mature enough to support mobile agents in a preferred way, for example, composite delegation and key distribution. Some of this will be developed for the needs of the mobile code, but the rest will have to be developed exclusively for agents. Security for mobile agents still has many unsolved problems and is an open area of research.

Lack of Infrastructure (naming, locating, controlling, communication). There is no widespread agent system that accepts agents. This problem may be alleviated only by providing plug-ins for existing widespread environments, such as Web browsers. Push technology is still not mature enough and the base of agent servers available throughout the world is too small for agents to become ubiquitous. Another limitation is the lack of meta-data, which prevents agents from making sense on the Web. This will improve with the introduction of XML.

Standards are important for application developers. Without common standards for a technology, individual application development requires high investments. Initial steps in developing agent standards are encouraging, but little experience from using reference implementations has been fed back. Standards and applications relate as a "chicken and egg" problem. Standards need to predate application development so that applications can be developed in a consistent manner, while the applications examples need to exist in order for standards to be developed. This leads to a spiral in which both applications and standards are being incrementally developed and improved.

13.2 Applications

Electronic commerce is one of the frequently mentioned areas for mobile agent deployment. The infrastructure for electronic commerce is emerging rapidly. Virtual stores that sell all kinds of goods are becoming available on the Internet. Standardized protocols, like the Secure Electronic Transactions (SET) internet payment specification [SET] developed by MasterCard and Visa, are used for transactions in order to provide security.

Software distribution is less frequently mentioned as an application, but it is one with a lot of opportunities for agents. Installing and maintaining software becomes significantly more difficult to achieve as the number of supported machines steadily grows. Agents can automate this process by customized installation of, on-site verification of versioning, and inventorying of software packages.

Information retrieval is another important application for mobile agents. Being able to visit the site with the information contributes to locality of reference and significantly improves performance. Filtering large amounts of information close to its source is an important benefit.

System administration. Administration is asynchronous and distributed in nature. Furthermore, it is frequently important to have a local view of the system in order to determine the cause or consequence of a problem. Mobile agents are suitable for such an environment because they can travel around large systems independently, and perform periodic management (backups, identifying unwanted *core* files, and other system-related operations); they can do initial assessment of problems on-site, prior to intervention by a human operator; they can be used to automate routine tasks; and they can periodically check systems and networks.

Network Management. Networks are inherently distributed and provide a suitable environment for deploying agents. Active networks [Tennenhouse, 1996; Wetherall, 1998] represent the lowest, physical level at which it is possible to deploy "active abstractions," whereas agents are meant for the higher levels of management, such as installing versions of software updates.

13.3 Myths and Facts

There are various myths related to mobile agents that prevent a clear understanding of what agents are suited for and what their limitations are. We describe some of the myths, trying to illustrate the background of each concern.

Myth: *Mobile agents are risky to use.* **Fact:** The largest concern of potential mobile agent users is related to security. Users are afraid to accept agents at their computers. The concern about accepting agents at servers is even higher. However, accepting mobile agents is not necessarily different from allowing remote access, as enabled by Java applets, or accepting e-mail that contains active entities, such as Word documents with macros. If agents were allowed to communicate with the servers through a set of limited, well-defined interfaces, and if they were contained within a clearly defined sandbox (as the Java Virtual Machine supports), the risks would be significantly reduced.

Myth: *The mobile agent paradigm needs a killer application to survive.* **Fact:** It is widely accepted that mobile agents do not have widely deployed applications. While this is true, there is also a belief that only the existence of a killer application can alleviate this problem. Mobile agents are a recent technique. It takes time for any technology to gain acceptance and even longer for it to be commonly adopted. No matter how much technically superior only few technologies end up having killer applications. For example, object-oriented programming took two decades to become widely used. It took over 15 years for the Internet to get the Web as its killer application. While it is certainly true that the appearance of a killer application could contribute to the success of mobile agents, it is unlikely that this will happen. Our belief is that a number of different applications, rather than one killer application, will push this technology further. The appearance of standards, such as the OMG's MASIF standard [Milojičić, 1998a] and the recently developed Foundation for Intelligent Physical Agents (FIPA) standard [FIPA], could provide common frameworks for mobile agent applications developers.

Myth: *Wide deployment of an agent environment is unlikely to emerge.* **Fact:** One of the key obstacles to the wider deployment of mobile agents is linked to the availability of agent environments that can accept and execute the agents. The existing mobile agent systems require agent environments that execute on the nodes that agents will visit. It is unlikely, critics say, that agent environments can be available on a base of computers worldwide large enough to achieve a critical mass required for real-world applications. There are at least two counter arguments. First, the development of push technology can improve the opportunities to execute uninvited agents on the servers that are willing to accept them. Second is the existence of ORBs and JVM in each browser. Most mobile agent systems are developed in Java, and the MASIF standard relies on CORBA-compliant interfaces. Mobile agent systems are typically small in size, making it easy and inexpensive to download them on the servers where they do not already exist.

Myth: *Mobile agents are too complex to implement and difficult to use.* **Fact:** This view is mostly inherited from the problems associated with process migration in the past. Mobile agents do not have similar transparency requirements, which makes the implementation of agent systems and agent applications much simpler. Certain applications can benefit from an analogy with the real world. For example, a traveling salesman or buyer who visits various stores in turn can be more easily modeled with mobile agents than with the client-server model.

Myth: *All that can be achieved with mobile agents can also be achieved with static agents.* **Fact:** This discussion is a new incarnation of the old argument between the advocates of process migration and remote execution, that is, dynamic and static load balancing [Eager, 1986a, b; 1988; Harchol-Balter, 1997; Krueger, 1988]. Whereas it is true that most of the functionality provided by mobile agents can be achieved by their static counterparts, it is the performance benefits that will drive the use of the former. For this to happen, mobile agent environments must become widely available, as discussed in the previous paragraph. A variation of this myth is the current gap between the intelligent and multi-agent community on one side and the mobile agent community on the other. The former are developing applications based on static agents that communicate, and the latter are based on the infrastructures and agent prototypes. In our view, the former can benefit by using the mobile agent system infrastructure, whereas in most cases the work of the latter assumes that the intelligent agents will actually migrate.

Myth: *Mobile agents can solve any problem.* **Fact:** Advocates of mobile agents believe that they can solve all problems with mobile agents. Complex problems are rarely resolved with a single technique. There is a subset of problems suitable for mobile agents. Out of this subset, sometimes it is still the client-server model that is more appropriate; at other times, it is mobile agents, but frequently, it is a combination of the two. After all, the mobility support in mobile agents consists of a series of client-server transactions.

Chapter 14 Mobile Agent Systems

Industry has invested a lot of effort in developing mobile agents. It has effectively led the development of mobile agent systems, with General Magic at the forefront with Telescript [White, 1996]. General Magic's effort was followed by other industrial development, such as IBM's Aglets [Lange, 1998a], Concordia [Walsh, 1998] and Voyager [Glass], and university research, such as Agent TCL [Kotz, 1997] and Tacoma [Johansen, 1995]. It is also interesting that a lot of research was pursued in Europe, with systems such as Mole [Baumann, 1998], Ara [Peine, 1997], TACOMA [Glass], and others.

Our original idea was to categorize papers by agent systems, applications, security, reliability, and standards. However, almost all of the papers qualify for the agent systems category. This is due to the relatively early stage in the development of mobile agents, and also because our selection is biased toward systems. Therefore, we have decided to abandon categorization. Instead, we have selected papers describing Telescript, Aglets, Agent TCL, Concordia, Mole, Tacoma, Sumatra, Ara, MOA, and Voyager. We conclude with the paper describing the MASIF standard.

Telescript was the first mobile agent system developed [Tardo, 1996; White, 1996; 1997]. It was targeted for a Magic Cap, a small hand-held device, similar to subsequent computers, such as the Newton and 3Com PalmPilot computers. In many ways, both Telescript and Magic Cap were ahead of their time. Telescript is similar to Java, in that it is interpreted and has a security model, even though it is targeted for a closed network. Telescript opened the commercial marketplace for mobile agents and was followed by many other systems, such as IBM Aglets, Concordia, and Voyager. It has introduced many new concepts, such as *place* and *permit* or mechanisms, such as *meet* or *go*. There was a Java incarnation of Telescript, called Tabriz, and a lot of Telescript experience was used for the MASIF implementation. In the "Telescript Technology: Mobile Agents" paper, Jim White introduces mobile agents and describes how they are programmed and used.

Jim White has provided an **Afterword** containing a retrospective of mobile agent development since the introduction of Telescript. He compares Java and Telescript virtual machines and the goals of their developers.

Aglets is one of the best known and most widespread mobile agent systems today [Lange, 1998a]. It was developed at the IBM Tokyo Research Lab. Aglets was one of the first agent systems to jump on the Java bandwagon. It has a stable interface and a large base of users. Despite its popularity, Aglets was never productized within IBM in the traditional way. Aglets is available for free, and the consulting is charged for, similar to other commercial mobile agent systems, such as Voyager or Concordia. Aglets is used on a commercial Web site in Japan (Travel agent site called TabiCan [Nakamura]). In the paper "Mobile Agents with Java: The Aglet API" Danny Lange and Mitsuru Oshima analyze the benefits and shortcomings of the Java programming language for mobile agents. They also describe the Aglets API and present an example of Aglets deployment in electronic commerce.

Agent TCL is also one of the early projects on mobile agents [Gray, 1995; 1996; 1997a, b; 1998; Kotz, 1996; 1997], developed at Dartmouth University. Agent-TCL started as a Tcl/Tk based system, but later was extended to support Java, Scheme, and C/C++, under the new name, d'Agents. Agent TCL is used for many research activities internally, such as those exploring security mechanisms, resource control (using economic models), navigation and resource discovery, and automatic indexing. Agent TCL was also used in DAIS [Hofmann, 1998], which uses mobile agents for information retrieval and dissemination in military intelligence applications (funded by DARPA). In the paper included in this book, David Kotz and others describe how Agent TCL satisfies the needs of mobile computers, such as the support for docking station, minimizing connection time, and inter-agent communication.

Concordia is a mobile agent system developed at the Mitsubishi Electric ITA Laboratory in Waltham, Massachusetts [Castillo, 1998; Walsh, 1998; Wong, 1997]. It is a Java-based system that addresses security and reliability concerns. Concordia deploys an identity-based system protection of agents and it also relies on hashing the agent code, thereby protecting access to its resources, and extensions to the Java security manager. Reliable network transmission is achieved using the message queueing subsystem based on two-phase-commit protocol. Concordia is publicly available in binary form. In the paper included, Walsh, and others, provide the details of security and reliability aspects of Concordia, such as agent and resource protection, queueing, persistence, and the use of proxies.

Mole is one of the first academic agent systems written in Java [Baumann, 1997; 1998a, b, c]. It was started back in 1994 at Stuttgart University, Germany. It has been used by industry (for example, Siemens, Tandem, and Daimler-Benz), and academia (University of Geneva bases a few projects on Mole). The topics addressed include groups of agents, agent termination, and security for protecting agents against malicious hosts. In the paper "Mole–Concepts of a Mobile Agent System," included in this book, Baumann and others describe the agent model, mobility and communication concepts, and security of the Mole system.

An **Afterword** provided by Baumann and the Mole development team elaborates on the future developments of the mobile agentry at the Stuttgart University.

TACOMA is a joint project between Tromso University in Norway and Cornell University in the USA [Johansen, 1995; 1996; Schneider, 1997]. The main research topics include security and reliability. One of the applications of the TACOMA mobile agent system is the *WeatherStorm* distributed application. As opposed to other mobile agent research that addresses programming language aspects of mobile agents, TACOMA mainly addresses operating system aspects. The paper on Tacoma "Operating System Support for Mobile Agents" is one of the early ones, which provides an overview of the system. Johansen and others describe abstractions, such as *briefcase* and *folders*, and mechanisms, such as *meet*. They also present services and charging for them, scheduling, fault tolerant support, and prototype implementation.

An **Afterword** provided by the TACOMA team summarizes the lessons learned while developing and using TACOMA.

Sumatra is a Java-based mobile agent system, developed at the University of Maryland [Ranganathan, 1996; 1997]. It is one of a few systems based on a JVM that has been modified to support the transparent migration of agents. Agents can suspend execution at any line of code, migrate, and resume execution at the remote node. The Sumatra developers have demonstrated the benefits of using such a mobile agent system in the case of an Internet Chat application *adaptalk*, which adapts the location of the server in order to optimize the distance between clients. In the paper "Network-aware Mobile Programs," Ranganathan and others describe the Sumatra system, the *adaptalk* application, resource monitoring, and performance evaluation.

Ara (Agents for Remote Action) is a mobile agent system developed at the University of Kaiserslautern [Peine, 1997; 1998]. It started with TCL/Tk and C/C++ implementations, but now it also supports Java. Ara runs on various UNIX platforms. Holger Peine, the principal developer of Ara, has explored the benefits of changing the JVM, thereby supporting transparent continuation of an agent that initiates migration at an arbitrary point of the code, as well as the management of physical resources, such as memory. In the paper "The Architecture of the Ara Platform for Mobile Agents," Peine and Stolpmann describe Ara's architecture and its support for mobility, communication, security, and fault tolerance.

MOA (Mobile Objects and Agents) is a recently-developed system at The Open Group Research Institute [Milojičić, 1998]. It is written in Java, and it resembles the Telescript object model with a few notable exceptions, such as the *place* semantics and the communication model. The major contributions of the MOA system are related to resource management, transparent maintenance of communication channels across migration, and compliance with the Java Beans component model. In the included paper, Milojičić, et al. describe the design and implementation of the MOA system and some early experience using the component model approach.

Voyager, developed by ObjectSpace, is one of the rare systems that has achieved wide deployment [Glass, 1998]. It is a software package that supports mobile objects and agents but this is only one of the features. In addition it has other communication mechanisms, such as remote method invocation, object request broker, and DCOM support. These features have made Voyager widely used. Mobile agent support is another attractive alternative that potentially could be used if deemed appropriate. It is orthogonal to most other systems where mobility is the core feature, and messaging and communication is supported to the extent that it is needed for mobile agents. Voyager also claims good performance and robustness [Glass, 1998]. In the "ObjectSpace Voyager Core Package, Technical Overview," Graham Glass introduces Voyager and its concepts and analyzes the product evolution.

MASIF is the first attempt to standardize mobile agent system interoperability [Chang, 1997; Milojičić, 1998a; OMG]. It has been developed by IBM, General Magic, GMD Fokus, Crystaliz, and The Open Group. MASIF standardizes interoperability between mobile agent systems, by specifying agent management, transfer, and naming. MASIF does not address the interfaces between agent applications and the agent system. The MASIF standard is written by a team of experienced developers of other agent systems, such as Aglets, Telescript, Grasshopper [IKV], MuBot [Vidhagris-waram], and MOA. MASIF has been accepted as an OMG technology and reference implementations are being pursued by its proposers as well as by other companies. In the paper included in the book [Milojičić, 1998a], Milojičić and others summarize the OMG MASIF standard document and the MASIF interfaces.

One of the first introductory papers on mobile agents is by Chess and others [Chess, 1994; 1995]. Other papers that describe mobile agent systems include those on Knowbots [Hylton, 1997], Messenger [Tschudin, 1997], and Grasshopper [IKV]. Other papers describing use of mobile agents or related technology include Migratory Applications [Bharat, 1995], Distributed Computing Using Autonomous Objects [Bic, 1996], and work by Stone [Stone, 1996]. Papers describing security in mobile agents include work by Farmer, and others [Farmer, 1996a, b], Haertig and Reuther [Haertig, 1997], Hohl [Hohl, 1998], Vigna [Vigna, 1997; 1998], Vitek, and others [Vitek, 1997], da Silva [da Silva, 1997], and many others. Additional work related to standards and industry includes the FIPA standard [FIPA], Guideware [Guideware], and most recently, JumpingBeans [Ad Astra Engineering].

14.1 White, J.E., Telescript Technology: Mobile Agents

White, J.E., "Telescript Technology: Mobile Agents," General Magic White Paper, Appeared in Bradshaw, J., *Software Agents,* AAAI/MIT Press, 1996.

Mobile Agents

James E. White

Introduction

New products

The economics of computing, the growth of public information networks, the convergence of computers and communication, and advances in graphical user interfaces enable powerful new electronics products to be placed in the hands of consumers. Personal intelligent communicators provide one example. Intelligent televisions, providing access to vast amounts of scheduled program material and video on demand, will offer another.

In principle, such products could put people in closer touch with one another (e.g., by means of electronic postcards), simplify their relationships (e.g., by helping them make and keep appointments), provide them with useful information (e.g., television schedules, traffic conditions, restaurant menus, and stock market results), and help them carry out financial transactions (e.g., purchase theater tickets, order flowers, and buy and sell stock).

New applications

New devices alone are insufficient to deliver to the consumer services like those above. Also needed is a new breed of computer software we call the *communicating application*. Unlike the standalone applications of today's personal computer (e.g., word processing programs), communicating applications will adeptly use the public networks to which they give access to find and interact with people and information on the consumer's behalf.

Communicating applications will have qualities of timeliness and effectiveness that today's uncommunicative applications do not. One such application might maintain my stock portfolio, buying and selling stock under conditions that I establish for it in advance. Another might arrange that every Friday evening a romantic comedy of its choosing is ready for viewing on my television, and that my favorite pizza is delivered to my door.

New networks

Today's networks pose a barrier to the development of communicating applications. The barrier stems from the need for such applications to physically distribute themselves, that is, to run not only on the computers dedicated to individual users, but also on the computers that users share, the servers. For example, a communicating application that is to provide a forum for buying and selling used cars necessarily has two parts. A user interface component in a user's personal communicator gathers information from an individual buyer or seller. A database component in a public server records the information and uses it to bring buyers and sellers together.

The barrier posed by a public network is insurmountable. A user of an enterprise network might well be allowed to install any application she wishes on her own desktop computer. She might even persuade her network administrator to install a new application on the departmental servers. However, a user of a commercial on-line service would find that no amount of persuasion would succeed in making her software a part of that service.

The success of the personal computer is due in large part to third-party software developers. On the platform provided by hardware and operating system manufacturers, developers built standalone applications that made personal computers indispensable tools for people in all walks of life. One might expect that the success of the information superhighway will depend on developers in a similar way. Unless public networks are platforms on which third-party developers can build communicating applications, the networks will respond much too slowly to new and varied requirements and so will languish. Unfortunately, today's networks are not platforms.

Scope of the chapter

This chapter explores the concept of a public network designed as a platform for application developers. It introduces a new communication paradigm, the *mobile agent*, which provides the organizing principle for such a network, and a new communication software technology, *Telescript technology*, which implements the concept in a commercial setting. The chapter also presents and explores the vision of an electronic marketplace that allows automated access by agents, as well as conventional, interactive access by people.

The chapter has three sections. The first, "Enabling mobile agents", introduces and motivates the concept of a mobile agent, explains how Telescript technology implements the concept, and describes an electronic marketplace based upon the technology. The second section, "Programming mobile agents", explains by example how a communicating application works. The example serves as an introduction to the Telescript language in which mobile agents are programmed. The third section, "Using mobile agents", explores the variety of applications that mobile agents make possible. These are scenes from the electronic marketplace of the future.

Enabling mobile agents

Mobile agent paradigm

The concept of a mobile agent sprang from a critical examination of how computers have communicated since the late 1970's. This section sketches the results of that examination and so presents the case for mobile agents.

Current approach

The central organizing principle of today's computer communication networks is *remote procedure calling (RPC)*. Conceived in the 1970's, the RPC paradigm views computer-to-computer communication as enabling one computer to call procedures in another. Each message that the network transports either requests or acknowledges a procedure's performance. A request includes data which are the procedure's arguments. The response includes data which are its results. The procedure itself is internal to the computer that performs it.

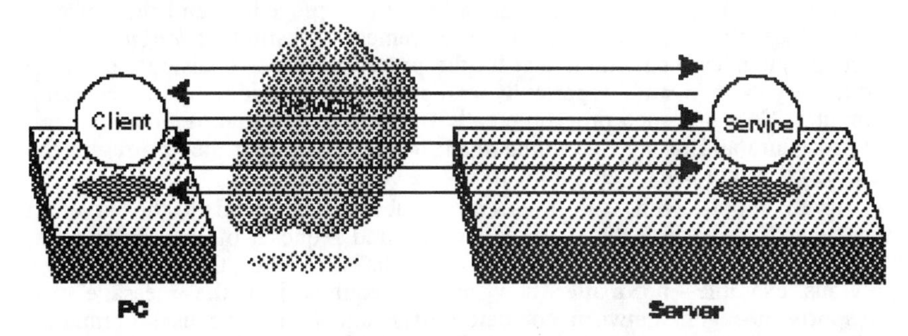

Two computers whose communication follows the RPC paradigm agree in advance upon the effects of each remotely accessible procedure and the types of its arguments and results. Their agreements constitute a *protocol*.

A user computer with work for a server to accomplish orchestrates the work with a series of remote procedure calls. Each call involves a request sent from user to server and a response sent from server to user. To delete from a file server, for example, all files at least two months old, a user computer might have to make one call to get the names and ages of the user's files and another for *each* file to be deleted. The analysis that decides which files are old enough to delete is done in the user computer. If it decides to delete n files, the user computer must send or receive a total of $2(n+1)$ messages.

The salient characteristic of remote procedure calling is that each interaction between the user computer and the server entails two acts of communication, one to ask the server to perform a procedure, another to acknowledge that the server did so. Thus *ongoing interaction requires ongoing communication*.

New approach

An alternative to remote procedure calling is *remote programming (RP)*. The RP paradigm views computer-to-computer communication as enabling one computer not only to call procedures in another, but also to supply the procedures to be performed. Each message that the network transports comprises a procedure which the receiving computer is to perform and data which are its arguments. In an important refinement, the procedure is one whose performance the sending computer began (or continued) but the receiving computer is to continue; the data are the procedure's current state.

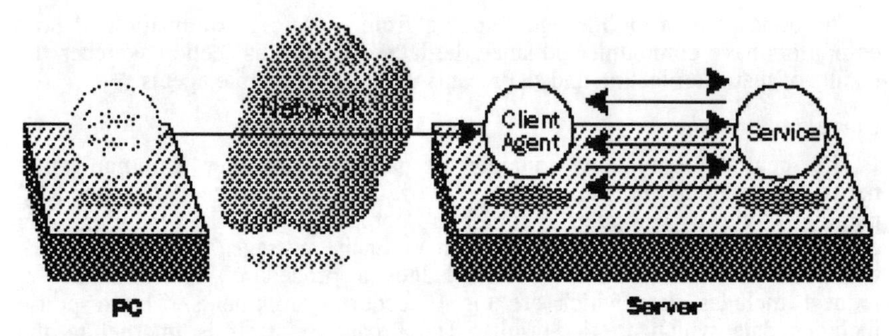

Two computers whose communication follows the RP paradigm agree in advance upon the instructions that are allowed in a procedure and the types of data that are allowed in its state. Their agreements constitute a *language*. The language includes instructions that let the procedure make decisions, examine and modify its state, and, importantly, call procedures provided by the receiving computer. Note that such procedure calls are local, rather than remote. We call the procedure and its state a *mobile agent* to emphasize that they represent the sending computer even while in the receiving computer.

A user computer with work for a server to accomplish[1] sends to the server an agent whose procedure there makes the required requests of the server (e.g., "delete") based upon its state (e.g., "two months"). Deleting the old files of the previous example—no matter how many—requires just the message that transports the agent between computers. The agent, not the user computer, orchestrates the work, deciding "on-site" which files should be deleted.

The salient characteristic of remote programming is that a user computer and a server can interact without using the network once the network has transported an agent between them. Thus *ongoing interaction does not require ongoing communication*. The implications of this are far-reaching.

---

[1] The opportunity for remote programming (like that for remote procedure calling) is bi-directional. The example depicts a user's agent visiting a server, but a server's agent can visit a user's computer as well. In an electronic marketplace, if the user's agent is a shopper, the server's agent is a door-to-door salesperson.

Tactical advantage

Remote programming has an important advantage over remote procedure calling. The advantage can be seen from two different perspectives. One perspective is quantitative and tactical, the other qualitative and strategic.

The tactical advantage of remote programming is *performance*. When a user computer has work for a server to do, rather than shout commands across a network, it sends an agent to the server and thereby directs the work locally, rather than remotely. The network is called upon to carry fewer messages. The more work to be done, the more messages remote programming avoids.

The performance advantage of remote programming depends in part upon the network: the lower its throughput or availability, or the higher its latency or cost, the greater the advantage. The public telephone network presents a greater opportunity for the new paradigm than does an Ethernet. Today's wireless networks present greater opportunities still. Remote programming is particularly well suited to personal communicators, whose networks are presently slower and more expensive than those of personal computers in an enterprise; and to personal computers in the home, whose one telephone line is largely dedicated to the placement and receipt of voice telephone calls.

A home computer is an example of a user computer that is connected to a network occasionally rather than permanently. Remote programming allows a user with such a computer to delegate a task—or a long sequence of tasks—to an agent. The computer must be connected to the network only long enough to send the agent on its way and, later, to welcome it home. The computer need not be connected while the agent carries out its assignment. Thus remote programming lets occasionally connected computers do things that remote procedure calling would make impractical.

Strategic advantage

The strategic advantage of remote programming is *customization*. Agents let manufacturers of user software extend the functionality offered by manufacturers of server software. Returning once again to the filing example, if the file server provides one procedure for listing a user's files and another for deleting a file by name, a user can effectively add to that repertoire a procedure that deletes all files of a specified age. The new procedure, which takes the form of an agent, customizes the server for that particular user.

The remote programming paradigm changes not only the division of labor among software manufacturers but also the ease of installing the software they produce. Unlike the standalone applications that popularized the personal computer, the communicating applications that will popularize the personal communicator have components that must reside in servers. The server components of an RPC-based application must be statically installed by the user. The server components of an RP-based application, on the other hand, are dynamically installed by the application itself. Each is an agent.

The advantage of remote programming is significant in an enterprise network but profound in a public network whose servers are owned and operated by public service providers (e.g., America Online™). Introducing a new RPC-based application requires a business decision on the part of the service provider. For an RP-based application, all that's required is a buying decision on the part of an individual user. Remote programming thus makes a public network, like a personal computer, a *platform*.

Mobile agent concepts

The first commercial implementation of the mobile agent concept is General Magic's *Telescript technology*™ which, by means of mobile agents, enables automated as well as interactive access to a network of computers. The commercial focus of Telescript technology is the *electronic marketplace*, a public network that will let providers and consumers of goods and services find one another and transact business electronically. Although the electronic marketplace doesn't exist today, its beginnings can be seen in the Internet.

Telescript technology implements the following principal concepts: places, agents, travel, meetings, connections, authorities, and permits. An overview of these concepts indicates how the remote programming paradigm provides the basis for a complete and cohesive remote programming technology.

Places

Telescript technology models a network of computers—however large—as a collection of places. A *place* offers a service to the mobile agents that enter it.

In the electronic marketplace, a mainframe computer might function as a shopping center. A very small shopping center (illustrated) might house a ticket place where agents can purchase tickets to theater and sporting events, a flower place where agents can order flowers, and a directory place where agents can learn about any place in the shopping center. The network might encompass many independently operated shopping centers, as well as many individually operated shops, many of the latter on personal computers.

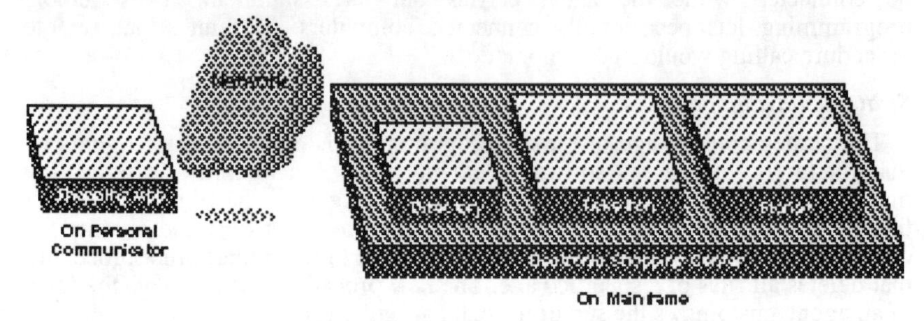

While servers provide some places, user computers provide others. A home place on a user's personal communicator, for example, might serve as the point of departure and return for agents that the user sends to server places.

Agents

Telescript technology models a communicating application as a collection of agents. Each *agent* occupies a particular place. However, an agent can move from one place to another, thus occupying different places at different times. Agents are independent in that their procedures are performed concurrently.

In the electronic marketplace, the typical place is permanently occupied by one, distinguished agent. This stationary agent represents the place and provides its service. The ticketing agent, for example, provides information about events and sells tickets to them, the flower agent provides information about floral arrangements and arranges for their delivery, and the directory agent provides information about other places, including how to reach them.

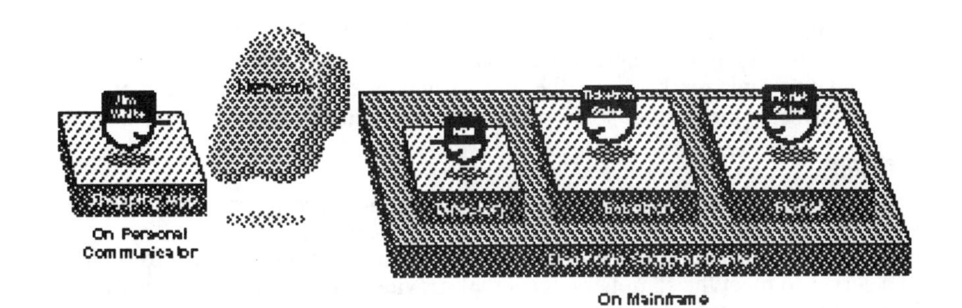

Travel

Telescript technology lets an agent *travel* from one place to another—however distant. Travel is the hallmark of a remote programming system.

Travel lets an agent obtain a service offered remotely and then return to its starting place. A user's agent, for example, might travel from home to a ticketing place to obtain orchestra seats for Phantom of the Opera. Later the agent might travel home to describe to its user the tickets it obtained.

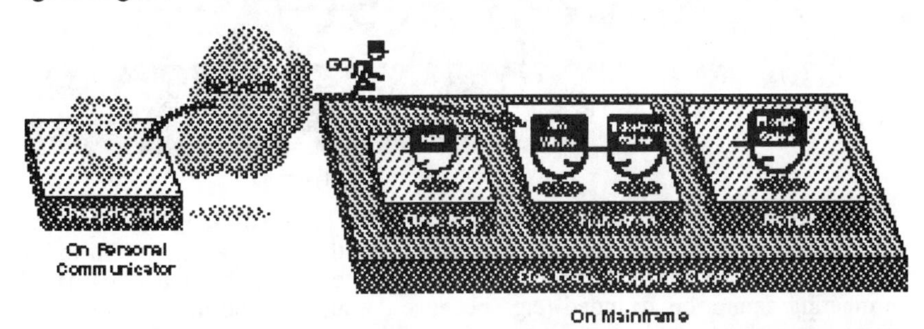

Moving software programs between computers by means of a network has been commonplace for 20 years or more. Using a local area network to download a program from the file server where it's stored to a personal computer where it must run is a familiar example. But moving programs *while* they run, rather than *before*, is unusual. A conventional program, written for example in C or C++, cannot be moved under these conditions because neither its procedure nor its state is portable. An agent can move from place to place throughout the performance of its procedure because the procedure is written in a language designed to permit this. The *Telescript language* in which agents are programmed lets a computer package an agent—its procedure and its state—so that it can be transported to another computer. The agent itself decides when such transportation is required.

To travel from one place to another an agent executes an instruction unique to the Telescript language, the `go` instruction. The instruction requires a *ticket*, data that specifies the agent's destination and the other terms of the trip (e.g., the means by which it must be made and the time by which it must be completed). If the trip cannot be made (e.g., because the means of travel cannot be provided or the trip takes too long) the `go` instruction fails; the agent handles the exception as it sees fit. However, if the trip succeeds, the agent finds that its next instruction is executed at its destination. Thus in effect the Telescript language reduces networking to a single instruction.

In the electronic marketplace, the `go` instruction lets the agents of buyers and sellers co-locate themselves so they can interact efficiently.

Meetings

Telescript technology lets two agents in the same place meet. A *meeting* lets agents in the same computer call one another's procedures.

Meetings are what motivate agents to travel. An agent might travel to a place in a server to meet the stationary agent that provides the service the place offers. The agent in pursuit of theater tickets, for example, travels to and then meets with the ticket agent. Alternatively, two agents might travel to the same place to meet each other. Such meetings might be the norm in a place intended as a venue for buying and selling used cars.

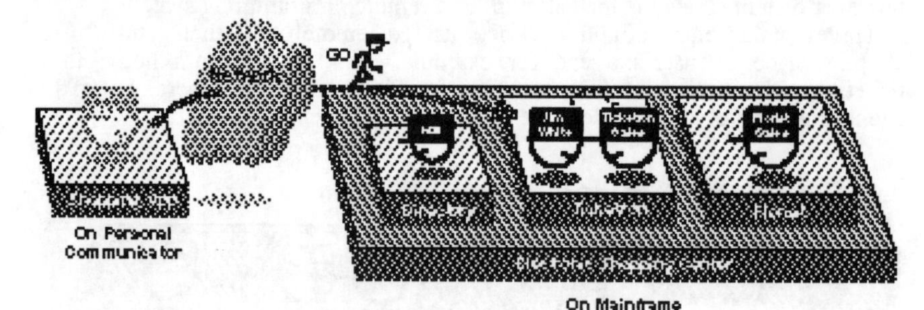

To meet a co-located agent an agent executes the Telescript language's `meet` instruction. The instruction requires a *petition*, data that specifies the agent to be met and the other terms of the meeting (e.g., the time by which it must begin). If the meeting cannot be arranged (e.g., because the agent to be met declines the meeting or arrives too late) the `meet` instruction fails; the agent handles the exception as it sees fit. However, if the meeting occurs, the two agents are placed in programmatic contact with one another.

In the electronic marketplace, the `meet` instruction lets the co-located agents of buyers and sellers exchange information and carry out transactions.

Connections

Telescript technology lets two agents in different places make a connection between them. A *connection* lets agents in different computers communicate.

Connections are often made for the benefit of human users of interactive applications. The agent that travels in search of theater tickets, for example, might send to an agent at home a diagram of the theater showing the seats available. The agent at home might present the floor plan to the user and send to the agent on the road the locations of the seats the user selects.

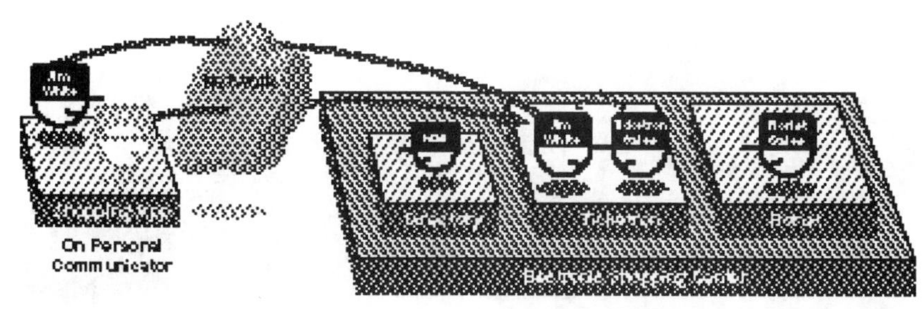

To make a connection to a distant agent an agent executes the Telescript language's `connect` instruction. The instruction requires a *target* and other data that specify the distant agent, the place where that agent resides, and the other terms of the connection (e.g., the time by which it must be made and the quality of service it must provide). If the connection cannot be made (e.g., because the distant agent declines the connection or is not found in time or the quality of service cannot be provided) the `connect` instruction fails; the agent handles the exception as it sees fit. However, if the connection is made, the two agents are granted access to their respective ends of it.

In the electronic marketplace, the `connect` instruction lets the agents of buyers and sellers exchange information at a distance. Sometimes, as in the theater layout phase of the ticking example, the two agents that make and use the connection are parts of the same communicating application. In such a situation, the protocol that governs the agents' use of the connection is of concern only to that one application's designer. It need not be standardized.

If agents are one of the newest communication paradigms, connections are one of the oldest. Telescript technology integrates the two.

Authorities

Telescript technology lets one agent or place discern the authority of another. The *authority* of an agent or place in the electronic world is the individual or organization in the physical world that it represents. An agent or place can discern but neither withhold nor falsify its authority. Anonymity is precluded.

Authority is important in any computer network. To control access to its files, a file server must know the authority of any procedure that instructs it to list or delete files. The need is the same whether the procedure is stationary or mobile. Telescript technology verifies the authority of an agent whenever it travels from one region of the network to another. A *region* is a collection of places provided by computers that are all operated by the same authority. Unless the source region can prove the agent's authority to the satisfaction of the destination region, the agent is denied entry to the latter. In some situations, highly reliable, cryptographic forms of proof may be demanded.

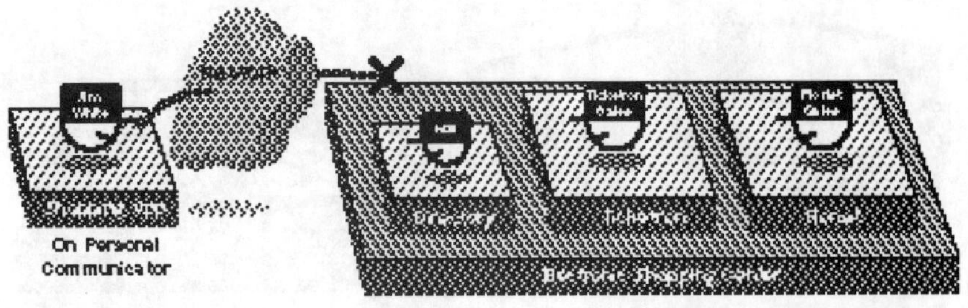

To determine an agent's or place's authority, an agent or place executes the Telescript language's `name` instruction. The instruction is applied to an agent or place within reach for one of the reasons listed below. The result of the instruction is a *telename*, data that denotes the entity's identity as well as its authority. *Identities* distinguish agents or places of the same authority.

Authorities lets agents and places interact with one another on the strength of their ties to the physical world. A place can discern the authority of any agent that attempts to enter it and can arrange to admit only agents of certain authorities. An agent can discern the authority of any place it visits and can arrange to visit only places of certain authorities. An agent can discern the authority of any agent with which it meets or to which it connects and can arrange to meet with or connect to only agents of certain authorities.

In the electronic marketplace, the `name` instruction lets programmatic transactions between agents and places stand for financial transactions between their authorities. A server agent's authority can bill a user agent's authority for services rendered. In addition, the server agent can provide personalized service to the user agent on the basis of its authority, or can deny it service altogether. More fundamentally, the lack of anonymity helps prevent viruses by denying agents that important characteristic of viruses.

Permits

Telescript technology lets authorities limit what agents and places can do by assigning permits to them. A *permit* is data that grants capabilities. An agent or place can discern its capabilities but cannot increase them.

Permits grant capabilities of two kinds. A permit can grant the right to execute a certain instruction. For example, an agent's permit can give it the right to create other agents.[1] An agent or place that tries to exceed one of these qualitative limits is simply prevented from doing so. A permit can also grant the right to use a certain resource in a certain amount. For example, an agent's permit can give it a maximum lifetime in seconds, a maximum size in bytes, or a maximum amount of computation, its *allowance*. An agent or place that tries to exceed one of these quantitative limits is destroyed.[2]

---

[1] An agent can grant any agents it creates only capabilities it has itself. Furthermore, it must share its allowance with them.

[2] An agent can impose temporary permits upon itself. The agent is notified, rather than destroyed, if it violates them. An agent can use this feature of the Telescript language to recover from its own misprogramming.

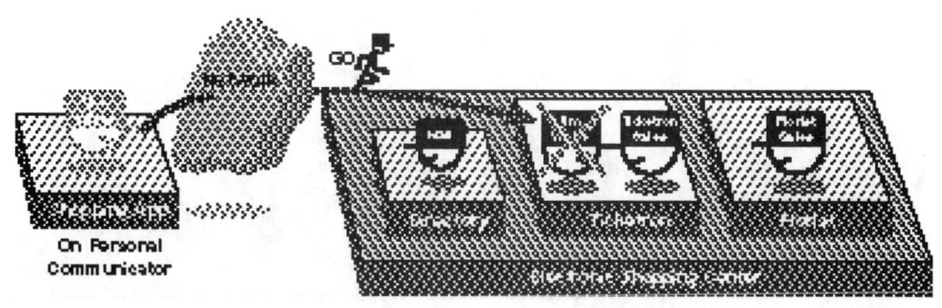

To determine an agent's or place's permit, an agent or place executes the Telescript language's `permit` instruction. The instruction is applied to an agent or place within reach for one of the reasons listed in "Authorities".

Permits protect authorities by limiting the effects of errant and malicious agents and places. Such an agent threatens not only its own authority but also those of the place and region it occupies. For this reason the technology lets each of these three authorities assign an agent a permit. The agent can exercise a particular capability only to the extent that all three of its permits grant that capability. Thus an agent's effective permit is renegotiated whenever the agent travels. To enter another place or region the agent must agree to its restrictions. When the agent exits that place or region, its restrictions are lifted but those of another place or region are imposed.

In the electronic marketplace, the `permit` instruction and the capabilities it documents help guard against the unbridled consumption of resources by illprogrammed or illintentioned agents. Such protection is important because agents typically operate unattended in servers rather than in user computers where their misdeeds might be more readily apparent to the human user.

Putting things together

An agent's travel is not restricted to a single round-trip. The power of mobile agents is fully apparent only when one considers an agent that travels to several places in succession. Using the basic services of the places it visits, such an agent can provide a higher-level, composite service.

In our example, traveling to the ticket place might be only the first of the agent's responsibilities. The second might be to travel to the flower place and there arrange for a dozen roses to be delivered to the user's companion on the day of the theater event. Note that the agent's interaction with the ticket agent can influence its interaction with the flower agent. Thus, for example, if instructed to get tickets for any evening for which tickets are available, the agent can order flowers for delivery on the day for which it obtains tickets.

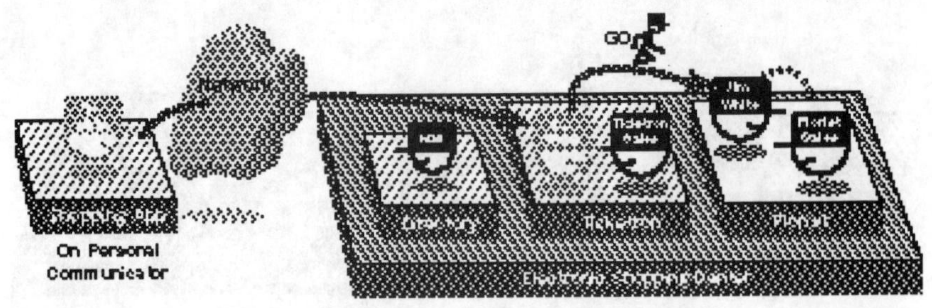

This simple example has far reaching implications. The agent fashions from the concepts of tickets and flowers the concept of special occasions. In the example as presented, the agent does this for the benefit of an individual user. In a variation of the example, however, the agent takes up residence in a server and offers its special occasion service to other agents. Thus agents can extend the functionality of the network. Thus the network is a platform.

Mobile agent technology

New communication paradigms beget new communication technologies. A technology for mobile agents is software. That software can ride atop a wide variety of computer and communication hardware, present and future.

Telescript technology implements the concepts of the previous section and others related to them. It has three major components: the language in which agents and places (or their "facades") are programmed; an engine, or interpreter, for that language; and communication protocols that let engines in different computers exchange agents in fulfillment of the go instruction.

Language

The Telescript programming language lets developers of communicating applications define the algorithms agents follow and the information agents carry as they travel the network[1]. It supplements systems programming languages such as C (or C++). Entire applications can be written in the Telescript language, but the typical application is written partly in C. The C parts include the stationary software in user computers that lets agents interact with users, and the stationary software in servers that lets places interact, for example, with databases. The agents and the "surfaces" of places to which they are exposed are written in the Telescript language.

---

[1] Despite its name, the Telescript language is not a scripting language. Its purpose is not to allow human users to create macros, or scripts, that direct applications written in "real" programming languages like C. Rather, it lets developers implement major components of communicating applications.

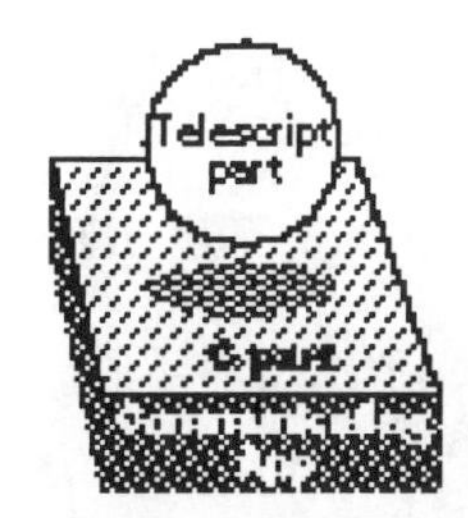

The Telescript language has a number of qualities that facilitate the development of communicating applications:

■ *Complete.* Any algorithm can be expressed in the language. An agent can be programmed to make decisions; to handle exceptional conditions; and to gather, organize, analyze, create, and modify information.

■ *Object-oriented.* The programmer defines classes of information, one class inheriting the features of others. Classes of a general nature (e.g., Agent) are predefined by the language. Classes of a specialized nature (e.g., Shopping Agent) are defined by communicating application developers.

■ *Dynamic.* An agent can carry an information object from a place in one computer to a place in another. Even if the object's class is unknown at the destination, the object continues to function: its class goes with it.

■ *Persistent.* Wherever it goes, an agent and the information it carries, even the program counter marking its next instruction, are safely stored in nonvolatile memory. Thus the agent persists despite computer failures.

■ *Portable and safe.* A computer executes an agent's instructions through a Telescript engine, not directly. An agent can execute in any computer in which an engine is installed, yet cannot access directly its processor, memory, file system, or peripheral devices. This helps prevent viruses.

■ *Communication-centric.* Certain instructions in the language, several of which have been discussed, let an agent carry out complex networking tasks: transportation, navigation, authentication, access control, etc.

A Telescript program takes different forms at different times. Developers deal with a high-level, compiled language not unlike C++. Engines deal with a lower-level, interpreted language. A compiler translates between the two.

Engine

The Telescript engine is a software program that implements the Telescript language by maintaining and executing places within its purview, as well as the agents occupying those places. An engine in a user computer might house only a few places and agents. The engine in a server might house thousands.

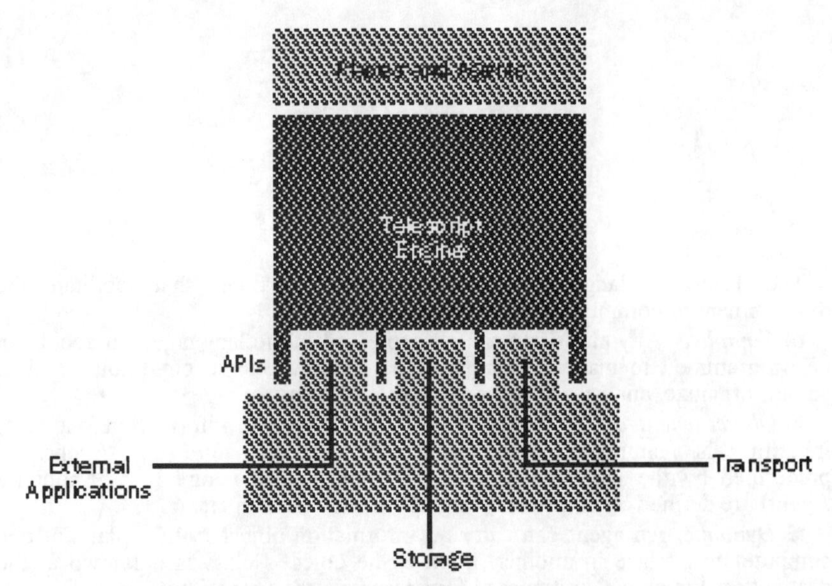

At least conceptually the engine draws upon the resources of its host computer through three application program interfaces (APIs). A storage API let the engine access the nonvolatile memory it requires to preserve places and agents in case of a computer failure. A transport API let the engine access the communication media it requires to transport agents to and from other engines. An external applications API lets the parts of an application written in the Telescript language interact with those written in C.

Protocols

The Telescript protocol suite enables two engines to communicate. Engines communicate in order to transport agents between them in response to the `go` instruction. The protocol suite can operate over a wide variety of transport networks including those based upon the TCP/IP protocols of the Internet, the X.25 interface of the telephone companies, or even electronic mail.

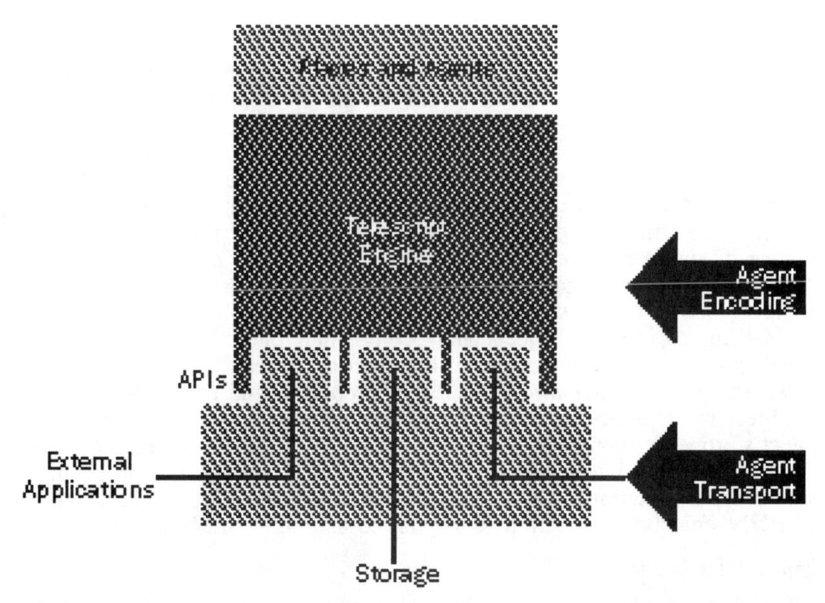

The Telescript protocols operate at two levels. The lower level governs the transport of agents, the higher level their encoding (and decoding). Loosely speaking, the higher level protocol occupies the presentation and application layers of the seven layer Open Systems Interconnection (OSI) model.[1]

The Telescript *encoding rules* specify how an engine encodes an agent—its procedure and its state—as binary data and sometimes omits portions of it as a performance optimization. Although free to maintain agents in different formats for execution, engines must employ a standard format for transport.

The Telescript *platform interconnect protocol* specifies how two engines first authenticate one another (e.g., using public key cryptography) and then transfer an agent's encoding from one to the other. The protocol is a thin veneer of functionality over that of the underlying transport network.

Programming mobile agents

Telescript object model

In this section we explain how a communicating application works. We do this by implementing, in the Telescript language, both an agent and a place. First we explain the low level concepts and terminology that underlie the example. In particular, we sketch the Telescript *object model* which governs how either an agent or a place is constructed from its component parts.

---

[1] The Telescript protocols are not OSI protocols. The OSI model is mentioned here merely to provide a frame of reference for readers acquainted with OSI.

Object structure

The Telescript language, like SmallTalk™, the first object-oriented language, treats every piece of information—however small—as an object. An *object* has both an external interface and an internal implementation.

An object's *interface* consists of attributes and operations. An *attribute*, an object itself, is one of an object's externally visible characteristics. An object can *get* or *set* its own attributes and the *public*, but not the *private*, attributes of other objects. An *operation* is a task that an object *performs*. An object can *request* its own operations and the *public*, but not the *private*, operations of other objects. An operation can accept objects as *arguments* and can *return* a single object as its *result*. An operation can *throw* an exception rather than return. The *exception*, an object, can be *caught* at a higher level of the agent's or place's procedure to which control is thereby transferred.

An object's *implementation* consists of properties and methods. A *property*, an object itself, is one of an object's internal characteristics. Collectively, an object's properties constitute its dynamic state. An object can directly get or set its own properties, but not those of other objects. A *method* is a procedure that performs an operation or that gets or sets an attribute. A method can have *variables*, objects that constitute the dynamic state of the method.

Object classification

The Telescript language, like many object-oriented programming languages, focuses on classes. A *class* is a "slice" of an object's interface combined with a related slice of its implementation. An object is an *instance* of a class.

The Telescript programmer defines his or her communicating application as a collection of classes. To support such *user-defined* classes the language provides many *predefined* classes a variety of which are used by every application. The example application presented in this section consists of several user-defined classes which use various predefined classes[1].

Classes form a hierarchy[2] whose root is Object, a predefined class. Classes other than the Object class inherit the interface and implementation slices of their superclasses. The *superclasses*[3] of a class are the root and the classes that stand between the class and the root. A class is a *subclass* of each of its superclasses. An object is a *member* of its class and each of its superclasses.

A class can both define an operation (or attribute) and provide a method for it. A subclass can provide an overriding method unless the class *seals* the operation. The overriding method can invoke the overridden method by

---

[1] The example uses the predefined Agent, Class, Class Name, Dictionary, Event Process, Exception, Integer, Meeting Place, Nil, Object, Part Event, Permit, Petition, Place, Resource, String, Teleaddress, Telename, Ticket, and Time classes and various subclasses of Exception (e.g., Key Invalid).

[2] The language permits a limited form of multiple inheritance by allowing other classes which extend the hierarchy to a directed graph.

[3] The language permits a class to have implementation superclasses that differ from its interface superclasses. Such classes are rare in practice.

escalating the operation using the language's "^" construct. The overriding method selects and supplies arguments to the overridden method.

One operation, which the Object class defines, is subject to a special escalation rule. The `initialize` operation is requested of each new object. Each method for the operation initializes the properties of the object defined by the class that provides the method. Each method for this operation must escalate it so that all methods are invoked and all properties are initialized.

Object manipulation

The Telescript language requires a method to have *references* to the objects it would manipulate. References serve the purpose of pointers in languages like C, but avoid the "dangling pointer" problem shared by such languages. References can be replicated, so there can be several references to an object.

A method receives references to the objects it creates, the arguments of the operation it implements, and the results of the operations it requests. It can also obtain references to the properties of the object it manipulates.

With a reference to an object in hand, a method can get one of the object's attributes or request one of the object's operations. It accomplishes these simple tasks with two frequently used language constructs, for example:

```
file.length
file.add("isEmployed", true)
```

The example application makes use of the predefined Dictionary class. A dictionary holds pairs of objects, its *keys* and *values*. Assuming `file` denotes a dictionary, the first program fragment above obtains the number of key-value pairs in that dictionary while the second adds a new pair to it. If these were fragments of a method provided by the Dictionary class itself, `file` would be replaced by "`*`", which denotes the object being manipulated.

References are of two kinds, *protected* and *unprotected*. A method cannot modify an object to which it has only a protected reference. The engine intervenes by throwing a member of the predefined Reference Protected class.

Programming a place

The agent and place of the example enable this scene from the electronic marketplace. A shopping agent, acting for a client, travels to a warehouse place, checks the price of a product of interest to its client, waits if necessary for the price to fall to a client-specified level, and returns when either the price reaches that level or a client-specified period of time has elapsed. Beyond the scope of the example are the construction of the warehouse place and the client's construction and eventual debriefing of the shopping agent.

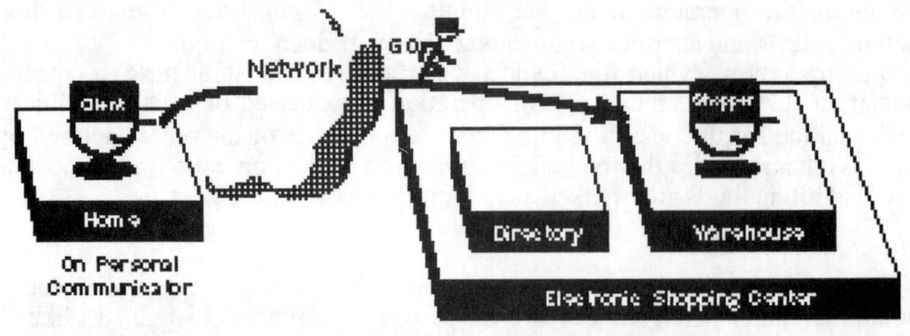

The warehouse place and its artifacts are implemented by three user-defined classes, each of which is presented and discussed below.

The Catalog Entry class

The user-defined Catalog Entry class implements each entry of the warehouse's catalog, which lists the products the warehouse place offers for sale. Implicitly below, this class is a subclass of the predefined Object class.

A catalog entry has two public attributes and two public operations. The `product` attribute is the name of the product the catalog entry describes, the `price` attribute its price. The two operations are discussed below.

```
CatalogEntry: class =
(
  public
    product: String;
    price: Integer; // cents
    see initialize
    see adjustPrice
  property
    lock: Resource;
);
```

The special `initialize` operation initializes the properties of a new catalog entry. There are three. The `product` and `price` properties, implicitly set to the operation's arguments, serve as the `product` and `price` attributes. The `lock` property, set by the method to a new resource, is discussed below.

```
initialize: op (
  product: String;
  price: Integer /* cents */ ) =
{
  ^();
  lock = Resource()
};
```

A catalog entry uses a resource to serialize price modifications made using its `adjustPrice` operation. A Telescript *resource* enables what some languages call critical conditional regions. Here the resource is used to prevent the warehouse place and an agent of the same authority, for example, from changing a product's price simultaneously and, as a consequence, incorrectly.

The public adjustPrice operation adjusts the product's price by the percentage supplied as the operation's argument. A positive percentage represents a price increase, a negative percentage a price decrease.

```
adjustPrice: op (percentage: Integer)
throws ReferenceProtected =
{
  use lock
  {
    price = price + (price*percentage).quotient(100)
  }
};
```

A catalog entry, as mentioned before, uses a resource to serialize price modifications. Here the language's use construct excludes one agent (or place) from the block of instructions in braces, so long as another is executing them.

The operation may throw an exception. If the catalog entry is accessed using a protected reference, the Engine throws a member of the predefined Reference Protected class. If the shopping agent, for example, rather than the warehouse place, tried to change the price, this would be the consequence.

The Warehouse class

The user-defined Warehouse class implements the warehouse place itself. This class is a subclass of the predefined Place and Event Process classes.

A warehouse has three public operations which are discussed below.

```
Warehouse: class (Place, EventProcess) =
(
  public
    see initialize
    see live
    see getCatalog
  property
    catalog: Dictionary[String, CatalogEntry];
);
```

The special initialize operation initializes the one property of a new warehouse place. The catalog property, implicitly set to the operation's argument, is the warehouse place's catalog. Each key of this dictionary is assumed to equal the product attribute of the associated catalog entry.

```
initialize: op (
  catalog: owned Dictionary[String, CatalogEntry]) =
{
  ^()
};
```

A region can prevent a place from being constructed in that region the same way it prevents an agent from traveling there (see "Permits"). Thus a region can either prevent or allow warehouse places and can control their number.

The special live operation operates the warehouse place on an ongoing basis. The operation is special because the engine itself requests it of each new place. The operation gives the place autonomy. The place *sponsors* the operation, that is, performs it under its authority and subject to its permit. The operation never finishes; if it did the engine would terminate the place.

```
live: sponsored op (cause: Exception|Nil) =
{
  loop {
    // await the first day of the month
    time: = Time();
    calendarTime: = time.asCalendarTime();
    calendarTime.month = calendarTime.month + 1;
    calendarTime.day = 1;
    *.wait(calendarTime.asTime().interval(time));

    // reduce all prices by 5%
    for product: String in catalog
    {
      try { catalog[product].adjustPrice(-5) }
      catch KeyInvalid { }
    };

    // make known the price reductions
    *.signalEvent(PriceReduction(), 'occupants)
  }
};
```

On the first of each month, unbeknownst to its customers, the warehouse place reduces by 5% the price of each product in its catalog. It signals this event to any agents present at the time. A Telescript *event* is an object with which one agent or place reports an incident or condition to another.

The public `getCatalog` operation gets the warehouse's catalog, that is, returns a reference to it. If the agent requesting the operation has the authority of the warehouse place itself, the reference is an unprotected reference. If the shopping agent, however, requests the operation, the reference is protected.

```
getCatalog: op () Dictionary[String, CatalogEntry] =
{
  if sponsor.name.authority == *.name.authority
{catalog}
  else {catalog.protect()@}
};
```

One agent or place, as mentioned before, can discern the authority of another. Using the language's `sponsor` construct, the warehouse place obtains a reference to the agent under whose authority the catalog is requested. The place decides whether to return to the agent a protected or unprotected reference to the catalog by comparing their `name` attributes.

The Price Reduction class

The user-defined Price Reduction class implements each event that the warehouse place might signal to notify its occupants of a reduction in a product's price. This class is a subclass of the predefined Event class.

```
PriceReduction: class (Event) = ();
```

Programming an agent

Once it opens its doors the warehouse needs customers. Shopping agents are implemented by the two user-defined classes presented and discussed below.

The Shopper class

The user-defined Shopper class implements any number of shopping agents. This class is a subclass of the predefined Agent and Event Process classes.

A shopping agent has four public operations and two private ones, all of which are discussed individually below.

```
Shopper: class (Agent, EventProcess) =
(
  public
    see initialize
    see live
    see meeting
    see getReport
  private
    see goShopping
    see goHome
  property
    clientName: Telename; // assigned
    desiredProduct: String;
    desiredPrice, actualPrice: Integer; // cents
    exception: Exception|Nil;
);
```

The special `initialize` operation initializes the five properties of a new shopping agent. The `clientName` property, set by the operation's method to the telename of the agent creating the shopping agent, identifies its client. The `desiredProduct` and `desiredPrice` properties, implicitly set to the operation's arguments, are the name of the desired product and its desired price. The `actualPrice` property is not set initially. If the shopping agent finds the desired product at an acceptable price, it will set this property to that price. The `exception` property is set by the method to a nil. If it fails in its mission the agent will set this property to the exception it encountered.

```
initialize: op (
  desiredProduct: owned String;
  desiredPrice: Integer) =
{
  ^();
  clientName = sponsor.name.copy()
};
```

A region can prevent an agent from being constructed in that region the same way it prevents one from traveling there (see "Permits"). Thus a region can either prevent or allow shopping agents and can control their number.

The special `live` operation operates the shopping agent on an ongoing basis. The engine requests the operation of each new agent. The new agent, like a new place, sponsors the operation and gains autonomy by virtue of it. When the agent finishes performing the operation, the engine terminates it.

```
live: sponsored op (cause: Exception|Nil) =
{
   // take note of home
   homeName: = here.name;
   homeAddress: = here.address;

   // arrange to get home
   permit: = Permit(
      (if *.permit.age      == nil {nil}
       else {(*.permit.age     *90).quotient(100)}),
      (if *.permit.charges == nil {nil}
       else {(*.permit.charges*90).quotient(100)})
   );

   // go shopping
   restrict permit
   {
      try { *.goShopping(Warehouse.name) }
      catch e: Exception { exception = e }
   }
   catch e: PermitViolated { exception = e };

   // go home
   try { *.goHome(homeName, homeAddress) }
   catch Exception { }
};
```

The shopping agent goes to the warehouse and later returns. The private `goShopping` and `goHome` operations make the two legs of the trip after the present operation records as variables the telename and teleaddress of the starting place. A *teleaddress* is data that denotes a place's network location.

Using the language's `restrict` construct the shopping agent limits itself to 90% of its allotted time and computation. It holds the remaining 10% in reserve so it can get back even if the trip takes more time or energy than it had anticipated. The agent catches and records exceptions, including the one that would indicate that it had exceeded its self-imposed permit.

The private `goShopping` operation is requested by the shopping agent itself. The operation takes the agent to the warehouse place, checks the price of the requested product, waits if necessary for the price to fall to the requested level, and returns either when that level is reached or after the specified time interval. If the actual price is acceptable to its client, the agent records it.

```
goShopping: op (warehouse: ClassName)
throws ProductUnavailable =
{
   // go to the warehouse
   *.go(Ticket(nil, nil, warehouse));

   // show an interest in prices
   *.enableEvents(PriceReduction(*.name));
   *.signalEvent(PriceReduction(), 'responder);
   *.enableEvents(PriceReduction(here.name));
```

```
  // wait for the desired price
  actualPrice = desiredPrice+1;
  while actualPrice > desiredPrice
  {
    *.getEvent(nil, PriceReduction());
    try
    {
      actualPrice =

  here@Warehouse.getCatalog()[desiredProduct].price
    }
    catch KeyInvalid { throw ProductUnavailable() }
  }
};
```

The shopping agent travels to the warehouse place using the go operation. Upon arrival there the agent expresses interest in the price reduction event which it knows the place will signal. Each time it sees a price reduction, the agent checks the product's price to see whether it was reduced and was reduced sufficiently. The agent contrives one such event to prompt an initial price check. If a price reduction is insufficient, the agent waits for another.

The agent provides the go operation with a ticket specifying the warehouse place's class but neither its telename nor its teleaddress. In an electronic marketplace of even moderate size this would not suffice. The agent would have to travel to a directory place to get the place's name, address, or both.

The operation may throw an exception. If the warehouse doesn't carry the product, a member of the user-defined Product Unavailable class is thrown.

The private goHome operation is requested by the shopping agent itself. The operation returns the agent to its starting place and initiates a meeting with its client. Before initiating the meeting, the agent asks to be signaled when it ends. After initiating the meeting, the agent just waits for it to end. During the meeting, the client is expected to request the getReport operation.

```
goHome: op (homeName: Telename; homeAddress:
Teleaddress) =
{
  // drop excess baggage
  *.disableEvents();
  *.clearEvents();

  // go home
  *.go(Ticket(homeName, homeAddress));

  // meet the client
  *.enableEvents(PartEvent(clientName));
  here@MeetingPlace.meet(Petition(clientName));

  // wait for the client to end the meeting
  *.getEvent(nil, PartEvent(clientName))
};
```

The shopping agent leaves the warehouse place using the `go` operation. Before leaving it retracts its interest in price reductions and, to lighten its load, discards any notices of price reductions it received but did not examine.

The agent provides the `go` operation with a ticket giving the telename and teleaddress of the agent's starting place, information it recorded previously.

The special `meeting` operation guards the agent's report by declining all requests to meet with the shopping agent. The agent itself initiates the one meeting in which it will participate. The operation is special because the engine itself requests it whenever a meeting is requested of an agent.

```
meeting: sponsored op (
  agent: protected Telename; // assigned
  _class: protected ClassName;
  petition: protected Petition) Object|Nil
throws MeetingDenied =
{
  throw MeetingDenied();
  nil
};
```

The operation may throw an exception. Indeed the shopping agent always throws a member of the predefined Meeting Denied class.

The public `getReport` operation returns the actual price of the desired product. The actual price is less than or equal to the desired price.

```
getReport: op () Integer // cents
throws Exception, FeatureUnavailable =
{
  if sponsor.name != clientName
    { throw FeatureUnavailable() };
  if exception != nil
    { throw exception };
  actualPrice
};
```

The operation may throw an exception. If the agent requesting the operation is not the shopping agent's client, the operation's method throws a member of the predefined Feature Unavailable class. If the shopping agent failed in its mission, the method throws a member of the predefined Exception class.

The Product Unavailable class

The user-defined Product Unavailable class implements each exception with which the shopping agent might notify its client that the warehouse doesn't carry the product. This class is a subclass of the predefined Exception class.

```
ProductUnavailable: class (Exception) = ();
```

Monitoring changing conditions

Having explained in the previous section how communicating applications work, we speculate in this final section about the myriad of applications that third-party developers could call into being. Each of three subsections adopts a theme, develops one variation on that theme, and sketches four others. These are the promised scenes from the electronic marketplace of the future.

The user experience

Two weeks from now, Chris must make a two-day business trip to Boston. He makes his airline reservations using his personal communicator. He's ready to go. Chris's schedule in Boston proves hectic. On the second day, he's running late. Two hours before his return flight is scheduled to leave, Chris's personal communicator informs him that the flight has been delayed an hour. That extra hour lets Chris avoid cutting short his last appointment.

The hour Chris saved was important to him. He could have called the airline to see whether his flight was on time, but he was extremely busy. Chris was startled—pleasantly so—when notice of his flight's delay hit the screen of his personal communicator. When he used his communicator to arrange his trip two weeks ago, Chris had no idea that this was part of the service.

How agents provide the experience

Chris can thank one mobile agent for booking his round-trip flight to Boston and another for monitoring his return flight and notifying him of its delay. The first of these two tasks was accomplished in the following steps:

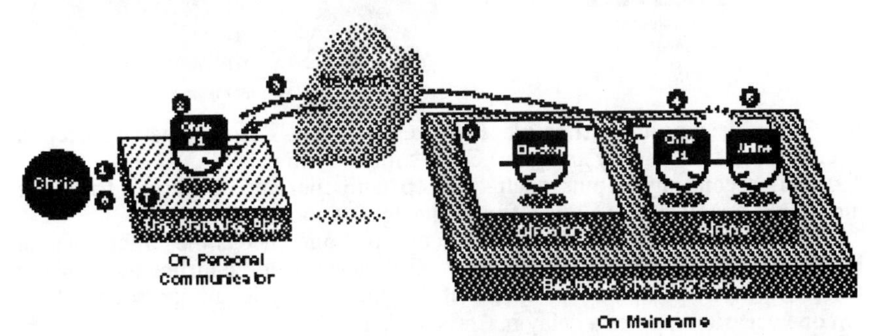

1. Chris gives to the trip planning application he bought for his personal communicator the dates of his trip, his means of payment (e.g., the number and expiration date on his Visa card), his choice of airline, etc. If he's used the application before, it has much of this information already.
2. The application creates an agent of Chris's authority and gives Chris's flight information to it. The part of the application written in C creates and interacts with the part written in the Telescript language, the agent, through the Telescript engine in Chris's personal communicator.

3. The agent travels from Chris's communicator to the airline place in the electronic marketplace. It does this using the `go` instruction and a ticket that designates the airline place by its authority and class.
4. The agent meets with the airline agent that resides in and provides the service of the airline place. It does this using the `meet` instruction and a petition that designates the airline agent by its authority and class.
5. The agent gives Chris's flight information to the airline agent, which compares the authority of Chris's agent to the name on Chris's Visa card and then books his flight, returning a confirmation number and itinerary.
6. The agent returns to its place in Chris's communicator. It does this using the `go` instruction and a ticket that designates that place by its telename and teleaddress which the agent noted before leaving there.
7. The agent gives the confirmation number and itinerary to the trip planning application. Its work complete, the agent terminates.
8. The application conveys to Chris the confirmation number and itinerary, perhaps making an entry in his electronic calendar as well.

The remaining task of monitoring Chris's return flight and informing him if it's delayed is carried out in the following additional steps:

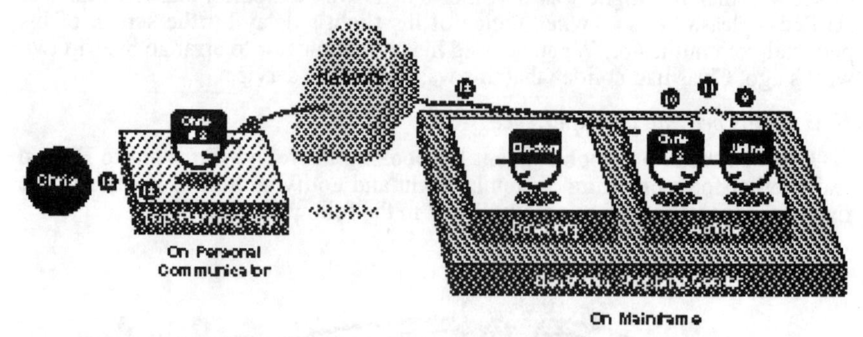

9. Before leaving the airline place (in step 6), Chris's agent creates a second agent of Chris's authority and gives Chris's itinerary to it.
10. This second agent puts itself to sleep until the day of Chris's trip. The airline place may charge Chris a fee for the agent's room and board.
11. On the day of Chris's flight, the agent arises and checks the flight once an hour throughout the day. On each occasion it meets with the airline agent using the `meet` instruction and a petition that designates it by its authority and class. On one occasion it notes a delay in Chris's flight.
12. The agent returns to Chris's personal communicator (as in step 6), the agent notifies the trip planning application of the delay in Chris's return flight and then terminates (as in step 7), and the application gives Chris the information that allows him to complete his meeting (as in step 8).

Variations on the theme

This first scenario demonstrates how mobile agents can monitor changing conditions in the electronic marketplace. There are many variations:

■ Chris learns by chance that the Grateful Dead are in town next month. He tries to get tickets but learns that the concert sold out in 12 hours. Thereafter Chris's agent monitors Ticketron every morning at 9 am. The next time a

Grateful Dead concert in his area is listed, the agent snaps up two tickets. If Chris can't go himself, he'll sell the tickets to a friend.

■ Chris buys a television from The Good Guys. Chris's agent monitors the local consumer electronics market for 30 days after the purchase. If it finds the same set for sale at a lower price, the agent notifies Chris so that he can exercise the low price guarantee of the store he patronized.

■ Chris invests in several publicly traded companies. Chris's agent monitors his portfolio, sending him biweekly reports and word of any sudden stock price change. The agent also monitors the wire services, sending Chris news stories about the companies whose stock he owns.

■ Mortgage rates continue to fall. Chris refinances his house at a more favorable rate. Thereafter Chris's agent monitors the local mortgage market and notifies him if rates drop 1% below his new rate. With banks forgoing closing costs, such a drop is Chris's signal to refinance again.

Doing time-consuming legwork

The user experience

John's in the market for a camera. He's read the equipment reviews in the photography magazines and in *Consumer Reports* and has visited his local camera store. He's buying a Canon EOS A2. The only remaining question is: from whom? John asks his personal communicator. In 15 minutes, he has the names, addresses, and phone numbers of the three shops in his area with the lowest prices. A camera store in San Jose, 15 miles away, offers the A2 at $70 below the price his local camera shop is asking.

The $70 that John saved, needless to say, was significant to him. John could have consulted the three telephone directories covering his vicinity, made a list of the 25 camera retailers within, say, 20 miles of his office, and called each to obtain its price for the EOS A2, but who has the time? John now considers his personal communicator to be an indispensable shopping tool.

How agents provide the experience

John can thank a mobile agent for finding the camera store in San Jose, a task that was accomplished in the following steps:

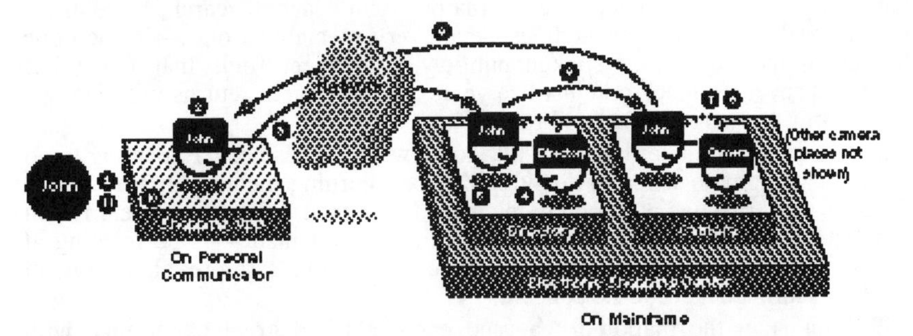

1. John gives to the shopping application he bought for his personal communicator the make and model of the camera he's selected. He also identifies the geographical area for which he wishes pricing information.

2. The application creates an agent of John's authority and gives John's shopping instructions to it. The C part of the application creates and interacts with the agent through the engine in John's communicator.
3. The agent travels from John's communicator to the directory place in the electronic marketplace. It uses the `go` instruction and a ticket that designates the directory place by its authority and class.
4. The agent meets with the directory agent that resides in and provides the service of the directory place. It uses the `meet` instruction and a petition that designates the directory agent by its authority and class.
5. The agent obtains from the directory agent the directory entries for all camera retailers about which the place has information. John's agent narrows the list to the retailers in the geographical area it is to explore.
6. The agent visits the electronic storefront of each retailer in turn. Each storefront is another place in the electronic marketplace. For each trip the agent uses the `go` instruction with a ticket that gives the telename and teleaddress that the agent found in the storefront's directory entry.
7. The agent meets with the camera agent it finds in each camera place it visits. It uses the `meet` instruction and a petition that designates the camera agent by its authority and class.
8. The agent gives to the camera agent the camera's make and model and is quoted a price. The agent retains information about this particular shop only if it proves a candidate for the agent's top-three list.
9. The agent eventually returns to its place in John's communicator. It does this using the `go` instruction and a ticket that designates that place by its telename and teleaddress which the agent noted before leaving there.
10. The agent makes its report to the shopping application. Its work complete, the agent terminates.
11. The application presents the report to John, perhaps making an entry in his electronic diary as a permanent record.

Variations on the theme

This second scenario demonstrates how mobile agents can find and analyze information in the electronic marketplace. There are many variations:

■ John hasn't talked to his college friend, Doug, in 20 years. He remembers that Doug was a computer science major. John's agent searches the trade journals and conference proceedings—even very specialized ones—in the hope that Doug has written or spoken publicly. The agent finds that Doug has published several papers, one just two years ago. The agent returns with Doug's address in LA where he has lived for 5 years.

■ It's Friday. John has been in Chicago all week on business. Expecting to go home today, John is asked to attend a Monday morning meeting in New York. He faces an unplanned weekend in midtown Manhattan. John's agent learns that Charles Aznavour—John has every recording he ever made—is performing at Radio City Music Hall on Saturday night. With John's approval, the agent purchases him a ticket for the concert.

■ John is in the market for a used car. He's had great experience with Toyota's. John's agent checks the classified sections of all 15 Bay Area newspapers and produces for John a tabular report that includes all used Toyota's on the market. The report lists the cars by model, year, and mileage, allowing John to easily make comparisons between them.

■ John yearns for a week in Hawaii. His agent voices his yearning in the electronic marketplace, giving details John has provided: a few days on Kauai, a few more on Maui, beach-front accommodations, piece and quiet. The agent returns with a dozen packages from American Airlines, Hilton Hotels, Aloha Condominiums, Ambassador Tours, etc.. Unlike the junk mail John receives by post, many of these offers are designed specifically for him. The marketplace is competing for his business.

Using services in combination

The user experience

Mary and Paul have been seeing each other for years. Both lead busy lives. They don't have enough time together. But Mary's seen to it that they're more likely than not to spend Friday evenings together. Using her personal communicator she's arranged that a romantic comedy is selected and ready for viewing on her television each Friday at 7 pm, that pizza for two is delivered to her door at the same time, and that she and Paul are reminded earlier in the day of their evening together and of the movie to be screened.

Paul and Mary recognize the need to live well-rounded lives but their demanding jobs make it difficult. Their personal communicators help them achieve their personal, as well as their professional objectives. And it's fun.

How agents provide the experience

Mary relies upon a mobile agent to orchestrate her Friday evenings. Born months ago, the agent waits in a quiet corner of the electronic marketplace for most of each week; each Friday at noon it takes the following steps:

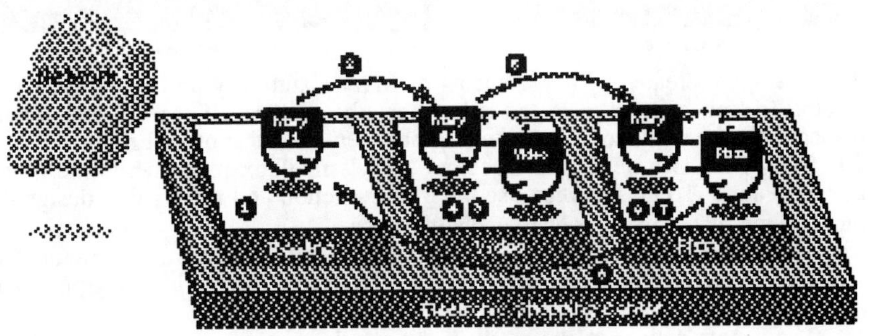

1. Mary's agent keeps a record of the films it selected on past occasions in order to avoid selecting one of those films again.
2. The agent travels from its place of repose to one of the many video places in the electronic marketplace. It uses the `go` instruction and a ticket that designates the video place by its authority and class.
3. The agent meets with the video agent that resides in and provides the service of the video place. It uses the `meet` instruction and a petition that designates the video agent by its authority and class.
4. The agent asks the video agent for the catalog listing for each romantic comedy in its inventory. The agent selects a film at random from among the

more recent comedies, but avoids the films it's selected before, whose catalog numbers it carries with it. The agent orders the selected film from the video agent, charges it to Mary's Visa card, and instructs the video agent to transmit the film to her home at 7 pm. The video agent compares the authority of Mary's agent to the name on the Visa card.

5. The agent goes next to Domino's pizza place. It uses the `go` instruction and a ticket that designates the pizza place by its authority and class.

6. The agent meets with the pizza agent that resides in and provides the service of the pizza place. It uses the `meet` instruction and a petition that designates the pizza agent by its authority and class.

7. The agent orders one medium-size Pepperoni pizza for home delivery at 6:45 pm. The agent charges the pizza, as it did the video, to Mary's Visa card. The pizza agent, like the video agent before it, compares the authority of Mary's agent to the name on the agent's Visa card.

8. Mary's agent returns to its designated resting place in the electronic marketplace. It uses the `go` instruction and a ticket that designates that place by its telename and teleaddress, which it noted previously.

What remains is for the agent to notify Mary and Paul of their evening appointment. This is accomplished in the following additional steps:

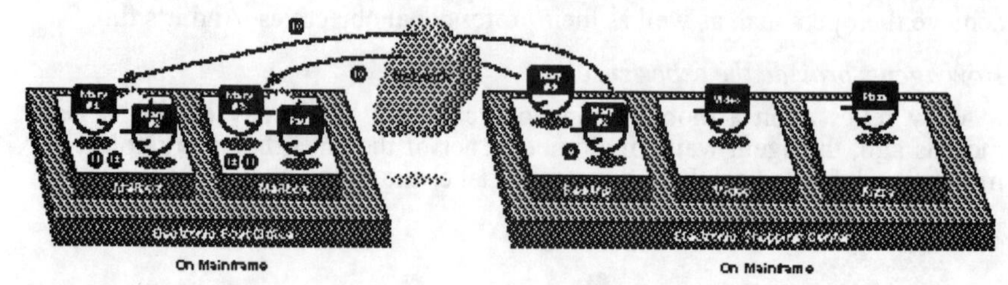

9. The agent creates two new agents of Mary's authority and gives each the catalog listing of the selected film and the names of Mary and Paul, respectively. Its work complete, the original agent awaits another Friday.

10. One of the two new agents goes to Mary's mailbox place while the other goes to Paul's. To do this they use the `go` instruction and tickets that designate the mailbox places by their class and authorities.

11. The agents meet with the mailbox agents that reside in and provide the services of the mailbox places. They use the `meet` instruction and petitions designating the mailbox agents by their class and authorities.

12. The agents deliver to the mailbox agents electronic messages that include the film's catalog listing and that remind Mary and Paul of their appointment with one another. The two agents terminate. The mailbox agents convey the reminders to Mary and Paul themselves.

Variations on the theme

This third scenario demonstrates how mobile agents can combine existing services to create new, more specialized services. There are many variations:

■ Mary plans to take Paul to see Phantom of the Opera next weekend. Her agent tries to book orchestra seats for either Saturday or Sunday, gets them for Sunday, reserves a table at a highly regarded Indian restaurant within walking distance of the theater, and orders a dozen roses for delivery to Paul's apartment that Sunday morning.

■ Mary's travel plans change unexpectedly. Rather than return home this evening as planned, she's off to Denver. Mary's agent alters her airline reservation, books her a non-smoking room at a Marriott Courtyard within 15 minutes of her meeting, reserves her a compact car, and provides her driving instructions to the hotel and the meeting. The agent also supplies Mary with a list of Indian restaurants in the vicinity.

■ Mary receives and pays her bills electronically. Her agent receives each bill as it arrives, verifies that Mary has authorized its payment, checks that it is in the expected range, and issues instructions to the bank if so. Mary's agent prepares for her a consolidated monthly report and, at tax time, sends to her accountant a report of her tax deductible expenses.

■ Mary takes a daily newspaper but it isn't her idea of news. At 7 am each day Mary's agent delivers to her a personalized newspaper. It includes synopses of the major national and international news stories of course, but it also reports the local news from her hometown in Virginia; the major events of yesterday in her field, physics; the market activity of the stocks in her portfolio; and Doonesbury which always gives her a chuckle.

Acknowledgments

A talented team of software engineers at General Magic led by Steve Schramm conceived Telescript technology and implemented it for servers. A second team at AT&T led by Alex Gillon used the technology to create a commercial service, AT&T's PersonaLink™. A third team at General Magic led by Darin Adler implemented the technology for user computers, initially Sony's MagicLink™ and Motorola's Envoy™ personal communicators. In July 1994 Nippon Telegraph and Telephone, AT&T, and Sony announced a joint venture to deploy a second mobile agent-based service in Japan.

General Magic licenses its technologies to members of a global alliance of computer, communication, and consumer electronics companies which include Apple, AT&T, Cable and Wireless, France Telecom, Fujitsu, Matsushita, Mitsubishi, Motorola, NorTel, NTT, Oki, Philips, Sanyo, Sony, and Toshiba.

Related work

Computer scientists have long explored how programming languages tailored for the purpose could ease the development of communicating applications. An overview of their work is found in [Bal, Steiner, and Tanenbaum 1989].

The RPC paradigm was conceived and first implemented in the mid-1970's [White 1976], was formalized and in other ways advanced in the mid-1980's [Birrell and Nelson 1984], and is today the basis for client-server computing.

The RP paradigm in its most basic form involves transporting procedures before they begin executing, not after (i.e., before they develop state). A persuasive argument for the basic paradigm is made in [Gifford and Stamos 1990]. The basic paradigm has been implemented in many settings. In the Internet the most recent effort is Java [Gosling and McGilton 1995].

The RP paradigm in the form described in this chapter involves transporting executing procedures. The case for this more advanced form is made in [Chess, Harrison, and Kershenbaum 1994]. An early noncommercial implementation is Emerald [Black, Hutchinson, Jul, and Levy 1988]. In the Internet the most recent undertaking is Obliq [Bharat and Cardelli 1995].

The RP paradigm arose not only in the field of programming languages but also in that of electronic mail, where procedures can be transported as the contents of messages. One of the earliest efforts is described in [Vittal 1981]. In the Internet the most recent work is Safe-Tcl [Ousterhout 1995].

If mobile agents actually become an important element of the electronic marketplace, standards for mobile agents will arise and technologies for mobile agents will one day be taken for granted. Attention will shift to higher-level matters, for example, agent strategies for effective negotiation. A foretaste of the work to come is found in [Rosenschein and Zlotkin 1994].

References

Bal, H. E.; Steiner, J. G.; and Tanenbaum, A. S. 1989. Programming Languages for Distributed Computing Systems, *ACM Computing Surveys* 21(3).

Bharat, K. A. and Cardelli, L. 1995. Migratory Applications, *Draft of Paper for Publication*. DEC Systems Research Center.

Birrell, A. D. and Nelson, B. J. 1984. Implementing remote procedure calls. *ACM Transactions on Computer Systems* 2(1): 39-59.

Black, A.; Hutchinson, N.; Jul, E.; and Levy, H. 1988. Fine-grained mobility in the Emerald System. *ACM Transactions on Computer Systems* 6(1): 109-133.

Chess, D. M.; Harrison, C. G. and Kershenbaum, A. 1994. Mobile Agents: Are they a good idea? *IBM Research Report*, RC 19887.

Gifford, D. K. and Stamos, J. W. 1990. Remote Evaluation. *ACM Transactions on Programming Languages and Systems* 12(4): 537-565.

Gosling, J. and McGilton, H. 1995. *The Java Language Environment: A White Paper*. Sun Microsystems.

Ousterhout, J. K. 1995. Scripts and Agents: The New Software High Ground, *Invited Talk at the Winter 1995 USENIX Conference*, New Orleans, LA.

Rosenschein, J. S. and Zlotkin, G. 1994. *Rules of Encounter*. MIT Press.

Vittal, J. 1981. Active Message Processing: Messages as Messengers. *Computer Message Systems*, North-Holland Publishing Company.

White, J. E. 1976. A high-level framework for network-based resource sharing. *AFIPS Conference Proceedings* 45: 561-570.

Language-Neutral Virtual Machines

The mobile agent idea first arose in 1988 in reaction to the complexity of OSI application protocols, such as X.400 and X.500. Rather than standardize a lot of functionality, why not allow clients to dynamically introduce the functionality they need? Our inspiration came from PostScript. We asked ourselves, "PostScript is to printing as ______ is to networking."

The essential requirement for mobile agents (and for mobile code in general) is not a universally accepted high-level language, but merely a universal instruction set, or virtual machine. This is where work on Telescript began. We developed a stack-oriented instruction set (Low Telescript) and only much later an expression-oriented language (High Telescript). The instruction set enabled the movement of programs and data and so was essential. The language merely facilitated program development.

The commercial introduction of Telescript in 1994 was quickly followed by the commercial introduction of Java, which won the title of The Network Programming Language. Like Telescript, Java enables code mobility by means of a virtual machine. Unlike Telescript, Java also aspires to be a general-purpose platform for application development by means of a so far open-ended collection of operating system-like libraries.

The Telescript and Java virtual machines share one important trait: they institutionalize a particular object model. In my view, this is a mistake -- technically, commercially, and culturally. A better approach is a virtual machine that is language-neutral -- for example, a virtual RISC processor. Technically, such a virtual machine would evenhandedly and compatibly enable any number of high-level languages for distributed computing. Commercially, it would avoid both the Object Model War and the Platform War in which Java is presently engaged. Culturally, such a virtual machine would honor the Internet's tradition of programming language neutrality in application protocol design.

The Telescript and Java virtual machines depart on one important point. Telescript has a strong notion of process, while Java has a strong notion of thread, but little or no notion of process. The process abstraction is a prerequisite for mobile agents (and, I would have thought, for servlets) in that it defines an agent's extent for purposes of capturing the agent's state whenever it moves, bounding its consumption of "computational" resources (processor time and storage space), and authorizing its use of "service" resources (the functionality offered by the machines agents visit, as well as the services that agents render to one another).

If I were developing a mobile agent system from scratch today, I'd employ a language-neutral virtual machine able to sustain any number of independent processes. For inter-process communication, I'd simply provide a means for the exchange of XML strings.

Jim White
Kaneohe, Hawaii
August 1998

14.2 Lange, D. and Oshima, M., Mobile Agents with Java: The Aglet API

This article is originally based on a chapter of the book by Lange and Oshima entitled "Programming and Deploying Mobile Agents with JAVA," Addison Wesley Longman 1998.

Lange, D. and Oshima, M., "Mobile Agents with Java: The Aglet API," *World Wide Web*, 1(3), September 1998.

Mobile Agents with Java: The Aglet API[1]

Danny B. Lange
General Magic Inc.
Sunnyvale, California, U.S.A.
danny@acm.org

Mitsuru Oshima
IBM Tokyo Research Laboratory
Yamato, Kanagawa, Japan
moshima@trl.ibm.co.jp

Abstract

Java, the C++-like language that changed the Web overnight, offers some unique capabilities that are fueling the development of mobile agent systems. In this article we will show what exactly it is that makes Java such a powerful tool for mobile agent development. We will also direct the attention to some shortcomings in Java language systems that have implications for the conceptual design and use of Java-based mobile agent systems. At last, but not least we will introduce the aglet – a Java-based agile agent. We will give an overview of the aglet, its application programming interface, and present a real-world example of its use in electronic commerce.

1 Introduction

Mobile agents are an emerging technology that promises to make it very much easier to design, implement, and maintain distributed systems [7]. We have in our research and development found that mobile agents reduce the network traffic, provide an effective means of overcoming network latency, and perhaps most importantly, through their ability to operate asynchronously and autonomously of the process that created them, help us to construct more robust and fault-tolerant systems.

This is not the place to examine the characteristics of the numerous agent systems made available to the public by many research and development labs. But if we looked at all these systems, we would find that a property shared by all agents is that fact that they live in some environment. They have the ability to interact with their execution environment, and to act asynchronously and autonomously upon it. No one is required either to deliver information to the agent or to consume any of its output. The agent simply acts continuously in pursuit of its own goals.

We define agents in the following way: An agent is a software object that

- is situated within an execution environment;

---

[1] This article is based on a chapter of a forthcoming book by Lange and Oshima entitled *Programming and Deploying Mobile Agents with Java* (Addison-Wesley.) [4]

- possesses the following mandatory properties:

  Reactive - senses changes in the environment and acts accordingly to those changes;
  Autonomous - has control over its own actions;
  Goal driven - is pro-active;
  Temporally continuous - is continuously executing;

- and may possess any of the following orthogonal properties:

  Communicative - able to communicate with other agents;
  Mobile - can travel from one host to another;
  Learning - adapts in accordance with previous experience;
  Believable - appears believable to the end-user.

Mobility is an orthogonal property of agents. That is, all agents do not necessarily *have* to be mobile. An agent can just sit there and communicate with the surroundings by conventional means. These means include various forms of remote procedure calling and messaging. We call agents that do not or cannot move *stationary agents.* In contrast, a *mobile* agent is not bound to the system where it begins execution. The mobile agent is free to travel among the hosts in the network. Created in one execution environment, it can transport its state and code with it to another execution environment in the network, where it resumes execution.

By the term "state," we typically understand the agent attribute values that help it determine what to do when it resumes execution at its destination. By the term "code," we understand, in an object-oriented context, the class code necessary for the agent to execute. We define agents in the following way: A mobile agent is not bound to the system where it begins execution. It has the unique ability to transport itself from one system in a network to another. The ability to travel, allows a mobile agent to move to a system that contains an object with which the agent wants to interact, and then to take advantage of being in the same host or network as the object.

So does mobile agent technology sound appealing? Our interest in mobile agents should not be motivated by the technology *per se*, but rather by the benefits they provide for the creation of distributed systems. So here are *seven good reasons* to start using mobile agents.

1. **They reduce the network load.** Distributed systems often rely on communications protocols that involve multiple interactions to accomplish a given task. This is especially true when security measures are enabled. The result is a lot of network traffic. Mobile agents allow us to package a conversation and dispatch it to a destination host where the interactions can take place locally. Mobile agents are also useful when it comes to reducing the flow of raw data in the network. When very large volumes of data are stored at remote hosts, these data should be processed in the locality of the data, rather that transferred over the network. The motto is simple: move the computations to the data rather than the data to the computations.

2. **They overcoming network latency.** Critical real-time systems such as robots in manufacturing processes need to respond to changes in their environments in real time. Controlling such systems through a factory network of a substantial size involves significant latencies. For critical real-time systems, such latencies are not acceptable.

Mobile agents offer a solution, since they can be dispatched from a central controller to act locally and directly execute the controller's directions.

3. **They encapsulate protocols.** When data are exchanged in a distributed system, each host owns the code that implements the protocols needed to properly code outgoing data and interpret incoming data, respectively. However, as protocols evolve to accommodate new efficiency or security requirements, it is a cumbersome if not impossible task to upgrade protocol code properly. The result is often that protocols become a legacy problem. Mobile agents, on the other hand, are able to move to remote hosts in order to establish "channels" based on proprietary protocols.

4. **They execute asynchronously and autonomously.** Often mobile devices have to rely on expensive or fragile network connections. That is, tasks that require a continuously open connection between a mobile device and a fixed network will most likely not be economically or technically feasible. Tasks can be embedded into mobile agents, which can then be dispatched into the network. After being dispatched, the mobile agents become independent of the creating process and can operate asynchronously and autonomously. The mobile device can reconnect at some later time to collect the agent.

5. **They adapt dynamically.** Mobile agents have the ability to sense their execution environment and react autonomously to changes. Multiple mobile agents possess the unique ability to distribute themselves among the hosts in the network in such a way as to maintain the optimal configuration for solving a particular problem.

6. **They are naturally heterogeneous.** Network computing is fundamentally heterogeneous, often from both hardware and software perspectives. As mobile agents are generally computer- and transport-layer-independent, and dependent only on their execution environment, they provide optimal conditions for seamless system integration.

7. **They are robust and fault-tolerant.** The ability of mobile agents to react dynamically to unfavorable situations and events makes it easier to build robust and fault-tolerant distributed systems. If a host is being shut down, all agents executing on that machine will be warned and given time to dispatch and continue their operation on another host in the network.

It would be an understatement to say that the Java programming language [2] has revolutionized the mobile agent field. The remainder of this paper is devoted to the application of Java for mobile agents. Most of our presentation is based on the experience gained from creating IBM's Aglets Workbench (`http://www.trl.ibm.co.jp/aglets`)- a system that provides an *applet-like* programming model for mobile agents [4].

This paper is structured as follows: in Section 2 we present the agent characteristics of the Java programming language as well some of its shortcomings; Section 3 and Section 4 introduce the Aglet programming model and the Aglet API; Section 5 reports on an application of aglets in electronic commerce; Section 6 gives a brief overview of contemporary Java-based agent systems; and Section 7 concludes the paper.

2 Agent Characteristics of Java

Java is an object-oriented network-savvy programming language. Some call it a *Better C++* that omits many rarely used, confusing features of C++. Others call it *the* language of the Internet. We think of as *jet fuel* for mobile agents. Let us view some of the properties of Java that make it a good language for mobile agent programming.

Platform-independence. Java is designed to operate in heterogeneous networks. To enable a Java application to execute anywhere on the network, the compiler generates architecture-neutral byte code, as opposed to non-portable native code. For this code to be executed on a given computer, the Java runtime system needs to be present. There are no platform-dependent aspects of the Java language. Primitive data types are rigorously specified and not dependent on the underlying processor or operating system. Even libraries are platform-independent parts of the system. For example, the window library provides a single interface for the GUI that is independent of the underlying operating system. It allows us to create a mobile agent without knowing the types of computers it is going to run on.

Secure execution. Java is intended for use on the Internet and Intranets. The demand for security has influenced the design in several ways. For example, Java has a pointer model that eliminates the possibility of overwriting memory and corrupting data. Java simply does not allow illegal type casting or any pointer arithmetic. Programs are no longer able to forge access to private data in objects that they do not have access to. This prohibits most activities of viruses. Even if someone tampers with the byte code, the Java runtime system ensures that the code will not be able to violate the basic semantics of Java. The security architecture of Java makes it reasonable safe to host an untrusted agent, because it cannot tamper with the host or access private information.

Dynamic class loading. This mechanism allows the virtual machine to load and define classes at runtime. It provides a protective name space for each agent, thus allowing agents to execute independently and safely from each other. The class-loading mechanism in extensible and enables classes to be loaded via the network.

Multithread programming. Agents are by definition autonomous. That is, an agent executes independently of other agents residing within the same place. Allowing each agent to execute in its own lightweight process, also called a thread of execution, is a way of enabling agents to behave autonomously. Fortunately, Java not only allows multithread programming, but it also supports a set of synchronization primitives that are built into the language. These primitives enable agent interaction.

Object serialization. A key feature of to mobile agents is that they can be serialized and deserialized. Java conveniently provides a built-in serialization mechanism that can represent the state of an object in a serialized form sufficiently detailed for the object to be reconstructed later. The serialized form of the object must be able to identify the Java class from which the object's state was saved, and to restore the state in a new instance. Objects often refer to other objects. Those other objects must be stored and retrieved at the same time, to maintain the object structure. When an object is stored, all the objects in the graph that are reachable from that object are stored as well.

Reflection. Java code can discover information about the fields, methods, and constructors of loaded classes, and can use reflected fields, methods, and constructors to operate on their underlying counterparts in objects, all within the security restrictions. Reflection accommodates the need for agents to be smart about themselves and other agents.

Although the Java language system is highly suitable for creating mobile agents, we should be aware of some signification shortcomings. Some of these shortcomings can be worked around, others are more serious and have implications for the overall conceptual design and deployment of Java-based mobile agent systems.

Inadequate support for resource control. The Java language system provides no means for us to control the resources consumed by a Java object. For example, an agent can start looping and waste processor cycles. The agent can also start consuming memory resources. These two examples relate to a specific type of security attack termed *denial of service.* This is the most feared type of attack by mobile agents: Agents swarm into a computer and take over all its resources, making it impossible for the owner to control his or her own computer. Unfortunately, Java provides no ways for the host to limit the processor and memory resources allocated by a given object or thread. A related issue is the ability of the agent to allocate resources external to the program, for example by opening files and sockets, and creating windows. The agent's allocation of these resources can be controlled but once the agent is disposed of or dispatched to another host, these resources must be released. However, it is difficult to do so, because there is no way of binding such resources to a specific object. So what we may see is a mobile agent that "forgets" its GUI and leaves behind an open window on our display when it leaves for another host.

No protection of references. A Java object's public methods are available to any other object that has a reference to it. Since there is no concept of a protected reference, some objects are allowed access to a larger set of public methods in the Java object's interface than others. This is important for the agent. There is no way that it directly can monitor and control which other agents are accessing its methods. We have found that a practical and powerful solution to this problem is to insert a *proxy* object between the caller and the callee to control access. It not only provides protection of references, it also offers a solution to the problem mentioned in the next item.and provides location transparency in general (explained later).

No object ownership of references. No one owns the references to a given object in Java. For an agent, this means that we can take its thread of execution away from it, but we cannot explicitly void the agent (object) itself. This is a task for the automated garbage collector. Now, the garbage collector will not reclaim any object until all references to the object have been voided. So if some other agent has a reference to our agent, it will unavoidably open a loophole for that agent to keep *our* agent alive against our will. All we can do is to repeat the warning against giving away direct references to agents in Java. Upcoming Java Development Kit (JDK) 1.2 has support of weak reference, which will solve the sketched problem.

No support for preservation and resumption of the execution state. It is currently impossible in Java to retrieve the full execution state of an object. Information such as the status of the program counter and frame stack is permanently forbidden territory for Java programs. Therefore, for a mobile agent to properly resume a computation on a remote host, it must rely on internal attribute values and external events to direct it. An embedded automaton can keep track of the agent's travels and ensure that computations are properly halted and properly resumed.

3 The Aglet Model

IBM's Aglets Workbench (Aglets), created by the authors of this article, mirrors the applet model in Java. The goal was to bring the flavor of mobility to the applet. The term *aglet* is indeed a portmanteau word combining *agent* and *applet*. We attempted to make Aglets an exercise in "clean design," and it is our hope that applet programmers will appreciate the many ways in which the aglet model reflects the applet model.

Basic Elements

Now let us take a closer look at the model underlying the Aglet API. This model was designed to benefit from the agent characteristics of Java while overcoming some of the above-mentioned deficiencies in the language system.

Most notably, the model defines a set of abstractions and the behavior needed to leverage mobile agent technology in Internet-like open wide-area networks. The key abstractions are aglet, proxy, context, message, future reply, and identifier:

Aglet. An aglet is a mobile Java object that visits aglet-enabled hosts in a computer network. It is autonomous, since it runs in its own thread of execution after arriving at a host, and reactive, because of its ability to respond to incoming messages.

Proxy. A proxy is a representative of an aglet. It serves as a shield for the aglet that protects the aglet from direct access to its public methods. The proxy also provides location transparency for the aglet. That is, it can hide the real location of the aglet.

Context. A context is an aglet's workplace. It is a stationary object that provides a means for maintaining and managing running aglets in a uniform execution environment where the host system is secured against malicious aglets. One node in a computer network may run multiple servers and each server may host multiple contexts. Contexts are named and can thus be located by the combination of their server's address and their name.

Message. A message is an object exchanged between aglets. It allows for synchronous as well as asynchronous message passing between aglets. Message passing can be used by aglets to collaborate and exchange information in a loosely coupled fashion.

Future reply. A future reply is used in asynchronous message-sending as a handler to receive the result later asynchronously.

Identifier. An identifier is bound to each aglet. This identifier is globally unique and immutable throughout the lifetime of the aglet.

Behavior supported by the aglet object model is based on a careful analysis of the "life and death" of mobile agents. There are basically only two ways to bring an aglet alive: either it is instantiated from scratch (creation) or it is copied from an existing aglet (cloning). To control the population of aglets one can of course destroy aglets (disposal of). Aglets are mobile in two different ways: active and passive. The first is characterized by an aglet pushing itself from its current host to a remote host (dispatching). A remote host pulling an aglet away from its current host (retracting) characterizes the passive way of aglet mobility. When aglets are well and running they take up resources. To economies with the resources aglets can go to sleep temporarily release their resources (deactivation) for being brought back to running mode (activation). Finally, multiple aglets may exchange information to accomplish a given task (messaging).

This is just about the minimal set of operations required to create and manage a distributed mobile agent environment. When creating the Aglet API, it was one of our goals to create a lightweight API that would be easy learn to use and yet applicable for real applications. It is tempting to call the Aglet API the "RISC" of mobile agents.

Below we summarize the aglets' fundamental operations such as creation, cloning, dispatching, retraction, deactivation, activation, disposal of, and messaging (see also Figure 1):

Creation. The creation of an aglet takes place in a context. The new aglet is assigned an identifier, inserted in the context, and initialized. The aglet starts executing as soon as it has successfully been initialized.

Cloning. The cloning of an aglet produces an almost identical copy of the original aglet in the same context. The only differences are the assigned identifier and the fact that execution restarts in the new aglet. Notice that execution threads are not cloned.

Dispatching. Dispatching an aglet from one context to another will remove it from its current context and insert it into the destination context, where it will restart execution (execution threads do not migrate). We say that the aglet has been pushed to its new context.

Retraction. The retraction of an aglet will pull (remove) it from its current context and insert it into the context from which the retraction was requested.

Activation and deactivation. The deactivation of an aglet is the ability to temporarily halt its execution and store its state in secondary storage. Activation of an aglet will restore it in a context.

Disposal of. The disposal of an aglet will halt its current execution and remove it from its current context.

Messaging. Messaging between aglets involves sending, receiving, and handling messages synchronously as well as asynchronously.

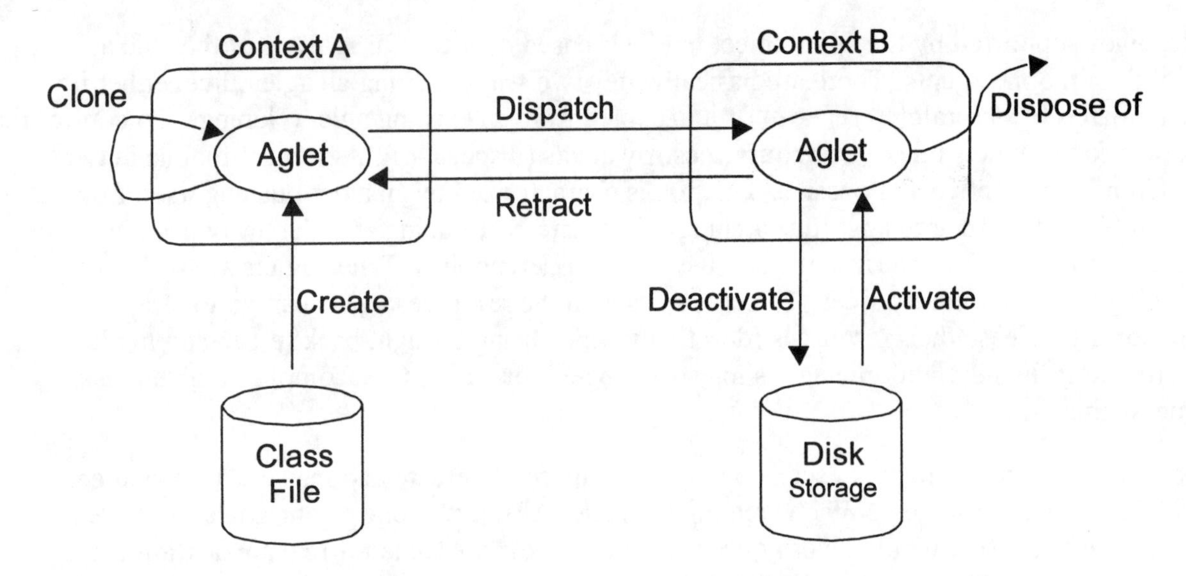

Figure 1 Aglet Life Cycle Model

The Aglet Event Model

The Aglet programming model is event-based. The model allows the programmer to "plug in" customized *listeners* into an aglet. Listeners will catch particular events in the lifecycle of an aglet and allow the programmer to take action when the aglet for instance is being dispatched.

Three kinds of listeners exist:

> **Clone Listener.** Listens for clone events. One can customize this listener to take specific actions when an aglet is about to be cloned, when the clone is actually created, and after the cloning has taken place.
>
> **Mobility Listener.** Listens for mobility events. One can use this listener to take action when an aglet is about to be dispatched to another context, when it is about to be retracted from a context, and when it actually arrives in a new context.
>
> **Persistence Listener.** Listens for persistent events. This listener allows the programmer to take action when an aglet is about to be deactivated and after it has been activated.

The `CloneLimiter` is an example of a reusable listener. It can be plugged into any aglet for which we want to limit the number of clones that can be made from it. The `CloneLimiter` listener implements three methods, `onCloning`, `onClone`, and `onCloned`, which are called in sequence when an application tries to clone the aglet that hosts the listener. Most notably, this implementation of `CloneLimiter` will at most allow five clones of the original aglet and none of the clones can be cloned.

```
public class CloneLimiter implements CloneListener() {
   final integer MAX = 5;
   boolean original = true;
   int number_of_clones = 0;
```

```
        // Called when the aglet is about to be cloned.
        public void onCloning(CloneEvent ev) {
           if (original == false)
              throw new SecurityException("Clone cannot create a clone");
           if (number_of_clones > MAX)
              throw new SecurityException("Exceeds the limit");
        }

        // Called in the cloned aglet.
        public void onClone(CloneEvent ev) {
           original = false;
        }

        // Called in the original aglet after the cloning.
        public void onCloned(CloneEvent ev) {
           number_of_clone++;
        }
     }
```

Agent Design Patterns

A very important part of the Aglet programming model is the idea of *agent design patterns.* [1] During the early work on the Aglet API, we recognized a number of recurrent patterns in the design of aglet applications. Several of these patterns were given intuitive meaningful names such as *Master-Slave*, *Messenger*, and *Notifier.* They were implemented in Java and included in the first release of the Aglets Workbench. These early patterns were found to be highly successfully for jump-starting users who were new to Aglets and the mobile agent paradigm.

This experience tells us that it is important to identify the elements of good and reusable designs for mobile agent applications and to start formalizing people's experience with these designs. This is the role of agent design patterns. The concept originated with software engineers and researchers in the object-oriented community, and has been recognized as one of the most significant innovations in the object-oriented field.

The patterns we have discovered so far can conceptually divided into three classes: *traveling, task,* and *interaction.* This classification scheme makes it easier to understand the domain and application of each pattern, to distinguish different patterns, and to discover new patterns.

4 A Tour of the Aglet API

Now, get ready for a tour of the core classes and interfaces of the aglet API. This is the API that the programmer will use to create and operate aglets. It contains methods for initializing an aglet, message handling, and dispatching, retracting, deactivating/activating, cloning, and disposing of the aglet. The aglet API is simple and yet flexible. Created in the spirit of Java and representing a lightweight pragmatic approach to mobile agents, the aglet API is a Java package (`aglet`) consisting of classes and interfaces, most notably: `Aglet`, `AgletProxy`, `AgletContext`, and `Message`.

`Aglet` Class

The Aglet class is the key class in the Aglet API. This is the abstract class that the aglet developer uses as base class when he or she creates customized aglets. The `Aglet` class defines methods for controlling its own life cycle, namely, methods for cloning, dispatching, deactivating, and disposing of itself. It also defines methods that are supposed to be overridden in its subclasses by the aglet programmer, and provides the necessary "hooks" to customize the behavior of the aglet. These methods are systematically invoked by the system when certain events take place in the life cycle of an aglet.

Let us describe some of the methods in the `Aglet` class. The `dispatch` method causes an aglet to move from the local host to the destination given as the argument. The `deactivate` method allows an aglet to be stored in secondary storage, and the `clone` method spawns a new instance of the aglet, which has the state of the original aglet.

The Aglet class is also used to access the attributes associated with an aglet. The `AgletInfo` object, which can be obtained by `getAgletInfo()`, contains an aglet's built-in attributes, such as its creation time and code base, as well as its dynamic attributes, such as its arrival time and the address of its current context.

Now, let us demonstrate how simple it is to create a customized aglet. Start by importing the `aglet` package, which contains all the definitions of the Aglet API. Next, define the `MyFirstAglet` class, which inherits from the `Aglet` class:

```
import aglet.*;
public class MyFirstAglet extends Aglet { ... }
```

For instance, if we want an aglet to perform some specific initialization when it is created, we just need to override its `onCreation` method:

```
public void onCreation(Object init) {
   // Do some initialization here...
}
```

When an aglet has been created or when it arrives in a new context, it is given its own thread of execution through a system invocation of its `run` method. This invocation can be seen as a means of giving the aglet a degree of autonomy. The `run` method is called every time the aglet arrives at or is activated in a new context. One can say that the `run` method becomes the main entry point for the aglet's thread of execution. By overriding this method, it is possible to customize an aglet's autonomous behavior.

```
public void run() {
   // Do something here...
}
```

For example, one can use `run()` to let the aglet dispatch itself to some remote context. This can be done by letting the aglet call its dispatch method with the Uniform Resource Locator (URL) of the remote host as the argument. This URL should specify the host and domain names of the destination context, and the protocol (`atp`) to be used for transferring the aglet over the network. The URL can also include the name of the remote context (if more than one context is supported at the remote server). If no context name is specified, the aglet will move into the default context.

```
dispatch(new URL("atp://some.host.com/context"));
```

So what exactly happens to an aglet when `dispatch()` is called? Basically the aglet will disappear from the current host machine and reappear in the same state at the specified destination. First, a special technique called object serialization is used to preserve the state information of the aglet. What it does is to make a sequential byte representation of the aglet. Next, it is passed to the underlying transfer layer that brings the aglet (byte code and state information) safely over the network. Finally, the transferred bytes are de-serialized to recreate the aglet's state:

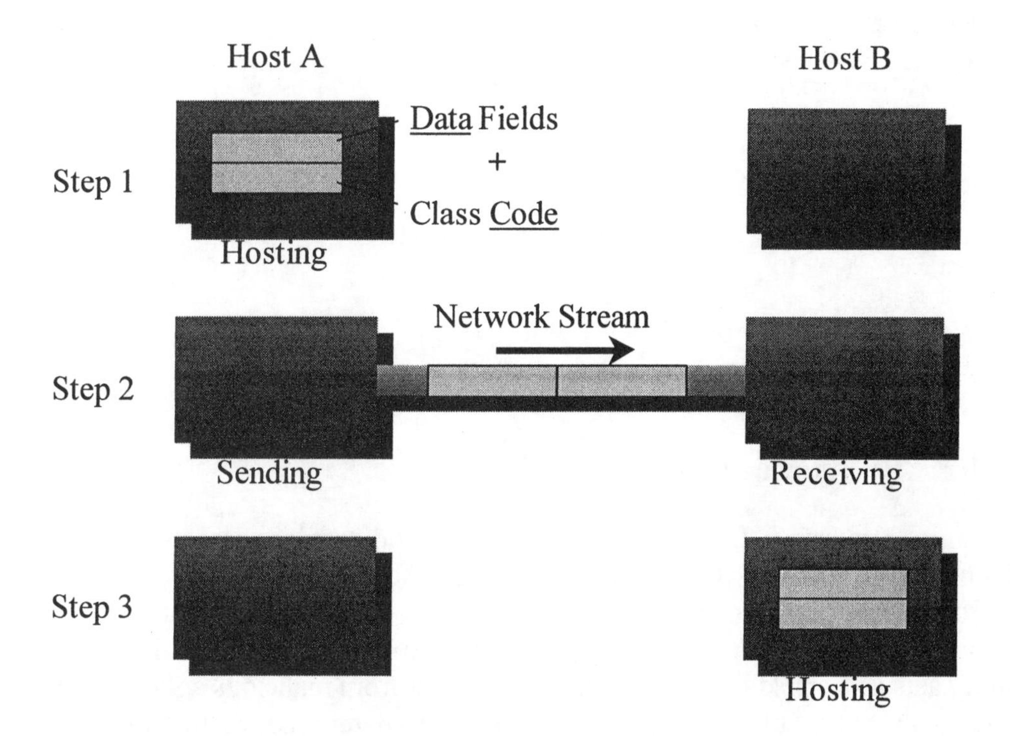

Figure 2 Transfer of an Aglet

Below is a tiny example in which we let the aglet named `DispatchingExample` dispatch itself. `DispatchingExample` extends the `Aglet` class and overrides two methods: `onCreation` and `run`. A *mobility listener* is defined as an inner class. This listener handles mobility related events. We use a Boolean field, `_theRemote`, to distinguish between the aglet before and after it has been dispatched. When this aglet is created and starts running (`run()`), it creates a URL for its destination. When the aglet has been dispatched all its threads will be killed. In other words, one should not expect the execution to return from a successful call to the `dispatch` method. When the aglet arrives at a new host, `onArrival` is called and the Boolean field is toggled. The aglet will now remain at this host until it is disposed of.

```
public class DispatchingExample extends Aglet {

   boolean _theRemote = false;

   public void onCreation(Object o) {
      addMobilityListener(
         new MobilityAdapter() {
            public void onDispatching(MobilityEvent e) {
               // Print to the console...
            }
            public void onArrival(MobilityEvent e) {
               _theRemote = true;  //-- Yes, I am the remote aglet.
```

```
                // Print to the console...
            }
        }
    );
}

public void run() {
    if (!_theRemote) {
        // The original aglet runs here
        try {
            dispatch(destination);
            // You should never get here!
        } catch (Exception e) {
            System.out.println(e.getMessage());
        }
    } else {
        // The remote aglet runs here...
    }
}
}
```

`AgletProxy` Interface

The `AgletProxy` interface acts as a handle of an aglet and provides a common way of accessing the aglet behind it. Since an aglet class has several public methods that should not be accessed directly from other aglets for security reasons, any aglet that wants to communicate with other aglets has to first obtain the proxy, and then interact through this interface. In other words, the aglet proxy acts as a shield object that protects an aglet from malicious aglets. When invoked, the `AgletProxy` object consults the *SecurityManager* [3] to determine whether the current execution context is permitted to perform the method. Another important role of the `AgletProxy` interface is to provide the aglet with location transparency. If the actual aglet resides at a remote host, the proxy forwards the requests to the remote host and returns the result to the local host.

Creating an aglet is one way to get a proxy. The `AgletContext.createAglet`[2] method will return the proxy of the newly created aglet. Other methods that return proxies include `AgletContext.retractAglet`, `AgletContext.activateAglet`, `AgletProxy.clone`, and `AgletProxy.dispatch`.

Proxies of existing aglets can also be obtained in the following ways:

- Retrieve an enumeration of proxies in a context by calling the `AgletContext.getAgletProxies` method.
- Get an aglet proxy for a given aglet identifier via the `AgletContext.getAgletProxy` method.
- Get an aglet proxy via message passing. An `AgletProxy` object can be put into a `Message` object as an argument, and sent to the aglet locally or remotely.

---

[2] We use the "dot-notation" (*class.method*) to denote the class that a given method belongs to.

- Put an `AgletProxy` object into a context property by the `AgletContext.setProperty` method and share the proxy object.

`AgletContext` Interface

An aglet spends most of its life in an aglet context. It is created in the context, it goes to sleep there, and it dies there. When it travels in a network, it really moves from context to context. In other words, the context is a uniform execution environment for aglets in an otherwise heterogeneous world.

The `AgletContext` interface is used by an aglet to get information about its environment and to send messages to the environment, including other aglets currently active in that environment. It provides means for maintaining and managing running aglets in an environment where the host system is secured against malicious aglets.

IBM's Aglets Workbench comes with a graphical user interface for the context. It is called Tahiti and enables the user to create, clone, deactivate/activate, dispose of, dispatch, and retract aglets. Tahiti allows the user to monitor aglets running in the local context. Tahiti is basically an application built on top of the Aglet API. Developers can build proprietary versions of Tahiti that serve specific purposes.

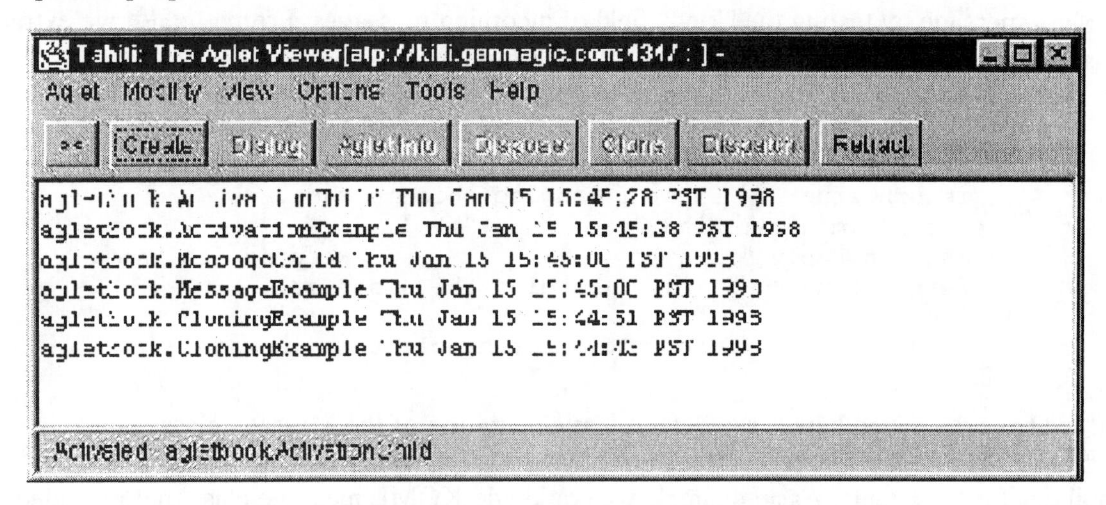

Figure 3 A Screendump of Tahiti

`Message` Class

Aglets communicate by exchanging objects of the `Message` class. A string field named "kind" distinguishes messages. This field is set when the message is created. The second parameter of the message constructor is an optional message argument:

```
Message myName = new Message("my name", "Jacob");
Message yourName = new Message("your name?");
```

Having created the message objects, we can send them to an aglet by invoking one of the following methods defined in the `AgletProxy` class:

- `Object sendMessage(Message msg)`

- `FutureReply sendFutureMessage(Message msg)`
- `void sendOnewayMessage(Message msg)`

The Message object is passed as an argument to the aglet's `handleMessage` method. It is now up to this method to handle the incoming messages. It should return true if a given message is handled; otherwise, it should return false. The sender will then know if the aglet actually handled the message. In this example, the aglet will recognize and respond accordingly to `"hello"` messages:

```
public boolean handleMessage(Message msg) {
    if (msg.sameKind("hello")) {
      doHello();      // Respond to the 'hello' message...
      return true;    // Yes, I handled this message.
    } else
      return false; // No, I did not handle this message.
}
```

Only from `yourName` do we expect to receive a return value:

```
proxy.sendMessage(myName);
String name = (String)proxy.sendMessage(yourName);
```

In the aglet's handleMessage method, we distinguish between the `myName` message and the `yourName` message by testing the "kind" field of incoming messages. From `myName` we extract the name argument, and for `yourName` we set the return value (`sendReply()`):

```
public boolean handleMessage(Message msg) {
   if (msg.sameKind("my name")) {
      String name = (String)msg.getArg();  // Gets the name...
      return true;  // Yes, I handled this message.
   } else if (msg.sameKind("your name?")) {
      msg.sendReply("Yina");                // Returns its name...
      return true;  // Yes, I handled this message.
   } else
      return false; // No, I did not handle this message.
}
```

What we have described here is the default implementation of the Message class. The programmer can create specialized message classes that provide more advanced messaging capabilities for the agents. As an example we could add KQML message class that provided some of the necessary abstractions required for agents to use KQML [8] for inter-agent communication.

5 Application of Aglets

Electronic commerce is one the fields that we expect to benefit from mobile agent technology. In this section we will describe an aglet-based framework that has been used to create a commercial service named Tabican.[3] Tabican is an *electronic marketplace* for air tickets and package tours

[3] The URL is `http://www.tabican.ne.jp`. Unfortunately, this service is available in Japanese language only.

(air + hotel) that has been designed to host thousands of agents. In the server, shop agents are waiting for requests from consumer agents. Different shop agents may implement different sales policies and consumer agents may implement different negotiation strategies. Users visiting this site can leave agents behind that search for deals for up to 24 hours.

Taking account of both business and technology backgrounds, an IBM team developed an electronic marketplace framework [5] on the basis of Aglets. An electronic marketplace is a multi-agent system in which selling and buying agents interact with each other. This architecture redefines the roles of participants, namely, marketplace owners, shop owners, and consumers. The role of marketplace owners is simply to manage system resources such as hardware and database systems. Users of marketplaces, that is, shop owners and consumers are responsible for maintaining applications that are implemented as software agents. The framework is developed on top of Aglets and is called Aglet Meeting Place Middleware (AMPM). An important construct of AMPM is a type database that stores information on the types of messages between agents. The type database enables unfamiliar agents, independently developed to interact with each other. See the figure in the below for an overview of the electronic marketplace.

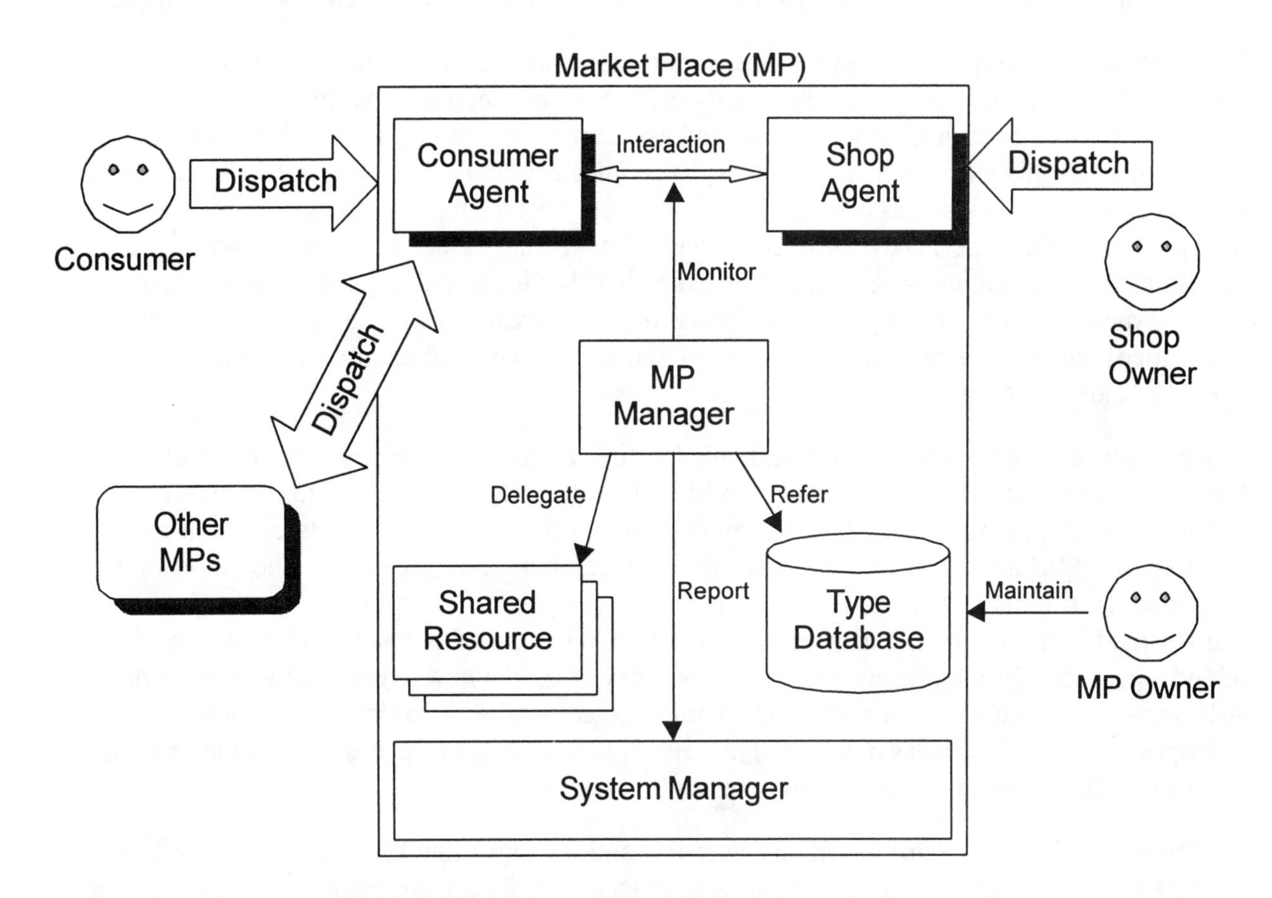

Figure 4 The Aglet Marketplace Architecture

From the perspective of electronic marketplace implementation, the key functions in Aglets are agent mobility and messaging. Mobility makes possible the following agent activities in marketplace:

- Shop agents go to a market from a shop owner's terminal.

- Customer agents travel around various markets to get more information.
- Market advertisers go to other markets to invite customer agents.

These activities were easily implemented by simply using the built-in mobility in Aglets. In the actual implementation, the aglet programming model was reported useful, since the marketplace agents' activities nicely fit into the programming model. The aglet programming model defines events such as creation, cloning, dispatching, retraction, disposal, deactivation, and activation. Then, on the basis of these events, listeners and associated methods are defined, for instance, `onCreation()`, `onDispatching()`, `onArrival()` and `onDeactvating()`.

A typical behavioral pattern of the customer agent is (1) load type information, (2) shout a request, (3) receive the results, and (4) go to another market. In (1), an agent obtains an interaction protocol in order to talk with unfamiliar agents: this activity is carried out in `onCreation()` and in `onArrival()`. (2) and (3) are ordinary activities within a marketplace. Finally, (4) causes a dispatching event and `onDispatching()` is invoked. Within this function, the customer agent is unregistered from the list of participants in the electronic marketplace.

For electronic marketplace, the key aglets constructs related to interaction are the `AgletProxy` and asynchronous messaging. An important purpose of the electronic marketplace is to develop agents that are self-serving. Aglet proxies and asynchronous messages directly contribute to the achievement of this purpose. An aglet proxy is an interface object for an aglet: aglets cannot talk to each other directly, but has to talk through a proxy. This means that an aglet cannot directly manipulate another aglet, and it provides a way of protecting aglets from each other. Asynchronous messaging is also important assuming that unfamiliar agents have to interact. By using asynchronous messaging between the agents, the possibility of two agent getting hung due to protocol mismatch is minimized. In general, the team developed agents so that they can be executed independently.

Since agents can be developed independently by different people, the agent system must be flexible in the sense that one can easily modify it by adding a new agent or updating an existing agent. For example, one can replace a shop or a customer agent with an updated agent as long as the new agent behaves in accordance with the defined interaction protocol. When developing electronic marketplace, the team first implemented a single market application, and then transformed it into a multiple market configuration. What they did was as follows: first, they added a protocol for advertising a market, next, they created a new agent, that is, the Market Advertiser, and finally, they updated the customer agent so that it could move to another market in response to an advertiser's message. The actual modification to the customer agent was the addition of the necessary code for going to another market.

On the basis of the team's observations we see promising signs that the use of agent technology such as Aglets ultimately will lead to a new approach to software development and maintenance. A trend of improved encapsulation and delegation that was started by object technology and perhaps now brought to new levels by agent technology.

6 Contemporary Mobile Agent Systems

So what kind of mobile agent systems are currently available? Fortunately, Java has generated a flood of experimental mobile agent systems. Numerous systems are currently under development, and most of them are available for evaluation on the Web.

The field is developing so dynamically and so fast that any attempt to map the agent systems will be outdated before this article goes to press. We will, however, mention a few interesting Java-based mobile agent systems: Odyssey, Concordia, and Voyager. Notice that we do not attempt to compare any of these systems to each other.

> **Odyssey.** General Magic Inc. invented the mobile agent and created the first commercial mobile agent system called Telescript. Being based on a proprietary language and network architecture, Telescript had a short life. In response to the popularity of the Internet and later the steamrollering success of the Java language, General Magic decided to re-implement the mobile agent paradigm in its Java-based Odyssey. This system effectively implements the Telescript concepts in the shape of Java classes. The result is a Java class library that enables developers to create their own mobile agent applications. More information available at `http://www.genmagic.com/html/agent_overview.html`.

> **Concordia.** Mitsubishi's Concordia is a framework for the development and management of mobile agent applications which extend to any system supporting Java. Concordia consists of multiple components, all written in Java, which are combined together to provide a complete environment for distributed applications. A Concordia system, at its simplest, is made up of a standard Java VM, a Server, and a set of agents. Concordia provides similar to Aglets a security manager that allows for secure execution of agents. It also supports a checkpoint mechanism for fault tolerance. Similar to aglets it provides event-based communication. More information available at `http://www.meitca.com/HSL/Projects/Concordia`.

> **Voyager.** ObjectSpace's Voyager is a platform for agent-enhanced ORB (Object Request Broker) for Java. While Voyager provides an extensive set of object messaging capabilities it also allows object to move as agents in the network. Voyager combines the properties of a Java-based object request broker with those of a mobile agent system. In this way Voyager allows Java programmers to create network applications using both traditional and agent-enhanced distributed programming techniques. Voyager provides no security management for untrusted agents. More information available at `http://www.objectspace.com/voyager`.

We would like to note that the Java-based mobile agent systems have a lot in common. Beside the programming language they all rely on standard versions of the Java virtual machine and Java's object serialization mechanism. A common server-based architecture permeates all the systems. Agent transport mechanisms and the support for interaction (messaging) varies a lot.

7 Conclusion

We have given an introduction to the world of mobile agents. In particular, we have given a quick tour of the essentials of the Aglet API and its underlying programming model. Aglets' event-based programming model emphasizes delegation and reuse of components such as listeners that efficiently define agent behavior. Code fragments of aglets have been shown in order to convince the reader that there is nothing mysterious about programming mobile agents in Java.

In addition we have reported on one of the first real Internet applications based on mobile agent technology, namely the Tabican electronic marketplace in Japan. This and future projects will demonstrate that Java-powered mobile Internet agents have a role to play outside the research labs. We envision a not too distant future were mobile agent technology is an integrated part of the network or perhaps even the operating system. This process will accelerate with the adoption of Java as a universal language system and with standardization efforts such as Object Management Group's Mobile Agent Facility (MAF) [6].

The Aglets project has been a very successful research project that has exceeded even our wildest imagination. We are confident that the entire field of mobile agents will continue to be an interesting and exciting place on the map in the years to come.

Acknowledgements

We wish in particular to thank the past and present members of the *Aglets Team* in the IBM, without whom aglets would not have made it. They include, in alphabetic order: Günter Karjoth, Kazuya Kosaka, Yoshiaki Mima, Dr. Yuichi Nakamura, Kouichi Ono, Hideaki Tai, Kazuyuki Tsuda, Gaku Yamamoto, Dr. Yariv Aridor, and Tsutomu Kamimura. We are also grateful to Mike McDonald of IBM Japan for checking the wording of this paper.

References

1. Aridor, Y. and Lange, D.B., Agent Design Patterns: Elements of Agent Application Design. To appear in *Second International Conference on Autonomous Agents* (Agents '98), Minneapolis/St. Paul, May 10-13, 1998.

2. Arnold, K. and Gosling, J. *The Java Programming Language.* Addison-Wesley, 1998.

3. Karjoth, G., Lange, D.B., and Oshima, M. A Security Model for Aglets. IEEE Internet Computing. Vol. 1, No. 4, pp. 68-77, 1997.

4. Lange, D. B. and Oshima, M. *Programming and Deploying Mobile Agents with Java.* Forthcoming book. Addison-Wesley, June 1998.

5. Nakamura, Y. and Yamamoto, G. *An Electronic Marketplace Framework Based on Mobile Agents.* Research Report, RT0224, IBM Research, Tokyo Research Laboratory, 1997.

6. Object Management Group. *The Mobile Agent Systems Interoperability Facility.* 1997.

7. White, J. Telescript Technology: Mobile Agents. In Bradshaw, J. (ed.), *Software Agents.* MIT Press, 1996.

8. DARPA. *Draft Specification of the KQML Agent-Communication Language.* 1993.

14.3 Kotz, D., Gray, R., Nog, S., Rus, D., Chawla. S., and Cybenko, G., AGENT TCL: Targeting the Needs of Mobile Computers

Kotz, D., Gray, R., Nog, S., Rus, D., Chawla. S., and Cybenko, G., "AGENT TCL: Targeting the Needs of Mobile Computers," *IEEE Internet Computing*, 1(4):58-67, July/August 1997.

AGENT TCL: Targeting the Needs of Mobile Computers

David Kotz, Robert Gray, Saurab Nog, Daniela Rus, Sumit Chawla, and George Cybenko
Dartmouth College

Agent Tcl accommodates mobile computers with features like laptop docking, which lets an agent return to a periodically disconnected machine.

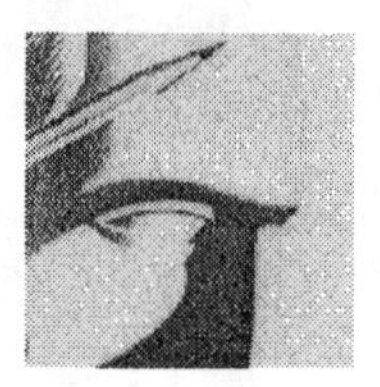

Mobile computers have become increasingly popular as users discover the benefits of having their electronic work available at all times. Using Internet resources from a mobile platform, however, is a major challenge. Mobile computers do not have a permanent network connection and are often disconnected for long periods. And when the computer *is* connected, the connection is often prone to sudden failure, such as when a physical obstruction blocks the signal from a cellular modem. In addition, the network connection often performs poorly and can vary dramatically from one session to the next, since the computer might use different transmission channels at different locations. Finally, depending on the transmission channel, the computer might be assigned a different network address each time it reconnects.

Mobile agents are one way to handle these unforgiving network conditions. A mobile agent is an autonomous program that can move from machine to machine in a heterogeneous network under its own control. It can suspend its execution at any point, transport itself to a new machine, and resume execution on the new machine from the point at which it left off. On each machine, it interacts with service agents and other resources to accomplish its task, returning to its home site with a final result when that task is finished. The sidebar "Why Mobile Agents?" describes the motivations for and benefits of these agents in more detail.

Agent Tcl is a mobile-agent system whose agents can be written in Tcl, Java, and Scheme, although the version available to the public supports only Tcl at present. Agent Tcl has extensive navigation and

WHY MOBILE AGENTS?

Mobile agents have several strengths. By migrating to the location of a needed resource, an agent can interact with the resource without transmitting any intermediate data across the network, significantly reducing bandwidth consumption in many applications. Similarly, by migrating to the location of a user, an agent can respond to user actions rapidly. In either case, the agent can continue its interaction with the resource or user even if the network connection goes down. These features make mobile agents particularly attractive for mobile-computing applications.

Mobile agents let traditional clients and servers offload work to each other, and *change* who offloads to whom according to machine capabilities and current loads. Similarly, mobile agents allow an application to dynamically deploy its components to arbitrary network sites, and to *re*deploy those components in response to varying network conditions.

Finally, most distributed applications fit naturally into the mobile-agent model, since mobile agents can migrate sequentially through a set of machines, send out a wave of child agents to visit multiple machines in parallel, remain stationary and interact with resources remotely, or perform any combination of these three extremes. Complex, efficient, and robust behaviors can be realized with surprisingly little code. Our own experience with undergraduate programmers at Dartmouth suggests that mobile agents are easier to understand than many other distributed computing paradigms.

Although each of these strengths is a reasonable argument for mobile agents, any specific application can be implemented just as efficiently and robustly with more traditional techniques, such as queued RPC, high-level query languages, dedicated proxies within the network, automatic installation facilities, and Java applets.[1] However, mobile agents eliminate the need for these other techniques, combining their strengths into a single, general, and convenient framework. Distributed applications can be implemented efficiently and easily—even if they must exhibit extremely flexible behavior in the face of changing network conditions. For example, a search application can migrate to a dynamically selected proxy site and do its merging and filtering there, while a server can continually migrate to new machines to minimize the average latency between itself and its clients.[2]

In short, the true strength of mobile agents is that they are a uniform paradigm for distributed applications. Thus, all existing systems—Agent Tcl, Telescript,[3] Odyssey,[4] IBM Aglets,[5] and Sumatra[2]—are intended for general applications, differing only in their languages, migration and security models, and support services. Agent Tcl distinguishes itself by combining a true jump instruction (one that automatically captures the entire program state); support for multiple languages; simple but effective security mechanisms; and significant navigation, communication, and debugging tools.

REFERENCES

1. D. Chess et al., "Itinerant Agents for Mobile Computing," IEEE Personal Comm., Oct. 1995, pp. 34-49.
2. M. Ranganathan et al., "Network-Aware Mobile Programs," Proc. Usenix Tech. Conf.,Usenix, Berkeley, Calif., 1997, pp. 91-104.
3. J.E. White, "Mobile Agents," in Software Agents, J.M. Bradshaw ed., MIT Press, Cambridge, Mass., 1997, pp. 437-472.
4. http://www.genmagic.com/agents/
5. http://www.trl.ibm.co.jp/aglets/

communication services, security mechanisms, and debugging and tracking tools. In this article we focus on Agent Tcl's architecture and security mechanisms, its RPC system, and its docking system, which lets an agent move transparently among mobile computers, regardless of when they are connected to the network.

Agent Tcl is being used in experimental projects at numerous academic and industrial research laboratories, including labs at Lockheed Martin, Siemens, Cornell University, and the University of Bordeaux, and has begun to find its way into production-quality applications as well. The public release provides migration, low-level communication, and security mechanisms for protecting a machine against malicious agents. The internal version includes the docking and RPC systems, the debugging tools, additional security features, and support for Java and Scheme agents. The new components in the internal version will be available in fall 1997. The current public release and all Agent Tcl publications are online (http://www.cs.dartmouth.edu/~agent).

OVERVIEW

Like all mobile-agent systems, the main component of Agent Tcl is a server that runs on each machine. When an agent wants to migrate to a new machine, it calls a single function, agent_jump, which automatically captures the complete state of the agent and sends this state information to the server on the destination machine. The destination server starts up an appropriate execution environment (for example, a Tcl interpreter for an agent written in Tcl), loads the state information into this execution environment, and restarts the agent from the exact point at which it left off. Now the agent is on the destination machine and can interact with that machine's resources without any further network communication. In addition to reducing migration to a single instruction, Agent Tcl has several important features:

- *Simple architecture.* The simple, layered architecture supports multiple languages and transport mechanisms. The main language is Tcl. The main transport mechanism is TCP/IP.

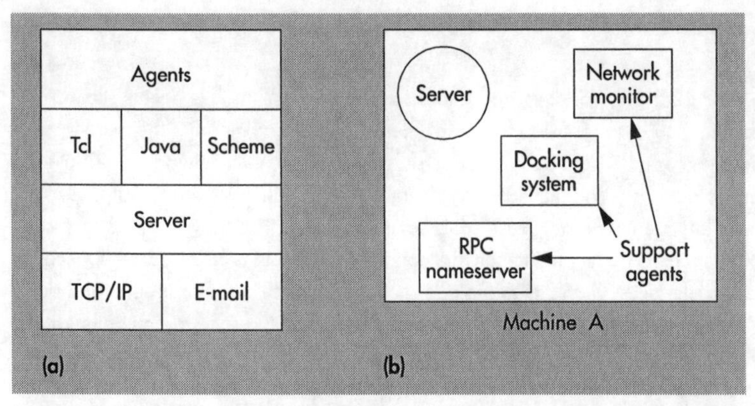

Figure 1. The architecture of Agent Tcl. (a) **The core system has four levels: transport mechanisms, a server that runs on each machine, an interpreter for each supported agent language, and the agents themselves.** (b) **Support agents provide navigation, communication, and resource management services to other agents.**

- *Security.* Agent Tcl protects individual machines against malicious agents—agents that try to access or destroy restricted information or consume too many machine resources. It also protects groups of machines controlled by a single organization.
- *Docking system.* The docking system lets an agent transparently jump off a partially connected computer (such as a mobile laptop) and return later, even if the computer is connected only briefly.
- *Interagent communication.* Agents communicate with either low-level mechanisms (message passing and streams) or high-level mechanisms (RPC) that are implemented at the agent level atop the lower level mechanisms. All communication mechanisms work the same whether or not the communicating agents are on the same machine.

ARCHITECTURE

As Figure 1 shows, Agent Tcl's architecture has a four-level core system and an agent-level support system.

Core System

At the lowest level of the core system, Figure 1a, is an interface to each available transport mechanism. The next level is the server that runs on each machine. The server keeps track of the agents running on its machine, provides the low-level interagent communication facilities (message passing and streams), receives and authenticates agents arriving from another host, and restarts an authenticated agent in an appropriate execution environment.

The third level of the architecture consists of the execution environments, one for each supported agent language. Agent Tcl supports Tcl, Java, and Scheme, so its execution environments are simply a Tcl interpreter (Tcl 7.5), a Scheme interpreter (Scheme 48), and the Java virtual machine (Sun JDK 1.2). For each incoming agent, the server starts up the appropriate interpreter.

The last level of the architecture comprises the agents themselves, which execute in the interpreters and use the facilities provided by the server to migrate from machine to machine and to communicate with other agents. Agents include both moving agents, which visit different machines to access needed resources, and stationary agents, which stay on a single machine and provide a specific service to either the user or other agents. From the system's point of view, there is no difference between the two kinds of agents, except that a stationary agent has authority to access more system resources.

To add a language to Agent Tcl, programmers simply extend the interpreter to provide

- state-capture routines that capture and restore the state of an executing program, and
- an interface to the agent servers.

In the Tcl interpreter, for example, the state-capture routines capture and restore all defined variables and procedures, the procedure-call stack, and the control stack. The interface to the servers is a set of Tcl commands, such as agent_begin and agent_jump, which are provided as a standard Tcl extension. The agent_jump command calls the state-capture routines to capture and restore the state of an executing Tcl agent. Similarly, in Java, the state-capture routines capture and restore the state of a single Java thread (including all accessible objects). The interface to the servers is a special Java class. Finally, in Scheme, the state-capture routines capture and restore the current continuation (the rest of the program), and the interface to the servers is a set of Scheme functions.

Most of the interface between the interpreters and the servers is implemented in a C/C++ library and shared among all interpreters. The language-specific portion is just a set of stubs that call into this library.

Agent-Level Support

The agent servers provide low-level functionality. As Figure 1b shows, all other services are provided at the agent level by dedicated service agents. Such services include navigation, high-level communication protocols, and resource management. Both the docking system and Agent RPC (described later) are implemented entirely at the agent level.

Sample Agent

Figure 2 shows a simple Agent Tcl agent written in Tcl. The agent's task is to make a list of all users logged onto some Dartmouth machines and then show this list to its owner. The agent has several important parts:

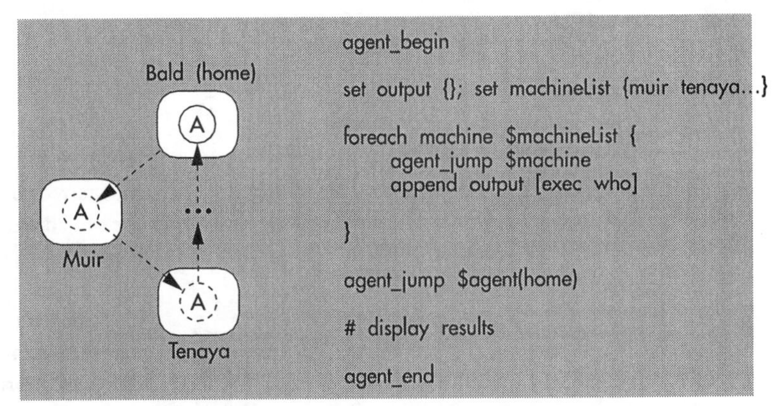

Figure 2. A simple Tcl agent that figures out which users are logged onto some set of machines. Bald, Muir, and Tenaya are machines at Dartmouth College. The agent starts on Bald.

- agent_begin. The agent registers with Agent Tcl through the server on its home machine (Bald).
- agent_jump $machine. The agent migrates sequentially through the machines of interest (Muir, Tenaya, and others not shown). It continues executing from the point of the jump on each successive machine. On each machine, the agent executes the Unix who command (exec who) to obtain the user list.
- agent_jump $agent(home). Once the agent has migrated through all the machines, it migrates one last time to return to Bald.
- # display results. Once on Bald, the agent displays the complete user list to its owner (not shown).
- agent_end. The agent tells Agent Tcl that it has finished.

Although this agent performs a simple task, any agent that migrates sequentially through one or more machines has the same general form. The exec who command can be replaced with any desired local processing. Some agents will need learning and reasoning capabilities. Agent Tcl does not provide such capabilities directly, but an agent is just a program written in Tcl, Scheme, or Java, so it can include and use any existing libraries.

SECURITY

Security, of course, is a critical issue in any mobile-code system. Agent Tcl currently protects machines from malicious agents (both individual machines and groups of machines under single administrative control), but does not protect agents from malicious machines. We give a brief summary of our current implementation here; a detailed description and our future plans are given elsewhere.[1,2]

The security mechanism has three major features:

- Agents and messages sent between machines are encrypted, which maintains agent privacy.
- Agents and messages sent between machines are signed, which authenticates the agent to the new host.
- Resource managers control access to system resources.

Each resource (CPU, memory, file system, screen, network, and so on) has a stationary agent that acts as a manager. For Tcl agents, each visiting agent is run in an *untrusted* Tcl interpreter, and all resource accesses are trapped into a separate, *trusted* interpreter. The trusted interpreter asks the appropriate resource manager to determine if the visiting agent should have access to the resource. For example, the memory manager might limit an agent to 100 Kbytes of memory. The trusted interpreter then enforces the manager's policy decision, either proceeding with the resource access or throwing a security exception back to the untrusted interpreter.

In addition to absolute limits on resource use, we plan to use a currency-based model in which agents purchase resources from the managers, thus limiting their total resource use even across administrative domains.[2] The prices will vary according to supply and demand and the changing priorities of agents and servers.

The same resource managers are used for all agents; only the enforcement mechanism differs among agent languages. In Java, for example, a special security manager class contacts the resource managers and enforces the policy decisions, rather than a separate trusted interpreter.

DOCKING SYSTEM

When a mobile agent tries to return to its home machine with final results, the machine might be disconnected. Thus, the agent must have some way to determine when the home machine reconnects. A simple approach is polling (try, timeout, sleep, try again, and so on), but polling wastes network resources and will fail outright if the home machine reconnects for only brief periods. For this reason, we devised a docking system, Figure 3, that pairs each laptop computer ("laptop" meaning any mobile device) with a permanently connected *dock* machine. When a mobile agent is unable to migrate to a laptop (laptopX), it waits at the laptop's dock machine (laptopX_dock). When the laptop reconnects, it notifies the dock machine of its new network address, and the dock machine forwards all waiting agents.

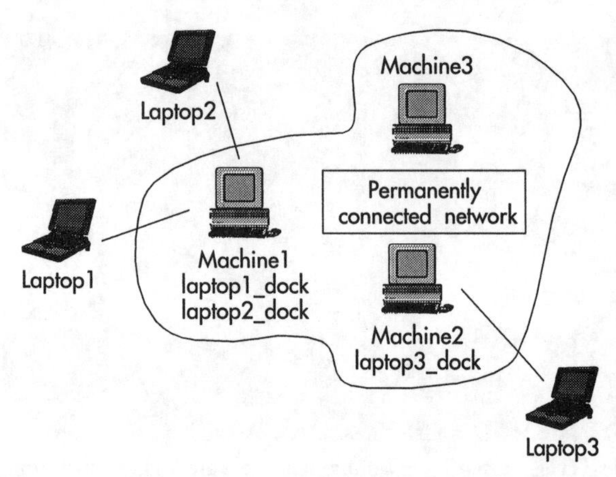

Figure 3. Docking system. Each laptop is paired with a permanently connected machine, where agents wait for the laptop to reconnect. Here "laptop" refers to any partially connected machine.

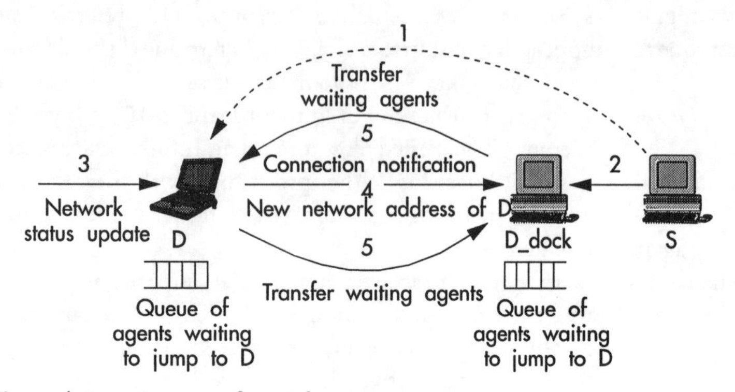

Figure 4. Jumping to or from a laptop.

Although only certain machines act as docks, all machines have a *dock master*, an agent that maintains a queue of waiting agents. The queue is always stored on disk rather than in memory. On the dock machine, the queue contains the agents waiting to visit the laptop; on the laptop itself, the queue contains the agents waiting to leave the laptop.

Application

Figure 4 illustrates how the docking system works in an application. The figure depicts the following sequence of events (numbers in parentheses correspond to the numbers in the figure). An agent wants to jump from a source machine *S* to a disconnected destination laptop *D*, so it executes the command agent_jump~D. The agent_jump command tries but fails to contact laptop *D* directly (1). Once the agent_jump command discovers that *D* is disconnected, it contacts the dock master on *D*'s dock machine *D*_dock and transfers the agent to this dock master (2). The dock master adds the agent to the queue of agents waiting to jump to *D*. When *D* reconnects to the network (3), the agent system on *D* detects the reconnection and notifies the dock master on *D*_dock (4). The dock master on *D*_dock transfers all waiting agents to *D*, where they resume execution (5). If *D* has changed its network address, the new address is included in the notification, so that waiting agents can be transferred to the new address. Agents trying to reach *D* at the old address will fail, jump to *D*_dock, and eventually reach *D* at its new address.

If the agent is trying to *leave* the disconnected laptop *D*, it again executes the agent_jump command, which detects that the laptop is disconnected, saves the state of the agent to disk, and informs the local dock master. The local dock master continually monitors network status, and when the laptop reconnects to the network, the dock master transfers the waiting agent to the desired destination.

A more complex case is when both the agent's source *S* and destination *D* are laptops. The two laptops might never be connected to the network at the same time. If *S* is disconnected, the dock master on *S* saves the agent's state on disk. When the dock master on *S* detects network reconnection, it tries to transfer the agent to *D*. If *D* is unreachable, it tries to transfer the agent to *D*'s dock. If *D*_dock is unreachable, perhaps because of a temporary problem on the Internet, the dock master on *S* tries to transfer the agent to *S*_dock. If *S*_dock is also unreachable, the dock master will try the entire process again at a later time. If *S*_dock can be reached, the agent is sent to *S*_dock, and the dock master on *S*_dock will periodically attempt to transfer the agent to either *D* or *D*_dock. The agent may reside at *D*_dock until *D* connects and notifies the dock master at *D*_dock of *D*'s new location. Once at *D*, the agent continues executing.

Multidestination Jumps

We are extending our docking system to support multidestination jumps, which are useful when an agent wants to visit multiple hosts (D_1, D_2, …, D_N) but in no particular order. The agent may be searching all sites for information or visiting one of a replicated set of servers. The dock master on *S* first tries to transfer the agent to one of the final destinations by trying each in order (D_1, D_2, …, D_N). If all destinations are unreachable, the *S* dock master transfers the agent to *S*_dock. The dock master at *S*_dock periodically tries to reach the destinations until one of the transfers succeeds. *S*_dock does not transfer the agent to a dock machine $D_K$dock, so that it does not prematurely commit to a destination that may rarely connect (although this issue is an open research topic). When the agent awakes (returns from its call to agent_jump), it checks its actual destination and proceeds with its task.

For agents that desire more control over the jumping process, we provide hooks to allow agents to query the status of the current machine's network connection, to request a failure notification rather than being blocked when a destination cannot be reached immediately, and to request that the jump go as far toward the destination as possible and then wake up the agent.

Performance

To determine the docking system's overhead, we measured the time needed for an agent to jump onto a laptop from a nearby host. The laptop was a 66-MHz Intel 486 running Red Hat Linux; the nearby host was a 100-MHz Intel Pentium running FreeBsd 2.1; the two machines were connected via a 10-Mbps Ethernet (with no intervening routers). In one set of experiments, the laptop was still connected, and the agent jumped directly onto it. In the second set of experiments, the laptop was still connected, but we forced the agent to go through the laptop's dock machine. Under normal operation, of course, the agent goes through the dock machine only if the laptop is disconnected, but forcing the agent to go through the dock was the easiest way to measure the docking overhead.

When the agent was carrying 8 Kbytes of code and data, a direct jump onto the laptop took 0.3 second, due mainly to the cost of starting up a new Tcl interpreter for the incoming agent. An indirect jump (agent going through the dock machine) took 1.6 seconds. The extra time came from the need to connect to the dock master, transfer the agent, save the agent to disk, and get the agent off the disk for transmission to its final destination. In addition, all agents are currently written in Tcl, which is slower than most other interpreted languages. Rewriting the dock masters in Java and providing a pool of ready interpreters will reduce these times significantly because a pool of ready interpreters eliminates the need to start a new interpreter for each incoming agent.

Benefits and Limitations

The docking system has several advantages. The agents depart from or arrive at the laptop as soon as possible and do not miss any transmission opportunities (because there is no polling). In addition, because waiting agents are saved on disk, they survive machine crashes and do not occupy precious memory and CPU time. Finally, all the state of a waiting agent has already been captured, so the agent is ready for transfer as soon as the network is connected.

On the down side, if an agent is *running* on a machine when the machine goes down, the agent is lost. If an agent is running on a machine, and the machine becomes disconnected from the network for a long period, the agent remains in exile. Finally, the dock for a given host named X.domain is the host named X_dock.domain. Although this allows immediate identification of a machine's dock, it also means that the machine must have a permanent name, even if the host gets a new network address at every machine restart. We are working to address these disadvantages.

INTERAGENT COMMUNICATION

Agent Tcl provides message passing and byte streams at its lowest level. Higher-level communication mechanisms, which make many applications much easier, are implemented at the agent level using message passing or streams. Our most important high-level mechanism, Agent RPC,[3] is similar to traditional RPC.[4] Agent RPC lets two agents communicate through the procedure-call abstraction. The agents can be on the same or different machines, but usually will be on the same machine, since most client agents jump to the same location as the desired service.

Programmers using Agent RPC begin by writing an interface in AIDL (Agent Interface Definition Language). The interface specifies the procedures a server agent provides to its clients. The programmer presents the AIDL specification to a stub compiler, which generates the Tcl procedures (called stubs) that let the client and server agents communicate. The client and server agents simply include these stubs with their application-specific code. In addition to accepting client requests, the server stubs register the server with a *nameserver agent*, which client agents consult when searching for a needed service.

Although this basic structure is no different from that of traditional RPC systems, Agent RPC offers two unique advantages:

- *Flexible interface language*. AIDL allows both default and position-independent parameters in interface definitions.
- *Client-server bindings*. Bindings (or connections between compatible client and server agents) are based on interface matching rather than on names. Thus, a client agent can obtain the desired service from any server that supports the appropriate interface, rather than only servers that have a particular service name and version. In addition, a client agent can have multiple, simultaneous bindings to the same or different servers. Finally, a server agent can accept or reject a bind request according to any security information the agent system provides, such as the authenticated identity of the client agent's owner.

AIDL and Bindings

AIDL aims to support extensibility and flexible matching. To illustrate, we present a running example that begins with the interface definition:

```
{constant {version 1} {service travel_agent}
                      {category airlines}}
{list_flights       {{Origin CityCode}
```

```
                  {Destination CityCode}}
                  {flights flight_list}}
{buy_ticket       {{payment_form sale_type credit_card}
                  {flight flight_number}}
                  {success boolean}}
{refund_ticket    {{flight flight_number}}
                  {success boolean}}
```

The constants convey the version number for this server (version 1), its service type (travel_agent), and its service specialization (airlines). The names of these constants are not meaningful to the stub compilers or the nameserver, but they are meaningful to the client and server agents.

Following the constants are three procedure definitions, list_flights, buy_ticket, and refund_ticket. Each procedure has a list of parameters and a return value. Each parameter and return value has a name, a type, and an optional default value.

The named parameters and the default values make interface matching more flexible. The process has the following steps:

- A server agent provides the AIDL description of its interface to the nameserver.
- A client agent provides an AIDL description of the interface it needs.
- The nameserver looks for a match between the client's desired interface and the interfaces the servers support. Two interfaces match if the server's AIDL matches all the functions and constants described in the client's AIDL. Two constants match if they have the same name and value (or if they have the same name and the server's value falls within the client's set of values). Two function descriptions match if they have the same function name, same parameter names and types (in any order), and same return name and type. If the server function provides a default value for a parameter not mentioned in the client's function description, the functions still match. If the server provides additional functions, the interfaces still match.

Thus, a client searching for the interface defined below will find all servers that support the interface defined earlier:

```
{constant {category airlines} {service travel_agent}}
{list_flights     {{Destination CityCode} {Origin CityCode}}
                  {flights flight_list}}
{buy_ticket       {{flight flight_number}}
                  {success boolean}}
```

In this example, the client does not care about the version number, uses the default value for the sale_type parameter to the buy_ticket function, and does not need the refund_ticket function. The client will find all servers that have a different version number, all servers whose buy_ticket function does not even have a sale_type parameter, and all servers that provide only the list_flights and buy_ticket functions. In other words, a client can find and use all servers that provide the functions it needs, regardless of whether those servers provide additional functions as well.

This flexibility is important in the dynamic world of the Internet, where clients and servers are not implemented by the same parties, where older or simpler clients must interact with newer or more complex servers, and so forth. It is easy to add more functions, constants, or parameters to a server and still support clients that expect an older, simpler interface.

The stub compiler compacts and sorts the interface definitions, so that the constants, the functions, and the parameters within each function are sorted by name. This sort lets the nameserver quickly compare two interfaces for a match. The generated client stubs pack the parameters (into a byte stream) in sorted order, and the generated server stubs unpack the parameters in sorted order. Thus, there is no explicit sorting step, which saves considerable time.

Performance

We measured the performance of Agent RPC using two machines: Bald, a 200-MHz Intel Pentium running Linux version 2.0, and q, a 100-MHz Intel Pentium running FreeBsd version 2.1. The two machines were connected with a 10 Mbps Ethernet and one intervening router. The server agent was always on Bald. The client agent was on either machine. For both client locations, we measured the end-to-end wall clock time for a remote procedure call that had a single parameter, an empty server procedure, and no return data. We repeated the experiments for various parameter sizes.

Table 1 presents the average timing results. The first column is the size of the single parameter. The second column is the total time needed for the RPC call. The third column shows the time needed for the client stub to pack the procedure parameters into a byte stream. The fourth column is the time needed for the server stub to unpack the parameters, invoke the actual procedure, and pack the (void) return value. The last column is the percentage of time actually spent transmitting data from one agent to the other. As expected, this percentage is significantly smaller when the two agents are on the same machine. The total RPC time is just the sum of the client stub, server stub, and communication times.

Communication time (last column) shows the time to make a local procedure call (one in the same program) with the same data. Because Tcl is inherently slow, this measure is a good baseline for evaluating results with remote procedure calls. When the client and server agents are on the same machine, and when the parameter size is zero, the remote procedure call takes 200 times longer than the local call. In all other cases, it takes 40 to 140 times longer than a local

Table 1. Agent RPC performance. All times are in milliseconds, and are averages of more than 1,000 trials. The processors passed the data bytes as a single parameter. Bald and q are the names of the two client machines in the experiment.

Data size (bytes)	RPC time		Client stub time		Server stub time	Communication time				Local call time	
	Same machine	Different machine	Bald	q		Same machine		Different machine		Bald	q
0	5.1	6.9	1.1	1.8	2.2	1.8	35.8%	3.0	43.0%	0.024	0.049
64	5.8	8.3	1.2	1.9	3.0	1.7	29.8%	3.5	42.3%	0.062	0.099
256	6.6	9.5	1.3	2.0	3.6	1.8	26.9%	4.0	41.7%	0.100	0.148
1,024	9.6	14.9	1.6	2.4	5.9	2.2	22.9%	6.6	44.6%	0.233	0.321

call. This ratio is fairly common in RPC systems. In Agent RPC, however, we plan to significantly reduce it by using a faster interprocess communication mechanism when the two agents are on the same machine and implementing the parameter packing routines in C instead of in Tcl. Communication was 23 to 43 per cent of the total time in all cases. Thus the overhead imposed by Tcl and our software is only two to four times greater—not unreasonable given Tcl's slow interpretation speed. Of course, Tcl will be too slow for certain applications; for those, the client and server agents could be written in Java or Scheme, either of which is 10 to 1,000 times faster than Tcl, depending on the application.

SEARCH APPLICATIONS

Agent Tcl is used primarily in distributed information retrieval applications. Our most full-featured application is a mobile Tcl agent that searches distributed collections of technical reports. The mobile agent starts on the user's home machine, typically a laptop, where it asks the user for a free-text query. The agent then travels to the network site of each collection, where it interacts with a dedicated information retrieval agent to retrieve relevant documents. As it travels, the agent merges and organizes the results from each collection.

The agent does not actually travel sequentially through the collection sites. Instead, it sends out child agents to search the collections in parallel. More specifically, the agent makes two decisions:

- If the home machine is connected to the network with an unreliable or low-bandwidth link, the agent first migrates to some dynamically selected proxy site within the network. This eliminates any use of the low-quality link except for the initial transmission of the agent and the final transmission of the merged query results.
- If the information retrieval agents provide a low-level interface to the collections, the agent sends a child agent to each collection. The child agents can perform a multistep query using only local communications; only the final query results are sent back to the main agent. On the other hand, if the information retrieval agents provide a high-level interface and the query requires only a single operation, the agent does not send out child agents. It simply interacts with the collections from across the network, avoiding the migration overhead.
- In either case, once the main agent has results from each collection, it merges and filters the results and carries the final list of relevant reports back to the home machine. If the home machine is disconnected, the agent goes through the docking system. Once the agent is back on its home machine, it displays the list of reports to the user. If the user wants to read a specific report, the agent retrieves the complete text.

Benefits and Limitations

Using a mobile agent in this and other search applications has several advantages:

- *Task continuation.* Because the agent migrates onto a proxy site, it can continue its task even if the home machine disconnects. For example, a user can launch the mobile agent from her laptop, disconnect the laptop, fly across the country (or walk down the hall to a conference room), and then have the agent immediately return with the final results when she reconnects.
- *Minimal connection use.* The agent merges and filters the documents at the proxy site, so the use of the connection between the network and the laptop is minimal. This is critical if the connection is a low-bandwidth wireless or modem link.
- *No application-specific support.* Neither the proxy sites nor the document collections need to provide any application-specific support. In fact, the document collections can provide an extremely low-level interface, such as a single operation that just returns the complete text of a specific technical report. By migrating to the collection, the agent can still perform its search efficiently, since all resource accesses are then local. Thus, once the agent system is installed at a site, developers can efficiently imple-

ment numerous distributed applications without any additional software support at the service sites, which makes life much easier for the service providers.

- *Straightforward code*. Even though the agent exhibits relatively complex behavior, it was extremely easy to write, since the communication mechanisms work the same regardless of whether two agents are on the same machines. Basically, the agent just asks the agent system about the current network link and then jumps to a proxy site if that network link has low bandwidth or is expected to go down. Then, using its knowledge of its query and the collection interfaces, the agent will either send out child agents or interact with the collections remotely. The code to perform the query is the same in both cases.

On the downside, because the agent is written in Tcl, it uses significant CPU time at each collection site. In addition, migration overhead in the current system is large. Thus, if the link between the home machine and network has high bandwidth and stays connected, the mobile agent takes more total time than a traditional implementation. With a high-quality network, even though the agent always eliminates intermediate network transmissions when performing a multistep query, the data amount is not large enough for the transmission time to outweigh the CPU and migration time. On the other hand, if the link between the home machine and network is going up and down or has low bandwidth, the agent takes less time, since it transmits minimal data across the link and can proceed even if the home machine is unavailable.

Reducing the migration time should make the agent competitive in all cases, and we are currently doing the necessary implementation and experimental work to verify this belief. Of course, agents that do a large amount of processing per resource access will need to be written in a faster language, such as Java.

Other Applications

Other information-retrieval applications of Agent Tcl include searching for products that meet a customer's needs, searching for a particular mechanical part (with a CAD drawing of the part provided as input), and searching for medical records that match given criteria. In addition, a joint project between Lockheed Martin and the US Army uses Agent Tcl (and a second, proprietary mobile agent system) to propagate tactical information between the battlefield and command headquarters (and to retrieve information relevant to the current situation from various online sources). Lockheed and the Army developed the application over the course of six Military Intelligence Brigade exercises. In all cases, the agents eliminated intermediate data transmission, continued with their task even when the home machine was disconnected, and performed efficiently without application-specific support at each network site.

Agent Tcl allows the rapid development of efficient, robust distributed applications, particularly when mobile computers are involved. Despite its current capabilities, we see several areas for future work. The two most important are network sensing and service directories. To best plan its route through the Internet, an agent needs information about the network's current state, such as its bandwidth, latency, and connectivity. We are developing sensing agents that glean this information from recent past communications with remote hosts.[5]

Once an agent can roam the network, it needs to know where to go to find relevant services. We are constructing a distributed "yellow pages" infrastructure in which agents can advertise their services and client agents can look for agents that meet their needs.[5] These "yellow pages" are similar to the RPC nameservers except that they are hierarchical and can contain arbitrary service descriptions. Eventually the RPC interface definitions will be included in the "yellow page" entries, eliminating the need for the separate RPC nameservers.

Other areas of future work include rewriting some of the service agents in the much faster Java language, making the agent servers more efficient, and extending the security model to protect agents from malicious machines. We are also investigating multidestination jump support, and are integrating our interagent message-passing with the docking system so that messages go through docks when necessary. We are adding a persistent store so that an agent may leave most of its data (such as the results of a database search) at one host, carry a small amount of its data along with it, and yet be able to remotely access the stored data if necessary. Finally, in cooperation with other groups, we are continuing to develop applications that demonstrate the effectiveness of mobile agents in different network environments. ■

ACKNOWLEDGMENTS

We thank the students who helped construct the support agents described in this article: Ting Cai, Saurab Nog, Vishesh Khemani, and Jun Shen built the dock system; Saurab Nog and Sumit Chawla built the RPC system; Katya Pelekhov implemented the technical report search agent; and Saurab Nog and David Hofer maintained the Agent 007 research lab. We also thank the students of CS88/188 (fall 1995) for their many discussions.

This research was supported by ONR contract N00014-95-1-1204, AFOSR contract F49620-93-1-0266, and Air Force Multidisciplinary University Research Initiative grant F49620-97-1-0382.

REFERENCES

1. R.S. Gray. "Agent Tcl: A Flexible and Secure Mobile-Agent System," *Proc. Tcl/Tk Workshop*, Usenix, Berkeley, Calif., 1996, pp. 9-23.
2. R.S. Gray, "Agent Tcl: A Flexible and Secure Mobile-Agent System," doctoral dissertation, CS Dept., Dartmouth College, Hanover, N.H., 1997.
3. S. Nog, S. Chawla, and D. Kotz, "An RPC Mechanism for Transportable Agents," Tech. Report PCS-TR96-280, CS Dept., Dartmouth College, Hanover, N.H., 1996.
4. A. Birrell and B. Nelson, "Implementing Remote Procedure Calls,"

ACM Trans. Computer Systems, Feb. 1984, pp. 39-59.

5. D. Rus, R. Gray, and D. Kotz, "Transportable Information Agents," in *Proc. Ann. Conf. Autonomous Agents*, ACM Press, New York, 1997, pp. 228-236.

David Kotz is an associate professor of computer science at Dartmouth College, where his research interests include parallel operating systems and architecture, multiprocessor file systems, transportable agents, parallel computer performance monitoring, and parallel computing in computer science education. Kotz received a PhD in computer science from Duke University. He is a member of the ACM, IEEE Computer Society, USENIX, and Computer Professionals for Social Responsibility.

Robert Gray is a research professor in the Thayer School of Engineering at Dartmouth College, where his main interests are the performance, security, and fault tolerance of mobile-agent systems. He is the lead programmer for the Agent Tcl project. Gray received a PhD in computer science from Dartmouth College.

Saurab Nog is a software design engineer at Microsoft, where his interests are mobile agents and their applications, operating systems, and computer networks. Nog received an MS in computer science from Dartmouth College and a BTech in computer science and engineering from the Indian Institute of Technology.

Daniela Rus is an assistant professor of computer science at Dartmouth College, where she cofounded and codirects the Transportable Agents Laboratory. She also founded and directs the Dartmouth Robotics Laboratory. Her research interests include mobile agents, robotics, and information capture and access. Rus received a PhD in computer science from Cornell University. She holds an NSF Career award and is a member of Phi Beta Kappa.

Sumit Chawla is a PhD candidate in computer science at Dartmouth College, where his research interests include image compression, fast medical image acquisition, and parallel computing. He received an MS in computer science from Dartmouth and a BA in computer science and mathematics from Knox College.

George Cybenko is the Dorothy and Walter Gramm Professor of Engineering at Dartmouth College, where his current interests are advanced information retrieval and management systems. He is principal investigator of the Transportable Agents and Wireless Computing project, which is part of the DoD's Multidisciplinary University Research Initiative. He is also editor-in-chief of *IEEE Computational Science and Engineering*. Cybenko received a PhD in electrical engineering from Princeton University.

Readers can contact Kotz at Dartmouth College, 6211 Sudikoff Laboratory, Hanover, NH 03755; dfk@cs.dartmouth.edu; or Gray at Dartmouth College, Thayer School of Eng., 8000 Cummings Hall, Hanover, NH 03755; rgray@cs.dartmouth.edu.

14.4 Walsh, T., Paciorek, N., and Wong, D., Security and Reliability in Concordia

Walsh, T., Paciorek, N., and Wong, D., "Security and Reliability in Concordia," *Proceedings of the 31st Hawaii International Conference on Systems Sciences,* VII:44-53, January 1998.

Security and Reliability in Concordia ™

Tom Walsh, Noemi Paciorek, David Wong
Mitsubishi Electric ITA
Horizon Systems Laboratory
300 Third Avenue
Waltham, MA 02154, USA
{walsh, noemi, wong}@meitca.com

Abstract

Concordia provides a robust and highly reliable framework for the development and execution of secure, mobile agent applications. Concordia incorporates many advanced security and reliability features beyond the basic functionality found in other mobile agent systems.

Concordia provides a rich security model that can be used to allow or deny access to system resources down to a very fine level of granularity and that protects agents and the information they carry from tampering or unauthorized access. The system utilizes transactional message queuing to provide reliable network transmissions. Further, Concordia uses proxy objects and a persistent object store to insulate applications from system or network failures. This paper discusses the design and implementation of these features.

1. Introduction

The popularity of the Internet and the World Wide Web has brought much attention to the world of distributed applications development. Now, more than ever, the network is being viewed as a platform for the development of cost-effective, mission-critical applications. While distributed applications represent a great potential future, today's reality of distributed applications development contains many pitfalls. The task of writing distributed applications which run over the Internet is wrought with problems of scalability, reliability, and security. Much interest exists in technologies that aid in the development of these applications.

Mobile agent technologies present an attractive option for distributed application development. The mobility of an agent hides the specifics of the network from the application developer and provides a familiar development paradigm. Further, this mobility allows an agent to inter-operate with existing systems in a manner not possible with other technologies. Also, an agent's mobility and autonomy enable it to perform its task without regards to the quality or reliability of the underlying network, allowing it to provide a solution applicable to WAN or wireless environments.

The scale of applications now being considered for network environments will require a level of security and reliability previously only available in large transaction processing systems. In order for agent systems to present a realistic development alternative for such applications, these systems must evolve to provide the highly secure and reliable environments needed.

Concordia is a framework for developing and executing mobile agents. It has been designed to provide flexible agent mobility within a highly robust and secure environment. Concordia provides a rich security model that allows access to server resources to be controlled based on the identity of the user on whose behalf the agent is executing. Concordia protects agents from tampering when travelling on the network or when stored on disk. Concordia uses proxies and a transactional message queuing system to shield applications from the reliability problems inherent in a distributed environment.

The remainder of this paper discusses the general architecture of Concordia as well as the architecture of the specific components that provide security and reliability.

2. System Overview

The Concordia infrastructure toolkit consists of a set of Java class libraries for server execution, agent application development, and agent activation. Each node in a Concordia system consists of a *Concordia Server* that executes on top of a Java virtual machine. The Concordia Server consists of a number of components as shown in Fig. 1.

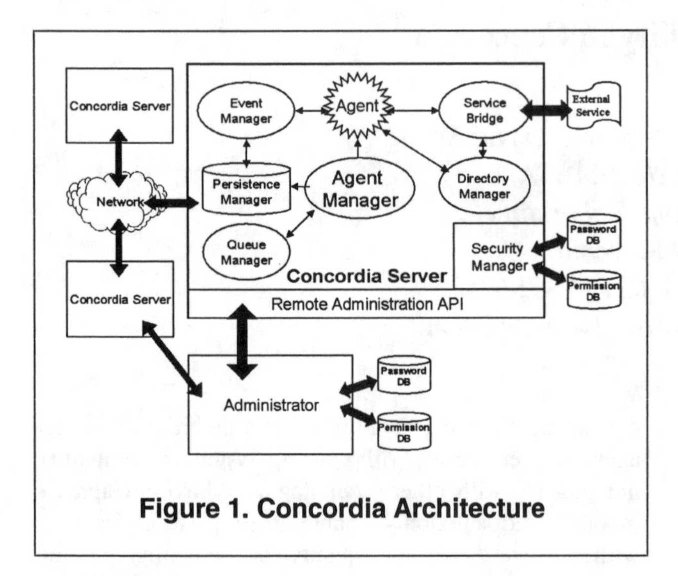

Figure 1. Concordia Architecture

Concordia, like a number of existing agent infrastructures and toolkits, supports the basic communication plumbing needed for agent mobility [19]. Within a Concordia Server, this propagation server is called the *Agent Manager* (formerly known as the Conduit Server in prior Concordia documents).

The Agent Manager provides the basic mobility support needed for agent transfer. It also provides the basic execution context for agents. Moreover, Concordia's agent mobility mechanism extends beyond the functionality provided in other Java-based agent systems. It offers a flexible scheme to dynamically invoke arbitrary method entry points within a common agent application. Details of this flexible scheme for agent mobility can be found in [19].

Concordia provides two forms of inter-agent communication: distributed events and agent collaboration. Distributed events are scheduled and managed by the *Event Manager*. Agent collaboration, which makes use of this event mechanism, allows agents to interact, modify external states (e.g., a database), as well as internal agent states. Details of these two forms of inter-agent communications can be found in [19].

Concordia system administration is handled by the *Administrator*. The Administrator, which resides on a separate Java virtual machine, starts up and shuts down the components in a Concordia Server. It also manages changes in the security profile of both agents and servers in a Concordia system and makes requests to the Security Manager on behalf of the agent or server. The Administrator also monitors the progress of agents throughout the network and maintains agent and system statistics.

The Concordia *Service Bridge* component makes it possible for an application/agent developer to add services to a Concordia Server to support agents when they travel to the server. Service Bridges may be managed remotely using the remote administration API. One major use of Service Bridges is to provide controlled access to native system services.

The *Directory Manager* enables mobile agents to locate the application servers they wish to interact with on each host. It maintains a registry of application services available on each host it manages. A Concordia distributed system may be configured to include one or more Directory Managers (at most one per host). Application servers export their services to mobile agents by registering them with the Directory Manager. Mobile agents then obtain references to application servers via the Directory Manager's lookup operation.

Concordia's security model provides support for three types of protection: (1) agent storage protection, (2) agent transmission protection, (3) server resource protection. Details of Concordia security features are discussed in Section 3.

Concordia agent mobility can extend to a number of local as well as wide area networks. To prevent potential performance and reliability problems associated with the transmission of agents across networks with different characteristics, Concordia also provides support for transactional queuing of agents between Agent Managers residing on different networks. The *Queue Manager* is discussed in more detail in Section 4.

Agent persistence is required to ensure that agents can recover successfully from system crashes. The *Persistent Store Manager* allows the internal state of agent objects to be persisted on disk. The Persistent Store Manager is discussed in more detail in Section 5.

Proxy objects increase the reliability of the Concordia system by shielding agents and other objects from the effects of server and system failures and this feature is discussed in Section 6.

3. Security

Security in mobile agent systems is generally accepted to encompass four distinct problems: (1) the secure network transfer of agents, (2) the protection of a host from attack or misuse by malicious agents, (3) the protection of an agent from attack by a another malicious agent, and (4) the protection of an agent from attack from a malicious host [3][6]. The Concordia security model provides the first three types of protection. There has been considerable debate as too whether the problem of protecting an agent from a malicious server is solvable [5]. The current Concordia security implementation does not support this type of protection. Within Concordia, once a server has been identified as a valid Concordia Server, it is considered to be a *trusted* environment and the Concordia security package does in fact take steps to ensure server integrity.

Within the Concordia system, the term *agent protection* is used to refer to the protection of agents from tampering during transmission or when stored in an on-disk representation. Protection of an agent during transmission is a generally recognized problem in agent systems. For added reliability, Concordia uses a persistent object store for periodically saving agent state. In case of system failure, this server uses the object store to reconstruct executing agents and resume agent travels (See Section 5 for details on Concordia persistence). Since the object store saves an agent and its state information to disk, it could also become a potential security risk. Concordia agent protection addresses securing this on-disk representation as well. Concordia does not attempt to protect an agent when it is present in memory, but relies on the protection offered by the operating system and the Java virtual machine.

Concordia uses the term *resource protection* to refer to the process of protecting server resources from unauthorized access. This area of protection addresses the problem of securing a host from attack or misuse by agents. In addition, this protection is applied to protect agents from attack by other agents.

3.1. Agent Protection

As previously mentioned, agent protection refers to the process of protecting an agent's contents from tampering or inspection during transmission across a network connection or when stored on-disk. Such protection exists to ensure the privacy and integrity of the agent and the potentially sensitive information it carries. This problem can be thought of as being composed of two sub-problems: (1) *transmission protection*, which deals with the network protection of an agent and (2) *storage protection* which protects an agent when represented on disk.

3.1.1. Transmission Protection

Concordia provides transmission protection using Secure Sockets Layer version 3 (SSLv3). SSL is a general-purpose network security protocol, which can provide authentication and encryption services for TCP connections. We chose SSL as the basis for Concordia transmission protection for several reasons. First, SSL itself is emerging as a standard for secure communication over the Internet. Due to its popularity, several implementations are available, including pure Java implementations.

In addition, since SSL plugs into the application at the socket layer, it can conveniently sit underneath other protocols such as HTTP, RPC, IIOP and RMI, in essence securing those protocols. The current implementation of Concordia makes use of Java RMI middleware [12] to provide network communications. RMI provides a Java specific distributed object layer on top of standard TCP/IP sockets. Providing for secure communications over RMI is simply a matter of replacing RMI's usage of standard sockets with secure sockets. SSL's ability to plug into our existing network infrastructure made it a very attractive solution.

Further, since SSL plugs in at the socket level, it doesn't require any special security knowledge or security related code at higher levels of the application, making it somewhat easier to implement than other solutions. Network security is encapsulated at a very low level and the details of that security remain hidden at that level.

SSL also allows for the mutual authentication of both sides of a network connection. This feature allows a server to protect itself from spoofing if an untrusted machine tries to masquerade itself as a Concordia server.

We investigated the possibility of ensuring transmission protection using a public-private key encryption scheme independent of the socket communications level. Similar mechanisms are used by the Agent/TCL and Itinerant Agents systems [2][6]. However, SSL provides a more attractive solution primarily because it encapsulates security mechanisms below the application level and can be made to work with our existing RMI infrastructure.

3.1.2. Storage Protection

Storage protection guards an agent from access or modification when stored in Concordia's persistent store. This type of protection is particularly useful when Concordia is running on an operating system that does not have a secure file system.

When an agent is stored, the agent's bytecodes, internal state and travel status are all written to disk. This data is referred to collectively as the *agent's information.* This agent's information is encrypted using a symmetric key encryption algorithm and a generated agent-specific symmetric key. Concordia takes advantage of the Java Cryptography Extension to allow any of the following symmetric algorithms to be used: IDEA, DES, RC4, RC5, Misty, or Triple DES.

The symmetric key used in the encryption is automatically generated by the Concordia Server when the Agent arrives. After the encrypted agent is written to disk, the symmetric key is then encrypted using a server-specific public key and then stored with the Agent. Decrypting an agent requires the server's private key. System administrators must ensure the security of this private key. This can be accomplished by storing the key on a secured filesystem or on removable media which can be physically secured by the administrator.

For further protection, the bytecodes for uninstantiated agents can optionally be stored in signed Java Archive (JAR) files [8]. Concordia can then use the signature on the JAR file to guarantee that the agent was in fact written by a *trusted* author and has not been tampered with since it was released by that author. This is similar to more traditional code signing done with web applets.

3.2. Resource Protection

With the popularity of the World Wide Web, Java applets, and other forms of downloadable executable content, the need to protect machine resources from unauthorized, potentially harmful, access has become apparent. This problem is no less evident in the world of mobile agents where downloadable executable content is replaced by autonomous mobile executable content. Ensuring protection of sensitive machine resources is one of the most important features of Concordia security. Concordia's approach to resource protection builds upon existing Java security mechanisms but greatly extends them to provide a useful solution for mobile agent systems.

Within the world of Java, traditional approaches to resource protection have focused either on the *sandbox* model or on the use of signed classes. Systems that make use of the sandbox approach classify Java objects into two categories: trusted and untrusted. Trusted objects are granted full access to machine resources while untrusted objects are given very limited access. In general, untrusted objects are not allowed to access the local file system and have very limited access to the network. Java makes the decision as to whether an object is trusted or untrusted by examining the way in which its class was loaded into the Java virtual machine. If a class is loaded from a local disk then it is assumed to be trusted, whereas a class loaded from a network source via a *ClassLoader* object [9], is assumed to be untrusted. Thus, classes that arrive from a network source receive very limited access to machine resources.

The sandbox approach, while providing reliable security, does have some drawbacks, which are especially evident in the mobile agent world. In general, the sandbox is too restrictive. The limits it places on Java classes makes it very difficult to write full-scale applications. Further, since any implementation of a mobile agent system involves the mobility of code around a network, virtually all agent and agent related classes executing on a particular machine are loaded from a network source and thus would be treated as untrusted. This fact would severely limit the usefulness of agents, since their access to machine resources would be so restricted.

The restrictions of the Java sandbox can be relaxed through the use of signed code. The Java Development Kit version 1.1 introduced code-signing support, which allows the author of a Java class to sign that class using a digital certificate. Users can then configure their systems to trust certain authors and can configure differing levels of trust for different authors. In essence, the user of the code trusts that the author of the code will not write malicious or damaging code or at least partially trusts the author not to do so.

While code signing can play a role in mobile agent systems, we feel it does not supply a suitable basis for the security of a mobile agent environment. In most cases, an owner of a machine resource wants to restrict access based on the identity of the user on whose behalf the agent is acting. For example, if you send a mobile agent to your desktop machine, you most likely would want to give it access to your personal files stored on that machine, whereas you would want to restrict access to agents sent by other users. The identity of the author of that code is not the deciding factor in the decision. Thus, in the Concordia philosophy, there is a level of trust between the system owner and the agent owner and not just between the system owner and the agent author. This approach is clearly superior to the one adopted by other mobile agent systems such as IBM's Aglets and General Magic's Odyssey [1][11]. These systems rely solely on the standard Java security model, which does not provide the proper degree of flexibility and extensibility for mobile agent systems, as the basis for their security.

3.2.1. User Identities

In order to identify an agent with a particular user, Concordia associates a *user identity* with agents executing in the system. This identity is a Java object and is composed of three pieces of information: (1) a user name, (2) a user group, and (3) a password. These are roughly equivalent to the user names, groups and password found in secure operating systems. Within the user identity, the password is always stored in a secure hashed form and is never represented in clear text. Construction of an agent requires supplying the clear text password. Thus knowledge of the hashed password is not sufficient for assigning a user identity to an agent.

The inclusion of user groups is primarily intended to ease administration, by allowing the security properties of multiple users to be administered simultaneously. In the current implementation, a user can only be a member of a single group. Allowing multiple group membership is being considered as a future enhancement.

The user identity is usually represented in a shorthand form which looks like the following: username@group. So if a user named *john* were a member of the group *accounting*, this user's identity would be represented as john@accounting.

As the agent travels around the network, it carries its identity with it. During an agent's travels, its identity is protected by the fact that the password is only stored in a hashed form and by the fact that during network travel or on disk storage, the entire agent is protected by Concordia's Agent Protection. Thus it would be very difficult for a third party to even obtain the hashed password moreover to discover the clear text password.

At each stop in the agent's travels, its identity is verified against a list of valid users of the system. Each server is configured with a list of users as well as corresponding resource access permissions allowed for the user. This approach is actually quite different than that adopted by the Agent/TCL system. Within Agent/TCL, an agent's identity is not re-verified at each stop in its travels [6]. Each server passes the agent's identity to the next server in an unverifiable form and this form is accepted because of an implied trust between servers. We decided to perform verification at each stop because of the added level of security this introduced as well as the further scalability of naming that is possible. In a large intranet or particularly in a large extranet, it is not unimaginable that two users could have the same user name and group name. Due to the expected uniqueness of their password, Concordia would identify these two users as different.

The server stores its list of valid users in a file referred to as the *password file*. This password file can either be stored on a local disk of the server, in a location that can be accessed through standard protocols via a URL, or in a JDBC database. In a single server setup or an installation of a small number of servers, the password file most likely would be stored locally on each server. In larger installations, the password file would be stored and administered centrally and accessed remotely by the servers using either JDBC or URLs for access. The network accessibility of the password file was introduced primarily in the expectation of such central administration of large installations.

The password file contains the usernames and hashed password of all the users of the system. Validation of an incoming agent is performed by comparing the hashed password value travelling with the agent to the value stored in the password file.

To prevent unauthorized modification, the contents of the password file are hashed and the hash value is stored with the file. Any modification to the file requires recomputing the hash value for the modified file. If the server discovers that the hash value is incorrect, it considers the password file to be corrupt, an error is logged, and all security requests are rejected. For further security, the hash value can then be signed using the DSA digital signature algorithm and the server's private key. Signing only of the hash value is a performance enhancement used commonly in the signing of documents [13].

We are investigating the use of personal digital certificates as a future mechanism for detecting the identity of agents. In such a system, an agent could be reliably detected to have originated from a particular user because it would have been signed using that user's personal certificate. This is very much like the approach used by the Itinerant Agents system [2].

Another option for user identity would be to map an agent into the identity of an operating system account.

For example a particular agent would be mapped to a particular UNIX or Windows NT user name and then restricted according to the operating system's security policy. This approach is attractive because it does not require the agent system's administrators to maintain multiple databases of user account information. We did not adopt this approach for Concordia because current implementations of Java do not provide a platform independent way to access operating system account information. Platform independence was one of the goals of the Concordia system and we did not want to introduce OS specific code unless absolutely necessary.

Once the server has identified and validated the user, it examines a list of *resource permissions* to see what level of system access is allowed to that user. Resource permissions are analogous to access control lists (ACLs) supported by operating systems. They can be assigned either to individual users or to groups of users.

3.2.2. Resource Permissions

Resource permissions can be used to allow or deny access to machine resources at a very low level of granularity. For example, a resource can be constructed to allow read access to the filesystem of the machine. Another could be constructed to deny such access. A third could specify read access only to a particular file on the machine. There is roughly a one-to-one mapping between Concordia resource permissions and the standard security checks built into Java. For example, resource permissions can be used to control access to files or network resources, the ability to create new threads or processes, the ability to access or change the Java virtual machine's operating properties, the ability to load non-Java code, and the ability to access to the system's console or graphical user interface. Concordia's resource permission mechanism is built on top of the standard Java security classes.

The standard Java security model makes use of a class called a SecurityManager [9]. The SecurityManager class contains methods, which are invoked internally by the Java runtime libraries whenever a potentially dangerous call has been made into the library. For example, if a Java class attempts to read from a file on the local filesystem, the *checkRead* method of the SecurityManager is called. The standard SecurityManager class attempts to determine if the class making the call is trusted or untrusted and then permits or denies access based on that fact.

Concordia security provides a subclass of the Java SecurityManager class. When a call is made into the Concordia Security Manager, it first determines whether the call was made by a class that originated from a network source or by a class loaded from the local machine. If a local class (also referred to as a *native class*) made the call, the Security Manager allows the access to be made. Local classes are always considered trusted and have full access to machine resources.

If the Security Manager determines that the call was made by network code, it then determines if an agent made the call. If this is the case, the manager retrieves that user identity associated with the agent from the agent's execution context. Then the manager validates the identity against the password file and then checks the resource permissions of the user to determine if it should allow the access.

If the Security Manager determines that the class making the call was loaded from a network source but is not an agent, the call is considered to have been made by an untrusted source and is only allowed to execute in a sandbox. Further, if the Security Manager determines an agent made the call, but is unable to verify the identity of the agent, the call also executes in the sandbox. The Concordia security system defines a special user identity known as *untrusted* which is used to control resource access given to these types of unknown or unverifiable agents. The resource permissions of the untrusted identity can be configured just like any other identity. Since the untrusted account defines Concordia's sandbox, this sandbox is configurable to any level of access desired.

The resource permissions for a server are stored in a file called the *permissions file*. The permissions file is hashed and signed in the same manner as the password file and also can be accessed either from the local filesystem, an URL or a JDBC database.

As mentioned before, Concordia's resource permissions build upon the standard Java SecurityManager class. The SecurityManager provides method callbacks for such things as file access, network access, and thread and process management. Concordia provides resource permissions for all of the resources protected by the standard Java SecurityManager. Concordia also provides resource permissions for higher-level procedures such as starting, suspending or stopping a server or an agent. These high level procedures are generally accessed remotely using the Concordia Administrator, a GUI administration tool. A request to administer a server is validated and regulated in the same way as a resource request by an agent.

Through the application of resource permissions, it is possible to protect an agent from attack by other agents. Since any request to administer an agent requires an authorized user identification, it is not possible for an unauthorized agent to stop, suspend, or alter the travel plans of another agent. Concordia provides no facilities that allow agents to obtain direct references to other agents nor make direct method calls on each other as is seen in Telescript meetings [17]. Agents usually communicate via messages or Concordia's group-oriented communications mechanisms. However, if an agent were to retrieve a reference to another agent through other means, it would be allowed to make public method calls on the other agent, but would not be allowed to interfere with its travels. When agents arrive from the network, their bytecodes are verified using the standard Java bytecode verifier. This guarantees that an agent cannot make an illegal call into the non-public interface of another agent. Thus, the only method of entry into an agent is its public interface. It is the responsibility of the agent developer to guarantee that the public interface is secure. For example, the public interface should not publish confidential information.

Concordia also takes steps to guarantee the integrity of the system. In the next release, Concordia's bytecodes will be shipped in a signed JAR file. During the startup procedure, Concordia will inspect the JAR file to guarantee that it has not been altered in any way. While this provides additional security, it does not completely secure an agent against the possibility of harm from a malicious server. Any code loaded from the local disk of a server will be considered trusted and will be given full access to the system. Further, the bytecodes of locally loaded classes are not verified on loading for performance reasons. Finally, given that a server manages the marshalling and unmarshalling of agents from a network, handles their writing to and reading from persistent storage, and controls their thread of execution, a high level of trust in the server is needed. In general, this is not a problem since the user decides where an agent is to travel, and can send highly sensitive agents only through trustworthy machines.

4. Queuing

The Concordia infrastructure provides support for reliable transmission of agents across the network via use of an underlying message queuing subsystem. Concordia's queuing support allows the Agent Manager to submit an agent enqueue request to the Queue Manager in an asynchronous manner. The agent message queue serves as a transmission buffer for the request in this manner. This feature of the message queuing subsystem is a natural fit to the disconnected operational mode of the mobile agent paradigm because it provides a "store and forward" mechanism. Agents can be stored on the message queue of a local server while a remote server is undergoing repair or is simply moved to a different physical location. When the remote server comes back online, the local server would then forward the agent to the returned server.

A message queuing subsystem provides additional reliability by maintaining a copy of the agent to be transmitted in an on-disk queue until the recipient of this agent transmission has acknowledged its receipt via the well-known handshaking protocol called the *Two Phase Commit* protocol. This is known in the industry as *transactional message queuing*. The Queue Manager communicates with its local Agent Manger and performs handshaking with other remote Queue Managers for reliable agent transmission.

The Queue Manager manages both an inbound and an outbound queue. The Agent Manager submits an agent enqueue request for the local Queue Manager's outbound queue. The local Queue Manager then proceeds to dequeue agents off its outbound queue and submits an enqueue request with a remote Queue Manager's inbound queue. The remote Queue Manager then proceeds to dequeue agents off its inbound queue and transmits the agent object to its local Agent Manager.

The Queue Manager is implemented as a multi-threaded Java thread object to allow for concurrent access to the underlying queue storage. The preservation of an object's class specification on disk is handled by the Java object serialization facilities while Queue Manager communication relies on the Java RMI package [12]. Reliable queuing of agent transmission appears to be an important feature lacking in other commercial mobile agent system offerings.

The Queue Manager's design goals included achieving: optimal disk space utilization, fast write operations, fast recovery from server failure, and reliable management of pending agent transmissions. Its on-disk queue storage architecture implementation borrows some ideas from the log-structured file systems research area [14][15] to employ a unique data architecture that ensures better overall performance over traditional message queuing system architectures [4][10].

Traditional message queuing architectures are generally not optimized for write operations without requiring extra hardware to work efficiently (e.g., utilization of low level RAID devices to cluster data blocks). Such systems appear to require special performance optimization in both hardware and software in order to handle workloads with high throughput and low message residence time, important characteristics of mobile agent systems.

Furthermore, traditional message queuing architectures employ separate queue data and log files, which introduce an extra level of *unreliability* since there are two points of potential file corruption and media failure. There is also usually no means for the message queuing systems administrator to predefine the amount of work needed to do recovery a priori. The utilization of automatic system checkpointing by the Queue Manager handles this problem efficiently.

In Concordia, the data architecture of the on-disk queue storage design consists of a combined on-disk file structure for the message queue data and log records. This architecture utilizes a circular queue and consists of a single flat file that is created when a Queue Manager is first initialized. Each entry in the queue data file contains the agent object in contiguous blocks on the disk. The queue data file is partitioned into a predefined number of logical segments. Each segment contains a predefined number of control blocks at the beginning of each segment. These control blocks contain control information for the queue entries and log record information. Queued agent objects and their log records are stored after the control blocks with potential mixing of agent objects and log records. When a new segment is reached in the queue data file, a new set of control blocks is written to disk at the beginning of the new segment. In this manner, the logical segments serve as forced *checkpoint* intervals on the state of the entire queuing system and the data that is stored in the queue file.

The Queue Manager design adheres to the Atomicity, Consistency, and Isolation properties of the well-known ACID properties. The Durability property to guard against hardware failures can be achieved by duplicating the combined queue data/log file on a separate disk. We can utilize existing RAID technology to do duplicate writes transparently. We made this architectural design decision in order to provide flexibility to Concordia application developers with respect to the possible tradeoffs between cost and levels of reliability required. For example, for deployment on mobile hand-held devices, one may wish to limit the hardware requirements on such systems at the expense of lack of media durability.

5. Persistence

Concordia's infrastructure incorporates highly- reliable servers that save their internal state to persistent storage and, during failure recovery, retrieve and reconstruct any data required to continue servicing their clients. Concordia also transparently saves agents' states to persistent storage, thereby enabling agents to recover from system and server failures and to continue execution unaffected. Applications and agents running on Concordia systems may also utilize persistent storage to checkpoint themselves and, after failure, restart from their latest checkpoint.

The *Persistent Store Manager* (PSM) is a general-purpose facility utilized by Concordia's servers to save their state and also that of mobile agents. Typically, each server or application owns an instance of the PSM and each instance utilizes a different file for persistent storage. The PSM exports methods to create, delete, update, and fetch objects from persistent storage.

The PSM manages saved objects by utilizing a random access facility built upon Java's object serialization package. The object serialization code ensures that when an object is stored, all objects reachable from it (i.e., all objects it refers to and all objects its references refer to, etc.) are also stored. Similarly, when an object is retrieved, all the objects reachable from it must also be retrieved.

The Agent Manager on each host is responsible for sending agents to remote hosts, receiving them from remote nodes, and providing an execution environment for agents. When the Agent Manager receives an agent, it writes its state to persistent storage before creating its main thread of execution. (For more details on the Agent Manager's operation, see [19].) After the agent finishes executing, the Agent Manager updates the agent's state in persistent storage before transmitting it to the Agent Manager at the agent's next destination. The agent remains in persistent storage until it has been successfully transferred to the next Agent Manager. When an agent finishes executing at its final destination, the Agent Manager deletes it from persistent storage.

After a system or server failure, the Agent Manager retrieves each agent's state from persistent storage and restarts it. A restarted agent may potentially repeat work it has already completed. Therefore this persistence scheme only guarantees correctness for agents performing idempotent operations. This criterion is sufficient for many mobile agent applications, which can often be classified as information retrieval and filtering.

However, the need for idempotency can be eliminated if agents utilize the PSM to checkpoint their internal state and restart their execution from their last checkpoint.

If the Agent Manager's recovery process fetches an agent that has already finished executing, it transfers the agent to its next destination (if one exists) and deletes it from the persistent store once the transfer is complete. This algorithm may result in duplicate agent transfers if a system or server failure occurs during or immediately after an agent transfer. This problem is detected and handled by the Queue Manager, which ensures that agent transfers complete reliably and that each time an agent travels to a new host, the Agent Manager receives exactly one copy of that agent. (See Section 4 for details on the Queue Manager.)

Many of Concordia's servers, in addition to the Agent Manager, employ the PSM. Both the Event Manager and Directory Manager save their registrations in persistent storage. Objects (e.g., agents) that receive distributed events, register their interest in specific events with the Event Manager. Each time the Event Manager receives a registration, it caches it and writes its updated registration information to persistent storage. If the Event Manager is restarted, its initialization process retrieves the registrations from its persistent store file. While the Event Manager is unavailable, some of its registrants may terminate. If so, the Event Manager will be unable to notify the defunct registrants of new events and will delete their registrations.

The Directory Manager utilizes persistent storage in a similar manner. As described in Section 2, Application Servers may export their services to mobile agents by registering with the Directory Manager. The Directory Manager caches registration information, saves it in persistent storage, and retrieves it when it restarts.

System administrators may optionally enable persistence for the Agent Manager, Event Manager, and Directory Manager. The Concordia approach is flexible – it allows administrators and developers to weigh the costs of persistence against the reliability requirements of the applications. This is in contrast to the Telescript paradigm, in which all objects are persistent [7]. Because the Telescript engine provides a garbage-collected persistent object store, it incurs a large cost in terms of resource utilization and performance[1]. Other recent mobile agent systems [18] do not provide a high degree of reliability, and therefore do not include support for persistence.

[1] General Magic's Telescript User's Guide recommends that systems running the Telescript engine have 96MB of memory.

6. Proxies

Proxies increase the reliability of the Concordia system by shielding agents and other objects from the effects of server and system failures [16]. Server proxies transparently attempt to re-establish connections when they are unable to communicate with their counterparts. Concordia provides proxies for servers that support potentially long-lived connections (currently only the Event Manager).

A distributed system with multiple nodes executing Concordia servers may be configured with one or more Event Managers. Agents may choose to maintain a reliable connection to an Event Manager by communicating via a proxy instead of directly with the server. (In applications that are more concerned about performance than reliability, agents may interact directly with the Event Manager.) An Event Manager Proxy is essentially an agent's local representative for a potentially remote Event Manager. An agent may communicate with the Event Manager on its local host, on the node where it was launched, or on any other host. Hence, as an agent travels, it often interacts with a remote Event Manager.

Each Event Manager registers itself with Java's RMI Registry and, potentially, the Directory Manager. The Event Manager Proxy is responsible for obtaining a reference to the Event Manager, via the Registry, and establishing a connection to it. If communication fails, the proxy also re-establishes a connection to the Event Manager. (Concordia's Administration Manager is responsible for restarting the Event Manager.)

If a mobile agent attempts to interact with a failed Event Manager, its proxy will be unable to communicate with the defunct server. After retrying its operation, the proxy requests a new reference to the Event Manager from the Registry. If the Event Manager has been restarted by this time, the proxy obtains a reference to the new instance; otherwise, the proxy throws an exception. This combination of persistent Event Manager state and server proxies comprises a highly reliable form of communication between mobile agents and a potentially remote Event Manager.

7. Conclusion

Concordia provides an environment for the development of highly secure and highly robust mobile agent applications. The infrastructure extends the standard security mechanisms of the Java language to provide an identity-based system where the rights given to an agent are determined from the identity of the user who launched the agent. Concordia also protects an agent and the information it carries from tampering when stored on disk or when transmitted over a network connection.

Through the use of proxies and object persistence, Concordia provides a highly robust environment where applications can gracefully recover from system or network failures. Using a transactional message queuing system, Concordia provides for reliable network transmission of agents even over unreliable network connections. As described in this paper, Concordia provides many features vital to the development of complex distributed applications.

Information on obtaining Concordia is available at the Mitsubishi Electric ITA Web site (URL= http://www.meitca.com/HSL/Projects/Concordia). Future extensions to the existing functionality may include support for transactional multi-agent applications and knowledge discovery for collaborating agents.

8. Acknowledgments

The authors wish to thank the members of the Concordia team for their contributions to this paper. In particular we would like to thank Joe DiCelie for his input into the section on Concordia security.

9. References

[1] *Aglets: Mobile Java Agents,* IBM Tokyo Research Lab, URL=http://www.trl.ibm.co.jp/aglets

[2] D. Chess, B. Grosof, C. Harrison, D. Levine, C. Parris, "Itinerant Agents for Mobile Computing", *IEEE Personal Communications Magazine*, 2(5), October 1995.

[3] M. Condict, D. Milojicic, F. Reynolds, D. Bolinger, "Towards a World-Wide Civilization of Objects", In *Seventh ACM SIGOPS European Workshop,* 1997.

[4] *Encina RQS Programmer's Guide*, Transarc Corporation, Pittsburgh, Pennsylvania, 1994.

[5] W. M. Farmer, J. D. Guttman, and V. Swarup, "Security for Mobile Agents: Issues and Requirements", In *19th National Information Systems Security Conference (NISSC 96)*, 1996.

[6] R. S. Gray, "Agent Tcl: A flexible and secure mobile-agent system", In *Proceedings of the Fourth Annual Tcl/Tk Workshop (TCL 96)*, Monterey, California, July 1996.

[7] "An Introduction to Safety and Security in Telescript", General Magic White Paper, 1996.

[8] "JAR - Java Archive", Javasoft Corporation, URL=http://www.javasoft.com/products/jdk/1.1/docs /guide/jar/index.html

[9] "Java ™ Platform 1.1.4 Core API", Javasoft Corporation, URL= http://www.javasoft.com/products/jdk/1.1/docs/api/p ackages.html

[10] *MQSeries: Message Queuing Interface Technical Reference*, IBM Corporation, Armonk, New York, 1994.

[11] *Odyssey*, General Magic, URL= http://www.genmagic.com/agents/odyssey.html

[12] "Remote Method Invocation for Java", Javasoft Corporation, URL= http://chatsubo.javasoft.com/current/rmi/index.html

[13] B Schneier, *Applied Cryptography*, p. 38, John Wiley & Sons, Inc., New York, NY, 1994.

[14] M. Seltzer, "Transaction Support in a Log-Structured File System", In *Proceedings of the Ninth International Conference on Data Engineering*, February, 1993.

[15] M. Seltzer, K. Bostic, M. McKusick, C. Staelin, "A Log-Structured File System for UNIX", In *Proceedings of the 1993 Winter Usenix Conference.*

[16] M. Shapiro, "Structure and Encapsulation in Distributed Systems: The Proxy Principle", *Proceedings of the 6th International Conference on Distributed Systems (ICDCS)*, IEEE Computer Society, 1986, pp. 198-205.

[17] J. E. White, "Telescript Technology: Mobile Agents", General Magic White Paper, 1996.

[18] J. E. White, "Mobile Agents – A Presentation by Jim White", General Magic, 1997.

[19] D. Wong, N. Paciorek, T. Walsh, J. DiCelie, M. Young, B. Peet, "Concordia: An Infrastructure for Collaborating Mobile Agents", In *Mobile Agents: First International Workshop*, Lecture Notes in Computer Science, Vol. 1219, Springer-Verlag, Berlin, Germany, 1997.

14.5 Baumann, J., Hohl, F., Rothermel, K., and Straßer, M., Mole - Concepts of a Mobile Agent System

Baumann, J., Hohl, F., Rothermel, K., and Straßer, M., "Mole-Concepts of a Mobile Agent System," *World Wide Web*, 1(3):123-137, September 1998.

MOLE - CONCEPTS OF A MOBILE AGENT SYSTEM

J. Baumann, F. Hohl, K. Rothermel, M. Straßer

IPVR (Institute for Parallel and Distributed High-Performance Computers)
University of Stuttgart
Breitwiesenstraße 20-22
D-70565 Stuttgart

Abstract

Due to its salient properties, mobile agent technology has received a rapidly growing attention over the last few years. Many developments of mobile agent systems are under way in both academic and industrial environments. In addition, there are already various efforts to standardize mobile agent facilities and architectures.

Mole is the first mobile agent system that has been developed in the Java language. The first version has been finished in 1995, and since then Mole has been constantly improved. Mole provides a stable environment for the development and usage of mobile agents in the area of distributed applications.

In this paper we describe the basic concepts of a mobile agent system, i.e. mobility, communication and security, discuss different implementation techniques, present the decisions made in Mole and give an overview of the system services implemented in Mole.

1. Introduction

Throughout the past years the concept of software agents has received a great deal of attention. Depending on the particular point of view, the term 'agent' is associated with different properties and functionalities, ranging from adaptive user interfaces or cooperating intelligent processes to mobile objects. Our particular interest lies in the exploration of mobile agents in context of the Internet and the key benefits provided by the application of this new technology (e.g. in the area of the WWW).

Mobile agents are defined as active objects (or clusters of objects) that have behaviour, state and location (see e.g. [Mobile Agent Facility 1997]). Mobile agents are *autonomous* because once they are invoked they will autonomously decide which location they will visit and what instructions they will perform. This behaviour is either defined implicitly through the agent code (see e.g. [Gray 1995]) or alternatively specified by an - at runtime modifiable - itinerary (see e.g. [Wong *et al.* 1997]). Mobile agents are mobile since they are able to migrate between locations that basically provide the environment for the agents' execution and represent an abstraction of the underlying network and operating system.

With the properties printed out above it has often been argued that mobile agents provide certain advantages compared to traditional approaches, e.g. the reduction of communication costs, better support of asynchronous interactions, or enhanced flexibility in the process of software distribution. The employment of mobile agents has been particularly promising in application domains like information retrieval in widely distributed heterogeneous open environments (e.g. the WWW), network management, electronic commerce, or mobile computing (for details please refer to [Harrison *et al.* 1995; Baumann *et al.* 1997c; Rothermel *et al.* 1997])

To support the paradigm of mobile agents, a system infrastructure is needed, that provides the functionality for the agents to move, to communicate with each other and to interact with the underlying computer system. Furthermore this infrastructure has to guarantee privacy and integrity of the agents and the underlying system, to prevent malicious agents attacking other agents or the computer system. At the same time the agents have to be protected against a potentially malicious system to avoid manipulations of the agents while they visit this system.

In this paper we describe the current state of our agent system infrastructure, the mobile agent system Mole V3. Mole builds on Java version 1.1 [Sun 1997] as the environment for the agent system and as the language for the implementation of the agents.

This paper is organized as follows: after a short introduction into our agent model in Section 2 we discuss possible mobility concepts for mobile agent in Section 3. In Section 4 we describe the communication concepts of Mole V3. In Section 5 security concerning mobile agent systems is examined. After giving an overview over the related work in Section 6 we present the Mole system in Section 7. Here we discuss some of the design decisions made in Mole and our experiences with the system. In Section 8 we summarize the paper and present our planned future work.

2. Our Agent Model

In this section we present our agent model. Our model of an agent-based system - as various other models - is mainly based on the concepts of agents and places. Places provide the environment for safely executing local as well as visiting agents. An agent system consists of a number of (abstract) places, being the home of various services. Agents are active entities, which may move from place to place to meet other agents and access the places' services. In our model, agents may be multi-threaded entities, whose state and code is transferred to the new place when agent migration takes place.

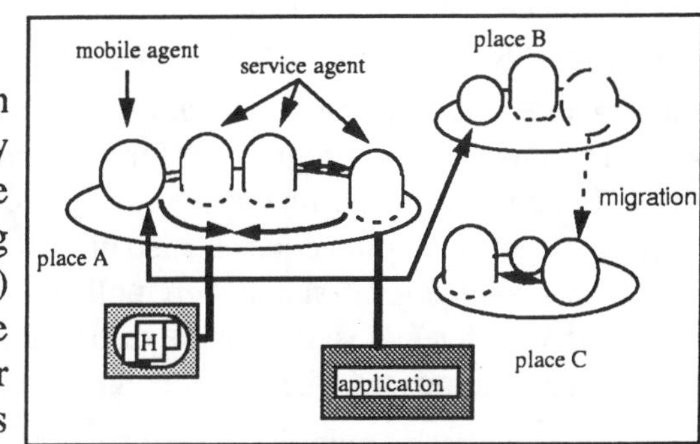

Figure 1: The Agent Model

Each agent is identified by a globally unique agent identifier. An agent's identifier is generated by the system at agent creation time. The creating place can be derived from this name. It is independent of the agent's current place, i.e. it does not change when the agent moves to a new place. In other words, the applied identifier scheme provides location transparency.

A place is entirely located at a single node of the underlying network, but multiple places may be implemented on a given node. For example, a node may provide a number of places, each one assigned to a certain agent community, allowing access to a certain set of services or implementing a certain prizing policy. Places are divided into two types, depending on the connectivity of the underlying system. If a system is connected to the network all the time (barring network failures and system crashes), a place on this system is called *connected.* If a system is only part-time connected to the network, e.g. a user's PDA (Personal Digital Assistant), the place is called *associated.*

Additionally we have a model of the lifecycle of an agent, which is particular to the Mole system. This will be presented in section 7.1.

3. Mobility Concepts for Mobile Agents

In this section we define a taxonomy of the different kinds of mobility, discuss the advantages of the different approaches and present the implementation of mobility in Mole.

3.1 Taxonomy of Mobility

Different degrees of mobility can be distinguished (see Figure 2). In the case of ***remote execution***, the agent program is transferred before its activation to some remote node, where it runs until its termination, i.e., an agent is transferred only once. The information transferred includes the agent code plus a set of parameters (although the transfer of code may not be necessary

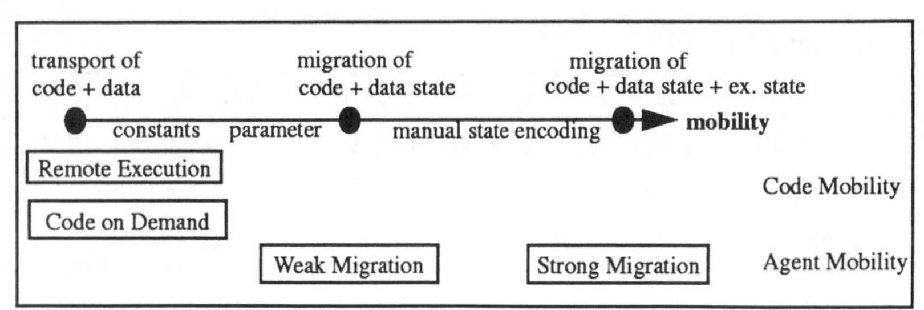

Figure 2: Degrees of Mobility

at runtime, compare e.g. [Hohl *et al.* 97] for a discussion of code transport issues). Once activated, an agent itself may use the *remote execution* mechanism to start the execution of other agents. A very similar approach - tailored to a client/server-style of interaction - is the *remote evaluation* scheme introduced by Stamos [Stamos 1990]. With this approach, an operation (e.g. a procedure plus parameters) is transferred to a remote site, where it is performed entirely. After executing the operation, the remote site returns the operation's results back to the issuer of the remote evaluation. The remote evaluation mechanism can be applied recursively, resulting in a tree-structured execution model.

With the above scheme, the destination of the agent program to be transferred is determined by the entity initiating the remote execution. In contrast, with ***code on demand***, the destination itself initiates the transfer of the program code. If the *code on demand* scheme is used in client/server settings, programs stored on server machines are downloaded to clients on demand. The currently most popular technologies supporting this type of mobility are ActiveX (see e.g. [Aaron and Aarron 1997]) and Java Applets (see e.g. [Sun 1994]).

Both *remote execution* and *code on demand* support 'code mobility' rather than 'agent mobility' as both schemes transfer agent programs before their activation. In the following two schemes, agents (i.e., executions of agent programs) may actively migrate from node to node in a computer network. Obviously, for migrating agents not only code but also the state of the agent has to be transferred to the destination. We will start with *strong migration* and then motivate the existence of *weak migration*. For our discussion we will assume that an agent's state consists of data state (i.e. the arbitrary content of the global or instance variables) and execution state (i.e. the content of the local variables and parameters and the executing threads).

The highest degree of mobility is *strong migration* [Ghezzi and Vigna 1997]. In this scheme, the underlying system captures the entire agent state (consisting of data and execution state) and transfers it together with the code to the next place. Once the agent is received at its new place, its state is automatically restored. From a programmer's perspective, this scheme is very attractive since capturing, transfer and restoration of the complete agent state is done transparently by the underlying system. On the other hand, providing this degree of transparency in heterogeneous environments at least requires a global model of agent state as well as a transfer syntax for this information. Moreover, an agent system must provide functions to externalize and internalize agent state. Only few languages allow to externalize state at such a high level, e.g., Facile [Knabe 1995] or Tycoon [Mathiske *et al.* 1997]. Since the complete agent state (including data and execution state) can be large - in particular for multi-threaded agents - strong migration can be a very time-consuming and expensive operation.

These difficulties have led to the development of the so-called ***weak migration*** scheme, where only **data** state information is transferred. The size of the transferred state information can be limited even more by letting the programmer select the variables making up the agent state. As a consequence, the programmer is responsible for encoding the agent's relevant execution states in program variables. Moreover, the programmer must provide some sort of a *start* method that decides, on the basis of the encoded state information, where to continue execution after migration. This method reduces substantially the amount of state to be transferred. But it changes the semantics of a migration, a fact that every agent programmer has to be aware of.

The following table classifies existing agent systems with regard to the degree of mobility supported.

Table 1: Mobile Agent Systems Classified

Type of Mobility	Systems		
Remote Execution	Java Servlets (push [Sun 1996])	Remote Evaluation [Stamos 1990]	Tacoma [Johansen *et al.* 1995]
Code on Demand	ActiveX [Aaron and Aaron 1997]	Java Applets [Sun 1994]	Java Servlets (pull [Sun 1996])
Weak Migration	Aglets [IBM 1996]	Mole [Mole 1997]	Odyssey [General Magic 1997]
Strong Migration	AgentTcl [Gray 1997]	Ara [Peine 1997]	Telescript [White 1997]

3.2 Advantages of Code and Agent Mobility

In the following, we will examine the advantages resulting from code and agent mobility. During our discussions, for each of the indicated advantages we will point out which degree of mobility is required. We have identified the following prime advantages: Software distribution on-demand, asynchronous operation of tasks, reduction of communication cost, scalability due to dynamic placement of functions.

3.2.1 Software-Distribution on Demand

In existing client-server systems, new code has to be installed manually by users or system operators. The installation is sometimes rather challenging and often requires detailed knowledge about the current state of the used computer system. Finally, software tends to depend on certain versions of other software packages and the installation varies on different machines or operating systems.

With the wide employment of code-on-demand systems (i.e. the success of Java-enabled web browsers) another, easier installation alternative showed up: the *Software-Distribution on Demand* which is able not only to transport code, but also to install packages automatically. For this purpose, code servers offer programs to clients, which include an environment to install these modules. The usage of a platform-independent language like Java allows the system to employ the same installation process on each system and hides differences at the execution level of the code. Since software-distribution on demand is a potential advantage of every mobile code system, mobile agent systems can offer it too. But since the latter uses a less transient model of applications (e.g. the active existence of Java Applets is bound to the existence of the invocation of a browser), code can be installed in a persistent way. What degree of mobility is needed here? If the client actively calls the code, *code on demand* is certainly the matching scheme, but also a kind of *remote execution* can be applied if the server e.g. periodically disseminates new versions of a program to registered clients. Software-distribution on demand only simplifies the management of an existing structure. The following advantages will allow to build up new structures that are better than the old ones in some aspects.

3.2.2 Reduction of Communication Costs

Using agent technology does not reduce communication cost per se. However, in certain situations mobile programs/agents may reduce this cost substantially. Two types of reductions can be distinguished:

- number of (remote) interactions (i.e., between entities residing on different nodes), and
- the amount of data communicated over the network

The first type of reduction can be achieved by bringing two entities that (heavily) interact with each other to the same place. For the second type of reduction consider a client/server relationship, where the client includes a function filtering the data retrieved from the server. The amount of data transferred from the server to the client can be reduced by moving the filter function (or the entire client) to the server. Then, filtering can be done before the data is communicated over the network.

Of course, moving agents is not for free. An overall cost reduction is only obtained if the performance gains exceed the extra overhead for transferring agents. A performance model taking into account an agent's itinary is described in [Straßer and Schwehm 1997], and an overview is given in section 7.8. Other models are given in [Carzaniga *et al.* 1997] and in [Chia and Kannapan 1997]. Communication cost reductions can already be achieved with the *remote execution* scheme. However, agent mobility provides more room for optimizations.

3.2.3 Asynchronous Tasks

Asynchronous communication mechanisms, such as asynchronous message queues (see e.g. [IBM 1997]), allow asynchronous processing of requests (Figure 3). While the individual requests of a task can be processed asynchronously, the client performing this task must be available to receive replies and react on them. Especially in the case of mobile clients or *associated* places this can be problematic. Keeping a mobile client up and connected while task processing is in progress is expensive at least and might even be impossible.

With agent technology, the client part of the application can be transferred from the mobile device to stationary servers in the network. From an end user's perspective, not only individual requests but the entire task is moved to the network, where it is performed asynchronously.

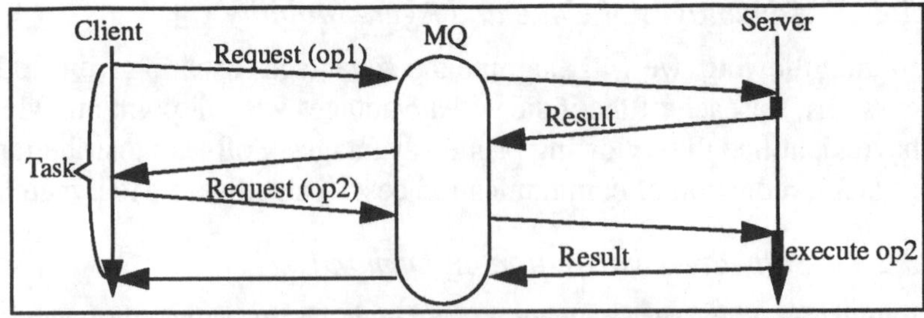

Figure 3: Asynchronous Tasks using Message Queues (MQ)

Clearly, once the task transfer is complete, the mobile device can be disconnected from the network. Later, after hours, days or even weeks, the device can be reconnected to receive the task's results. It is important to notice that the hidden assumption of those scenarios is that the underlying system guarantees 'exactly once' semantics of agents, i.e. when accepting an agent, the network guarantees that the agent is not lost and is performed exactly once, independent of communication and node failures. Unfortunately, none of the current agent systems supports this level of fault-tolerance.

Now, the question is what degree of mobility is required for those scenarios. *remote execution* is sufficient for moving the client program to a *connected* place. This *connected* place (see section 2) is assumed to be an infrastructure component, whose purpose is to host clients downloaded to the network. Downloaded clients run on these places and access remote services to perform their tasks. In general, common servers cannot be used to do this hosting job, e.g., a Lufthansa server is certainly not willing to host an agent booking a flight with a remote British Airways server. With a *weak* or *strong migration* scheme, hosting agents is not needed anymore. An agent moves from server to server to achieve its task. In our example above, the mobile agent just goes on to the British Airways server if the Lufthansa server cannot offer the desired flight.

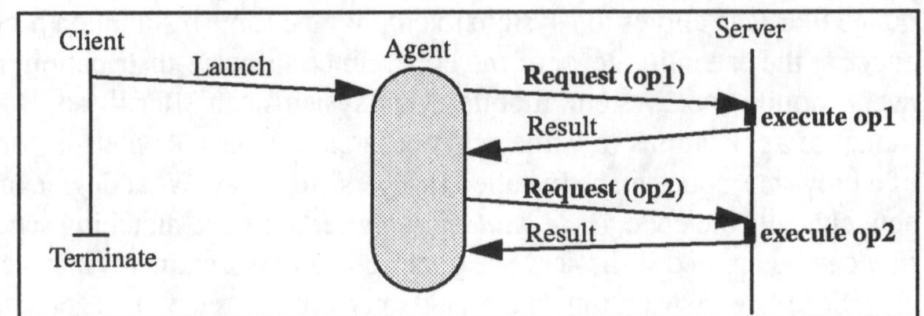

Figure 4: Asynchronous Tasks using Mobile Agents

3.2.4 Scalability Due to Dynamic Deployment

Dynamic deployment of agent programs allows for more scalable applications. Assume, for example, a search application that accesses a large number of globally distributed data sources. Assume documents are retrieved from the data sources and selected (or indexed) based on a content-based filtering function. In a pure client/server setting, a client would access the remote data sources, and all retrieved documents would be transferred to the client. The final filtering would be performed at the client site. If accessing the data sources is performed in parallel, the client as well as (parts of) the network may become a bottleneck.

With mobile agents, a hierarchy of filter agents can be set up. Filter agents not only perform content-based filtering but also get rid of redundantly retrieved documents. The structure of the hierarchy and the placement of the individual filter agents mainly depends on the set of data sources accessed. Both placement and structure can change if new data sources are detected while the search operation is in progress. Obviously, this setting is more scalable since filtering is distributed and can be performed close to the data sources. Moreover, redundant information can be detected early and thus does not have to be transported all the way to the client.

What degree of mobility is needed here? *remote execution* is certainly sufficient if the placement of (filter) agents is static for a given search operation. If, however, the placement is changed and (filter) agents maintain context information (e.g., in order to detect redundant documents), *strong* or *weak migration* would simplify the implementation. We have now discussed the different mobility concepts possible for mobile agent system. In section 7.2 we will present the design decision made for the Mole system.

4. Communication Concepts for Mobile Agents

In this section, we will address the various types of communication suitable for agents and discuss their use in Mole. An in-depth discussion of communication paradigms suitable for agents can be found in [Baumann *et al.* 1997a].

A fundamental question tightly related to communication is how mobile agents are identified. There is certainly a need for globally unique agent identifiers (which we will discuss in section 7.5). Identifier schemes that provide code migration transparency are well-understood today. However, such a scheme might be too inflexible in agent-based systems. Assume for example, that a group of agents cooperatively performs a user-defined task. Assume further that one group member wants to meet another member of this group at a particular place for the purpose of cooperation. In this case, the member should be identified by a tuple consisting of place and group identifier. If the agent to be met additionally is expected to play a particular role in this group, the identifier would have the form *(placeId, groupId, roleId)*. For supporting those application-specific naming schemes we propose the concept of badges (see section 7.3).

A number of the currently existing agent systems are purely based on an RPC-style communication. While this type of communication is mostly appropriate for interactions with service agents, i.e. those agents that represent services in the agents' world, it has its limitations if agents interact like peers. For the purpose of cooperation mobile agents must 'meet' and establish communication relationships from time to time. For this purpose, we introduce the concept of a session, which is an generalization of Telescript's meeting metaphor [White 1997]. Therefore, we support both message passing and remote method invocations (with or without session context).

In the general case, a group of agents performing a common task may be arbitrarily structured and highly dynamic (see [Baumann and Radouniklis 1997] for details). In those environments, we can not assume that an agent that wants to synchronize on an event (e.g., the completion of some subtask this agent depends upon) knows a priori which agent or agent subgroup is responsible for generating this event.

4.1 *Types of Agent Communication*

Considering inter-agent interaction, we distinguish between following types of communication:

- mobile agent/service agent interaction
  Since service agents are the representatives of services in the agent world, the style of interaction is typically client/server. Consequently, services are requested by issuing requests, results are reported by responses. To simplify the development of agent software, an RPC-like communication mechanism should be provided.
- mobile agent/mobile agent interaction
  This type of interaction differs significantly from the previous one. The role of the communication partners are peer-to-peer rather than client/server. Each mobile agent has its own agenda and hence initiates and controls its interactions according to its needs and goals. Furthermore, the communication patterns that may occur in this type of interaction might not be limited to request/response only. Assume e.g. a mobile agent passing a form to another agent and terminating afterwards. The receiving agent would fill out that form by using various services and finally would deliver the filled out form to another agent waiting at some previously specified place. The required degree of flexibility for those interactions is provided by a message passing scheme. Even higher-layer cooperation protocols, such as KQML/KIF [Finin *et al.* 1994], are based on message passing.
- Anonymous agent group interaction
  In the previous two types, we have assumed that the communication partners know each other, i.e. the sender of a message or RPC is able to identify the recipient(s). However, there are situations, where a sender does not know the identities of the agents that are interested in the message sent. Assume, for example, a given task is performed by a group of agents, each agent taking over a subtask. In order to perform their subtasks, agents themselves may dynamically create subgroups of agents. In other words, the member set of the agent group responsible for performing the original task is highly dynamic. Of course, the same holds true for each of the subgroups involved in this task. Now assume that some agent

wants to terminate the entire group or some subgroup. In general, the agent that has to send out the *terminate* request does not know the individual members of the group to be terminated. Therefore, communication has to be anonymous, i.e., the sender does not identify the recipients. This type of communication is supported by group communication protocols (e.g., see [Birman and van Renesse 1994; Kaashoek and Tanenbaum 1991]), the concept of tuple spaces [Carriero and Gelernter 1989], as well as sophisticated event managers. In the latter approach, senders send out event messages anonymously, and receivers explicitly register for those events they are interested in. A group model using such a distributed event service for the coordination has been presented in [Baumann and Radouniklis 1997].

- User/Agent Interaction
  Although a very interesting area of research, the interaction between human users and software agents is beyond the scope of this article. For a discussion of this type of communication the reader is referred to e.g. [Maes 1994].

Let us briefly summarize our findings. Different types of communication schemes are needed in agent-based systems. Besides anonymous communication for group interactions, message passing and an RPC-style of communication is needed. In our model, message passing and RPC is session-oriented, which means that agents wanting to communicate have to establish a session before they can send and receive data. In section 7.3 we describe how session-oriented communication has been designed in Mole.

5. Security for Mobile Agent Systems

The vision of mobile agents as the key technology for future electronic commerce applications can only become reality if all security issues are well understood and the corresponding mechanisms are available. As illustrated by Figure 5, four security areas within mobile agent systems can be identified, namely (1) inter-agent security, (2) agent-host security, (3) inter-host security, and (4) security between hosts and unauthorized third parties. Existing cryptographic technology seems to be applicable to areas (1), (3) and (4), but area (2) is specific to mobile code systems (see [Hohl 1997] for an overview).

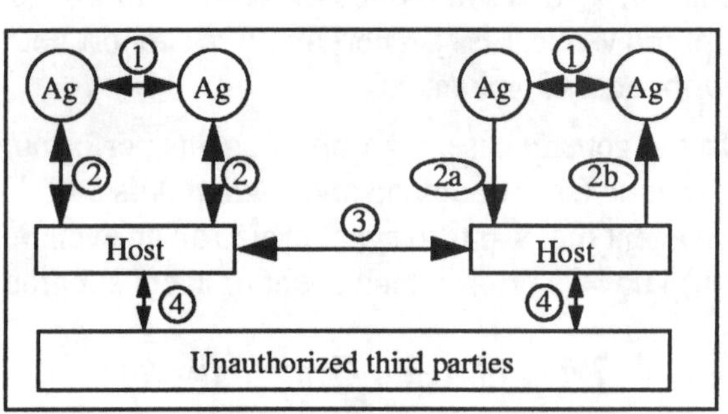

Figure 5: Security Areas of Mobile Agent Systems

The security between host and agent is twofold: on the one hand, hosts have to be protected against malicious agents, on the other hand, agents have also to be protected against malicious hosts. The first direction, protection against agents, can be solved using existing technology known from Java Applets and SafeTcl programs, since there the same problem - the execution of unknown programs - has to be addressed. Both systems use an approach, the so-called *Sandbox* security model, where all potentially dangerous procedure calls are restricted by special security control components that decide which programs can use these procedures and which not.

The other direction, the protection of agents against malicious hosts, is specific to mobile agents, and ongoing research efforts try to provide approaches in this field. Currently four research directions exist: the organizational approach (as in [White 1997]) eliminates the problem by allowing only trustworthy institutions to run mobile agent systems (and does, therefore not allow open systems). The trust / reputation approach (see [Farmer *et al.* 1996] or [Rasmusson and Jansson 1996]) allows agents to migrate only to trusted hosts or such with good reputation, but trust/reputation are problematic terms or they restrict the openness of the system. The manipulation detection approach [Vigna 1997] offers mechanisms to detect manipulations of agent data or the execution of code, but does not protect against read attacks. The blackbox protection approach [Hohl 1997] tries to generate a 'blackbox' out of agent code by using code obfuscating techniques. Since an attacker needs time to analyse the blackbox code before it can attack the code, the agent is protected for a certain time interval. After this 'expiration interval', the agent and the data it transports become invalid. All of these approaches are subject of ongoing work. None of them is currently used in real-world application, but the blackbox approach has been implemented for Mole, and is currently tested in regard to performance, message overhead and possible attacks.

6. Existing Mobile Agent Systems

In Table 2 an overview of available mobile agent systems is given. Many of these agent systems are research prototypes, and only a few of these have users outside their own university or research institute. We have already classified most of these systems regarding mobility support (see Table 1), now we will examine the systems on the subject of communication support.

Table 2: Mobile Agent Systems and their Availability

Name of the System	Supported Languages	Company	Availability
ARA	Tcl, C, Java	University of Kaiserslautern, Germany	free
ffMAIN	Tcl, Perl, Java	University of Frankfurt, Germany	no
Tacoma	Tcl, C, Python, Scheme, Perl	Cornell (USA), Tromso, Norway	free
AgentTcl	Tcl	Dartmouth College, USA	free
Aglets	Java	IBM, Japan	binary only
Concordia	Java	Mitsubishi, USA	binary only
CyberAgents	Java	FTP Software, Inc., USA	no longer
Java-2-go	Java	University of California at Berkeley, USA	free
Kafka	Java	Fujitsu, Japan	binary only
Messengers	M0	University of Zurich, Switzerland	free
MOA	Java	The Open Group, USA	no
Mole	Java	University of Stuttgart, Germany	free
MonJa	Java	Mitsubishi, Japan	binary only
Odyssey	Java	General Magic, USA	binary only
Telescript	Telescript	General Magic, USA	binary only
Voyager	Java	ObjectSpace, Inc., USA	binary only

All of these systems for mobile agents employ many communication mechanisms such as messages, local and remote procedure calls or sockets, but, to our knowledge, no system provides a global event service. There are "events" in AgentTcl [Gray *et al.* 1996], but they are simply (local) messages plus a numerical tag.

Although the use of sessions offers certain advantages as shown above, existing agent systems barely provide session support. Telescript [White 1997], for example, which introduced a kind of sessions by using the term *meeting* for mobile agent processing, offers only local meetings, that allow the agents only to exchange local agent references. The meet command is asymmetric, i.e. there is an active meeting requester, the "petitioner" and a passive meeting accepter, the "petitionee". The petitionee can accept or reject a meeting, but only the petitioner gets a reference to the petitionee. Agents communicate after opening a meeting by calling procedures of each other (i.e. the petitioner can call procedures of the petitionee). As there is no possibility during the execution of a procedure to obtain information about an enclosing meeting, agents cannot access session context data. Furthermore, an agent can open only one meeting per agent as a petitioner. Finally, agents may migrate during meetings, and if an agent takes shared objects with it, the other agent will not see this until it tries to access a shared object and gets a "Reference void" exception. To summarize the Telescript meeting, we can say, that it is not a session according to our definition. There are also "meetings" in ARA [Peine 1996] and in AgentTcl. Meetings in ARA build up communication relations between two agents over which (string) messages can be exchanged, meetings are local and the only supported "specification method" is anonymous addressing via meeting names. Meetings in AgentTcl are just a mechanism that opens a socket between two agents.

7. The Mole System

Until now we have discussed mobile agent systems in general. In this section we focus on the Mole system developed at the university of Stuttgart. We cannot present all of the functionality of the system, we will instead concentrate on the important and/or unique features of Mole. We present the lifecycle of an agent, show the implementation of mobility in Mole and discuss session-oriented communication as implemented in Mole. We give an overview over the Mole security model, show how agent naming is done in Mole, present the directory services and resource management for Mole, and conclude with experiences made using the Mole system.

7.1 The Lifecycle of an Agent

After an agent is created, its initialization routine *init()* is processed (see Figure 6). The arguments given to the agent at creation time are passed to this routine. The programmer can set up the internal state and initialize the agent attributes.

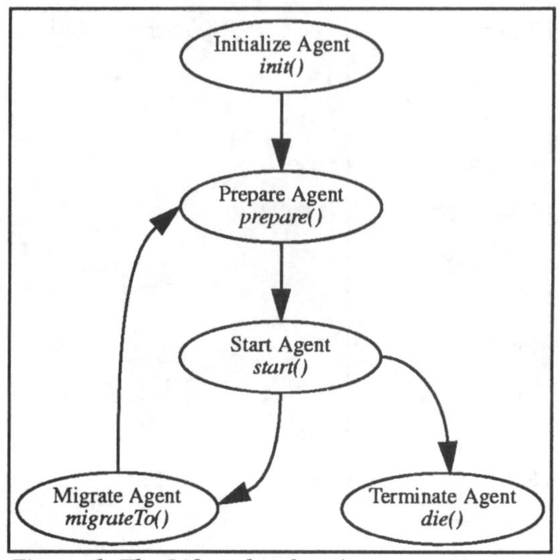

Figure 6: The Lifecycle of an Agent

At this time the agent is still outside any place. Now the agent system injects the agent into the system as if it had just arrived after a migration. First the agent is made known to the place, but other agents are not yet allowed to communicate with it. The *prepare()*-method is called, allowing the agent to do its place-specific setup, e.g. identifying local services. Finally the agent is started by calling the *start()*-method. After that normal processing takes place, the agent can start its own threads, use local services, and can communicate locally or remotely. The agent stays on the place until it decides to either migrate or to terminate.

If the agent wants to migrate it calls the method *migrateTo()* (described in more detail in the next section). The system suspends the agent's threads, serializes the agent and sends it to the target place. If the target place accepts the agent, the agent is injected into the system and started again via the methods *prepare()* and *start()*. Now the target place sends an acknowledgement back to the source place which removes the suspended agent from the system. If the target place does not accept the agent, an error message is sent back and the agent resumes its work on the original place. It receives an exception as the result of the failed migration, and can react e.g. by trying to migrate to another place.

If the agent has reached the end of its life, it calls the method *die()*. The system now stops all threads of the agent, removes it from the place, and deletes the agent.

Additionally the system supports periodic operation. This provides a simple mechanism for the programmer to implement recurring tasks, e.g. checking a database for changes if no trigger mechanism exists. If an agent implements the interface *Periodical*, then the system, in addition to calling the *init()*-method, executes the method *heartbeatInit()*. Now as soon as the *start()*-method has been called for the first time, the system begins executing a method name *heartbeat()* in regular intervals. This continues until the *die()*-method is called.

7.2 Mobility in Mole

As has been discussed in section 3, *migration*, be it *weak* or *strong*, has many advantages. *Remote execution*, while sufficient for many applications, provides neither the flexibility nor the simplicity in use, that *weak* and *strong migration* supply.

The difference between *weak* and *strong migration* is a change in semantics, but not in expressive power. We have decided not to implement *strong migration*, but to choose *weak migration* instead. Why? One of the design goals of Mole is the ability to run out of the box on every Java virtual machine (VM). A normal Java VM doesn't support capturing the state of a thread, which would be a prerequisite for capturing the execution state. Thus our decision was to choose the changed semantics of weak migration and with it the ability to run Mole on unchanged Java interpreters. This includes that, while agents in Mole can be multi-threaded, after a migration only one thread is started. If more threads are necessary the agent has to start them explicitly.

Weak migration in Mole is implemented by using a part of the Remote Method Invocation package RMI, the object serialization provided as part of Java 1.1. After an agent thread calls the *migrateTo()*-method, all threads belonging to the agent are suspended (not stopped). No new messages and calls (RPC) to the agent are accepted. After all pending messages to the agent have been delivered, the agent is removed from the list of active agents. Now the agent is serialized using the object serialization. The object serialization computes the transitive closure of all objects belonging to the agent (ignoring transient objects and threads), and creates a system-independent representation of the agent. This *serialized* version of the agent is sent to the target place that reinstantiates the agent. If any of the Java classes needed are not available locally, the target place requests these classes either from a code server [Hohl *et al.* 1997], or from the source place. Now the agent is reinstantiated. One new agent thread is started. First the *prepare()*-method is called to initialize the place-specific attributes. Then this thread begins its work at the *start()*-method. As soon as the thread assumes control of the agent, a success message is sent back to the source place. The source place now terminates all threads pertaining to the agent and removes it from the system. If at any stage of the migration an error occurs, the migration is stopped and the agent threads at the source place are resumed. The control flow continues after the *migrateTo()* statement, where error handling can be implemented (an exception is thrown in the case of failure).

```
prepare()
{
    ...
}
start()
{
    // here starts the agent thread
    // after migration or at instantiation
    ...
    try
    {
        migrateTo(targetplace);
    }
    catch(Event e)
    {
        // migration failed if control flow
        // executes the following statements
    ...
}
```

Figure 7: Migration of an Agent

Interestingly, experience showed that the semantics of *weak migration* are well understood and easily used even by inexperienced agent programmers. After working with *weak migration* for over two years we no longer deem *strong migration* strictly necessary, and a large fraction of agent system builders concurs [Baumann *et al.* 1997b].

7.3 Session-Oriented Communication in Mole

As will be seen below, a session between agents can be established only if the agents can identify each other. In our model, there are basically two ways how agents can be identified, the unique agent identifiers (called Agent Ids) - comparable to object Ids - and the so-called badges. Agent Ids are well-suited for identifying service agents, as long as there exists a directory system, that maps user-defined service names to service agent Ids. Note, however, that the directory service is not part of our base system, i.e., we clearly separate the mechanism for identifying services from the one for finding services. As a consequence, different naming schemes and directory systems can be used on top of this system. We will present the directory service Mole provides in section 7.6.

In the case of mobile agents the concept of agent Ids is not always sufficient. Assume for example, that an agent wants to meet some other agent participating in the same task at a given place. If only agent Ids were available, both agents would have to know each others ids. Actually, for identification it would be sufficient to say „At place XYZ I would like to meet an agent participating in task ABC". This type of identification is supported by the concept of badges. A badge is an application-generated identifier, such as „task ABC", which agents can „pin on" and „pin off". This badge does not necessarily have to be unique, it simply represents a role of an agent at a given time. As long as the agent provides the functionality associated with this role it wears the badge.

An agent may have several badges pinned on at the same time. Badges may be copied and passed on from agent to agent, and hence multiple agents can wear the same badge. For example, all agents participating in a subtask may wear a badge for the subtask and another one for the overall task. The agent that carries the result of the subtask may have an additional badge saying „CarryResult". Using badges, an agent is identified by a (*place_Id, badge predicate*)-tuple, which identifies all agents fulfilling the *badge predicate* at the place identified by *place_Id*. A badge predicate is a logical expression, such as („task ABC" AND („CarryResult" OR „Coordinator")). Obviously, this is a very flexible naming scheme, which allows to assign any number of application-specific names to agents. To change the name assignments two functions are provided, *PinOnBadge(badge)* and *PinOffBadge(badge)*.

Now let us take a closer look at sessions. A session defines a communication relationship between a pair of agents. Agents that want to communicate with each other, must establish a session before the actual communication can be started. After session setup, the agents can interact by remote method invocation or by message passing. When all information has been communicated, the session is terminated. Sessions have the following characteristics:

- Sessions may be intra-place as well as inter-place communication relationships, i.e., two agents participating in a session are not required to reside at the same place. Limiting sessions to intra-place relationships seems to be too restrictive. There are many situations, where it is more efficient to communicate from place to place (i.e., generally over the network) than migrating the caller to the place where the callee lives. Consequently, we feel that the mobility of agents cannot replace the remote communication in all cases.
- In order to preserve the autonomy of agents, each session peer must explicitly agree to participate in the session. Further, an agent may unilaterally terminate the sessions it is involved in at any point in time. Consequently, agents cannot be "trapped" in sessions.
- While an agent is involved in a session, it is not supposed to move to another place. However, if it decides to move anyway, the session is terminated implicitly. The main reason for this property is to simplify the underlying communication mechanism, e.g., to avoid the need for message forwarding.

The question may arise, why sessions are needed at all. There are basically two reasons: Firstly, the concept of a session can be used to synchronize agents that want to 'meet' for cooperation. Note that the first property stated above allows agents to 'meet' even if they stay on different places. The concept of a session is introduced to allow agents to specify which other agents they are interested to meet at which places. Furthermore, it allows agents to wait until the desired cooperation partner arrives at the place and indicates its willingness to participate. Secondly, we want to support both "stateless" and "stateful" interactions. In contrast to the first, the latter maintain state information for a sequence of requests. Obviously, if they encapsulate "stateful" servers, service agents have to be "stateful" also. A prerequisite for building "stateful" entities are explicit communication relationships, such as sessions.

7.3.1 Session Establishment

In order to set up sessions two operations are offered, *PassiveSetUp()* and *ActiveSetUp()* (see Figure 8). The first operation is non-blocking and is used by agents to express that they are willing to participate in a session. In contrast, *ActiveSetUp()* is used to issue a synchronous setup request, i.e., the caller is blocked until either the session is successfully established or a timeout occurs.

```
void PassiveSetUp(PeerQualifier , PlaceId)
SessionObject ActiveSetUp(PeerQualifier, PlaceId, Timeout)
boolean SetUp(SessionObject)
void SessionObject.Terminate()
```

Figure 8: Session Methods

If *ActiveSetUp()* succeeds, it returns the reference of the newly created session object to the caller. Input parameter *PlaceId* identifies the place, where the desired session peer is expected, and *PeerQualifier* qualifies the peer at the specified place. A *PeerQualifier* is either an agent Id or a badge predicate. Note that at most one agent qualifies in the case of a single agent Id, while several agents may qualify if a single badge predicate is specified. In that case a randomly picked agent is chosen. To avoid infinite blocking, the parameter *TimeOut* can be used to specify a timeout interval. The operation blocks until the session is established or a timeout occurs, whatever happens first.

The parameters *PeerQualifier* and *Place_Id* of the operation *PassiveSetUp()* are optional. If neither of both parameters is specified, the caller expresses its willingness to establish a session with any agent residing at any place. By specifying *Place_Id* and / or *PeerQualifier* the calling agent may limit the group of potential peers. For example, a group may be limited to all agents wearing the badge "Stuttgart University" and / or that are located at the caller's place.

As pointed out above, before a session is established both participants must agree explicitly. An agreement for session setup is achieved if both agents issue matching setup requests. Two setup requests, say R_A and R_B of agents A resp. B, match if

- *Place_Id* in R_A and R_B identifies the current place of B and A, respectively, and
- *PeerQualifier* in R_A and R_B qualifies B and A, respectively.

If a setup request issued by an agent matches more than one setup request, one request is chosen randomly and a session is established with the corresponding agent. A combination of *PassiveSetUp()* and *ActiveSetup()* allows a client / server style of communication (see Figure 9). The agent playing the server role once issues *PassiveSetUp()* when it is ready to receive requests. When an agent playing the client role invokes *ActiveSetup()*, this causes the *SetUp()* method of the server side to be invoked implicitly. *SetUp()* implicitly establishes a session with the caller and assigns a thread for handling this session. Therefore, once the server agent has called *PassiveSetup()*, any number of sessions can be established in parallel, where session establishment is purely client driven.

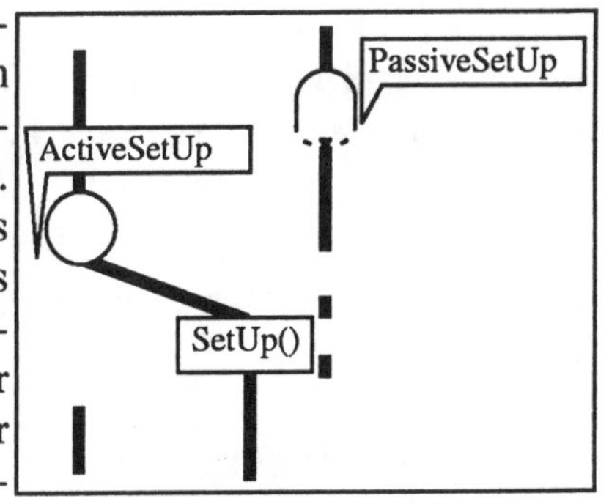

Figure 9: Client / Server Interaction

If both agents issue (matching) *ActiveSetUp()* requests this corresponds to a rendezvous. Both requesters are blocked until the session is established or timeout occurs (see Figure 10). This type of session establishment is suited for agents that want to establish peer-to-peer communication relationships with other agents. Communication between agents is peer-to-peer if both have their own "agenda" in terms of communication, i.e., both decide - depending on their individual goals - when they want to interact with whom in which way.

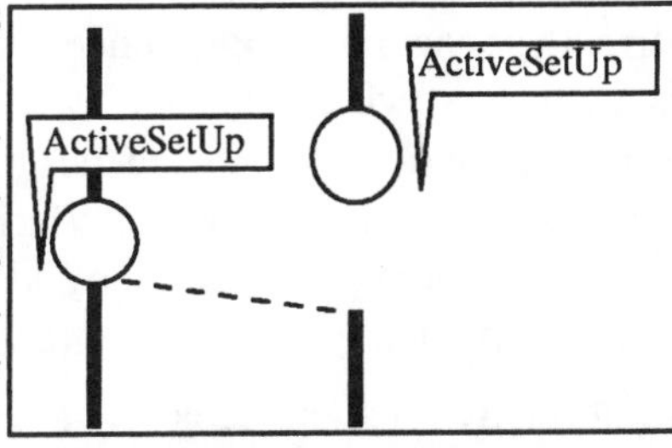

Figure 10: Peer to Peer Interaction

7.3.2 Session Communication

As pointed out above, Remote Method Invocation (RMI), the object-oriented equivalent to RPC, seems to be the most appropriate communication paradigm for a client / server style of interaction, while message passing is required to support peer-to-peer communication patterns. The available communication mechanisms are realized by so-called *com* objects. Currently, there are two types of *com* objects, RMI objects and messaging objects.

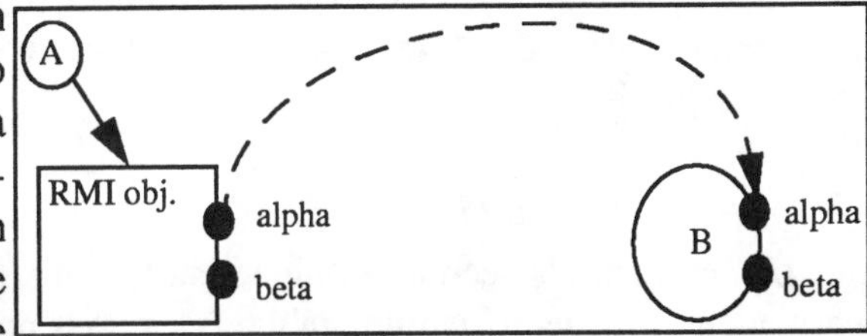

Figure 11: RMI Object

Com objects are associated with sessions. Each session may have an RMI object, a messaging object, or both. Each session object offers a method for creating *com* objects associated with this session. With the ***RMI object*** the methods exported by the session peer can be invoked. It can be compared with a proxy object known from distributed object-oriented systems. Figure 11 shows the RMI object enabling access to methods *alpha* and *beta* of object B.

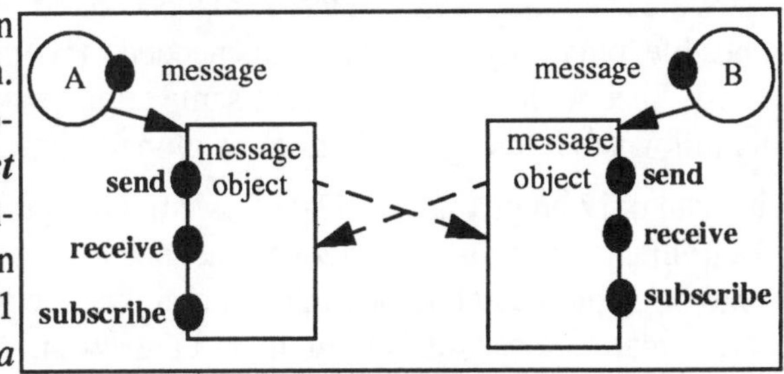

Figure 12: Message Object

With the ***Messaging object***, messages can be conveyed asynchronously between the participants of a session (see Figure 12). Messages are sent by calling the *send()* method. For receiving messages the *receive()* and *subscribe()*-methods are provided. The *receive()*-method blocks until a message is received or timeout occurs, whatever happens first. If the *subscribe()*-method is invoked instead, the incoming messages are handed over by calling the *message()*-method of the recipient and passing the message as method parameter.

The advantage of having the concept of *com* objects is twofold. Firstly, only those communication mechanisms have to be initiated that are actually needed during a session. Secondly, other mechanisms, such as streams, can be added to the system transparently. The latter advantage enhances the extensibility of the system.

7.3.3 Session Termination

At any time, a session can be terminated unilaterally by each of the both session participants, either explicitly or implicitly. A session is terminated explicitly by calling *SessionObject.Terminate()* (see Figure 8), and implicitly when a session participant moves to another place. When a session is terminated, this is indicated by calling the *SessionTerminated()*-method exported by agents. Moreover, all resources associated with the terminated session are released.

We want to mention, that for easier programming, we still allow the programmer to use "traditional" RMIs or messages without the need of a session overhead, giving them the opportunity to issue single communication acts.

7.4 Security in Mole

In Mole, the *Sandbox* security model (as mentioned above) is enforced by implementing a simple concept. In section 2 we presented our agent model, and with it mobile and service agents. Mobile agents are the normal user agents, programmed and employed by the user. They have absolutely no access to the underlying system. Service agents are agents with access to system resources, providing controlled, secure abstractions of these resources inside the agent system. Furthermore, service agents may offer access to legacy software, using the native code interface offered by Java. This does not cause any security problems, because the service agents are immobile and may be started only by the administrator of the place. User agents may only communicate with other agents and have no direct access to system resources.

Additionally it can be decided on a per-place basis which types of agents to allow on a place. Only agents that are derived from a specific type can migrate to a place. This mechanism can be used to implement access restrictions. Take e.g. a place that allows only agents of a very specific type. These can only be created at one other, open place. Then every agent wanting to access a service on the first, closed place has to migrate to the open place and request a service. This service then creates one of the specific agents that migrate to the closed place.

7.5 Agent Identifiers

In Mole, an agent is seen as a unique entity. This view is supported by using a globally unique name for every agent. This name is immutable, i.e. it does not change when the agent migrates. The uniqueness can only be guaranteed if the system creates the names used. If the system creates the agent ids, then these ids should be devisable without global knowledge. Additionally it is of advantage to be able to derive the site where the agent has been created from the agent id. Why do we place such constraints on the agent id? First, to be able to identify an agent (this is needed for communication, termination etc.), its name must be unique locally. Second, to be able to do the same after an agent has migrated, the name has to be immutable. From this follows that the agent id has to be globally unique.

This can only be guaranteed if the system itself provides a service to create agent ids conforming to these requirements. If global knowledge is needed to create this agent name, then either an expensive mechanism has to be implemented to obtain the global knowledge, or a single point of failure is introduced if e.g. an id server creating these ids is brought into the system (see e.g. the Amoeba sequencer in [Tanenbaum 1995]). The ability to derive the site where the agent has been created is of advantage e.g for localization algorithms utilizing the *home location registry* approach. This approach is used in GSM (*Global System for Mobile Communications*, see e.g. [Mouly and Pautet 1992]), where the id of the user (his telephone number) leads to a designated place (the *home location registry*) that contains the information how this user can be reached.

The agent id in Mole is created from information that can be obtained locally. Table 3 contains the components of the agent id. The internet protocol version 6 address of the underlying system together with the port number of the engine allows to identify the engine on which the agent has been created. The uniqueness of the name is guaranteed by using a combination of a normal counter that is set to 0 at the start of the engine, and a so-called crash counter, that is incremented every time the engine is started. If more than 2^{32} agents are started the crash counter is incremented also. 2 more bytes are reserved for future use, giving a total of 24 bytes.

Table 3: Components of the Agent Id

# Bytes	Meaning
4	Dynamic Counter, incremented for every new agent id
4	Crash counter, incremented every time the system is started. Also incremented if dynamic counter overflows.
12	IP Version 6 address of the system on which the engine runs
2	The port number of the engine
2	Reserved for future use (set to 0)

7.6 Directory Service

A directory service is an electronic database that contains information on entities. An example for a full-fledged directory service is X.500 (see e.g. [Chadwick 1994]). In our Mole system we simply provide a local directory service. It supplies information on agents providing a service denoted by a string. This local directory service exists on every place. An agent can register itself locally if it provides a service by submitting a string identifying the service to the directory service. Another agent wanting to use this service first asks the directory service. The directory service returns a list containing all agents providing the service. This list is either empty, or contains one or more agent ids. The agent now chooses one of the agent ids and contacts the agent.

7.7 Resource Management

Resource management is necessary for two purposes. One is accounting, the other is resource control. Accounting is a prerequisite for commercial applications with agents, and resource control is necessary to prevent e.g. denial-of-service attacks.

In Mole the following resources are managed:

- CPU time
- local network communication
- communication with remote networks
- number of created children
- total time at the local place

The CPU time used is calculated by counting the time slices given to threads of an agent. Mole has a central scheduler, the MCP (Master Control Process), that schedules all threads in the Mole system. We decided to implement our own scheduler, when problems with Java 1.0 led to the conclusion that the Java scheduler of the Solaris implementation had problems with concurrent execution of more than 4 threads of the same priority. This results from the Java specification being imprecise in this respect. The network communication is an important cost factor. Thus it is important for both accounting and resource control. Because all agent communication has to use the mechanisms provided by the agent system, control is done here. When an agent arrives at one place the arrival time is logged. This way the total time at the local place can be computed without problems. One other resource is not managed, the memory consumption of an agent. While in principle extremely important, it can not be implemented with acceptable costs without modifying the Java virtual machine.

7.8 *Experience*

Since the first public release of Mole nearly two years have passed, but even before we had internal versions we worked with. In this section we will describe our experience with the mobile agent paradigm regarding the efficiency and the needed infrastructure (hardware and software) to use Mole.

7.8.1 *Efficiency of the Agent Paradigm*

It is often argued that the advantage of agent migration lies in the reduction of (expensive) global communication costs by moving the computation to the data ([White 1997; Harrison *et al.* 1995]). Although this argument is understandable from an intuitive point of view, not much work has yet been done to evaluate the performance of migration on a quantitative basis. A simple performance model was investigated in [Carzaniga *et al.* 1997]. The evaluation of three scenarios was done in [Chia and Kannapan 1997].

Interaction between entities (objects, agents) in distributed systems which are located at different sites can take place in many different ways ([Baumann *et al.* 1997a]). In [Straßer and Schwehm 1997] a performance model has been developed where agents can alternatively use remote procedure calls or agent migration for the interaction with partners on different places. The model was first used to identify situations where a single agent migration has advantages compared to a single remote procedure call. This is basically the case when the amount of data to be processed is large compared to the size of the agent and if the selectivity of the agent, i.e. the ability of the agent to reduce the size of the reply by remote processing, is high. The model was then extended to describe a sequence of interactions.

From this model the conclusion can be drawn that an alternating sequence of remote procedure calls and agent migrations normally performs better than a pure sequence of remote procedure calls or a sequence of agent migrations. This result has been confirmed by measurements in Mole.

The experiments defined a sequence of interactions which were tested in the following way: an 'RPC-Only'-agent remained on the originating place L_0 and used remote procedure calls for each interaction. An 'Migrate-Always'-agent always migrated to the next interaction partner and then used local procedure calls. The 'Optimized'-agent used the performance model to decide when and to which place it should migrate and when it should communicate via RPC instead.

As an example one of the interaction sequences is shown in Table 4 (slightly simplified). The i-th interaction is described by a tuple $I_i = \{R_i, m_i, B_{req}, B_{rep}\}$ where R_i is the remote place with which the communication should take place. Each communication consists of m_i (local or remote) method calls with request size B_{req} and reply size B_{rep}:

Table 4: Sequence of Interactions

i	R_i	m_i	B_{req} [Bytes]	B_{rep} [Bytes]
1	L_1	1	700	3000
2	L_2	5	700	3000
3	L_3	1	700	3000
4	L_4	5	700	3000
5	L_0	1	3000	700

The execution times for the three mobility strategies averaged over 50 runs of the mobile agent are listed in Table 5.

Table 5: Measurements for the interaction sequence

mobility strategy	average time [ms]	standard deviation [ms]
'RPC-Only'	19127	1516
'Migrate-Always'	11394	1414
'Optimized'	10953	1341

This example hints that there is not one paradigm that is clearly the better one, but that in many cases a combination of 'classical' RPC-like communication and mobile agent migration yields the superior results.

7.8.2 Requirements for Mole

Mole is undemanding regarding hardware and software requirements, especially compared to Telescript [White 1997]. This is due to the design focusing on the use of an unmodified Java Virtual Machine and existing hardware. While this didn't allow e.g. strong migration or the control of the memory consumption of agents, it allows Mole to run on every hardware platform that runs a Java Virtual Machine version 1.1 or higher.

We have tested the system on various computer types and operating systems. Normal PC's with a Intel Pentium (we have not tested systems with Intel 486) or compatible CPU with main memory ranging from 16 to 128 MBytes under Windows 95, Windows NT V3.51 or V4.0, OS/2 or Linux runs the system as well as Sun Sparc with Solaris, IBM RS/6000 with AIX, or HP workstations with HPUX.

7.8.3 The Future of Mole

We are in the process of porting Mole to a browser environment, i.e. the Mole system runs wholly in the browser. At the moment this doesn't allow Mole system agents to access the system. But with JDK 1.2 new and more flexible security mechanisms will be implemented, and then it will be possible to run a full-fledged Mole system simply by following a URL. This means that in the future Mole will run on every computer that has a Java-compliant browser. This could possibly be even a small notebook running Windows CE or a PDA.

At the same time Mole can be integrated with WWW servers (e.g. Jigsaw [W3C 1997]) or provide an HTTP server as a system agent (see e.g. [Konstantas *et al.* 1996]). This allows to integrate Mole tightly with the existing WWW infrastructure, providing a safe and smooth migration path for WWW sites toward a mobile agent environment on top of the existing infrastructure.

Mobile agents provide a promising abstraction especially in the area of electronic commerce. The rapid growth of information applications, particularly on the Internet, suggests that public computer networks have the potential to establish a new kind of electronic marketplace for services and goods. So-called virtual malls or electronic storefronts will provide their goods and services to a huge number of customers. Such an electronic marketplace will be only vital if basically everybody can offer and exploit services with minimal investment. In other words, service deployment and distribution of client code must be an efficient, secure and simple process. In addition electronic storefronts must guarantee the same level of availability, scalability and reliability as offered by today's transaction processing systems. Those properties are a prerequisite for merchants and customers to rely on their services.

It is widely believed that mobile agents in conjunction with WWW technology will play an important role for future electronic commerce. One strong indicator for this is the OMG specification for a mobile agent facility [Mobile Agent Facility 1997]. Mobile agents, which can migrate over the network to remote sites to access services (e.g. entering electronic malls), are performed securely in an environment protecting the receiving system form malicious agent behaviour. It is this property of agents that allows for easy and secure service deployment and client code distribution. Moreover, agent technology may be used by customers to delegate complex tasks to rather autonomous agents, which perform them asynchronously. While performing a task, an agent may migrate from system to system. It delivers the result to the customer as soon as its tasks are finished, the customer is reachable and ready for receipt.

By only needing a connection between the systems hosting agents in case of communication or migration, network failures (e.g. partitioning) and even the loss of single systems (e.g. through crashs) is not fatal. In the worst case the agent that resided in the unreachable part of the network or on a crashed computer is reinitiated. This also allows for the better scalability of the overall distributed system, first by making the system more independent on the existence of a specific node in the network and second by lowering the number of existing connections compared e.g. to classical Client / Server interactions.

8. Conclusion and Future Work

In this paper we presented the mobile agent system Mole and the design decisions that led to the existing implementation of Mole V3. Different kinds of mobility have been presented, remote execution, code-on-demand, weak migration and strong migration, and their advantages and disadvantages have been discussed. Possible communication concepts for mobile agents and security problems in mobile agent systems have been reviewed. Other mobile agent systems have been presented. We then have concentrated on the Mole system. First we showed the lifecycle of an agent. Then we gave the reasons for choosing weak migration, presented the advanced communication concepts implemented in Mole, namely sessions and badges. We discussed the security model of Mole, the structure of agent Ids as used in Mole, directory service and the resource management. Finally we presented some of the experiences made with the Mole system in the last few years.

Mole is used e.g as the prototypical infrastructure for an electronic document system [Konstantas *et al.* 1996], as a simulation environment for distributed network management, as an environment for an enhanced WWW server, as an execution environment for Tandem server classes [Straßer *et al.* 1997], and in a distributed variant of a Multi-User Dungeon (MUD), in which players can use mobile agents as artificial teammates. Since its beginning the system had very active external users, giving us feedback (i.e. bug reports, new features etc.) which greatly helped and helps to improve the system. These are Siemens, Tandem Computers, University of Freiburg, University of Zürich, and University of Geneva. Apart from these active users over 400 different persons downloaded the version 1.0 of Mole, and in the first two months after the release of version 2 nearly 200 downloads have been counted. Mole is available as source code. Further informations about the Mole project can be found at
http://www.informatik.uni-stuttgart.de/ipvr/vs/projekte/mole.html.

We will continue to develop Mole, and research in this area follows many directions. A considerable part of today's commercial applications require a high degree of robustness. An *exactly-once* semantics for task processing will be a prerequisite for a majority of future net-based applications. Those semantics require a tight integration of agent technology and transaction management. There are two challenges to achieve this integration. Firstly, due to their asynchronous nature, agents are best suited to be engaged in long-lived activities. It must be investigated which transaction models meet these requirements, and Contracts [Wächter and Reuter 1992] and Sagas [Garcia-Molina *et al.* 1991] seem to be a good starting point. Secondly, at least parts of the agent state must be made recoverable. Current approaches to integrate mobile entities and transactions, such as Java Database Connection (JDBC, see [Hamilton *et al.* 1998]) and Java Transaction Service (JTS, see [JavaSoft 1997]), only consider server state to be recoverable. If operations performed on mobile state are part of transactions, existing protocols for transaction management, such as commit protocols, must be modified. We will continue to study this challenging area of mobile agent systems. Furthermore we will investigate agent group models in more detail, agent security issues and advanced communication concepts.

Acknowledgements: A system as large as this cannot be implemented with only a few researchers. We wish to thank the many students that have implemented parts of the system as their student or diploma thesis. Furthermore we thank the nameless reviewers for their helpful comments.

A References

[Aaron and Aaron 1997] Aaron, B. and Aaron A. (1997), *ActiveX Technical Reference*, Prima Publishing, Rocklin, CA 95765, USA.

[Baumann *et al.* 1997a] Baumann, J. and Hohl, F. and Radouniklis, N. and Rothermel, K. and Straßer, M. (1997), "Communication Concepts for Mobile Agent Systems", in *Proceedings of the First International Workshop on Mobile Agents '97*, Lecture Notes in Computer Science 1219, Springer Verlag, Berlin, Germany, pp. 123 - 135.

[Baumann *et al.* 1997b] Baumann, J. and Shapiro, M. and Tschudin, C. and Vitek, J. (1997), "Mobile Object Systems: Workshop Summary", in *Proceedings for the third ECOOP Workshop on Mobile Object Systems*, to appear.

[Baumann *et al.* 1997c] Baumann, J. and Tschudin, C. and Vitek, J. (1997), "Mobile Object Systems: Workshop Summary", in *Special Issues in Object-Oriented Programming, Workshop Reader of the ECOOP '96*, M. Mühlhäuser, Ed., dpunkt.verlag, Heidelberg, Germany, pp. 301 - 308.

[Baumann 1997] Baumann, J. (1997), "A Protocol for Orphan Detection and Termination in Mobile Agent Systems", Tech. Report Nr. 1997/09, Faculty of Computer Science, U. of Stuttgart, Germany.

[Baumann and Radouniklis 1997] Baumann, J. and Radouniklis, N. (1997), "Agent Groups for Mobile Agent Systems", in *Distributed Applications and Interoperable Systems*, H. König, K. Geihs and T. Preuß, Eds., Chapman & Hall, London, UK, pp. 74 - 85.

[Carzaniga *et al.* 1997] Carzaniga, A. and Picco, G. and Vigna, G. (1997), "Designing Distributed Applications with Mobile Code Paradigms", in *Proceedings 19th International Conference on Software Engineering*, ACM Press, New York, USA, pp. 22 - 32.

[Chadwick 1994] Chadwick, D. (1994), *Understanding the X.500 Directory*, Chapman & Hall, London, UK.

[Chia and Kannapan 1997] Chia, T.-H. and Kannapan, S. (1997), "Strategically Mobile Agents", in *Proceedings of the First International Workshop on Mobile Agents '97*, Lecture Notes in Computer Science 1219, Springer Verlag, Berlin, Germany, pp. 149 - 161.

[Carriero and Gelernter 1989] Carriero, N. and Gelernter, D. (1984), "Linda in Context", *CACM 32*, 4, pp. 444 - 458.

[Farmer *et al.* 1996] Farmer, W. and Guttmann, J. and Swarup, V. (1996), "Security for Mobile Agents: Authentication and State Appraisal", in *Proc. of the European Symposium on Research in Computer Security (ESORICS)*, E. Bertino, Ed., Springer Verlag, Berlin, Germany, pp. 118 - 130.

[Finin *et al.* 1994] Finin, T. and Fritzson, R. and McKay, D. and R. McEntire, R. (1994), "KQML as an Agent Communication Language", in *Proceedings of the third Conferenc on Information and Knowledge Management*, ACM Press, pp. 456-463.

[Garcia-Molina *et al.* 1991] Garcia-Molina, H. and Gawlick, D. and Klein, J. and Kleissner, K. and Salem, K. (1991), "Modeling Long-Running Activities as Nested Sagas", *Data Engineering Bulletin 14*, 1, pp. 14-18.

[General Magic 1997] General Magic (1997), "Odyssey Web Site", web page, URL: http://www.genmagic.com/agents/

[Ghezzi and Vigna 1997] Ghezzi, C. and G. Vigna, G. (1997),"Mobile Code Paradigms and Technologies: A Case Study", in *Proceedings of the First International Workshop on Mobile Agents '97*, Lecture Notes in Computer Science 1219, Springer Verlag, Berlin, Germany, pp. 39 - 49.

[Goscinski 1991] Goscinski, A. (1991), *Distributed Operating Systems - The Logical Design*, Addison-Wesley.

[Gray *et al.* 1996] Gray, R. and Cybenko, G. and Kotz, D. and Rus, D. (1996), "Agent Tcl", in *Mobile Agents: Explanations and Examples with CD-ROM*, W. Cockayne and M. Zyda, Ed., Manning Publishing, pp. 58 - 95.

[Gray 1997] Gray, R. S. (1997), "AgentTcl: A flexible and secure mobile-agent system", *Dr. Dobbs Journal 22*, 3, San Mateo, USA, pp. 18 - 27.

[Hamilton *et al.* 1998] Hamilton, G. and Cattell, R. and Fisher, M. (1998), *JDBC Database Access with Java*, JavaSoft Press, Addison-Wesley, to appear.

[Hammer and Shipman 1980] Hammer, M. and Shipman, D. (1980), "Reliability Mechanisms for SDD-1: A System for Distributed Databases", in *ACM Transactions on Database Systems 5*, 4, pp. 1 - 17.

[Harrison *et al.* 1995] Harrison, C. and Chess, D. and Kershenbaum, A. (1995), "Mobile Agents: Are they a good idea?", IBM Research Report, IBM T.J. Watson Research Center, Westchester County, USA.

[Hohl 1997] Hohl, F. (1997), "An approach to solve the problem of malicious hosts", Technical Report Nr. 1997/03, Faculty of Computer Science, University of Stuttgart.

[Hohl *et al.* 1997] Hohl, F. and Klar, P. and Baumann, J. (1997), "Efficient Code Migration for Modular Mobile Agents", in *Proceedings for the third ECOOP Workshop on Mobile Object Systems*, dpunkt-Verlag, Heidelberg, Germany, to appear.

[IBM 1997] IBM Corporation (1997), "Messaging and Queuing Technical Reference", web page, URL: http://www.software.ibm.com/ts/mqseries/

[IBM 1996] IBM Tokyo Research Labs (1996), "Aglets Workbench: Programming Mobile Agents in Java", web page, URL: http://www.trl.ibm.co.jp/aglets

[IONA 1996] IONA Technologies Ltd. (1996), "OrbixTalk Programming Guide", IONA Technologies Inc., Cambridge, MA, USA.

[JavaSoft 1997] JavaSoft, Inc. (1997), "The Java Transaction Service API.", web page, URL: http://java.sun.com/marketing/enterprise/jts.html

[Johansen *et al.* 1995] Johansen, D. and van Renesse, R. and Schneider, F. (1995), "An Introduction to the TACOMA Distributed System - Version 1.0", Technical Report TR-95-23, University of Tromsø, Norway.

[Kaashoek and Tanenbaum 1991] Kaashoek, M. F. and Tanenbaum, A. S. (1991), "Group Communication in the Amoeba Distributed Operating System.", in *Distributed Computing Systems Engineering 1*, 6, pp. 48 - 58.

[Knabe 1995] Knabe, F. (1995), "Language Support for Mobile Agents", PhD Dissertation, School of Computer Science, Carnegie Mellon University.

[Konstantas *et al.* 1996] Konstantas, D. and Morin, J. H. and Vitek, J. (1996), "MEDIA: A Platform for The Commercialization of Electronic Documents", in *Object Applications*, Dennis Tsichritzis, Ed., University of Geneva, pp. 7 - 18.

[Maes 1994] Maes, P. (1994), "Agents that Reduce Work and Information Overload", in *CACM 37*, 7, pp. 31 - 40.

[Mobile Agent Facility 1997] Object Management Group (1997), "Mobile Agent Facility (MAF) specification", URL: http://www.omg.org/library/schedule/Mobile_Agents_Facility_RFP.htm

[Mathiske *et al.* 1997] Mathiske, B. and Matthes, F. and Schmidt, J. (1997), "On Migrating Threads", *Journal of Intelligent Information Systems 8*, 2, pp. 167 - 191.

[Mouly and Pautet 1992] Mouly, M. and Pautet, M. (1992), *The GSM System for mobile Communication*, Europe Media Publications S. A., ETSI, Palaiseau, France.

[Mole 1997] "Mole Project Pages" (1997), University of Stuttgart, web page, URL: http://www.informatik.uni-stuttgart.de/ipvr/vs/projekte/mole.html

[OMG 1994] OMG (1994), "Common Object Services Specification", Volume 1, OMG Document Number 94-1-1, Object Management Group, Framingham, MA, USA.

[Peine 1996] Peine, H. (1996), "Ara: Agents for Remote Action." In *Mobile Agents: Explanations and Examples with CD-ROM*, W. Cockayne and M. Zyda, Ed., Manning PublishingCo., Greenwich, CT, USA, pp. 96 - 164.

[Rasmusson and Jansson 1996] Rasmusson, L. and Jansson, S. (1996), "Simulated Social Control for Secure Internet Commerce", Accepted Position Paper for the *New Security Paradigms '96 Workshop*, URL: http://www.sics.se/~lra/nsp96/nsp96.html

[Rothermel 1997] Rothermel, K. and Hohl, F. and Radouniklis, N. (1997), "Mobile Agent Systems: What is Missing?", in *Distributed Applications and Interoperable Systems*, H. König, K. Geihs and T. Preuß, Eds., Chapman & Hall, London, UK, pp. 111 - 124.

[Stamos 1990] Stamos, J. W. and Gifford, D. K. (1990), "Remote Evaluation", *ACM Transactions Programmming Languages and Systems 12*, 4, pp. 537 - 565.

[Straßer *et al.* 1996] Straßer, M. and Baumann, J. and Hohl, F. (1996), "Mole - A Java Based Mobile Agent System", in *Special Issues in Object-Oriented Programming, Workshop Reader of the ECOOP '96*, M. Mühlhäuser, Ed., dpunkt-Verlag, Heidelberg, Germany, pp. 327 - 334.

[Straßer *et al.* 1997] Straßer, M. and Baumann, J. and Hohl, F. and Radouniklis, N. and Rothermel, K. and Schwehm, M. (1997), "ATOMAS: A Transaction-oriented Open Multi Agent-System. Annual Report", Technical Report Nr. 1997/14, Faculty of Computer Science, University of Stuttgart.

[Straßer and Schwehm 1997] Straßer, M. and Schwehm, M. (1997), "A Performance Model for Mobile Agent Systems", in *Proceedings of the International Conference on Parallel and Distributed Processing Techniques and Applications PDPTA'97, Volume II*, H. R. Arabnia, Ed., Computer Science Research, Education, and Applications Technology (CSREA), pp. 1132 - 1140.

[Sun 1994] Sun Microsystems (1994), "The Java Language: A White Paper", Technical Report, Sun Microsystems, Palo Alto, CA, USA.

[Sun 1996] Sun Microsystems (1996), "Solaris NEO", web page, URL: http://www.sun.com/solaris/neo/

[Sun 1997] Sun Microsystems (1997), "The Java Web Pages", web page, URL: http://www.javasoft.com

[Tacoma 1997] "Tacoma Project Pages" (1997), web page, URL: http://www.cs.uit.no/DOS/Tacoma/index.html

[Tanenbaum 1995] Tanenbaum, A. (1995), *Distributed Operating Systems*, Prentice Hall.

[Vigna 1997] Vigna, G. (1997), "Protecting Mobile Agents through Tracing", in *Proceedings for the third ECOOP Workshop on Mobile Object Systems*, dpunkt-Verlag, Heidelberg, Germany, to appear.

[W3C 1997] World Wide Web Consortium (1997), "Jigsaw Overview", web page, URL: http://www.w3.org/Jigsaw/

[Wächter and Reuter 1992] Wächter, H. and Reuter, A. (1992), "The ConTract Model", in *Transaction Models*, A. Elmagarmid, Ed., Morgan Kaufmann, San Francisco, CA, USA, pp. 219 - 263.

[Walter 1982] Walter, B. (1982), "A Robust and Efficient Protocol for Checking the Availability of Remote Sites", in *Proceedings of the 6th Berkeley Workshop on Distributed Data Management and Computer Networks*, Technical Information Department, Lawrence Berkeley Laboratory, University of California, Berkeley, CA, USA, pp. 45 - 67.

[White 1997] White, J. E. (1997), "Telescript", In *Mobile Agents: Explanations and Examples with CD-ROM*, W. Cockayne and M. Zyda, Ed., Manning Publishing, Greenwich, CT, USA, pp. 37 - 57.

[Wong *et al.* 1997] Wong, D. and Paciorek, N. and Walsh, T. (1997), "Concordia: An Infrastructure for Collaborating Mobile Agents", in *Proceedings of the First International Workshop on Mobile Agents '97*, Lecture Notes in Computer Science 1219, Springer Verlag, Berlin, Germany, pp. 86 - 97.

THE FUTURE OF MOLE

J. Baumann, F. Hohl, K. Rothermel, M. Schwehm, M. Straßer, W. Theilmann

IPVR (Institute for Parallel and Distributed High-Performance Computers)
University of Stuttgart, Breitwiesenstraße 20-22, D-70565 Stuttgart

In 1994, General Magic presented its all new technology, Telescript, and with it, the new agent paradigm that has interested researchers since. While most of the underlying technology (e.g. process migration) was known for a long time, the combination proved to be ingenious. We at the IPVR were quite excited about the possibilities of this new approach. But as a research institute we were simply not able to pay the high license fees. Because no other mobile agent system was available we decided to create an alternative ourselves. This alternative we called Mole. In mid-1995 we finished the first version that was used subsequently for in-house and selected external projects. A year later we released the first public version of Mole. Since then we worked on the concepts and on the implementation of Mole. Mole now has reached version number 3, and is in use in many research institutes around the world, for research as well as for teaching. We believe that Mole could be called a success.

One of the most interesting application areas for mobile agents is *electronic commerce* [Rothermel *et al.* 1997]. To support this application area a high level of security has to be guaranteed. One of the most problematic questions in that area that has not yet been answered by the research community is the question, how a mobile agent could be protected against a malicious host. Other security-related problems can be solved with well-known cryptographical means, but this specific problem is hard to solve. However, a viable answer has to be presented to allow the use of mobile agent technology in commercial environments. At the IPVR, a first solution has been developed. "Time limited blackbox protection" (see [Hohl 1998; Hohl 1998] for more details) transforms agents into a form that makes it impossible to deduce the functionality of the code in a given time interval. By limiting the validity of this specific form to the time interval, an agent can only be understood, and thus attacked, after its momentary form has already expired. After expiration a new transformation takes place, which makes it impossible for a potential attacker to correlate the different forms. This in combination with techniques that allow to compensate the effects of attacks after the expiration date, agents can be protected against privacy, modification and manipulation attacks. Mole will integrate this approach in a security framework that will also include:

- authentication of agents, places and users.
- authentication of agent code.
- encryption of the communication between places on different computers.
- resource control both quantitatively (usage of CPU, memory and other resources) and qualitatively by using authorisation techniques and security policies.
- usage of standard key certification mechanisms.

Additional support needed for electronic commerce applications encompasses the reliable execution of agents. Due to the asynchronous execution of mobile agents, a reliable execution environment is an important precondition for the widespread use of mobile agent systems. If a user launches a mobile agent, he expects that the agent either fulfils its task or that he gets a message stating why the agent wasn't able to as told. To provide this functionality it must be ensured that an agent does not get lost during migration or due to a node crash. Additionally it has to be guaranteed that an agent performs its task exactly once. If an agent buys a flight ticket on a place and the place crashes before the agent migrates to the next place, it would be all but convenient for the user to end up with 2 tickets after the recovery of the place and the restarting of the agent. To enable agents to be executed as fast as possible even in the presence of long lasting place crashes and network partitionings, fault-tolerant mechanisms have to be provided. In [Rothermel and Straßer 1997; Straßer and Rothermel 1998], we presented a fault-tolerant mechanism to provide exactly-once execution of mobile agents based on transactional message queues and distributed transactions. This mechanism will be integrated in future Mole releases as an optional feature. Additional research will be invested to extend the mechanism, e.g. to allow a partial rollback of an agent execution.

Equally important for applications are control and orphan detection protocols for agents and agent groups. If a user sends out an agent, he does not want to lose the control over it, e.g. to terminate it in case he decides not to need the results the agent could provide. Furthermore, orphan detection is a very important functionality to guarantee that superfluous agents are removed from the system. Protocols supporting orphan detection and termination have been presented in [Baumann 1997; Baumann and Rothermel 1998] and will be developed further. A model for agent groups has already been developed in [Baumann and Radouniklis 1997], and simple groups of agents (AND and OR groups) have already be designed. But the model is able to support much more complex groups, and we try to find out which types of groups are the most important from an application's viewpoint.

The other important application area besides electronic commerce is information retrieval in large heterogeneous networks (e.g. the Internet). Here it often becomes necessary to analyse remote documents or document collections concerning their relevancy according to a user's query. Mobile agents are supposed to perform this task much more efficient (in terms of network bandwidth) than traditional techniques. They can go to remote sites in order to examine a great amount of data and return with a small precise result. In this context it is very important to be able to calculate the possible performance advantages that can be reached by the use of mobile agents. [Straßer and Schwehm 1997] proposed a performance model that deals with single interactions (remote procedure call, migration, remote execution) and sequences of them. In the context of information retrieval the problem is more complicated. Possibly interesting documents can be spread all over the world. So groups of mobile agents are needed to process these documents in an acceptable time period. And for the complex interaction patterns in groups of agents no performance model exists.

While the basic system infrastructure of Mole is quite stable and used by a large community, the support for commercial applications, and for applications of high complexity, does not yet exist. But we are working on it and hope to provide the necessary functionality in the near future.

References

[Baumann 1997] Baumann, J. (1997), "A Protocol for Orphan Detection and Termination in Mobile Agent Systems", Technical Report (Fakultätsbericht) Nr. 1997/09, Faculty of Computer Science, U. of Stuttgart, Germany.

[Baumann and Radouniklis 1997] Baumann, J. and Radouniklis, N. (1997), "Agent Groups for Mobile Agent Systems", in *Distributed Applications and Interoperable Systems*, H. König, K. Geihs and T. Preuß, Eds., Chapman & Hall, London, UK, pp. 74 - 85.

[Baumann and Rothermel 1998] Baumann, J. and Rothermel, K. (1998), "The Shadow Approach: An Orphan Detection Protocol for Mobile Agents", submitted to the Mobile Agents '98.

[Hohl 1997] Hohl, F. (1997), "An approach to solve the problem of malicious hosts", Technical Report (Fakultätsbericht) Nr. 1997/03, Faculty of Computer Science, University of Stuttgart.

[Hohl 1998] Hohl, F (1998), "Time Limited Blackbox Security: Protecting Mobile Agents From Malicious Hosts", in *Mobile Agents and Security*, G.Vigna, Ed., Springer-Verlag, Berlin, Germany.

[Rothermel *et al.* 1997] Rothermel, K. and Hohl, F. and Radouniklis, N. (1997), "Mobile Agent Systems: What is Missing?", in *Distributed Applications and Interoperable Systems*, H. König, K. Geihs and T. Preuß, Eds., Chapman & Hall, London, UK, pp. 111 - 124.

[Rothermel and Straßer 1997] Rothermel, K.; Straßer, M.: "A Protocol for Preserving the Exactly-Once Property of Mobile Agents". Technical Report (Fakultätsbericht) 1997/18, Faculty of Information Science, University of Stuttgart, 1997.

[Straßer *et al.* 1996] Straßer, M. and Baumann, J. and Hohl, F. (1996), "Mole - A Java Based Mobile Agent System", in *Special Issues in Object-Oriented Programming, Workshop Reader of the ECOOP '96*, M. Mühlhäuser, Ed., dpunkt-Verlag, Heidelberg, Germany, pp. 327 - 334.

[Straßer and Rothermel 1998] Straßer, M., and Rothermel, K. (1998), "Providing Reliable Agents for Electronic Commerce." in *Proceedings GI/IFIP Conference Trends in Electronic Commerce '98*, Hamburg, to appear.

[Straßer and Schwehm 1997] Straßer, M. and Schwehm, M. (1997), "A Performance Model for Mobile Agent Systems", in *Proceedings of the International Conference on Parallel and Distributed Processing Techniques and Applications PDPTA'97, Volume II*, H. R. Arabnia, Ed., Computer Science Research, Education, and Applications Technology (CSREA), pp. 1132 - 1140.

14.6 Johansen, D., van Renesse, R., and Schneider, F.B., Operating System Support for Mobile Agents

Johansen, D., van Renesse, R., and Schneider, F.B., "Operating system support for mobile agents," *Proceedings of the 5th Workshop on Hot Topics in Operating Systems*, pp. 42-45, May 1995.

Operating System Support for Mobile Agents

Position paper for 5th IEEE Workshop on Hot Topics in Operating Systems

Dag Johansen[*] Robbert van Renesse[**] Fred B. Schneider[**]

1. Introduction

An *agent* is a process that may migrate through a computer network in order to satisfy requests made by its clients. Agents implement a computational metaphor that is analogous to how most people conduct business in their daily lives: visit a place, use a service (perhaps after some negotiation), and then move on. Thus, for the computer illiterate, agents are an attractive way to describe network-wide computations.

Agents are also useful abstractions for programmers who must implement distributed applications. This is because in the agent metaphor, the processor or *place* the computation is performed is not hidden from the programmer, but the communications channels are. Contrast this with the traditional approach of employing a client at one site that communicates with servers at other sites. Communication is not hidden and must be programmed explicitly. Moreover, pieces of a computation performed at different sites must be coordinated, placing an added burden on the programmer.

By structuring a system in terms of agents, applications can be constructed in which communication-network bandwidth is conserved. Data may be accessed only by an agent executing at the same site as the data resides. An agent typically will filter or otherwise reduce the data it reads, carrying with it only the relevant information as it roams the network; there is rarely a need to transmit raw data from one site to another. In contrast, when an application is built using a client and servers, raw data may have to be sent from one site to another if, for example, the client obtains its computing cycles from a different site than it obtains its data.

Most current research on agents has focused on language design (e.g. [W94]) and application issues (e.g. [R94]). The TACOMA project (Tromsø And COrnell Moving Agents) has, instead, focused on operating system support for agents and how agents can be used to solve problems traditionally addressed by operating systems. We have implemented prototype systems to support agents using UNIX and using Tcl/Tk [O94] on top of Horus [vRHB94].

---

[*] Department of Computer Science, University of Tromsø, N-9037 Tromsø, Norway. Johansen is supported by grant No. 100413/410 from the Norwegian Science Foundation.

[**] Department of Computer Science, Cornell University, Ithaca, New York 14853, U.S.A. Van Renesse is supported by ARPA/ONR grant N00014-92-J-1866. Schneider is supported by the ARPA/NSF Grant No. CCR-9014363, NASA/ARPA grant NAG-2-893, and AFOSR grant F49620-94-1-0198.

The remainder of this paper outlines insights and questions based on that experience. In section 2, we briefly discuss abstractions needed by an operating system to support agents. Section 3 discusses some problems that arise in connection with electronic commerce involving agents. How to schedule agents by using other agents is the subject of section 4. Some preliminary thoughts on implementing fault tolerance are given in section 5. Section 6 discusses the status of our implementations.

2. Abstractions and Mechanisms for Agents

An agent must be accompanied by data in order for its future actions to depend on its past ones. For this reason, our implementations associate with each agent a *briefcase*, which contains a collection of named *folders*. A folder is a list of elements, each of which is an uninterpreted sequence of bits. Because it is a list, it can be treated as a stack or a queue. This makes folders reminiscent of the familiar objects used to group documents. Unlike files in a traditional operating system, folders must be easy to transfer from one computing system to another, since this operation occurs frequently. Thus, elaborate index structures are not suitable for implementing the folders that accompany agents.

It is also important that agents be able to read and write folders that are bound and local to a site executing the agent. Site-local folders allow more efficient use of network bandwidth. If an agent requires certain information only when it is executing at a given site, then it is inefficient to carry along that information to every site the agent visits. A site-local folder allows an agent to leave such information behind. In addition, site-local folders allow communication between agents that are not simultaneously resident at a given site. For example, consider a flooding algorithm to deliver a message at all sites in a network. One implementation would have each agent deliver the message and then create a clone of itself at every adjacent site. Unfortunately, here the number of agents increases without bound. If, instead, an agent also records its visit in a site-local folder, then an agent can simply terminate—rather than clone—when it finds itself at a site that has already been visited.

Just as an agent's folders are grouped into briefcases, we have found it useful to group site-local folders. We refer to such a grouping as a *file cabinet*. File cabinets support the same operations as briefcases, but we expect these operations to be implemented differently. In particular, since it is rare to move a file cabinet from site to site, file cabinets can be implemented using techniques that optimize access times even if this increases the cost of moving the file cabinet from one site to another.

One agent causes another to execute using the **meet** operation, where a briefcase allows information to be exchanged between the two agents. The **meet** operation is thus analogous to a procedure call, and the specified briefcase is analogous to an argument list (with each folder containing the value of one argument). For example, an agent *A* executing

meet *B* **with** *bc*

causes agent *B* to be executed at the current site with briefcase *bc*; *A* continues executing only after *B* terminates the **meet** operation. Note that after the **meet** terminates, *B* may continue executing concurrently with *A*.

Surprisingly, no additional abstractions are required to implement our basic computational metaphor. Services for agents—communication, synchronization, and so on—are provided directly by other agents. For example, an agent moves from one site to another by **meet**ing with the local *rexec* agent. The *rexec* agent expects to find two folders in the briefcase with which it is invoked: a *HOST* folder names the site where execution is to be moved and a *CONTACT* folder names the agent to be executed at that site. The *CONTACT* folder might contain the name of an agent that is a shell or a compiler. Such an agent would expect to find a *CODE* folder in the briefcase, which it would then translate and execute. Since the contents of this *CODE* folder might be the source code for the agent that originally met with *rexec*, it is possible for an agent to travel from one site to another. Note that this scheme allows an agent to move to a destination site having a completely different machine language.

Given an *rexec* agent, it is not difficult to program a *courier* agent, which transfers a folder to a specified agent on a specified machine. This allows agents to communicate without having to **meet** (on a common machine). It is also not difficult to program our *diffusion* agent, which executes a specified agent locally and then creates a clone of itself at every site that appears in the set difference of the site-local *SITES* folder and the briefcase *SITES* folder.

3. Obtaining and Paying for Services

Once agents are employed for commerce—as some proponents [W94] of the metaphor intend—support for a negotiable instrument becomes necessary. We, therefore, decided to explore the implementation and use of *electronic cash*. Electronic cash is nothing more than an unforgeable and untraceable capability that enables its owner to obtain goods and services. By implementing an electronic analogue to a well understood concept, we hoped to produce a system that remained understandable to the computer illiterate. We also hoped that electronic cash would provide a mechanism for controlling run-away agents. Specifically, charging for services would limit possible damage by a run-away agent.

Even as simple an operation as transferring electronic cash from one agent to another turned out to be surprisingly subtle to implement. With the familiar physical form of cash, money transfers are achieved by moving physical objects (coins or pieces of paper). This works only because it is difficult to manufacture copies of such objects. In a computer system, however, "copy" is a cheap operation. The usual solution to this problem would be to employ indirection and store all electronic cash in a single trusted agent. One agent could then transfer money to another by invoking an operation provided by this trusted agent.

We must reject solutions based on indirection because they necessarily violate our untraceability requirement for funds transfers. Following [C92], the solution we adopted was to implement each unit of electronic cash (ECU) as a record containing an amount and a large random number. Only certain of these random numbers appear on the records for valid ECUs. Each agent stores records for the ECUs it owns. An agent transfers funds by placing these records in a briefcase that is then passed to the intended recipient of those funds.

The recipient of such a briefcase, however, has no guarantee that the sending agent has not already spent (a copy of) the ECUs being transferred. To solve this problem, a trusted *validation* agent is employed. This agent can check whether a record it is shown corresponds to a valid ECU. If it is valid, then a record for an equivalent ECU is returned, but this record has a new random number (effectively retiring an old bill and replacing it by a new one). An attempt by an agent to spend retired or copied ECUs will be foiled if a validation agent is always consulted before any service is rendered. Notice that using a validation agent supports our untraceability requirement, since the validation agent does not require knowledge of the source or destination of a transfer.

A second problem that we encountered in supporting electronic cash concerned implementing the exchange of funds for services. It must not be possible to obtain a service without paying for it or to pay without obtaining the service. This precludes the obvious two-step protocols, because as long as electronic cash is untraceable either party might cheat the other. For example, the customer might claim to have paid when it has not, or the service-provider might claim not to have been paid when it has. What would seem to be required is support for transactions, so that we are guaranteed that both actions ("paying" and "providing the service") occur or that neither action occurs.

We rejected adding support for transactions to our system for two reasons:

(1) Having such a mechanism would impact performance and would be effective only if it were trusted.

(2) Such a mechanism would be alien to the computer illiterate, because such a mechanism does not exist in current business practice.

Our solution was to employ the threat of audits, a scheme that is well-known in current business practice.

- Participants document their actions so that a third party (a court, in real life) can perform an audit to find violations of a contract.
- An aggrieved agent requests an audit.

Documenting actions sometimes requires the presence of a third agent and the use of cryptographic protocols—we omit the details here. Having to interact with such a third agent will be familiar to computer-illiterate users (at least, to those who have purchased a house).

4. Scheduling

In our prototypes, scheduling allows the enforcement of policies that govern when and where an agent is executed. Sites in a computer network are presumed to be autonomous, so facilities must be provided for system administrators to control the resources comprising a site. Agents are also presumed to be autonomous, though. Thus, implementing support for scheduling requires mechanisms to match the needs of agents with the providers of services while, at the same time, respecting constraints imposed by system administrators.

Scheduling is implemented by *broker* agents, which are ordinary agents whose names are well known. Some broker agents maintain databases of service providers; these brokers serve as matchmakers. An agent that requires a given service consults a broker to identify which agents provide that service. Brokers are expected to communicate among themselves and with the service providers, so that requests can be distributed amongst service providers based on load and capacity. The problem of maintaining the requisite state information and intelligently distributing service requests seems to be equivalent to that of routing in a wide-area network. We do not yet have experience with various routing protocols to know how they can be adapted to this new setting, but this is a topic under investigation.

Another use of broker agents is to enforce some *protected* agent's policies with regard to **meet**ing other agents. This is accomplished by keeping the name of the protected agent secret from all but its broker. The broker, then, provides the only way to **meet** with the protected agent. To do this, the broker maintains a folder for each agent that has requested a **meet**ing with the protected agent. This folder contains the agent that has requested the meeting (along with its briefcase). Notice that this scheme is possible only because folders are uninterpreted and typeless and, therefore, can themselves store agents and sets of folders.

5. Fault-tolerance

It is to be expected that sites in a computer network will fail. When such a failure occurs, agents at that site are no longer able to continue executing. To deal with this problem, we have been investigating ways to ensure that a computation can proceed, even though one or more of its agents is the victim of a site failure. The solutions we have studied involve leaving a *rear guard* agent behind whenever execution moves from one site to another. This rear guard is responsible for (i) launching a new agent should a failure cause an agent to vanish and (ii) terminating itself when its function is no longer necessary (because the agent it protects is itself ready to terminate). The details of implementing rear guards efficiently are complex, because the sites traversed by an agent computation may be cyclic and because a single agent may clone itself and fan out through a network.

6. Prototype Implementations

Our most recent version of TACOMA is based on Tcl [O94]. Each site in our system runs a Tcl interpreter, which provides the place where agents execute. An agent is implemented by a Tcl procedure; the text of the procedure is stored in the agent's *CODE* folder. Folders, briefcases, and file cabinets are Tcl data structures. File cabinets can be flushed to disk when permanence is required.

A collection of system agents provides a variety of support functions. The most basic of these is *ag_tcl*, which pops a Tcl procedure from the *CODE* folder and executes that procedure. Currently, two implementations exist for the *rexec* agent. The first uses the UNIX rsh command to start a Tcl interpreter on the remote host. The second uses Tcl/TCP, an extension to Tcl that allows Tcl processes to set up TCP communication channels. We are now completing a third implementation based on Tcl/Horus, a version of Tcl that uses Horus [vRHB94] to support group communication and fault-tolerance.

In our first prototype of TACOMA, we implemented the electronic cash of section 3. The implementation used the security mechanisms provided by UNIX; this simplified our implementation, but relies on UNIX for security. We are now investigating alternatives.

Our TACOMA prototype currently supports a scheduling service that assigns to processors based on load. It uses four different agents to implement a scheme like that outlined in section 4. One of these agents is the broker, another is responsible for monitoring the status of a site and reporting that to the brokers, one is a courier, and one issues tickets to allow access to the service.

To evaluate the metaphor we are using our prototype to construct a variety of distributed applications. First, we are reimplementing StormCast [J93], which uses a set of expert systems to predict severe storms in the Arctic based on weather data obtained from a distributed network of sensors. Second, we have started to build an interactive mail system where messages are implemented by agents.

References

[C92] Chaum, D. Achieving Electronic Privacy. *Scientific American* 267,2 (Aug 1992), 96-101.

[J93] Johansen, Dag. StormCast: Yet another exercise in distributed computing. *Distributed Open Systems* F.M.T. Brazier and D. Johansen, eds. *IEEE Computer Society Press*, California (Oct 1993), 152-174.

[O94] Ousterhout, John K. *Tcl and the Tk Toolkit* Addison Wesley, Reading, Massachusetts, 1994.

[R94] Riecken, D. (guest editor). Intelligent Agents. *Commun. of the ACM* 37,7 (July 1994), 19-21.

[vRHB94] Van Renesse, Robbert, Takako M. Hickey, and Kenneth P. Birman. Design and Performance of Horus: A Lightweight Group Communications System. Technical Report TR 94-1442, Department of Computer Science, Cornell University, Aug 1994.

[W94] White, J.E. Telescript Technology: The Foundation for the Electronic Marketplace. General Magic White Paper, General Magic Inc., 1994.

What TACOMA Taught Us

Dag Johansen* Fred B. Schneider† and Robbert van Renesse†

1 Introduction

TACOMA is a system for supporting processes—so called agents—whose execution moves from processor to processor. Our first prototype was completed in March 1994; the version documented elsewhere in this volume was up and running the following year. There have since been four major system releases. The current versions of TACOMA provide support for agents written in C, C++, ML, Perl, Python, Scheme, and Visual Basic. They run on most flavors of UNIX, the Win32 API (including Windows 95, Windows NT, and Windows CE platforms), and the Palm Pilot (from US Robotics). Our practice has been to build and discard prototypes; we try to learn from what worked and what didn't.[1]

2 Some Lessons Learned

Two basic low-level mechanisms are characteristic of TACOMA.

- *Folders* enable an agent to transfer uninterpreted strings of bits from one processor to another.
- A *meet* operation enables a program to be started on a specific host by a program running on the same or another host.

*Department of Computer Science, University of Tromsø, Norway. Supported by NSF (Norway) grant No. 17543/410 and 112578/431.

†Department of Computer Science Cornell University, Ithaca, New York 14853. F. B. Schneider is supported in part by ARPA/RADC grant F30602-96-1-0317 and AFOSR grant F49620-94-1-0198. R. van Renesse is supported by ARPA/ONR grant N00014-92-J-1866. The views and conclusions contained herein are those of the authors and should not be interpreted as necessarily representing the official policies or endorsements, either expressed or implied, of these organizations or the U.S. Government.

[1]The Tacoma Narrows suspension bridge is a notable example of learning from mistakes. And our project name, TACOMA, was chosen with this in mind.

We provided only low-level mechanisms in TACOMA, fearing that higher-level abstractions would preclude one or another programming model from being supported. With no real experience in how to structure systems using agents, constraining the programming model seemed unwise.

In retrospect, our choice of mechanisms had fortuitous consequences. One consequence was that we completely avoided having to solve the "state capture" problem within the TACOMA system itself. With TACOMA, the only way to move an agent's state from one processor to another is by explicitly storing that state in one or more folders. TACOMA programmers must be cognizant of what state to capture and move; they must program these actions explicitly. Awkward as this may seem, the problem of automatically performing the state capture is now understood to be quite complex. By our choice of mechanisms, we managed to avoid confronting it. Moreover, in higher-level programming models where state is invisible to the programmer, automatic state capture becomes a necessity. The cost of moving an agent from one processor then cannot be predicted, and designing applications to meet performance goals becomes difficult.

The generality of our folder and meet mechanisms decouples TACOMA from the choice of language used in writing individual agents. A program in any language can be stored in a folder and moved from host to host. And, any language that a given host supports can be used to program the portion of an agent executed on that host. This generality is particularly useful in using agents for system integration. Existing applications do not have to be rewritten; COTS components can be accommodated. Two non-trivial applications we have developed make extensive use of this flexibility:

- **StormCast** is a system to support weather prediction and environmental monitoring in the Arctic. The current version is over 50K lines of code spanning multiple programming languages.

- **Tacoma Image Server** is a system to retrieve satellite images from a large database. This system employs agents to link together a database system (PostgreSQL) and an HTML forms-based user interface.

Our experience building these and other applications revealed some non-obvious benefits associated with the use of agents for structuring systems. The most surprising was that efficient and flexible service interfaces become practical when agents implement a client-server architecture. We use an agent to carry a service request to the processor where a server is executing, and the agent invokes server operations there using (local) procedure calls. Since the overhead of invoking a server operation is low, server interfaces can comprise more-primitive operations. Sequences of these operations would be invoked by the agent in order to service any given client request. In effect, the agent dynamically defines its own high-level server operations—high-level operations that can be both efficient and well suited for the task at hand.

3 Still to Learn

Despite our experiences, we, along with others in the field, have yet to demonstrate applications that depend on agents in an essential way. We have found agents to be a convenient form of glue for system integration. And, we have gotten leverage from employing agents as remote server-interfaces that can conform to the needs and capabilities of a client/host that makes use of that server. But, we have rarely had use for agents that visit a series of processors to complete a task—something that vexes us.

Two oft-advertised benefits of using agents to structure systems are:

1. the potential they offer for reducing the communications bandwidth required by a computation and
2. the potential they offer for tolerating intermittent communications outages between hosts that participate in a computation.

By moving an agent to the data, data can be processed locally, so less data needs to be transmitted between hosts. And, by having programs move from host to host, fewer hosts need to be communicating during each stage of a computation.

In networks today, however, bandwidth is ample and outages are infrequent. Cellular telephony and portable devices with limited power budgets is about to change that. So, as the need increases for distributed systems in which hosts are smart cellular telephones or other small hand-held devices (e.g. PDA's) the utility of agents may also increase.

As our project continues, we are now focusing more on security and fault-tolerance. Our quest to understand application architectures and agent-based system management issues continues. Interested readers can visit the TACOMA web site[2] for updates on our progress.

---

[2] http://www.cs.uit.no/DOS/Tacoma/index.html

14.7 Ranganathan, M., Acharya, A., Sharma, D.D., and Saltz, J., Network-aware Mobile Programs

Ranganathan, M., Acharya, A., Sharma, D.D., and Saltz, J., "Network-aware Mobile Programs," *Proceedings of the USENIX 1997 Conference*, pp. 91-103, January 1997.

Network-aware Mobile Programs*

M.Ranganathan, Anurag Acharya,†Shamik D. Sharma and Joel Saltz
Department of Computer Science
University of Maryland
College Park, MD 20740

Abstract

In this paper, we investigate network-aware *mobile programs, programs that can use mobility as a tool to adapt to variations in network characteristics. We present infrastructural support for mobility and network monitoring and show how* adaptalk, *a Java-based mobile Internet chat application, can take advantage of this support to dynamically place the chat server so as to minimize response time. Our conclusion was that on-line network monitoring and adaptive placement of shared data-structures can significantly improve performance of distributed applications on the Internet.*

1 Introduction

Mobile programs can move an active thread of control from one site to another during execution. This flexibility has many potential advantages. For example, a program that searches distributed data repositories can improve its performance by migrating to the repositories and performing the search on-site instead of fetching all the data to its current location. Similarly, an Internet video-conferencing application can minimize overall response time by positioning its server based on the location of its users. Applications running on mobile platforms can react to a drop in network bandwidth by moving network-intensive computations to a proxy host on the static network. The primary advantage of mobility in these scenarios is that it can be used as a tool to adapt to variations in the operating environment. Applications can use online information about their operating environment and knowledge of their own resource requirements to make judicious decisions about placement of computation and data.

For different applications, different resource constraints are likely to govern the decision to migrate, e.g. network latency, network bandwidth, memory availability, server availability. In this paper, we investigate *network-aware* mobile programs, i.e. programs that position themselves based on their knowledge of network characteristics. Whether the potential performance benefits of network-aware mobility are realized in practice depend on answers to three questions. First, how should programs be structured to utilize mobility to adapt to variations in network characteristics? In particular, what policies are suitable for making mobility decisions? Second, is the variation in network characteristics such that adapting to them can be profitable? Finally, can adequate network information be provided to mobile applications at an acceptable cost?

In order to adapt to network variations, mobile programs must be able to decide when to move, what to move and where to move. There are three types of network variations which may be cause for migration: (1) *population* variations, which represent changes in the distribution of users on the network, as sites join or leave an ongoing distributed computation; (2) *spatial* variations, i.e. stable differences between in the quality of different links, which are primarily due to the hosts' connectivity to the Internet; and (3) *temporal* variations, i.e. changes in the quality of a link over a period of time, caused presumably by changes in cross-traffic patterns and end-point load. Spatial variations can be handled by a *one-time placement* based on the information available at the beginning of a run. Adapting to temporal and population variations requires *dynamic placement* which needs a periodic cost-benefit analysis of current and alternative placements of computation and data. Dynamic placement decisions have two partially conflicting goals: maximize the performance improvement from mobility and minimize the cost of mobility. If an opportunity for im-

*This paper was originally published in the USENIX 1997 Annual Technical Conference.

†Now at the Department of Computer Science, University of California, Santa Barbara CA 93106

proving performance presents itself, it should be capitalized upon; however, reacting too rapidly to changes in the network characteristics can lead to performance degradation as the performance gain may not offset the mobility cost.

We investigate these issues in the context of *Sumatra*, an extension of the *Java*[1] programming environment [11] that provides a flexible substrate for adaptive mobile programs. Since mobile programs are scarce, we developed a mobile chat server for our experiments. This application, called `adaptalk`, monitors the latencies between all participants and locates the chat server so as to minimize the maximum response time. We selected this application since it is highly interactive and requires fine-grain communication. If such an application is able to take advantage of information about network characteristics, we expect that many other distributed applications over the Internet would be similarly successful. The resource that governs the migration decisions of `adaptalk` is network latency. To provide latency information, we have developed *Komodo*, a distributed network latency monitor.

To evaluate if mobile applications can take advantage of network-awareness, we examined the performance of `adaptalk` with and without mobility. Our evaluation had two main goals: (1) to determine the performance benefits, if any, of network-aware placement of the central chat server over a network-oblivious placement; and (2) to determine if dynamic placement based on online network monitoring provides significant performance gains over a one-time placement based on initial information. Our results are encouraging - they indicate that on-line monitoring and dynamic placement can significantly improve performance of distributed applications on the Internet.

The paper is organized as follows. Section 2 describes Sumatra and the programming model that it provides. Section 3 describes the design and implementation of Komodo. Section 4 describes the `adaptalk` application and the policy it uses to make mobility decisions. Section 5 describes our experiments and presents the results. Section 6 discusses the results and their implications. Section 7 describes related work and Section 8 provides our conclusions and plans for future work.

[1] *Java* is a registered trademark of Sun Microsystems.

2 Sumatra: a Java that walks

Sumatra is an extension of the Java programming environment that supports adaptive mobile programs. Platform-independence was the primary rationale for choosing Java as the base for our effort. In the design of Sumatra, we have not altered the Java language. Sumatra can run all legal Java programs without modification. All added functionality was provided by extending the Java class library and by modifying the Java interpreter without affecting the virtual machine interface.

Our design philosophy for Sumatra was to provide the mechanisms to build adaptive mobile programs. Policy decisions concerning when, where and what to move are left to the application. The main feature that distinguishes Sumatra from previous systems [4, 12, 14, 25] that support mobile programs is that *all* communication and migration happens under application control. Furthermore, combination of distributed objects and thread migration allows applications the flexibility to dynamically choose between moving data or moving computation. The high degree of application control allows us to easily explore different policy alternatives for resource monitoring and for adapting to variations in resources. We believe that the space of design choices for adaptive mobile programs is yet to be mapped out and such flexibility is important to help explore this space.

Sumatra adds two programming abstractions to Java: *object-groups* and *execution engines*. An object-group is a dynamically created group of objects. Objects can be added to or removed from object-groups. All objects within an object-group are treated as a unit for mobility-related operations. This allows the programmer to customize the granularity of movement and to amortize the cost of moving and tracking individual objects. This is particularly important in languages like Java because every data structure is an object and moving the state one object at a time can be prohibitively expensive. An execution-engine is the abstraction of a location in a distributed environment. In concrete terms, it corresponds to an interpreter executing on a host. Sumatra allows object-groups to be moved between execution-engines. An execution-engine may also host active threads of control. Multiple threads on the same engine are scheduled in a *run-to-completion* manner. The rest of this section provides a brief description of Sumatra. Further details about Sumatra are presented in [1].

The principal new operations provided by Sumatra are:

Object-group migration: Object-groups can be moved between engines on application request. As mentioned earlier, all objects within an object-group are treated as a unit for mobility-related operations. Objects in an object-group are automatically marshalled using type-information stored in their class templates. When an object-group is moved, all local references to objects in the group (stack references and references from other objects) are converted into *proxy references* which record the new location of the object. Some objects, such as I/O objects, are tightly bound to local resources and cannot be moved. References to such objects are reset and must be reinitialized at the new site. The class template for an object (and the associated bytecode) can be downloaded into an execution-engine on application request.

Remote method invocation: Method invocations on proxy objects are translated into calls at the remote site. Type information stored in class-templates is used to achieve RPC functionality without a stub compiler. Exceptions generated at the called site are forwarded to the caller. Sumatra does not automatically track mobile objects. Requesting a remote method invocation on an object that is no longer at the called site results in an *object-moved* exception at the calling site. To facilitate application-level tracking, the exception carries with it a forwarding address. The caller can handle the exception as it deems fit (e.g., re-issue the request to the new location, migrate to the new location, raise a further exception and so on). This mechanism allows applications to locate mobile objects lazily, paying the cost of tracking only if they need to. It also allows applications to abort tracking if need be and pursue an alternative course of action.

Thread migration: Sumatra allows explicit thread migration using a `engine.go()` function that bundles up the stack and the program counter and moves the thread to the specified execution-engine. Execution is resumed at the first instruction after the call to go. To automatically marshal the stack, the Sumatra interpreter maintains a type stack parallel to the value stack, which keeps track of the types of all values on the stack. When a thread migrates, Sumatra transports with it all local objects that are referenced by the stack but do not belong to any object-group. Objects that belong to an object-group move only when that object-group is moved. Stack references to the objects that are left behind (i.e were part of some object-group) are converted to proxy references. After the thread is moved to the target site, it is possible that its stack contains proxy references that point to objects that used to be remote but are now local. These references are converted back to local references before the call to go returns.

Remote execution: A new thread of control can be created by *rexec*'ing the `main` method of a class existing on a remote engine. The arguments for new thread are copied and moved to the remote site. Unlike remote method invocation, remote execution is non-blocking; the calling thread resumes immediately after the `main` method call is sent to the remote engine. Remote execution is different from thread migration as it creates a new thread at the remote site that runs concurrently with the original thread; thread migration moves the current thread to the remote site without creating a new thread. Concurrent threads communicate using calls to shared objects. The thread initiating a remote execution can share objects with the new thread by passing it references to these objects as arguments to `main`.

Resource monitoring: Sumatra provides a resource-monitoring interface which can be used by applications to register monitoring requests and to determine current values of specific resources. This interface is similar to an object-oriented version of the Unix `ioctl()` interface. When an application makes a monitoring request, Sumatra forwards the request to the local resource monitor. If the monitor does not support the requested operation, an exception is delivered to the application.

Signal handlers: Sumatra allows applications to register handlers for a subset of Unix signals. Signals can be used by the external environment (the operating system or some other administrative process) to inform the application about urgent asynchronous events, in particular resource revocation. Using a handler, the application can take appropriate action including moving away from the current execution site.

2.1 Example

In this section, we provide a feel for the Sumatra programming model using a simple example. The task is to scan through a database of X-ray images stored at a remote site for images that show lung cancer. This task can be performed in two steps. In the first step, a computationally cheap pruning algorithm is used to quickly identify lungs that might have cancer. A compute-intensive cancer-detection algorithm

```
.....
filter_object = new Lung_filter();
cancer_object = new Lung_checker(filter_object);
myengine = System.rpc.myEngine();

// Create a engine at the xray database site.
remote_engine = new Engine("xrays.gov");
// Send the lung_filter class to the remote engine
remote_engine.downloadClass("Lung_filter");
// Create a new object group.
objgroup = new ObjGroup("lung_filter_group");
// Add the lung_filter_object to the object group
objgroup.checkIn(filter_object);
// Move the object group to the database site
objgroup.moveTo(remote_engine);

// a remote method call selects interesting xrays
size = filter_object.query(db,"DarkLungs");

// Are there too many images to bring over?
if ( size > too_many_images ) {
  // Migrate thread, process images and return.
  remote_engine.go();
  result = cancer_object.detect_cancer();
  myengine.go();
}
else {
  // there are only a few interesting xrays. Fetch them
  // and process locally.
  objgroup.moveTo(myengine);
  result = cancer_object.detect_cancer();
}

// display result locally
System.display(result);
```

Figure 1: Excerpt of a Sumatra program that adaptively migrates to reduce its network bandwidth requirements

is then used to identify images that actually show cancer.

One way to write a program for this task would be to download all lung images from the image server and do all the processing locally. If the absence of cancer in most lung images can be cheaply established, this scheme wastes network resources as it moves all lung images to the destination site. Another approach would be to send the selection procedure to the site of the image database and to send only the "interesting" images back to the main program. If the selection procedure is able to filter out most of the images, this approach would significantly reduce network requirements. A third, and even more flexible, approach would allow the shipped selection procedure to extract all the interesting images from the database but return only the *size* of the extracted images to the main program. If the size is too big, the program may choose to move itself to the database site and perform the cancer-detection computation there rather than downloading all the data. This avoids downloading most images at the cost of (possibly) slower processing at the server. On the other hand if the size of the images is small, the data can be shipped over and processed locally. Figure 1 shows code for the third approach. This program makes its decision to migrate in a rudimentary fashion; a more realistic version of this application would also take network bandwidth and the processing power available on both machines into consideration.

Sumatra assumes that a local resource monitor is available which can be queried for information about the environment. In the next section, we describe one such monitor which allows `Sumatra` applications to request information about network latency between any pair of sites that run the monitor.

3 Komodo: a distributed network latency monitor

Komodo[2] is a distributed network latency monitor. The design principles of Komodo are: low-cost active monitoring and fault-tolerance. Active monitoring uses separate messages for monitoring; passive monitoring generates no new messages and piggybacks monitoring information on existing messages. An active monitoring approach is needed for `adaptalk` (described in the next section) as passive monitoring cannot provide information about links that are not used in the current placement but could be used in alternative placements. It is our working hypothesis that effective mobility decisions can be based on medium-term (30sec-few minutes) and long-term (hours) variations. At these resolutions, we believe that active monitoring can be achieved at an acceptable cost. This section briefly describes the design and implementation of Komodo. Further details about Komodo are presented in [19].

Komodo allows applications to initiate monitoring of network latency between any pair of hosts running

[2]Komodo dragons are a species of *monitor* lizards found on the island of Komodo which is close to both Java and Sumatra.

the monitor; the application need not be resident on either of the hosts. Komodo is implemented as a user-level daemon that runs on every host participating in the computation. Applications pass monitoring requests to their local Komodo daemon. If the requested link includes the current host, the local daemon handles the request. Otherwise, it forwards the request to the daemon on the appropriate host. Daemons determine network latency by sending 32-byte UDP packets to each other. If an echo is not received within an expected interval, (the maximum of the ping period or five times the current round trip time estimate) the packet is retransmitted. Using UDP for communication may, occasionally, lead to loss of messages. Message loss can lead only to a short-term loss of efficiency. If monitoring requirements are coarse-grained as we expect, the effect of packet loss should be small.

Applications that initiate a monitoring request can specify the frequency with which Komodo *pings* a link. Komodo enforces an upper bound on this frequency to keep the monitoring cost at an acceptable level. Applications need to refresh requests periodically to keep them alive; Komodo deactivates requests that have not been refreshed for longer than its *request-timeout* period.

Latency measures acquired by Komodo are passed through a filter before being provided to applications. This filter eliminates singleton impulses as well as noise within a jitter threshold (we use a jitter threshold of 10 ms, which is the resolution of most Unix timers). If the measure changes rapidly, a moving window average is generated. This filter was designed on the basis of our study of a large number of Internet latency traces (see Section 5.1) which revealed that: (1) there is a lot of short-term jitter in the latency measures but in most cases, the jitter is small; (2) there are occasional sharp jumps in latency that appear only for short time intervals; (3) occasionally, the latency measure fluctuates rapidly; (4) for time windows of 10 seconds or larger, the mode value (with a 10 ms jitter threshold) dominates. To elaborate the last point, in most time windows, 70-90% of the latency values fall within a jitter threshold of the most common value. Our filter attempts to find the mode for a recent time window. If there is no stable mode (as happens occasionally), it returns the mean. Figure 2(a) illustrates the operation of the filter.

Each daemon maintains a cache of current latency estimates for all links it is currently monitoring. This cache is maintained in a well-known shared memory segment and can be efficiently read by all Sumatra applications executing on the same machine. Cooperating Komodo daemons forward latency information in response to persistent remote requests. A latency estimate for a request received from another host is forwarded only when a new filtered estimate (different from the previous filtered estimate) is generated and is piggybacked onto a ping reply if possible. Currently, Komodo is implemented in C.

To address concerns about the cost of active monitoring, we measured the CPU utilization of Komodo for varying number of links. Results in Figure 2 (b) show that the maximum CPU utilization for up to sixteen links is about 0.5 %. The amount of data transferred is 512 bytes/second. This experiment was conducted on Sparc 5 machines (110MHz,32 MB of memory) running SunOS Release 5.5.

4 `Adaptalk`: An adaptive internet chat application

`Adaptalk` is a relatively simple network chat application built using Sumatra and Komodo. It allows multiple users to have an online conversation; new participants can join an ongoing conversation at any point; multiple independent conversations can be held. To ensure that all participants see the same conversation and that new participants can join ongoing conversations, a central server is used to serialize and broadcast the contributions.

`Adaptalk` is divided into three modules: handling keyboard events, managing the chat screen and coordinating the communication between participants. Each component is implemented by a separate object-group. Each host participating in the conversation runs two execution-engines, one houses the `screen` object-group and the other houses the `keyboard` object-group. The central server is implemented as a separate shared object-group, the `msgboard`, which can be placed on any host participating in the conversation. Each message issued by a participant starts from a keyboard object which invokes a remote method on the msgboard. The msgboard serializes incoming messages and issues a series of remote-execution requests, one per participant, which update the screen objects on all participants. In this case, remote execution is preferred to remote method invocation as there is no useful return value and remote execution allows fast one-way communication.

Individual messages in `adaptalk`, and most other chat applications, consist of single lines of characters, usually no more than 50-60 characters. The goal of a chat application is to provide a short response-time to

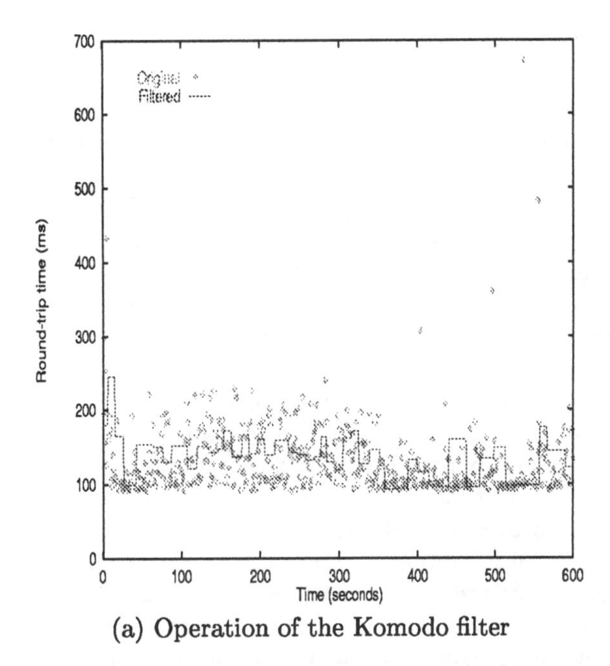

(a) Operation of the Komodo filter

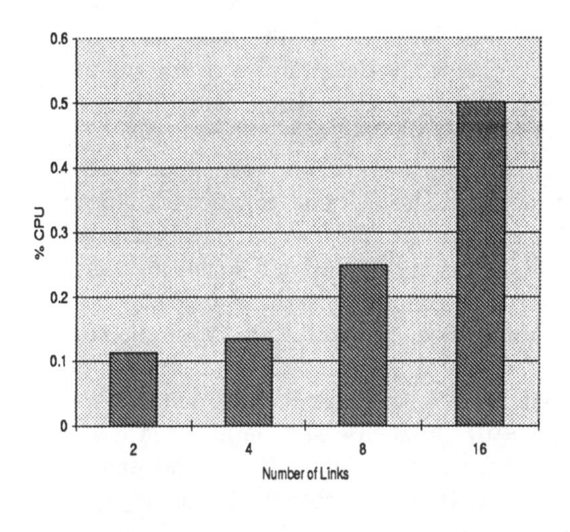

(b) CPU utilization of Komodo

Figure 2: (a) The input to the filter is a 10-minute trace of one-per-second latency measures between `baekdoo.cs.umd.edu` and `lanl.gov`. Note that the four single-ping impulses towards the right end have been eliminated. (b) The CPU utilization is computed by dividing the (user+system) time by the total running time. Each experiment was run for 1000 seconds with one ping per second for all links.

all participants so that a conversation can make quick progress. The response-time for a particular participant depends on the latency between it and the central server. Given the latencies of all the links, the primary knob that `adaptalk` can turn to maintain a low response-time for all participants is the position of the central server.

4.1 Mobility policy

There are two main features of the `adaptalk` mobility policy. (1) continuous tracking of the instantaneously *most-suitable-site* and (2) deferral of server-motion till the potential for a significant and *stable* performance advantage has been seen. The first feature allows it to quickly take advantage of opportunities for optimization; the seconds helps ensure the gain is greater than the cost. The goal of `adaptalk` is to minimize the maximum response-time seen by any participant. The suitability of a participating machine as the location of the msgboard is characterized by the maximum network latency between it and all other participants. The machine that achieves the lowest measure is designated the most-suitable-site.

`Adaptalk`'s migration policy is shown in pseudocode in Figure 3. This algorithm is run at the location that hosts the msgboard and recomputes the most-suitable-site each time a new message is posted by any participant. The msgboard maintains an array of counters, one for each potential location, which keep track of the number of times each location is found to be the most-suitable-site. The msgboard moves whenever: (1) the current site receives a very low score (< loss_threshold) over a given period (the decision_cycle); or (2) a different site receives more than a threshold score (the win_threshold). The first condition is used to quickly move away from locations that provide poor performance; the second condition is used to move the msgboard to locations that consistently promise better performance. The counters are reset whenever the msgboard moves, the decision_cycle completes or a participant enters or leaves the conversation.

We expect three types of variations in the network characteristics which may be cause for migration: (1) *population* variations, which represent changes in the distribution of users on the network, as participants join or leave an ongoing conversation; (2) *spatial* variations, i.e. stable differences between latencies of different links; and (3) *temporal* variations, i.e. changes in the latency of a link over a period of time. `Adaptalk`'s migration policy can adapt to all three types of variations. Consider the case with a fixed number of participants with significant spatial variation in network latency and little temporal variation. In this case, the migration algorithm rapidly recognizes the best location for the msgboard, but waits until this choice has been ratified over some period of time (count[newloc] > win_threshold) before moving it. As shown in Section 5, this policy allows `adaptalk` to effectively insure itself against poor ***initial placement.*** Once a good location has been found, the msgboard does not move, unless temporal variations or changes in population distribution cause another node to become a substantially better location (i.e. count[newloc] > win_threshold) or the current host to become a substantially bad choice (i.e. count[curr_engine] < loss_threshold && rounds % decision_cycle == 0). In such cases, the msgboard will move during the conversation. After initial experiments with `adaptalk`, we set the win_threshold to be $25 \times n$, the loss_threshold to be $12 \times n$ and the decision_cycle to be $50 \times n$. Here, n is the number of participants. The length of the decision_cycle was set large enough to amortize the cost of movement in cases where large temporal variations or fluctuations in population distribution cause frequent repositioning.

```
........
Get the all to all latency map from Komodo;
Find the site s that would minimize the max
  latency for messages posted to msgboard;
count[s] = count[s] + 1; rounds++;
let w be the site with the largest count;
let curr_engine be the engine which
   currently houses msgboard;
// Found a clear cut winner.
if (count[w] > win_threshold) return w;
else if (rounds % decision_cycle == 0) {
// Is the current engine an ok location ?
if (count[curr_engine] > loss_threshold) {
    clear count for each host;
    return curr_engine;
 } else {
    // Current engine is a bad location.
    set new_host to the host with the
         maximum count;
    clear count for each host;
    return new_host;
 }
} else return null; // cycle not yet over.
```

Figure 3: Decision Algorithm for *msgboard* placement used in Adaptalk. This algorithm is run at the location where the *msgboard* resides each time a message is posted.

5 Evaluation

To evaluate the performance impact of network-aware adaptation on the Internet, we performed two sets of experiments. First, we monitored round-trip times for 32-byte ICMP packets between a large set of host-pairs over several days. The goal of these experiments was to study the spatial and temporal variation in network latency on the Internet. Results from this study are presented in section 5.1.

Second, we measured the performance of three versions of `adaptalk` over long-haul networks, using traces collected during the Internet study. Our evaluation had two main goals: (1) to determine if network-aware placement of components of an application distributed over multiple hosts on the Internet provides significant performance gains over a network-oblivious placement; and (2) to determine if dynamic placement based on online network monitoring provides significant performance gains over a one-time placement based on initial information. Results from this study are presented in section 5.3.

5.1 Variations in Internet latency

We selected 45 hosts: 15 popular `.com` web-sites (US), 15 popular `.edu` web sites (US) and 15 well-known non-US hosts. These host were pinged from four different locations in the US. The study was conducted over several weekdays, each host-pair being monitored for at least 48 hours. We used the commonly available `ping` program and sent one ping per second. This resolution was acceptable as our goal was to discover medium-term (30sec/minutes) and long-term (hours) variations.

The conclusions of our study, briefly, are: (1) there is large spatial variation in Internet latency (the per-hour mean latency varied between 15 ms and 863 ms for US hosts and between 84 ms and 4000 ms for non-US hosts); (2) there is a large and stable variation in the latency of a single host-pair over the period of a day (maximum daily variation in per-hour mean latency for US hosts was 550 ms and for non-US hosts was 5750 ms); (3) There is a lot of jitter in the latency measures but in most cases, the jitter is small; (4) there are isolated peaks in latency that appear only for a single time interval; (5) for time windows of 10 seconds or larger, the mode value (with a 10 ms jitter threshold) dominates (in most time windows, 70-90% of the latency values fall within a jitter threshold of the most common value); (6) the moving-window mode changes quite slowly.

5.2 Experimental Setup

Having established that there are significant spatial and temporal variations in network latency on the Internet, we examined how well `adaptalk` could adapt to these variations.

To simulate the characteristics of long-haul networks, we decided to run our experiments over a low-latency LAN and delay all packets based on the ICMP `ping` traces described above (see Figure 4 (a)). This approach also allowed us to perform repeatable experiments. To ensure that delaying packets instead of using a real network does not skew the latency measures, we performed a simple test. Free-running Komodo monitors were installed at `bookworm.cs.umd.edu` and `jarlsberg.cs.wisc.edu` and were used to collect UDP latency measures between this host-pair. In parallel, a trace of ICMP ping times between these two hosts over the same period (5000 sec) was collected. This trace was later fed into trace-driven Komodo monitors running on two hosts on our LAN. The latency measures reported by the trace-driven monitors matched quite well with the actual latency measures reported by free-running monitors. The average of the actual latency measures was 128 ms (std dev = 64); the average of the values reported by the trace-driven monitors was 144 ms (std dev = 68).

We performed all our experiments on four Solaris machines on our LAN. We picked six trace-segments from the Internet study and used them to delay packets between the machines. All these segments were over the noon-2pm EDT period. We selected this period since noon is the approximate beginning of the daily latency peak for US networks as well as the approximate end of the daily latency peak for many non-US networks. These traces were selected to approximate the network latency spectrum observed in the Internet study. Hosts participating in the selected traces include: `java.sun.com`, `home.netscape.com`, `www.opentext.com`, `cesdis.gsfc.nasa.gov`, `www.monash.edu.au` and `www.ac.il`. This setup makes the four local machines behave like four far-flung machines on the Internet. Figure 4 (b) shows the configuration used for the experiments.

5.3 Experiments

We performed a series of experiments to evaluate the benefits of adapting to various types of network variations. The experiments consisted of running three different versions of the chat server. The first version, called static-placement, had no migration support and no network-awareness. The location of the msgboard was chosen in a network-oblivious fashion. The second version was a stripped-down version of ***adaptalk***, called one-shot-placement. It used network information from Komodo to find the best initial placement for the msgboard, and used mobility support to move it there. After initial placement, migration decisions and network-awareness were turned off. The third version, called dynamic-placement, was the full-fledged `adaptalk`, as described in section 4. It used on-line monitoring and dynamic placement to position the msgboard.

The performance of static-placement depends on the location of the msgboard. If static-placement chooses the same location as one-shot-placement, both would have the same performance. On the other hand, since static-placement is network-oblivious, it is just as likely to place the msgboard at the worst possible location. As the performance of one-shot-placement already provides a rough upper-bound on the performance of static-placement, we deliberately chose the worst initial placement when running the static-placement version.

Adapting to Population Variation: To evaluate the effect of changing user distribution we used the following workload: A conversation was initiated between hosts C and D. Host B joins the conversation after 15 minutes, and host A joins 15 minutes thereafter. Each host sends a sequence of 70-character sentences with a 5-second think time between sentences. With only two hosts initiating the conversation, there is no difference between the best and worst initial placements for the msgboard and both static-placement and one-shot-placement perform identically (both place the msgboard on host D). Figure 5 (a) plots the maximum latency over all hosts for the one-shot-placement version. Note that even after new hosts join the conversation there is no noticeable difference in maximum latency. continues to In contrast, dynamic-placement adapts to the changing population. Soon after host B joins the conversation, the adaptive placement policy moves the msgboard there, causing a drop in the maximum latency. After host A joins the conversation, the msgboard moves between hosts A and B in response to temporal fluctuations. This can be seen from the variation in latency for host B in Figure 5 (b). These movements help keep the maximum latency steady even in the presence of temporal fluctuations.

Adapting to Temporal and Spatial Variation: In this case the client population is assumed to be stable. The workload consists of all 4 hosts jointly

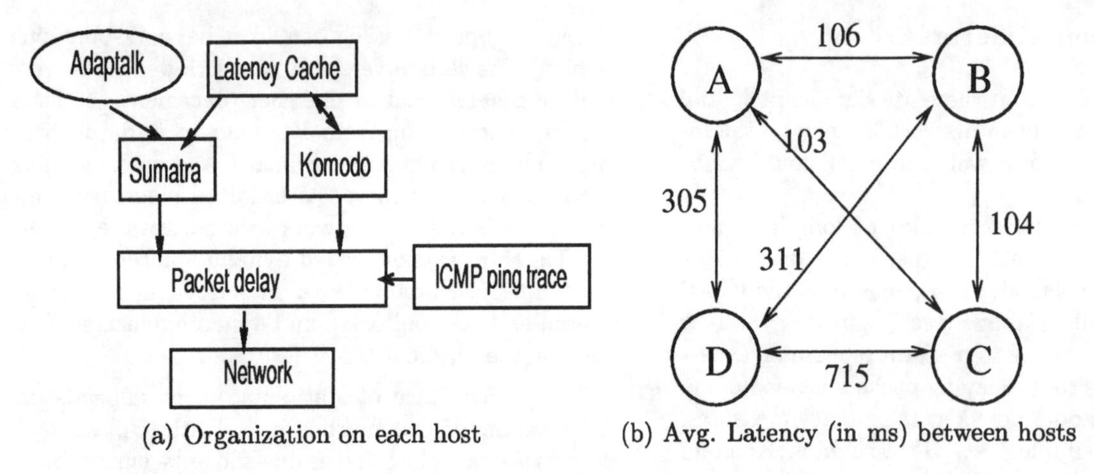

(a) Organization on each host (b) Avg. Latency (in ms) between hosts

Figure 4: Experimental Setup. Four local machines on a LAN were used to simulate four remote machines on the Internet by adding delays to packets. ICMP ping traces between real Internet hosts were used to generate the delays, so as to capture real-life temporal variations in latencies.

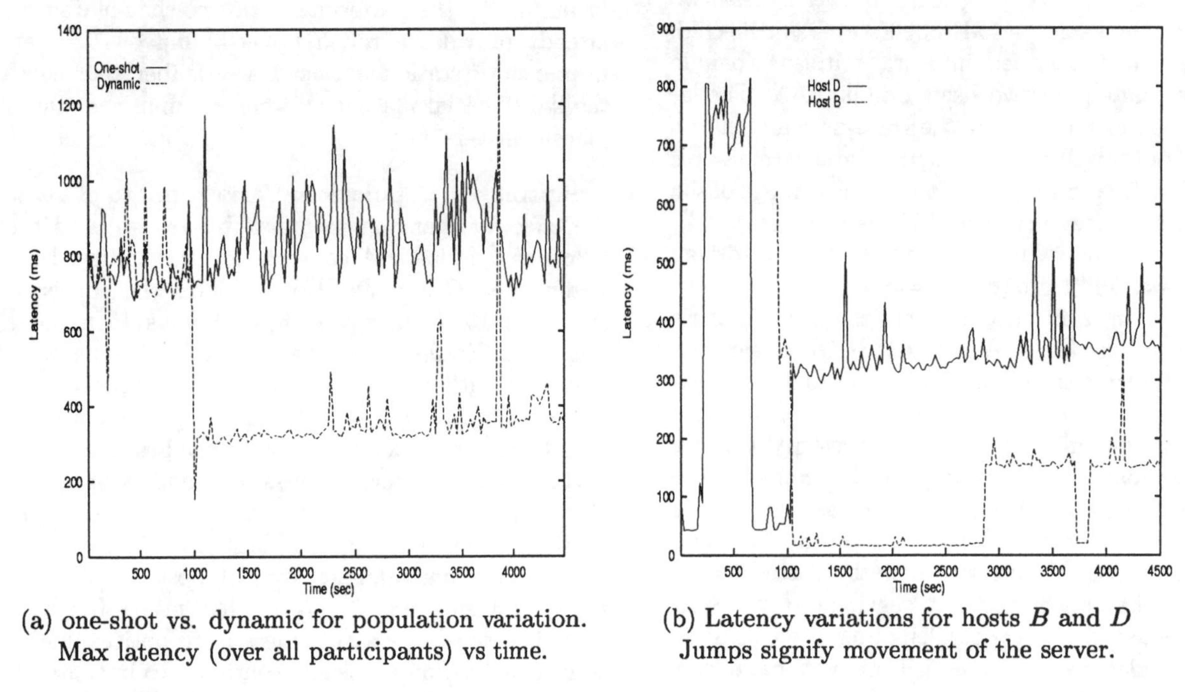

(a) one-shot vs. dynamic for population variation. Max latency (over all participants) vs time.

(b) Latency variations for hosts B and D Jumps signify movement of the server.

Figure 5: Adapting to population variation. Hosts C and D initiate the conversation. Host B joins after 900 seconds and host A joins 1800 seconds after the beginning. The one-shot-placement version places the chat server at host D. The dynamic-placement version migrates the server when new hosts join.

initiating a conversation which runs for 75 minutes. As before, each host generates a new sentence every 5 seconds. In this case, the network-oblivious (static-placement) version places the chat server on host D. The network-aware (one-shot-placement) version uses latency information provided by Komodo to determine that host B is a much better placement. For the dynamic-placement version, initial placement is less important as it should be able to recover from a bad initial placement. For this version, we place the msgboard at host D, the worst-possible location.

To avoid clutter, Figure 6 shows the performance of these three versions in two different graphs. Figure 6 (a) compares the maximum latency (over all participants) for the dynamic-placement and static-placement versions. As seen from the sharp drop on the left end of the graph, the dynamic-placement version is successfully able to move the msgboard away from its bad initial placement to a more suitable location. Figure 6 (b) compares the average maximum latency (over all participants) for the dynamic-placement and one-shot-placement versions. It shows that once the dynamic-placement version moves the server to a more suitable location, the performance of the two versions is largely equivalent. This implies that adapting to short-term temporal variations in a steady population workload does not provide much performance advantage over one-shot network-aware placement. It may, however, still be advantageous to adapt to long-term temporal variations. Note that at the far right of graph Figure 6 (b), temporal variation in the link latencies do allow the dynamic-placement version to do better than the one-shot-placement version.

6 Discussion

In the introduction, we raised three questions with respect to network-aware mobility. First, how should programs be structured to utilize mobility to adapt to variations in network characteristics? Second, is the variation in network characteristics such that adapting to them proves profitable? Finally, can adequate network information be provided to mobile applications at an acceptable cost?

Our experience with Sumatra and `adaptalk` provides some early insights about application structure suitable for adaptive mobile programs. First, the migration policy should be cheap so that applications don't have to analyze the tradeoffs of the migration decision itself. An easy-to-compute policy allows frequent decisions and rapid adaptation to changes in the environment. We believe that an easy-to-compute migration policy was key to `adaptalk`'s ability to quickly find good locations for the chat server. Second, good modularization helps an application take advantage of mobility. Modularization is important for all distributed applications but it is more so for mobile programs as they have to make online decisions about the placements of different components. Third, to be resource-aware, remote accesses should be split-phase; the first phase delivers an *abbreviation*, a small and cheaply computed metric of the data (for example, size, number of data items, thumbnail sketch etc) and the second phase actually accesses the data. This allows the application to change its data access modality for the second phase (retrieve remotely, request filtering, move to data location) based on the value of the abbreviation and knowledge of its own requirements. This insight comes from our experience with writing other applications in Sumatra; `adaptalk` does not benefit from this as the size of all messages is small.

An important question that needs further investigation is where the control for mobility decisions should be placed – whether mobility decisions should be made by a central controller that keeps track of the state of all links or by multiple local controllers that use information only from a small subset of the links. Centralized decisions are likely to be more expensive than distributed decisions (the latter need less information and less synchronization) but could yield better performance (as they use global information).

To answer the second question, we evaluated the profitability of adapting to changes in the user-distribution as well as spatial and temporal variations in network latency. Adapting to changes in user-distribution led to significant gains allowing `adaptalk` to find better placements as more users came online. Support for mobility allows applications built around a central data-structure to recover from a poor initial placement of this structure by repositioning it to a more suitable location. Adapting to temporal variations alone did not not lead to significant benefits over the period of an hour. In light of this experience, we expect that a simpler migration policy for `adaptalk` for short periods would consider migration only when users join or leave the conversation, rather than on every message as is currently done. Since long-term variation of latency could be as large as 550 ms (US hosts) and 5750 ms (non-US-hosts), longer conversations could still benefit from adapting to temporal variations.

Our experiments with Komodo illustrate that cheap active monitoring can provide network information that can be profitably exploited. Though it would be best to use Komodo as a stand-alone system supplying network information to many distributed applications, its cost is so low that one can contemplate rolling Komodo into individual applications such as `adaptalk` without overloading the network. Active monitoring was needed for `adaptalk` as it needed information about links that are not used in the current placement but could be used in alternative placements. Other applications that change the location of computation but do not change the pattern of communication would not need active monitoring as they could piggyback monitoring information on existing messages. An example of such an application would be an information access program on a mobile platform which moves primarily between this platform and a proxy host on the static network. Active monitoring, as implemented in Komodo, will not be as cheap for applications that are

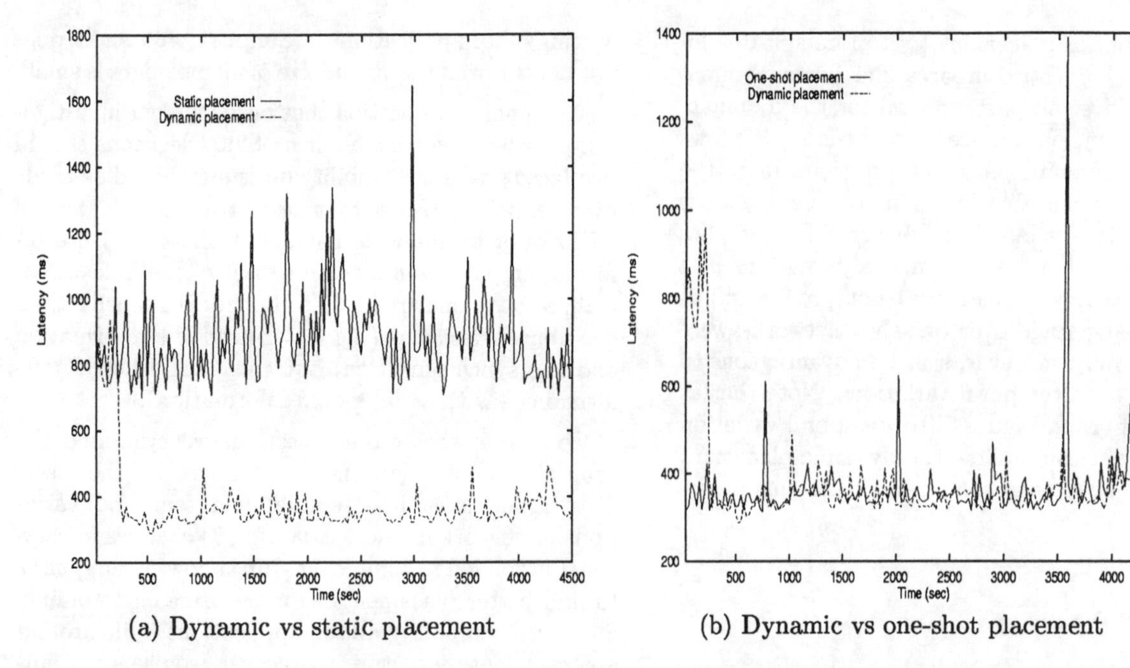

(a) Dynamic vs static placement

(b) Dynamic vs one-shot placement

Figure 6: Maximum latency (over all participants) vs time in **adaptalk**. The one-shot-placement and the static-placement are computed based on latency information available when the conversation is initiated. The client population is stable throughout the experiment.

bandwidth-sensitive and not latency-sensitive. We are currently investigating methods to cheaply estimate Internet bandwidth.

In this paper, we have considered Internet hosts that are static. If the platform is mobile and is able to switch between multiple wireless networks [15], the temporal variation in latency could be greater and more abrupt. In these cases, adapting to short-term temporal variations could provide a significant benefit even for latency-sensitive applications.

System stability is a potential concern for programs whose components are mobile. We believe that system stability is a property of the application and not the underlying system support. Accordingly, Sumatra does not provide automatic tracking. Instead, it provides support (in the form of *object-moved* exceptions) that allows applications to track mobile objects (see section 2 for details). We did not encounter stability problems in any of our applications.

Finally, we would like to argue the need for mobility as an adaptation mechanism. An alternative adaptation mechanism, which places replicated servers at all suitable points in the network, could adapt to spatial, temporal and population variation by handing off control between servers and by using dynamically created hierarchies of servers. It is quite likely that for any particular application, such a strategy would be able to achieve the performance achieved by programs that use program mobility as the adaptation tool. The advantage of mobility-based strategies is that it allows small groups of users to rapidly set up private communities on-demand without requiring extensive server placement.

7 Related work

Process migration and remote execution have been proposed, and have been successfully used, as mechanisms for adapting to changes in host availability [6, 8, 16, 23, 26]. Remote execution has also been proposed for efficient execution of computation that requires multiple remote accesses [7, 9, 24] and for efficient execution of graphical user interfaces which need to interact closely with the client [3]. Both these application scenarios use remote execution as a way to avoid using the network. Most proposed uses of Java [11] also use remote execution to avoid repeated client-server interaction. In these applications, decisions about the placement of computation are hard-coded. To the best of our knowledge, Sumatra (together with Komodo) is the first system that allows distributed applications to *monitor* the network state and *dynamically* place computation and data in response to changes in the network state. We also believe that our experiment with **adaptalk** is the first attempt to determine if the variation in Internet characteristics is such that it is profitable for applications to adapt to them.

Network-awareness is particularly important to applications running on mobile platforms which can see rapid changes in network quality. Various forms of network-awareness have been proposed for such applications. Application-transparent or system-level adaptation to variations in network bandwidth has been successfully used by the designers of the Coda file system [18] to improve the performance of applications. The Odyssey project on mobile information access plans to provide support for application-specific resource monitoring and adaptation. The primary adaptation mechanism under consideration is change in data fidelity [22]. Athan and Duchamp [2] propose the use of remote execution for reducing the communication between a mobile machine and the static network. In all these systems, location of the various computation modules is fixed; adaptation is achieved by changing the way in which the network is used.

Several systems have been built which permit an executing program to move while it is in execution - for example Obliq [4], Agent TCL [12], Emerald [14], Telescript [25] and TACOMA [13].The primary distinction between these systems and Sumatra is that in Sumatra, ***all*** communication and migration happens under application control. Complete application control allows us to easily explore different policy alternatives for resource monitoring and for adapting to variations in resources.

Several studies have been performed to determine end-to-end Internet performance. Sanghi et al [21] and Mukherjee [17] have studied network latency. Their observations show that while round trip times show significant variability with sharp peaks, there exist dominant low frequency components. This is consistent with our observations that in a time window of reasonable size, the mode value usually dominates and that the mode value changes slowly.

Golding [10] and Carter and Crovella [5] have studied mechanisms to estimate end-to-end Internet bandwidth. Golding's results indicate that attempts to predict bandwidth using previous observations alone is unlikely to work well. Carter and Crovella propose the use of round trip times for short packets to estimate network congestion. They propose to use the network congestion information to estimate changes in network bandwidth (assuming the inherent bandwidth of the link has been previously computed by flooding the link). Their results indicate that it might be possible to estimate the *change* in network bandwidth using information about the *change* in network latency.

8 Conclusions and Follow-up

This paper is a first step in demonstrating that distributed programs can use mobility as a tool to adapt to variations in their operating environment. Our exploration of network-aware mobile programs lead us to the following conclusions. First, network-aware placement of components of a distributed application can provide significant performance gains over a network-oblivious placement. For short term applications (applications that run for an hour or so), exploiting spatial variations as well as variations in the number and location of the clients achieves most of the gains. For longer-running applications, exploiting temporal variations might be worthwhile. Second, effective mobility decisions can be based on coarse-grained monitoring. This allows cheap active monitoring without losing effectiveness. Finally, there is significant spatial and temporal variation in Internet latency which can be effectively adapted to by mobile programs.

Since the publication of this paper, we have extended our work to examine the utility of changing the location of combination operators as a technique to adapt to variations in wide-area network bandwidth. We tried to answer the following questions. First, does relocation of operators provide a significant performance improvement? Second, is on-line relocation useful or does a one-time positioning at start-up time provide most if not all the benefits? If on-line relocation *is* useful, how frequently should it be done?

We addressed these questions in an empirical manner. We developed three algorithms that use bandwidth information to adapt data combination plans. We evaluated the performance of these algorithms in the context of a specific task – composition of satellite images from geographically distributed sites. In addition to comparing the end-to-end performance of these algorithms, we tried to determine how frequently should the adaptation take place for on-line algorithms. To ensure realistic network conditions, we conducted a multi-day study of Internet bandwidth for a large number of host-pairs. This study included US hosts (east coast, west coast, midwest and south), European hosts (in Spain, France and Austria) and one host in Brazil. We used the bandwidth traces acquired in this study to emulate realistic network conditions.

We found that all relocation algorithms significantly outperform the strategy of downloading all data to the client and that while an initial bandwidth-aware placement achieves large gains, on-line relocation provides a consistent and substantial additional improvement. For the set of traces that we used, a 5-10 minute relocation period provides the best performance. Details about this work can be found in [20]. We believe that this experience provides additional evidence for our conclusion that distributed programs can successfully use mobility as a tool to adapt to variations in their operating environment.

Acknowledgments

We would like to thank Mustafa Uysal, Manuel Ujaldon and anonymous referees for their suggestions. We would also like to thank John Kohl, our shepherd.

References

[1] A. Acharya, M. Ranganathan, and J. Saltz. *Sumatra: A Language for Resource-Aware Mobile Programs*, chapter II, pages 111–30. Springer Verlag, Lecture Notes in Computer Science, 1997.

[2] A. Athan and D. Duchamp. Agent-mediated Message Passing for Constrained Environments. In *Proceedings of the USENIX Mobile and Location-independent Computing Symposium*, pages 103–7, Aug 1993.

[3] K. Bharat and L. Cardelli. Migratory Applications. In *Proceedings of the Eighth ACM Symposium on User Interface Software and Technology*, pages 133–42, Nov 1995.

[4] L. Cardelli. A Language with Distributed Scope. In *Proceedings of the* 22nd *ACM SIGPLAN-SIGACT Symposium on Principles of Programming Languages*, January 1995.

[5] R. Carter and M. Crovella. Dynamic Server Selection using Bandwidth Probing in Wide-Area networks. Technical Report BU-CS-96-007, Boston University, March 1996.

[6] J. Casas, D. Clark, R. Konuru, S. Otto, and R. Prouty. MPVM: A Migration Transparent Version of PVM. *Computing Systems*, 8(2):171–216, Spring 1995.

[7] S. Clamen, L. Leibengood, S. Nettles, and J. Wing. Reliable Distributed Computing with Avalon/Common Lisp. In *Proceedings of the International Conference on Computer Languages*, pages 169–79, 1990.

[8] F. Douglis and J. Ousterhout. Transparent Process Migration: Design Alternatives and the Sprite Implementation. *Software - Practice and Experience*, 21(8):757–85, Aug 1991.

[9] J. Falcone. A Programmable Interface Language for Heterogeneous Systems. *ACM Transactions on Computer Systems*, 5(4):330–51, November 1987.

[10] R. Golding. End-to-end performance prediction for the Internet (Work In Progress). Technical Report UCSC-CRL-92-26, University of California at Santa Cruz, June 1992.

[11] J. Gosling and H. McGilton. The Java Language Environment White Paper, 1995.

[12] R. Gray. Agent TCL: A Flexible and Secure Mobile-agent System. In *Proceedings of the Fourth Annual Tcl/Tk Workshop (TCL 96)*, July 1996.

[13] D. Johansen, R. van Renesse, and F. Schneider. An Introduction to the TACOMA Distributed System Version 1.0. Technical Report 95-23, University of Tromso, 1995.

[14] E. Jul, H. Levy, N. Hutchinson, and A. Black. Fine-Grained Mobility in the Emerald System. *ACM Transactions on Computer Systems*, 6(2):109–33, February 1988.

[15] R. Katz. The Case for Wireless Overlay Networks. Invited talk at the ACM Federated Computer Science Research Conferences, Philadelphia, 1996.

[16] M. Litzkow and M. Livny. Experiences with the Condor Distributed Batch System. In *Proceedings of the IEEE Workshop on Experimental Distributed Systems*, Huntsville, Al., 1990.

[17] A. Mukherjee. On the dynamics and significance of low frequency components of Internet load. *Internetworking: Research and Experience*, 5(4):163–205, Dec 1994.

[18] L. Mummert, M. Ebling, and M. Satyanarayanan. Exploiting Weak Connectivity for Mobile File Access. In *Proceedings of the Fifteenth ACM Symposium on Operating System Principles*, December 1995.

[19] M. Ranganathan, A. Acharya, and J. Saltz. Distributed Resource Monitors for Mobile Objects.

In *Proceedings of the Fifth International Workshop on Operating System Support for Object Oriented Systems*, pages 19–23, October 1996.

[20] M. Ranganathan, A. Acharya, and J. Saltz. Adapting to Bandwidth Variations in Wide-Area Data Combination. In *Proceedings of the International Conference on Distributed Computing Systems*, 1998. To appear.

[21] D. Sanghi, A.K. Agrawala, O. Gudmundsson, and B.N. Jain. Experimental Assessment of End-to End Behavior on Internet. Technical Report CS-TR-2909, University of Maryland, June 1992.

[22] M. Satyanarayanan, B. Noble, P. Kumar, and M. Price. Application-aware adaptation for mobile computing. *Operating Systems Review*, 29(1):52–5, Jan 1995.

[23] J. Smith. A Survey of Process Migration Mechanisms. *Operating Systems Review*, 22(3):28–40, July 1988.

[24] J. Stamos and D. Glifford. Implementing Remote Evaluation. *IEEE Transactions on Software Engineering*, 16(7):710–22, July 1990.

[25] J. White. Telescript Technology: Mobile Agents. *http://www.genmagic.com/Telescript/Whitepapers.*

[26] E. Zayas. Attacking the Process Migration Bottleneck. In *Proceedings of the Eleventh ACM Symposium on Operating System Principles*, pages 13–24, November 1987.

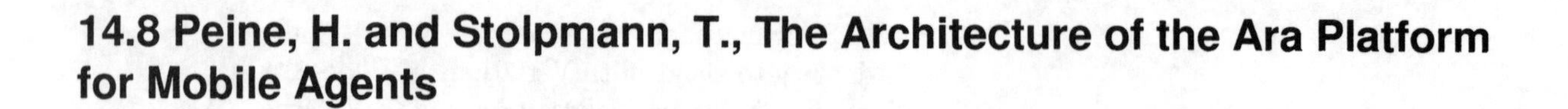

14.8 Peine, H. and Stolpmann, T., The Architecture of the Ara Platform for Mobile Agents

Peine, H. and Stolpmann, T., "The Architecture of the Ara Platform for Mobile Agents," *Proceedings of the First International Workshop on Mobile Agents (MA'97)*, Berline, Springer Verlag, LNCS 1219, pp. 50-61, April 1997.

Published in
Kurt Rothermel, Radu Popescu-Zeletin (Eds.): Proceedings of the
First International Workshop on Mobile Agents, MA'97, April 7-8th 1997, Berlin, Germany.
Lecture Notes in Computer Science Nr. 1219, Springer Verlag 1997. ISBN: 3-540-62803-7

The Architecture of the Ara Platform for Mobile Agents

Holger Peine and Torsten Stolpmann
Dept. of Computer Science
University of Kaiserslautern, Germany
{peine, stolp}@informatik.uni-kl.de

Abstract: We describe a platform for the portable and secure execution of mobile agents written in various interpreted languages on top of a common run-time core. Agents may migrate at any point in their execution, fully preserving their state, and may exchange messages with other agents. One system may contain many virtual places, each establishing a domain of logically related services under a common security policy governing all agents at this place. Agents are equipped with allowances limiting their resource accesses, both globally per agent lifetime and locally per place. We discuss aspects of this architecture and report about ongoing work.

Keywords: migration, multi-language, interpreter, Tcl, C, byte code, Java, persistence, authentication, security domain.

1. Introduction

Mobile agents have raised considerable interest as a new concept for networked computing, and numerous software platforms for various forms of mobile code have recently appeared and are still appearing [CGH95, CMR+96, GRA96, HMD+96, LAN96, LDD95, JRS95, RAS+97, SBH96]. While there seems to have emerged a wide agreement about the general requirements for such systems, most notably portability and security of agent execution, many issues are still debated, as witnessed by the numerous approaches exploring diverging solutions. Prominent issues here include the right balance between necessary functionality and incurred complexity, and the degree of compatibility with existing models, languages, and software.

The Ara[1] system is a mobile agent platform under development at the University of Kaiserslautern. Its design rationale is to *add* mobility to the well-developed world of programming, rather than attempt to build a new realm of "mobile programming". Mobility should be integrated as comfortably and unintrusively as possible with existing programming concepts — algorithms, languages, and programs. A mobile agent in Ara is a program able to move at its own choice and without interfering with its execution, utilizing various established programming languages. Complementing this, the platform provides facilities for access to system resources and agent communication under the characteristic security and portability requirements for mobile agents in heterogeneous networks.

The rest of this paper is structured as follows: The subsequent main section will describe the system architecture of Ara, featuring agent execution, mobility, communication, security, and fault tolerance. This is followed by a section discussing selected

1. "Agents for Remote Action"

aspects of mobile agent architecture. A subsequent section gives an account of the ongoing work with Ara, and the paper closes with a conclusion section. An extensive description of the Ara system will appear in [PEI97].

2. The Ara Architecture

The programming model of Ara consists of agents moving between and staying at places, where they use certain services, provided by the host or other agents, to do their job. A place is physically located on some host machine, and may impose specific security restrictions on the agents staying at that place. Keeping this in mind, agents are programmed much like conventional programs in all other respects, i.e. they work with a file system, user interface and network interface. Corresponding to the rationale stated above, the Ara architecture deliberately abstains both from high-level agent-specific concepts, such as support for intelligent interaction patterns, and from complex distributed services, such as found in distributed operating systems.

2.1 System Core and Interpreters

Portability and security of agent execution are the most fundamental requirements for mobile agent platforms, portability being an issue because mobile agents should be able to move in heterogeneous networks to be really useful, and security being at stake because the agent's host effectively hands over control to a foreign program of basically unknown effect[1]. Most existing platforms, while differing considerably in the realization, use the same basic solution for portability and security: They do not run the agents on the real machine of processor, memory and operating system, but on some virtual one, usually an interpreter and a run-time system, which both hides the details of the host system architecture as well as confines the actions of the agents to that restricted environment.

This is also the approach adopted in Ara: Mobile agents are programmed in some interpreted language and executed within an interpreter for this language, using a special run-time system for agents, called the *core* in Ara terms. The relation between core and interpreters is characteristic here: Isolate the language-specific issues (e.g. how to capture the Tcl-specific state of an agent programmed in the Tcl programming language) in the interpreter, while concentrating all language-independent functionality (e.g. how to capture the general state of an Ara agent and use that for moving the agent) in the core. To support compatibility with existing programming models and software, Ara does not prescribe an agent programming language, but instead provides an interface to attach existing languages. In contrast to most other systems, such as Telescript [GEM95] or Java [ARG96], this separation of concerns makes it possible to employ several interpreters for different programming languages at the same time on

1. There is also the reverse problem of the agent's security against undue actions of the host. There is, however, no general solution for this problem; see section 2.4, "Security" for a discussion of this.

top of the common, generic core, which makes its services, e.g. agent mobility or communication, uniformly available to agents in all languages.

Since part of an agent's execution state is inevitably contained in its interpreter, a given interpreter necessarily has to be extended by state capturing functions if the full transfer of the executing agent is desired. Currently, interpreters for the *Tcl* scripting language as well as for *C/C++*, the latter by means of precompilation to an efficiently interpretable byte code [STO95], have been adapted to the Ara core, opening up a wide spectrum of applications. An adaption of the *Java* language is on the way, and other languages such as Pascal and Lisp are being considered.

The functionality of the system core is kept to the necessary minimum, with higher-level services provided by dedicated server agents. The complete ensemble of agents, interpreters and core runs as a single application process on top of an unmodified host operating system. Fig. 1 shows this relation of agents, core, and interpreters for languages called 𝒜 and ℬ.

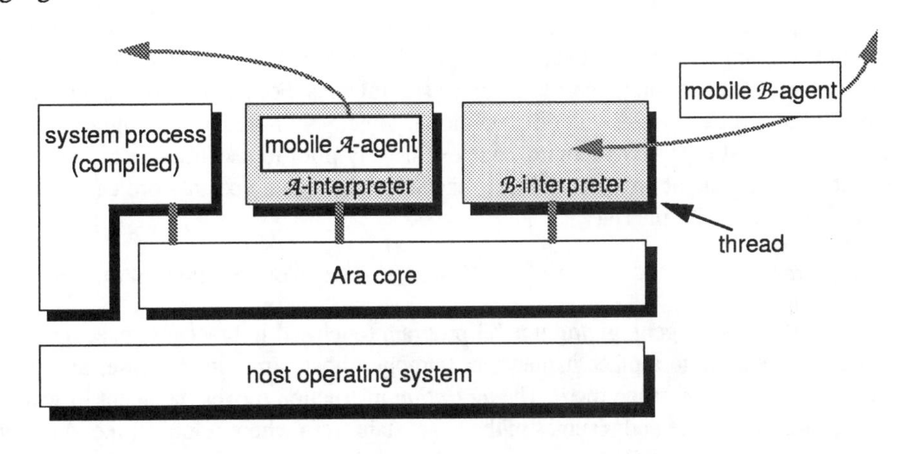

Fig. 1. High-level view of the Ara system architecture

Ara agents are executed as parallel processes, using a fast *thread* package, and are transparently transformed into a portable representation whenever they choose to move. The system also employs processes for certain internal tasks (*"system processes"*) in order to modularize the architecture[1]. Employing threads as opposed to host operating system processes keeps the agent management completely under control of the core and achieves superior performance. The use of multiple threads in a common address space does not induce a memory protection problem here, as protection is already ensured on a higher level by the interpreters (see below), independent of hardware facilities such as privileged processor modes or page protection.

1. If such processes are trusted and not mobile, they may also be compiled to native machine code for undiminished performance.

Adapting a given interpreter for some programming language to the Ara core is a clearly defined procedure. First, it requires the definition of calling interfaces (*stubs*) in this language for the functions of the core API, and conversely the provision of functions for interpreter management (*upcalls*) to the core. The job of the stubs is mostly a matter of data format conversions and similar interface translations. Regarding the interpreter upcalls, the most prominent functions are those for the extraction of an executing interpreter's state as it is necessary to transfer the agent being interpreted, and conversely for the restoration of such a state on arrival of a migrated agent. Further, during execution the interpreter must ensure that the agent program will not call illegal code or access illegal memory locations; interpreters for languages without physical memory access such as Tcl or Java will ensure this anyway. Finally, the interpreter has to assist the core in the preemptive execution of the agent programs by performing regular calls to a core function for time slice surveillance.

2.2 Mobility

Many applications require agents to be moved not only once from their source to a destination site, but to move further, based upon their intermediate results and perceived environment, and continue their task across several sites. For such purposes a moving agent needs to carry its execution state along, effectively making it a migrating process. In contrast to systems moving code exclusively prior to execution, e.g. Java, Ara agents can migrate at any point in their execution through a special core call, named `ara_go` in Ara's Tcl interface[1]:

```
ara_agent {puts "Going to ida"; ara_go ida; puts "Hello at ida!"}
```

This creates a new agent, giving it a Tcl program (enclosed in braces) to execute. The agent will migrate to a place named `ida` (simply a host name, in this case) and then print the greeting message there. The *migration* instruction moves the agent in whole to the indicated place and resumes in the exact state from where it left off, i.e. directly after this instruction, while hiding the complexity of extracting the agent from the local system, marshaling it to another, possibly heterogeneous, machine and reinstalling it there. Furthermore, the act of migration does not affect the agent's flow of execution nor its set of data (including local variables), allowing the programmer to make the agent migrate whenever it seems appropriate, without having to deal with preparation or reinstallation measures.

Note that while the internal state of a moving agent is transferred transparently, this does not hold for its "external state", i.e. its relations to other, stationary system objects and resources like files or communication end points. It might be tempting to add a software layer over such stationary resources making them appear as mobile, effectively creating a distributed operating system. However, Ara opted against this, since the complex protocols and tight coupling involved with this approach do not seem well adapted to the low-bandwidth and heterogeneous networks targeted by mobile agents.

1. The same could be achieved in a C agent a by calling a C function `Ara_Go()` etc.

Ara agents move between *places*, which are both an obvious association to physical location and a concept of the architecture. Places are virtual locations within an Ara system, which in turn is running on a certain machine. In fact an agent is always staying at some place, except when in the process of moving between two of them. In practice, a place might be run by an individual or an organization, presenting its services. Service points (see subsequent section), for instance, are always located at a specific place. More importantly besides structuring, however, places also exercise control over the agents they admit and host (see section 2.4, "Security").

Places have names which make them uniquely identifiable and serve as the destination of a migration. A place name in Ara is, in the most general case, a list of *URLs*, corresponding to the different transport protocols[1] by which the site hosting the designated place might be reached, e.g. MIME mail, HTTP or raw TCP. On migration, the system will try the indicated protocols until one of them succeeds. Apart from designating a protocol and site, place name URLs will contain a local name of meaning to the targeted Ara system only. This local name will identify the specific place, using a simple hierarchical name space.

Agents bear names as well, consisting of a globally unique id, an identification of their principal, and an optional symbolic name from a hierarchical name space (disjoint from the place name space).

2.3 Communication

It can be argued whether agent communication should be remote or restricted to agents on the same machine. Considering that one of the main motivations for mobile agents was to avoid remote communication in the first place, Ara emphasizes local agent interaction. This is not to say that agents should be barred from network access (which depends on the policy of the hosting place, see section 2.4). Rather, the system encourages local communication. There are various options for this, including disk files, more or less structured shared memory areas ("tuple space", "blackboard"), direct message exchange, or special procedure calls, each entailing different ways of access and addressing. For reasons of efficiency and simplicity, Ara chose a variant of message exchange between agents, providing client/server style interaction. The core provides the concept of a *service point* for this. This is a meeting point with a well-known name where agents located at a specific place can interact as clients and servers through an *n:1* exchange of synchronous request and reply messages. Each request is stamped with the name of the client agent, and the server may use that in deciding on the reply.

Service points provide a simple and efficient mechanism for interaction between heterogeneous agents. However, for a widely deployed real-world mobile agent system an integration with existing, more structured service interfaces such as CORBA [OMG96] would certainly be a preferable alternative.

In spite of the emphasis on local interaction, a simple asynchronous remote *messaging*

1. Currently, only raw TCP is supported.

facility between agents will be added for pragmatic reasons, appropriate e.g. for simple status reports, error messages or acknowledgments which do not reward the overhead of sending an agent. However, to avoid remote coupling, the messaging facility will not involve itself in any guarantees against message losses. Messages will be addressed to an agent at a place, named as explained above. A message will be delivered to all agents at the indicated place whose names are subordinates of the indicated recipient name in the sense of the hierarchical agent name space. This addressing scheme may be used to send place-wide multicast messages or implement application-level transparent message forwarding by installing a subordinate proxy agent.

Quite apart from programming the agents' actions, the term "agent language" is sometimes also applied to the language interacting agents, in particular "intelligent" ones, use for mutual communication. However, there is no set of agreed basic functionality for such languages, and it is a current issue of research to find powerful, yet general patterns of agent communication (see [MLF95] for an example). Ara, in particular, leaves the choice of communication language open, offering only a general data exchange mechanism; applications may implement their own customized interaction scheme on top of this.

2.4 Security

The most basic layer of security in the Ara architecture is the memory protection through the interpreters as described in section 2.1. Besides this fundamental and undiscriminating protection, the different places existing on an Ara system play the central role in the Ara security concept. An Ara place establishes a *domain* of logically related services under a common security policy governing all agents at that place.

Allowances to Limit Resource Access

The central function of a place is to decide on the conditions of admission, if at all, of an agent applying to enter. These conditions are expressed in the form of an *allowance* conceded to the agent for the time of its stay at this place. An allowance is a vector of access rights to various system resources, such as files, CPU time, memory, or disk space. The elements of such a vector constitute resource access limits, which may be quantitative (e.g. for CPU time) or qualitative (e.g. for the network domains to where connection is allowed). An agent migrating to a place specifies the allowance it desires for its task there, and the place in turn decides what allowance to actually concede to the applicant and imposes this on the entering agent. The system core will ensure that an agent never oversteps its allowance.

Besides the local allowance conceded by an agent's current place, every agent may also be equipped with a global allowance at the time of creation. The global allowance puts overall limits to an agent's actions throughout its lifetime, effectively limiting its principal's liability. The system core ensures that a place will never concede a local allowance to an agent which exceeds the agent's global one. Agents may inquire about their current global and local allowance at any time, and may transfer amounts of it among each other under certain conditions. Agents may also form groups sharing a common allowance.

Entering a Place

Places may be created dynamically, by specifying a name and an *admission function.* The admission function has a predefined interface, receiving the agent's name and *authentication* status (i.e. the strength, if any, of its authentication), along with its desired local allowance as input parameters, possibly accompanied by further security attributes such as the agent's past itinerary record. The admission function returns either the local allowance to be imposed on this agent, or a denial of admission. Each place may thus implement its own specific security policy, discriminating between individual agents, principals, or source domains, and controlling resource access with the appropriate granularity.

When an agent resumes after a successful migration and admission procedure, it may check its local allowance, discovering to what extent the place has honored its desires. This enables the agent to decide on its own what to do if it finds the conceded local allowance insufficient. An agent which has been denied access to the destination place of a desired migration is sent back to its source place, there to discover the failure in the form of an error return from its migration call.

General resource access restrictions as imposed by allowances are an adequate mechanism for securing common accesses like allocating memory, writing a file or sending to a certain network location. However, certain higher-level security requirements such as enforcing that only data of a specific format are sent, or that consistency conditions across several files are preserved, require correspondingly higher-level access restrictions. This may be achieved by using service points as controlled outlets of the security domain, served by a trusted agent maintaining those high-level requirements. In particular with respect to such outlets, the security domain concept of Ara places is somewhat similar to the *padded cell* security model of Safe-Tcl [OLW96], but realized independent of a specific language and also somewhat more comprehensive, regarding allowances for CPU time and memory consumption.

Open Problems

The implementation of authentication will be based on digital signatures using public key cryptography. However, since a mobile agent usually changes during its itinerary, it cannot be signed in whole by its principal, which makes it difficult to authenticate the changing parts of the agent. It seems most desirable that the agent's code should be signed by the principal; this, however, would preclude dynamically generated code as it is common e.g. in the Tcl programming language. Other security-relevant components of a mobile agent might be authenticated by dedicated schemes, e.g. its itinerary record can be incrementally signed by the nodes the agent has passed through. In addition to authentication, public key cryptography will also be used to optionally *encrypt* Ara agents during migration to protect against eavesdropping.

As with any cryptographic scheme, the question of key distribution must be resolved. As this is a general problem not specific to mobile agents in any way, Ara does not define specific support for this, but assumes the existence of a well-known trusted public key server.

Quite apart from the security of the host system against malfunctioning or malicious agents, which is indispensable for any mobile agent platform to be practically accepted, there is also the reverse problem of the agent's security against undue actions of the host, e.g. spying on the agent's content or modifying it to an harmful effect. It is fortunate that the agent's security requirements are not as severe in practice as those of the host, since there is no general solution for the problem of agent security. The Ara system will provide certain measures, such as protecting immutable parts of the agent (e.g. its code) against tampering by a digital signature of its principal; other threats, however, such as spying on the agent's content, cannot usually be solved by technical means.

2.5 Fault Tolerance

When moving through a large and unreliable network such as the Internet, mobile agents may fall a prey to manifold accidents, e.g. host crashes or line breakdowns. Rather than trying to anticipate all potential pitfalls, Ara offers a basic means of recovery from such accidents: An agent can create a *checkpoint*, i.e. a complete record of its current internal state, at any time in its execution. Checkpoints are stored on some persistent media (usually a disk), and can be used to later restore the agent to its state at the time of checkpointing. The obvious application for this scheme is for an agent to leave a checkpoint behind as a "back-up copy" before undertaking a risky operation. Applications may build their own fault tolerance schemes upon this. The system will, however, provide a facility to implicitly checkpoint all locally existing agents in the event of an emergency shutdown.

3. Discussion

Most of the technical problems involved with mobile agents appear solvable in principle. However, considerable work is still needed to arrive at solutions which strike a satisfactory balance between conflicting requirements, such as necessary functionality vs. incurred complexity, security vs. flexibility and performance, or conceptual purity vs. compatibility with existing models, languages, and software. This section discusses three selected issues of debate and makes a case for Ara's decisions.

Language Integration

Mobile agent systems are often discussed from point of view of programming languages, suggested by prominent examples [GEM95, ARG96]. However, integrating concept and language blurs the differences between both and raises the hurdle for widespread use by requiring new skills and tools and hindering the interoperation with existing software. Analogous experience from distributed programming rather suggests to employ libraries and run-time systems instead of enhanced programming languages, interfaced from whatever languages seem appropriate for the application. Experience has shown here that distribution handling is not intertwined so intimately with the local processing as to require language support very strongly[1], relative to the disadvantages of changing the language. It is remarkable in this respect that even the seminal Telescript system, a typical example of the integration of language and system, has recently been suggested by its creator to play the role of one of several language environments on top of a common platform [WHI96].

Location Transparence

Distributed object systems and distributed operating systems often strive towards the goal of location transparence, i.e. the property of a logical object that its physical location is neither discernible nor important. It might be argued that a mobile agent platform seek such transparency, too, for maximum convenience. However, the wide area networks targeted by mobile agents tend to make distributed objects unwieldy to use. Moreover, there is a conceptual mismatch between location transparency and the principle of mobile agents to explicitly move between locations. Both problems are rooted in the different underlying network assumptions, since hiding distances is only practicable assuming reasonable network bandwidth and reliability; otherwise it seems sensible to admit the distance and deal with it. Accordingly, the distributed functions in a mobile agent system should be kept to a minimum.

Performance

Performance has not been as much in the focus of mobile agent systems as, say, operating systems. This may stem from the idea of an agent performing relatively few and high-level operations, such that its performance is mostly determined by that of the underlying host system. While this may be true for an individual agent, the performance overhead of an agent platform on a server executing hundreds of agents may be crucial. Analogous experience from WWW servers strongly suggests the use of threads instead of operating system processes. Using threads in a common address space allows highly efficient context creation and switching without sacrificing protection, since the latter may conveniently be ensured by the agent interpreters. Moreover, the threads may be scheduled non-preemptively while preserving sufficiently fine-grained preemption semantics from point of view of the agents, achievable by performing time slice checking synchronously in the run-time system (as opposed to an asynchronous interrupt handler). Non-preemptive thread scheduling enables parallelism without synchronization within the run-time system, further benefiting performance.

4. Ongoing Work

Both the core mobile agent platform functionality as well as tools and applications building on top of this are active areas of work. Most components of the Ara platform have been implemented, including the larger part of the core providing agent execution, service points, checkpointing, and migration; the same holds for the Tcl and C interpreters.

1. Parallelism, as opposed to distribution, constitutes an instructive counter example: Parallelizing languages and compilers are well-established in high-performance computing. This can be attributed to the fact that parallelism appears and can be realistically exploited in a more fine-grained form than distribution.

The focus of current work at the platform is on the security implementation. Agents will be able to create places with programmable admission policies implemented by application code; at the moment, however, there is only one implicit default place supported per system. Accordingly, place names currently reduce more or less to machine names[1]. The default place has a fixed behavior; it admits all arriving agents and fully honors their desires for local allowance. Consequently, there is no authentication of agents yet. However, allowance enforcement is implemented, and the set of resources currently controlled by allowances (CPU time and memory consumption) will be enlarged by files, network connections, disk space, bandwidth, and visited places.

Apart from the core system functionality, two other areas of work are tools and applications. We are developing a visual on-line monitoring and control tool for a set of Ara systems distributed across a network, which will include control and debugging of remote agents. As a first application based on mobile agents, we are implementing a service for searching and retrieving Usenet news articles [HOA87], a class of application we consider typical for mobile agents. Usenet is a network of servers exchanging news articles, where each server possesses only a constantly changing subset of all articles. Mobile agents visit servers in search for interesting articles, adapting their search objective and itinerary based on the contents of articles they already found, by means of exploiting meta information in the article headers such as article propagation path or cross references.

5. Conclusion

Ara is a system platform trying to provide mobile processes in heterogeneous networks in an efficient and secure way while retaining as much as possible of established programming models and languages. This paper has laid out the architecture of the system, based on a run-time core, on top of which mobile agents are executed inside interpreters to support portability and security. The system offers a clear interface to adapt interpreters for established programming languages to the core, demonstrated by the adaption of interpreters for such diverse languages as C/C++ and Tcl. Ara offers full migration of agents, i.e. orthogonal to the conventional program execution, which relieves the programmer of all details involved with remote communication and state transfer.

The security model of Ara is flexible in that domains of protected resources can be dynamically created in the form of places, and that the admission of agents to such a domain, as well as their actual rights at that place, can be controlled in a fine grained manner down to individual agents and resources.

However, the described architecture is still lacking in the area of structured agent interoperation. Further, supportive services for distributed resource discovery will be needed for real world applications.

---

1. To be precise, a place name currently designates one (of possibly several) specific Ara systems on a specific machine.

A usable development snapshot of the Ara platform is expected to be available in full source code from the Ara WWW pages[1] by the time of this publication. The system has been ported to the Solaris, SunOS and Linux operating systems so far.

References

[ARG96] ARNOLD, K. and GOSLING, J. (1996) *The Java Programming Language*, Addison-Wesley, Reading (MA), USA.

[CGH95] CHESS, D., GROSOF, B. and HARRISON, C (1995) *Itinerant Agents for Mobile Computing*, Research Report RC-20010, IBM Th. J. Watson Research Center. http://www.research.ibm.com:8080/main-cgi-bin/gunzip_paper.pl?/PS/172.ps.gz

[CMR+96] CONDICT, M., MILOJICIC, D., REYNOLDS, F. and BOLINGER, D. (1996) *Towards a World-Wide Civilization of Objects*, Proc. of the 7th ACM SIGOPS European Workshop, September 9-11th, Connemara, Ireland. http://www.osf.org/RI/DMO/WebOs.ps.

[GEM95] GENERAL MAGIC, Inc. (1995) *The Telescript Language Reference*, Sunnyvale (CA), USA. http://cnn.genmagic.com/Telescript/TDE/TDEDOCS_HTML/telescript.html

[GRA96] GRAY, R. (1996) *Agent-Tcl: A Flexible and Secure Mobile Agent system*, Proc. of the 4th annual Tcl/Tk workshop (ed. by M. Diekhans and M. Roseman), July, Monterey, CA, USA. http://www.cs.dartmouth.edu/~agent/papers/tcl96.ps.Z

[HMD+96] HYLTON, J., MANHEIMER, K., DRAKE, F., WARSAW, B., MASSE, R., and VAN ROSSUM, G. (1996) *Knowbot Programming: System support for mobile agents*, Proceedings of the Fifth IEEE International Workshop on Object Orientation in Operating Systems, Oct. 27-28, Seattle, WA, USA. http://the-tech.mit.edu/~jeremy/iwooos.ps.gz

[HOA87] HORTON, M.R. and ADAMS, R. (1987) *Standard for interchange of USENET messages*, Internet RFC 1036, AT&T Bell Laboratories and Center for Seismic Studies, December. http://ds.internic.net/rfc/rfc1036.txt.

[JRS95] JOHANSEN, D., van RENESSE, R. and SCHNEIDER, F. B. (1995) *An Introduction to the TACOMA Distributed System*, Technical Report 95-23, Dept. of Computer Science, University of Tromsø, Norway. http://www.cs.uit.no/Lokalt/Rapporter/Reports/9523.html.

[LAN96] LANGE, D. (1996) *Programming Mobile Agents in Java - A White Paper*, IBM Corp. http://www.ibm.co.jp/trl/aglets/whitepaper.htm

1. http://www.uni-kl.de/AG-Nehmer/Ara/

[LDD95] LINGNAU, A. DROBNIK, O. and DÖMEL, P. (1995) *An HTTP-based Infrastructure for Mobile Agents,* Proc. of the 4th International WWW Conference, December, Boston (MA), USA. http://www.w3.org/pub/Conferences/WWW4/Papers/150/.

[MLF95] MAYFIELD, J., LABROU, Y. and FININ, T. (1995) *Desiderata for Agent Communication Languages,* Proc. of the AAAI Symposium on Information Gathering from Heterogeneous, Distributed Environments, AAAI-95 Spring Symposium, Stanford University, Stanford (CA). March 27-29, 1995. http://www.cs.umbc.edu/kqml/papers/desiderata-acl/root.html.

[OMG96] OBJECT MANAGEMENT GROUP (1996) *CORBA 2.0 specification*, OMG document ptc/96-03-04, http://www.omg.org/docs/ptc/96-03-04.ps.

[OLW96] OUSTERHOUT, J. K., LEVY, J., and WELCH, B. (1996) *The Safe-Tcl Security Model*, draft, Sun Microsystems Labs, Mountain View, CA, USA. http://www.sunlabs.com/research/tcl/safeTcl.ps

[PEI97] PEINE, H. (1997) *Ara – Agents for Remote Action*, in *Itinerant Agents: Explanations and Examples with CD-ROM*, ed. by W. Cockayne and M. Zyda, Manning/Prentice Hall. To appear[1].

[RAS+97] RANGANATHAN, M., ACHARYA, A., SHARMA, S., and SALTZ, J. (1997) *Network-Aware Mobile Programs*, Dept. of Computer Science, University of Maryland, MD, USA. To appear in USENIX'97. http://www.cs.umd.edu/~acha/papers/usenix97-submitted.html

[SBH96] STRASSER, M., BAUMANN, J. and HOHL, F. (1996) *Mole – A Java Based Mobile Agent System*, Proc. of the 2nd ECOOP Workshop on Mobile Object Systems, University of Linz, Austria, July 8-9. http://www.informatik.uni-stuttgart.de/ipvr/vs/Publications/1996-strasser-01.ps.gz

[STO95] STOLPMANN, T. (1995) MACE - *Eine abstrakte Maschine als Basis mobiler Anwendungen*, diploma thesis, Department of Computer Science, University of Kaiserslautern, Germany. German text and English abstract at http://www.uni-kl.de/AG-Nehmer/Ara/mace.html.

[WHI96] WHITE, J. (1996) *A Common Agent Platform*, position paper for the Joint WWW Consortium / OMG Workshop on Distributed Objects and Mobile Code, June 24-25, Boston, MA, USA. http://www.genmagic.com/internet/cap/w3c-paper.htm.

1. A preprint can be obtained from the author (see cover page of this paper for address).

14.9 Milojičić, D.S., Chauhan, D., LaForge, W., Mobile Objects and Agents (MOA)

Milojicic, D.S., Chauhan, D., and laForge, W., "Mobile Objects and Agents (MOA), Design, Implementation and Lessons Learned," *Proceedings of the 4th USENIX Conference on Object-Oriented Technologies (COOTS)*, pp. 179-194, April 1998. Also appeared in *IEE Distributed Systems Engineering*, 5:1-14, 1998.

Mobile Objects and Agents (MOA)

Dejan S. Milojičić, William LaForge, and Deepika Chauhan

The Open Group Research Institute

[dejan, laforge, dchauhan]@opengroup.org

Abstract

This paper describes the design and implementation of the Mobile Objects and Agents (MOA) project at the Open Group Research Institute. MOA was designed to support migration, communication and control of agents. It was implemented on top of the Java Virtual Machine, without any modifications to it. The initial project goals were to support communication across agent migration, as a means for collaborative work; and to provide extensive resource control, as a basic support for countering denial of service attacks. In the course of the project we added two further goals: compliance with the Java Beans component model which provides for additional configurability and customization of agent system and agent applications; and interoperability which allows cooperation with other agent systems.

This paper analyzes the architecture of MOA, in particular the support for mobility, naming and locating, communication, and resource management. Object and component models of MOA are discussed and some implementation details described. We summarize the lessons learned while developing and implementing MOA and compare it to related work.

1. Introduction

Mobility has always attracted researchers in computer science. This interest spans from general observations, such as *"if it weren't for mobility, we would still be trees"* [10], and the analogies with the real world *"migrating birds and nomadic tribes moving due to the lack of resources"*, to purely technical reasons, such as improving locality of reference and difference between local and remote semantics.

One of the first incarnations of software mobile entities. is worms [30], which could spread across nodes and arbitrarily clone. Unrestricted implementations of worms and viruses have received negative connotations, due to security breaches and denial of service attacks [13].

The next generation of mobile entities, known as process migration, were implemented at the operating system (OS) level. There were many implementations of process and object migration [3, 12, 19, 31], but none has achieved wide acceptance. Due to inherent complexity, it was hard to introduce process migration without impacting the stability and robustness of the underlying OS.

Mobile objects and agents have attracted significant attention recently. In addition to mobile code (such as applets), agents consist of data and non-transient system state that can travel between the nodes in a distributed system (intranet or Internet). Compared to mobile objects, mobile agents also represent someone; they can perform autonomous actions on behalf of a user or another agent. A number of academic systems (such as Agent Tcl [20], Mole [4], Ara [27] and Tacoma [18]) and industrial systems (such as Telescript [34], Aglets [1], Concordia [9] and Voyager [33]) exist. The products using mobile agents have started to appear, such as Guideware [16]. The government is interested in funding work on agents [11]. A patent has been approved on mobile agents [35]. A standard has been adopted (OMG MASIF [26]), and reference implementations are in progress. A couple of books have been published on agents [6, 8] and a few more are in progress [21, 24].

This paper describes the Mobile Objects and Agents (MOA) project at the Open Group Research Institute. The obvious question is why yet another mobile agent system? There were a few reasons. None of the existing systems at the time of starting the project were mature enough to be used as a starting point for our work. We found it easier to develop another system that would suit our needs from the beginning. Additionally, some areas of our interest, such as communication and resource control, are deeply involved in the design decisions of any system, making it very hard to add them as an afterthought. Finally, we were interested in interoperability between the systems, and therefore supporting another implementation was a good idea.

At the beginning of the project we were interested in the first two of the four features listed below, and during the course of development we added the last two:

Collaboration. Frequently, agents need to collaborate during their execution either with other agents or their

This work was supported in part by the Advanced Research Projects Agency and the Rome Laboratory of the Air Force Materiel Command.

user. For agent collaboration, it is required to support naming, locating and communication among agents.

Denial of service attacks. Agents, as well as hosts, are vulnerable to mutual attacks, either over a network or locally. In order to prevent denial of service attacks, it is required to maintain resource control of agents and agent systems, and to impose security and resource policies.

Configurability and customization. It is increasingly difficult to configure and customize software. In the case of mobile agents, this applies both to agent applications, as well as to agent systems. Being compliant with a component model, such as Java Beans, allows for a standardized way to access and change component properties.

Interoperability. Agents, as well as agent systems, need to interoperate. In the case of agent systems, interoperability leads to a larger base that agents can visit. We were active in the OMG Mobile Agent Facility proposal which addresses mobile agents systems interoperability [26].

More details on how these goals have been achieved is described in Sections 4.3, 4.6, 4.2 and 4.10 respectively.

The rest of this paper is organized as follows. In Section 2 we provide a background on mobile agents and component-based computing. Section 3 describes Java's suitability for mobile agents and for component-based computing. Section 4 presents the MOA design and implementation. Section 5 discusses MOA current status. Section 6 describes some MOA applications. In Section 7 we present lessons learned while designing and implementing MOA. MOA is compared to related work in Section 8. Finally conclusions and future work are presented in Section 9.

2. Background

In this section, we provide background on mobile agents and component-based computing.

2.1 Mobile Agents

Among the benefits of mobile agents we would like to underline the following. **Improving locality of reference** is achieved by moving the action towards the source of data or other end point of communication, resulting in substantial performance improvement. **Survivability:** similar to nomadic tribes or migratory birds, agents can survive if moved closer to resources, or away from partially failed nodes. **Analogy to the real world** helps some programmers to better understand programming paradigms expressed in terms of mobile agents. Examples are travelling salesman, shoppers and workflow management systems. **Customization** of software can be achieved using mobile agents, for example, by adjusting the search according to a user-specific criteria, or by performing an action specific to a remote site. **Autonomicity** represents agent's independence from its owner. A user can start an agent to act on his behalf and disconnect. When the user reconnects, the agent returns or otherwise provides results.

Agents have various areas of deployment. One is **slow and unreliable links**, such as radio communication, where locality of reference improves performance, and avoids potential loss while transferring large amounts of data. **Software distribution** becomes increasingly hard. Mobile code has provided a revolutionary breakthrough, by allowing downloading code for heterogeneous environment. Mobile agents makes this effort even easier, by associating actions and state with each distributed version and copy of a particular software. **Network management**: agents migrate both code and data, making them useful for automating control and configuration in large scale environments, such as networks [15]. **Electronic commerce** deploys mobile agents by modeling travelling salesmen or shoppers visiting stores in an electronic mall. **Data mining** is a convenient application for mobile agents due to locality of reference: agents optimize a search by wandering from site to site with large volumes of information. (See [7] for additional benefits.)

Nevertheless, mobile agents still haven't achieved wide acceptance. Some of the reasons include the following. **Lack of applications**: mobile agents have achieved a reputation of "the solution searching for the problem". Many systems have been developed but few applications exist. **Security:** the problems caused by mobile code are frequently reported. Mobile agents push the security problems even further. **Lack of infrastructure** adapted for mobile agents, such as name servers, messaging systems, and management, is still not widely deployed. **Survivability** is both a benefit and a challenge for mobile agents. Mobile agents are inherently survivable, but this does not come free; they need to be designed and implemented for survivability. In particular, they should minimize residual dependencies on previously visited nodes, or servers.

2.2 Component-Based Computing

Component-based programming, including OpenDoc, VBX, and ActiveX, has been quite successful in speeding the development of GUI applications. Java Beans (components written in Java), are promising for non-GUI component programming. The runtime behavior of a Java Bean is defined by an ordinary Java class. The difference between a bean and other objects is the metadata used for configuration. It is provided by an associated BeanInfo class, or it is derived from the runtime class.

Component-based programming enhances object oriented benefits, such as flexibility, and code reuse with two new characteristics: independence and configuration.

Independence. The source code defining a component does not directly reference any other component; instead, relationships between components are created at runtime. The relationships may be established by the container holding the components, or even by the component itself. This has several benefits:

- a "building block" approach: programs are constructed from existing components by defining relationships
- each component can be individually tested
- components are more easily reused; there is a minimum of interdependency between components
- a program can be restructured for new requirements without impacting the logic of individual components
- updating a program with the latest version of 3rd party components is simplified

Configuration. A component is constructed by a general configuration tool. The component participates in its own configuration. Application programs are assembled from pre-configured components. The implementation specifics of a component are separated from other elements of the program. Separating the configuration of components from an application program facilitates the use of alternative implementations and component upgrades are backward compatible. However, this impacts the development cycle, as changes made to a component's source code will often invalidate its configuration. The edit, compile, and test of the development cycle now becomes edit, compile, configure, and test.

3. Java

We have chosen Java because it seemed to be the mainstream programming language, but also because of the features that make it suitable for mobile agents and components.

3.1 Java and Mobile Agents

Java offers advantages for mobile agents, as well as some disadvantages. Advantages consist of the support for mobile code, heterogeneity, language safety, object serialization [28], reflection, dynamic class loading, and multithreading.

Disadvantages consist of inadequate support for resource management (e.g. memory and disk limits), no support for preserving the thread execution context, limited support for versioning, no ownership of objects and fine grained protection at the object granularity [21].

Figure 1. MOA Configurations: *front-end supports starting and controlling agents and other MOA components. Home node is the node where agent was originally started and where agent-related state is maintained. Remote node is one of the nodes where agent currently executes.*

3.2 Java and Components

Components written in Java are lightweight and little code is required for conformance to the component model. Java supports a number of key features of component programming:

- A component may have several interfaces. Java provides for the implementation of multiple interfaces, unrestricted casting, and an instanceof operator to determine if a component supports an interface.
- Several components may be aggregated into a single component. The JDK 1.1 methods Beans.isInstanceOf and Beans.getInstanceOf can be used in place of the instanceof operator and casting, allowing for the future use of aggregation in JDK 1.2.
- The life of a component may span more than one program. JDK's provision for serializing components allows converting an object into a form which can be written to a disk file or passed across a network.
- A component is configured by modifying its properties identified by examining the method signatures of the component, e.g. the class is recognized as having the property slices, if the methods getSlices and setSlices exist.
- The component's properties are accessed using the class Introspector.

4. MOA Design and Implementation

The MOA architecture is presented in Figure 1. There are three types of nodes that run MOA system: front-end node allows users to control and monitor agents; home node is used as a repository for agent's data; and remote node is where agents typically run throughout their life time. The MOA system has a Telescript-like model (although sufficient difference avoids infringing their patent). Agents travel visiting places held by Agent Environments (AE). Places accept agents, and store information. Agent environments host various objects. A name server tracks the location of agents and other objects, whereas a monitor serves for controlling and monitoring objects. These and other objects in the MOA architecture are described in more detail in this section.

4.1 Object Model

The MOA objects on remote nodes can be classified as agent- or system-related (see Figure 2). Agent-related objects are circled; they have migratory state. Agent and place belong to the user trust domain (see Section 4.7 on more details related to security), whereas other components belong to the MOA trust domain.

An agent and place are application extended classes. **Agent** is the first class MOA object. It is a template class extended by agent applications. Agents are named (see Section 4.4), and they can communicate (see Section 4.3). An agent can *move* to an agent environment (or a place within it), it can request to *meet* other agents at a certain place or agent environment, *openChannel* to another agent, or *sendMessages* to it. An agent always executes within a place (see below). There is a one-to-one mapping between an agent and a place within an agent system. However, an agent can leave places behind when it moves. Therefore there is one-to-many mapping between an agent and places on different agent systems.

Place is the second class MOA object. The main difference from agent is that place is a stationary object, and therefore it can not *move* or *meet*. However, places can communicate with other places and agents; they can be active, i.e. they can have threads running. Place also serves the container-proxy role. They are proxies because they can remain after an agent leaves and represent it there (be proxy). They serve the container role for security and resources of an agent.

Agent Control, Agent Properties, Family, LogService and Bucket are agent-related classes that belong to the MOA trust domain. **Agent control** is an internal class that represents an interface between agent system and agent/place. It manages agent system resources (communication channels, agent properties, etc.). These objects can be accessed by an agent/place, but they cannot be changed. **AgentProperties** contains the properties that characterize agents and are transferred across the nodes. Examples are owner, home Agent Environment, and locatingStrategy. **Family** is used for monitoring agent activities. Each agent has its own Family object which it carries across migrations. **LogService** manages local logs. **Bucket** holds the contents of a JAR file (JAva aRchive). Each bucket implements a classloader for the dynamic loading of the jar file. A hashtable contained within a bucket enables the client of a bucket to efficiently index into the contents of a jar file. During migration,

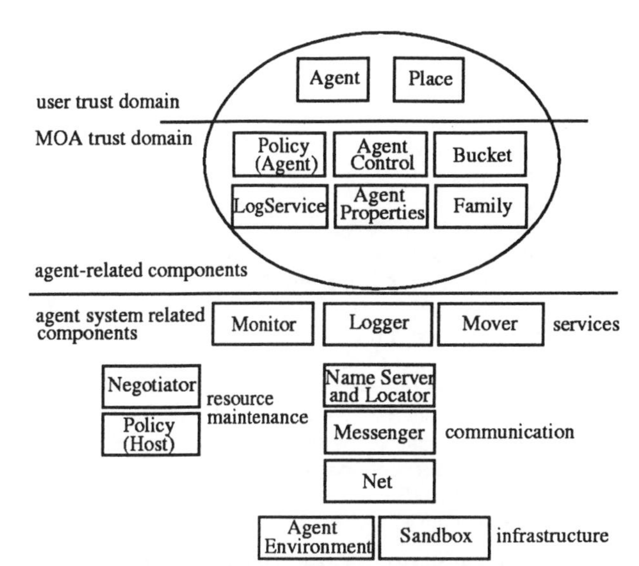

Figure 2. MOA Objects: *consist of objects in user trust domain, and MOA objects in MOA trust domain. Agent environment represents container for all MOA objects.*

components circled in Figure 2 (except for Place and Agent Control) are serialized, put into the bucket and sent to destination node.

Policy and Negotiator objects maintain and manage information about resources. **Policy** is a placeholder for properties describing the policy of an agent arriving at a node (agentPolicy), and a host receiving the agent (hostPolicy), such as agent's maximum lifetime, maximum number of channels and maximum threads. **Negotiator** performs negotiation between the agent and the receiving agent environment prior to agent's visit. Agent movement is subject to resource requirements and security arrangements between the two entities.

Sandbox and Agent Environment provide basic infrastructure. **Sandbox** class separates the agent application from the agent system state. It switches from the agent system thread to application thread when there is a different protection domain; it also serves to switch from synchronous to asynchronous communication when going across the network. Resource usage and limits are tracked on a per sandbox basis. **AgentEnvironment** (AE) is the container for agents and their related objects at an agent system. There is one agent environment per Java Virtual Machine (JVM), but there can be many per a node. Each agent has its home AE and alternate home AE in case the home is not accessible.

Net, Messenger and Name server comprise the MOA communication model. **Net** provides the basic communication support for establishing and maintaining communication channels between components on remote and local JVMs. Communication channels are established by specifying the component name, host and port number.

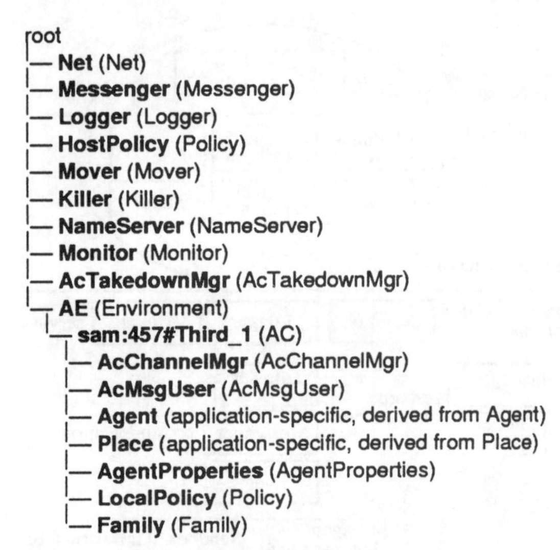

Figure 3. Agent System Object Tree *defines the names of MOA internal objects. Parenthesis contain the class names from which the objects are derived. Internal object names are important when communication between various objects on different MOA systems is established, and for initialization.*

The **Messenger** layer uses the services of Net to support one- and two-way messages between components. Components are addressed using destination agent system and the component name. **Name server** tracks agent locations. The name server clients can *(un)register* and *lookup* an agent location. Name server clients are user (monitor) and agents that *lookup* locations of other agents in order to communicate. Information about the agent location (or how to find it) is cached at agent systems that the agent visited or communicated with. Name server also plays the **Locator** role.

Mover, Monitor and Logger provide MOA services. **Mover** supports agent movement. It negotiates migration, captures the agent state, and transfers it. **Monitor** provides a user interface to control and monitor applications (e.g. agent's movement, communication and resource usage). **Logger** logs events in an Agent System to persistent media.

MoaApplet, BatchDriver and User classes support the interface to the MOA system and its applications. **MoaApplet** is an applet-based interface enabling users to interactively monitor and debug agents, to launch them, to snapshot the agent's state, and to query the logged data. **BatchDriver** is a script-based user interface to provide the services of MoaApplet; typically it is used for testing purposes. **User** class serves as an interface between the user applet and the agent. It launches the preconfigured agents, tracks the agents, and maintains information of interest to a user.

```
class sandbox.environment.EnvironmentChild
  class moa.ac.AcTakedownMgr
  class sandbox.environment.Environment
    class moa.ac.AC
  class sandbox.environment.Killer
  class moa.ac.LogService
  class sandbox.message.MessageUser
    class moa.nameserver.LocatorUser
      class moa.ac.AcMsgUser
      class moa.nameserver.NameServer
        class moa.nameserver.HNameServer
      class moa.user.User
    class moa.logger.Logger
    class moa.monitor.Monitor
  class moa.mover.Mover
  class sandbox.net.Net
  class sandbox.net.NetUser
    class moa.ac.AcChannelMgr
    class sandbox.message.ChannelManager
      class sandbox.message.Messenger
  class moa.api.Service
```

Figure 4. Inheritance tree of EnvironmentChild Class: *the EnvironmentChild class supports accessing other components within the object tree (see Figure 3). For example, the Mover needs access to NameServer and Messenger and therefore has to be wired. Inheritance tree indicates objects accessible by inheritance, for example, Mover has access to Messenger by being a subclass of MessageUser, and needs not be wired.*

4.2 Component Model

MOA components are configured using the MOAbatch tool (see Section 5). The components configuration defines the system object tree. The MOA system is loaded by first loading the root component. Each component then successively loads the component below it in the tree. Components are locally organized into a labeled tree (see Figure 3) used to dynamically establish relationship between the components, in contrast to static binding, typical of OO programming.

After the components have been loaded, they can locate other components of the agent system by name in order to establish dynamic binding. Non-leaf elements of the object tree subclass the environmentChild class (see Figure 4) which provides methods for locating a tree element given a relative or full pathname. For example, any component in Figure 4 can access the Mover object by calling the method:

```
getEnvironment().getInstanceOf("/Mover",Mover.class)
```

Coupling of components to a certain extent negates the benefits of component programming, and as such it has been kept to a minimum. For example, the net component is made known to other components, such as those that subclass NetUser: AC, Messenger and arbitrary applications using Net.

We have successfully used components for preconfiguring agent applications, as well as the agent system. For agent applications, we can easily preconfigure an agent's itinerary, policy, types of logging, debugging, etc. Agent system configuration specifies which components will

Debug (List of components being debugged)
AE (agent application components)
Messenger (netBean,timeout)
Monitor (messengerBean,timeout)
Mover (messengerBean,timeout)
Net (retryDelays(Increment,max,init),runSrv,host,port,exitSrvOnError)
HomeService (nameSrvBean,messengerBean)
Logger (fileSuffix,logPath,retentionPeriod,messengerBean,timeOut)
NameServer (nameSrvBean,messengerBean)
Users (userPath,nameSrvBean,passwd,messengerBean,HSretryDelay)
UE (listOfUserBeans)
HostPolicy (maxLifeTime,timeRemain,maxChnl,remainPlaces,maxThrd)
Killer (killerTimeout)
Root (AE,debugger,net,messenger,logger,hostPolicy,mover,nameSrv,
UE,homeServer,monitor,ACTakeDownMgr,killer)
ACChannelManager (timeout, netBean)
AcMsgUser (nameSrvBean,messengerBean,timeout)
AC (AcChannelMgr,AcMsgUser)

Figure 5. MOA Components, *described with the list of properties (configuration of the remote MOA system).*

be integrated into the tree. For example, homeService is not present on all agent systems, and agent environment is not needed on the front-end and on the home node. Debugging can be specified as a part of the agent system configuration at class granularity. Message timeouts can be configured on a component basis. For the Net component we specify its port number; for each agent, we specify the number of service threads for each agent; Killer component's property includes time when it will take the system down; for each user, we specify the password, login id and login time; for each system, we specify the host policy for negotiating with agents.

Components can also be organized using subcomponents. For example, AC is an aggregate which includes AcChannelMgr and AcMsgUser. Only AC is configured into the whole system. This way configuration is simplified using a hierarchical structure. The MOA components of a remote system and their properties are presented in Figure 5.

4.3 Communication

The MOA communication is built on top of JVM sockets. It provides a higher level of abstraction, such as the channels and messaging between MOA objects (agents, places and servers). The communication channels support object streams. Messaging provides for passing objects of arbitrary type specific to the application.

The Net package supports opening of channels with automated retry. The destination system can optionally reject a request for the channel subject to resource limitations. This can happen at the agent system, as well as at the application level. When opening a channel to an agent, only the agent name needs to be specified. The agent system resolves the actual agent location with the help of Locator object in a distributed manner.

The Message package is able to pass application specific objects by delaying deserialization until the name space of destination is identified. Messages can be synchronous (RPC-like), or asynchronous (one-way messages).

Locator (AcMsgUser, NameServer, User)
Messenger (LocatorUser, Monitor, Mover, Logger)
Net (AcChannelMgr, ChannelMgr)

Figure 6. Stacked Communications Layers: *Net supports stream based communication, messenger messaging and locator introduces transparent locating of migratory objects.*

Messaging is built on top of the Net package, using a common pool of channels dedicated to message passing. The destination of messages can either be a specific location (destination name, host and port), or a logical name (e.g. agent name) in the case where the destination is a migrateable object. This is reflected in the implementation, where a layered approach is applied by building the Locator on top of the Message layer which builds on top of the Net (see Figure 6). The Locator handles transparent routing of messages when a location is not specified. It enables the agents tc transparently communicate and collaborate with each other by using the name of the agent. The Locator relies on the locating strategies described in Section 4.4 to find new agent location.

In the case of two way messages, responses are routed back to the originating thread, which is suspended pending either a response or a time out. A response can arrive from the node other than the original destination, if the destination agent moved.

While moving from one node to another, the agent does not notice that its channels have been closed and reopened on the remote node. When the transfer is initiated, the channel migration process is performed first. During this process, the agent informs its collaborating partners of its intent. From this point onwards, the data received on the channels is not passed to the application, but is stored in a Vector of unread Objects. Upon learning about the move, the agent channel manager on the other end of the channel replies back an acknowledgment and closes the channel socket without informing the application. For the migrating agent, when the acknowledgment is received on a channel, the Reader thread for the channel is stopped, and the socket is closed. The agent transports itself to a new location, only when the migration process is completed for all open channels.

During channel migration, though the socket is closed, the information regarding the other communicating agents/places is still maintained. When the agent moves, it carries along the information about the channels, and uses it to reopen channels at the new location. Prior to reopening channels it first sends all the unread objects to

AEname *(ae)*:	$h{:}p$	*h* - host name
AgentName *(a)*:	$ae_{home}\#f\_l.g$	*p* - port number
PlaceName:	$a_{owner}\%ae_{residing}$	*f* - family name *l* -launch number
ServerName:	$h{:}p$	*g* - generation number

Figure 7. MOA Named Objects: *Agent Environment, Agent, Place and Server*

the application. Reopening of channels can be done either eagerly or lazily, depending on the type of the application. If there are many channels not used often, they are re-opened lazily. If there are a few channels likely to be used after migration, then they are reopened eagerly.

MOA uses remote method invocation within a number of its components. It would have been a reasonable choice to rely on Javasoft's RMI [36] instead, but we have not made such a decision. There is nothing to prevent us from using it in the future (and most likely we shall switch to using it), but for the initial implementation we did not want to adopt yet another technology (in addition to Java Beans) that would introduce a learning curve. Furthermore the unclear situation with CORBA v. RMI was another contributing factor. Instead of using RMI, we simulate its functionality by sending objects across the network; based on the object type we invoke appropriate methods. The actual implementation is trivial. There are a small number of uses of remote method invocation and these are confined to limited scope of the MOA implementation, whereas MOA applications can use RMI. Overall, it was more a political- rather than a technical-decision not to use RMI for the initial implementation.

Communication and resource management are deeply involved in the design decisions in MOA. This is reflected in many MOA layers and components. For example, communication is involved in the communication stack (Net, Message, NameServer and Locator), but also in the sandbox, AC, and agent/place interfaces. Similar applies to resource management. The AC and sandbox components were shaped to enable resource tracking. It would have been hard to add this support as an afterthought. Our earlier experience with Mach task migration [23] indicates that to a certain extent it is possible to add resource management or to make communication modifications as an afterthought, however, any significant support needs to be well elaborated in advance.

4.4 Naming and Locating

The following objects are named in a MOA system: agent environment, agent, place and servers. Name syntax is presented in Figure 7. An agent environment is named after the node and port on which they perform communication. A server name has similar syntax.

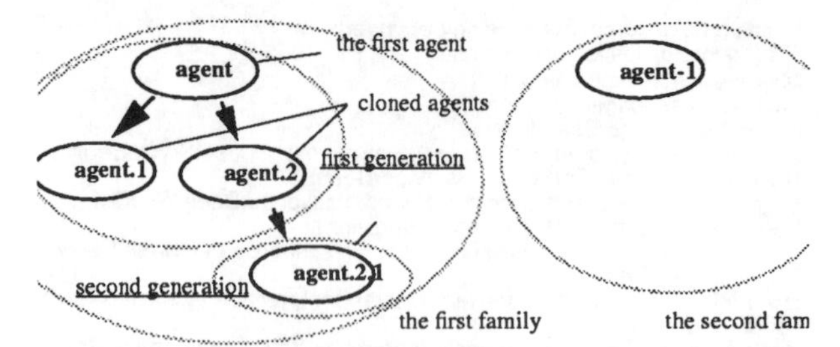

Figure 8. Naming of agents: *agent names are organized around agent families and generations. Cloned agents always carry the name of its ancestor as a part of their name.*

Agents are named after the AE where they were created if they are first generation. If cloned, agents are named after the AE of their first ancestor extended with the generation number (see Figure 8), irrespective of the AE where they were actually cloned. In other words, each agent bears the sign of the original site responsible for initiating the agent family. This is used as an ultimate source of information on the current agent location: as a last resort, the home AE can be queried for current location information. The place name consists of the name of an agent it belongs to, extended with the AE name where place currently resides.

The agent location may be needed by its owner explicitly in order to track the agent location, or implicitly in order to be able to control it (*kill, suspend, resume*, etc.). It is also needed by agents in order to be able to communicate to (open channels) or synchronize (propose meeting) with other agents.

Locating agents is performed through name servers. Application requests to name servers (*lookup, register, unregister*) are issued through the agent environment which performs security and consistency verifications. When migrating the agent, the Mover object makes a local request to the Name Server. When opening channels or sending messages, NetUser and Messenger interact with the Name Server. Name servers on multiple nodes then cooperate to satisfy these requests.

The location object contains either the current location of an agent, or sufficient information to obtain it. In particular, it contains some or all of the following: the name of the residing agent system (if known), the type of the strategy to locate the agent (discussed below), the list of the nodes where the agent may reside (*itinerary*), the lifetime of this location object. Even though the lifetime of an agent is limited by its owner, because of the delays in transferring the agent over network, it is not possible to assure its accuracy.

The location object is cached at each node the agent visited, or where there was a channel opened with the agent. When searching for an agent, the location object is first looked up at a local name server. If not found, it is looked up at name servers higher in the hierarchy, if any exists (name servers may be organized in a tree-like hierarchy). If the agent is still not found, then the agent's home node is approached. The home agent environment is the ultimate source of information of agent location. The location of the home agent environment is implicitly known from the agent's name.

When locating agents, different location schemes are used, similarly to those used in distributed operating systems, such as Charlotte [3], V kernel [31], Sprite [12] and Mach [23]. The MOA system supports: a) **updating** the home after agent moves, b) **registering** at a predefined name server, c) **searching** based on predefined itinerary and d) **forwarding** based on the trails left after migration (see Figure 9). Locating scheme is selected subject to:

- destinations (a local or a far away region; limited set of destination hosts or unknown),
- security aspects (if agent crosses security domain)
- type of migration (burst v. sporadic; frequent v. rare; random v. cyclic).

For example, updating the home node is suitable for an agent that moves within a local region. It is not suitable for agents that visit distant nodes. Registering is suitable for an agent that migrates within a far away region; in the case of a large number of nodes, registering nodes are organized in a hierarchical manner; it is not suitable for a large number of migrations. Searching scheme is suitable for agents that visit a small number of known hosts; it is not suitable for destinations not known in advance and for large number of nodes visited. Forwarding is suitable for a small number of migrations; it is not appropriate for long chains.

In many cases, locating agents is application specific. Even if the agent is successfully located, it might migrate further away by the time the location is reported back to the requesting node and communication or delivery of a control message is attempted. This is especially critical in cases of control messages such as *meet*, *suspend* or *kill*. It is required to perform optimizations, such as to batch a locating request with the control message. This eliminates the delay between the time the agent is located and the control message is delivered to the visiting agent system. This may not be sufficient for highly dynamic agent applications or heavy loaded nodes in the case of forwarding locating strategy. Instead, the updating and/or registering strategy needs to be used, combined with trapping the agent when registering/updating its location. This way agent movement is delayed until communication/control messages are delivered.

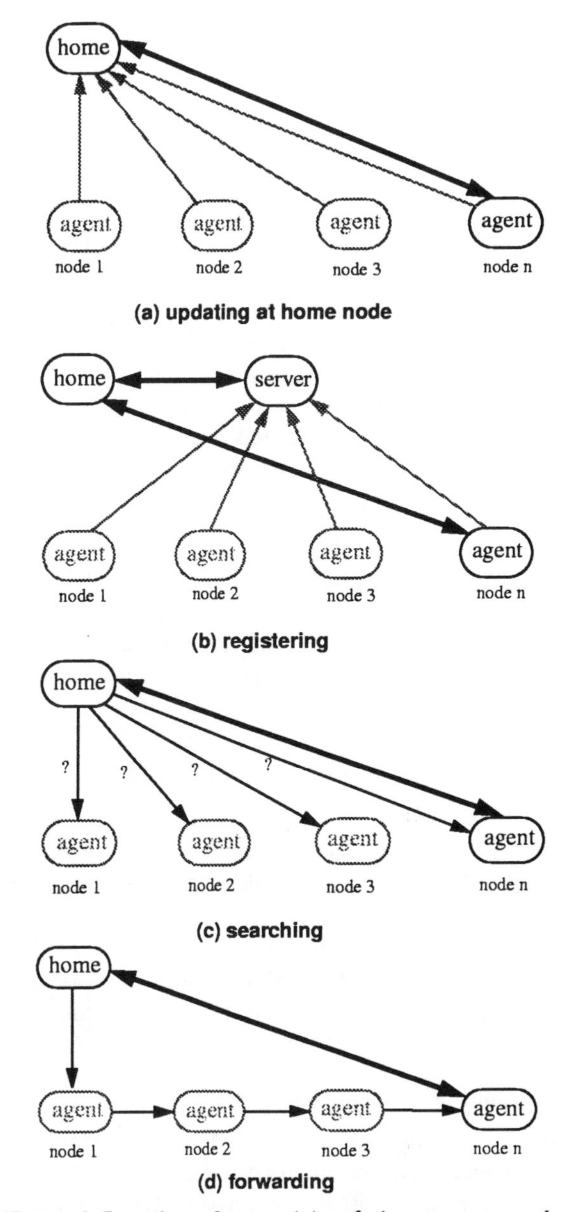

Figure 9. Locating schemes: *(a)* ***updating****: an agent updates its location with the home node name server; (b)* ***registering****: agent registers at a predefined name server; (c)* ***searching****: based on available itinerary, the sites are searched for agent's location; (d)* ***forwarding*** *based on trails agent left behind.*

4.5 Mobility

When an agent migrates, its state is extracted from the source agent system and transferred to its destination where it is restored into a new instance of an agent object. During transfer, only site-independent information is transferred. In the case of communication channels, this information consists of the agent names with which the migrating agent had opened channels as well as their current location. The state relevant to each particular node is transient, i.e., it is discarded. For example the sockets maintained in the agent control object are closed

AgentProperties (owner,familyName,home/alternateHomeAE,lifeTime)
Agentpolicy (maxLifeTime,timeRemaining,maxChannel,
remainingPlaces,maxThreads)
Family (typesOfLog,tracing,watches,limits)
Agent (applicationSpecific)
Place (applicationSpecific)

Figure 10. MOA agent applications Component Model: *in parenthesis we present configurable properties.*

and then reopened in the remote agent control object. Figure 10 describes the transferred agent state. The state extraction starts at the application level, where the application state is serialized (non-transient data), then the state of the agent control is serialized (agent resources, such as agent limits and logging data). This state is then transferred to the remote node through the cooperation of Mover objects in the source and destination agent system. The Mover objects involve negotiation based on the agentPolicy and the destination node hostPolicy.

Mobility is based on messaging, where the message object is the bucket containing the agent and related resources. When an agent arrives at a node, the Mover creates a new instance of the AC object (unless there already is one for that agent - the agent is returning to a place it left). All other agent-related objects are instantiated from the serialized versions in the bucket. Objects are loaded using the class loader associated with the bucket.

We do not provide for sharing of objects remotely, i.e. as an agent migrates to another node, it should not maintain any references to an object on the source node. Our experience is that distributed shared state is very hard to support at the system level [5], it is more appropriate to rely on distributed shared memory packages for such needs.

4.6 Resource Management

One of the initial goals of the MOA project was to support extensive resource control of various MOA resources. The following limits are enforced on MOA resources:

- agent: lifetime, places, hops, open channels, clones
- place: lifetime, nested places, open channels, agents
- agent environment: agents, places, channels

These limits are verified upon each MOA function that can impact the values, such as moving, or opening a channel. Should the limits be exceeded, the function is interrupted and the appropriate exception is thrown to the component that invoked the function.

Prior to being accepted at a node, the agent negotiates which and how many MOA resources it can utilize at the visiting MOA system.This is achieved by calculating local policy from the agent policy and host policy. The agent local policy is enforced during its lifetime at the visiting MOA system.

We did not address resource management not supported by JVM, such as the size of VM, the amount of processing, and communication. Whereas it would be possible to enforce some of them by making modifications to the JVM, we refrained from any deviation from *de facto* standard solutions. Imposing resource limits has impacted the design and implementation of the MOA system.

4.7 Security

The first MOA release is fully compatible with the JDK 1.1 security manager; however, no security manager has actually been implemented. Many security features were left open for the next release, such as the work on authentication, and authorization of agents. We have actually implemented only the following features.

Thread switching was employed to allow conformance with the Java security model. Services are provided by threads containing only trusted classes. When a MOA system thread has to switch the trust boundaries (for example in the case of an incoming message, or opening a channel), the request is passed to a system queue serviced by a pool of application threads allocated for that specific trust domain. A thread from the pool processes requests by calling the application specific methods. The request resumes either upon receipt of the response, or upon the timeout, whichever happens first. This way, the application is prevented from stalling the system by thread exhaustion, or by impacting performance through overusing system threads. In addition, resource usage is tracked on a per sandbox basis.

Each agent has its own name space as defined by the bucket in which it is transported. A name space consists primarily of bytecodes and serialized objects. One complication arises when an agent returns to a place that it had left. In this case, the name space is a combination of the original bytecodes and the returning objects. This is achieved by nesting the returning bucket (with the meaning of classloader in this context) within the bucket of the remaining place.

We are using the standard JAR file format for passing agents. This format has provision for digital signatures, allowing for authentication. However, we have not addressed authentication at the moment. It is left as an open issue, even though we have considered its deployment during design and implementation. For example, the agent's authenticity will be maintained as a part of the agent's name object.

4.8 User Interface

MOA's User Interface provides users with various types of interactive monitoring and debugging services. An applet-based and a script-based User Interface are provided. The script-based interface is primarily used for testing purposes whereas the applet-based interface is used by the user to interact with the MOA system.

The User Interface is a component in the front-end of the system. It supports interaction with both the home and remote system. A user can log into the system from any remote location. The login information is verified at the home agent system and a list of all the agents (preconfigured applications available for launching and already launched applications) is returned to the user as a result of a successful login. The user can select any preconfigured application agent and launch it to a remote or local destination.

Users can launch a preconfigured agent, send a command to suspend or kill a selected agent and monitor agent-related activities. An agent can be queried by specifying its start time, duration (to determine which log records to access) and a query pattern. The home of an agent maintains a cyclic array of agent snapshots (captures of agent's state at different times). The user can fetch a snapshot and use it to start a new application agent. The interface accepts various types of messages pertaining to an agent's movement, notices, system statistics and log items. All these messages are displayable.

4.9 MOA Tools

We also developed the tools for manipulation of Java Beans. Even though many new tools are becoming available commercially, or will be developed soon, we needed some functionality that was not available at the time of development. In particular we developed MoaBatch program for instantiating Beans and MOAJar for manipulating JAR files.

MoaBatch is a simple script program (726 lines of code) which lets you instantiate beans (saving them to disk as serialized objects) and edit the properties. It can not use the property editors which do not support text. MoaBatch works with all of the property editors provided with the BDK except for the font editor. MoaBatch fully supports indexed properties. Source and executables are available from http://www.camb.opengroup.org/~laforge/java/moabatch/. Some of the commands that MoaBatch supports for manipulation of beans are included below:

- **Instantiate X** - create a bean X.ser.
- **Properties X** - list the properties of bean X.
- **Limit N** - limit display of elements of an indexed property
- **Set X Y Z** - set property Y of bean X to Z.
- **SetAt X Y I Z** - set property Y at I of bean X to Z.

MOAJar is a GUI utility for editing JAR files layered on top of an API for manipulating the JAR file contents and manifest. MoaJar supports:

- Add, remove, extract, or rename a file in the JAR.
- Edit the name/value attributes in the JAR manifest.
- Serialize object from a class in JAR or on CLASSPATH.
- Edit the properties of a serialized object in the JAR.

4.10 Interoperability

During the development of MOA, we participated in the OMG Mobile Agent Systems Interoperability Facility (MASIF) standard [26]. MASIF is an attempt by General Magic, IBM, Crystaliz, GMD and the Open Group to establish a standard for mobile agents using CORBA. It standardizes agent control, locating, and migration. It does not address communication among the agents. This participation was intertwined with the development of MOA. For example, our experience with MOA has impacted some of the choices in MASIF, and conversely, some of the MASIF specification choices have impacted MOA. In particular, our experience with locating contributed to standardization. MASIF impacted our selection of interfaces for the name server, as well as for the naming in future versions of MOA. OMG MASIF is important for enlarging the base of agent systems that can accept visiting agents.

5. MOA Current Status and Performance

MOA has been delivered to SECOM, utilizing funding provided by MITI. The project started in the summer of 1996. On average the project had four people working full time. Approximately 6 staff years were invested in the effort. The system is at the advanced research prototype stage. It has been tested for a number of scenarios, and we are currently conducting robustness and performance tests. MOA has been adopted as a base technology for a follow-up project ANIMA [2] in The Open Group Grenoble Research Institute. Three other sites have been using MOA: SECOM, University of Denver (further development of the Rent-a-Soft application), and INRIA (for security work).

MOA was developed on Windows NT, PC-based machines. We were mainly using the bare JDK for development, although throughout various phases of the project and for various purposes, developers have also been using other tools, such as Symantec Cafe, J++ and Java Studio. The main reason for using JDK was due to the relatively slow response of the industry to Java Beans development.

For development, we have used ATRIA's ClearCase. While we highly regard this tool in general and especially on UNIX machines (we had other development on HP/UX), the match between Clearcase and the JDK on NT was not a good experience. The problems consisted of the use of upper case letters for file names, very poor response time and interference with debugging. By the

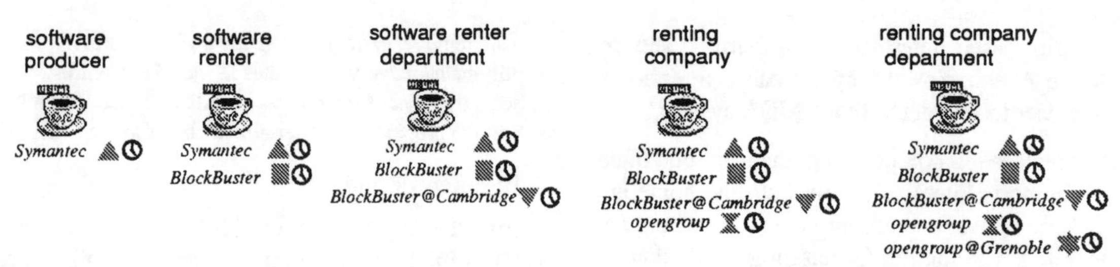

Figure 11. Rent-A-Soft: *each level (producer, renter, renting company, and their departments) encapsulates its data and supporting code for maintaining the renting process, upgrades, inventory, etc.*

very end, we had found ways around all these problems, but they mostly consisted of *ad hoc* solutions, and by staying away from ClearCase as much as possible. The bare minimum consisted of maintaining the source code control, and Clearcase was indeed very suitable for this.

At the moment of writing this paper, the MOA system consists of approximately 30,000 lines of code (including comments), organized in 21 packages, 200 classes, and 10 interfaces. This does not include test programs, developed as unit tests for most packages, and 25 test scenarios exercising various aspects of the system. The tests, including the configuration files, represent additional 11,500 lines of code. The footprint (accumulative size of classes) of the whole MOA system, along with the test programs, is approximately 730KB.

We have only now started working on performance and robustness. We eliminated a few obvious performance bottlenecks and are improving it further. Because of this, and because all measurements were done using interpreted Java (JIT for 1.2 will be available only for the final release on NT platform), results should be taken with the grain of salt. Measurements were conducted between two 100MHz pentium PCs connected in a separate LAN (10Mb ethernet), running NT 4.0 and JVM 1.2. All measurements are an average of 5 runs, which in turn consist of 1,000 RPCs or of 100 moves, subject to measurement.

The RPC with a null message between two agents running on two different nodes takes approximately 25 ms. Note that even though the context of the message is null, the message itself is not null, since it contains destination and source fields. Serializing this object incurs additional cost. Out of 25ms, approximately 3ms is part of the MOA code before the message is passed to the JVM stream, and it takes 3ms since it is read from the JVM stream and until it is delivered to receiving agent. This is when the agent and MOA system are collocated in the same sandbox. If they reside in different sandboxes, it takes 47ms (the corresponding times on the write and read path are 11ms and 6ms). For comparison, the null RPC using RMI on the same platform was measured at 5 ms.

We have measured the move time of a simple agent to be 1177ms. It takes 88ms to serialize the agent and its properties, and 783ms to deserialize it. Transfer of the JAR file over network takes approximately 98ms. For comparison, it takes 45ms to transfer the same JAR file using RMI. The higher costs bring in return a higher flexibility, such as forwarding of messages. Nevertheless, we hope that using JIT and further improving the MOA communication will significantly improve the performance.

6. Applications

"Rent-A-Soft" is a demo program, presented at Uniforum. The idea is to use agents to help out with distributing and renting software packages. This is applicable for relatively expensive packages, or for cheaper software (such as games) assuming large quantities. A chain of participants is envisioned, such as producer, wholesale renter, renting departments, the company which rents and its own departments which sub-rent and occasionally exchange software. By encapsulating information about the renting source, duration and usage of the package, an otherwise complex inventory tracking process is replaced. Each encapsulated layer can associate an agent responsible for specific activities, such as revoking the rent, statistics maintenance, etc. Communication in the presence of mobility can be used to revoke the rent within the same security domains. For example, the company which rents may revoke certain copies of the package within its own departments, it can install new versions of software, and dynamically monitor software use. The Rent-a-soft application is presented in Figure 11.

Radio Communication. The Grenoble facility of the Open Group Research Institute is planning to use the MOA project for a police force application over radio connections [2]. This is a type of application suited for mobile agents. Radio communication has slow and unreliable links, allowing mobile agents to exploit the locality of reference. Another important feature that prevents alternative solutions based on the client server model to be used as effectively as possible is the unpredictable availability of connections. It can easily happen that connections with the centralized server are broken (e.g. a police car enters a tunnel). In such a case, mobile agents can still continue to be functional, e.g. by cooperating with, or visiting other available locations (e.g. another police car that is also in the tunnel).

Security. Even though mobile agents are considered to suffer from (still) inadequate security, they can help to solve some security problems. In certain cases it is not allowed to give access to the actual data stored in a security domain, whereas it is allowed to provide some attributes of the data or some other information about the data. For example, Swiss banks do not allow WWW access to its old accounts since World War II, but they allow anyone checking if the user (or someone from his family, based on the last name) had an account opened. This type of application is a relatively simple query. Imagine a more complex application, where queries need to do complex, arbitrary searches across the whole security domain. Such an application can be easily achieved using mobile agents. A mobile agent would be allowed to enter the security domain. While in the security domain, it can do activities within certain limitations. Before leaving the security domain, agent data could be inspected to see what data it takes out; alternatively only certain data could be allowed to leave. There is an opportunity for establishing covert channels, but depending on how secret the data is and how much information is allowed to leave, this solution could be acceptable in a range of applications. The Open Group RI has an ongoing proposal for the use of mobile agents for improving security.

In all three applications, communication, resource management and interoperability requirements are very important. Our belief is that MOA satisfies these requirements well. For example, communication channels can be temporarily suspended or disabled during the application lifetime, and mobile applications need to reconnect from various sites, requiring the MOA migrateable channel support. For all applications, and especially for security, resource management plays one of the most important roles. Being able to track and limit resources is invaluable for Web server applications. Finally, interoperability is one of the key requirements for many application nowadays, particularly for mobile agents.

7. Lessons Learned

In this section we summarize lessons learned while developing MOA.

Operating System Support vs. Middleware. Recently, work in the development of operating systems has significantly reduced. The current trend is toward using NT on lower-end systems and some version of UNIX on high-end servers. Linux is a dominating freely available version of an OS, and there are also a few real-time executives. In many cases, operating system modules are being replaced with middleware solutions (true for the MOA project).

Nevertheless, many operating systems techniques can be applied in the development of middleware systems. Again, the same applies in the development of MOA. We have drawn on substantial experience in the area of operating systems, such as communication channels and messaging protocols; locating and naming of mobile agents; resource management; negotiation policies, synchronization among agents, etc.

Transparency in communication (maintaining channels across migration) was more complex to support than we originally thought. We were aware that this is a hard goal to achieve, but we hoped that relaxing assumptions would make it simpler to implement.

Resource management was straightforward to design and implement. We believe that its extensive usage will demonstrate its ultimate benefits even more. We strongly recommend that resource management be initially planned for the development of agent systems. We shall heavily rely on it for some of the future work related to policies for management of agent based systems.

Component-based computing has somewhat slowed us down during the development. Compliance with the component model does not come for free. There are costs both in terms of development effort, as well as run-time. The learning curve was high to get accustomed to Java Beans; we had to provide additional methods to inspect/set properties; we had to take care that all classes are serializable; we had to create jar files for both the agent application and system; it is required to link (or wire) components once they are loaded. Nevertheless, we feel that the benefits have at least returned the investment so far, and that the benefits will significantly outweigh the investment once we start using and especially configuring the MOA system and MOA applications.

Immediate benefits of complying with the component model were stronger enforcement of component boundaries than is the case with object boundaries. The components are loaded instead of constructed and component boundaries enforced careful design of what is serialized, particularly useful for application development.

In the future, we expect even higher benefits from the component model, allowing for inspection of visiting agents, reconfiguring agent applications, and agent systems. Evolvement of the MOA system will be easier, since changes will be isolated to single components.

Interoperability. It is too early elaborate on the benefits of participating in the **OMG MASIF** proposal. It was a useful experience to collaborate with implementors of other mobile agent systems. We were solving similar problems, sometimes finding different solutions. Because of the different underlying infrastructure, the current compliance is still a future goal, because we need to come up with a reference implementation first. At the moment, we have taken care that nothing stands in the way of the MOA design to prevent us from switching to different communication infrastructure.

8. Related Work

There are three classes of related work to MOA. The first class consists of process migration, the second of distributed systems on the Web and the third of mobile agents.

Charlotte process migration [3] dealt with the interprocess communication among the migrating processes and introduced forwarding as a locating scheme. Process migration in the **Sprite** operating system supported the notion of a home node [12]. In the **V Kernel** process migration [31], migrating processes are located by searching them. **Emerald** supports fine grain mobility on a small scale network, addressing mobility at the language level [19]. In Mach task migration, transparency of communication and resource maintenance is achieved at the microkernel level [23]. A comprehensive survey of process migration is available at [22]. A theoretical description of mobility in form of Actors is presented in [1].

Two distributed object-based systems on the Web explore similar issues as MOA. **Legion** is an object-based, meta-systems software project, developed at University of Virginia [14] that provides a single, coherent virtual machine and that addresses issues of scalability, fault tolerance, site autonomicity, and security. **Globe** is an object-based wide-area distributed system constructed as a middleware layer on top of existing networks and operating systems [17]. It is based on the concept of a distributed shared object whose state can be physically distributed and that encapsulates implementation aspects (communication, replication, and migration).

Telescript is the first commercial implementation of the mobile agent concept [34]. Recently, it was discontinued and re-implemented in Java, under the name Odyssey. **AgentTcl** is a mobile agent system implemented in the Tcl language [20]. It has two components: a special Tcl interpreter that executes the Tcl agents, and a server that runs on each machine to which agents can be sent. It uses the SafeTcl model for security. **Aglets** is one of the first mobile agent systems written in Java [1]. It supports rich communication semantics (location independency, synchronous, asynchronous and multicast). **Mole** project at University of Stuttgart was one of the first academic efforts in mobile agents in Java [4]. It collaborates with a few industrial partners, such as Siemens and Tandem. **Concordia** supports agent persistence and recovery [9]. Collaborative work is based on event manager and two forms of asynchronous distributed events: selected and group-oriented. **Ara** is a Java-based agent system that applied some changes to the JVM in exchange for increased functionality, such as maintaining thread execution context and imposing limits on memory usage [27]. **Tacoma** and its descendent T2 address fault tolerance and security issues [18]. **Voyager** is a Java-based system for developing distributed applications using mobile objects and agents. It includes an ORB with support for migration, services for persistence, scalable group communication, and basic directory services.

Of the systems presented, the most elaborate schemes for maintaining communication channels across migration were implemented in process migration. This was achieved at the cost of complexity introduced in the operating system [12, 23, 24]. Voyager also supports communication with the migrated away agent, but it relies only on the forwarding strategy. Even though this strategy may appear superior to others (see Section 4.4), it is really the combination of different strategies that offers most benefits to an application writer.

Almost all the systems described provide some support for resource management. None of them, to our knowledge, have made an elaborate use of resource information to pursue negotiation and control.

None of the systems that we described is compliant with the component model. MOA was developed later than most of the agent systems, allowing it to overlap in a timely fashion with the development of Java Beans. This is one of the rare cases when being late happened to be an advantage. Voyager is integrated with the Java Beans event model, but the Voyager system is not built from components.

Of the agent systems we described, Aglets is the only other agent system that plans to pursue a MASIF reference implementation. MOA and Aglets are currently similar with regards to MASIF compliance, i.e. both are Java-, rather than CORBA-oriented. It will be required to adapt security and communication models to adjust to MASIF requirements.

In summary, the MOA system is different from other agent systems in the following unique aspects. The MOA system and applications are Java Beans compliant. The place in MOA can be retained after an agent leaves. Agent naming supports families and generations of agents that can be managed. Agents are tracked using four, per-agent, configurable locating schemes. Communication channels are migrateable.

9. Conclusion and Future Work

In this paper, we described the design and implementation of the MOA project. In particular we presented the MOA object and component models and described its components, such as communication, naming and locating, mobility, and resource management. We also discussed some lessons learned during its development and presented some preliminary performance measurements.

MOA contributions consist of: supporting agent collaboration by maintaining communication channels across migration; providing basic support for denial of service attacks by extensive resource management and negotiation policies; compliance with the Java beans component model, leading to better configurability; and complying with the OMG MASIF standard.

There are many mobile agent systems available nowadays, both from academia and from industry. Even though MOA represents yet another new mobile object system among many research vehicles today, we believe that we have distinguished it sufficiently enough to justify its development. In particular, we believe that it was easier to achieve compliance with the component model while designing the system; similar reasoning applies to managing resources, and to maintaining communication channels.

The lessons we learned range from resemblance of middleware solutions to experience with operating systems. Our experience with the component model distinguishes between costs and benefits of complying with the component model. We strongly believe the latter will outweigh the former already for moderate requirements for configurability. We introduced a lot of complexity by maintaining communication transparency. Resource maintenance proved to be very useful with expectations to significant benefits in the future. Finally, OMG MASIF standard impacted only the design decisions of MOA so far. We expect to learn more about MASIF as we pursue the reference implementations.

The future work consists of four areas. First, we plan to extensively improve security. In particular, we plan to include authentication, authorization, integrity checking, and the trust model of MOA. The second area consists of applications, which we plan to support a few. The third area addresses improvements to the current implementation, in particular related to performance and robustness. Finally, we plan to demonstrate interoperability in practice, by interoperating with another OMG MASIF reference platform, such as Aglets.

Acknowledgments

We are grateful to Shai Guday and Holger Peine for reviewing this paper. They significantly improved its presentation and contents. Rosemary Hudson and Jackie Clark undertook the impossible task to insert all missing articles and to eliminate the superfluous ones that an author, a non-native English speaker, introduced.

Availability

The MOA project is available for scientific and research purposes under a Collaborative Research Agreement from The Open Group. The URL of the project is: http://www.camb.opengroup.org/RI/Techno/OS/moa.html.

References

[1] Agha, G., "A Model of Concurrent Computation in Distributed Systems", *MIT Press,* Cambridge, MA, 1987.

[2] ANIMA Project, http://www.gr.opengroup.org/anima.

[3] Artsy, Y. and Finkel, R. (September 1989). Designing a Process Migration Facility: The Charlotte Experience. *IEEE Computer,* pp 47–56.

[4] Baumann, J., Hohl, F., Rothermel, K., Strasser, M., "Mole, Concepts of a Mobile Agent System", to appear in the WWW Journal, Special Issue on Applications and Techniques of Web Agents, *Baltzer Science Publishers.*

[5] Black, D., Milojicic, D., Langerman, A., Dominijanni, M., Dean, R., Sears, S., "Distributed Memory Management", accepted for publication, Software Practice & Experience, 1997.

[6] Bradshaw, J., "Software Agents", *AAAI/MIT Press,* 1996.

[7] Chess, D., Grossof B., Harrison, C., Levine, D., Parris, C., Tsudik, G., "Itinerant Agents for Mobile Computing", *IEEE Personal Communications Magazine,* October 1995.

[8] Cockayne, W., and Zyda, M., "Mobile Agents: Explanations and Examples", *Manning,* 1997.

[9] Concordia: "Concordia: An Infrastructure for Collaborating Mobile Agents" *Proc. of Workshop on Mobile Agents MA'97,* Berlin, April 7-8th. LNCS 1219, Springer Verlag.

[10] Cybenko, G., Spontaneous comment during Transportable Agents Workshop, September 1997.

[11] DARPA Broad Agency Announcement, 98-01, "Agent-Based Systems", http://ballston.prc.com/baa9801/abspipv1.htm.

[12] Douglis, F. and Ousterhout, J. (August 1991). Transparent Process Migration: Design Alternatives and the Sprite Implementation. *Software-Practice and Experience,* 21(8):757–785.

[13] Ford, W., Baum, M., "Secure Electronic Commerce", *Prentice Hall,* New Jersey, 1997.

[14] Grimshaw, A., et al. "The Legion Vision of a Worldwide Virtual Computer", CACM, vol 40, no 1, Jan. 1997, pp 39-45.

[15] Goldszmidt, G., Yemini, Y., "Distributed Management by Delegating Mobile Agents", Proc. of the 15th ICDCS, Vancouver, British Columbia, June 1995.

[16] Guideware Corporation, http://www.guideware.com.

[17] Homburg, P., van Steen, M., and Tanenbaum A., "An Architecture for A Wide Area Distributed System", Proc. of the Seventh SIGOPS European Workshop, Connemara Ireland, September 1986, pp 75-82.

[18] Johansen, D., van Renesse, R., and Schneider, F., "Operating system support for mobile agents", *Proc. of the 5th. IEEE HOTOS Workshop*, Orcas Island, USA (4th-5th May, 1995),.

[19] Jul, E., Levy, H., Hutchinson, N., Black, A., "Fine-Grained Mobility in the Emerald System", *ACM TOCS*, vol 6, no 1, February 1988, pp 109-133.

[20] Kotz, D., et al., "Mobile Agents for Mobile Internet Computing", July/August 1997, IEEE Internet Computing, vol 1, no 4, pp 58-67.

[21] Lange, D., Oshima, M., "Java Agent API: Programming and Deploying Aglets with Java", Addison Wesley, expected publication date, Winter 1998. (Aglets Web Page: http://www.ibm.co.jp/trl/projects/aglets/).

[22] Milojicic, D., Douglis, F., Paindaveine, Y., Wheeler, R., Zhou, S., "Process Migration Survey", *The Open Group Research Institute, Collected Papers, vol. 5, March 1997.*

[23] Milojicic, D., Zint, W., Dangel, A., Giese, P., "Task Migration on the top of the Mach Microkernel", Proc. of the third USENIX Mach Symposium, April 1993, pp 273-290, Santa Fe, New Mexico.

[24] Milojicic, D., Douglis, F., Wheeler, R., Guday, S., "Mobility, and Edited Collection", and "Mobility in Distributed Systems", Addison Wesley, expected dates of publication, Winter 1998 and Fall 1999 respectively.

[25] Odyssey Web Page, http://www.genmagic.com/agents/odyssey.html.

[26] OMG Mobile Agent Systems Interoperability Facilities Specification (MASIF), OMG TC Document ORBOS/97-10-05, also available from http://www.opengroup.org/~dejan/maf/draft10.

[27] Peine, H., and Stolpmann, T., "The Architecture of the Ara Platform for Mobile Agents", *Proc of the First Intl Workshop on Mobile Agents MA'97, Berlin,* April 7-8, Springer Verlag, http://www.uni-kl.de/AG-Nehmer/Ara

[28] Riggs, R., et al.,"Pickling State in the Java System," *Proc. of the USENIX 1996 Conference on Object-Oriented Technologies (COOTS)*, pp 241-250.

[29] Ranganathan, M., Acharya, A., Sharma, S., Saltz. J.,., "Network-aware Mobile Programs", *Proceedings of the Annual Usenix 1997 Conf.*, January 6-10, Anaheim, California, USA.

[30] Shoch, J., Hupp., J., "The Worms Programs - Early Experience with Distributed Computing", Communications of the ACM, 25 (3), pp 172-180, March 1982.

[31] Theimer, M., Lantz, K., and Cheriton, D. (December 1985). Preemptable Remote Execution Facilities for the V System. *Proc. of the 10th ACM SOSP*, pp 2–12.

[32] Vitek, J., and Tschudin, C., "Mobile Objects Systems: Towards the Programmable Internet", *Springer Verlag*, April 1997.

[33] Voyager Technical Overview, ObjectSpace, http://www.objectspace.com/voyager.

[34] White, J., "Telescript Technology: Mobile Agents", General Magic White Paper (http://www.genmagic.com/Telescript/Whitepapers/wp4/whitepaper-4.html).

[35] White, J., et al., "System and Method for Distributed Computation Based upon the Movement, Execution, and Interaction of Processes in a Network", *United States Patent*, no 5603031, February 1997.

[36] Wollrath, A., et al., "A Distributed Object Model for the Java System," *Proc. USENIX 1996 Conf. on Object-Oriented Technologies (COOTS)*, pp. 219-231.

14.10 Glass, G., ObjectSpace Voyager Core Package Technical Overview

ObjectSpace Voyager Core Package Technical Overview

The ObjectSpace Voyager™ Core Technology (Voyager) contains the core features and architecture of the ObjectSpace Voyager platform, including a full-featured, intuitive ORB with support for mobile objects and autonomous agents. Also in the core package are services for persistence, scalable group communication, and basic directory services. Voyager can be downloaded for free commercial use from www.objectspace.com and is everything you need to get started building high-impact systems in Java™ today.

This text presents a high-level overview of Version 1.0[†] of the Voyager Core Technology. It first presents Voyager concepts, then an example of Voyager in action.

> *There are three models of distributed computing: client/server, peer-based, and agent-based. ObjectSpace's Voyager is the only product that offers the ability to build applications that mix all three models. This makes Voyager a leading tool in distributed application development.*
>
> J.P. Morgenthal, President, New Horizon Computing Corp. Leading analyst on Java

1 Introduction

ObjectSpace Voyager is the ObjectSpace product line designed to help developers produce high-impact distributed systems quickly. Voyager is 100% Java and is designed to use the Java language object model. Voyager allows you to use regular message syntax to construct remote objects, send them messages, and move them between programs. This reduces learning curves, minimizes maintenance, and, most importantly, speeds your time to market for new, advanced systems. Voyager's architecture is designed to provide developers full flexibility and powerful expansion paths.

The root of the Voyager product line is the ObjectSpace Voyager Core Technology. This product contains the core features and architecture of the platform, including a full-featured, intuitive object request broker (ORB) with support for mobile objects and autonomous agents. Also in the core package are services for persistence, scalable group communication, and basic directory services. The ObjectSpace Voyager Core Technology is everything you need to get started building high-impact systems in Java today.

As the industry evolves, other companies providing distributed technologies struggle as they try to adapt to the new Java language. These companies are required to adapt older object models to fit Java. This results in a series of compromises that together have a dramatic impact on time to market and development costs. Voyager, on the other hand, is developed to use the Java language as its fundamental interface.

One of Java's primary distinctions is the ability to load classes into a virtual machine at run time. This capability enables infrastructures to use mobile objects and autonomous agents as another tool for building distributed systems. Adding this capability to older distributed technologies is often impractical and results in difficult-to-use infrastructures. Voyager provides seamless support for mobile objects and autonomous agents.

---

†. For information about subsequent versions of Voyager, see the ObjectSpace Web site at www.objectspace.com.

1.1 CORBA Integration

Complete bidirectional CORBA integration was included in the 2.0 Beta 1 enhancement of Voyager, released early December 1997. This additional Java package allows Voyager to be used as a CORBA 2 client or server. You can generate a Voyager remote interface from any IDL file. You can use this interface to communicate with any Voyager or CORBA server. Without modifying the code, you can export any Java class as a CORBA server in seconds, automatically generating IDL for use by CORBA implementations.

As part of the 2.0 Beta 1 Voyager Core Technology, the CORBA integration is also free for most commercial use. For more information, read the *ObjectSpace Voyager CORBA Integration Technical Overview* paper or Part 4 of the *ObjectSpace Voyager Core Technology User Guide*, Version 2.0 Beta 1, on www.objectspace.com.

1.2 Developing with Voyager

Voyager was designed from the ground up to solve problems encountered in the development of distributed systems in Java. As the premier Java distributed systems architecture, Voyager is a technology that enables developers to solve these problems quickly and efficiently.

Consider the following issues.

- **Problem.** Time to market is crucial and development time is expensive. Extra months spent in development mean extra months for competitors to gain market share.

  Solution. Voyager is the easiest way to build distributed systems in Java. Previous technologies require a tedious, clumsy, and error-prone multistep process to prepare a class for remote programming. A single Voyager command replaces this hassle and automatically enables any class for distributed computing and persistence in just seconds. Voyager does not require Java classes to be altered. You can remotely construct and communicate with any Java class, even third-party libraries, without accessing the source code. You can remotely persist any serializable object. Other technologies typically require the use of .idl files, interface definitions, and modifications to the original class, all of which consume development time and couple your domain classes tightly to a particular ORB or database technology.

- **Problem.** Enterprise networks are usually composed of many hardware and operating system platforms. Systems built with legacy ORBs often require separate binaries for each platform. This increases developer load and system complexity and complicates system maintenance.

  Solution. Voyager is 100% Java. Voyager applications can be written once and run anywhere Java 1.1 is supported.

- **Problem.** Resources in a distributed system need to be used wisely. When a machine is being overused, the load should be shifted to other, less used machines. Existing ORBs do not help developers solve this problem.

  Solution. Voyager is dynamic. Mobile objects and agents can be used to encapsulate processing and can migrate through the network, carrying their workloads with them. Instead of being limited to running only static processes on a given virtual machine, developers can now exploit the natural connection between agents and processing. The result is effortless, dynamic load balancing.

- **Problem.** Information access needs vary. Sometimes information should be broadcast across the enterprise; sometimes it should be filtered based on the needs of the user. Sometimes information is transient, while at other times it is stored for future use.

Solution. Voyager is comprehensive. It supports development of high-performance push systems using the built-in publish/subscribe technology. Voyager also supports distributed persistence, multicast messaging, and a rich set of message types.

- **Problem.** Developers need to leverage JavaBeans™ components in a distributed context, but cannot afford to modify the architecture with wrapper code or spend time developing complex glue logic.

  Solution. Voyager is JavaBeans-enabled. It provides support for distributed JavaBeans events without requiring any modifications to the beans. No other ORB has such a seamless beans event distribution model.

- **Problem.** Large systems and congested networks often result in sluggish software. Today's high-performance needs require responsive software.

  Solution. Voyager is fast. Remote messages with Voyager are as fast as other CORBA ORBs. Messages delivered by mobile agents can be up to 1,000,000 times quicker.

- **Problem.** Today's embedded systems require small run-time footprints. Similarly, Web applets must be small to minimize download times.

  Solution. Voyager is compact. The entire Voyager system is less than 300KB, not including the JDK classes it uses. Voyager is a fully functional, agent-enhanced ORB and does not require any additional software beyond JDK 1.1.

2 Concepts

This section describes the concepts behind the ObjectSpace Voyager Core Technology architecture, using a mix of text, example code, and drawings.

2.1 Objects

Objects are the building blocks of all Voyager programs. An object is a software component that has a well-defined set of public functions and encapsulates data. The following object is an instance of the class Store with a public function to accept new stock.

```
Store store = new Store();
store.stock( "widget", 43 );
```

2.2 Voyager-Enabled Programs

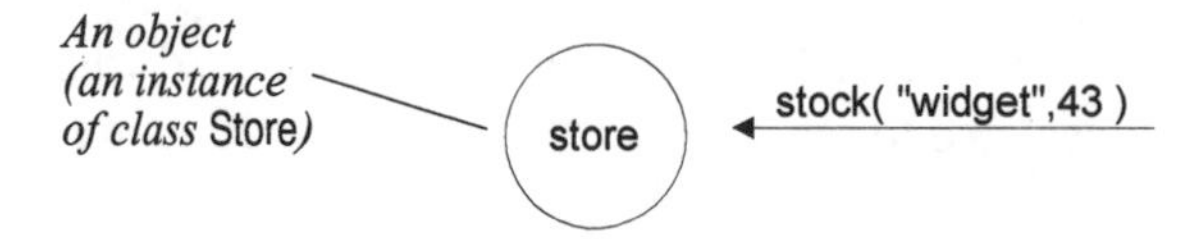

When a Voyager-enabled program starts, it automatically spawns threads that provide timing services, perform distributed garbage collection, and accept network traffic. Each Voyager-enabled program has a network address consisting of its host name and a communications port number, which is an integer unique to the host.

Port numbers are usually randomly allocated to programs. This is sufficient for clients communicating with remote objects and for creating and launching agents into a network. However, if a program will be addressed by other programs, you can assign a well-known port number to the program at startup.

```
Voyager.startup( 7000 ); // assign port number 7000 to this program
```

```
Store store = new Store();
```

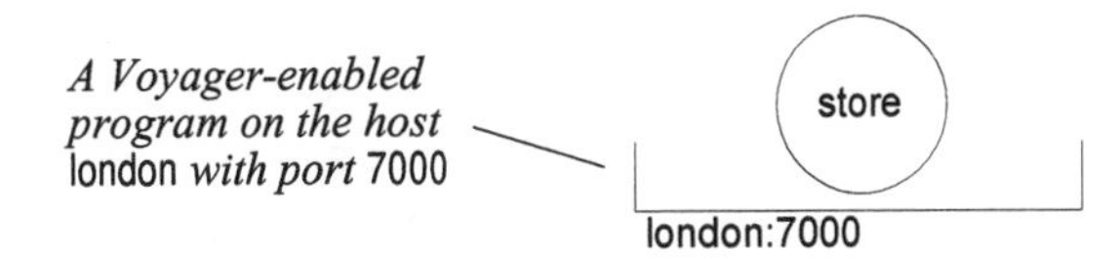

2.3 Remote-Enabled Classes and Virtual References

A class is remote-enabled if its instances can be created outside the local address space of a program and if these instances can receive messages as if they were local. Voyager allows an object to communicate with an instance of a remote-enabled class via a special object called a *virtual reference*. When messages are sent to a virtual reference, the virtual reference forwards the messages to the instance of the remote-enabled class. If a message has a return value, the target object sends the return value to the virtual reference, which returns this message to the sender.

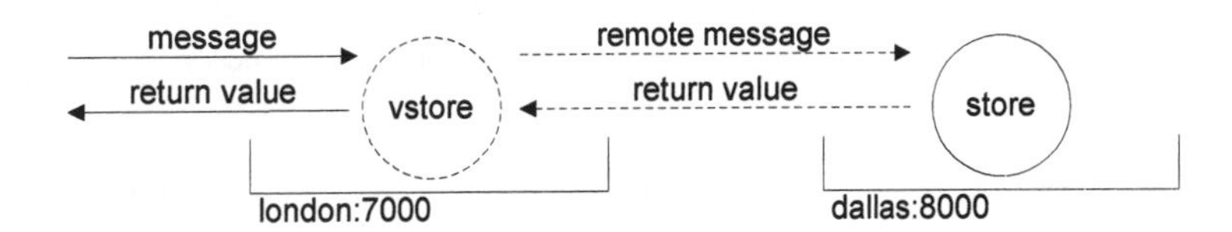

After remote-enabling a class, you can:

- Construct instances remotely, even if the class code does not exist on the remote machine.
- Send messages to remote instances using regular Java syntax.
- Connect to existing remote instances in other programs.
- Move remote instances to other programs, even if the class code is not already in the destination program.
- Make remote instances persistent.

2.4 Generating a Remote-Enabled Class

Use Voyager's vcc utility to generate a remote-enabled class from an existing class. The vcc utility reads a .class or .java file and generates a new *virtual class*. The virtual class contains a superset of the original class functions and allows function calls to occur even when objects are remote or moving.

The virtual class name is V plus the original class name. For example, if the file Store.java contains the source code for class Store, the compiled class file is Store.class. You can remote-enable the Store class by running vcc on either Store.java or Store.class to create a new, virtual class named VStore.

For more detailed information about remote enabling, refer to Chapter 5, "Fundamental ORB Features," of the *ObjectSpace Voyager Core Technology User Guide*, Version 1.0.

Agent technology will be as important for the Internet as the Internet has been for personal computing. Voyager is the most powerful and easy-to-use solution for agent-enabled distributed computing I have seen.

John Nordstrom, Sabre Decision Technologies

2.5 Constructing a Remote Object

After remote-enabling a class, you can use the class constructors of the resulting virtual class to create a remote instance of the original class. The remote instance can reside in your current program or a different program, and a virtual reference to the remote instance is created in your current program.

To construct a remote instance of a class, give the virtual class constructor the address of the destination program where the remote instance will reside. If the original class code for the remote instance does not exist in the destination program, the Voyager network class loader automatically loads the original class code into the destination program.

```
Voyager.startup( 7000 );
VStore vstore = new VStore( "dallas:8000/Acme" ); // alias is Acme
```

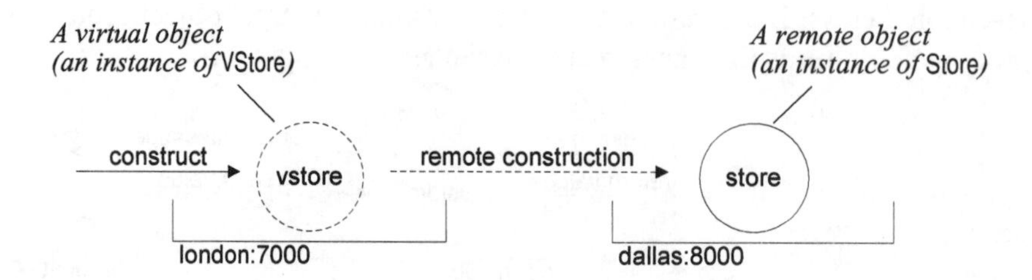

When a remote object is constructed, it is automatically assigned a 16-byte globally unique identifier (GUID), which uniquely identifies the object across all programs worldwide. Optionally, you can assign an alias to an object during construction. The GUID or the optional alias can be used to locate or connect to the object at a later point in time. This directory service is a basic Voyager feature. Voyager also includes an advanced federated directory service for more complex directory requirements. Refer to Chapter 16, "Federated Directory Service," of the *ObjectSpace Voyager Core Technology User Guide*, Version 1.0, for more information.

2.6 Sending a Message to a Remote Object

When a message is sent to a virtual reference, the virtual reference forwards the message to its associated remote object. If the message requires a return value, the remote object passes the return value to the virtual reference, which forwards it to the sender. Similarly, if the remote object throws an exception, the exception is caught and passed back to the virtual reference, which throws it to the caller.

```
vstore.stock( "widget", 43 );
```

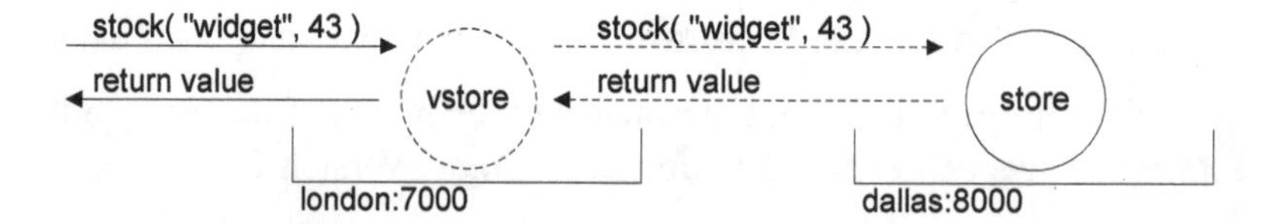

2.7 Connecting to an Existing Remote Object

A remote object can be referenced by any number of virtual references. To create a new virtual reference and associate it with an existing remote object, supply the address of the program where the existing remote object currently resides and the alias of the remote object to the static VObject.forObjectAt() method.

```
// connect using alias
Voyager.startup( 9000 );
VStore vstore2 = (VStore) VObject.forObjectAt( "dallas:8000/Acme" );
int price = vstore2.buy( "widget" );
```

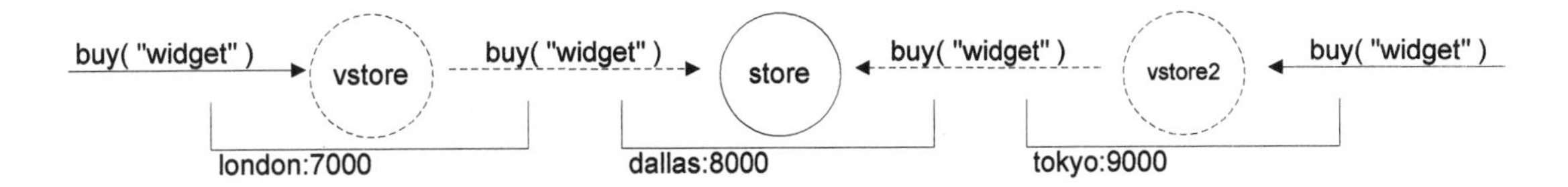

2.8 Mobility

You can move any serializable object from one program to another by sending the moveTo() message to the object via its virtual reference. Supply the address of the destination program as a parameter.

```
vstore.moveTo( "tokyo:9000" );
```

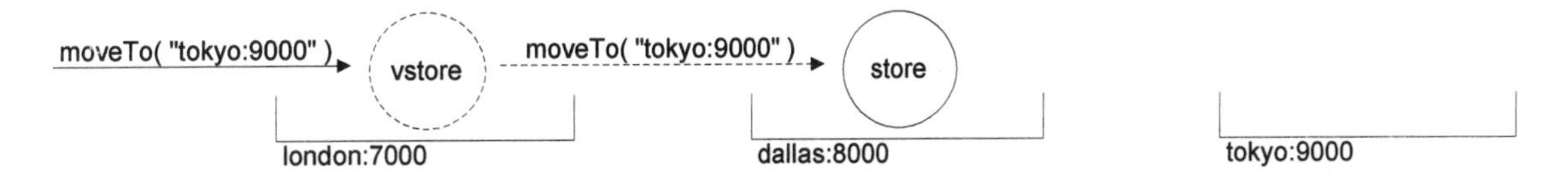

The object waits until all pending messages are processed and then moves to the specified program, leaving behind a forwarder to forward messages and future connection requests.

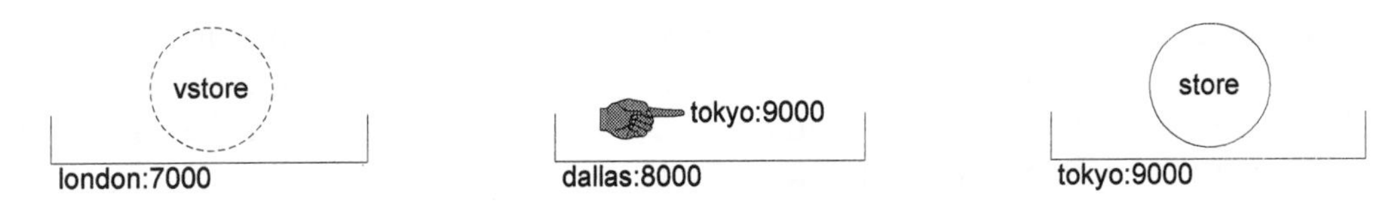

You can send a message to an object even if the object has moved from one program to another. Simply send the message to the object at its last known address. When the message cannot locate its target object, the message searches for a forwarder. If the message locates a forwarder representing the object, the forwarder sends the message to the object's new location.

```
int price = vstore.buy( "widget" );
```

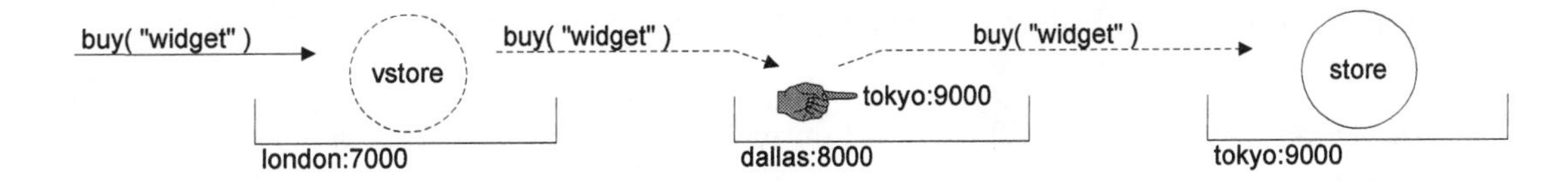

The return value is tagged with the remote object's new location, so the virtual reference can update its knowledge of the remote object's location.

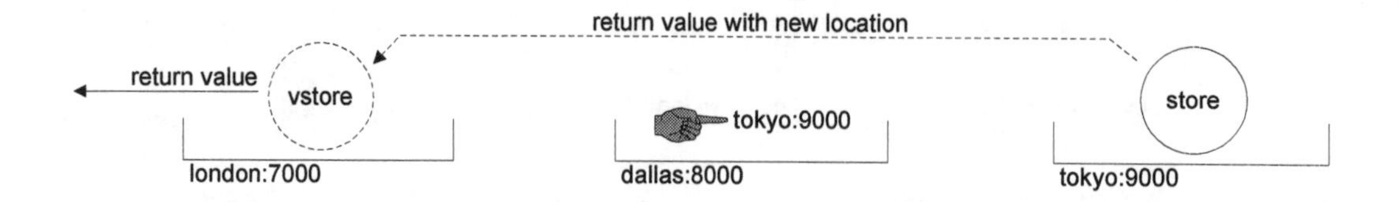

Subsequent messages are sent directly to the remote object at its new location, bypassing the forwarder.

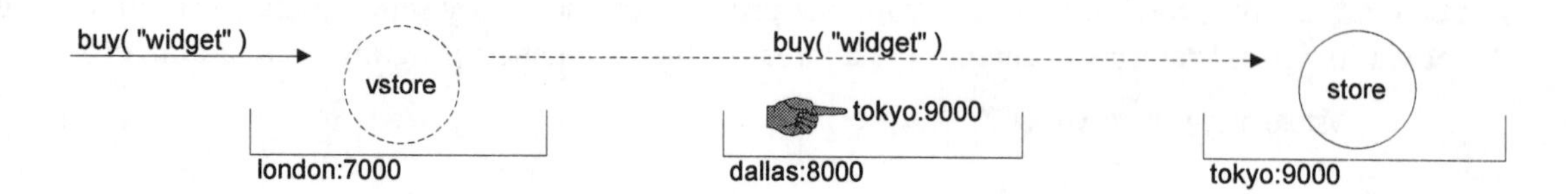

2.9 Agents

An agent is a special object type. Although there is no single definition of an agent, all definitions agree that an agent has autonomy. An autonomous object can be programmed to satisfy one or more goals, even if the object moves and loses contact with its creator.

Some definitions state that an agent has mobility as well as autonomy. Mobility is the ability to move independently from one device to another on a network. Voyager agents are both autonomous and mobile. They have all the same features as simple objects—they can be assigned aliases, have virtual references, communicate with remote objects, and so on.

To create an agent, extend the base class COM.objectspace.voyager.Agent and then use Voyager's vcc utility to remote-enable the agent's class. Use the resulting virtual class to instantiate an agent object and use virtual references to communicate with this object even if it moves.

Like all objects, an agent can be moved from one program to another. However, unlike simple objects, an agent can move itself autonomously. An agent can move to other programs, allowing the execution of distributed itineraries, or an agent can move to other objects, allowing communication using high-speed, local messaging.

An agent can move to another program and continue to execute when it arrives by sending itself moveTo() with the address of the destination program and the name of the member function that should be executed on arrival.

For example, an agent in dallas:8000 is told to travel. The agent sends itself a moveTo() message with two parameters: dallas:9000, the destination address, and atTokyo, the name of the callback function.

```
public void travel() // defined in Shopper
  {
  moveTo( "tokyo:9000", "atTokyo" );
```

```
}
```

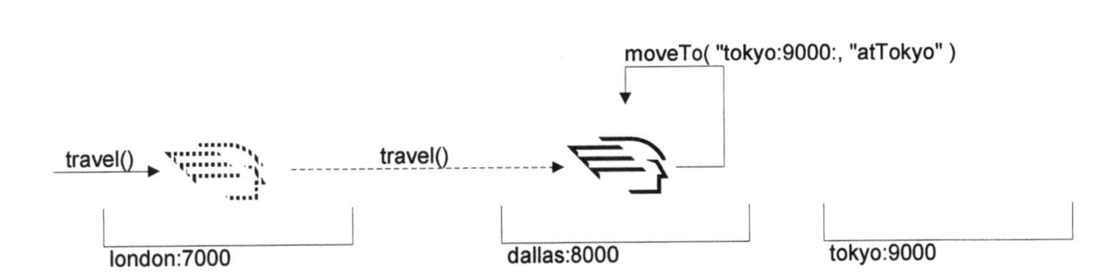

The agent then moves to tokyo:9000, leaving behind a forwarder to forward messages.

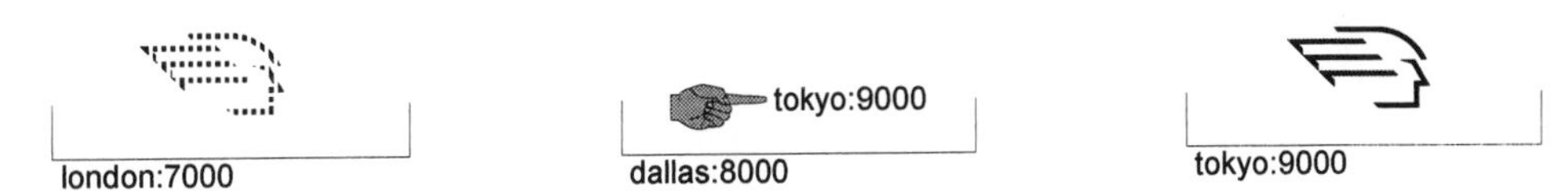

After arriving at its new location, the agent automatically receives the atTokyo() message.

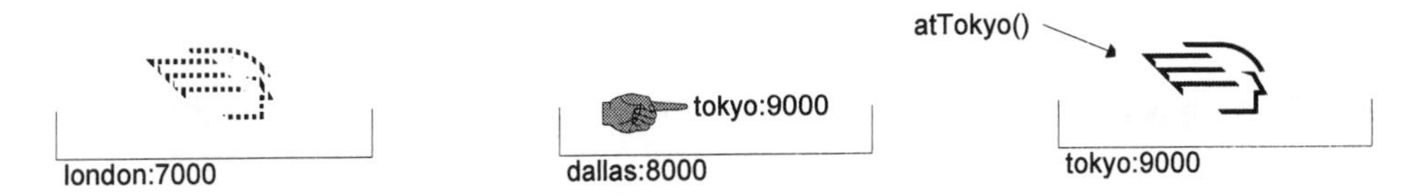

The following code in the agent is then executed.

```
public void atTokyo() // defined in Shopper
  {
  // this code is executed when I move successfully to tokyo:9000.
  }
```

If an agent wants to have a high-speed conversation with a remote object, the agent can move to the object and then send it local Java messages. The easiest way for an agent to move to an object is by sending itself a variation of moveTo() that specifies both a virtual reference to the destination object and a callback parameter.

For example, an agent in dallas:8000 is told to buy from a store object. The agent sends itself a moveTo() message with two parameters: vstore, a virtual reference to the remote store object, and shop, the name of a callback function.

```
public void buyFrom( VStore vstore ) // defined in Shopper
  {
  moveTo( vstore, "shop" );
  }
```

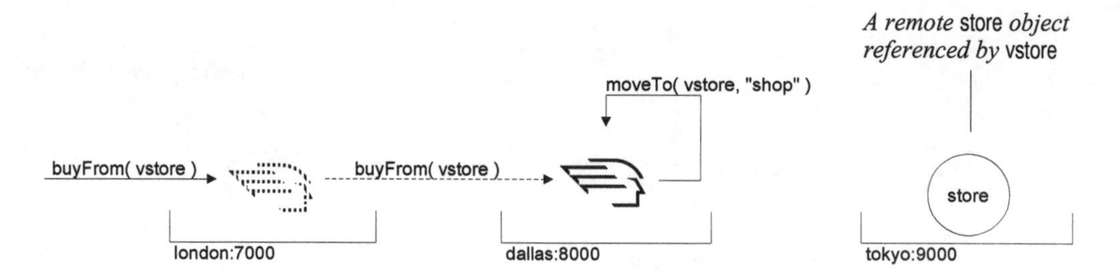

After leaving behind a forwarder and moving to tokyo:9000, the agent receives the callback message shop() with a local native Java reference to the object store.

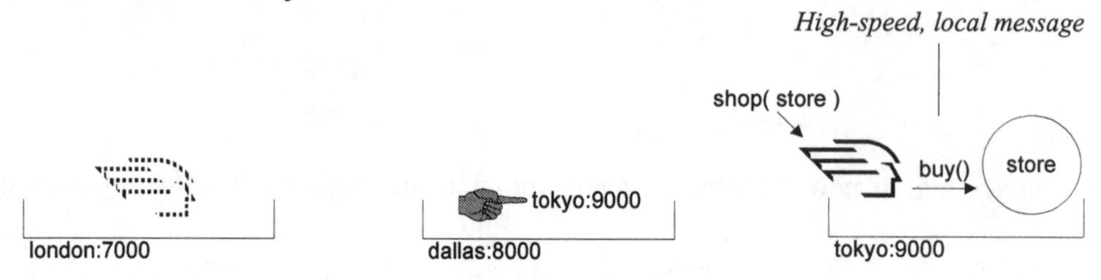

The following code in the agent is then executed.

```
public void shop( Store store ) // defined in Shopper
  {
  // this code is executed when I successfully move to the store
  // note that store is a regular Java reference to the store
  int price = store.buy( "widget" );
  }
```

2.10 JavaBeans Integration

Voyager is designed to integrate with the JavaBeans component model. Existing JavaBeans can be used in a Voyager system. Voyager extends the beans delegation event model by allowing all events to be transmitted across the network. This is possible without modifying the bean or event classes in any way.

Voyager also uses the beans event model for object- and system-level monitoring. Every remote Voyager object is automatically a source of events. Objects can listen to remote objects and monitor every aspect of the remote object's behavior. In particular, listeners are notified when the remote object receives messages, when it moves, when it is saved to or loaded from a database, and when the remote object dies.

Voyager allows system-level monitoring with the beans event model as well. Listeners can monitor when the system garbage-collects remote objects, when classes are loaded into the system, when messages are sent and received, and when agents and mobile objects arrive and depart.

Voyager extends the beans event model further by introducing persistent listeners. Typically, developers use standard beans listeners for transient listening. However, more complex systems often require listeners that can move with objects, or listeners that can automatically be stored to and retrieved from databases with the source objects. Voyager adds this critical piece of functionality to all listeners of Voyager events.

Voyager's integration with the JavaBeans event model allows developers to apply their bean knowledge and experience directly to their Voyager systems. The event system provides a wealth of information useful for monitoring, auditing, logging, and other higher-level, application-specific actions.

2.11 Dynamic Properties

Voyager allows developers to attach key value *properties* to remote objects. These properties are dynamically attached to any object without requiring modification to the object's source. This property mechanism is used by the publish/subscribe system to allow objects to specify what subjects they are interested in and can also be used to attach application-specific information to an object at run time.

```
myObject.addProperty( Subscription.SUBSCRIBE, "sports.basketball.*" );
```

2.12 Database-Independent Persistence

A persistent object has a backup copy in a database. A persistent object is automatically recovered if its program is unexpectedly terminated or if it is flushed from memory to the database to make room for other objects. Voyager includes seamless support for object persistence. In many cases, you can make an object persistent without modifying its source.

Each Voyager program can be associated with a database. The type of database can vary from program to program and is transparent to a Voyager programmer. Voyager includes a high-performance object storage system called VoyagerDb, but provides an interface layer to allow developers to drop in their own custom bindings to other popular relational and object databases.

To save an object to the program's database, send saveNow() to the object. This method writes a copy of the object to the database, overwriting any previous copy. If the program is shut down and then restarted, the persistent objects are left in the database. Any attempt to communicate with a persistent object causes the object to be reloaded from the database.

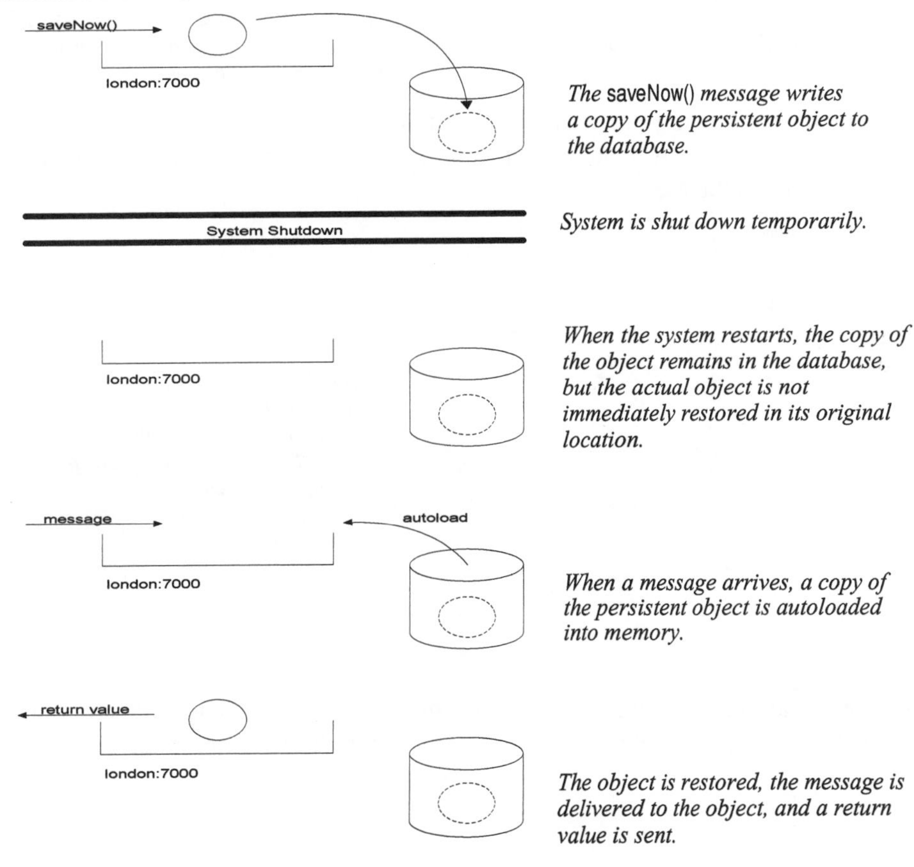

The saveNow() *message writes a copy of the persistent object to the database.*

System is shut down temporarily.

When the system restarts, the copy of the object remains in the database, but the actual object is not immediately restored in its original location.

When a message arrives, a copy of the persistent object is autoloaded into memory.

The object is restored, the message is delivered to the object, and a return value is sent.

See Chapter 14 of the *ObjectSpace Voyager Core Technology User Guide*, Version 1.0, for more details about Voyager persistence.

If a persistent object is moved from one program to another, the copy of the object is automatically removed from the source program's database and added to the destination program's database.

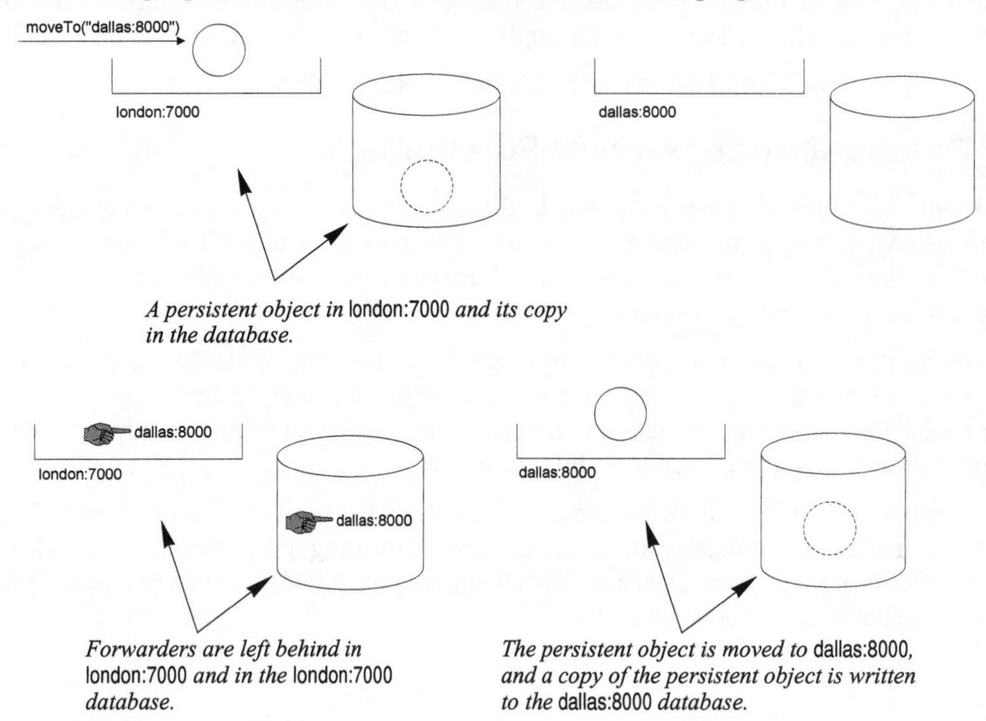

A persistent object in london:7000 *and its copy in the database.*

Forwarders are left behind in london:7000 *and in the* london:7000 *database.*

The persistent object is moved to dallas:8000, *and a copy of the persistent object is written to the* dallas:8000 *database.*

You can conserve memory by using one of the flush() family of methods to remove a persistent object from memory and store it in a database. Any subsequent attempt to communicate with a flushed persistent object reloads the object from the database.

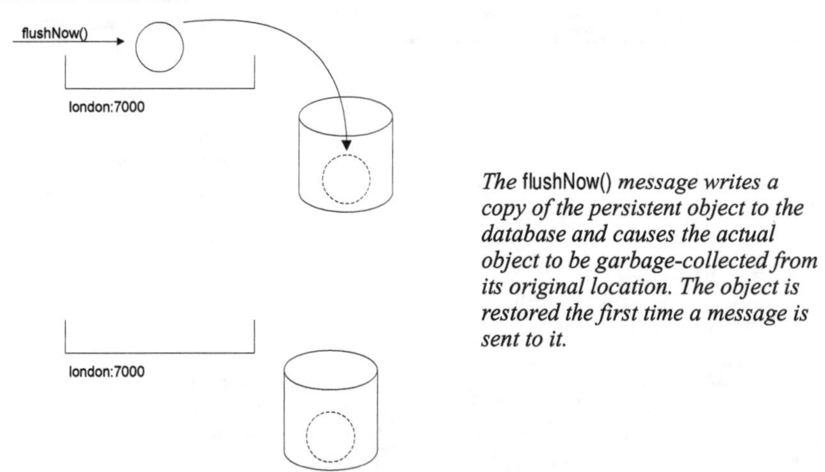

The flushNow() *message writes a copy of the persistent object to the database and causes the actual object to be garbage-collected from its original location. The object is restored the first time a message is sent to it.*

By default, Voyager's database system automatically makes Java classes loaded into a program across a network persistent, thus avoiding a reload of these classes when the program is restarted.

2.13 Space – Scalable Group Communication

Many distributed systems require features for communicating with groups of objects. For example:

- Stock quote systems use a distributed event feature to send stock price events to customers around the world.
- A voting system uses a distributed messaging feature (multicast) to send messages around the world to voters, asking their views on a particular matter.
- News services use a distributed publish/subscribe feature so that broadcasts are received only by readers interested the broadcast topic.

Most traditional systems use a single *repeater* object to replicate the message or event to each object in the target group.

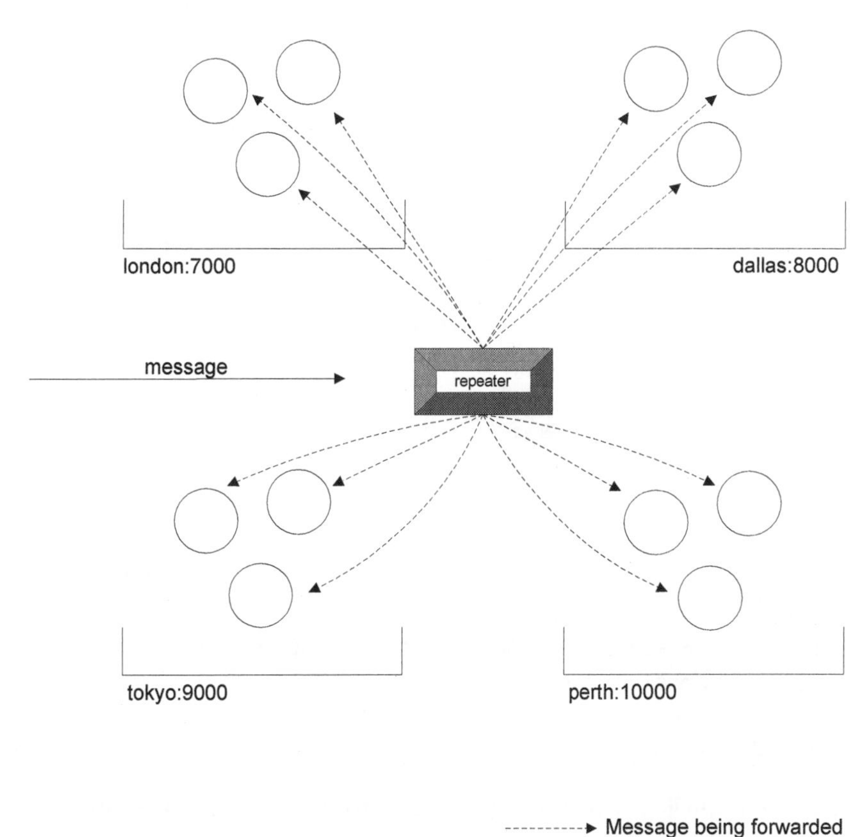

This traditional approach works well if the number of objects in the target group is small, but does not scale well when large numbers of objects are involved.

Voyager uses a different and innovative architecture for message/event replication called Space™ that can scale to global proportions. Clusters of objects in the target group are stored in local groups called *subspaces*. Subspaces are linked to form a larger logical group called a *Space*. When a message or event is sent into one of the subspaces, the message or event is cloned to each neighboring subspace before being delivered to every object in the local subspace. This process results in a rapid parallel fanout of the message or event to every object in the Space. A special mechanism in each subspace ensures that no message or event is accidentally processed more than once, regardless of how the subspaces are linked.

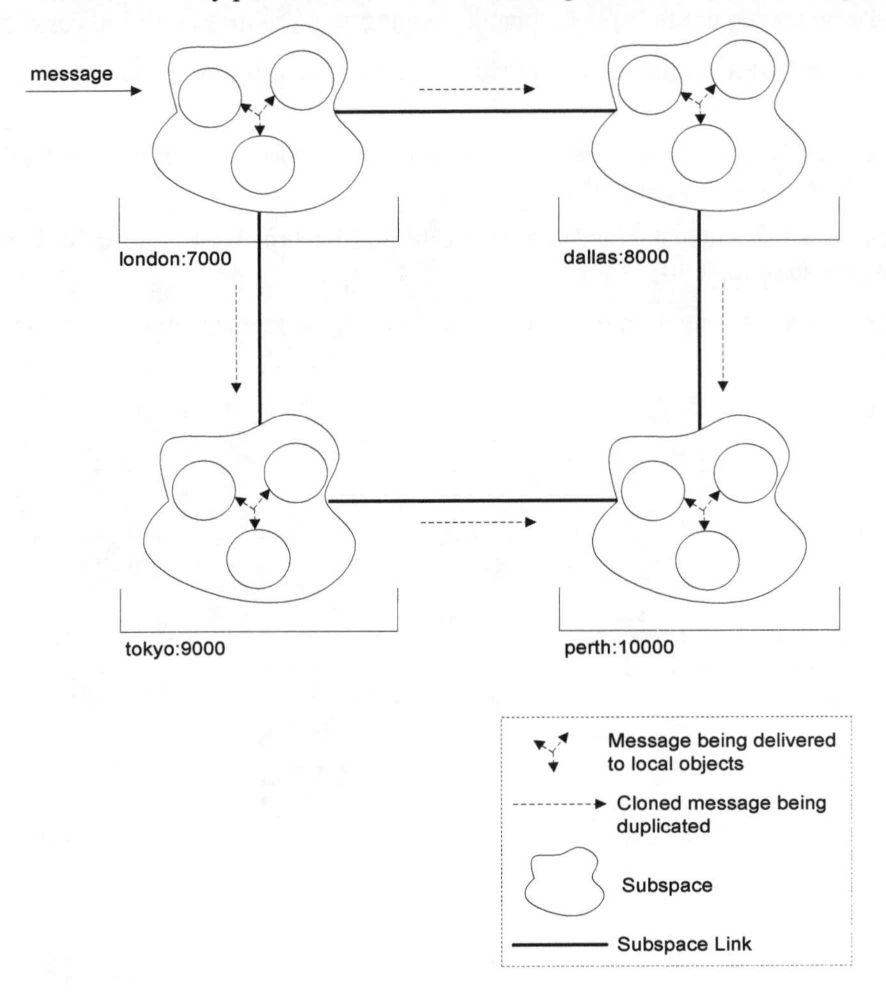

Voyager's multicast messaging, distributed events, and publish/subscribe features all use and benefit from the same underlying Space architecture.

2.14 Message Types

Unlike traditional ORBs, which use a simple, on-the-wire message protocol, Voyager messages are delivered by lightweight agents called *messengers*. Voyager has four predefined message types.

Synchronous Messages

By default, Voyager messages are synchronous. When a caller sends a synchronous message, the caller blocks until the message completes and the return value is received. You can use regular Java syntax to send a synchronous message to an object. Arguments are automatically encoded on the sender side and decoded on the receiver side.

```
int price = vstore.buy( "Killer Rabbits" );
```

One-Way Messages

Although messages are synchronous by default, Voyager supports one-way messages as well. One-way messages do not return a value. When a caller sends a one-way message, the caller does not block while the message completes.

```
vstore.buy( "Killer Rabbits", new OneWay() ); // no return
:
```

Future Messages

Voyager also supports future messages. When a caller sends a future message, the caller does not block while the message completes. The caller receives a placeholder that can be used to retrieve the return value later by polling, blocking, or waiting for a callback.

```
Result result = vstore.buy( "Killer Rabbits", new Future() );
:
int price = result.readInt(); // Block for price.
```

One-Way Multicast Messages

One-way multicast messages can be used to send one-way messages to all objects in a Space using a single operation.

```
VStore stores = new VStore( space ); // gateway into space
stores.stock( "video", 25 ); // send stock() to all stores in space
```

To send a one-way message to only certain objects in a Space, use a one-way multicast message with a selector.

Selective Multicast Messages

Multicast messages can be selectively broadcast to a subset of objects in a Space. For instance, Voyager supports traditional publish/subscribe multicasting where objects are selected based on whether or not they are subscribed to given subjects (defined as hierarchical strings). However, Voyager also supports a more general selection mechanism in that messages can be multicast to objects that meet any user-defined criterion.

```
VAccount accounts = new VAccount( space ); // gateway into space
Selector selector = new DelinquentSelector(); // select if delinquent
accounts.close( selector ); // close account if delinquent in payment
```

2.15 Federated Directory Service

Voyager provides a directory service for remote object lookup. Using the directory service, an object can get a reference to a remote or mobile object without advance knowledge of its location. Voyager's directory service avoids the single-server bottleneck/point-of-failure associated with monolithic directory services by allowing distributed directory services to be linked together to form a single, federated directory service.

All directories are completely integrated with Voyager's persistence mechanism, and like any object, can be saved to a database with a single command.

2.16 Dynamic Messaging

Voyager supports dynamic message construction at run time. The following code creates a synchronous message at run time using the Java virtual machine syntax for signature definition.

```
// dynamically create and execute a synchronous message
Sync sync = new Sync();
sync.setSignature( "buy( Ljava.lang.String; )I" );
sync.writeObject( "Killer Rabbits" );
Result result = vstore.send( sync );
int price = result.readInt(); // price
```

2.17 Life Spans and Garbage Collection

Each instance of a remote-enabled class has a life span. When an object reaches the end of its life span, the object dies and is garbage-collected. Garbage collection destroys an object, freeing the object's memory for reclamation by the Java virtual machine.

Voyager includes a distributed garbage collector that supports a variety of life spans.

- An object can live forever.
- An object can live until there are no more local or virtual references to it. By default, an instance of any class that does not extend Agent has this kind of reference-based life span.
- An object can live for a specified length of time. By default, an instance of any class that extends Agent lives for one day.
- An object can live until a particular point in time.

You can change an object's life span at any time.

Voyager leverages Java's run-anywhere code mobility to provide true agent-based computing as well as traditional distributed object communication. The ability of agents to move seamlessly about the new network provides a significant advantage for multitier, client/server, and peer-to-peer architectures.

Michael Greenspon, Sequential Interface, Inc.

3 Product Evolution

Many people ask us, "If Voyager is free, then how will you make money?" We believe that, years from now, companies will not make much money by selling basic middleware. DCOM will be embedded and distributed everywhere. CORBA price points are already plummeting. In the not-too-distant future, the bulk of the features currently in the Voyager Core Technology will be freely available in several forms and locations. Your cost is in development time, and your revenues are increasingly dependent on time to market. ObjectSpace believes Voyager's Java-centric binding, advanced mobile object features, and innovative services provide the best basis for rapid development of distributed systems in Java.

As the industry changes, we will continue to develop and sell partnerships, support, and other services, but we will also begin to unveil more and more next-generation, add-on features for the ObjectSpace Voyager platform. These add-ons will progress in areas of security, group communication, and persistence concurrency and will deliver the same time-to-market and rapid development advantages found in the Voyager Core Technology today. Unlike the Voyager Core Technology, these add-ons will not be free.

ObjectSpace is also pursuing several partnerships for the creation of technology integrations and enhanced development tools. Our relationships, based on the deployment of JGL, will enable the rapid distribution, adoption, and integration of the ObjectSpace Voyager platform.

As you look further into the future, you will see ObjectSpace using the Voyager technology base as the platform for its own next-generation product lines. As definite product release dates approach, we will announce these longer-term projects. Be assured that we will leverage Voyager's advantages, such as agent technology, to deliver products that, until now, you have only speculated about.

For additional information on Voyager, visit the ObjectSpace Web site at www.objectspace.com. You will find several additional white papers, customer stories, and of course, the complete Voyager Core Technology download. This download includes a comprehensive user guide that covers additional details on the Voyager 1.0 feature set. The Version 2.0 Beta 2 is also available for download.

ObjectSpace also offers several packaged services to help you in the evaluation, adoption, and use of ObjectSpace Voyager.

- ObjectSpace Voyager Core Technology Support
- ObjectSpace Voyager Core Technology Training
- ObjectSpace Voyager Platinum Partners Program
- ObjectSpace Voyager QuickStart Adoption Program

14.11 Milojičić, D.S., Breugst, B., Busse, I., Campbell, J., Covaci, S., Friedman, B., Kosaka, K., Lange, D., Ono, K., Oshima, M., Tham, C., Virdhagriswaran, S., and White, J., MASIF, The OMG Mobile Agent System Interoperability Facility

---

Milojicic, D.S., Breugst, B., Busse, I., Campbell, J., Covaci, S., Friedman, B., Kosaka, K., Lange, D., Ono, K., Oshima, M., Tham, C., Virdhagriswaran, S., and White, J., "MASIF, The OMG Mobile Agent System Interoperability Facility," *Proceedings of the Second International Workshop on Mobile Agents*, pp. 50-67, September 1998. Also appeared in *Personal Technologies*, 2:117-129, 1998.

MASIF
The OMG Mobile Agent System Interoperability Facility

Dejan Milojicic[5], Markus Breugst[3], Ingo Busse[3], John Campbell[5†], Stefan Covaci[3], Barry Friedman[2‡], Kazuya Kosaka[4], Danny Lange[2††], Kouichi Ono[4], Mitsuru Oshima[4], Cynthia Tham[2‡‡], Sankar Virdhagriswaran[1] and Jim White[2]

Crystaliz[1], General Magic, Inc.[2], GMD Fokus[3], IBM[4], Open Group[5]

† John Campbell is currently with the Sun Microsystems, East
‡ Barry Friedman is currently with Cisco
†† Danny Lange was with IBM during the initial part of MASIF
‡‡ Cynthia Tham is currently with Freegate

Abstract

MASIF is a standard for mobile agent systems which has been adopted as an OMG technology. It is an early attempt to standardize an area of industry that, even though very popular in the recent past, still has not caught on. In its short history MASIF has raised a lot of interest in both industry and academia. There are already a number of projects that are pursuing MASIF reference implementation or will be in the near future. MASIF addresses the interfaces between agent systems, not between agent applications and the agent system. Even though the former seem to be more relevant for application developers, it is the latter that impacts interoperability between different agent systems and therefore applications. This paper describes two sets of interfaces that constitute MASIF: MAFAgentSystem and MAFFinder (the acronym MAF is used for historical reasons). MASIF extensively addresses security. The paper provides a brief, but complete, description of MASIF and its interfaces, data types and data structures.

Keywords: Mobile Agents, Standards, OMG, Security, Interoperability, CORBA, and Java.

1. Introduction

Mobile agents are a relatively new technology, but there are already a number of implementations, such as MuBot [5], AgentTcl [6], Aglets [4], MOA [8], Grasshopper [13], and Odyssey [7]. These systems differ widely in architecture and implementation, thereby impeding interoperability, rapid proliferation of agent technology and growth of the industry. To promote interoperability and system diversity, some aspects of mobile agent technology must be standardized. MASIF [1] is a collection of definitions and interfaces that provides an interoperable interface for mobile agent systems. It is as simple and generic as possible to allow for future advances in mobile agent systems. MASIF specifies two interfaces: **MAFAgentSystem** (for agent transfer and management) and **MAFFinder** (for naming and locating MASIF objects).

Interoperability in this document is not about language interoperability. MASIF is about interoperability between agent systems written in the same language (potentially by different vendors) expected to go through revisions within the life time of an agent. Language interoperability for active objects that carry "continuations" around is difficult, and it is not addressed by MASIF. Furthermore, MASIF does not standardize local agent operations such as agent interpretation, serialization/deserialization, and execution. In order to address interoperability, concerns, the interfaces have been defined at the agent system rather than at the agent level. MASIF standardizes:

- **Agent Management.** One can envision a system administrator managing agent systems of different types via standard operations in a standard way: create an agent, suspend it, resume, and terminate.
- **Agent Transfer.** It is desirable that agent applications can freely move among agent systems of different types, resulting in a common infrastructure, and a larger base of available system agents can visit.
- **Agent and Agent System Names.** Standardized syntax and semantics of agent and agent system names allows agent systems and agents to identify each other, as well as clients to identify agents and agent systems.
- **Agent System Type and Location Syntax.** The agent transfer can not happen unless the agent system type can support the agent. The location syntax is standardized so that the agent systems can locate each other.

Table 1. describes the types of interoperability MASIF addresses, and estimates the complexity of agent systems required to support it. **Agent management** allows agent systems to control agents of another agent system. Management is addressed by interfaces for suspending, resuming, and terminating agents. This is straightforward to implement. **Agent tracking** supports locating agents registered with MAFFinders (naming service) of different agent systems. This is also straightforward to implement. **Agent communication** is outside the scope of MASIF, and it is extensively addressed by CORBA [2]. **Agent transport** defines methods for receiving agents and fetching their classes. This requires cooperation between different agent systems, and it is complex to achieve.

Function	Addressed by MASIF	Complexity
agent management	yes	straightforward
agent tracking	yes	straightforward
agent communication	no	n/a
agent transport	yes	complex

Table 1. Interoperability addressed by MASIF and associated complexity to implement it.

There are other aspects that should be standardized when the industry is more mature. The security issues become complex when an agent makes a multi-hop between security domains. Most security systems today deal only with single-hop transfer. Standardizing multi-hop security should be delayed until security systems can handle the problem. Today's mobile agent systems use different languages (e.g. Tcl and Java). The effort to convert between encodings is too complex. When the code and serialization formats are similar, it should be possible to build standard bridges between different agent system types.

2. Basic Concepts

An **agent** is a computer program that acts autonomously on behalf of a person or organization. Most agents are programmed in an interpreted language for portability. Each agent has its own thread of execution, so tasks can be performed on its own initiative. A **mobile agent** is not bound to the system where it begins execution. It has ability to transport itself from one system in a network to another. **Agent state** (execution state and the attributes) and code are transported while agent travels. An **agent's authority** identifies the person or organization for whom the agent acts. **Agent names** are required for identification, management operations, and locating. Agents are named by their authority, identity, and agent system type, whose combination has a unique value. An agent's identity is a unique value within the scope of the authority that identifies a particular agent instance.

An **agent system** is a platform that can create, interpret, execute, transfer and terminate agents. Like an agent, an agent system is associated with an authority that identifies the person or organization for whom the agent system acts. An agent system is uniquely identified by its name and address. A host can contain one or more agent systems. An **agent system type** describes the profile of an agent. For example, if the agent system type is Aglet, the agent system is implemented by IBM, supports Java as the Agent Language, uses Itinerary for travel, and uses Java Object Serialization. MASIF recognizes agent system types that support multiple languages and serialization methods. A client requesting an agent system function must specify the agent profile (agent system type, language, and serialization method) to uniquely identify the requested functionality.

An agent transfers itself between **places.** A place is a context in which an agent executes. It is associated with a location, which consists of the place name and the address of the agent system where the place resides. An agent system can contain one or more places and a place can contain one or more agents. If an agent system does not implement places, then it acts as a default place. When a client requests the location of an agent, it receives the address of the place where the agent is executing.

A **region** is a set of agent systems of the same authority, but not necessarily of the same agent system type. Regions allow more than one agent system to represent the same person or organization. A region may grant a richer set of privileges to one agent than to another agent with a different authority, e.g. an agent with the same authority as the region may be granted administrative privileges. A region can be same as an identity domain of CORBA security if its authority equals domain's identity. Region can be regarded as a security domain.

Serialization is the process of storing the agent in a serialized form, sufficient to reconstruct the agent. Deserialization is inverted process. The serialized form must be able to identify and verify the classes from which the fields were saved. **Codebase** specifies the location of the classes used by an agent. It can be an agent system or non CORBA object such as Web servers. In **remote agent creation**, a client program interacts with the destination agent system to request that an agent of a particular class be created. A client can be an agent or a program in a non-agent system The client authenticates itself to the destination agent system, establishing the authority and credentials that the new agent will possess. Then it supplies initialization arguments and, if necessary, the class needed to instantiate the agent.

During **agent transfer** the destination place and the quality of communication service are identified. The latter is not specified by MASIF, it is left open to agent system implementors. When the destination agent system agrees to the transfer, the source agent's state, authority, security credentials, and (if necessary) its code are transferred to the destination. The destination agent system reinstantiates the agent, and resumes its execution. **Remote method invocation** on another agent or object needs to be authorized and requires a reference to the object and specification of required level of quality of service (not covered by MASIF). When an agent invokes a method, the security information supplied to the communications infrastructure executing the method invocation is the agent's authority. Most distributed object systems support this.

3. Functions of an Agent System

a) Initiating agent transfer (executed on the sender side).

1. Suspend the agent (halt the agent's execution thread).
2. Identify transferable pieces of the agent's state.
3. Serialize the instance of the Agent class and state.
4. Encode it for the chosen transport protocol.
5. Authenticate client.
6. Transfer the agent.

b) Receiving an agent (executed on the receiver side).

1. Authenticate client.
2. Decode the agent.
3. Deserialize the Agent class and state.
4. Instantiate the agent.
5. Restore the agent state.
6. Resume agent execution.

Figure 1. Algorithms for agent transfer.

Transferring an Agent. The mobile agent requests the source agent system for a transfer to the destination agent system, as a part of an internal API. When the destination agent system receives the transfer request it executes the algorithm described in Figure 1.a. Before an agent is received into a destination agent system, the destination agent system must determine whether it can interpret the agent. If the agent system can interpret the agent, it accepts the agent, an executes the algorithm presented in Figure 1.b. There are three cases when class transfer is necessary:

1. **Agent instantiation (remote agent creation).** When an agent is created remotely by invoking a create operation at the agent system, the Agent class is needed to instantiate the agent. If the class does not exist there, it must be transferred from the source agent system.
2. **Agent instantiation (agent transfer).** After an agent travels to another agent system, the Agent class is needed to instantiate the agent. If it does not exist there, it is transferred from the source.
3. **Agent execution after instantiation.** After an agent is instantiated due to 1) or 2), the agent often creates other objects. If these objects' classes are not available at the destination, they are transferred from the source.

Class transfer. The common conceptual model is flexible enough to support variations of class transfer so that implementors have more than one method available. Specifically, the model supports:

1. **Automatic transfer of classes.** The source agent system (the class provider or the agent sender) sends all classes needed to execute the agent with each remote agent creation or transfer request. This eliminates the need for the destination agent system to request classes. However, it consumes unnecessary bandwidth if the classes are already cached at the destination.
2. **Automatic transfer of the Agent class, transfer on demand of other classes.** The source agent system sends the class needed to instantiate the agent with each remote agent creation or transfer request. If more classes are needed after instantiation the agent, the destination issues requests to the class provider. If the class provider is not accessible from the destination, the destination agent system issues the request to the sender agent system by calling *fetch_class* method with the codebase as parameter. The sender locates the requested classes either by using the codebase information, or by sending a further request to another agent system associated with the codebase. The sender may have cached the classes. This approach does not require the source to determine all possible classes necessary before creating or transferring an agent, and it is more

efficient as more classes are cached at the destination. However, the agent creation or transfer request fails if the destination agent system cannot access the source agent system to transfer the necessary classes, e.g. if the source agent system is a portable computer that disconnected since the agent creation or transfer.

3. **Variations of the previous two: a)** Case 2) when transferring an agent and case 1) when creating an agent remotely. When a remote agent creation is launched by a client that is not always connected, all classes are automatically transferred for remote agent creation operations. **b)** The source sends a list of class names that includes all the classes necessary to perform the specific agent operation. The destination requests only the classes that have not been cached. This approach is efficient, but it still requires the source agent system to know which classes the agent needs before making the agent creation or transfer request.

1. Start a thread for the agent.
2. Instantiate the Agent class.
3. Generate assign a globally unique agent name.
4. Start execution of the agent within its thread.

Figure 2. Algorithm for Agent Creation.

Creating an Agent. To create an agent, an agent system creates an instance of the Agent class within a default place or a place the client specifies. The Agent class specifies the interface and the implementation of the agent. It executes on its own thread. A simplified algorithm for agent creation is presented in Figure 2.. An agent system must generate a unique name for itself, and the agents and places it creates. When an agent wants to communicate with another agent, it must be able to find the destination agent system to establish communication. The ability to locate a particular mobile agent is also important for agent management.

4. Security

It is imperative for agent systems to identify and screen incoming agents. An agent system must protect resources including its operating system, file system, disks, CPU, memory, other agents, and local programs. To ensure the safety of system resources, an agent system must identify and verify the authority that sent the agent. The agent system must know the access the authority is allowed. The ability to identify the agent authority enables access control and agent authentication. Also, an agent might want to keep its activities confidential.

Threats and Attacks. Agent systems may be vulnerable to security threats due to weaknesses in the communications infrastructure and programming languages. MASIF is mainly concerned with communications security threats, such as: denial of service, unauthorized access through agent or agent system; unauthorized modification or corruption of data, spamming, spoofing or masquerade, trojan horse, replay, and eavesdropping.

Countering Threats and Attacks. To ensure that agents act responsibly, sets of rules are created and defined as security policies that govern agent activities. The security and safety services that the underlying communications infrastructure and the programming language provide enforce the rules. Both agents and agent systems can have multiple security policies. The authority that the agent or agent system represents sets the policies. The particular policy is determined based on the authenticity of the communicating parties credentials, agent class, agent authority, and/or other factors. Security policies contain rules for restricting or granting agent capabilities, setting agent resource consumption limits, and restricting or granting access.

Agent System Authentication. Agent systems are typically co-located with the information that their authority uses to authenticate itself. Authentication services normally available in secure communications infrastructures include this functionality. Agent systems use communication transport calls (e.g. RPC) to transfer agents between systems. To satisfy the destination agent system's security policies, mutual authentication of agent systems may be required. Agent systems operate without human supervision (e.g. without entering a password).

Agent authentication and delegation. Agents cannot carry their encryption key with them when they travel. Instead, agent authentication uses authenticators. An authenticator is an algorithm that determines an agent's authenticity. An authenticator uses information such as the authenticity of the source agent system or launching client, the authorities of the agent and agent system involved, and possibly information about trusted authorities that can authenticate an agent. If an agent is migrating to destination agent system, the agent credentials are transferred along. The credentials may be weakened depending on the authentication. If the client (or the server on its behalf) invokes a remote method, the client agent credentials are passed for charging or auditing.

Authenticators are divided into one-hop and multi-hop. It is currently possible to specify the behavior of and requirements for a one-hop authenticator. A one-hop authenticator can authenticate an agent traveling one hop from its source agent system. For example, an agent of authority A is executing on a source agent system of Authority A, then migrates to a destination agent system of authority B. If destination can successfully authenticate the source, the agent retains its authenticity on destination. If destination cannot authenticate the source, the agent is not defined as authenticated. MASIF currently does not address multi-hop authentication.

Authentication of clients for remote agent creation. Security services must provide for the authentication of non-agent system client applications. This might be done using passwords or smart cards. Authenticating a client establishes the credentials of agents that the client launches determines security policy.

Agent and agent system security policies. The agent and its agent system must set and enforce the access controls. If the access controls are self-defined and self-enforced, the source agent's credentials must be available to the destination agent system.

Level of Network Communications Security. For any communication, the requestor must be able to specify its integrity, confidentiality, replay detection, and authentication requirements. The communications infrastructure must honor these requirements, or return a failure indication to the requestor.

5. CORBA Services

This chapter contains brief descriptions of the CORBA services related to mobile agent technology (see Figure 3.): naming service, lifecycle service, externalization service and security service

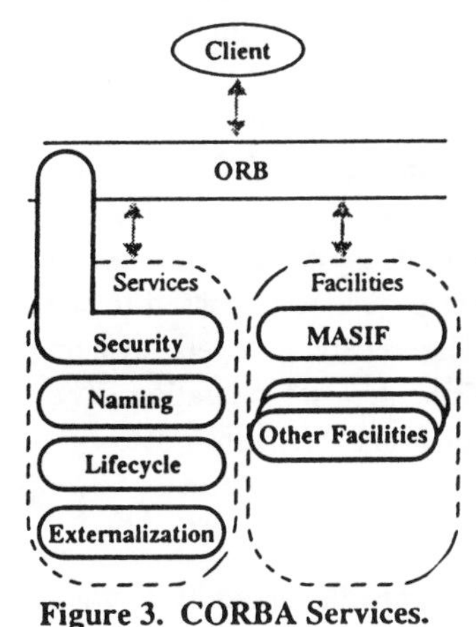

Figure 3. CORBA Services.

5.1 Naming Service

The CORBA Naming Service binds names to CORBA objects. The resulting name-to-object association is called a *name binding*, which is related to a *naming context*. A naming context is an object containing a set of name bindings in which each name is unique. Naming contexts can be combined into a *naming graph*. A specific object can be addressed by a sequence of names that builds a path in the naming graph. Applications use the Naming Service to publish named objects, or to find an object given the name. An application typically bootstraps a reference to a naming context using the ORB::resolve_initial_references operation.

MASIF describes two CORBA object interfaces: MAFAgentSystem and MAFFinder. These objects may be published in the Naming Service. Agents acting as CORBA objects may publish themselves, and thereby allow applications to dynamically obtain object references to remote agents and interact with them using CORBA RPC. Since a CORBA object reference (IOR) comprises, among others, the name of the host on which an object resides and the corresponding port number, a mobile agent gets a new IOR after each migration, and the IOR kept by the accessing application becomes invalid. This can be solved in different ways:

ORB itself is responsible for keeping the IOR of moving objects constant. The mapping of the original IOR to the actual IOR of the migrated agent is managed by a corresponding proxy object maintained by the ORB. This is not a mandatory feature in CORBA, and MASIF does not rely on this.

Update the name binding after each migration, i.e. supply the Naming Service with the new IOR. This is achieved either by the agent systems involved in the migration or by the migrating agent. This way, the Naming Service maintains the actual IOR during agent's lifetime. If an application tries to access the agent after it migrated, it receives an exception, and has to contact the Naming Service to obtain the new agent IOR.

The original instance (proxy) remains at home and forwards subsequent access. A disadvantage is that the proxy must be contacted by the agent after each migration in order to provide the new IOR, and that the home system must be accessible at any time. If the home system is terminated, the agent cannot be accessed.

In the context of the CORBA Naming Service, each of the components is represented by one CosName.Name object. The MAFFinder object is independent of specific authorities. The identification of such an object is managed by means of a single CosName.Name object corresponding to the CORBA Naming Service.

5.2 Lifecycle Service

The CORBA Life Cycle Service defines services and conventions for managing CORBA objects. The CORBA objects can be created, deleted, copied and moved using the Life Cycle Service. Since it is necessary to transfer the agent state, the Life Cycle Service must be combined with the CORBA Externalization Service. The Life Cycle Service can only be used for CORBA objects. MASIF does not require agents to be CORBA objects. In order to provide a uniform interface for both CORBA- and non-CORBA-based agents, new operations have been introduced. The *create_agent* and *terminate_agent* operations of the MAFAgentSystem interface (See Appendix A) can use the Life Cycle Service internally for CORBA-based agents.

5.3 Externalization Service

```
#include <CosExternalization.idl>
typedef sequence<octet> OctetString;
// MemoryStream externalizes objects to an in-memory octet sequence.
// After calling externalize() and flush(), the octet sequence
// representation may be accessed by calling get_octets().
interface MemoryStream : CosExternalization::Stream{
      OctetString get_octets();};
// Use MemoryStreamFactory to create a MemoryStream object. Call
// create() to make an empty MemoryStream for object externalization.
// Call create_from_octets() to make a MemoryStream that can
// internalize the objects from the supplied octet sequence.
interface MemoryStreamFactory {
      MemoryStream create();
      MemoryStream create_from_octets(OctetString octets);};
```

Figure 4. Externalization and Internalizations Interfaces.

The CORBA Externalization Service provides a standardized mechanism for recording an object's state onto, and for restoring it from a data stream. However, the implementor is free to choose other methods, e.g. Java Object Serialization. By using the Externalization Service to serialize an agent, the agent's state must be represented by a CORBA object that implements the Streamable interface. The agent system should also implement a MemoryStream object that has two purposes: 1) output an in-memory octet sequence when externalizing the agent and 2) read from an in-memory octet sequence when internalizing an agent. A MemoryStreamFactory interface allows for the creation of MemoryStream objects. A suggested set of interface definitions is presented inFigure 4.. Once an agent is externalized, the octet sequence is passed to the remote agent system's *receive_agent()* operation to transmit the agent's state. The receiving agent system constructs a MemoryStream from the received octet sequence using the *create_from_octets()* operation. The receiving agent system then calls the MemoryStream's *internalize()* operation to reconstruct the agent's state.

5.4 Security Service

This section describes how CORBA implementations fulfill the agent security requirements discussed in Section 4. The security capabilities of current CORBA implementations can be categorized as follows:

- **No security services.** The implementation includes neither proprietary nor standardized security interfaces. This type of implementation is limited to secure environments (physically or firewall protected), or to applications that contain no data or services worth protecting. Intranet applications are a typical example.
- **Proprietary security services.** The implementation includes a vendor-defined set of security capabilities such as authentication and access control. These services may be transparent to the application, or may be accessed via vendor-defined interfaces.They do not involve the ORB, and do not provide an acceptable level of safety.
- **Conforming to CORBA security services** (see [10] for more information about CORBA-defined security services). The implementation includes security services that conform to CSI level 0, 1 or 2 as defined in [11], and interfaces defined in [10]. The rest of the paper is related to this type of security service.

Agent Naming. The destination agent system must identify the principal on whose behalf an agent is acting. This is true even when that principal is not authenticated, because certain applications may find it acceptable to use application-defined heuristics to evaluate authenticity. An agent system provides the following information to an authorized user about an agent it is hosting (CORBA security uses the term principal instead of authority):

- The agent's name (principal and identity)
- Whether or not the principal has been authenticated (authenticity)
- The authenticator (algorithm) used to evaluate the agent's authenticity

Secure ORBs exchange security information about principals when remote operations are invoked. This information is available as a Credential object. If an ORB does not support security services, or a principal is not authenticated, the principal identity information is not available. An agent system exchanges principal information when agents are transferred. If available, the information in the Credential may be used to evaluate the authenticity of the exchanged information.

Client Authentication for Remote Agent Creation. CORBA security services offer client authentication services via the PrincipalAuthenticator object. The client invokes the authenticate operation to establish its credentials. When the client makes a request to create an agent, it makes the credentials (obtained via PrincipalAuthenticator object) available to the destination agent system. The principal for the new agent is then obtained via this credentials. The agent system uses this information to find and apply the appropriate security policies. A non-secure ORB does not provide client authentication. If a client creates an agent in such an environment, the client may supply a name for the agent, but the agent will be marked as "not authenticated".

Mutual Authentication of Agent Systems. CORBA security services allow administrators to require the mutual authentication of agent systems by setting the association options for agent systems. Specifically, both the EstablishTrustInClient and EstablishTrustInTarget association options are required. Both the source and destination agent systems transfer credentials before an agent transfer occurs, making it possible to apply security policy before transferring the agent and protecting agents from illegitimate agent systems and agent systems from illegitimate agents. A non-secure ORB does not provide mutual authentication of agent systems. An agent initially marked as authenticated is marked as "not authenticated" if it visits a non-authenticated agent system.

Access to Authentication Results and Credentials. At the destination of an agent transfer, CORBA security services provide access to the credentials of the source via the SecureCurrent interface. The get_credentials operation may be used to obtain a reference to a Credentials object. The Credentials object includes the sender's principal if the sender was authenticated. The receiver of an agent transfer request may evaluate the sender's credentials to determine the identity and authenticity of the sender. If an agent invokes operations on CORBA objects, the agent needs to have the credentials for its principal for secure invocations, even if the agent is defined as non CORBA object. In this case, the credentials object of the agent should be available at the destination. Therefore, the credentials of both the agent systems and the agents must be available at the destination.

If a secure ORB supports CSI level 2 with composite delegation, credentials of both the agent's principal and the sender agent system's principal can be made available at the receiver side. These credentials are obtained using the SecurityLevel2:Current interface (not possible if composite delegation is not supported). It is possible for a sender agent system to set and use the agent system's credentials for the agent transfer using SecurityLevel2:Current interface or may use the override_default_credentials on the reference of the target agent system. If the agent system's credentials are used for the agent transfer, the destination agent system can evaluate the sender agent system's principal to determine the identity and authenticity of the sender. However, the invocation credentials for the agent may become those of the agent system that is hosting the agent. This makes object invocations by any agent appear to have the authority of the agent system's principal.

Agent systems without having the CSI level 2, may choose to use the agent's credentials for the agent transfer. The receiver agent system can then use that credentials for the agent's secure invocations. In this case, the sender's credentials are not available and the receiver cannot evaluate the sender agent system's principal. Note that the secure agent system can be built on top of CSI level 0 or 1 if an agent does not invoke operations on CORBA objects. In a non-secure ORB, all agent transfers and agent operation invocations are anonymous. The only identifying information available is the unauthenticated principal value that an agent system may include during an agent transfer. The ORB does not transfer or support access to credentials.

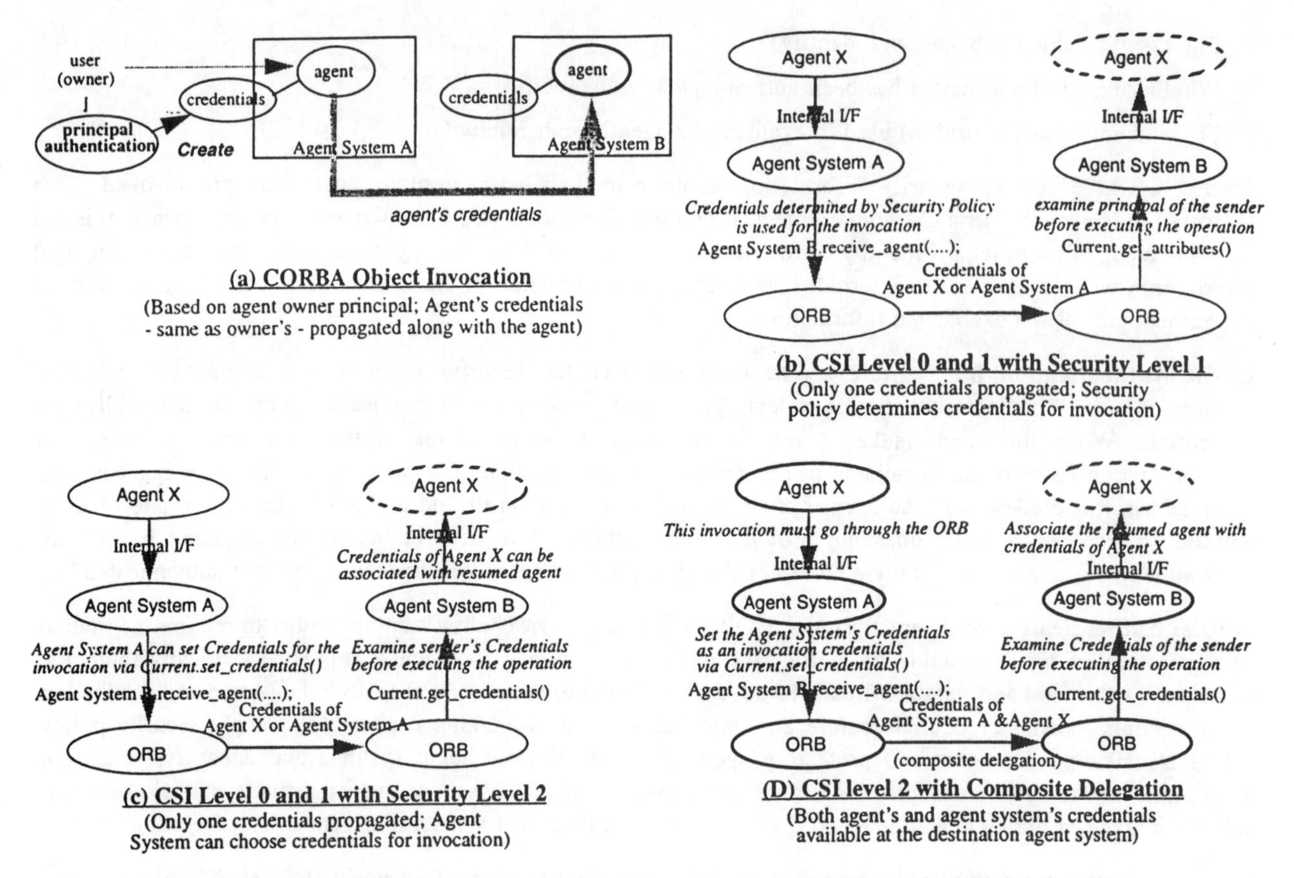

(a) CORBA Object Invocation
(Based on agent owner principal; Agent's credentials - same as owner's - propagated along with the agent)

(b) CSI Level 0 and 1 with Security Level 1
(Only one credentials propagated; Security policy determines credentials for invocation)

(c) CSI Level 0 and 1 with Security Level 2
(Only one credentials propagated; Agent System can choose credentials for invocation)

(D) CSI level 2 with Composite Delegation
(Both agent's and agent system's credentials available at the destination agent system)

Figure 5. Delegation of credentials in case of remote method invocation (a) and migration (b, c, d)

Agent Authentication and Delegation. When possible, it is desirable that secure ORB implementations propagate the agent's credentials along with the agent as it moves between agent systems. This may only be possible using composite delegation, which involves both parties in the transfer request, and propagates the credentials of the agent and the sending agent system. Upon receiving an agent's credentials, the receiving agent system should establish the agent's credentials as the invocation credentials of the agent. By doing so, operations invoked by the agent are subject to the policies associated with the agent's principal. This approach ensures the propagation of the agent's credentials during subsequent transfers. If an agent system receives an agent from an untrusted agent system, the agent system may choose to weaken the agent's credentials.

The propagation of both agent's credentials and agent system's credentials is only possible with composite delegation, which is only available with ORB implementations that conform to CSI level 2. It is not known whether ORB implementations will support delegation of credentials to application-created threads of execution. Delegation of credentials is needed to identify an agent's principals when an agent invokes a method on CORBA objects. In CSI level 0 and 1 implementations, only one of the credentials of the agent or the agent systems can be transmitted. If mutual authentication between agent systems is not required (e.g. in a trusted environment), the agent's credentials may be propagated to the destination agent system in lieu of the agent system's credentials. In non-secure ORB implementations, an agent's credentials are not propagated between agent systems.

Agent and Agent System Defined Security Policies. A CORBA object implementation may refuse to service a request. Secure ORB implementations (CSI levels 0,1, or 2) can provide the object with the credentials of the requestor, allowing objects to make their own access decisions. Typically, when a CORBA object throws exception CORBA::NO_PERMISSION of a type that indicates that a security violation was attempted and refused. Object implementations based on non-secure ORBs do not have the requestor's credentials available. They may refuse a request based on other criteria (e.g. values of the parameters).

Security Features. Secure ORB implementations allow applications to specify the quality of security service when they invoke operations. To specify the security level, set the security features of the invoker's credentials, or set the quality of protection in an object reference. Security features set via the invoker's credentials include: integrity, confidentiality, replay detection, misordering detection, and target authentication (establish trust). Security features set via the quality of protection in an object reference include: integrity and confidentiality.

6. MASIF Naming and Locating

The CORBA services are designed for static objects, CORBA naming services applied to mobile agents may not handle all cases well. Therefore, MASIF defines a MAFFinder interface as a naming service. A client can obtain the object reference to the MAFFinder using either the CORBA Naming Service or the method *AgentSystem.get_MAFFinder()*. The MAFFinder interface provides methods for maintaining a database of agents, places, and agent systems, and it defines operations to register, unregister, and locate these objects. The interface does not dictate the method a client uses to find an agent. Instead, it provides a range of location techniques:

- **Brute force search.** Find every agent system in the region, then check it to find the agent.
- **Logging.** Whenever an agent leaves an agent system, it leaves a mark that says where it is going. An agent system can follow the logs to locate that agent. The logs are garbage-collected after the agent dies.
- **Agent registration.** Every agent registers its current location in a database, which always has the latest information available about agents' locations. This can add an overhead to the agent *go()* operation.
- **Agent advertisement.** Register the places only. An agent's location is registered only when the agent advertises itself. To find a non-advertised agent, the agent system can use a brute force search or logging.

In the MASIF module, **Name** is defined as a structure that consists of three attributes: *authority, identity,* and *agent_system_type*. These attributes represent a globally-unique name for an agent or agent system. When Name is an agent name, the agent_system_type is the type of agent system that generated the identity of the agent. Authority defines the person or organization the agent or agent system represents. The authority of the agent must be equivalent to the principal of the agent's credentials if the CORBA security is used. Agent systems of different types may use different mechanisms to generate identities. Therefore, it is possible that two agent systems of different types might generate the same authority and identifier. The responsibility for naming an agent may also differ for each agent system type. The client may be responsible for naming in some agent system, while the agent system generates a name for the agent in others.

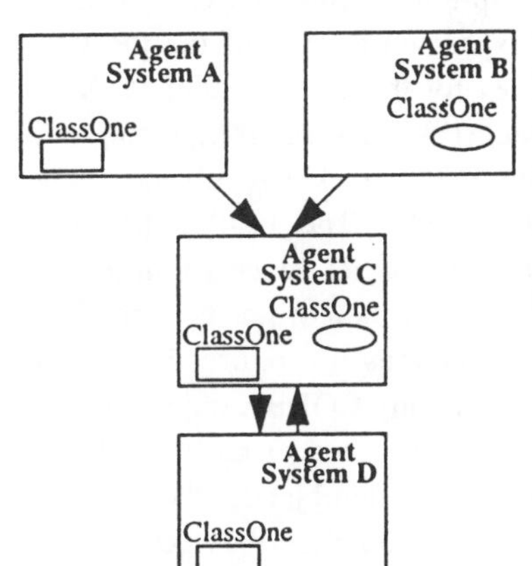

Figure 6. Class name uniqueness.

The **ClassName** structure defines the syntax for a class name. A class name has a human-readable name and an octet string that ensures uniqueness within the scope. MASIF does not specify mechanisms to make class names globally unique. MASIF implementors are responsible for ensuring that class names are unique within the scope of the source agent system.

Figure 6. illustrates the minimum requirement for class name uniqueness, (uniqueness within an agent system). When Agent System C requests ClassOne from Agent System A, it should be unique within the scope of A. Similarly, when C requests ClassOne from B, it should be unique within the scope of B. C must distinguish between the two versions. This is necessary, for example, if D needs ClassOne to create an instance of an A type of agent. Suppose C wants to create an agent on D. If a ClassOne is involved in this creation, C uses the class_names parameter in Create_Agent to specify which ClassOne is necessary. Once D receives the class, it can rename it. The difference between names for the same version on the two systems can cause an unnecessary class transfer. For example, if Agent System D later attempts to transfer an agent that uses the A version of ClassOne to C, C might not recognize that the ClassOne specified in the class list for the call is the same as the class A:ClassOne that it already has. If the region administrators of the communicating agent systems agree on a globally-unique class naming scheme, the problem of duplicate names for the same class can be avoided. For example, if class names were globally unique in Figure 6., Agent System C would not encounter two classes with the same name.

In the MAFFinder interface, **Location** specifies the path to an agent system based on the name of an agent system, agent, or place. For example, when *MAFFinder.lookup_agent()* is called using an agent name, a Location specifying the agent system that contains the agent is returned. Once the client gets the Location (a String) of an agent system, it must convert it to the object reference of the agent system to invoke the operations offered. The Location is in one of two forms: a) a URI containing a CORBA name; b) a URL containing an Internet address. The advantage of using the CORBA Naming Service is that it is protocol independent. The advantage of using an internet address is that it is better suited to mobile agents and the Internet. To determine the format of the Location, the client parses the string up to the first colon (:). If the characters preceding the colon are "CosNaming", the string is a CORBA name; if they are "mafiiop", the string is an Internet address.

```
mafuri     :=scheme":"location
scheme     :="CosNaming"
location   :=components I "/"location
components:=component I component"/"components
component:=id"!"kind
id         :=xpalphas
kind       :=xpalphas
```

Figure 7. Syntax of COSNaming Format.

COSNaming Location String Format. When the Location is in a CORBA name format, the client must convert the URI to the syntax of a CosNaming.Name (see RFC 1630 for URI). Once the Location is converted to a CosNaming.Name, the client uses it as the key for a search that returns the object reference. The format of a CosNaming. Name is an ordered sequence of components, consisting of two attributes: the identifier and the kind (both are strings). The location for an agent system or a place can be written in URI form using the syntax presented in Figure 7. (see RFC 1630 for a definition of xpalphas). Conversion of a CosNaming URI to a CosNaming.Name is a straightforward mapping from one to the other. In a CosNaming URI, the components are separated with slashes (/), and the identifier and kind attributes of each component are separated with an exclamation mark (!). For example, the Location containing *URI: CosNaming:/user!domain/user_name!u3* can be converted to the following CosNaming.Name: *{{"user", "domain"}, {"u3", "user_name"}}.*

```
mafurl        := scheme ":" location
scheme        := "mafiiop"
location      := "//" [hostport "/"] agentsystem ["/" place]*
hostport      := host ":" port
host          := hostname I hostnumber
port          := digit+
agentsystem   := uchar+
place         := uchar+
components    := component ["&" components]
component     := tagname "=" tagvalue
tagname       :=    "TAG_ORB_TYPE" I
                    "TAG_CODE_SETS" I
                    "TAG_SEC_NAME" I
                    "TAG_ASSOCIATION_OPTIONS"I
                    "TAG_GENERIC_SEC_MECH"
tagvalue      := uchar+
```

Figure 8. Syntax of Internet-Specific naming format.

Internet-Specific (MAFIIOP) Location Conversion Method. If IIOP is used, an IOR for an agent system in another domain may be constructed directly from the Location. The requirements for an IIOP IOR are host name, port number, and an object key (an octet string defined in the CORBA Interoperation description of IIOP IOR). The host and port number might be expressed as part of the location information, and the object key for the agent system can be a string value (e.g. "AgentSystem1"). Such references may not be actual object references, since CORBA objects may migrate and thus change their IOR. The IIOP redirection capability is used to map a received reference to an actual reference. The client's ORB caches the correct version. The requesting client object is oblivious to redirection. This works for obtaining a reference to any CORBA object, and is not MASIF specific. However, CORBA does not mandate redirection. Therefore, clients also have to obtain actual IORs, e.g. by contacting a Naming Service supplied with the actual IORs of mobile objects. The location of an agent system or a place can be expressed in URL form (see Figure 8. RFC1738 describes component definitions). URLs of this type can define locations accessible via IP-based networks such as the Internet.

Even though a place is addressable via this scheme, MASIF does not mandate places as first class CORBA objects. To get a reference to an agent system, a client creates an IOR using the host, port, and agent system components of URL. If there is a place component (a path separated by one or more slashes), it is passed as an argument to operations. If there are components (equations separated by "&"), then it denotes tagged components included in IIOP 1.1 IOR. Because locations can be strings, no special data type is required for them.

Additional Location Conversion Method. For non-IP networks that do not use the CORBA Naming Service, other URIs (see RFC1630) could be developed. Those URIs are distinguished from the URL defined above by choosing different scheme tags. The location specification and how it can be mapped to an IOR will be defined.

OMG has agreed to become the naming authority for mobile agent technology. Having a naming authority benefits the interoperability and unambiguous exchange of information between different MASIF applications. The identifiers assigned to parameters such as agent system type and authenticator should be unique across all implementations of MASIF.

Agents and agent systems provide application specific properties. A client may specify the properties in order to restrict the scope of a search operation while looking for a specific agent or agent system using the corresponding lookup method of the MAFFinder interface. In order to specify a property, a client must support the application specific format of the value component of the property. The semantics and syntax of the value are identified by the name component of the property.

7. Conclusions and Future Work

The MASIF goals are to accelerate the industry and promote a higher level of interoperability between different agent systems. It is a result of the significant effort of a team comprised of experienced agent developers. Experience accumulated in developing Telescript, Odyssey, Aglets, Mubot, Magna and MOA was incorporated into MASIF. A number of presentations given at five OMG meetings (Cape Cod, Tampa, Montreal, Ireland and East Brunswick) in front of different audiences, such as working groups (ORBOS and Common Facilities), Security SIG, Architecture Board, Technical Committee, as well as many individual OMG members who have contributed to improving the technical contents and presentation of the specification.

According to the OMG rules, after each technology submission is made, a reference implementation should follow within a year of submission acceptance. This applies only to official submitters (IBM and GMD) and not to supporters (General Magic, Crystaliz and Open Group). This is the only difference between these two different levels of participation. The latter can, and are encouraged to, pursue reference implementation as well. The Open Group, for example, plans to pursue reference implementation within the project ANIMA [9].

The immediate feedback on MASIF is encouraging. A number of European projects are interested in adopting MASIF, for example MARINE at Italtel, MIAMI at GMD Fokus, MONTAGE at INTRACOM, Greece, MARINER at Teltec Ireland [12]. FIPA [14] has tentative plans to adopt OMG technology to avoid duplication of standardizations and repeating efforts. We believe that after the reference implementations are introduced that MASIF will increase its relevance. The developers hope that MASIF will achieve a higher level of interoperability than is the case with ORBs, i.e. that the reference implementations by different developers will interoperate. This is the ultimate goal of MASIF.

Acknowledgment

The MASIF proposal (originally MAF) is a result of a collaborative work of many contributors. It started more than three years ago as OMG issued a Request for Proposals [3] on the Mobile Agents Facility. Sankar Virdhagriswaran from Crystaliz was instrumental in creating and maintaining the CFP. The original submissions were made by Crystaliz and IBM [15]. After the first presentations, a joint proposal was made. At that time, The Open Group, GMD Berlin and General Magic joined the team. The joint proposal was soon replaced with a new one, which was eventually presented at OMG and (many times) voted upon. The co-authors of this paper contributed through various phases of project. Dan Chang and Danny Lange of IBM and Sankar Virdhagriswaran wrote the original proposals. Dan Chang was instrumental in the first joint proposal. The second joint proposal was a true joint effort, and it was led by Cynthia Tham. The major contributors were Barry Friedman and Jim White of GMI, Danny Lange of IBM, Stefan Covaci and Ingo Busse of GMD, Sankar Virdhagriswaran of Crystaliz and Dejan Milojicic and John Campbell of The Open Group. Larry Smith of IBM was driving the last phases of the project. He successfully took us through the "politicking" waters of OMG. Dejan Milojicic, Mitsuru Oshima and Markus Breugst of GMD Fokus technically led the last phases of submission. According to an approximate estimate, the invested effort in MASIF by co-submitter companies is 7.5 engineer years. We are indebted to various reviewers of MASIF and of this paper who are too numerous to be mentioned here.

References

[1] OMG MASIF, OMG TC Document ORBOS/97-10-05, also available from http://www.opengroup.org/~dejan/maf/draft10.

[2] OMG, The Common Object Request Broker: Architecture and Specification, Revision 2.0, July 1995.

[3] OMG, Common Facilities RFP 3, OMG TC Document 95-11-3, November 3, 1995.

[4] Aglets Workbench (http://www.trl.ibm.co.jp/aglets).

[5] Mubot: (http://www.crystaliz.com).

[6] Agent-Tcl, http://www.cs.dartmouth.edu/~agent/.

[7] Odyssey: (http://www.genmagic.com/agents/).

[8] Mobile Objects and Agents, to appear in the Proc. of the COOTS'98 (http://www.opengroup.org/RI/java/moa/).

[9] ANIMA, http://www.gr.opengroup.org/anima/.

[10] CORBA Security Services Specification.

[11] Common Secure Interoperability Specification (CSI).

[12] ACTS Domain 5, Agent Cluster Baseline Document, editor T. Magedanz, January 1998.

[13] Grasshopper, http://www.ikv.de/products/grasshopper.html.

[14] FIPA http://drogo.cselt.it/fipa/.

[15] Chang, D., Covaci, S., "The OMG Mobile Agent Facility: A Submission", Proc. of MA'97, April 1997, pp 98-100.

Appendix. Interfaces, Data Structures, and Data Types

```
interface MAFFinder {
    void register_agent (in Name agent_name,   in Location agent_location,   in AgentProfile agent_profile,);
    void register_agent_system (in Name agent_system_name, in Location agent_system_location,
        in AgentSystemInfo agent_system_info);
    void register_place (in string place_name, in Location place_location);
    Locations lookup_agent (in Name agent_name, in AgentProfile agent_profile);
    Locations lookup_agent_system (in Name agent_system_name, in AgentSystemInfo agent_system_info);
    Location lookup_place (in string place_name);
    void unregister_agent (in Name agent_name);
    void unregister_agent_system (in Name agent_system_name);
    void unregister_place (in string place_name);};
interface MAFAgentSystem {
    Name create_agent (in Name agent_name, in AgentProfile agent_profile, in OctetString agent, in string place_name,
        in Arguments arguments, in ClassNameList class_names, in string code_base, in MAFAgentSystem class_provider);
    OctetStrings fetch_class(in ClassNameList class_name_list, in string code_base, in AgentProfile agent_profile);
    Location find_nearby_agent_system_of_profile (in AgentProfile profile);
    AgentStatus get_agent_status(in Name agent_name);
    AgentSystemInfo get_agent_system_info();
    AuthInfo get_authinfo(in Name agent_name);
    MAFFinder get_MAFFinder();
    NameList list_all_agents();
    NameList list_all_agents_of_authority(in Authority authority);
    Locations list_all_places();
    void receive_agent(in Name agent_name, in AgentProfile agent_profile, in OctetString agent, in string place_name,
        in ClassNameList class_names, in string code_base, in MAFAgentSystem agent_sender);
    void resume_agent(in Name agent_name);
    void suspend_agent(in Name agent_name);
    void terminate_agent(in Name agent_name);};
```

MASIF does not dictate agent system types, languages, serialization mechanisms, and authentication methods must be used to accommodate new systems. The OMG naming authority should begin with these initial values:

Language: NotSpecified (0), Java (1), Tcl (2), Scheme (3), Perl (4)
Agent system types: NonAgentSystem (0), Aglets (1), MOA (2), AgentTcl (3)
Authenticator types: none (1), one-hop (2)
Encoding mechanisms: SerializationNotSpecified (0), Java Object Serialization (1)

```
typedef short               AgentSystemType;
typedef sequence<octet>OctetString;
struct ClassName{
      string                name;
      OctetString           discriminator;};
typedef sequence<ClassName> ClassNameList;
typedef OctetString         Authority;
typedef OctetString         Identity;
struct Name{
      Authority             authority;
      Identity              identity;
      AgentSystemType       agent_system_type;};
typedef string              Location;
typedef short               LanguageID;
typedef short               AgentSystemType;
typedef short               Authenticator;
typedef short               SerializationID;
typedef sequence<SerializationID> SerializationIDList;
typedef any                 Property;
typedef sequence<Property>PropertyList;
struct LanguageMap {
      LanguageID            language_id;
      SerializationIDList   serializations;};
typedef sequence<LanguageMap>LanguageMapList;
struct AgentSystemInfo {
      Name                  system_name;
      AgentSystemType       system_type;
      LanguageMapList       language_maps;
      string                system_description;
      short                 major_version;
      short                 minor_version;
      PropertyList          properties;};
struct AuthInfo {
      boolean               is_auth;
      Authenticator         authenticator;};
struct AgentProfile {
      LanguageID            language_id;
      AgentSystemType       agent_system_type;
      string                agent_system_description;
      short                 major_version;
      short                 minor_version;
      SerializationID       serialization;
      PropertyList          properties;};
```

Chapter 15 Other Sources of Information

There are many sources of information on mobile agents. In this section we offer some pointers to books, mailing lists, standards, conferences, journals, sources of funding, and useful sites on mobile agents.

Books. There are many useful books on agents in general and on mobile agents in particular. Recently, both themes have often been represented in the same books, conferences, and mailing lists. For example, Bradshaw's book on software agents [Bradshaw, 1997] and "Readings in Agents" by Huhns and Singh [Huhns, 1997], both contain sections on mobile agents. The books exclusively dedicated to mobile agents are *Mobile Agents* [Cockayne, 1997] and *Programming Mobile Agents in Java™—With the Java Aglet API* [Lange, 1998b]. There are also a few proceedings of mobile agent-related workshops, such as "Mobile Object Systems" by Vitek and Tschudin [Vitek, 1997b] (in addition it contains some other related materials), and the Proceedings of the First and Second International Workshop on Mobile Agents.

Mailing lists. The main source of information on agents is the agents' mailing list, maintained by Tim Finin at University of Maryland at Baltimore County. This mailing list, and especially Tim Finin, deserve credit for keeping the agent-related discussions alive. The address of the mailing list is agents@cs.umbc.edu, and archives are available at http://www.csee.umbc.edu/agentslist/archive/, along with the instructions on how to join. The mailing list dedicated exclusively to mobile agents is a new mobility mailing list maintained at MIT. To subscribe, send a "subscribe mobility" to majordomo@media.mit.edu. There is an archive and associated Web site maintained at http://mobility.lboro.ac.uk. Various projects maintain their project-related mailing lists, such as Aglets (aglets-request@javalounge.com), and Voyager (voyager-interest@objectspace.com).

Standards. There are a few standard bodies that deal with mobile agents. The most active one so far has been OMG MASIF, which became an official OMG technology (see http://www.omg.org/news/pr98/8_10.htm). The spec is available at http://www.omg.org/library/schedule/Technology_Adoptions.htm, and it is also described in a paper included in this book [Milojičić, 1998a]. FIPA (http://drogo.cset.it/fipa) recently expressed interest in merging or overlapping activities with the OMG MASIF effort. Finally, the Agent Society (http://www.agent.org) was a good source of various information and activities in the past, but it became inactive recently. Some other bodies also deal with agents peripherally, but their work is of great importance for the agents. Such bodies include the W3C (http://www.w3c.org), which is relevant for development of various Web technologies, the IETF (http://www.ietf.org) for its development of Internet technologies; and the Active Group (http://www.activex.org), which was introduced for opening up the Active X specification.

Conferences and Workshops. There are many events either entirely dedicated to mobile agents or that have sessions, workshops, or minitracks on mobile agents:

The Dartmouth Workshop on Transportable Agents (DWTA, http://www.cs.dartmouth.edu/~agent/workshop/) has been held for two consecutive years. In 1999, DWTA will be co-located with another workshop, "Agent Systems and Applications" (ASA), sponsored by the IEEE Task Force on Internetworking, and co-sponsored by USENIX.

The International Workshop on Mobile Agents has also been held for two consecutive years and produced a number of quality papers exclusively related to mobile agents (http://www.fokus.gmd.de/ice/events/MA_97.html, http://www.informatik.uni-stuttgart.de/ipvr/vs/ws/ma98/ma98.html). Mobile Agents '99 is also planned to integrate with ASA.

Workshops on Mobile Object Systems (http://cuiwww.unige.ch/~ecoopws/) have been held in conjunction with ECOOP for four consecutive years. They feature many mobile agent papers.

HICSS (http://www.cba.hawaii.edu/hicss/) held minitracks on agents in two consecutive years ("Agent Mobility and Communication" in 1998, http://www.camb.opengroup.org/RI/HICSS31-agents/, and "Software Agents" in 1999, http://www.ca.sandia.gov/~carmen/hicss.html), which address both mobile and static agents.

WET ICE (http://www.cerc.wvu.edu/WETICE/) organized an agent related workshop (Collaboration in Presence of Mobility, http://www.camb.opengroup.org/RI/WETICE) that addresses various aspects of mobility with respect to collaboration.

The Third International Conference and Exhibition, Practical Applications of Intelligent Agents and Multi-Agents (PAAM97) (http://www.demon.co.uk/ar/PAAM98), features demonstrations of applications of various forms of agents.

The International Conference on Autonomous Agents is sponsored by the ACM, ISI, and others, and was last held in Minneapolis/St. Paul MN, USA, 10–13 May 1998 (http://www.cssi.udel.edu/~agents98).

Dagstuhl organized a Seminar on Mobile Software Agents at Schloss Dagstuhl, Germany, in October 1997. Abstracts, slides, and reports are available at http://www.informatik.tu-darmstadt.de/VS/mobags97/.

Journals and magazines. *IEEE Internet* and *IEEE Concurrency* magazines have regular columns, departments, and occasional special issues on agents and on mobile agents.

Sources of funding. Mobile agents have been funded by various governments around the world. In the US, DARPA has recently started the Agent Based Systems (ABS) program [DARPA] (http://www.darpa.mil/iso). NSF has mostly funded intelligent agents. In Europe, there are two programs that fund agents related research. One is ESPRIT and the other is ACTS. MITI in Japan has also funded work on mobile agents.

Useful sites. The following sites contain extensive sources of mobile agent related information, such as links to various projects, papers, bibliographies, tutorials, and FAQs:

- **UMBC AgentWeb:** http://www.cs.umbc.edu/agents/

- **Agent Society:** http://www.agent.org

- **Cetus Links:** http://www.cetus-links.org/oo_mobile_agents.html

Part V

Summary

Chapter 16 Summary

We have presented an overview of mobility in distributed systems by providing a selection of relevant papers in three different areas: process migration, mobile computing, and mobile agents. The first and the third areas are about logical migration, migrating bits (code and data); the second area is about physical migration or the migration of the hardware. These three areas have been introduced in the chronological order of their appearance.

Each area is classified thematically. Process migration papers are grouped by early systems, kernel-based systems, user-space implementations, and migration policies. Mobile computing papers are grouped into those that discuss limits on connectivity, mobile IP, and ubiquitous computing. The mobile agent those represent ten different mobile agent systems and a standard specification. For each of these areas, an introduction, summary, and other sources of information are provided.

This part of the book summarizes how each of the areas presented addresses the common elements of mobile systems. A summary of the benefits of mobility, deployment challenges, and technical issues is presented in Section 16.1. In Section 16.2, we speculate about the future of each type of mobility. Finally, we summarize the book in Section 16.3.

16.1 Comparison

This section compares the three fields in a tabular format. Table V.1 presents a concise summary of the benefits of each of the three areas. Similarly, Table V.2 summarizes how the deployment challenges are addressed, and Table V.3 concludes with a presentation of the technical issues.

Benefits

As discussed in Part I, the benefits of mobility include locality of reference, the use of resources while moving, and flexibility. Locality of reference has been demonstrated to improve the performance of applications that take advantage of process migration, such as in the case of MOSIX [Barak, 1985a], Mach [Milojičić, 1993b], and Emerald [Jul, 1988]. MOSIX demonstrates the performance benefits of accessing resources locally by comparing the performance of local and remote access for various system functions, throughput for accessing data, and overall performance benefits achieved by using process migration. Mach task migration presents performance trade-offs for local and remote execution of programs for different migration scenarios. Emerald compares performance of call-by-move, call-by-visit, and remote-reference. In each case, it is locality of reference that provides dramatic performance gains when it is applied appropriately.

Sprite [Douglis, 1989] and Accent [Zayas, 1987a] demonstrate how resources can be used while processes are moving. The two papers on load distributed scheduling, by Cabrera [Cabrera, 1986] and Harchol-Balter and Downey [Harchol-Balter, 1996], argue that it is reasonable to use process migration for improved load distribution. Sprite presents how under-utilized computers can be used to off-load the computation intensive applications and how to use migration to evict the processes once they are reclaimed.

The Worm, Charlotte, and Condor papers discuss the flexibility offered by migration. Worm programs can locate available computers, clone them, and migrate to them [Shoch, 1982]. Condor supports user level migration and checkpointing [Litzkow, 1992].

Table V.1 Benefits

	Process migration	Mobile computing	Mobile agents
Locality of reference, moving toward resources	move toward other processes, special devices, or data sources	user and computer move where needed	search databases remotely administer computers locally
Using computer resources while moving	use resources of other computers (memory, processor, etc.)	wireless networking	migratory applications
Flexibility	leave partially failed or about to be administered (for example, shut-down) host	move computers around as needed	leave partially failed domains, reconfigure system, adapt

Deployment Challenges

There are only a few commercial systems that support process migration, such as Utopia [Zhou, 1993]. There is no widespread infrastructure for mobile computing. There are demo applications, such as those presented for Rover [Joseph, 1997b], ubiquitous computing [Weiser, 1993b], and other papers, but few if any of these applications have been adopted commercially. Finally, mobile agents have many potential applications but only a few commercial systems have evolved.

Mobile IP [Perkins, 1998b] is still pretty rare in the commercial world, but there are other aspects to mobile technology that have some commercial success. For example, the "briefcase" metaphor in Microsoft Windows 95 allows one to synchronize the files on a laptop computer and with those on a desktop computer. Support for palmtop computers, like the PalmPilot, also supports this kind of synchronization between mobile and non-mobile systems.

Although most of the work in mobility revolves around the development of infrastructure, no common infrastructure is in common use. This is especially true for process migration where the technically superior distributed operating systems (for example, Sprite [Ousterhout, 1988]) has not reached wide acceptance. Lacking a popular base operating system with access to system source code, process migration mechanisms have been largely left to the research community. In the case of mobile computing and more recently in the case of mobile agents, standardization efforts are promising, but this also slows down development and innovation. Wide adoption is a prerequisite to global scalability. There is still a lot of effort needed to improve the scalability of mobile systems.

Table V.2 Deployment Challenges

	Process migration	Mobile computing	Mobile agents
Lack of applications	mainly relies on remote execution; parallelized apps are rare	mostly used for nomadic computing, rare use	huge potential, but few commercial apps available
Lack of a widespread infrastructure	support for (near) single system image: distributed IPC, DFS, naming, and so forth	heterogeneous networks, performance of mobile IP	docking station that would accept agents
Global scalability	heterogeneous and wide area network process migration	different countries have different schemes (for example, GMS in Europe)	WWW scale

Technical Issues

Technical issues best show the similarity between the various areas of mobility. Mobile agents and mobile computing both need to resolve issues dealing with authenticating mobile entities and allowing them access to their own data. From the perspective of security, process migration is mostly concerned with preserving the integrity of the migrated process. Most process migration systems assume a single trust domain.

Reliability and autonomous behavior use some similar solutions, such as avoiding residual dependencies and relying on various fault tolerant solutions, including checkpointing state and using transactional queues. Caching in Sprite's process migration mechanism [Nelson, 1988] is similar to the Coda hoarding scheme [Kistler, 1992] for mobile computing and the way that MOA packs agent state for mobile agents [Milojičić, 1998b].

Finally, various complexities in locating, naming, and communicating are resolved in a similar manner. An address in mobile IP [Perkins, 1998a] is similar to a home address in Sprite [Ousterhout, 1994] or MOSIX [Artsy, 1989]. Forwarding communication is similar to mobile IP and Charlotte's forwarding scheme [Artsy, 1989]. Various locating strategies, such as those in V Kernel [Theimer, 1985], Sprite [Douglis, 1989], Charlotte [Artsy, 1989], and MOSIX [Barak, 1989b], are equally applicable to systems such as Aglets [Lange, 1998a], MOA [Milojičić, 1998b], and the MASIF standard [Milojičić, 1998b]. Finally, maintaining communication channels is approached in a similar way by the various process migration implementations, such as Demos/MP [Powell, 1983] and Sprite, and by mobile agent systems, such as Voyager [Glass] and MOA [Milojičić, 1998b].

Table V.3 Technical Issues

	Process migration	Mobile computing	Mobile agents
Security	relies on host OS, typically single security domain; preserve integrity and resources of visiting host	authentication of the mobile computer, encryption of the data	protecting agents and hosts privacy; authentication and authorization
Reliability and autonomous, disconnected behavior	checkpointing; avoid residual dependency on the "home" and intermediate nodes	dependency on the forwarding computer, inconsistent data; hoarding reconciliation	relying on FT systems, transactional queues, checkpointing; encapsulating all agent state
Complexity in locating, naming, and communication	underlying distributed OS support and locating strategies (register, home, search, forward)	care/of addresses (update forwarding location)	similar to process migration, extended for Internet (for example, LDAP, RMI)

16.2 Looking into the Future

Process migration has been implemented successfully in a number of systems. Despite the technical advantages migration provides, actual deployment of migration has been a rare occurrence due to a lack of a common, single distributed operating system and the reluctance of users to share access to their computers with users from other computing nodes in a local cluster. Despite this, today there are several promising markets for process migration systems. In traditional UNIX systems, commercial vendors of user-level, checkpoint/restart migration schemes have gained popularity with users of long-running scientific or other heavy applications. In addition, the increasingly common nature of free UNIX systems, like Linux or FreeBSD, and the ever-growing presence of Microsoft's NT operating system might just provide the common base required for a traditional, kernel implementation of migration technologies.

Mobile computing is impacted by the dramatic changes that are taking place today. There is ongoing integration between the cable and networking, televisions and computers, telephones and computers, and the Web and public broadcasting. Satellite communication is competition for cable and phone lines. Wireless networking is slowly but steadily gaining momentum, while the integration of mobile phone with the computer will further increase wireless deployment. Ubiquitous connectivity in mobile computing will become an everyday need for people: continuous access to one's home, office, and to the Web wherever one is and whenever one wants.

The prospects of the acceptance of mobile agents is mainly based on the following three factors. First, as the numbers of computers continues to increase in everyday life, there will be more pressure for automated control and administration. Some form of agents will be most suitable for such tasks. Mobility will be a useful feature because of its local access and dynamic reconfiguration which prevents client server models from being equally well deployed. Second, with increasing amounts of available information, agents will likely become more common. There will be a need for some automated, probably intelligent, way to reduce the amount of information available to users into manageable amounts. Agents represent an obvious candidate for such a task. Third, due to the different network characteristics, agents will be used to improve the locality of reference, or even to perform the role of network configuration and protocols, such as envisioned in the active networks paradigm [Tennenhouse, 1996].

16.3 Summary

The goals of this book are threefold: first, we want to gather relevant papers on mobility in distributed systems and offer them as a unique collection to our readers. Our second goal is to illustrate how similar these areas are by drawing comparisons between the different forms of mobility based on the selected papers and our own experience in building and using mobile systems. Finally, we attempt to extrapolate the future based on these two previous goals.

Clearly, it is not possible to make a perfect selection of papers in any single area. However, we believe that the papers presented here give a broad and appropriate representation of each field. This is especially true for process migration, which is a well established and researched area.

Mobile computing has matured rapidly as a field of computer science. While research on mobile computing appears in many forums, the existence of an annual international conference specifically devoted to mobile computing is an indication of its growing importance. In addition, support within the commercial sector for mobile computing—particularly tools for file synchronization and for wireless networking—indicate the relative maturity of the field.

Mobile agents are the most recent form of mobility addressed in this book. This is recognized by the type of papers presented: with some notable exceptions, the papers mostly appeared at workshops and conferences as opposed to the journal papers predominantly selected for the first two areas.

The second goal is much harder to achieve. Our intuition, which led us to organize the book around these three forms of mobility in the first place, suggests that there are obvious similarities between the various areas of mobility despite the different deployment issues. This book only begins to address this goal: we plan to present a much more comprehensive treatment of this in a companion textbook. Nevertheless, the various aspects of each area that justify our intuition are at least outlined in this book.

Predicting the future is difficult: the development and popularity of any technology is unpredictable and does not depend solely on the quality or utility of the technology. On the other hand, this also makes predicting some part of the future easy: clearly, many possible predictions might just come true. We have presented our view of how each of the technologies might evolve. We hope that this book and its companion book will enhance the chances for successful exploitation of the experiences gathered here for each of the fields of mobility and in each of the fields discussed.

References

References

[Accetta 1986] Accetta, M., Baron, R., Bolosky, W., Golub, D., Rashid, R., Tevanian, A., and Young, M., "Mach: A New Kernel Foundation for UNIX Development," *Proceedings of the Summer USENIX Conference*, pp. 93–112, July 1986.

[Acharya 1995] Acharya, S., Franklin, M.J., and Zdonik, S., "Dissemination-based Data Delivery Using Broadcast Disks," *IEEE Personal Communications*, 2(6):50–60, December 1995.

[Acharya 1996] Acharya, S., Franklin, M.J., and Zdonik, S., "Disseminating Updates on Broadcast Disks," *Proceedings of the 22nd International Conference on Very Large Databases,* pp. 354–365, September 1996.

[Ad Astra Engineering] Ad Astra Engineering, Incorporated, JUMPING BEANS,™ http://www.JumpingBeans.com.

[Agha 1986] Agha, G., *Actors: A Model of Concurrent Computation in Distributed Systems*, Artificial Intelligence Series, MIT Press, Cambridge, MA, 1986.

[Almeroth 1998] Almeroth, K.C., Ammar, M.H., and Fei, Z., "Scalable Delivery of Web Pages Using Cyclic Best-Effort (UDP) Multicast," *Proceedings of IEEE INFOCOM'98*, March 1998.

[Alon 1987] Alon, N., Barak, A., and Manber, U., "On Disseminating Information Reliably without Broadcasting," *Proceedings of the Seventh International Conference on Distributed Computing Systems*, pp. 74–81, September 1987.

[Alonso 1988] Alonso, R. and Kyrimis, K., "A Process Migration Implementation for a UNIX System," *Proceedings of the USENIX Winter Conference,* pp. 365–372, February 1988.

[Amaral 1992] Amaral, P., Jacqemot, C., Jensen, P., Lea, R., and Mirowski, A., "Transparent Object Migration in COOL-2," *Proceedings of the ECOOP,* pp. 72–77, June 1992.

[Amir 1995] Amir, E., Balakrishnan, H., Seshan, S., and Katz, R., "Efficient TCP over Networks with Wireless Links," *Proceedings of the Fifth Workshop on Hot Topics in Operating Systems*, pp. 35–40, May 1995.

[Artsy 1987] Artsy, Y., Chang, H.-Y., and Finkel, R., "Interprocess Communication in Charlotte," *IEEE Software*, 4(1):22–28, January 1987.

[Artsy 1989] Artsy, Y. and Finkel, R., "Designing a Process Migration Facility: The Charlotte Experience," *Computer*, 22(9):47–56, September 1989. Reprinted in this book.

[Baker 1996] Baker, M., Zhao, X., Cheshire, S., and Stone, J., "Supporting Mobility in MosquitoNet," *Proceedings of the USENIX 1996 Annual Technical Conference,* pp. 127–139, January 1996.

[Balakrishnan 1995] Balakrishnan, H., Seshan, S., Amir, E., and Katz, R.H., "Improving TCP/IP Performance over Wireless Networks," *Proceedings of the First Annual International Conference on Mobile Computing and Networking,* pp. 2–11, November 1995.

[Balakrishnan 1997a] Balakrishnan, H., Padmanabhan, V.N., and Katz, R.H., "The Effects of Asymmetry on TCP Performance," *Proceedings of the Third Annual ACM/IEEE International Conference on Mobile Computing and Networking,* pp. 77–89, September 1997.

[Balakrishnan 1997b] Balakrishnan, H., Padmanabhan, V.N., Seshan, S., and Katz, R.H., "A Comparison of Mechanisms for Improving TCP Performance over Wireless Links," *IEEE/ACM Transactions on Networking,* 5(6):756–769, December 1997. Reprinted in this book.

[Banga 1997] Banga, G., Douglis, F., and Rabinovich, M., "Optimistic Deltas for WWW Latency Reduction," *Proceedings of the 1997 USENIX Technical Conference,* pp. 289–303, January 1997.

[Barak 1985a] Barak, A. and Litman, A., "MOS: a Multicomputer Distributed Operating System," *Software–Practice and Experience,* 15(8):725–737, August 1985.

[Barak 1985b] Barak, A. and Shiloh, A., "A Distributed Load-Balancing Policy for a Multicomputer," *Software–Practice and Experience,* 15(9):901–913, September 1985.

[Barak 1989a] Barak, A., Shiloh, A., and Wheeler, R., "Flood Prevention in the MOSIX Load-Balancing Scheme," *IEEE Technical Committee on Operating Systems Newsletter,* 3(1):24–27, Winter 1989.

[Barak 1989] Barak, A. and Wheeler, R., "MOSIX: An Integrated Multiprocessor UNIX," *Proceedings of the USENIX Winter 1989 Technical Conference,* pp. 101–112, February 1989. Reprinted in this book.

[Barak 1993] Barak, A., Guday, S., and Wheeler, R.G., *The MOSIX Distributed Operating System, Load Balancing for UNIX,* Springer Verlag, 1993.

[Barak 1998] Barak, A. and La'adan, O., "The MOSIX Multicomputer Operating System for High Performance Cluster Computing," *Journal of Future Generation Computer Systems,* 13(4–5):361–372, March 1998.

[Baumann 1997] Baumann, J., Hohl, F., Radouniklis, N., Rothermel, K. and Straßer, M., "Communication Concepts for Mobile Agent Systems," *Proceedings of the First International Workshop on Mobile Agents, MA'97,* Springer Verlag, pp. 123–135, April 1997.

[Baumann 1998a] Baumann, J. and Rothermel, K., "The Shadow Approach: An Orphan Detection Protocol for Mobile Agents," *Proceedings of the Second International Workshop on Mobile Agents (MA'98),* pp. 2–13, September 1998.

[Baumann 1998b] Baumann, J., Hohl, F., Rothermel, K., and Straßer, M., "Mole–Concepts of a Mobile Agent System," *World Wide Web* 1(3):123–137, September 1998. Reprinted in this book.

[Baumann 1998c] Baumann, J., Hohl, F., Rothermel, K., Schwehm, M., and Straßer, M., "Mole 3.0: A Middleware for Java-Based Mobile Software Agents," *Proceedings of the IFIP International Conference on Distributed Systems Platforms and Open Distributed Processing (Middleware '98),* pp. 355–370, September 1998.

[Beguelin 1993] Beguelin, A., Dongarra, J., Geist, A., Manchek, R., Otto, S., and Walpole, J., "PVM: Experiences, Current Status and Future Directions," *Proceedings of Supercomputing 1993*, pp. 765–766, November 1993.

[Bhagwat 1993] Bhagwat, P. and Perkins, C.E., "A Mobile Networking System Based on Internet Protocol (IP)," *Proceedings of the USENIX Mobile and Location-Independent Computing Symposium*, pp. 69–82, August 1993.

[Bharat 1995] Bharat, K.A. and Cardelli, L. "Migratory Applications," *Proceedings of the Eight Annual ACM Symposium on User Interface Software Technology*, pp. 133–142, November 1995.

[Bic 1996] Bic, L.F., Fukuda, M., and Dillencourt, M.B., "Distributed Computing Using Autonomous Objects," *Computer*, 29(8):55–61, August 1996.

[Blackwell 1994] Blackwell, T., Chan, K., Chang, K., Charuhas, T., Gwertzman, J., Karp, B., Kung, H., Li, D., Lin, D., Morris, R., Polansky, R., Tang, D., and Young, C., "Secure Redirects in Mobile IP," *Proceedings of the USENIX Summer 1994 Conference*, pp. 305–316, June 1994.

[Bradshaw 1997] Bradshaw, J., *Software Agents*, AAAI Press / MIT Press, Cambridge, MA, 1997.

[Brooks 1995] Brooks, C., Mazer, M.S., Meeks, S., and Miller, J., "Application-Specific Proxy Servers as HTTP Stream Transducers," *Proceedings of the Fourth International WWW Conference*, pp. 539–548, December 1995.

[Bryant 1981] Bryant, R. M. and Finkel, R. A., "A Stable Distributed Scheduling Algorithm," *Proceedings of the Second International Conference on Distributed Computing Systems, pp.* 314–323, April 1981.

[Bryant 1995] Bryant, B., "Design of AD 2, a Distributed UNIX Operating System," *OSF Research Institute*, December 1995.

[Cabrera 1986] Cabrera, L.-F., "The Influence of Workload on Load Balancing Strategies," *Proceedings of the USENIX Summer Conference,* pp. 446–458, June 1986. Reprinted in this book.

[Cáceres 1995] Cáceres, R. and Iftode, L. "Improving the Performance of Reliable Transport Protocols in Mobile Computing Environments," *IEEE Journal on Selected Areas in Communications* 13(5):850–857, June 1995.

[Casas 1995] Casas, J., Clark, D.L., Conuru, R., Otto, S.W., Prouty, R.M., and Walpole, J., "MPVM: A Migration Transparent Version of PVM," *Computing Systems*, 8(2):171–216, Spring 1995.

[Castillo 1998] Castillo, A., Kawaguchi, M., Paciorek, N., and Wong, D., "Concordia™ as Enabling Technology for Cooperative Information Gathering," *Proceedings of the Japanese Society for Artificial Intelligence Conference,* pp. 280–283, June 1998.

[Chang 1997] Chang, D. and Covaci, S., "The OMG Mobile Agent Facility: A Submission," *Proceedings of the First International Workshop on Mobile Agents*, Lecture Notes in Computer Science No. 1219, Springer Verlag, pp. 98–110, April 1997.

[Chang 1997] Chang, H., Tait, C., Cohen, N., Shapiro, M., Mastrianni, S., Floyd, R., Housel, B., and Lindquist, D., "Web Browsing in a Wireless Environment: Disconnected and Asynchronous Operation in ARTour Web Express," *Proceedings of the Third Annual ACM/IEEE International Conference on Mobile Computing and Networking,* pp. 260–269, September 1997.

[Chawathe 1998] Chawathe, Y., Fink, S., McCanne, S., and Brewer, E.A., "A Proxy Architecture for Reliable Multicast in Heterogeneous Environments," *Proceedings of ACM Multimedia '98*, pp. 151–159, September 1998.

[Cheriton 1988] Cheriton, D.R., "The V Distributed System," *Communications of the ACM*, 31(3):314–333, March 1988.

[Cheriton 1990] Cheriton, D.R., "Binary Emulation of UNIX Using the V Kernel," *Proceedings of the USENIX Summer Conference*, pp. 73–86, June 1990.

[Cheshire 1996] Cheshire, S. and Baker, M., "Internet Mobility 4x4," *Proceedings of the ACM SIGCOMM'96 Conference on Applications, Technologies, Architectures, and Protocols for Computer Communications,* pp. 318–329, August 1996. Reprinted in this book.

[Chess 1994] Chess, D.M., Grossof B., Harrison, C., Levine, D., Parris, C., and Tsudik, G., "Mobile Agents: Are They a Good Idea," IBM Research Report, RC 19887, October 1994.

[Chess 1995] Chess, D.M., Grossof B., Harrison, C., Levine, D., Parris, C., and Tsudik, G., "Itinerant Agents for Mobile Computing," *IEEE Personal Communications*, 2(5):34–49, October 1995.

[Cockayne 1997] Cockayne, W. and Zyda, M., *Mobile Agents*, Manning, Greenwich, CT, USA, 1997.

[Cohn 1989] Cohn, D.L., Delaney, W.P., and Tracey, K.M., "Arcade: A Platform for Distributed Operating Systems," *Proceedings of the USENIX Workshop on Experiences with Distributed and Multiprocessor Systems,* pp. 373–390, October 1989.

[Dannenberg 1982] Dannenberg, R.B., *Resource Sharing in a Network of Personal Computers*, Ph.D. thesis, Technical Report CMU-CS-82-152, Carnegie Mellon University, December 1982.

[Dannenberg 1985] Dannenberg, R.B. and Hibbard, P.G., "A Butler Process for Resource Sharing on a Spice Machine," *IEEE Transactions on Office Information Systems,* 3(3):234–252, July 1985.

[DARPA] DARPA Broad Agency Announcement, 98-01, "Agent-Based Systems," http://ballston. prc.com/baa9801/abspipv1.htm.

[Dediu 1992] Dediu, H., Chang, C.H., and Azzam, H., "Heavy Weight Process Migration," *Proceedings of the Third Workshop on Future Trends of Distributed Computing Systems*, pp. 221–225, April 1992.

[Demers 1994] Demers, A., Petersen, K., Spreitzer, M., Terry, D., Theimer, M., and Welch, B., "The Bayou Architecture: Support for Data Sharing Among Mobile Users," *Proceedings of the Workshop on Mobile Computing Systems and Applications,* pp. 2–7, December 1994.

[Douglis 1987] Douglis, F. and Ousterhout, J., "Process Migration in the Sprite Operating System," *Proceedings of the Seventh International Conference on Distributed Computing Systems*, pp. 18–25, September 1987.

[Douglis 1989] Douglis, F., "Experience with Process Migration in Sprite," *Proceedings of the USENIX Workshop on Experiences with Distributed and Multiprocessor Systems*, pp. 59–72, October 1989.

[Douglis 1990] Douglis, F., *Transparent Process Migration in the Sprite Operating System*, Ph.D. thesis, Technical Report UCB/CSD 90/598, CSD(EECS), University of California, Berkeley, September 1990.

[Douglis 1991] Douglis, F. and Ousterhout, J., "Transparent Process Migration: Design Alternatives and the Sprite Implementation," *Software—Practice and Experience*, August 1991, 21(8):757–785. Reprinted in this book.

[Dubach 1989] Dubach, B., "Process-Originated Migration in a Heterogeneous Environment," *Proceedings of the 17th ACM Annual Computer Science Conference,* pp. 98–102, 1989.

[Eager 1986] Eager, D., Lazowska, E., and Zahorjan, J., "A Comparison of Receiver-Initiated and Sender-Initiated Adaptive Load Sharing," *Performance Evaluation*, 6(1):53–68, April 1986.

[Eager 1986] Eager, D., Lazowska, E., and Zahorjan, J. "Dynamic Load Sharing in Homogeneous Distributed Systems," *IEEE Transactions on Software Engineering*, 12(5):662–675, May 1986.

[Eager 1988] Eager, D., Lazowska, E., and Zahorjan, J., "The Limited Performance Benefits of Migrating Active Processes for Load Sharing," *Proceedings of the 1988 ACM SIGMETRICS Conference on Measurement and Modeling of Computer Systems, Performance Evaluation Review,* 16(1):63–72, May 1988.

[Ebling 1998] Ebling, M., *Translucent Cache Management for Mobile Computing*, Ph.D. thesis, Carnegie Mellon University, 1998.

[Eskicioglu 1990] Eskicioglu, M.R., "Design Issues of Process Migration Facilities in Distributed Systems," *IEEE Technical Committee on Operating Systems Newsletter,* 4(2):3–13, 1990.

[Farber 1973] Farber, D., Feldman, J., Heinrich, F.R., Hopwood, M.D., Larson, K.C., Loomis, D.C., and Rowe, L.A., "The Distributed Computing System," *Proceedings of the Spring COMPCON'73*, pp. 31–34, 1973.

[Farmer 1996] Farmer, W.M., Guttman, J.D., and Swarup, V., "Security for Mobile Agents: Authentication and State Appraisal," *Proceedings of the Fourth European Symposium on Research in Computer Security,* Springer Verlag Lecture Notes in Computer Science No. 1146, pp. 118–130, September 1996.

[Farmer 1996] Farmer, W.M., Guttman, J.D., and Swarup, V., "Security for Mobile Agents: Issues and Requirements," *Proceedings of the National Information Systems Security Conference,* pp. 591–597, 1996.

[FIPA] FIPA http://drogo.cselt.it/fipa/.

[Forman 1994] Forman, G.H. and Zahorjan, J., "The Challenges of Mobile Computing," *Computer,* 27(4):38–47, April 1994. Reprinted in this book.

[Fox 1996] Fox, A. and Brewer, E.A., "Reducing WWW Latency and Bandwidth Requirements via Real-Time Distillation," *Proceedings of the Fifth International World Wide Web Conference*, pp. 1445–1456, May 1996.

[Fox 1996a] Fox, A., Gribble, S.D., Brewer, E.A., and Amir, E., "Adapting to Network and Client Variability via On-Demand Dynamic Distillation," *Proceedings of the Seventh International Conference on Architectural Support for Programming Languages and Operating Systems*, pp. 160–170, October 1996.

[Fox 1996b] Fox, A. and Gribble, S.D., "Security On the Move: Indirect Authentication Using Kerberos," *Proceedings of the Second ACM Conference on Mobile Computing*, pp. 155–164, November 1996.

[Fox 1997] Fox, A., Gribble, S.D., Chawathe, Y., Brewer, E.A., and Gauthier, P., "Cluster-Based Scalable Network Services," *Proceedings of the 16th ACM Symposium on Operating Systems Principles*, pp. 78–91, October 1997.

[Fox 1998a] Fox, A., Gribble, S.D., Chawathe, Y., and Brewer, E.A., "Adapting to Network and Client Variation Using Active Proxies: Lessons and Perspectives," *IEEE Personal Communications,* 5(4):10–19, August 1998. Reprinted in this book.

[Fox 1998b] Fox, A., Goldberg, I., Gribble, S.D., Lee, D.C., Polito, A., and Brewer, E.A., "Experience With Top Gun Wingman: A Proxy-Based Graphical Web Browser for the USR PalmPilot," *Proceedings of the IFIP International Conference on Distributed Systems Platforms and Open Distributed Processing (Middleware '98)*, pp. 407–424, September 1998.

[Freedman 1991] Freedman, D., "Experience Building a Process Migration Subsystem for UNIX," *Proceedings of the USENIX Winter Conference*, pp. 349–355, January 1991.

[Glass] Glass, G., "ObjectSpace Voyager Core Package Technical Overview," ObjectSpace, White Paper. Reprinted in this book.

[Glass 1998] Glass, G., Personal Communication, April 1998.

[Goodman 1997] Goodman, D.J., *Wireless Personal Communications Systems*, Addison-Wesley, Reading, MA, 1997.

[Gray 1995] Gray, R., "Agent Tcl: A Transportable Agent System," *Proceedings of the CIKM Workshop on Intelligent Information Agents, Fourth International Conference on Information and Knowledge Management (CIKM 95),* December 1995.

[Gray 1996] Gray, R., "Agent Tcl: A Flexible and Secure Mobile-Agent System," *Proceedings of the 1996 Tcl/Tk Workshop*, pp. 9–23, July 1996.

[Gray 1997a] Gray, R., *Agent Tcl: A Flexible and Secure Mobile-Agent System*, Ph.D. thesis, Technical Report TR98-327, Dept. of Computer Science, Dartmouth College, June 1997.

[Gray 1997b] Gray, R., Kotz, D., Nog, S., Rus, D., and Cybenko, G., "Mobile Agents: The Next Generation in Distributed Computing," *Proceedings of the Second Aizu International Symposium on Parallel Algorithms/Architectures Synthesis (pAs '97),* pp. 8–24, March 1997.

[Gray 1998] Gray, R., Kotz, D., Cybenko, G., and Rus, D., "D'Agents: Security in a Multiple-Language, Mobile-Agent System," In Giovanni Vigna, editor, *Mobile Agent Security,* Lecture Notes in Computer Science, Springer Verlag, 1998.

[Grimshaw 1997] Grimshaw, A., Wulf, W.A., and the Legion Team, "The Legion Vision of a Worldwide Virtual Computer," *Communications of the ACM*, 40(1):39–45, January 1997.

[Gruber 1994] Gruber, R., Kaashoek, M.F., Liskov, B., and Shrira, L., "Disconnected Operation in the Thor Object-Oriented Database System," *Proceedings of the Workshop on Mobile Computing Systems and Applications*, pp. 51–56, December 1994.

[Guideware Corporation] Guideware Corporation, http://www.guideware.com.

[Hac 1986] Hac, A., "Concurrency Control in a Distributed System and Its Performance for File Migration and Process Migration," *Proceedings of the Sixth International Conference on Distributed Computing Systems*, pp. 429–434, May 1986.

[Haertig 1997] Haertig, H. and Reuther, L., "Encapsulating Mobile Objects," *Proceedings of the 17th International Conference on Distributed Computing Systems,* pp. 355–362, May 1997. Reprinted in this book.

[Harchol-Balter 1996] Harchol-Balter, M. and Downey, A.B., "Exploiting Process Lifetime Distributions for Dynamic Load Balancing," *Proceedings of the 1996 ACM SIGMETRICS Conference on Measurement and Modeling of Computer Systems*, pp. 13–24, May 1996.

[Harchol-Balter 1997] Harchol-Balter, M. and Downey, A.B., "Exploiting Process Lifetime Distributions for Dynamic Load Balancing," *ACM Transactions on Computer Systems,* 15(3):253–285, August 1997. Reprinted in this book.

[Hewitt 1977] Hewitt, C., "Viewing Control Structures as Patterns of Passing Messages," *Journal of Artificial Intelligence,* 8(3), pp. 323–364, June 1977.

[Hofmann 1998] Hofmann, M.O., McGovern, A., and Whitebread, K., "Mobile Agents on the Digital Battlefield," *Proceedings of the Autonomous Agents '98,* pp. 219–225, May 1998.

[Hohl 1998] Hohl, F., "A Model of Attacks of Malicious Hosts Against Mobile Agents," *Proceedings of the Fourth Workshop on Mobile Objects Systems*, INRIA Technical Report, pp. 105–120, July 1998.

[Homburg 1996] Homburg, P., van Steen, M., and Tanenbaum A.S., "An Architecture for A Wide Area Distributed System," *Proceedings of the Seventh SIGOPS European Workshop,* pp. 75–82, September 1996.

[Housel 1996] Housel, B.C. and Lindquist, D.B., "WebExpress: A System for Optimizing Web Browsing in a Wireless Environment," *Proceedings of the Second Annual International Conference on Mobile Computing and Networking*, pp. 108–116, November 1996. Reprinted in this book.

[Howard 1988] Howard, J., Kazar, M., Menees, S., Nichols, D., Satyanarayanan, M., Sidebotham, R., and Sidebotham, R., "Scale and Performance in a Distributed File System," *ACM Transactions on Computer Systems*, 6(1):51–81, February 1988.

[Huhns 1997] Huhns, M. and Singh, M. (eds.), *Readings in Agents,* Morgan Kaufmann Publishers, October 1997.

[Huizinga 1994] Huizinga, D. and Heflinger, K., "Experience with Connected and Disconnected Operation of Portable Notebook Computers in Distributed Systems," *Proceedings of the Workshop on Mobile Computing Systems and Applications,* pp. 119–123, December 1994.

[Huston 1993] Huston, L. and Honeyman, P., "Disconnected Operation for AFS," *Proceedings of the USENIX Mobile and Location-Independent Computing Symposium,* pp. 1–10, August 1993.

[Hylton 1997] Hylton, J. and van Rossum, G., "Using the Knowbots Operating Environment in a Wide-Area Network," *Presented at the Third ECOOP Workshop on Mobile Object Systems,* June 1997.

[IKV++ Gmbh] IKV++ Gmbh, Grasshopper, http://www.ikv.de/products/grasshopper.html.

[Imieliński 1996] Imieliński, T. and Korth, H.F. (eds.), *Mobile Computing*, Kluwer Academic Publishers, Norwell, MA, USA, 1996.

[Ioannidis 1991] Ioannidis, J., Duchamp, D., and Maguire, G.Q., Jr., "IP-Based protocols for Mobile Internetworking," *Proceedings of the ACM SIGCOMM'91 Conference on Communications Architectures and Protocols*, pp. 235–245, September 1991.

[Ioannidis 1993] Ioannidis, J., *Protocols for Mobile Internetworking*, Ph.D. thesis, Columbia University in New York, 1993.

[Ioannidis 1993] Ioannidis, J. and Maguire, G.Q. Jr., "The Design and Implementation of a Mobile Internetworking Architecture," *Proceedings of the USENIX Winter 1993 Technical Conference,* pp. 491–502, January 1993. Reprinted in this book, page TBD.

[Johansen 1995] Johansen, D., van Renesse, R., and Schneider, F.B., "Operating System Support for Mobile Agents," *Proceedings of the Fifth Workshop on Hot Topics in Operating Systems*, pp. 42–45, May 1995. Reprinted in this book.

[Johansen 1996] Johansen, D., van Renesse, R., and Schneider, F.B., "Supporting Broad Internet Access to TACOMA," *Proceedings of the Seventh ACM SIGOPS European Workshop*, pp. 55–58, September 1996.

[Joseph 1995] Joseph, A.D., deLespinasse, A., Tauber, J., Gifford, D., and Kaashoek, M.F., "Rover: A Toolkit for Mobile Information Access," *Proceedings of the 15th ACM Symposium on Operating Systems Principles*, pp. 156–171, December 1995.

[Joseph 1997] Joseph, A.D. and Kaashoek, M.F., "Building Reliable Mobile-Aware Applications Using the Rover Toolkit," *Wireless Networks*, 3(5):405–419, October 1997.

[Joseph 1997] Joseph, A.D., Tauber, J.A., and Kaashoek, M.F., "Mobile Computing with the Rover Toolkit," *IEEE Transactions on Computers,* 46(3):337–352, March 1997. Reprinted in this book.

[Jul 1988] Jul, E., Levy, H., Hutchinson, N., and Black, A., "Fine-Grained Mobility in the Emerald System," *ACM Transactions on Computer Systems*, 6(1):109–133, February 1988. Reprinted in this book.

[Jul 1989] Jul, E., "Migration of Light-weight Processes in Emerald," *IEEE Technical Committee on Operating Systems Newsletter*, 3(1):20–23, Winter 1989.

[Kaufman 1995] Kaufman, C., Perlman, R., and Speciner, M., *Network Security: Private Communication in a Public World*, Prentice Hall, Englewood Cliffs, NJ, 1995.

[Kistler 1992] Kistler, J.J. and M. Satyanarayanan. "Disconnected Operation in the Coda File System," *ACM Transactions on Computer Systems,* 10(1):3–25, February 1992. Reprinted in this book.

[Kistler 1993] Kistler, J.J., *Disconnected Operation in a Distributed File System*, Ph.D. thesis, Technical report CMU–CS–93–156, Carnegie Mellon University, May 1993.

[Kotz 1996] Kotz, D., Gray, R., and Rus, D., "Transportable Agents Support Worldwide Applications," *Proceedings of the Seventh ACM SIGOPS European Workshop,* pp. 41–48, September 1996.

[Kotz 1997] Kotz, D., Gray., R., Nog, S., Rus, D., Chawla, S., and Cybenko, G., "AGENT TCL: Targeting the Needs of Mobile Computers," *IEEE Internet Computing*, 1(4):58–67, July/August 1997. Reprinted in this book.

[Krueger 1988] Krueger, P. and Livny, M., "A Comparison of Preemptive and Non-Preemptive Load Balancing," *Proceedings of the Eighth International Conference on Distributed Computing Systems,* pp. 123–130, June 1988.

[Kuenning 1997] Kuenning, G. and Popek, G., "Automated Hoarding for Mobile Computers," *Proceedings of the 16th ACM Symposium on Operating Systems Principles,* pp. 264–275, October 1997.

[Kumar 1993] Kumar, P. and Satyanarayanan, M., "Log-Based Directory Resolution in the Coda File System," *Proceedings of the Second International Conference on Parallel and Distributed Information Systems,* pp. 202–213, 1993.

[Kumar 1993] Kumar, P. and Satyanarayanan, M., "Supporting Application-Specific Resolution in an Optimistically Replicated File System," *Proceedings of the Fourth Workshop on Workstation Operating Systems,* pp. 66–70, October 1993.

[Kunz 1991] Kunz, T., "The Influence of Different Workload Descriptions on a Heuristic Load Balancing Scheme," *IEEE Transactions on Software Engineering,* 17(7):725–730. July 1991.

[Lange 1998] Lange, D. and Oshima, M., "Mobile Agents with Java: The Aglet API," *World Wide Web*, 1(3), September 1998. Reprinted in this book.

[Lange 1998] Lange, D. and Oshima, M., *Programming and Deploying Java^{TM} Mobile Agents With Aglets^{TM}* Addison Wesley Longman, Reading, MA, September 1998.

[Le 1994] Le, M.T., Seshan, S., Burghardt, F., and Rabaey, J., "Software Architecture of the Infopad System," *Proceedings of the Mobidata Workshop on Mobile and Wireless Information Systems*, October 1994.

[Leland 1986] Leland, W. and Ott, T., "Load Balancing Heuristics and Process Behavior," *Proceedings of the 1986 ACM SIGMETRICS Conference on Measurement and Modeling of Computer Systems,* pp. 54–69, May 1986.

[Litzkow 1987] Litzkow, M., "Remote UNIX–Turning Idle Workstations into Cycle Servers," *Proceedings of the USENIX Summer Conference*, pp. 381–384, June 1987.

[Litzkow 1988] Litzkow, M., Livny, M., and Mutka, M., "Condor—A Hunter of Idle Workstations," *Proceedings of the Eighth International Conference on Distributed Computing Systems*, pp. 104–111, June 1988.

[Litzkow 1992] Litzkow, M. and Solomon, M., "Supporting Checkpointing and Process Migration Outside the UNIX Kernel," *Proceedings of the USENIX Winter Conference*, pp. 283–290, January 1992. Reprinted in this book.

[Louboutin 1991] Louboutin, S., "An Implementation of a Process Migration Mechanism Using Minix," *Proceedings of the 1991 European Autumn Conference,* pp. 213–224, September 1991.

[LoVerso 1997] LoVerso, J. and Mazer, M., "Caubweb: Detaching the Web with Tcl," *Proceedings of the Fifth Annual Tcl/Tk Workshop,* July 1997.

[Lux 1993] Lux, W., Haertig, H., and Kuehnhauser, W.E., "Migrating Multi-Threaded, Shared Objects," *Proceedings of the 26th Hawaii International Conference on Systems Sciences,* II:642–649, January 1993.

[Mandelberg 1988] Mandelberg, K. and Sunderam, V., "Process Migration in Unix Networks," *Proceedings of the USENIX Winter Conference,* pp. 357–363, February 1988.

[Mazer 1994] Mazer, M. and Tardo, J., "A Client-Side-Only Approach to Disconnected File Access," *Proceedings of the Workshop on Mobile Computing Systems and Applications,* pp. 104–110, December 1994.

[Metcalfe 1976] Metcalfe, R.E. and Boggs, D. "Ethernet: Distributed Packet Switching for Local Computer Networks," *Communications of the ACM,* 19(7):395–404, July 1976.

[Metricom Corporation] Metricom Corporation. Ricochet wireless modem. http://www.ricochet.net/netoverview.html.

[Miller 1981] Miller, B. and Presotto, D., "XOS: an Operating System for the XTREE Architecture," *Operating Systems Review,* 2(15):21–32, 1981.

[Milojičić 1993] Milojičić, D.S., Giese, P., and Zint, W., "Experiences with Load Distribution on Top of the Mach Microkernel," *Proceedings of the USENIX Symposium on Experiences with Distributed and Multiprocessor Systems (SEDMS IV),* pp. 19–36, September 1993.

[Milojičić 1993] Milojičić, D.S., Zint, W., Dangel, A., and Giese, P., "Task Migration on the Top of the Mach Microkernel," *Proceedings of the Third USENIX Mach Symposium*, pp. 273–290, April 1993. Reprinted in this book.

[Milojičić 1994] Milojičić, D.S., *Load Distribution, Implementation for the Mach Microkernel*, Vieweg, Wiesbaden, Germany, 1994.

[Milojičić 1998] Milojičić, D.S., Breugst, B., Busse, I., Campbell, J., Covaci, S., Friedman, B., Kosaka, K., Lange, D., Ono, K., Oshima, M., Tham, C., Virdhagriswaran, S., and White, J., "MASIF, The OMG Mobile Agent System Interoperability Facility," *Proceedings of the Second International Workshop on Mobile Agents*, pp. 50–67, September 1998. Reprinted in this book. Also appeared in *Personal Technologies*, 2:117–129, 1998.

[Milojičić 1998a] Milojičić, D.S., Chauhan, D., and laForge, W., "Mobile Objects and Agents (MOA), Design, Implementation and Lessons Learned," *Proceedings of the Fourth USENIX Conference on Object-Oriented Technologies (COOTS),* pp. 179–194, April 1998. Reprinted in this book. Also appeared in *IEE Distributed Systems Engineering*, 5:1–14, 1998.

[Mogul 1997] Mogul, J., Douglis, F., Feldmann, A., and Krishnamurthy, B., "Potential Benefits of Delta-encoding and Data Compression for HTTP," *Proceedings of the ACM SIGCOMM'97 Conference on Applications, Technologies, Architectures, and Protocols for Computer Communications*, pp. 181–194, September 1997.

[Mummert 1994] Mummert, L. and Satyanarayanan, M., "Large Granularity Cache Coherence for Intermittent Connectivity," *Proceedings of the 1994 USENIX Summer Conference,* pp. 279–289, June 1994.

[Mummert 1995] Mummert, L., Ebling, M., and Satyanarayanan, M., "Exploiting Weak Connectivity for Mobile File Access," *Proceedings of the 15th ACM Symposium on Operating System Principles,* pp. 143–155, December 1995. Reprinted in this book.

[Mummert 1996] Mummert, L., *Exploiting Weak Connectivity in a Distributed File System,* Ph.D. thesis, Technical Report CMU-CS-96-195, Carnegie Mellon University, December 1996.

[Myles 1993] Myles, A. and Skellern, D., "Comparing Four IP-Based Mobile Host Protocols," *Journal of Computer Networks and ISDN Systems,* 26(3):349–355, November 1993.

[Nakamura] Nakamura, Y. and Yamamoto, G., "Aglets-based e-Marketplace: Concept, Architecture and Applications," *IBM Research, Tokyo Research Laboratory, White Paper,* http://aglets.trl.ibm.co.jp/RT0253/RT0253.html.

[Nelson 1988] Nelson, M.N., Welch, B.B., and Ousterhout, J.K., "Caching in the Sprite Network File System," *ACM Transactions on Computer Systems,* 6(1):134–154, February 1988.

[Nichols 1987] Nichols, D., "Using Idle Workstations in a Shared Computing Environment," *Proceedings of the 11th ACM Symposium on Operating Systems Principles,* pp. 5–12, November 1987.

[Noble 1997] Noble, B.D., Satyanarayanan, M., Narayanan, D., Tilton, J.E., Flinn, J., and Walker, K.R., "Agile Application-Aware Adaptation for Mobility," *Proceedings of the 16th ACM Symposium on Operating Systems Principles,* pp. 276–287, October 1997.

[Noble 1998] Noble, B.D., *Mobile Data Access*, Ph.D. thesis, Carnegie Mellon University, 1998.

[Nuttal 1994] Nuttal, M., "Survey of Systems Providing Process or Object Migration," *Operating Systems Review,* 28(4):64–79, October 1994.

[OMG 1995] OMG, "The Common Object Request Broker: Architecture and Specification," OMG Document Revision 2.0, July 1995.

[OMG 1997] OMG Mobile Agent Facility Interoperability Facilities Specification (MAF), OMG TC Document ORBOS/1997-10-05.

[Ousterhout 1988] Ousterhout, J.K., Cherenson, A.R., Douglis, F., Nelson, M.N., and Welch, B.B., "The Sprite Network Operating System," *Computer*, 21(2):23–36, February 1988.

[Ousterhout 1994] Ousterhout, J.K., *Tcl and the Tk Toolkit*, Addison-Wesley, Reading, MA, 1994.

[Paindaveine 1996] Paindaveine, Y. and Milojicic, D.S., "Process v. Task Migration," *Proceedings of the 29th Annual Hawaii International Conference on System Sciences,* pp. 636–645, January 1996.

[Peine 1997] Peine, H. and Stolpmann, T., "The Architecture of the Ara Platform for Mobile Agents," In Kurt Rothermel, Radu Popescu-Zeletin (eds.): *Proceedings of the First International Workshop on Mobile Agents (MA'97)*, April 1997. Lecture Notes in Computer Science No. 1219, Springer Verlag, pp. 50–61. Reprinted in this book.

[Peine 1998] Peine, H., "Security Concepts and Implementation for the Ara Mobile Agent System," *Proceedings of the Seventh IEEE Workshop on Enabling Technologies: Infrastructure for Collaborative Enterprises,* pp. 236–242, June 1998.

[Perkins 1996] Perkins, C.E., *IP Mobility Support,* RFC 2002, October 1996.

[Perkins 1998] Perkins, C.E., *Mobile IP: Design Principles and Practices*, Addison Wesley, Reading, MA, 1998.

[Perkins 1998] Perkins, C.E., "Mobile Networking with Mobile IP," *IEEE Internet Computing*, 2(1):58–69, January/February 1998. Reprinted in this book.

[Petri 1995] Petri, S. and Langendorfer, H., "Load Balancing and Fault Tolerance in Workstation Clusters Migrating Groups of Communicating Processes," *Operating Systems Review,* 29(4):25–36, October 1995.

[Philippe 1993] Philippe, L., *Contribution à l'étude et la réalisation d'un système d'exploitation à image unique pour multicalculateur*, Ph.D. thesis, Technical Report 308, Université de Franche-comté, 1993.

[Platform Computing 1996] Platform Computing, "LSF User's and Administrator's Guides," Version 2.2, Platform Computing Corporation, February 1996.

[Popek 1981] Popek, G., Walker, B.J., Chow, J., Edwards, D., Kline, C., Rudisin, G., and Thiel, G., "Locus: a Network-Transparent, High Reliability Distributed System," *Proceedings of the Eighth ACM Symposium on Operating System Principles*, pp. 169–177, December 1981.

[Popek 1985] Popek, G. and Walker, B., *The Locus Distributed System Architecture*, MIT Press, Cambridge, MA, 1985.

[Powell 1983] Powell, M. and Miller, B., "Process Migration in DEMOS/MP," *Proceedings of the Ninth ACM Symposium on Operating Systems Principles,* pp. 110–119, October 1983. Reprinted in this book.

[Python Home Page 1996] Python Home Page, http://www.python.org. Internet programming with Python, Aaron Watters, Guido van Rossum, James Ahlston, M&T Books, 1996.

[Ranganathan 1996] Ranganathan, M., Acharya, A., and Saltz, J., "Distributed Resource Monitors for Mobile Objects," *Proceedings of the Fifth International Workshop on Operating System Support for Object Oriented Systems,* pp. 19–23, October 1996.

[Ranganathan 1997] Ranganathan, M., Acharya, A., Sharma, S.D., and Saltz, J., "Network-aware Mobile Programs," *Proceedings of the USENIX 1997 Annual Technical Conference,* pp. 91–103, January 1997. Reprinted in this book.

[Rashid 1981] Rashid, R. and Robertson, G., "Accent: a Communication Oriented Network Operating System Kernel," *Proceedings of the Eighth ACM Symposium on Operating System Principles*, pp. 64–75, December 1981.

[Rashid 1986] Rashid, R., "From RIG to Accent to Mach: The Evolution of a Network Operating System," *Proceedings of the ACM/IEEE Computer Society Fall Joint Computer Conference*, pp. 1128–1137, November 1986.

[van Renesse 1996] van Renesse, R., Birman, K.P., and Maffeis, S., "Horus: A Flexible Group Communication System," *Communications of the ACM*, 39(4):76–85, April 1996.

[Rousch 1996] Rousch, E. and Campbell, R., "Fast Dynamic Process Migration," *Proceedings of the 16th International Conference on Distributed Computing Systems,* pp. 637–645, May 1996.

[Sandberg 1985] Sandberg, R., Goldberg, D., Kleiman, S., Walsh, S., and Lyon, B. "Design and Implementation of the Sun Network Filesystem," *Proceedings of the USENIX 1985 Summer Conference,* pp. 119–130, June 1985.

[Satyanarayanan 1990] Satyanarayanan, M., Kistler, J., Kumar, P., Okasaki, M., Siegel, E., and Steere, D., "Coda: A Highly Available File System for a Distributed Workstation Environment," *IEEE Transactions on Computers,* 39(4):447–459, April 1990.

[Satyanarayanan 1993] Satyanarayanan, M., Kistler, J., Mummert, L., Ebling, M., Kumar, P., and Lu, Q., "Experience with Disconnected Operation in a Mobile Environment," *Proceedings of the USENIX Symposium on Mobile and Location Independent Computing*, August 1993.

[Schill 1993] Schill, A. and Mock, M., "DC++: Distributed Object Oriented System Support on Top of OSF DCE," *Distributed Systems Engineering*, 1(2):112–125, December 1993.

[Schneider 1997] Schneider, F.B., "Towards Fault-Tolerant and Secure Agentry," *Proceedings of the Eleventh International Workshop on Distributed Algorithms,* September 1997.

[Seshan 1997] Seshan, S., Balakrishnan, H., and Katz, R., "Handoffs in Cellular Wireless Networks: The Daedalus Implementation and Experience," *Wireless Personal Communications,* 4(2):141–162, Kluwer Academic Publishers, March 1997.

[SET] SET, http://www.setco.org.

[Shapiro 1989] Shapiro, M., Gautron, P., and Mosseri, L., "Persistence and Migration for C++ Objects," *Proceedings of the ECOOP '89 European Conference on Object-Oriented Programming*, pp. 191–204, July 1989.

[Shivaratri 1992] Shivaratri, N., Krueger, P., and Singhal, M., "Load Distributing for Locally Distributed Systems," *Computer*, 25(12):33–44, December 1992.

[Shoch 1982] Shoch, J. and Hupp, J., "The Worm Programs—Early Experience with Distributed Computing," *Communications of the ACM*, 25(3):172–180, March 1982. Reprinted in this book.

[Shoham 1997] Shoham, Y., "An Overview of Agent-oriented Programming," in J.M. Bradshaw, editor, *Software Agents*, pp. 271–290. MIT Press, 1997.

[Shub 1990] Shub, C., "Native Code Process-Originated Migration in a Heterogeneous Environment," *Proceedings of the 18th ACM Annual Computer Science Conference,* pp. 266–270, February 1990.

[da Silva 1997] da Silva, M.M., "Mobility and Persistence," In *Mobile Object Systems: Towards the Programmable Internet*, Lecture Notes in Computer Science No. 1222, Springer Verlag, pp. 157–176, April 1997.

[Sinha 1991] Sinha, P., Maekawa, M., Shimuzu, K., Jia, X., Ashihara, Utsunomiya, N., Park, and Nakano, H., "The Galaxy Distributed Operating System," *Computer*, 24(8):34–40, August 1991.

[Skordos 1995] Skordos, P., "Parallel Simulation of Subsonic Fluid Dynamics on a Cluster of Workstations," *Proceedings of the Fourth IEEE International Symposium on High Performance Distributed Computing*, August 1995.

[Smith 1998] Smith, P. and Hutchinson, N.C., "Heterogeneous Process Migration: The Tui System," *Software—Practice and Experience*, 28(6):611–639, May 1998. Reprinted in this book.

[Solomon 1998] Solomon, J.D. *Mobile IP*, Prentice Hall, Englewood Cliffs, NJ,1998.

[Spreitzer 1993] Spreitzer, M. and Theimer, M., "Providing Location Information in a Ubiquitous Computing Environment," *Proceedings of the 14th ACM Symposium on Operating System Principles,* pp. 270–283, December 1993.

[Steensgaard 1995] Steensgaard, B. and Jul, E., "Object and Native Code Thread Mobility" *Proceedings of the 15th ACM Symposium on Operating Systems Principles,* pp. 68–78, December 1995.

[Steketee 1994] Steketee, C., Zhu, W., and Moseley, P., "Implementation of Process Migration in Amoeba," *Proceedings of the 14th International Conference on Distributed Computer Systems,* pp. 194–203, June 1994.

[Stone 1996] Stone, S., Zyda, M., Brutzman, D., and Falby, J., "Mobile Agents and Smart Networks for Distributed Simulation," *Proceedings of the 14th Distributed Simulations Conference*, March 1996.

[Tait 1995] Tait, C., Lei, H., Acharya, S., and Chang, H., "Intelligent File Hoarding for Mobile Computers," *Proceedings of the First Annual International Conference on Mobile Computing and Networking*, pp. 119–125, November 1995.

[Tanenbaum 1990] Tanenbaum, A.S., "Experiences with the Amoeba Distributed Operating System for the 1990s," *Communication of the ACM*, 33(12):46–63, December 1990.

[Tanenbaum 1992] Tanenbaum, A.S., *Modern Operating Systems*, Prentice Hall, Englewood Cliffs, NJ, 1992.

[Tardo 1996] Tardo, J. and Valente, L., "Mobile Agent Security and Telescript," *Proceedings of COMPCON'96*, pp. 52–63, February 1996.

[Taylor 1997] Taylor, M.S., Waung, W., and Banan, M., *Internetwork Mobility: The CDPD Approach*. Prentice Hall Series in Computer Networking and Distributed Systems, 1997.

[Tennenhouse 1996] Tennenhouse, D. and Wetherall, D., "Towards an Active Network Architecture," *Computer Communication Review*, 26(2):5–18, April 1996.

[Terry 1994] Terry, D., Demers, A., Petersen, K., Spreitzer, M., Theimer, M., and Welch, B., "Session Guarantees for Weakly Consistent Replicated Data," *Proceedings of the International Conference on Parallel and Distributed Information Systems,* pp. 140–149, September 1994.

[Terry 1995] Terry, D., Theimer, M., Petersen, K., Demers, A., Spreitzer, M., and Hauser, C., "Managing Update Conflicts in a Weakly Connected Replicated Storage System," *Proceedings of the 15th ACM Symposium on Operating System Principles*, pp. 173–183, December 1995. Reprinted in this book.

[Theimer 1985] Theimer, M., Lantz, K., and Cheriton, D., "Preemptable Remote Execution Facilities for the V System," *Proceedings of the 10th ACM Symposium on Operating System Principles,* pp. 2–12, December 1985. Reprinted in this book.

[Tracey 1991] Tracey, K. M., *Processor Sharing for Cooperative Multi-task Applications*, Ph.D. thesis, Technical Report, Department of Electrical Engineering, University of Notre Dame, April 1991.

[Tschudin 1997] Tschudin, C., "The Messenger Environment M0—a Condensed Description," In *Mobile Object Systems: Towards the Programmable Internet*, Lecture Notes in Computer Science No. 1222, Springer Verlag, pp. 149–156, April 1997.

[Tsichritzis 1987] Tsichritzis, D., Fiume, A., Gibbs, S., Nierstrasz, O., "KNOS: Knowledge Acquisition, Dissemination, and Manipulation Objects," *ACM Transactions on Office Information Systems*, 5(1):96–112, January 1987.

[Tso 1993] Tso, M., "Using Property Specifications to Achieve Graceful Disconnected Operation in an Intermittent Mobile Computing Environment," Technical report CSL-93-8, Xerox Palo Alto Research Center, 1993.

[Vigna 1998] Vigna, G., editor, *Mobile Agents Security*, Lecture Notes in Computer Science, Springer Verlag, 1998.

[Vigna 1997] Vigna, G., "Protecting Mobile Agents through Tracing," *Presented at the Third ECOOP Workshop on Mobile Object Systems*, June 1997.

[Virdhagriswaran] Virdhagriswaran, S., "Mobile Unstructured Business Object (MuBot)™ Technology," Crystaliz, Inc. White Paper, http://www.crystaliz.com/logicware/mubot.html.

[Vitek 1997] Vitek, I. and Tschudin, C., "Mobile Object Systems," *Proceedings of the Second International Workshop on Mobile Object Systems, MOS '96*, July 1996, Springer Verlag Lecture Notes in Computer Science No. 1222, April 1997.

[Vitek 1997] Vitek, I., Serrano, M., and Thanos, D., "Security and Communication in Mobile Object Systems," *In Mobile Object Systems: Towards the Programmable Internet,* Lecture Notes in Computer Science No. 1222, Springer Verlag, pp. 177–200, April 1997.

[Walker 1983] Walker, B., Popek, G., English, R., Kline, C., and Thiel, G., "The LOCUS Distributed Operating System," *Proceedings of the Ninth ACM Symposium on Operating Systems Principles*, pp. 49–70, October 1983.

[Walker 1989] Walker, B.J. and Mathews, R.M., "Process Migration in AIX's Transparent Computing Facility (TCF)," *IEEE Technical Committee on Operating Systems Newsletter*, 3(1):5–7, Winter 1989.

[Walsh 1998] Walsh, T., Paciorek, N., and Wong, D., "Security and Reliability in Concordia™," *Proceedings of the 31st Hawaii International Conference on Systems Sciences,* VII:44–53, January 1998. Reprinted in this book.

[Want 1992] Want, R., Hopper, A., Falcao, V., and Gibbons, J., "The Active Badge Location System," *ACM Transactions on Information Systems,* 10(1):91–102, 1992.

[Want 1992] Want, R. and Hopper, A., "Active Badges and Personal Interactive Computing Objects," *IEEE Transactions on Consumer Electronics,* 38(1):10–20, February 1992.

[Want 1995] Want, R., Schilit, B., Adams, N., Gold, R., Petersen, K., Goldberg, D., Ellis, J., and Weiser, M., "An Overview of the ParcTab Ubiquitous Computing Experiment," *IEEE Personal Communications,* 2(6):28–43, December 1995.

[Want 1996] Want, R., Schilit, B., Adams, N., Gold, R., Petersen, K., Goldberg, D., Ellis, J., and Weiser, M., "The ParcTab Ubiquitous Computing Experiment," in *Mobile Computing*, Tomasz Imieliński and Henry F. Korth, editors, pp. 45–101, 1996.

[Weiser 1991] Weiser, M., "The Computer for the Twenty-First Century," *Scientific American,* 265(3):94–104, September 1991.

[Weiser 1993] Weiser, M., "Hot Topic: Ubiquitous Computing," *Computer*, 26(10):71–72, October 1993.

[Weiser 1993] Weiser, M., "Some Computer Science Issues in Ubiquitous Computing," *Communications of the ACM*, 36(7):74–84, July 1993. Reprinted in this book.

[Weiser 1994] Weiser, M., Welch, B., Demers, A., and Shenker, S., "Scheduling for Reduced CPU Energy," *Proceedings of the First Symposium on Operating Systems Design and Implementation (OSDI)*, pp. 13–23, November 1994.

[Wetherall 1998] Wetherall, D., Guttag, J., and Tennenhouse, D.L., "ANTS: A Toolkit for Building and Dynamically Deploying Network Protocols," *Proceedings of the First IEEE Conference on Open Architectures and Network Programming,* April 1998.

[White 1996] White, J.E., "Telescript Technology: Mobile Agents," General Magic White Paper, Appeared in Bradshaw, J., *Software Agents*, AAAI/MIT Press, 1996. Reprinted in this book.

[White 1997] White, J.E., Helgeson, S., and Steedman, D.A., "System and Method for Distributed Computation Based upon the Movement, Execution, and Interaction of Processes in a Network," *United States Patent* no. 5603031, February 1997.

[White 1998] White, J.E., personal communication, August 1998.

[Wireless Application Protocol Forum, Ltd] Wireless Application Protocol Forum, Ltd., Wireless Application Protocol Architecture Specification, Version 30-Apr-1998. http://www.wapforum.org/docs/technical/arch-30-apr-98.pdf.

[Wollrath 1996] Wollrath, A., Riggs, R., and Waldo, J., "A Distributed Object Model for the Java System," *Proceedings of the Second Conference on Object-Oriented Technologies (COOTS)*, pp. 219–231, June 1996.

[Wong 1997] Wong, R., Walsh, T., and Paciorek, N., "Concordia: An Infrastructure for Collaborating Mobile Agents," *Proceedings of the First International Workshop on Mobile Agents*, Lecture Notes in Computer Science No. 1219, Springer Verlag, pp. 86–97, April 1997.

[Zajcew 1993] Zajcew, R., Roy, P., Black, D., Peak, C., Guedes, P., Kemp, B., LoVerso, J., Leibensperger, M., Barnett, M., Rabii, F., and Netterwala, D., "An OSF/1 UNIX for Massively Parallel Multicomputers," *Proceedings of the USENIX Winter 1993 Technical Conference,* pp. 449–468, January 1993.

[Zayas 1987] Zayas, E., "Attacking the Process Migration Bottleneck," *Proceedings of the 11th ACM Symposium on Operating Systems Principles*, pp. 13–24, November 1987. Reprinted in this book.

[Zayas 1987] Zayas, E., *The Use of Copy-on-Reference in a Process Migration System*, Ph.D. thesis, Technical Report CMU-CS-87-121, Carnegie Mellon University, April 1987.

[Zhou 1988] Zhou, S. and Ferrari, D., "A Trace-Driven Simulation Study of Dynamic Load Balancing," *IEEE Transactions on Software Engineering,* 14(9):1327–1341, September 1988.

[Zhou 1993] Zhou, S., Zheng, X., Wang, J., and Delisle, P., "Utopia: A Load Sharing Facility for Large, Heterogeneous Distributed Computer Systems," *Software—Practice and Experience*, 23(12):1305–1336, December 1993.

[Zhu 1992] Zhu, W., *The Development of an Environment to Study Load Balancing Algorithms, Process migration and Load Data Collection*, Ph.D. thesis, Technical Report, University of New South Wales, March 1992.

[Zhu 1995] Zhu, W., Steketee, C., and Muilwijk, B., "Load Balancing and Workstation Autonomy on Amoeba," *Australian Computer Science Communications (ACSC'95),* 17(1):588–597, 1995.

Index

Index